VOICE OF LABOR

WCFL

Surveys 1965-1976

John Celarek

Printed in the United States of America.

Library of Congress Control Number: 2022948209

ISBN	Paperback	978-1-68536-896-8
	eBook	978-1-68536-897-5

Westwood Books Publishing LLC
Atlanta Financial Center
3343 Peachtree Rd NE Ste 145-725
Atlanta, GA 30326

www.westwoodbookspublishing.com

This book is dedicated to Tom Konard who was responsible for putting the WCFL surveys together from 1971 to 1975. Without his help this book would not be possible.

INTRODUCTION

What you have in your hands is a great resource of WCFL information from 1965-1976. I was a big fan of this radio station from 1971-76. I first started listening to WCFL while listening to Jerry Kay, Dick Biondi and Larry O'Brien when I came home from school in 1971. I really loved the slogans "Big 10 WCFL" and "All Hit Music." The excitement of the radio station continued in 1972 when Larry Lujack came aboard and the station called itself "Super CFL." Big Ron O'Brien was a big inspiration to me in the nighttime as was Dr. Brock and Johnny Driscoll. It was one of the saddest days of my life when WCFL changed from rock to beautiful music on March 15, 1976. I was really wondering at the time why the radio station had to change formats as I thought Super CFL was still a very good station. But now as I look back it was truly one of the greatest stations that ever existed and really gave WLS a run for their money. And now please enjoy this information and look back at a truly great era of Chicago radio history!

John Celarek

CONTENTS

WCFL DISCJOCKEYS: 1965-1976

TED ANTHONY – He worked weekends at WCFL from the summer of 1971 until the summer of 1972. He presented the "Golden Weekend."

DICK BIONDI – He started working at WCFL on October 3, 1967. He worked the overnight shift from 12-5 AM until around the beginning of 1969. Then he worked the 1-4 PM shift until the spring of 1971 when he started working the 4-7 shift until July 4, 1972 when he was fired for no apparent reason by John Rook.

JERRY G. BISHOP – Jerry started working at WCFL in the spring of 1967 and he did the morning shift from 5-9 AM. Then he was sent to the overnight shift from 1-6 AM around the beginning of 1969 then he left CFL shortly thereafter.

SCOTTY BRINK – Basically only worked at WCFL in the summer of 1970 on the 4-7 PM shift.

RON BRITAIN – One of the original men from ten, started working at WCFL in the summer of 1965 and worked the 6-9 PM shift until the summer of 1968 when he was placed on the 10 PM to 1 AM shift. He also hosted the "Subterranean Circus" on Sunday nights. He stayed at WCFL until the winter of 1971.

DR. BROCK – He was called the ugliest mother in rock and roll and chief crazy legs doctory. He started working at WCFL in the spring of 1974 on the 6-10 PM shift and stayed there until the fall of 1974. He was famous for the "ugly contest" where listeners would try to guess what Dr. Brock looks like.

DEX CARD – He was the overnight discjockey on WCFL in the summer of 1967. He also filled in for various WCFL discjockeys including Jerry Kay in the fall of 1971 when he was out with pneumonia.

BUDDY CARR – He was a weekend discjockey on WCFL in 1969.

PAUL CHRISTY – He worked as a weekend discjockey on WCFL from the winter of 1970 until March 11, 1971. Interesting story about him: He recorded a 1 hour show on a Thursday to be broadcast the following Monday at sign-on. But Paul was fired that Thursday afternoon so the show never aired!

DOUG DAHLGREN – He started working at WCFL in the fall of 1974 and teamed up with Dick Sainte to form the "Dick 'N Doug" morning show from 6-10 AM. This continued until March 2, 1976 when Bob Gould called up Lew Witz on the air and Dick 'N Doug were taken off the air.

BOB DEARBORN – He started at WCFL in the fall on 1970 and began by doing the overnight shift from 1-6 AM. Then in the spring of 1971 he was put on the 10 PM to 1 AM shift. He stayed there until the spring of 1972 when he was put on the 10 AM to 2 PM shift. This continued until the fall of 1972 when he was teamed up with Larry O'Brein in the morning from 6-10 AM. Later that fall Larry left and Bob was doing mornings by himself. In the winter of 1973 Bob was teamed up with Tom Murphy to do a morning show. Then starting in the spring of 1973 Bob did the mid-day shift from 10 AM to 2 PM. This continued until his final show on WCFL on March 15, 1976. So Bob had the longest tenure of any WCFL discjockey at 5.5 years.

JOHNNY DRISCOLL – He started on WCFL in the fall of 1974 on the 10 PM to 2 AM shift. Shortly thereafter Dr. Brock left and John was doing the 6-10 PM shift. This continued until July 7, 1975 when he was put on the 10 PM to 2 AM shift. This continued until January 1, 1976 when Big Ron O'Brien left. Then he was put on the 6-10 PM shift until March 2, 1976 when he took over for Dick 'N Doug in the morning from 6-10 AM and he stayed there until March 15, 1976.

GARY GEARS – He worked the 4-7 PM afternoon shift from the fall of 1970 to the spring of 1971. He also filled in for Dick Sainte at times in 1973 and 1974.

DON GIRARDEAUX – He started working at WCFL in the fall of 1974 on weekends and continued there until July 6, 1975.

JIM BERNARD HEBEL – Big Ron O'Brien hired him to do weekends starting on December 26, 1975 and he continued until March 2, 1976. Then he worked the 2 to 6 am shift until March 13, 1976.

JERRY KAY – He started on WCFL in the spring on 1971 and did the 1-4 PM shift. He continued there until the spring of 1972. He is my favorite WCFL discjockey!

TIM KELLY – Started working at WCFL in September of 1975 on weekends there until January 1, 1976. Then he worked the 10 PM to 2 AM shift until March 2, 1976. Then he worked the 6 PM to 10 PM shift until March 13, 1976.

PAUL KIRBY – Started working at WCFL on weekends in the summer of 1972 and continued there until the fall of 1973. He did a lot of promotional work for CFL.

CHUCK KNAPP – He only worked at WCFL in the spring of 1973 and he did the 10 PM to 2 AM shift.

ROBERT E. LEE – He started working at WCFL in the winter of 1970 and he did the 7 to 10 PM nighttime shift. Then in the spring of 1970 he was put on the 10 AM to 1 PM shift and stayed there until the spring of 1972.

LARRY LUJACK – Superjock started his first tour of duty with WCFL in the spring of 1967 doing the overnight shift from 12-5 AM. He continued there June of 1967 when he left for WLS. He came back in resounding fashion on July 5, 1972 and did the afternoon shift from 2-6 PM until March 15, 1976. He was the heart and soul of super CFL and was the man that made the station big.

SID MCCOY – He did the all-night jazz show on WCFL from 12-6 AM from the summer of 1965 until the winter of 1967.

HOWARD MILLER – He worked the morning shift on WCFL in 1969 from 6-10 AM.

TOM MURPHY – He started working at WCFL in the winter of 1973 as he teamed up with Bob Dearborn on the morning shift from 6-10 AM. Then in the spring of 1973 he did the morning show himself until the summer of 1974. Larry Lujack was responsible for bringing him to CFL. Called himself world famous.

LARRY O'BRIEN – He started working at WCFL in the winter on 1970 and did the overnight shift from 1-6 AM. In the spring of 1970 he started working the 7-10 PM nighttime shift. This continued until the summer of 1972 when he started to do the 10 PM to 2 AM shift. In the fall of 1972 he teamed up with Bob Dearborn to do a morning show from 6-10 AM. He then left CFL soon thereafter.

BIG RON O'BRIEN – He started working at WCFL on June 7, 1972 and he did the 6-10 PM shift until December 21, 1973. Then Big Ron returned to WCFL on July 7, 1975 and worked the 6-10 PM shift. He was also the program director at this time. He worked the second time around until January 1, 1976.

BARNEY PIP – One of the original men from ten. He was famous for the slogan "turn into peanut butter" and horn antics. He started at WCFL in the summer of 1965 and worked the 9 PM to Midnight shift. This continued until the winter of 1970 when he was doing weekends. Then in the spring of 1970 he started the overnight shift from 1-6 AM until July 26, 1970.

GARY PRICE – He basically filled in for Larry Lujack on Saturdays in 1974 and 1975. He did a lot of promotional work for the station.

RON RILEY - Worked as a weekend jock on WCFL in 1969.

ART ROBERTS – He started working at WCFL in the summer of 1971 in the morning from 6-10 AM and this continued until the summer of 1972. In the fall of 1972 he switched to mid-days from 10 AM to 2 PM until the winter of 1973. A work of Art.

JIM RUNYON – Worked on WCFL from the summer of 1965 to 1967.

DICK SAINTE – He started working at WCFL in the summer of 1973 on the 10 PM to 2 AM shift. Then in the fall of 1974 he teamed up with Doug Dahlgren to do the Dick 'N Doug morning show from 6-10 AM. This continued until March 2, 1976 when Dick 'N Doug were yanked off the air by Lew Witz.

MICHAEL SCOTT – He started on WCFL when Dick 'N Doug were fired on March 2, 1976 on weekends until March 14, 1976.

JOEL SEBASTIAN – One of the original men from ten. He started in 1965 and did the 9 AM to noon shift until the summer of 1968. Then he worked the 10 AM to 1 PM shift until the winter of 1970.

DICK SHANNON – He started working at WCFL in the summer of 1972 doing the overnight shift from 2-6 AM and continued there until March 2, 1976. Then he worked the 10 PM to 2 AM shift until March 14, 1976.

JIMMY P. STAGG – One of the original men from ten, he started working at WCFL in the summer of 1965 and did the 3-6 PM shift until the end of 1969. Then in the winter of 1970 he did the 4-7 PM shift until the summer of 1970. He was famous for his Beatles interviews.

KRIS ERIK STEVENS – He took over for Larry O'Brien in the fall of 1972 and worked the 10 PM to 2 AM shift until the winter of 1973.

BILL TAYLOR – He started working at WCFL in the winter of 1974 and did weekends. This continued until the fall of 1974 when he did the 10 PM to 2AM shift. This continued until the summer of 1975 when he was switched to weekends.

CLARK WEBER – He started working at WCFL in 1969 and did the morning shift from 6-10 AM until the spring of 1971.

DICK WILLIAMSON – One of the original men from ten, he started working on WCFL in the summer of 1965 and did the noon to 3 PM shift until the end of 1969.

STEVE YORK – He started working at WCFL on the overnight shift from 1-6 AM in the spring of 1971 and continued there until the spring of 1972.

The following are discjockeys that worked at WCFL and WLS:

1. Dick Biondi
2. Scotty Brink
3. Dex Card
4. Buddy Carr
5. Gary Gears
6. Jerry Kay
7. Chuck Knapp
8. Larry Lujack
9. Ron Riley
10. Art Roberts
11. Dick Sainte
12. Joel Sebastian
13. Kris Erik Stevens
14. Clark Weber

WCFL TOP 1000 - 1971-1976

	YEAR	PEAK	WEEKS
1. I HONESTLY LOVE YOU - OLIVIA NEWTON-JOHN	1974	1(3)	18
2. JIVE TALKIN' - BEE GEES	1975	1	20
3. YOU'RE SO VAIN - CARLY SIMON	1973	1(3)	16
4. SEASON'S IN THE SUN - TERRY JACKS	1974	1(3)	16
5. LOVE WILL KEEP US TOGETHER - CAPTAIN & TENILLE	1975	1(3)	15
6. RHINESTONE COWBOY - GLEN CAMPBELL	1975	1	21
7. JOY TO THE WORLD - THREE DOG NIGHT	1971	1(4)	12
8. MORNING AFTER - MAUREEN MCGOVERN	1973	1(2)	15
9. HAVE YOU NEVER BEEN MELLOW - OLIVIA NEWTON-JOHN	1975	1	16
10. AMERICAN PIE - DON MCLEAN	1972	1(3)	13
11. A COWBOY'S WORK IS NEVER DONE - SONNY & CHER	1972	1(2)	14
12. KNOCK THREE TIMES - DAWN	1971	1(2)	12
13. IT'S TOO LATE - CAROLE KING	1971	1(3)	14
14. UNCLE ALBERT/ADMIRAL HALSEY - PAUL MCCARTNEY	1971	1(4)	11
15. DOWN BY THE LAZY RIVER - OSMONDS	1972	1(2)	13
16. ALONE AGAIN (NATURALLY) - GILBERT O'SULLIVAN	1972	1(2)	14
17. MY MELODY OF LOVE - BOBBY VINTON	1974	1(2)	14
18. PHILADELPHIA FREEDOM - ELTON JOHN	1975	1(2)	17
19. BRAND NEW KEY - MELANIE	1971	1(3)	15
20. BAND ON THE RUN - PAUL MCCARTNEY & WINGS	1974	1(2)	16
21. BILLY, DON'T BE A HERO - BO DONALDSON/HEYWOODS	1974	2(2)	16
22. BRANDY - LOOKING GLASS	1972	1(2)	14
23. HALF-BREED - CHER	1973	1(2)	15
24. FALLIN' IN LOVE - HAMILTON, JOE FRANK & REYNOLDS	1975	1	18
25. LITTLE WILLY - SWEET	1973	1	15
26. THE WAY WE WERE - BARBARA STREISAND	1974	1(3)	15
27. LISTEN TO WHAT THE MAN SAID - PAUL MCCARTNEY	1975	1(2)	16
28. THE RIGHT THING TO DO - CARLY SIMON	1973	1	13
29. (YOU'RE) HAVING MY BABY - PAUL ANKA	1974	1(2)	14
30. JUST MY IMAGINATION - TEMPTATIONS	1971	1	15
31. GET DOWN TONIGHT - KC AND THE SUNSHINE BAND	1975	1(2)	16
32. SHE'S A LADY - TOM JONES	1971	1(2)	14
33. HEARTBEAT, IT'S A LOVEBEAT - DEFRANCO FAMILY	1973	1	16
34. MAGGIE MAY - ROD STEWART	1971	1(3)	10
35. THE FIRST TIME EVER I SAW YOUR FACE - ROBERTA FLACK	1972	1(2)	12
36. THE HUSTLE - VAN MCCOY	1975	3	18
37. YOU DON'T MESS AROUND WITH JIM - JIM CROCE	1972	1(2)	13
38. CROCODILE ROCK - ELTON JOHN	1973	1(2)	13
39. CAT'S IN THE CRADLE - HARRY CHAPIN	1974	1	12

40. HEART OF GOLD - NEIL YOUNG	1972	1	14
41. HOOKED ON A FEELING - BLUE SWEDE	1974	1(2)	16
42. SUNDOWN - GORDON LIGHTFOOT	1974	1	17
43. MONSTER MASH - BOBBY “BORIS” PICKETT	1973	2(2)	14
44. LEAN ON ME - BILL WITHERS	1972	1(3)	11
45. FRANKENSTEIN - EDGAR WINTER GROUP	1973	1	13
46. MY LOVE - PAUL MCCARTNEY	1973	1	14
47. THE LOCO-MOTION - GRAND FUNK RAILROAD	1974	2(3)	16
48. ROCK THE BOAT - HUES CORPORATION	1974	1	15
49. DOCTOR MY EYES - JACKSON BROWNE	1972	1	15
50. WHY CAN’T WE BE FRIENDS? - WAR	1975	1	17
51. HEY, WON’T YOU PLAY - B.J. THOMAS	1975	1(2)	17
52. BAD BLOOD - NEIL SEDAKA	1975	1(2)	17
53. MY SWEET LORD - GEORGE HARRISON	1971	1(3)	9
54. LET’S GET IT ON - MARVIN GAYE	1973	1	15
55. MAGIC - PILOT	1975	2	17
56. BENNIE & THE JETS - ELTON JOHN	1974	16	2(2)
57. DELTA DAWN - HELEN REDDY	1973	1(2)	14
58. TUMBLING DICE - ROLLING STONES	1972	1	14
59. NICE TO BE WITH YOU - GALLERY	1972	1	15
60. ANNIE’S SONG - JOHN DENVER	1974	1(2)	16
61. TOP OF THE WORLD - CARPENTERS	1973	1(2)	17
62. STAY WITH ME - FACES	1972	1(2)	11
63. DOESN’T SOMEBODY WANT TO BE WANTED - PARTRIDGE FAM.	1971	1(2)	11
64. GAMES PEOPLE PLAY - SPINNERS	1975	3	18
65. BALLROOM BLITZ - SWEET	1975	1(2)	16
66. SUPERSTAR - CARPENTERS	1971	1	10
67. LADY MARMALADE - LABELLE	1975	1	14
68. ANGIE - ROLLING STONES	1973	1(2)	14
69. MY EYES ADORED YOU - FRANKIE VALLI	1975	1	17
70. I HEAR YOU KNOCKING - DAVE EDMUNDS	1971	1(2)	10
71. I’D LOVE YOU TO WANT ME - LOBO	1972	1	15
72. SWEET HOME ALABAMA - LYNYRD SKYNYRD	1974	1(2)	15
73. MAMA’S PEARL - JACKSON FIVE	1971	1(2)	10
74. THE ENTERTAINER - MARVIN HAMLISCH	1974	2	16
75. SOMEONE SAVED MY LIFE TONIGHT - ELTON JOHN	1975	1	16
76. WE’RE AN AMERICAN BAND - GRAND FUNK RAILROAD	1973	3	13
77. STAY AWHILE - BELLS	1971	1(2)	11
78. PILLOW TALK - SYLVIA	1973	1(2)	12
79. STAR BABY - GUESS WHO	1974	3	16
80. NEVER ENDING SONG OF LOVE - DELANEY & BONNIE	1971	1	15
81. I CAN SEE CLEARLY NOW - JOHNNY NASH	1972	2(2)	13
82. ONE BAD APPLE - OSMONDS	1971	1(2)	9
83. JOY - APOLLO 100	1972	1	12

84. BEACH BABY - FIRST CLASS	1974	2	15
85. HOCUS POCUS - FOCUS	1973	1	13
86. I CAN HELP - BILLY SWAN	1974	1	13
87. WANT ADS - HONEY CONE	1971	1(2)	11
88. BACK STABBERS - O'JAYS	1972	1	13
89. KILLING ME SOFTLY - ROBERTA FLACK	1973	1(3)	11
90. THE STREAK - RAY STEVENS	1974	1(3)	12
91. DON'T PULL YOUR LOVE - HAMILTON, JOE FRANK & REYNOLDS	1971	1	13
92. FAME - DAVID BOWIE	1975	1(2)	15
93. MOCKINGBIRD - CARLY SIMON & JAMES TAYLOR	1974	1(3)	13
94. BEST OF MY LOVE - EAGLES	1975	1(2)	14
95. TIE A YELLOW RIBBON - DAWN	1973	3	13
96. THEME FROM "SHAFT" - ISAAC HAYES	1971	1(2)	11
97. RUN, RUN, RUN - JO JO GUNNE	1972	2	14
98. BACK OFF BOOGALOO - RINGO STARR	1972	1	12
99. BLACK WATER - DOOBIE BROTHERS	1975	1(2)	12
100. MANDY - BARRY MANILOW	1975	1	12
101. THE NIGHT THE LIGHTS WENT OUT IN GA.-VICKI LAWRENCE	1973	1(3)	11
102. SUNSHINE ON MY SHOULDERS - JOHN DENVER	1974	2(3)	15
103. SISTER GOLDEN HAIR - AMERICA	1975	1(2)	17
104. HORSE WITH NO NAME - AMERICA	1972	1	13
105. ISLAND GIRL - ELTON JOHN	1975	1	13
106. WILDFIRE - MICHAEL MURPHY	1975	1	16
107. SONG SUNG BLUE - NEIL DIAMOND	1972	2	14
108. IF YOU COULD READ MY MIND - GORDON LIGHTFOOT	1971	1	11
109. GO AWAY LITTLE GIRL - DONNY OSMOND	1971	1	12
110. LAST SONG - EDWARD BEAR	1973	2(2)	14
111. KILLER QUEEN - QUEEN	1975	1(2)	18
112. YOU'VE GOT A FRIEND - JAMES TAYLOR	1971	1	12
113. SING - CARPENTERS	1973	1	13
114. IF YOU LOVE ME (LET ME KNOW) - OLIVIA NEWTON-JOHN	1974	1(2)	12
115. YESTERDAY ONCE MORE - CARPENTERS	1973	1(2)	13
116. BURNING LOVE - ELVIS PRESLEY	1972	2	12
117. ROCK ON - DAVID ESSEX	1974	3	14
118. MIND GAMES - JOHN LENNON	1973	1	13
119. THE NIGHT CHICAGO DIED - PAPER LACE	1974	1	15
120. LUCY IN THE SKY WITH DIAMONDS - ELTON JOHN	1975	1(2)	13
121. HOW CAN YOU MEND A BROKEN HEART - BEE GEES	1971	1	12
122. LOVES ME LIKE A ROCK - PAUL SIMON	1973	1	13
123. KODACHROME - PAUL SIMON	1973	1	15
124. SMOKE ON THE WATER - DEEP PURPLE	1973	3	15
125. BEN - MICHAEL JACKSON	1972	2	12
126. SMILING FACES SOMETIMES - UNDISPUTED TRUTH	1971	1	11
127. GYPSYS, TRAMPS & THIEVES - CHER	1971	1(2)	11

128. PLEASE MR. POSTMAN - CARPENTERS	1975	1	12
129. NOTHIN' FROM NOTHIN' - BILLY PRESTON	1974	2(2)	16
130. WHO LOVES YOU - FOUR SEASONS	1975	1(2)	13
131. ROCKIN' ROBIN - MICHAEL JACKSON	1972	2(2)	12
132. MIDNIGHT AT THE OASIS - MARIA MULDAUR	1974	1	15
133. DANCING MACHINE - JACKSON FIVE	1974	6	15
134. GOODBYE TO LOVE - CARPENTERS	1972	2	13
135. I AM WOMAN - HELEN REDDY	1972	1	13
136. WEDDING SONG (THERE IS LOVE) - PAUL STOOKEY	1971	2(2)	13
137. NIGHTS IN WHITE SATIN - MOODY BLUES	1972	1	13
138. HOLD YOUR HEAD UP - ARGENT	1972	2	13
139. AT SEVENTEEN - JANIS IAN	1975	2	16
140. RAINY DAYS & MONDAYS - CARPENTERS	1971	2(3)	13
141. PAPA WAS A ROLLING STONE - TEMPTATIONS	1972	1(2)	13
142. I'LL BE AROUND - SPINNERS	1972	1(2)	13
143. THE JOKER - STEVE MILLER BAND	1974	1(2)	14
144. GIVE ME LOVE - GEORGE HARRISON	1973	1(2)	13
145. COME AND GET YOUR LOVE - REDBONE	1974	1(3)	12
146. INDIAN RESERVATION - RAIDERS	1971	1	9
147. THE MOST BEAUTIFUL GIRL - CHARLIE RICH	1974	5	14
148. NEVER CAN SAY GOODBYE - JACKSON FIVE	1972	2	12
149. OUTA-SPACE - BILLY PRESTON	1972	1	11
150. FEELIN' STRONGER EVERY DAY - CHICAGO	1973	2	12
151. DANNY'S SONG - ANNE MURRAY	1973	2	13
152. FREE RIDE - EDGAR WINTER GROUP	1973	3	15
153. LOVIN' YOU - MINNIE RIPPERTON	1975	1(2)	13
154. I SHOT THE SHERIFF - ERIC CLAPTON	1974	1	15
155. PLAYGROUND IN MY MIND - CLINT HOLMES	1973	2(2)	10
156. WITHOUT YOU - NILLSON	1972	1	11
157. BLACK & WHITE - THREE DOG NIGHT	1972	1	11
158. LAUGHTER IN THE RAIN - NEIL SEDAKA	1975	1	13
159. ROCK YOUR BABY - GEORGE MCCRAE	1974	1	13
160. THE NIGHT THEY DROVE OLD DIXIE DOWN - JOAN BAEZ	1971	2(2)	11
161. THE AIR THAT I BREATHE - HOLLIES	1974	2	12
162. DANCE WITH ME - ORLEANS	1975	2(3)	15
163. TAKIN' CARE OF BUSINESS - BACHMAN-TURNER OVERDRIVE	1974	5	14
164. MIRACLES - JEFFERSON STARSHIP	1975	2	13
165. TUSH - ZZ TOP	1975	2(2)	14
166. DON'T LET THE SUN GO DOWN ON ME - ELTON JOHN	1974	1	16
167. L.A. GOODBYE - IDES OF MARCH	1971	2	12
168. SWEET MARY - WADSWORTH MANSION	1971	2	11
169. DANCING IN THE MOONLIGHT - KING HARVEST	1973	1	12
170. CLAIR - GILBERT O'SULLIVAN	1972	1(2)	12

171. CAN'T GET ENOUGH - BAD COMPANY	1974	1	14
172. LADY - STYX	1975	3	11
173. DRAGGIN' THE LINE - TOMMY JAMES	1971	2	13
174. LONG TRAIN RUNNIN' - DOOBIE BROTHERS	1973	3	15
175. SAUVECITO - MALO	1972	1	11
176. I'M SORRY - JOHN DENVER	1975	1	14
177. KUNG FU FIGHTING - CARL DOUGLAS	1974	1	16
178. SUMMER BREEZE - SEALS & CROFTS	1972	2	11
179. TIN MAN - AMERICA	1974	2(2)	11
180. YOU ARE SO BEAUTIFUL - JOE COCKER	1975	2	15
181. YOU'RE SIXTEEN - RINGO STARR	1974	1	13
182. YO-YO - OSMONDS	1971	1	14
183. CHEVY VAN - SAMMY JOHNS	1975	1	17
184. MOTHER & CHILD REUNION - PAUL SIMON	1972	2	12
185. RIKKI, DON'T LOSE THAT NUMBER - STEELY DAN	1974	4	14
186. IT'S A MIRACLE - BARRY MANILOW	1975	1	14
187. MIDNIGHT TRAIN TO GEORGIA - GLADYS KNIGHT/PIPS	1973	1(2)	14
188. I WRITE THE SONGS - BARRY MANILOW	1976	1	16
189. ONLY YESTERDAY - CARPENTERS	1975	2	16
190. MORNING HAS BROKEN - CAT STEVENS	1972	1	11
191. PHOTOGRAPH - RINGO STARR	1973	1	12
192. GET DOWN - GILBERT O'SULLIVAN	1973	1	13
193. BROTHER LOUIE - STORIES	1973	1	13
194. THAT'S THE WAY I LIKE IT - KC & THE SUNSHINE BAND	1975	1	13
195. FIRE - OHIO PLAYERS	1975	1	12
196. SWEET & INNOCENT - DONNY OSMOND	1971	1(2)	13
197. PEACE TRAIN - CAT STEVENS	1971	1	13
198. BEAUTIFUL SUNDAY - DANIEL BOONE	1972	1	13
199. LA GRANGE - ZZ TOP	1974	2(2)	15
200. STONED LOVE - SUPREMES	1971	2	9
201. PINBALL WIZARD - ELTON JOHN	1975	-	-
202. MY MARIA - B.W. STEVENSON	1973	3	13
203. LOVE'S THEME - LOVE UNLIMITED ORCHESTRA	1974	2	14
204. HE DON'T LOVE YOU - TONY ORLANDO & DAWN	1975	2(2)	15
205. ONE OF THESE NIGHTS - EAGLES	1975	3	17
206. YOU AIN'T SEEN NOTHIN' YET - BACHMAN-TURNER OVERDRIV 1974	1(2)	12	
207. THE BITCH IS BACK - ELTON JOHN	1974	1	13
208. HELLO IT'S ME - TODD RUNDGREN	1973	1	13
209. FOX ON THE RUN - SWEET	1975	1	14
210. WATERLOO - ABBA	1974	2	15
211. ROCK 'N ROLL PART II - GARY GLITTER	1972	1	13
212. YOU WON'T SEE ME - ANNE MURRAY	1974	2	15
213. ROCKIN' CHAIR - GWEN MCCRAE	1975	3	16

214. IT DON'T COME EASY - RINGO STARR	1971	2	10
215. RIDERS ON THE STORM - DOORS	1971	2	11
216. BLACK DOG - LED ZEPELLIN	1972	2(2)	11
217. SUNSHINE - JONATHON EDWARDS	1971	1	11
218. FEEL LIKE MAKIN' LOVE - ROBERTA FLACK	1974	1	12
219. WHEN WILL I BE LOVED - LINDA RONSTADT	1975	2(2)	15
220. SKY HIGH - JIGSAW	1975	2	14
221. EVERYBODY PLAYS THE FOOL - MAIN INGREDIENT	1972	1	12
222. PUPPY LOVE - DONNY OSMOND	1972	1	12
223. THE WAY I WANT TO TOUCH YOU - CAPTAIN & TENILLE	1975	2(2)	13
224. SYLVIA'S MOTHER - DR. HOOK & MEDICINE SHOW	1972	1	12
225. BOOGIE WOOGIE BUGLE BOY - BETTE MIDLER	1973	1	12
226. ONE TIN SOLDIER - COVEN	1974	1	11
227. CANDY MAN - SAMMY DAVIS JR.	1972	2	9
228. GOODBYE YELLOW BRICK ROAD - ELTON JOHN	1973	3	13
229. FOR ALL WE KNOW - CARPENTERS	1971	3	13
230. FEELINGS - MORRIS ALBERT	1975	1	11
231. ME & MRS. JONES - BILLY PAUL	1973	1	11
232. THEN CAME YOU - DIONNE WARWICKE/SPINNERS	1974	3	14
233. CONVOY - C.W. MCCALL	1976	1(2)	11
234. RAMBLIN' MAN - ALLMAN BROTHERS BAND	1973	3	14
235. THE SHOW MUST GO ON - THREE DOG NIGHT	1974	3	15
236. ROCKIN' PNEUMONIA - JOHNNY RIVERS	1972	2	14
237. NEVER CAN SAY GOODBYE - GLORIA GAYNOR	1975	2(2)	12
238. LONG TALL GLASSES - LEO SAYER	1975	2	13
239. JUST DON'T WANT TO BE LONELY - MAIN INGREDIENT	1974	3	15
240. EIGHTEEN WITH A BULLET - PETE WINGFIELD	1975	3	15
241. WINNERS & LOSERS - HAMILTON, JOE FRANK & REYNOLDS	1975	2	14
242. KEEP ON TRUCKIN' - EDDIE KENDRICKS	1973	3	14
243. LOVE ROLLERCOASTER - OHIO PLAYERS	1976	1	15
244. DARK LADY - CHER	1974	2	13
245. DUELING BANJOES - DELIVERANCE	1973	2(2)	12
246. HURTING EACH OTHER - CARPENTERS	1972	1	11
247. ALSO SRPACH ZARATHUSTRA (2001) - DEODATO	1973	1(2)	9
248. PLEASE MR. PLEASE - OLIVIA NEWTON-JOHN	1975	2	13
249. JACKIE BLUE - OZARK MOUNTAIN DAREDEVILS	1975	3	15
250. NO NO SONG - RINGO STARR	1975	3	15
251. COULD IT BE I'M FALLING IN LOVE - SPINNERS	1973	3	13
252. I'VE FOUND SOMEONE OF MY OWN - FREE MOVEMENT	1971	2	13
253. GARDEN PARTY - RICK NELSON	1972	2	13
254. ANOTHER SATURDAY NIGHT - CAT STEVENS	1974	2	13
255. ALL I EVER NEED IS YOU - SONNY & CHER	1971	2	13
256. TUBULAR BELLS - MIKE OLDFIELD	1974	2	14
257. LAST NIGHT (I DIDN'T GET TO SLEEP AT ALL) - 5TH DIMENSION	1972	2(2)	12

258. I GOT A NAME - JIM CROCE	1973	2	13
259. KEEPER OF THE CASTLE - FOUR TOPS	1973	2	13
260. THE HURT - CAT STEVENS	1973	1	11
261. KISSING MY LOVE - BILL WITHERS	1973	1	11
262. BROWN SUGAR - ROLLING STONES	1971	2	10
263. LOVE HER MADLY - DOORS	1971	3	13
264. ROCK ME GENTLY - ANDY KIM	1974	2	12
265. SATURDAY NIGHT - BAY CITY ROLLERS	1975	2	14
266. SMOKIN' IN THE BOYS ROOM - BROWNSVILLE STATION	1974	2	13
267. I DON'T WANT TO BE RIGHT - LUTHER INGRAM	1972	2	13
268. TELL ME SOMETHING GOOD - RUFUS	1974	3	13
269. SPACE RACE - BILLY PRESTON	1973	3	13
270. BAD TIME - GRAND FUNK RAILROAD	1975	3	14
271. GO ALL THE WAY - RASPBERRIES	1972	2	13
272. LET ME BE THERE - OLIVIA NEWTON-JOHN	1974	2	13
273. TROGLODYTE (CAVE MAN) - JIMMY CASTOR BUNCH	1972	1	10
274. RIGHT PLACE, WRONG TIME - DR. JOHN	1973	3	14
275. STOP AND SMELL THE ROSES - MAC DAVIS	1974	3	14
276. TAKE IT EASY - EAGLES	1972	2	13
277. CHINA GROVE - DOOBIE BROTHERS	1973	2	13
278. WHAT ARE YOU DOING SUNDAY? - DAWN	1971	3	14
279. RING THE LIVING BELL - MELANIE	1972	3	11
280. MY DING-A-LING - CHUCK BERRY	1972	1(2)	8
281. PICK UP THE PIECES - AVERAGE WHITE BAND	1975	3	13
282. LOVE TRAIN - O'JAYS	1973	2	11
283. WHAT'S GOING ON - MARVIN GAYE	1971	3	12
284. LONG COOL WOMAN - HOLLIES	1972	2	11
285. JAZZMAN - CAROLE KING	1974	3	12
286. THE LION SLEEPS TONIGHT - ROBERT JOHN	1972	2	12
287. SATURDAY IN THE PARK - CHICAGO	1972	3	12
288. DAY AFTER DAY - BADFINGER	1972	3	12
289. IT NEVER RAINS IN SOUTHERN CALIFORNIA-ALBERT HAMMOND	1972	3	12
290. WHERE IS THE LOVE - FLACK/HATHAWAY	1972	2	11
291. EVERYTHING I OWN - BREAD	1972	2	12
292. DAY DREAMING - ARETHA FRANKLIN	1972	2	11
293. AMOS MOSES - JERRY REED	1971	3	11
294. DON'T EXPECT ME TO BE YOU'RE FRIEND - LOBO	1973	2	11
295. SWEET CITY WOMAN - STAMPEDERS	1971	2	11
296. ROCK 'N ROLL ALL NIGHT - KISS	1976	2	15
297. PEACEFUL - HELEN REDDY	1973	2	12
298. MAYBE TOMORROW - JACKSON FIVE	1971	2(2)	10
299. ABSOLUTELY RIGHT - FIVE MAN ELECTRICAL BAND	1971	2	12
300. BEFORE THE NEXT TEAR DROP FALLS - FREDDY FENDER	1975	3	12
301. TOO LATE TO TURN BACK NOW - CORNELIUS BROS.	1972	2	11

302. TAKE ME HOME, COUNTRY ROADS - JOHN DENVER	1972	2	12
303. PUT YOUR HAND IN THE HAND - OCEAN	1971	3	11
304. SPIDERS & SNAKES - JIM STAFFORD	1974	3	11
305. WHY CAN'T WE LIVE TOGETHER - TIMMY THOMAS	1973	2	11
306. JUNIOR'S FARM - PAUL MCCARTNEY	1974	3	12
307. IMAGINE - JOHN LENNON	1971	2	11
308. PROUD MARY - IKE & TINA TURNER	1971	2	10
309. OLD FASHIONED LOVE SONG - THREE DOG NIGHT	1971	3	12
310. STUCK IN THE MIDDLE WITH YOU - STEELERS WHEEL	1973	2	12
311. SCHOOL'S OUT - ALICE COOPER	1972	2	11
312. HEY GIRL - DONNY OSMOND	1971	2	11
313. REELING IN THE YEARS - STEELY DAN	1973	3	12
314. LIVING IN THE PAST - JETHRO TULL	1973	2	11
315. I'M JUST A SINGER - MOODY BLUES	1973	3	12
316. FLOY JOY - SUPREMES	1972	2	11
317. WALK ON THE WILD SIDE - LOU REED	1963	3	11
318. PIECES OF APRIL - THREE DOG NIGHT	1973	2	11
319. SISTER MARY ELEPHANT - CHEECH & CHONG	1974	2(2)	10
320. ABRA-CA-DABRA - DEFRANCO FAMILY	1974	3	12
321. THEME FROM MAHOGHANY - DIANA ROSS	1976	3	15
322. MR. BIG STUFF - JEAN KNIGHT	1971	3	11
323. DIAMOND GIRL - SEALS & CROFTS	1973	2	10
324. DO YOU KNOW WHAT I MEAN - LEE MICHAELS	1971	2	10
325. TWELTH OF NEVER - DONNY OSMOND	1973	3	11
326. HOT ROD LINCOLN - COMMANDER CODY	1972	2	10
327. I WOKE UP IN LOVE THIS MORNING - PARTRIDGE FAMILY	1971	3	11
328. BASKETBALL JONES - CHEECH & CHONG	1973	3	11
329. CONQUISTADOR - PROCOL HAREM	1972	3	11
330. THE WITCH QUEEN OF NEW ORLEANS - REDBONE	1972	3	11
331. SOFTLY WHISPERING I LOVE YOU - ENGLISH CONGREGATION	1972	3	11
332. CHARITY BALL - FANNY	1971	3	11
333. NOTHIN' TO HIDE - TOMMY JAMES	1972	2	10
334. 50 WAYS TO LEAVE YOUR LOVER - PAUL SIMON	1976	1(2)	6
335. SUPERSTITION - STEVIE WONDER	1973	3	11
336. YOU'RE THE 1ST, LAST, MY EVERYTHING - BARRY WHITE	1975	3	11
337. HAVE YOU SEEN HER - CHI-LITES	1971	2	10
338. BABY I'M-A-WANT YOU - BREAD	1971	2	10
339. THE LORD'S PRAYER - SISTER JANET MEAD	1974	3	11
340. FREDDIE'S DEAD - CURTIS MAYFIELD	1972	3	10
341. CHERISH - DAVID CASSIDY	1971	3	10
342. LISTEN TO THE MUSIC - DOOBIE BROTHERS	1972	3	11
343. DON'T KNOCK MY LOVE - WILSON PICKETT	1971	3	11
344. IMMIGRANT SONG - LED ZEPELLIN	1971	2	9
345. STONEY END - BARBARA STREISAND	1971	2	6

346. SPANISH HARLEM - ARETHA FRANKLIN	1971	3	10
347. BREAKING UP IS HARD TO DO - NEIL SEDAKA	1976	2	8
348. THAT'S THE WAY I'VE ALWAYS HEARD - CARLY SIMON	1971	3	10
349. SUGAR DADDY - JACKSON FIVE	1972	2	9
350. TOO YOUNG - DONNY OSMOND	1972	3	10
351. WHERE DID OUR LOVE GO - DONNIE ELBERT	1972	3	11
352. RINGS - CYMARRON	1971	3	10
353. I DON'T BLAME YOU AT ALL - MIRACLES	1971	3	11
354. VAHEVELLA - LOGGINS & MESSINA	1972	3	11
355. FAMILY AFFAIR - SLY & THE FAMILY STONE	1971	3	9
356. LOVE TO LOVE YOU BABY - DONNA SUMMER	1976	2	11
357. STORY IN YOUR EYES - MOODY BLUES	1971	3	10
358. IF NOT FOR YOU - OLIVIA NEWTON-JOHN	1971	3	10
359. LOVE THE ONE YOU'RE WITH - STEPHEN STILLS	1971	3	7
360. GAMES - REDEYE	1971	3	8
361. DEC. '63 (OH, WHAT A NIGHT) - FOUR SEASONS	1976	3	7
362. I'M NOT IN LOVE - 10 CC	1975	4	16
363. BEST THING THAT EVER HAPPENED TO ME-GLADYS KNIGHT	1974	4	16
364. FEEL LIKE MAKIN' LOVE - BAD COMPANY	1975	4	16
365. NIGHTS ON BROADWAY - BEE GEES	1975	4	15
366. WILDFLOWER - SKYLARK	1973	4	15
367. TAKE ME IN YOUR ARMS (ROCK ME) - DOOBIE BROTHERS	1975	4	15
368. S.O.S. - ABBA	1975	4	15
369. FLY, ROBIN, FLY - SILVER CONVENTION	1975	4	14
370. YOU'RE NO GOOD - LINDA RONSTADT	1975	4	14
371. LYIN' EYES - EAGLES	1975	4	14
372. JUST YOU 'N ME - CHICAGO	1973	4	14
373. AIN'T NO WAY TO TREAT A LADY - HELEN REDDY	1975	4	14
374. DON'T CALL US, WE'LL CALL YOU - SUGARLOAF	1975	4	14
375. I'LL HAVE TO SAY I LOVE YOU IN A SONG - JIM CROCE	1974	4	14
376. ERES TU (TOUCH THE WIND) - MOCEDADES	1974	4	14
377. YOU LITTLE TRUSTMAKER - TYMES	1974	4	14
378. WILL IT GO ROUND IN CIRCLES - BILLY PRESTON	1973	4	13
379. THANK GOD I'M A COUNTRY BOY - JOHN DENVER	1975	4	13
380. I'M GONNA LOVE YOU - BARRY WHITE	1973	4	13
381. HELP ME - JONI MITCHELL	1974	4	13
382. ROCKY MOUNTAIN HIGH - JOHN DENVER	1973	4	13
383. SATURDAY NIGHTS ALLRIGHT FOR FIGHTING - ELTON JOHN	1973	4	13
384. HERE COMES THAT RAINY DAY FEELING - FORTUNES	1971	4	13
385. D'YER MAKER - LED ZEPPELIN	1973	4	13
386. EIGHTEEN - ALICE COOPER	1971	4	13
387. BABY DON'T GET HOOKED ON ME - MAC DAVIS	1972	4	12
388. WHEN WILL I SEE YOU AGAIN - 3 DEGREES	1974	4	12
389. DADDY DON'T YOU WALK SO FAST - WAYNE NEWTON	1972	4	12

390. IF - BREAD	1971	4	12
391. ANOTHER DAY - PAUL MCCARTNEY	1971	4	12
392. LONELY PEOPLE - AMERICA	1975	4	12
393. LIFE IS A ROCK (BUT CFL ROLLED ME) - REUNION	1974	4	12
394. ALL I KNOW - ART GARFUNKLE	1973	4	12
395. DYNOMITE - BAZUKA	1975	4	12
396. RADAR LOVE - GOLDEN EARRING	1974	4	12
397. ROLL ON DOWN THE HIGHWAY - B.T.O.	1975	4	12
398. BABY BLUE - BADFINGER	1972	4	12
399. WON'T GET FOOLED AGAIN - WHO	1971	4	12
400. THUNDER & LIGHTNING - CHI COLTRANE	1972	4	12
401. I DON'T KNOW HOW TO LOVE HIM - YVONNE ELLIMAN	1971	4	12
402. WHITE LIES, BLUE EYES - BULLET	1972	4	12
403. PAINTED LADIES - IAN THOMAS	1974	4	12
404. LET'S PRETEND - RASPBERRIES	1973	4	12
405. YOUR MOVE - YES	1971	4	12
406. LIVE & LET DIE - PAUL MCCARTNEY & WINGS	1973	4	11
407. DANIEL - ELTON JOHN	1973	4	11
408. BOOGIE ON REGGAE WOMAN - STEVIE WONDER	1975	4	11
409. AIN'T NO WOMAN (LIKE THE ONE I'VE GOT) - FOUR TOPS	1973	4	11
410. PLAY ME - NEIL DIAMOND	1972	4	11
411. SHAME, SHAME, SHAME - SHIRLEY (AND COMPANY)	1975	4	11
412. ROUNDABOUT - YES	1972	4	11
413. FUNKY NASSAU - BEGINNING OF THE END	1971	4	11
414. LIVING IN A HOUSE DIVIDED - CHER	1972	4	11
415. I'M COMIN' HOME - TOMMY JAMES	1971	4	11
416. MARIANNE - STEPHEN STILLS	1971	4	11
417. GOT TO BE THERE - MICHAEL JACKSON	1971	4	10
418. BLACK MAGIC WOMAN - SANTANA	1971	4	10
419. SCORPIO - DENNIS COFFEY	1972	4	10
420. I JUST WANT TO CELEBRATE - RARE EARTH	1971	4	10
421. HAVE YOU EVER SEEN THE RAIN - C.C.R.	1971	4	10
422. DON'T SAY YOU DON'T REMEMBER - BEVERLY BREMERS	1972	4	10
423. HUMMINGBIRD - SEALS & CROFTS	1973	4	10
424. CHIRPY CHIRPY CHEEP CHEEP - MAC & KATIE KISSOON	1971	4	10
425. HOW DO YOU DO? - MOUTH & MCNEIL	1972	4	9
426. DOUBLE LOVIN' - OSMONDS	1971	4	9
427. REMEMBER ME - DIANA ROSS	1971	4	9
428. PRECIOUS & FEW - CLIMAX	1972	4	8
429. EVIL WOMAN - ELECTRIC LIGHT ORCHESTRA	1976	4	8
430. HANG ON TO YOUR LIFE - GUESS WHO	1971	4	7
431. HOW LONG - ACE	1975	5	17
432. BAD, BAD LEROY BROWN - JIM CROCE	1973	5	15
433. ONE MAN WOMAN - PAUL ANKA	1975	5	15

434. LONGFELLOW SERENADE - NEIL DIAMOND	1974	5	14
435. LADY BLUE - LEON RUSSELL	1975	5	14
436. WILD THING - FANCY	1974	5	14
437. I AM LOVE - JACKSON FIVE	1975	5	14
438. TSOP - MFSB	1974	5	13
439. SHOW & TELL - AL WILSON	1974	5	13
440. BOOGIE DOWN - EDDIE KENDRICKS	1974	5	13
441. SOME KIND OF WONDERFUL - GRAND FUNK	1975	5	13
442. THE COVER OF "ROLLING STONE" - DR. HOOK	1973	5	13
443. JET - PAUL MCCARTNEY & WINGS	1974	5	13
444. I'LL MEET YOU HALFWAY - PARTRIDGE FAMILY	1971	5	13
445. IF I COULD REACH YOU - 5TH DIMENSION	1972	5	13
446. I DON'T WANT TO BE LONELY - JAMES TAYLOR	1973	5	13
447. JIM DANDY - BLACK OAK ARKANSAS	1974	5	13
448. LAST KISS - WEDNESDAY	1974	5	13
449. OH GIRL - CHI-LITES	1972	5	12
450. CAN'T GET ENOUGH OF YOUR LOVE - BARRY WHITE	1974	5	12
451. NEVER, NEVER GONNA GIVE YA UP - BARRY WHITE	1974	5	12
452. VENTURA HIGHWAY - AMERICA	1972	5	12
453. JAMBALAYA - BLUE RIDGE RANGERS	1973	5	12
454. JIMMY LOVES MARYANNE - LOOKING GLASS	1973	5	12
455. ME & BOBBY MCGEE - JANIS JOPLIN	1971	5	11
456. TOUCH ME IN THE MORNING - DIANA ROSS	1973	5	11
457. DO IT 'TILL YOU'RE SATISFIED - B.T. EXPRESS	1975	5	11
458. FIGHT THE POWER - ISLEY BROTHERS	1975	5	11
459. IN THE RAIN - DRAMATICS	1972	5	11
460. WILDWOOD WEED - JIM STAFFORD	1974	5	11
461. I NEED YOU - AMERICA	1972	5	11
462. STIR IT UP - JOHNNY NASH	1973	5	11
463. RESPECT YOURSELF - STAPLE SINGERS	1972	5	11
464. NATHAN JONES - SUPREMES	1971	5	11
465. CISCO KID - WAR	1973	5	10
466. MR. JAWS - DICKIE GOODMAN	1975	5	10
467. THE WAY OF LOVE - CHER	1972	5	10
468. TIRED OF BEING ALONE - AL GREEN	1971	5	10
469. SUPERSTAR - MURRAY HEAD	1971	5	10
470. COULD IT BE FOREVER - DAVID CASSIDY	1972	5	10
471. 500 MILES - HEAVEN BOUND	1972	5	10
472. AIN'T NO SUNSHINE - BILL WITHERS	1971	5	9
473. THE WORLD IS A GHETTO - WAR	1973	5	9
474. DIARY - BREAD	1972	5	9
475. TWO DIVIDED BY LOVE - GRASS ROOTS	1971	5	9
476. JUNGLE FEVER - CHACKACHAS	1972	5	8
477. BORN TO WANDER - RARE EARTH	1971	5	8

478. CONVENTION '72 - DELEGATES	1972	5	7
479. DRIFT AWAY - DOBIE GRAY	1973	6	15
480. SHINING STAR - EARTH, WIND & FIRE	1975	6	14
481. ONLY WOMEN - ALICE COOPER	1975	6	14
482. LOVE SONG - ANNE MURRAY	1974	6	14
483. I'M NOT LISA - JESSI COLTER	1975	6	13
484. MIDNIGHT BLUE - MELISSA MANCHESTER	1975	6	13
485. GOOD TIME CHARLIE'S GOT THE BLUES - DANNY O'KEEFE	1972	6	13
486. SITTING - CAT STEVENS	1973	6	13
487. OOH BABY - GILBERT O'SULLIVAN	1973	6	13
488. ROLL OVER BEETHOVEN - ELECTRIC LIGHT ORCHESTRA	1973	6	13
489. TIME IN A BOTTLE - JIM CROCE	1974	6	12
490. JUNGLE BOOGIE - KOOL & THE GANG	1974	6	12
491. MERCY, MERCY ME - MARVIN GAYE	1971	6	12
492. EARACHE MY EYE - CHEECH & CHONG	1974	6	12
493. ROCK 'N ROLL LULLABY - B.J. THOMAS	1972	6	12
494. BLACK SUPERMAN - JOHNNY WAKELIN	1975	6	12
495. NO MORE MR. NICE GUY - ALICE COOPER	1973	6	12
496. AUTOMATICALLY SUNSHINE - SUPREMES	1972	6	12
497. YOU ARE THE ONE - SUGAR BEARS	1972	6	12
498. I'VE GOT TO USE MY IMAGINATION - GLADYS KNIGHT/PIPS	1974	6	11
499. IF YOU REALLY LOVE ME - STEVIE WONDER	1971	6	11
500. FOR THE LOVE OF MONEY - O'JAYS	1974	6	11
501. WITCHY WOMAN - EAGLES	1972	6	11
502. MY LITTLE TOWN - SIMON & GARFUNKLE	1976	6	11
503. ANTICIPATION - CARLY SIMON	1972	6	11
504. I WANNA BE WHERE YOU ARE - MICHAEL JACKSON	1972	6	11
505. ONLY YOU KNOW AND I KNOW - DELANEY & BONNIE	1971	6	11
506. LET'S DO IT AGAIN - STAPLE SINGERS	1975	6	10
507. WHATEVER GETS YOU THRU THE NIGHT - JOHN LENNON	1974	6	10
508. ME & YOU & A DOG NAMED BOO - LOBO	1971	6	10
509. I WANNA BE WITH YOU - RASPBERRIES	1973	6	10
510. BIRDS OF A FEATHER - RAIDERS	1971	6	10
511. BABY LET ME TAKE YOU - DETROIT EMERALDS	1972	6	10
512. HEAVY MAKES YOU HAPPY - STAPLE SINGERS	1971	6	10
513. USE ME - BILL WITHERS	1972	6	9
514. ONLY YOU - RINGO STARR	1975	6	9
515. BEGINNINGS - CHICAGO	1971	6	9
516. BUNGLE IN THE JUNGLE - JETHRO TULL	1975	6	9
517. IT'S ONE OF THOSE NIGHTS - PATRIDGE FAMILY	1972	6	9
518. ALL BY MYSELF - ERIC CARMEN	1976	6	8
519. LONELY DAYS - BEE GEES	1971	6	8
520. SHAMBALA - THREE DOG NIGHT	1973	7	15
521. SEARCHIN' SO LONG - CHICAGO	1974	7	15

522. CHICK-A-BOOM - DADDY DEWDROP	1971	7	15
523. HEY YOU - BACHMAN-TURNER OVERDRIVE	1975	7	15
524. AFTER THE GOLDRUSH - PRELUDE	1974	7	15
525. ANGIE BABY - HELEN REDDY	1974	7	14
526. IT ONLY TAKES A MINUTE - TAVARES	1975	7	14
527. GIVE IT TO ME - J. GEILS BAND	1973	7	13
528. RUBBER BULLETS - 10 CC	1973	7	13
529. RUN, JOEY, RUN - DAVID GEDDES	1975	7	12
530. ON AND ON - GLADYS KNIGHT/PIPS	1974	7	12
531. CAN'T GET IT OUT OF MY HEAD - ELECTRIC LIGHT ORCHESTRA	1975	7	12
532. HERE COMES THE SUN - RICHIE HAVENS	1971	7	12
533. WILD NIGHT - VAN MORRISON	1971	7	12
534. MOONSHADOW - CAT STEVENS	1971	7	12
535. WHY - DONNY OSMOND	1972	7	11
536. MY WORLD - BEE GEES	1972	7	11
537. BIG CITY MISS RUTH ANN - GALLERY	1973	7	11
538. YOU COULD HAVE BEEN A LADY - APRIL WINE	1972	7	11
539. GLORY BOUND - GRASS ROOTS	1972	7	11
540. RAINY JANE - DAVY JONES	1971	7	11
541. MAMMY BLUE - STORIES	1973	7	11
542. LET'S STAY TOGETHER - AL GREEN	1972	7	10
543. IF YOU DON'T KNOW ME BY NOW - HAROLD MELVIN	1973	7	10
544. YOUR SONG - ELTON JOHN	1971	7	10
545. IT'S GONNA TAKE SOME TIME - CARPENTERS	1972	7	10
546. OPERATOR - JIM CROCE	1972	7	10
547. THE DRUM - BOBBY SHERMAN	1971	7	10
548. MORNING SIDE OF THE MOUNTAIN - DONNY & MARIE OSMOND	1975	7	9
549. STONES - NEIL DIAMOND	1971	7	9
550. HAPPY - ROLLING STONES	1972	7	9
551. I LOVE MUSIC – O'JAYS	1976	7	9
552. FANNY – BEE GEES	1976	7	9
553. TIMOTHY – BUOYS	1971	8	15
554. THE WAY WE WERE (MEDLEY) – GLADYS KNIGHT/PIPS	1975	8	14
555. MONEY – PINK FLOYD	1973	8	14
556. LEAVE ME ALONE (RUBY RED DRESS) – HELEN REDDY	1974	8	13
557. SAY, HAS ANYBODY SEEN MY SWEET GYPSY ROSE – DAWN	1973	8	13
558. YOUR MAMA DON'T DANCE – LOGGINS & MESSINA	1973	8	13
559. HIGHER GROUND – STEVIE WONDER	1973	8	13
560. PAPER ROSES – MARIE OSMOND	1973	8	13
561. DO IT AGAIN – STEELY DAN	1973	8	13
562. YOU AND ME AGAINST THE WORLD – HELEN REDDY	1974	8	13
563. I WON'T LAST A DAY WITHOUT YOU – CARPENTERS	1974	8	13
564. MY GIRL BILL – JIM STAFFORD	1974	8	13
565. RUNAWAY – CHARLIE KULIS	1975	8	13

566. THIS WILL BE – NATALIE COLE	1975	8	12
567. MOTORCYCLE MAMA – SAILCAT	1972	8	12
568. WE CAN WORK IT OUT – STEVIE WONDER	1971	8	12
569. HOLD HER TIGHT – OSMONDS	1972	8	12
570. DEAD SKUNK – LOUDON WAINWRIGHT III	1973	8	12
571. LOVE JONES – BRIGHTER SIDE OF DARKNESS	1973	8	12
572. I BELIEVE IN MUSIC – GALLERY	1972	8	12
573. CHERRY, CHERRY – NEIL DIAMOND	1973	8	12
574. POPCORN – HOT BUTTER	1972	8	11
575. INNER CITY BLUES – MARVIN GAYE	1971	8	11
576. DOCTOR'S ORDERS – CAROL DOUGLAS	1975	8	11
577. LET ME SERENADE YOU – THREE DOG NIGHT	1973	8	11
578. WATCHA SEE IS WATCHA GET – DRAMATICS	1971	8	10
579. WILD WORLD – CAT STEVENS	1971	8	10
580. DIALOGUE – CHICAGO	1971	8	10
581. REACH OUT…I'LL BE THERE – DIANA ROSS	1971	8	10
582. RESURRECTION SHUFFLE – ASHTON, GARDNER & DYKE	1971	8	10
583. DO I LOVE YOU – PAUL ANKA	1971	8	10
584. ISN'T LIFE STRANGE – MOODY BLUES	1972	8	9
585. POWER TO THE PEOPLE – JOHN LENNON	1971	8	8
586. CORNER OF THE SKY – JACKSON FIVE	1972	8	8
587. PUPPET MAN – TOM JONES	1971	8	8
588. AMERICANS – BYRON MCGREGOR	1974	8	6
589. OVER MY HEAD – FLEETWOOD MAC	1976	8	6
590. APEMAN – KINKS	1971	8	6
591. EMMA – HOT CHOCOLATE	1975	9	15
592. I'M LEAVING IT UP TO YOU – DONNY & MARIE OSMOND	1974	9	14
593. I DON'T LIKE TO SLEEP ALONE – PAUL ANKA	1975	9	14
594. YOU SEXY THING – HOT CHOCOLATE	1975	9	13
595. HEAT WAVE – LINDA RONSTADT	1975	9	13
596. LOVE ME FOR A REASON – OSMONDS	1974	9	13
597. OLD DAYS – CHICAGO	1975	9	12
598. I'M STONE IN LOVE WITH YOU – STYLISTICS	1973	9	12
599. YES WE CAN CAN – POINTER SISTERS	1973	9	12
600. ONE OF A KIND (LOVE AFFAIR) – SPINNERS	1973	9	12
601. WE MAY NEVER PASS THIS WAY (AGAIN) – SEALS & CROFTS	1973	9	12
602. EXPRESS – B.T. EXPRESS	1975	9	11
603. ARE YOU MAN ENOUGH – FOUR TOPS	1973	9	11
604. KISS AN ANGEL GOOD MORNIN' – CHARLEY PRIDE	1972	9	11
605. STAGGER LEE – TOMMY ROE	1971	9	11
606. SON OF MY FATHER – CHICORY	1972	9	11
607. LOOK WHAT YOU'VE DONE FOR ME – AL GREEN	1972	9	10
608. BACK HOME AGAIN – JOHN DENVER	1974	9	10
609. SWEET HITCH-HIKER – CREEDENCE CLEARWATER REVIVAL	1971	9	10

610. LOVES, LINES, ANGLES & RHYMES – 5TH DIMENSION	1971	9	10
611. MISTER CAN'T YOU SEE – BUFFY SAINTE-MARIE	1972	9	10
612. I HEAR THOSE CHURCH BELLS RINGING – DUSK	1971	9	10
613. TEENAGE LAMENT '74 – ALICE COOPER	1974	9	10
614. I'M STILL IN LOVE WITH YOU – AL GREEN	1972	9	9
615. WHAT IS LIFE – GEORGE HARRISON	1971	9	9
616. CHEER – POTLIQUOR	1972	9	9
617. STICK-UP – HONEY CONE	1971	9	8
618. BLUE MONEY – VAN MORRISON	1971	9	8
619. BURNING BRIDGES – MIKE CURB CONGREGATION	1971	9	8
620. 1900 YESTERDAY – LIZ DAMON'S ORIENT EXPRESS	1971	9	7
621. SQUEEZE BOX – THE WHO	1976	9	6
622. THIRD RATE ROMANCE – AMAZING RHYTHM ACES	1975	10	14
623. EVERLASTING LOVE – CARL CARLTON	1974	10	13
624. UNEASY RIDER – CHARLIE DANIELS	1973	10	13
625. ROCKFORD FILES – MIKE POST	1975	10	13
626. I'M THE LEADER OF THE GANG – BROWNSVILLE STATION	1974	10	13
627. LITTLE BITTY PRETTY ONE – JACKSON FIVE	1972	10	12
628. WHO DO YOU THINK YOU ARE – BO DONALDSON/HEYWOODS	1974	10	12
629. ROLL ON – NEW COLONY SIX	1971	10	12
630. LIVING FOR THE CITY – STEVIE WONDER	1974	10	11
631. I BELIEVE IN YOU – JOHNNY TAYLOR	1973	10	11
632. GUITAR MAN – BREAD	1972	10	11
633. SWEET SURRENDER – BREAD	1973	10	11
634. IT'S ONLY ROCK 'N ROLL – ROLLING STONES	1974	10	11
635. COOL AID – PAUL HUMPHREY	1971	10	11
636. PART OF THE PLAN – DAN FOGELBERG	1975	10	11
637. ANOTHER PARK, ANOTHER SUNDAY – DOOBIE BROTHERS	1974	10	11
638. YOU ARE THE SUNSHINE OF MY LIFE – STEVIE WONDER	1973	10	10
639. THAT LADY – ISLEY BROTHERS	1973	10	10
640. LET'S PUT IT ALL TOGETHER – STYLISTICS	1974	10	10
641. TREAT HER LIKE A LADY – CORNELIUS BROTHERS	1971	10	9
642. YOU GOT THE LOVE – RUFUS	1975	10	9
643. AMAZING GRACE – ROYAL SCOTS DRAGOON GUARDS	1972	10	9
644. THINKING OF YOU – LOGGINS & MESSINA	1973	10	9
645. HIGH TIME WE WENT – JOE COCKER	1971	10	9
646. OYE COMO VA – SANTANA	1971	10	8
647. RUNWAY – GRASS ROOTS	1972	10	8
648. IF I WERE YOUR WOMAN – GLADYS KNIGHT/PIPS	1971	10	7
649. DO YOU KNOW WHAT TIME IT IS? – P-NUT GALLERY	1971	10	6
650. THEME FROM "S.W.A.T." – RHYTHM HERITAGE	1976	10	5
651. CLAP FOR THE WOLFMAN – GUESS WHO	1974	11	12
652. DON'T YOU WORRY 'BOUT A THING – STEVIE WONDER	1974	11	12
653. VIRGINIA – BILL AMESBURY	1974	11	12

654. SWEARIN' TO GOD – FRANKIE VALLI	1975	11	11
655. LIAR – THREE DOG NIGHT	1971	11	11
656. IF YOU'RE READY (COME GO WITH ME) – STAPLE SINGERS	1973	11	11
657. SURE AS I'M SITTING HERE – THREE DOG NIGHT	1974	11	11
658. HOW LONG – POINTER SISTERS	1975	11	11
659. YOU'RE STILL A YOUNG MAN – TOWER OF POWER	1972	11	11
660. NEVER MY LOVE – BLUE SWEDE	1974	11	10
661. TIGHT ROPE – LEON RUSSELL	1972	11	10
662. SEXY MAMA – MOMENTS	1974	11	10
663. SMOKE GETS IN YOUR EYES – BLUE HAZE	1973	11	10
664. ONE MAN BAND – RONNIE DYSON	1973	11	10
665. WORKING AT THE CAR WASH BLUES – JIM CROCE	1974	11	10
666. SUPERFLY – CURTIS MAYFIELD	1973	11	9
667. SIMPLE SONG OF FREEDOM – BUCKWHEAT	1972	11	9
668. WATCHING SCOTTY GROW – BOBBY GOLDSBORO	1971	11	8
669. RIVER DEEP, MOUNTAIN HIGH – SUPREMES & FOUR TOPS	1971	11	7
670. COUNTRY ROAD – JAMES TAYLOR	1971	11	7
671. STANDING AT THE END OF THE LINE – LOBO	1974	12	13
672. ROCK 'N ROLL HEAVEN – RIGHTEOUS BROTHERS	1974	12	12
673. ROCK 'N ROLL BABY – STYLISTICS	1974	12	12
674. AUTOBAHN – KRAFTWERK	1975	12	12
675. I'LL TAKE YOU THERE – STAPLE SINGERS	1972	12	11
676. THE NEED TO BE – JIM WEATHERLY	1974	12	11
677. SOMETHING'S WRONG WITH ME – AUSTIN ROBERTS	1973	12	11
678. BEHIND CLOSED DOORS – CHARLIE RICH	1973	12	11
679. KEEP ON SINGING – HELEN REDDY	1974	12	11
680. WOODSTOCK – MATTHEWS SOUTHERN COMFORT	1971	12	11
681. I'M ON FIRE – 5,000 VOLTS	1975	12	11
682. YOU HAVEN'T DONE NOTHIN' – STEVIE WONDER	1974	12	10
683. DAY BY DAY – GODSPELL	1972	12	10
684. UP IN A PUFF IN SMOKE – POLLY BROWN	1975	12	10
685. MOTHER FREEDOM – BREAD	1971	12	10
686. LOVE – LETTERMEN	1971	12	10
687. SHA-LA-LA – AL GREEN	1974	12	9
688. RIGHT ON THE TIP OF MY TONGUE – BRENDA/TABULATIONS	1971	12	9
689. TAURUS – DENNIS COFFEY	1972	12	8
690. MACHINE GUN – COMMODORES	1974	12	8
691. NAMES, TAGS, NUMBERS & LABELS – ASSOCIATION	1973	12	8
692. ENERGY CRISIS '74 – DICKIE GOODMAN	1974	12	7
693. SWEET CHARLIE BABE – JACKIE MOORE	1973	13	13
694. LOVE WON'T LET ME WAIT – MAJOR HARRIS	1975	13	12
695. OH MY MY – RINGO STARR	1974	13	12
696. WILD ABOUT MY LOVIN' – ADRIAN SMITH	1973	13	12
697. HANG ON IN THERE BABY – JOHNNY BRISTOL	1974	13	11

698. GYPSY MAN – WAR	1973	13	11
699. ROCK STEADY – ARETHA FRANKLIN	1971	13	11
700. BRAZIL – RITCHIE FAMILY	1975	13	11
701. I'M DOIN' FINE NOW – NEW YORK CITY	1973	13	11
702. HOW CAN I TELL HER – LOBO	1973	13	11
703. SUMMER SAND – DAWN	1971	13	11
704. WALK AWAY FROM LOVE – DAVID RUFFIN	1976	13	10
705. #9 DREAM – JOHN LENNON	1975	13	10
706. I'VE GOT THE MUSIC IN ME – KIKI DEE BAND	1974	13	10
707. FAIRYTALE – POINTER SISTERS	1975	13	10
708. LAST TIME I SAW HIM – DIANA ROSS	1974	13	10
709. YOU TURN ME ON, I'M A RADIO – JONI MITCHELL	1973	13	10
710. BACK WHEN MY HAIR WAS SHORT – GUNHILL ROAD	1973	13	10
711. LONG TIME TO BE ALONE – NEW COLONY SIX	1972	13	10
712. NEITHER ONE OF US – GLADYS KNIGHT/PIPS	1973	13	9
713. VENUS & MARS ROCK SHOW – PAUL MCCARTNEY & WINGS	1975	13	9
714. MOVIN' ON – BAD COMPANY	1975	13	9
715. TRAIN OF THOUGHT – CHER	1974	13	9
716. COUNTRY BOY – GLEN CAMPBELL	1976	13	8
717. EASY LOVIN' – FREDDIE HART	1971	13	8
718. EVERYDAY OF MY LIFE – BOBBY VINTON	1972	13	7
719. COUNTRY WINE – RAIDERS	1972	13	7
720. LOVE MACHINE – MIRACLES	1976	13	5
721. I AM WHAT I AM – LOIS FLETCHER	1974	14	12
722. FUNNY FACE – DONNA FARGO	1973	14	11
723. DADDY'S HOME – JERMAINE JACKSON	1973	14	11
724. KEEP ON SMILIN' – WET WILLIE	1974	14	11
725. HELEN WHEELS – PAUL MCCARTNEY & WINGS	1973	14	11
726. LOVING YOU JUST CROSSED MY MIND – SAM NEELY	1972	14	11
727. LET ME IN – OSMONDS	1973	14	11
728. UNTIL YOU COME BACK TO ME – ARETHA FRANKLIN	1974	14	10
729. HARRY TRUMAN – CHICAGO	1975	14	10
730. 'TIL THE WORLD ENDS – THREE DOG NIGHT	1975	14	10
731. STANDING OVATION – DELLS	1973	14	10
732. YOU OUGHT TO BE WITH ME – AL GREEN	1972	14	9
733. WHAT AM I GONNA DO WITH YOU – BARRY WHITE	1975	14	9
734. THE FAMILY OF MAN – THREE DOG NIGHT	1972	14	9
735. I PLAY & SING – DAWN	1971	14	9
736. RENDEZVOUS – HUDSON BROTHERS	1975	14	9
737. SURE TOOK A LONG, LONG TIME – LOBO	1973	14	9
738. ALREADY GONE – EAGLES	1974	14	9
739. 13 QUESTIONS – SEATRAIN	1971	14	9
740. YOU GOT ME ANYWAY – SUTHERLAND BROTHERS & QUIVER	1973	14	9
741. STEPPIN' OUT (GONNA BOOGIE TONIGHT) – DAWN	1974	14	8

742. CRAZY HORSES – OSMONDS	1972	14	8
743. FUNKY WORM – OHIO PLAYERS	1973	14	8
744. FROM THE BEGINNING – EMERSON, LAKE & PALMER	1972	14	8
745. SING A SONG – EARTH, WIND & FIRE	1976	14	7
746. SUPERSTAR – TEMPTATIONS	1971	14	7
747. GET DANCIN' – DISCO TEX & THE SEX-O-LETTES	1975	15	11
748. NO LOVE AT ALL – BJ THOMAS	1971	15	11
749. CAN'T FIND THE TIME – ROSE COLORED GLASS	1971	15	11
750. HOLLYWOOD SWINGING – KOOL & THE GANG	1974	15	10
751. WISHING YOU WERE HERE – CHICAGO	1974	15	10
752. SKIN TIGHT – OHIO PLAYERS	1974	15	10
753. I WANNA DO SOMETHING FREAKY TO YOU – LEON HAYWOOD	1975	15	10
754. THE LAST FAREWELL – ROGER WHITTAKER	1975	15	10
755. SIDESHOW – BLUE MAGIC	1974	15	9
756. MORNIN' BEAUTIFUL – TONY ORLANDO & DAWN	1975	15	9
757. MIDNIGHT RIDER – GREGG ALLMAN	1974	15	9
758. WALK IN THE NIGHT – JUNIOR WALKER & THE ALL STARS	1972	15	9
759. POWER OF LOVE – JOE SIMON	1972	15	8
760. FREE – CHICAGO	1971	15	8
761. BANGLA-DESH – GEORGE HARRISON	1971	15	8
762. POP THAT THANG – ISLEY BROTHERS	1972	15	8
763. HOT LOVE – T. REX	1971	15	8
764. TAKE ME GIRL, I'M READY – JUNIOR WALKER/ALL STARS	1971	15	8
765. BRIDGE OVER TROUBLED WATER – ARETHA FRANKLIN	1971	15	7
766. HOW CAN I BE SURE – DAVID CASSIDY	1972	15	7
767. IT'S UP TO YOU, PETULA – EDISON LIGHTHOUSE	1971	15	5
767A. WHEN I'M DEAD AND GONE – BOB SUMMERS	1971	15	5
768. DREAM WEAVER – GARY WRIGHT	1976	15	4
769. YOU MAKE ME FEEL BRAND NEW – STYLISTICS	1974	16	11
770. MUST OF GOT LOST – J. GEILS BAND	1974	16	11
771. DOO DOO DOO DOO DOO (HEARTBREAKER) – ROLLING STONES	1974	16	11
772. POETRY MAN – PHOEBE SNOW	1975	16	10
773. CALL ME (COME BACK HOME) – AL GREEN	1973	16	10
774. KNOCKIN' ON HEAVEN'S DOOR – BOB DYLAN	1973	16	10
775. DISCO QUEEN – HOT CHOCOLATE	1975	16	10
776. THEME FROM "JAWS" – ORIGINAL SOUNDTRACK	1975	16	10
777. BE – NEIL DIAMOND	1973	16	10
778. I AM…I SAID – NEIL DIAMOND	1971	16	9
779. HAPPIEST GIRL IN THE WHOLE U.S.A. – DONNA FARGO	1972	16	9
780. YOU WEAR IT WELL – ROD STEWART	1972	16	9
781. WALK LIKE A MAN – GRAND FUNK	1974	16	9
782. I FEEL A SONG (IN MY HEART) – GLADYS KNIGHT/PIPS	1974	16	9
783. OPERATOR – MANHATTAN TRANSFER	1975	16	9
784. MIGHTY CLOUDS OF JOY – B.J. THOMAS	1971	16	9

785. WELCOME TO MY NIGHTMARE – ALICE COOPER	1975	16	9
786. WALKIN' IN THE RAIN – LOVE UNLIMITED	1972	16	8
787. RUN TO ME – BEE GEES	1972	16	8
788. SO YOU ARE A STAR – HUDSON BROTHERS	1974	16	8
789. A MILLION TO ONE – DONNY OSMOND	1973	16	8
790. LOOKIN' FOR A LOVE – J. GEILS BAND	1972	16	8
791. IF IT'S ALLRIGHT WITH YOU – ROSE COLORED GLASS	1971	16	8
792. HE'D RATHER HAVE THE RAIN – HEAVEN BOUND	1971	16	8
793. PROMISED LAND – ELVIS PRESLEY	1975	16	6
794. OH VERY YOUNG – CAT STEVENS	1974	17	11
795. OH BABE, WHAT WOULD YOU SAY – HURRICANE SMITH	1973	17	10
796. THIS HEART – GENE REDDING	1974	17	10
797. ONCE YOU GET STARTED – RUFUS	1975	17	9
798. ASK ME WHAT YOU WANT – MILLIE JACKSON	1972	17	9
799. FIRE, BABY I'M ON FIRE – ANDY KIM	1974	17	9
800. AIN'T GOT TIME ANYMORE – GLASS BOTTLE	1971	17	9
801. JESAHEL – ENGLISH CONGREGATION	1972	17	9
802. WHAT A DIFFERENCE A DAY MAKES – ESTHER PHILLIPS	1975	17	8
803. EVE – JIM CAPALDI	1972	17	8
804. DON'T WANNA LIVE INSIDE MYSELF – BEE GEES	1971	17	8
805. NO – BULLDOG	1973	17	8
806. LIFE IS A CARNIVAL – BAND	1971	17	8
807. WALK AWAY – JAMES GANG	1971	17	7
808. OVER THE HILLS & FAR AWAY – LED ZEPELLIN	1973	17	7
809. TAKE IT TO THE LIMIT – EAGLES	1976	17	6
810. NATURAL HIGH – BLOODSTONE	1973	18	10
811. THAT'S THE WAY OF THE WORLD – EARTH, WIND & FIRE	1975	18	9
812. JOIN TOGETHER – THE WHO	1972	18	9
813. THE SOUTH'S GONNA DO IT – CHARLIE DANIEL'S	1975	18	9
814. BREAK UP TO MAKE UP – STYLISTICS	1973	18	8
815. TROUBLE MAN – MARVIN GAYE	1973	18	8
816. HONKY CAT – ELTON JOHN	1972	18	8
817. ROCK 'N ROLL SOUL – GRAND FUNK RAILROAD	1972	18	8
818. CAN THIS BE REAL – NATURAL FOUR	1974	18	8
819. KINGS OF THE PARTY – BROWNSVILLE STATION	1974	18	8
820. LAYLA – DEREK & THE DOMINOES	1971	18	6
821. I'D LIKE TO TEACH THE WORLD TO SING – HILLSIDE SINGERS	1972	18	6
822. I DIDN'T KNOW I LOVED YOU – GARY GLITTER	1972	19	10
823. BERTHA BUTT BOOGIE – JIMMY CASTOR BUNCH	1975	19	9
824. SAVE THE LAST DANCE FOR ME – DEFRANCO FAMILY	1974	19	9
825. REMEMBER WHAT I TOLD YOU TO FORGET – TAVARES	1975	19	9
826. LIVING TOGETHER, GROWING TOGETHER – 5TH DIMENSION	1973	19	9
827. MISTY – RAY STEVENS	1975	19	8
828. STARTING ALL OVER AGAIN – MEL & TIM	1972	19	8

829. SHADY LADY – SHEPSTONE & DIBBINS	1973	19	8
830. BANG-A-GONG (GET IT ON) – T.REX	1972	19	7
831. ROBERTA – BONES	1972	19	7
832. WHEN A CHILD IS BORN – MICHAEL HOLM	1975	19	7
833. YOU CAN'T ALWAYS GET WHAT YOU WANT – ROLLING STONES	1973	19	7
834. TIMES OF YOUR LIFE – PAUL ANKA	1976	19	5
835. THEME FROM LOVE STORY – HENRI MANCINI	1971	19	5
836. STOP THE WAR NOW – EDWIN STARR	1971	19	5
837. LONELY NIGHTS (ANGEL FACE) – CAPTAIN & TENILLE	1976	19	3
838. AMIE – PURE PRAIRIE LEAGUE	1975	20	11
839. MASTERPIECE – TEMPTATIONS	1973	20	9
840. WASTED DAYS & WASTED NIGHTS – FREDDY FENDER	1975	20	9
841. CRIED LIKE A BABY – BOBBY SHERMAN	1971	20	9
842. IN THE MIDNIGHT HOUR – CROSS COUNTRY	1973	20	9
843. CAPTAIN HOWDY – SIMON STOKES	1974	20	9
844. GOIN' DOWN (ON THE ROAD TO L.A.) – BLACK/WARD	1972	20	9
845. SWEET SURRENDER – JOHN DENVER	1975	20	8
846. I LOVE YOU FOR ALL SEASONS – THE FUZZ	1971	20	8
847. WHO'S IN THE STRAWBERRY PATCH WITH SALLY – DAWN	1973	20	8
848. I LIKE TO LIVE THE LOVE – B.B. KING	1974	20	8
849. TELL HER SHE'S LOVELY – EL CHICANO	1973	20	8
850. I MISS YOU – DELLS	1974	20	8
851. THE LAST GAME OF THE SEASON – DAVID GEDDES	1975	20	7
852. STRAIGHT SHOOTIN' WOMAN – STEPPENWOLF	1974	20	7
853. PLAY SOMETHING SWEET – THREE DOG NIGHT	1974	20	7
854. IT COULD HAVE BEEN ME – SAMI JO	1974	20	7
855. SUMMER SUN – JAMESTOWN MASSACRE	1972	20	7
856. I CAN FEEL YOU – ADDRISSI BROTHERS	1972	20	7
857. FOR THE GOOD TIMES – RAY PRICE	1971	20	6
858. FLY AWAY – JOHN DENVER	1976	20	5
858A. AMAZING GRACE - JUDY COLLINS	1971	20	5
859. DREAM ON – AEROSMITH	1976	20	3
860. WE'RE ON OUR WAY – CHRIS HODGE	1972	21	10
861. SON OF SAGITTARIUS – EDDIE KENDRICKS	1974	21	9
862. LET IT RIDE – BACHMAN-TURNER OVERDRIVE	1974	21	8
863. HALLELUJAH – SWEATHOG	1971	21	7
864. ROCK ME ON THE WATER – LINDA RONSTADT	1972	21	7
865. DONE TOO SOON – NEIL DIAMOND	1971	21	7
866. ROSE GARDEN – LYNN ANDERSON	1971	21	6
867. DO IT BABY – MIRACLES	1974	21	6
868. PUT IT WHERE YOU WANT IT – CRUSADERS	1972	21	6
869. ONLY SIXTEEN – DR. HOOK	1976	21	4
870. LOOKIN' FOR A LOVE – BOBBY WOMACK	1974	22	9
871. THERE'S NOTHING STRONGER THAN OUR LOVE – PAUL ANKA	1975	22	9

872. SOMEONE, SOMETIME – NEW COLONY SIX	1972	22	9
873. ROCKY MOUNTAIN WAY – JOE WALSH	1973	22	8
874. ELECTED – ALICE COOPER	1972	22	8
875. IT'S ALLRIGHT – JIM CAPALDI	1975	22	8
876. A PIECE OF PAPER – GLADSTONE	1972	22	8
877. PEACEFUL EASY FEELING – EAGLES	1973	22	7
878. BLOCKBUSTER – SWEET	1973	22	7
879. AIN'T TOO PROUD TO BET – ROLLING STONES	1974	22	6
880. WILD HORSES – ROLLING STONES	1971	22	6
881. I'M COMIN' HOME – STORIES	1972	22	5
882. MOTHER – JOHN LENNON	1971	22	5
883. DON'T TELL ME GOODNIGHT – LOBO	1975	23	10
884. I'M ON FIRE – DWIGHT TWILLEY BAND	1975	23	9
885. TOAST & MARMALADE FOR TEA – TIN TIN	1971	23	8
886. BRING THE BOYS HOME – FREDA PAYNE	1971	23	7
887. GIVE IT TO THE PEOPLE – RIGHTEOUS BROTHERS	1974	23	7
888. SMALL BEGINNINGS – FLASH	1972	23	7
889. FOOL – ELVIS PRESLEY	1973	23	7
890. LOVE ME (LOVE THE LIFE I LEAD) – FANTASTICS	1972	23	7
891. NEVER MY LOVE – 5TH DIMENSION	1971	23	5
892. ONE FINE MORNING – LIGHTHOUSE	1971	23	5
893. SO VERY HARD TO GO – TOWER OF POWER	1973	24	9
894. ANGEL – ARETHA FRANKLIN	1973	24	8
895. I'M A TRAIN – ALBERT HAMMOND	1974	24	8
896. TRYING TO HOLD ON TO MY WOMAN – LAMONT DOZIER	1974	24	7
897. MCARTHUR PARK – FOUR TOPS	1974	24	7
898. ROCK 'N ROLL – KEVIN JOHNSON	1973	24	7
899. IN THE MOOD – BETTE MIDLER	1974	24	7
900. DON'T LET THE GREEN GRASS FOOL YOU – WILSON PICKETT	1971	24	6
901. STAND BY ME – JOHN LENNON	1975	24	6
902. ROCK 'N ROLL HOOCHIE KOO – RICK DERRINGER	1974	24	5
903. I SAW THE LIGHT – TODD RUNDGREN	1972	25	9
904. FINALLY GOT MYSELF TOGETHER – IMPRESSIONS	1974	25	9
905. TO THE DOOR OF THE SUN – AL MARTINO	1975	25	8
906. DON'T EVER BE LONELY – CORNELIUS BROTHERS	1972	25	8
907. TELL 'EM WILLIE BOYS'A COMIN' – TOMMY JAMES	1972	25	7
908. IKO IKO – DR. JOHN	1972	25	7
909. SIMONE – HENRY GROSS	1974	25	7
910. MAKE YOU, BREAK YOU – GERRY RAFFERTY	1972	25	7
911. HERE I AM – AL GREEN	1973	25	6
912. ME & BOBBY MCGEE – JERRY LEE LEWIS	1971	25	6
913. TOUCH A HAND, MAKE A FRIEND – STAPLE SINGERS	1974	25	6
914. TALK IT OVER IN THE MORNING – ANNE MURRAY	1971	25	6
915. I'VE BEEN LONELY FOR SO LONG – FREDRICK KNIGHT	1972	25	5

916. AMERICA – YES	1972	25	5
917. LOVE MAKES THE WORLD GO ROUND – ODDS & ENDS	1971	25	5
918. JODY'S GOT YOUR GIRL & GONE – JOHNNY TAYLOR	1971	25	4
918A. DO IT - NEIL DIAMOND	1971	20	5
919. LOVE IS THE DRUG – ROXY MUSIC	1976	25	3
920. I LOVE – TOM T. HALL	1974	26	9
921. WHO'S SORRY NOW – MARIE OSMOND	1975	26	7
922. THERE'S NO ME WITHOUT YOU – MANHATTANS	1973	26	7
923. IN A BROKEN DREAM – PYTHON LEE JACKSON	1972	26	7
924. GROOVE ME – KING FLOYD	1971	26	6
925. SECRET LOVE – FREDDY FENDER	1975	26	6
926. DON'T CROSS THE RIVER – AMERICA	1973	26	6
927. I WANNA DANCE WIT'CHOO – DISCO TEX & THE SEX-O-LETTES	1975	26	5
928. THEME FROM CLEOPATRA JONES – JOE SIMON	1973	26	4
929. ROCK ME ON THE WATER – JACKSON BROWNE	1972	26	4
930. AFTER MIDNIGHT – J.J. CALE	1972	26	4
931. HIJACK – HERBIE MANN	1975	27	8
932. DADDY COULD SWEAR – GLADYS KNIGHT/PIPS	1973	27	7
933. L-O-V-E (LOVE) – AL GREEN	1975	27	6
934. FOOL ME – JOE SOUTH	1971	27	6
935. I LOVE YOU LADY DAWN – BELLS	1971	27	6
936. BLUE COLLAR – BACHMAN-TURNER OVERDRIVE	1973	27	6
937. YOU'RE ALL I NEED – ARETHA FRANKLIN	1971	27	5
938. LITTLE BIT O'SOUL – BULLET	1972	27	5
939. ANGEL BABY – DUSK	1971	27	5
940. DROWNING IN THE SEA OF LOVE – JOE SIMON	1972	27	4
941. HEY BIG BROTHER – RARE EARTH	1971	27	4
942. RENEGADE – MICHAEL MURPHY	1976	27	4
943. BLACK FRIDAY – STEELY DAN	1975	28	8
944. YOU'RE A SPECIAL PART OF ME – DIANA ROSS/MARVIN GAYE	1973	28	6
945. BORN TO RUN – BRUCE SPRINGSTEEN	1975	28	6
946. JESSE – ROBERA FLACK	1973	28	6
947. MERCEDES BENZ – GOOSE CREEK SYMPHONY	1972	28	6
948. BLESS YOU – MARTHA REEVES/VANDELLAS	1971	28	6
949. TEDDY BEAR SONG – BARBARA FAIRCHILD	1973	28	5
950. I GOT STONED AND I MISSED IT – JIM STAFFORD	1975	28	5
951. DRINKING WINE SPO-DEE-O-DEE – JERRY LEE LEWIS	1973	28	5
952. MR. BOJANGLES – NITTY GRITTY DIRT BAND	1971	28	4
953. WHEN THERE'S NO YOU – ENGLEBERT HUMPERDINCK	1971	28	4
954. MONEY HONEY – BAY CITY ROLLERS	1976	28	2
955. RIVER ROAD – UNCLE DOG	1973	29	6
956. SUPERNATURAL THING – BEN E. KING	1975	29	5
957, SHOESHINE BOY – EDDIE KENDRICKS	1975	29	5
958. CRAZY MAMA – J.J. CALE	1972	29	5

959. YOUR BULLDOG DRINKS CHAMPAGNE – JIM STAFFORD	1975	29	5
960. PALOMA BLANCA – GEORGE BAKER SELECTION	1976	29	5
961. ROCK 'N ROLL – MITCH RYDER & DETROIT	1972	29	4
962. HI, HI, HI – PAUL MCCARTNEY & WINGS	1973	30	5
963. SPEAK TO THE SKY – RICK SPRINGFIELD	1972	30	5
964. CAN THE CAN – SUZY QUATRO	1976	30	4
965. OVER & OVER – DELPHONICS	1971	30	4
966. ANYTIME SUNSHINE – CRAZY PAVING	1971	30	4
967. CHERISH WHAT IS DEAR TO YOU – FREDA PAYNE	1971	30	4
968. MOST OF ALL – B.J. THOMAS	1971	30	3
969. LOVE HURTS – NAZARETH	1976	30	1
970. IT'S ALL UP TO YOU – JIM CAPALDI	1975	31	5
971. DREAM BABY – GLEN CAMPBELL	1971	31	4
972. TEMPTATION EYES – GRASS ROOTS	1971	31	3
973. SILVER MOON – MICHAEL NESMITH/1ST NATIONAL BAND	1971	31	3
974. TRY SOME, BUY SOME – RONNIE SPECTOR	1971	31	3
975. BOHEMIAN RHAPSODY – QUEEN	1976	31	2
976. WE GOTTA GET YOU A WOMAN – RUNT	1971	32	3
977. YESTERDAY'S HERO – JOHN PAUL YOUNG	1976	32	3
978. THE GLORY OF LOVE – DELLS	1971	32	2
979. HAPPY XMAS – JOHN & YOKO/PLASTIC ONO BAND	1971	32	2
980. I'LL STILL LOVE YOU – JIM WEATHERLY	1975	32	2
981. DO ME RIGHT – DETROIT EMERALDS	1971	32	2
982. JUNK FOOD JUNKIE – LARRY GROCE	1976	32	1
983. ACTION – SWEET	1976	33	2
984. SUGAR MOUNTAIN – NEIL YOUNG	1971	34	3
985. DEEP PURPLE – DONNY & MARIE OSMOND	1976	34	1
986. ADRIENNE – TOMMY JAMES	1971	35	3
987. BRIDGET THE MIDGET – RAY STEVENS	1971	35	3
988. SLOW RIDE – FOGHAT	1976	35	1
989. PRECIOUS, PRECIOUS – JACKIE MOORE	1971	35	1
990. I REALLY DON'T WANT TO KNOW – ELVIS PRESLEY	1971	36	3
991. SWEET THING – RUFUS	1976	36	1
992. ONE TOKE OVER THE LINE – BREWER & SHIPLEY	1971	36	1
993. RIGHT BACK WHERE WE STARTED FROM – MAX NIGHTINGALE	1976	37	1
994. I DO, I DO, I DO, I DO, I DO – ABBA	1976	38	1
995. BOOGIE FEVER – SYLVERS	1976	-	-
996. YOUNG BLOOD – BAD COMPANY	1976	-	-
997. ONLY LOVE IS REAL – CAROLE KING	1976	-	-
998. SHOUT IT OUT LOUD – KISS	1976	-	-
999. HIGH FLY – JOHN MILES	1976	-	-
1000. LOCOMOTIVE BREATH – JETHRO TULL	1976	-	-

NON-WCFL SONGS

WCFL had a lot of great songs. However, there are many songs that did not appear on the WCFL survey but were played on other stations such as WLS and the FM stations. The following is a list of songs that did not appear on the WCFL survey from 1970-75 and were played on other stations:

1. A Natural Man – Lou Rawls (71)
2. All Day Music – War (71)
3. All The Young Dudes – Mott The Hoople (72)
4. Angry Eyes – Loggins & Messina (73)
5. Animal Trainer & The Toad – Mountain (71)
6. Aqualung – Jethro Tull (71)
7. Are You My Woman – Chi-Lites (71)
8. Attitude Dancing – Carly Simon (75)
9. Aubrey – Bread (73)
10. Bad Company – Bad Company (74)
11. Bad Luck – Harold Melvin & The Blue-Notes (75)
12. Be Thankful For What You Got – William DeVaughn (74)
13. Been To Canaan – Carole King (73)
14. Behind Blue Eyes – Who (71)
15. Believe In Humanity – Carole King (73)
16. Bell Bottom Blues – Eric Clapton (71)
17. Bells – Originals (70)
18. Betcha By Golly Wow – Stylistics (72)
19. Big Yellow Taxi – Neighborhood (70)
20. Billion Dollar Babies – Alice Cooper (73)
21. Bitch – Rolling Stones (71)
22. Bloody Well Right – Supertramp (75)
23. Bluebird – Paul McCartney (73)

24. Bus Rider – Guess Who (70)
25. Call On Me – Chicago (74)
26. Calypso – John Denver (75)
27. Carry On – Crosby, Stills & Nash (70)
28. Changes – David Bowie (72)
29. Chestnut Mare – Byrds (70)
30. Cinnamon Girl – Neil Young (70)
31. City Of New Orleans – Arlo Guthrie (72)
32. Clean-Up Woman – Betty Wright (72)
33. Coconut – Harry Nilsson (72)
34. Cold Ethyl – Alice Cooper (75)
35. Colour My World – Chicago (71)
36. Come Monday – Jimmy Buffett (73)
37. Could It Be Magic – Barry Manilow (75)
38. Cross-Eyed Mary – Jethro Tull (71)
39. Crunchy Granola Suite – Neil Diamond (71)
40. Cut The Cake – AWB (75)
41. Daisy A Day – Jud Strunk (73)
42. Dancin' Fool – Guess Who (75)
43. Danny's Song – Loggins & Messina (72)
44. Dark Horse – George Harrison (75)
45. Department Of Youth – Alice Cooper (75)
46. Double Barrel – Doug & Ansil Collins (71)
47. Dreamer – Supertramp (74)
48. Easy Livin' – Uriah Heep (75)
49. Every Picture Tells A Story – Rod Stewart (71)
50. Everybody's Everything – Santana (71)
51. Express Yourself – Watts 103rd St. Band (70)
52. Feelin' Allright – Joe Cocker (72)

53. Fire On The Mountain – Marshall Tucker Band (75)
54. Footstompin' Music – Grand Funk Railroad (71)
55. Francene – ZZ Top (74)
56. Free Electric Band – Albert Hammond (73)
57. Free Man In Paris – Joni Mitchell (74)
58. Funk #48 & 49 – James Gang (70)
59. Funky Stuff – Kool & The Gang (73)
60. Get It On – Chase (71)
61. Geronimo's Cadillac – Michael Murphy (72)
62. Gimme Your Money Please – B.T.O. (75)
63. Give More Power To The People – Chi-Lites (71)
64. Good Lovin' Gone Bad – Bad Company (75)
65. Good Foot – James Brown (72)
66. Grey Seal – Elton John (73)
67. Happy Music – Blackbyrds (75)
68. Happy People – Temptations (75)
69. Haven't Got Time For The Pain – Carly Simon (74)
70. He's Armed & Extremely Dangerous – First Choice (73)
71. Help Me Make It Through The Night – Sammi Smith (71)
72. Hey Tonight – C.C.R. (71)
73. Holdin' On To Yesterday – Ambrosia (75)
74. Hot Pants – James Brown (71)
75. House At Pooh Corner – Loggins & Messina (72)
76. House At Pooh Corner – Nitty Gritty Dirt Band (71)
77. How Sweet It Is – James Taylor (75)
78. Hurts So Good – Millie Jackson (72)
79. I Almost Cut My Hair – Crosby, Stills, Nash & Young (70)
80. I Can't Stand The Rain – Ann Pebbles (73)
81. I Don't Know How To Love Him – Helen Reddy (71)

82. I Don't Need No Doctor – Humble Pie (72)
83. I Feel The Earth Move – Carole King (71)
84. I Gotcha' – Joe Tex (72)
85. I Know I'm Losing You – Rod Stewart (71)
86. I'd Like To Teach The World To Sing – New Seekers (71)
87. I'd Love To Change The World – Ten Years After (71)
88. If There's A Hell Below – Curtis Mayfield (71)
89. If You Wanna Get To Heaven – Ozark Mountain Daredevils (74)
90. If You Want Me To Stay – Sly & The Family Stone (73)
91. I'll Play For You – Seals & Crofts (75)
92. I'm Bad, I'm Nationwide – ZZ Top (74)
93. I'm Comin' Home – Spinners (74)
94. Iron Man – Black Sabbath (70)
95. Jessica – Allman Brothers (72)
96. Jesus Is Just Allright – Doobie Brothers (72)
97. Judy Mae – Boomer Castleman (75)
98. Jump Into The Fire – Nilsson (72)
99. Kentucky Woman – Deep Purple (73)
100. L.A. Woman – Doors (71)
101. Lake Shore Drive – Alliotta, Haynes & Jeremiah (75)
102. Landslide – Fleetwood Mac (75)
103. Let It Grow – Eric Clapton (73)
104. Let It Rain – Eric Clapton (72)
105. Let Your Love Go – Bread (71)
106. Levon – Elton John (70)
107. Livin' For The Weekend – O'Jays (74)
108. Long Distance Runaround – Yes (72)
109. Long, Long Time – Linda Ronstadt (70)
110. Love Don't Love Nobody – Spinners (74)

111. Love I Lost – Harold Melvin & The Blue Notes (74)

112. Love Is A Rose – Linda Ronstadt (75)

113. Love The One You're With – Isley Brothers (71)

114. Love's Made A Fool Of You – Cochise (71)

115. Love We Had – Dells (71)

116. Low Rider – War (75)

117. Lucky Man – Emerson, Lake & Palmer (70)

118. Makusa – Mandingo (73)

119. Mama Weer All Crazee Now – Slade (72)

120. Mandolin Wind – Rod Stewart (71)

121. Me & Baby Brother – War (74)

122. Me & Julio Down On The Schoolyard – Paul Simon (72)

123. Medicine Jar – Paul McCartney (75)

124. Melissa – Allman Brothers Band (72)

125. Midnight Rider – Joe Cocker (72)

126. Mighty Love – Spinners (74)

127. Misty Mountain Hop – Led Zeppelin (71)

128. Monday Morning – Fleetwood Mac (75)

129. Money Back Guarantee – Five Man Electrical Band (72)

130. Molena – C.C.R. (71)

131. My Music – Loggins & Messina (73)

132. My Old School – Steely Dan (73)

133. Never Been To Spain – Three Dog Night (72)

134. Nightingale – Carole King (75)

135. 1985 – Paul McCartney (73)

136. Ohio – Crosby, Stills, Nash & Young (70)

137. Old Man – Neil Young (72)

138. One Monkey Don't Stop No Show – Honey Cone (71)

139. One More Chance – Jackson Five (70)

140. Our House – Crosby, Stills & Nash (70)
141. Overnite Sensation – Raspberries (74)
142. Panama Red – New Riders Of The Purple Sage (73)
143. Panic In Detroit – David Bowie (73)
144. Paranoid – Black Sabbath (70)
145. Piano Man – Billy Joel (74)
146. Play On Love – Jefferson Starship (75)
147. Please Come To Boston – Dave Loggins (74)
148. Questions 67 & 68 – Chicago (71)
149. Rain Dance – Guess Who (71)
150. Rainy Day People – Gordon Lightfoot (75)
151. Ready For Love – Bad Company (74)
152. Reason To Believe – Rod Stewart (71)
153. Rebel Rebel – David Bowie (74)
154. Ride The Tiger – Jefferson Starship (75)
155. Rings – Lobo (74)
156. Road Food – Guess Who (74)
157. Roadhouse Blues – Doors (70)
158. Rock 'N Roll – Led Zeppelin (72)
159. Rock 'N Roll – Lou Reed (74)
160. Rock 'N Roll Crazies – Stephen Stills (72)
161. Rock Steady – Bad Company (74)
162. Rocket Man – Elton John (72)
163. Rocky – Austin Roberts (75)
164. Run Thru The Jungle – C.C.R. (70)
165. Sandman – America (72)
166. Saturday Morning Confusion – Bobby Russell (71)
167. Saturday Night Special – Lynyrd Skynyrd (75)
168. Sealed With A Kiss – Bobby Vinton (72)

169. Shaving Cream – Benny Bell (75)
170. Shinin' On – Grand Funk (74)
171. Shooting Star – Bad Company (75)
172. Show Biz Kids – Steely Dan (73)
173. Showdown – E.L.O (74)
174. Shower The People – James Taylor (75)
175. Signs – Five Man Electrical Band (71)
176. Silly Milly – Blue Swede (74)
177. Slippin' Into Darkness – War (72)
178. Smackwater Jack – Carole King (71)
179. So Far Away – Carole King (71)
180. Somebody's Watching You – Little Sister (70)
181. Someday Never Comes – C.C.R. (72)
182. Go Down Gamblin' – Blood, Sweat & Tears (72)
183. Son Of My Father – Giorgio (72)
184. Sooner Or Later – Grass Roots (71)
185. Southern Man – Neil Yound (70)
186. Spaceman – Nilsson (72)
187. Space Oddity – David Bowie (73)
188. Space Truckin' – Deep Purple (73)
189. Speak To Me – Pink Floyd (73)
190. Stealin' – Uriah Heep (70)
191. Summer (The 1st Time) – Bobby Goldsboro (73)
192. Superstition – Jeff Beck (73)
193. Surrender – Diana Ross (75)
194. Sweet Emotion – Aerosmith (75)
195. Sweet Seasons – Carole King (72)
196. Sweeter Than Honey – Jefferson Starship (75)
197. Tangled Up In Blue – Bob Dylan (75)

198. Tarkio Road – Brewer & Shipley (71)
199. Taxi – Harry Chapin (72)
200. Teenage Love Affair – Rick Derringer (74)
201. 10th Ave. Freezeout – Bruce Springsteen (75)
202. Tequila Sunrise – Eagles (73)
203. There Goes Another Love Song – Outlaws (75)
204. Thick As A Brick – Jethro Tull (72)
205. Thin Line Between Love And Hate – Persuaders (71)
206.30 Days In The Hole – Humble Pie (72)
207. Toys In The Attic – Aerosmith (75)
208. Trampled Under Foot – Led Zeppelin (75)
209. Truckin' – Grateful Dead (71)
210. Turn The Page – Bob Segar (75)
211. Two Hangmen – Mason Profit (71)
212. Undun – Guess Who (70)
213. Us And Them – Pink Floyd (75)
214. Vincent – Don McLean (72)
215. Castles In the Air – Don McLean (72)
216. That's Where The Music Takes Me – Neil Sedaka (75)
217. Walkin' In Rhythm – Blackbyrds (75)
218. War Pigs – Black Sabbath (70)
219. Welcome To Machine – Pink Floyd (75)
220. Westbound #9 – Flaming Ember (70)
221. We've All Got To Be Loved – Doobie Brothers (75)
222. We've Got To Get It On Again – Adrissi Brothers (72)
223. When You're Hot, You're Hot – Jerry Reed (71)
224. Willie & The Hand Jive – Eric Clapton (74)
225. Wish You Were Here – Pink Floyd (75)
226. Woman From Tokyo – Deep Purple (73)

227. Woodstock – Crosby, Stills, Nash & Young (70)
228. You – George Harrison (75)
229. You Are Everything – Stylistics (72)
230. You Need Love – Styx (75)
231. Young Americans – David Bowie (74)
232. Your Lovin' Is The Best Thing – Styx (72)
233. Ziggy Stardust – David Bowie (73)
234. Give Ireland Back To The Irish – Paul McCartney (72)
235. Mary Had A Little Lamb – Paul McCartney (72)

1965 - 1966

WEEK ENDING DECEMBER 23, 1965

1	Let's Hang On	(1)	**Four Seasons**
2	Five O'Clock World	(2)	**The Vogues**
3	A Taste Of Honey	(3)	Herb Alpert
4	We Can Work It Out	(17)	The Beatles
5	Over And Over	(4)	**Dave Clark 5**
6	1-2-3	(6)	**Len Barry**
7	It's Good News Week		**Hedgehoppers Anonymous**
8	Keep On Dancing	(5)	The Gentrys
9	The Sounds Of Silence	(11)	Simon & Garfunkel
10	Il Silenzio	(12)	**Nini Rosso**
11	Don't Think Twice	(13)	**The "Wonder Who"**
12	I'm A Man	(10)	The Yardbirds
13	Turn Turn Turn	(8)	The Byrds
14	Just Like Me	(18)	Paul Revere & Raiders
15	Sunday And Me	(15)	Jay & Americans
16	It's My Life	(19)	The Animals
17	Harlem Nocturne	(-)	The Viscounts
18	The Little Girl I Once Knew	(-)	The Beach Boys
19	England Swings	(20)	**Roger Miller**
20	Day Tripper	(-)	The Beatles

VERY IMPORTANT PREMIERS

- I See The Light — Five Americans
- Recovery — Fontella Bass
- Big Bright Eyes — Danny Hutton
- Little Boy — Four Seasons

WEEK ENDING DECEMBER 30, 1965

1	Five O'Clock World	**The Vogues**	(2)
2	We Can Work It Out/Day Tripper	**The Beatles**	(4)
3	Let's Hang On	Four Seasons	(1)
4	The Sounds Of Silence	Simon & Garfunkel	(9)
5	It's Good News Week	**Hedgehoppers Anonymous**	(7)
6	Over And Over	**Dave Clark 5**	(5)
7	A Taste Of Honey	**Herb Alpert**	(3)
8	A Must To Avoid	Herman's Hermits	(-)
9	1-2-3	Len Barry	(6)
10	Just Like Me	**Paul Revere & Raiders**	(14)
11	It's My Life	**The Animals**	(16)
12	Turn Turn Turn	The Byrds	(13)
13	I'm A Man	The Yardbirds	(12)
14	Il Silenzio	Nini Rosso	(10)
15	Don't Think Twice	The "Wonder Who"	(11)
16	She's Just My Style	Gary Lewis	(-)
17	Harlem Nocturne	The Viscounts	(17)
18	You Didn't Have To Be So Nice	Lovin' Spoonful	(-)
19	Ebb Tide	**Righteous Brothers**	(-)
20	The Little Girl I Once Knew	Beach Boys	(18)

(Number in parentheses indicates last week's position)

VERY IMPORTANT PREMIERS

Going To A Go-Go	The Miracles
Crying Time	Ray Charles

WEEK ENDING JANUARY 6, 1966

	Title	Artist	Last Week
1	The Sounds Of Silence	**Simon & Garfunkel**	(4)
2	We Can Work It Out/Day Tripper	**The Beatles**	(2)
3	Let's Hang On	Four Seasons	(3)
4	Five O'Clock World	The Vogues	(1)
5	It's Good News Week	**Hedgehoppers Anonymous**	(5)
6	A Must To Avoid	**Herman's Hermits**	(8)
7	Over And Over	**Dave Clark 5**	(6)
8	Just Like Me	Paul Revere & Raiders	(10)
9	A Taste Of Honey	Herb Alpert	(7)
10	It's My Life	**The Animals**	(11)
11	I'm A Man	**The Yardbirds**	(13)
12	I Confess	New Colony Six	(-)
13	You Didn't Have To Be So Nice	Lovin' Spoonful	(18)
14	Don't Think Twice	The "Wonder Who"	(15)
15	She's Just My Style	Gary Lewis	(16)
16	Harlem Nocturne	The Viscounts	(17)
17	The Men In My Little Girl's Life	Mike Douglas	(-)
18	A Well Respected Man	The Kinks	(-)
19	Ebb Tide	**Righteous Brothers**	(19)
20	Barbara Ann	Beach Boys	(-)

(Number in parentheses indicates last week's position)

VERY IMPORTANT PREMIERS

The Answer To My Prayer	Neil Sedaka
Take Me For What I'm Worth	The Searchers

WEEK ENDING JANUARY 13, 1966

	Title	Artist	Last Week
1	We Can Work It Out/Day Tripper	**The Beatles**	(2)
2	The Sounds Of Silence	**Simon & Garfunkel**	(1)
3	Five O'Clock World	The Vogues	(4)
4	Let's Hang On	Four Seasons	(3)
5	A Must To Avoid	**Herman's Hermit**	(6)
6	It's Good News Week	**Hedgehoppers Anonymous**	(5)
7	I Confess	**New Colony Six**	(12)
8	Just Like Me	Paul Revere & Raiders	(8)
9	It's My Life	The Animals	(10)
10	You Didn't Have To Be So Nice	**Lovin' Spoonful**	(13)
11	She's Just My Style	**Gary Lewis**	(15)
12	Over And Over	Dave Clark 5	(7)
13	Barbara Ann	Beach Boys	(20)
14	No Matter What Shape	The T-Bones	(-)
15	I'm A Man	The Yardbirds	(11)
16	Don't Think Twice	The "Wonder Who"	(14)
17	A Well Respected Man	The Kinks	(18)
18	The Men In My Little Girl's Life	Mike Douglas	(17)
19	As Tears Go By	**Rolling Stones**	(-)
20	Sandy	Ronny Daytonas	(-)

(Number in parentheses indicates last week's position)

VERY IMPORTANT PREMIERS

My World Is Empty	The Supremes
Working My Way	Four Seasons
Breaking Up Is Breaking My Heart	Roy Orbison
A Hard Days Night	Ramsey Lewis
Set You Free This Time Time	The Byrds
When Liking Turns To Loving	Ronny Dove

THE WCFL "SOUND 10 SURVEY" REPRESENTS THE RELATIVE POPULARITY OF RECORDS IN THE CHICAGO AREA CONDITIONED BY SUCH FACTORS AS INDUSTRY TABULATIONS, LOCAL RECORD SALES AND THE JUDGMENT OF WCFL'S RECORD SELECTION COMMITTEE. WCFL RADIO, THE VOICE OF LABOR, IS O

WEEK ENDING JANUARY 20, 1966

	Title	Artist	Last Week
1	We Can Work It Out/Day Tripper	**The Beatles**	(1)
2	The Sounds Of Silence	**Simon & Garfunkel**	(2)
3	Barbara Ann	Beach Boys	(13)
4	I Confess	New Colony Six	(7)
5	Five O'Clock World	**The Vogues**	(3)
6	It's Good News Week	**Hedgehoppers Anonymous**	(6)
7	A Must To Avoid	**Herman's Hermits**	(5)
8	It's My Life	The Animals	(9)
9	She's Just My Style	Gary Lewis	(11)
10	Let's Hang On	**Four Seasons**	(4)
11	No Matter What Shape	**The T-Bones**	(14)
12	You Didn't Have To Be So Nice	Lovin' Spoonful	(10)
13	The Men In My Little Girl's Life	Mike Douglas	(18)
14	Sandy	Ronny & Daytonas	(20)
15	A Well Respected Man	The Kinks	(17)
16	Lies	The Knickerbockers	(-)
17	I Ain't Gonna Eat My Heart Out	Young Rascals	(-)
18	Zorba The Greek	Herb Alpert	(-)
19	My Love	**Petula Clark**	(-)
20	As Tears Go By	Rolling Stones	(19)

(Number in parentheses indicates last week's position)

VERY IMPORTANT PREMIERS

My Generation	The "Who"
Night Time	The Strangeloves
Lightnin' Strikes	Lou Christie

WEEK ENDING JANUARY 27, 1966

1	We Can Work It Out/Day Tripper	**The Beatles**	(1)	
2	The Sounds Of Silence	**Simon & Garfunkel**	(2)	
3	Barbara Ann	Beach Boys	(3)	
4	I Confess	New Colony Six	(4)	
5	No Matter What Shape	**The T-Bones**	(11)	
6	Lies	**The Knickerbockers**	(16)	
7	It's Good News Week	**Hedgehoppers Anonymous**	(6)	
8	Five O'Clock World	The Vogues	(5)	
9	You Didn't Have To Be So Nice	Lovin' Spoonful	(12)	
10	Lightnin' Strikes	**Lou Christie**	(-)	
11	A Must To Avoid	**Herman's Hermits**	(7)	
12	Sandy	Ronny & Daytonas	(14)	
13	She's Just My Style	Gary Lewis	(9)	
14	It's My Lie	The Animals	(8)	
15	A Well Respected Man	The Kinks	(15)	
16	The Man In My Little Girl's Life	Mike Douglas	(13)	
17	As Tears Go By	Rolling Stones	(20)	
18	Zorba The Greek	Herb Alprt	(18)	
19	I Fought The Law	**Bobby Fuller**	(-)	
20	Spanish Eyes	Al Martino	(-)	

(Number in parentheses indicates last week's position)

VERY IMPORTANT PREMIERS

At The Scene	Dave Clark 5
Long Live Our Love	Shangri-La's
Homeward Bound	Simon & Garfunkel

WEEK ENDING FEBRUARY 3, 1966

1	We Can Work It Out/Day Tripper	**The Beatles**	(1)
2	Sounds Of Silence	**Simon & Garfunkel**	(2)
3	Barbara Ann	Beach Boys	(3)
4	Lightnin' Strikes	Lou Christie	(10)
5	I Confess	**New Colony Six**	(4)
6	Lies	**The Knickerbockers**	(6)
7	No Matter What Shape	**The T-Bones**	(5)
8	Sandy	Ronny & Daytonas	(12)
9	Five O'Clock World	The Vogues	(8)
10	You Didn't Have To Be So Nice	**Lovin' Spoonful**	(9)
11	I Fought The Law	**Bobby Fuller**	(19)
12	A Must To Avoid	Herman's Hermits	(11)
13	Zorba The Greek	Herb Alpert	(18)
14	It's My Life	The Animals	(14)
15	My Love	Petula Clark	(-)
16	I See The Light	Five Americans	(-)
17	Jenny Take A Ride	Mitch Ryder	(-)
18	The Men In My Little Girls' Life	Mike Douglas	(16)
19	Andrea	**The Sunrays**	(-)
20	California Dreamin'	The Mama's & Papa's	(-)

(Number in parentheses indicates last week's position)

VERY IMPORTANT PREMIERS

Husbands And Wives	Roger Miller
The Dedication Song	Freddy Cannon
Up And Down	The McCoys

WEEK ENDING FEBRUARY 10, 1966

1	Lightnin' Strikes	(4) (5)	**Lou Christie**
2	We Can Work It Out/Day Tripper	(1) (10)	**The Beatles**
3	Barbara Ann	(3) (7)	Beach Boys
4	Lies	(6) (11)	The Knickerbockers
5	The Sounds Of Silence	(2) (12)	**Simon & Garfunkel**
6	No Matter What Shape	(7) (10)	**The T-Bones**
7	I Fought The Law	(11) (6)	**Bobby Fuller**
8	I Confess	(5) (6)	New Colony Six
9	Sandy	(8) (9)	Ronny & Daytonas
10	These Boots Are Made For Walkin'	(-) (4)	**Nancy Sinatra**
11	My Love	(15) (8)	**Petula Clark**
12	Jenny Take A Rider	(17) (10)	Mitch Ryder
13	Zorba The Greek	(13) (8)	Herb Alpert
14	California Dreamin'	(20) (5)	The Mama's & The Papa's
15	Andrea	(19) (5)	The Sunrays
16	You Didn't Have To Be So Nice	(10) (12)	Lovin' Spoonful
17	I See The Light	(16) (8)	Five Americans
18	My World Is Empty Without You	(-) (5)	The Supremes
19	Georgiana	(-) (5)	**The Princetons**
20	Second Hand Rose	(-) (9)	Barbra Streisand

(Number in parentheses indicates last week's position)
NOTE—2nd figure in parentheses denotes weeks on list

S0UND 10 STAIRWAY

Answer To My Prayer	Neil Sedaka
At The Scene	Dave Clark 5
Ballad Of The Green Beret	S/Sgt. Barry Sadler
Batman	The Marketts
Big Bright Eyes	Danny Hutton
Homeward Bound	Simon & Garfunkel
Listen People	Herman's Hermits
Long Live Our Love	Shangri-La's
What Now My Love	Sonny & Cher
Workin' My Way Back To You	Four Seasons

CHICAGO PREMIERS

It's Too Late	Bobby Goldsboro
Georgia On My Mind	Righteous Brothers
Why Can't You Bring Me Home	Jay & Americas

WEEK ENDING FEBRUARY 17, 1966

1	Lightnin' Strikes	**Lou Christie**	(1/6)
2	Barbara Ann	**Beach Boys**	(3/8)
3	Boots Are Made For Walkin'	Nancy Sinatra	(10/5)
4	I Confess	New Colony Six	(8/7)
5	We Can Work It Out/Day Tripper	**The Beatles**	(2/11)
6	I Fought The Law	**Bobby Fuller**	(7/7)
7	California Dreamin'	**Mama's & Papa's**	(14/6)
8	Listen People	Herman's Hermits	(-/6)
9	Sounds Of Silence	Simon & Garfunkel	(5/13)
10	Lies	**The Knickerbockers**	(4/12)
11	My Love	**Petula Clark**	(11/9)
12	Ballad Of The Green Berets	S/Sgt. Barry Sadler	(-/3)
13	Andrea	The Sunrays	(15/6)
14	Zorba The Greek	Herb Alpert	(13/9)
15	I See The Light	Five Americans	(17/9)
16	Georgianna	The Princetons	(19/6)
17	My World Is Empty	The Supremes	(18/6)
18	It Was A Very Good Year	Frank Sinatra	(-/4)
19	At The Scene	**Dave Clark 5**	(-/4)
20	Homeward Bound	Simon & Garfunkel	(-/4)

(Number in parentheses indicates last week's position / Weeks heard on WCFL)

S0UND 10 STAIRWAY

Answer To My Prayer	Neil Sedaka
Batman	Jan & Dean
Call Me	Chris Montez
Crying Time	Ray Charles
Elusive Butterfly	Bob Lind
Gloria	Shadows Of Knight
Long Live Our Love	Shangri-La's
Uptight	Stevie Wonder
What Now My Love	Sonny & Cher
When Liking Turns To Loving	Ronnie Dove

VERY IMPORTANT PREMIERS

		Projection
Bossy Boss	Gene Simmons	50%
One More Heartache	Marvin Gaye	75%
Shake Me, Wake Me	Four Tops	77%

WEEK ENDING FEBRUARY 24, 1966

#	Title	Artist	(Last Week/Weeks)
1	Lightnin' Strikes	**Lou Christie**	(1/7)
2	California Dreamin'	**Mana's & Papa's**	(7/7)
3	Boots Are Made For Walkin'	Nancy Sinatra	(3/6)
4	Listen People	Herman's Hermits	(8/7)
5	Ballad Of Green Berets	**S/Sgt. Barry Sadler**	(12/4)
6	Barbara Ann	**Beach Boys**	(2/9)
7	I Fought The Law	**Bobby Fuller**	(6/8)
8	Work It Out/Day Tripper	The Beatles	(5/12)
9	I Confess	New Colony Six	(4/8)
10	Georgianna	**The Princetons**	(16/7)
11	Homeward Bound	**Simon & Garfunkel**	(20/5)
12	Andrea	The Sunrays	(13/7)
13	My Love	Petula Clark	(11/10)
14	I See The Light	Five Americans	(15/10)
15	At The Scene	Dave Clark 5	(19/5)
16	Gloria	Shadows Of Knight	(-/2)
17	Elusive Butterfly	Bob Lind	(-/4)
18	Bat Man	Marketts	(-/5)
19	My World Is Empty	**The Supremes**	(17/7)
20	19th Nervous Breakdown	Rolling Stones	(-/2)

(Number in parentheses indicates last week's position / Weeks heard on WCFL)

SOUND 10 STAIRWAY

Title	Artist
Answer To My Prayer	Neil Sedaka
Dedication Song	Freddy Cannon
It Won't Be Wrong	The Byrds
Long Live Our Love	Shangri-La's
Love Makes The World Go 'Round	Deon Jackson
Magic Town	The Vogues
What Now My Love	Sonny & Cher
Woman	Peter & Gordon
Workin' My Way Back To You	Four Seasons
Uptight	Stevie Wonder

CHICAGO PREMIERS

Title	Artist	*Projection*
Daydream	Lovin' Spoonful	93%
Don't Push Me	Hedgehoppers Anonymous	79%
Latin Lupe Lu	Mitch Ryder & Wheels	60%

WEEK ENDING MARCH 3, 1966

1	California Dreamin'	**Mama's & Papa's**	(2/8)
2	Lightnin' Strikes	**Lou Christie**	(1/8)
3	Boots Are Made For Walkin'	Nancy Sinatra	(3/7)
4	Listen People	Herman's Hermits	(4/8)
5	Ballad Of Green Berets	**S/Sgt. Barry Sadler**	(5/5)
6	I Fought The Law	**Bobby Fuller**	(7/9)
7	Gloria	**Shadows Of Knight**	(16/3)
8	Elusive Butterfly	Bob Lind	(17/5)
9	Homeward Bound	Simon & Garfunkel	(11/6)
10	Barbara Ann	**Beach Boys**	(6/10)
11	19th Nervous Breakdown	**Rolling Stones**	(20/3)
12	Georgianna	The Princetons	(10/8)
13	Nowhere Man	The Beatles	(-/2)
14	Magic Town	The Vogues	(-/4)
15	My Love	Petula Clark	(13/11)
16	At The Scene	Dave Clark 5	(15/6)
17	Workin' My Way Back	Four Seasons	(-/8)
18	My World Is Empty	The Supremes	(19/8)
19	Bat Man	**The Marketts**	(18/6)
20	What Now My Love	Sonny & Cher	(-/7)

(Number in parentheses indicates last week's position / Weeks heard on WCFL)

S0UND 10 STAIRWAY

Bat Man	Jan & Dean
Call Me	Chris Montez
Chester	Bob Kuban
Daydream	Lovin' Spoonful
Dedication Song	Freddy Cannon
I Want To Go With You	Eddy Arnold
Love Makes World Go 'Round	Deon Jackson
Phoenix Love Theme	Brass Ring
Woman	Peter & Gordon
You Baby	The Turtles

CHICAGO PREMIERS

		Projection
Lullaby Of Love	The Poppies	52%
Sure Gonna Miss Her	Gary Lewis	95%
Walkin' My Cat	Norma Tanega	50%

WEEK ENDING MARCH 10, 1966

1	California Dreamin'	**Mama's & Papa's**	(1/9)
2	Boots Made For Walkin'	**Nancy Sinatra**	(3/8)
3	Gloria	Shadow Of Knight	(7/4)
4	Ballad Of Green Berets	S/Sgt. Barry Sadler	(5/6)
5	Lightnin' Strikes	**Lou Christie**	(2/9)
6	19th Nervous Breakdown	**Rolling Stones**	(11/4)
7	Listen People	**Herman's Hermits**	(4/9)
8	Nowhere Man/What Goes On	The Beatles	(13/3)
9	Elusive Butterfly	Bob Lind	(8/6)
10	I Fought The Law	**Bobby Fuller**	(8/10)
11	Homeward Bound	**Simon & Garfunkel**	(9/7)
12	Magic Town	The Vogues	(14/5)
13	At The Scene	Dave Clark 5	(16/7)
14	Daydream	Lovin' Spoonful	(-/3)
15	Workin' My Way Back	Four Seasons	(17/9)
16	You Baby	The Turtles	(-/8)
17	My Love	Petula Clark	(15/12)
18	Woman	Peter & Gordon	(-/6)
19	Love Makes World Go Round	**Deon Jackson**	(-/5)
20	What Now My Love	Sonny & Cher	(20/8)

(Number in parentheses indicates last week's position / Weeks heard on WCFL)

S
0
U 10
N
D

Title	Artist
634-5789	Wilson Pickett
Call Me	Chris Montez
Cheater	Bob Kuban
I Can't Let You Go	The Hollies
I'm Ready	Little Boy Blues

S
T
A
I
R
W
A
Y

Title	Artist
It's Too Late	Bobby Goldsboro
Phoenix Love Theme	Brass Ring
Baby Scratch My Back	Slim Harpo
Soul & Inspiration	Righteous Bros.
Sure Gonna Miss Her	Gary Lewis

CHICAGO PREMIERS

Title	Artist	*Projection*
Kicks	Paul Revere & Raiders	72%
One Track Mind	Knickerbockers	80%
Shape Of Things	Yardbirds	87%

WEEK ENDING MARCH 17, 1966

	Title	Last Week/Weeks	Artist
1	California Dreamin'	(1/10)	**Mama's & Papa's**
2	Gloria	(3/5)	**Shadows Of Knight**
3	Boots Made For Walkin'	(2/9)	Nancy Sinatra
4	Ballad Of Green Berets	(4/7)	S/Sgt. Barry Sadler
5	Nowhere Man/What Goes On	(8/4)	**The Beatles**
6	19th Nervous Breakdown	(6/5)	**Rolling Stones**
7	Daydream	(14/4)	**Lovin' Spoonful**
8	Elusive Butterfly	(9/7)	Bob Lind
9	Magic Town	(12/6)	The Vogues
10	Listen People	(7/10)	**Herman's Hermits**
11	Homeward Bound	(11/8)	**Simon & Garfunkel**
12	Workin' My Way Back	(15/10)	Four Seasons
13	You Baby	(16/9)	The Turtles
14	Woman	(18/7)	Peter & Gordon
15	Shape Of Things	(-/3)	Yardbirds
16	Sure Gonna Miss Her	(-/3)	Yardbirds
17	I Can't Let You Go	(-/3)	The Hollies
18	Call Me	(-8)	Chris Montez
19	Phoenix Love Theme	(-/4)	**Brass Ring**
20	Baby Scratch My Back	(-/1)	Slim Harpo

(Number in parentheses indicates last week's position / Weeks heard on WCFL)

SOUND 10 STAIRWAY

Title	Artist
634-5789	Wilson Pickett
Bang Bang	Cher
Don't Push Me	Hedgehoppers Annon.
I'm So Lonesome I Could Cry	B.J. Thomas
Inside Looking Out	Animals
It's Too Late	Bobby Goldsboro
Kicks	Paul Revere
Little Latin Lupe Lu	Mitch Ryder
Outside Of Heaven	Lou Christie
Soul & Inspiration	Righteous Bros.

CHICAGO PREMIERS

Title	Artist	*Projection*
A Sign Of The Times	Petula Clark	78%
Frankie & Johnny	Elvis Presley	54%
Rhapsody In The Rain	Lou Christie	

WEEK ENDING MARCH 24, 1966

	Title	Artist	(Last/Weeks)
1	California Dreamin'	**Mama's & Papa's**	(1/11)
2	Gloria	**Shadows Of Knight**	(2/6)
3	Boots Made For Walkin'	Nancy Sinatra	(3/10)
4	Ballad Of Green Berets	S/Sgt. Barry Sadler	(4/8)
5	Nowhere Man/This Side Only	**Beatles**	(5/5)
6	19th Nervous Breakdown	**Rolling Stones**	(6/6)
7	Elusive Butterfly	**Bob Lind**	(8/8)
8	Daydream	Lovin' Spoonful	(7/5)
9	Magic Town	The Vogues	(9/7)
10	You Baby	**The Turtles**	(13/10)
11	Shapes Of Things	**Yardbirds**	(15/4)
12	Homeward Bound	Simon & Garfunkel	(11/9)
13	Woman	Peter & Gordon	(14/8)
14	Workin' My Way Back	Four Seasons	(12/11)
15	Sure Gonna Miss Her	Gary Lewis	(16/4)
16	Kicks	Paul Revere	(-/3)
17	I'm So Lonesome I Could Cry	B.J. Thomas	(-/3)
18	Baby Scratch My Back	Slim Harpo	(20/2)
19	Soul And Inspiration	**Righteous Bros.**	(-/4)
20	Bang Bang	Cher	(-/4)

(Number in parentheses indicates last week's position / Weeks heard on WCFL)

SOUND 10 STAIRWAY

Caroline No	Brian Wilson
Don't Push Me	Hedgehopper Annon.
Don't You Know	Keith Everett
Get Ready	Temptations
Good Lovin'	Young Rascals
I Can't Let You Go	The Hollies
Outside Gates Of Heaven	Lou Christie
Wait A Minute	Tim Tam
What Now My Love	Herb Alpert
634-5789	Wilson Pickett

CHICAGO PREMIERS

Title	Artist
Love's Made A Fool Of You	Bobby Fuller
Sloop John B	Beach Boys
Try Too Hard	Dave Clark 5

WEEK ENDING MARCH 31, 1966

#	Title	Artist	(Last/Weeks)
1	Gloria	**Shadows Of Knight**	(2/7)
2	California Dreamin'	**Mama's & Papa's**	(1/12)
3	Shapes Of Things	Yardbirds	(11/5)
4	Nowhere Man	Beatles	(5/6)
5	Daydream	**Lovin' Spoonful**	(8/6)
6	19th Nervous Breakdown	**Rolling Stones**	(6/7)
7	Boots Made For Walkin	**Nancy Sinatra**	(3/11)
8	Ballad Of Green Berets	S/Sgt. Barry Sadler	(4/9)
9	Bang Bang	Cher	(20/5)
10	Kicks	**Paul Revere**	(16/4)
11	You Baby	**The Turtles**	(10/11)
12	Elusive Butterfly	Bob Lind	(7/9)
13	Sure Gonna Miss Her	Gary Lewis	(15/5)
14	Magic Town	The Vogues	(9/8)
15	Soul And Inspiration	Righteous Bros.	(19/5)
16	I'm So Lonesome	B.J. Thomas	(17/4)
17	Secret Agent Man	Johnny Rivers	(-/2)
18	I Can't Let Go	The Hollies	(-/5)
19	Caroline No	**Brian Wilson**	(-/4)
20	Good Lovin'	Young Rascals	(-/5)

(Number in parentheses indicates last week's position / Weeks heard on WCFL)

SOUND 10 STAIRWAY

A Sign Of Our Times	Petula Clark
Don't You Know	Keith Everet
It's Too Late	Bobby Goldsboro
Little Latin Lupe Lu	Mitch Ryder
Outside Gates Of Heaven	Lou Christie
Rhapsody In Rain	Lou Christie
Till The End Of The Day	Kinks
Time Won't Let Me	Outsiders
Try Too Hard	Dave Clark 5
Walkin' My Cat Named Dog	Normal Tanega

CHICAGO PREMIERS

Title	Artist	*Projection*
EVOL-Not Love	5 Americans	65%
Got My Mojo Working	Jimmy Smith	42%
Monday Monday	Mama's & Papa's	82%

WEEK ENDING APRIL 7, 1966

#	Title	Artist	Last/Weeks
1	Gloria	**Shadows Of Knight**	(1/8)
2	California Dreamin'	**Mama's & Papa's**	(2/13)
3	Shapes Of Things	Yardbirds	(3/6)
4	You Baby	The Turtles	(11/12)
5	Kicks	**Paul Revere**	(10/5)
6	Daydream	**Lovin' Spoonful**	(5/7)
7	Nowhere Man	**Beatles**	(4/7)
8	Bang Bang	Cher	(9/6)
9	19th Nervous Breakdown	Rolling Stones	(6/8)
10	Secret Agent Man	**Johnny Rivers**	(17/3)
11	Good Lovin'	**Young Rascals**	(20/6)
12	I'm So Lomesome	B.J. Thomas	(16/5)
13	I Can't Let Go	The Hollies	(18/6)
14	Ballad Of Green Berets	S/Sgt. Barry Sadler	(8/10)
15	Soul And Inspiration	Righteous Bros.	(15/6)
16	Caroline No	Brian Wilson	(19/5)
17	Boots Made For Walkin'	Nancy Sinatra	(7/12)
18	Try Too Hard	Dave Clark 5	(-/3)
19	Time Won't Let Me	**Outsiders**	(-/4)
20	Rhapsody In The Rain	Lou Christie	(-/4)

(Number in parentheses indicates last week's position / Weeks heard on WCFL)

Title	Artist
A Sign Of Our Times	Petula Clark
Don't You Know	Keith Everett
I Lie Awake	New Colony Six
I'll Go Crazy	The Buckinghams
Inside Looking Out	Animals
One Track Mind	Knickerbockers
Sloop John B	Beach Boys
Till End Of The Day	Kinks
Walkin' My Cat	Norma Tanega
What Now My Love	Herb Alpert

S 0 U 10 N D

STAIRWAY

CHICAGO PREMIERS

Title	Artist	*Projection*
A Groovy Kind Of Love	Mindbenders	71%
Backstage	Gene Pitney	71%
Remember The Rain	Bob Lind	58%

WEEK ENDING APRIL 14, 1966

	Title	Artist	(Last/Weeks)
1	Gloria	**Shadows Of Knight**	(1/9)
2	Kicks	**Paul Revere**	(5/6)
3	Shapes Of Things	Yardbirds	(3/7)
4	Good Lovin'	Young Rascals	(11/7)
5	Daydream	**Lovin' Spoonful**	(6/8)
6	California Dreamin'	**Mama's & Papa's**	(2/14)
7	Secret Agent Man	**Johnny Rivers**	(10/4)
8	Bang Bang	Cher	(8/7)
9	You Baby	The Turtles	(4/13)
10	I Can't Let Go	**The Hollies**	(13/7)
11	I'm So Lonesome	**B.J. Thomas**	(12/6)
12	Time Won't Le ME	Outsiders	(19/5)
13	Nowhere Man	Beatles	(7/8)
14	Try Too Hard	Dave Clark 5	(18/4)
15	Soul And Inspiration	Righteous Bros.	(15/7)
16	Rhapsody In The Rain	Lou Christie	(20/5)
17	Sloop John B	Beach Boys	(-/4)
18	Don't You Know	Keith Everett	(-/3)
19	Monday, Monday	**Mama's & Papa's**	(-/3)
20	Caroline No	Brian Wilson	(16/6)

(Number in parentheses indicates last week's position / Weeks heard on WCFL)

SOUND 10 STAIRWAY

Sign Of The Times	Petula Clark
Eight Miles High	The Byrds
History Repeats Itself	Buddy Starcher
I Hear Trumpets	Tokens
I Lie Awake	New Colony Six
I'll Go Crazy	Buckinghams
Lara's Theme	Roger Williams
Leaning On Lamppost	Herman's Hermits
Rainy Day Women	Bob Dylan
Till End Of The Day	The Kinks

CHICAGO PREMIERS

		Projection
I Fall To You	Bob Morrison	83%
Pin The Tail On the Donkey	Paul Peek	42%
Running 'Round In Circles	Ivy League	56%

WEEK ENDING APRIL 21, 1966

#	Title	Artist	(LW/Wks)
1	Kicks	**Paul Revere**	(2/7)
2	Secret Agent Man	**Johnny Rivers**	(7/5)
3	Gloria	Shadows Of Knight	(1/10)
4	Good Lovin'	Young Rascals	(4/8)
5	Monday, Monday	**Mama's & Papa's**	(19/4)
6	Time Won't Let Me	**Outsiders**	(12/6)
7	Shapes Of Things	**Yardbirds**	(3/8)
8	Bang Bang	Cher	(8/8)
9	I'm So Lonesome	B.J. Thomas	(11/7)
10	You Baby	**The Turtles**	(9/14)
11	I Can't Let Go	**The Hollies**	(10/8)
12	Don't You Know	Keith Everett	(18/4)
13	Sloop John B	Beach Boys	(17/5)
14	Try Too Hard	Dave Clark 5	(14/5)
15	Soul And Inspiration	Righteous Bros.	(15/8)
16	Rhapsody In The Rain	Lou Christie	(16/6)
17	Caroline No	Brian Wilson	(20/7)
18	A Groovy Kind Of Love	Mindbenders	(-/3)
19	I Lie Awake	**New Colony Six**	(-/5)
20	Rainy Day Woman #12 & 35	Bob Dylan	(-/2)

(Number in parentheses indicates last week's position / Weeks heard on WCFL)

SOUND 10 STAIRWAY

A Sign Of The Times	Petula Clark
Eight Miles High	The Byrds
Frankie & Johnny	Elvis Presley
History Repeats Itself	Buddy Starcher
How Does That Grab You	Nancy Sinatra
I Hear Trumpets Blow	Tokens
I'll Go Crazy	Buckinghams
Lara's Theme	Roger Williams
Leaning On Lamppost	Herman's Hermits
Love's Made A Fool Of You	Bobby Fuller 4

CHICAGO PREMIERS

Title	Artist	*Projection*
Marble Breaks & Iron Bends	Drafi	60%
Solitary Man	Neil Diamond	65%
You're Ready Now	Frankie Valli	70%

WEEK ENDING APRIL 28, 1966

1	Kicks	**Paul Revere**	(1/8)
2	Monday, Monday	**Mama's & Papa's**	(5/5)
3	Secret Agent Man	Johnny Rivers	(2/6)
4	Good Lovin'	Young Rascals	(4/9)
5	Rainy Day Woman #12 & 35	**Bob Dylan**	(20/3)
6	Time Won't Let Me	**Outsiders**	(6/7)
7	Gloria	**Shadows of Knight**	(3/11)
8	Bang Bang	Cher	(8/9)
9	Soul And Inspiration	Righteous Bros.	(15/9)
10	Shapes Of Things	**Yardbirds**	(7/9)
11	A Groovy Kind Of Love	**Mindbenders**	(18/4)
12	I Can't Let Go	The Hollies	(11/9)
13	Leaning On Lamppost	Herman's Hermits	(-/7)
14	Sloop John B	Beach Boys	(13/6)
15	Don't You Know	Keith Everett	(12/5)
16	Rhapsody In The Rain	Lou Christie	(16/7)
17	Try Too Hard	Dave Clark 5	(14/6)
18	I'm So Lonesome	B.J. Thomas	(9/8)
19	History Repeats Itself	**Buddy Starcher**	(-/4)
20	Eight Miles High	Byrds	(-/5)

(Number in parentheses indicates last week's position / Weeks heard on WCFL)

S 0 U N D 10 S T A I R W A Y

A Sign Of The Times	Petula Clark
Backstage	Gene Pitney
EVOL-Not Love	5 Americans
Got My Mojo Working	Jimmy Smith
How Does That Grab You	Nancy Sinatra
I Go Crazy	Buckinghams
I Hear Trumpets Blow	Tokens
I'm Coming Home Cindy	Trini Lopez
Lara's Theme	Roger Williams
Love's Made A Fool Of You	Bobby Fuller 4

CHICAGO PREMIERS

		Projection
Daddy You Got To Let Him In	The Satisfactions	60%
Girl In Love	The Outsiders	80%
Younger Girl	The Hondells	75%

WEEK ENDING MAY 5, 1966

	Title	Artist	
1	Monday, Monday	**Mama's & Papa's**	(2/6)
2	Kicks	**Paul Revere**	(1/9)
3	Good Lovin'	Young Rascals	(4/10)
4	Secret Agent Man	Johnny Rivers	(3/7)
5	A Groovy Kind Of Love	**Mindbenders**	(11/5)
6	Rainy Day Woman #12 & 35	**Bob Dylan**	(5/4)
7	Time Won't Let Me	**Outsiders**	(6/8)
8	Soul & Inspiration	Righteous Bros.	(9/10)
9	Shapes Of Things	Yardbirds	(10/10)
10	Don't You Know	**Keith Everett**	(15/6)
11	Bang Bang	**Cher**	(8/10)
12	Leaning On Lamppost	Herman's Hermits	(13/8)
13	Sloop John B	Beach Boys	(14/7)
14	Try Too Hard	Dave Clark 5	(17/7)
15	Eight Miles High	Byrds	(20/6)
16	Till End Of The Day	Kinks	(-/8)
17	Lara's Theme	Roger Williams	(-/5)
18	Rhapsody In The Rain	Lou Christie	(16/8)
19	I Go Crazy	**Buckinghams**	(-/5)
20	How Does That Grab you	Nancy Sinatra	(-/4)

(Number in parentheses indicates last week's position / Weeks heard on WCFL)

SOUND 10 STAIRWAY

Title	Artist
Backstage	Gene Pitney
Did You Ever Have To Make Up Your Mind	Lovin' Spoonful
Dirty Water	Standells
EVOL-Not Love	5 Americans
I Hear Trumpets Blow	Tokens
Let's Start All Over Again	Ronnie Dove
Message To Michael	Dionne Warwick
Sun Ain't Gonna Shine	Walker Bros.
Twinkle Toes	Roy Orbison
Killer Joe	Kingsmen

CHICAGO PREMIERS

Title	Artist	*Projection*
David's Mood	Jack Eely	82%
Mama	B.J. Thomas	77%
My Little Red Book	Love	74%

WEEK ENDING MAY 12, 1966

#	Title	Artist	(Last Week/Weeks)
1	Monday, Monday	**Mamas & Papas**	(1/7)
2	Kicks	**Paul Revere & The Raiders**	(2/10)
3	Rainy Day Women #12 & 35	Bob Dylan	(6/5)
4	A Groovy Kind Of Love	Mindbenders	(5/6)
5	Good Lovin'	**Young Rascals**	(3/11)
6	Secret Agent Man	**Johnny Rivers**	(4/8)
7	Time Won't Let Me	**Outsiders**	(7/9)
8	Soul & Inspiration	Righteous Bros.	(8/11)
9	Eight Miles High	Byrds	(15/7)
10	Sloop John B	**Beach Boys**	(13/8)
11	Leaning On Lamppost	**Herman's Hermits**	(12/9)
12	Bang Bang	Cher	(11/11)
13	Don't You Know	Keith Everett	(10/7)
14	Rhapsody In The Rain	Lou Christie	(18/9)
15	Did You Ever Have To Make Up Your Mind	Lovin' Spoonful	(-/4)
16	Backstage	Gene Pitney	(-/6)
17	EVOL-Not Love	Five Americans	(-/6)
18	How Does That Grab you	Nancy Sinatra	(20/5)
19	Sun Ain't Gonna Shine	**Walker Bros.**	(-/4)
20	I Go Crazy	Buckinghams	(19/6)

(Number in parentheses indicates last week's position / Weeks heard on WCFL)

S0UND 10		STAIRWAY	
Dirty Water	The Standells	Message To Michael	Dionne Warwick
Green Grass	Gary Lewis	The "A" Team	S/Sgt. Barry Sadler
I Am A Rock	Simon & Garfunkel	Twinkle Toes	Roy Orbison
Killer Joe	Kingsmen	When A Man Loves A Woman	Percy Sledge
Let's Start All Over Again	Ronnie Dove	Love Is Like An Itching	Supremes

CHICAGO PREMIERS

		Projection
Come On Let's Go	The McCoys	72%
You Wouldn't Listen	The Ides Of March	88%

WEEK ENDING MAY 19, 1966

1	Monday, Monday	**Mamas & Papas**	(1/8)
2	Kicks	**Paul Revere**	(2/11)
3	A Groovy Kind Of Love	Mindbenders	(4/7)
4	Rainy Day Woman	Bob Dylan	(3/6)
5	Good Lovin'	**Young Rascals**	(5/12)
6	Make Up Your Mind	**Lovin' Spoonful**	(15/5)
7	Secret Agent Man	**Johnny Rivers**	(6/9)
8	Eight Miles High	Byrds	(9/8)
9	Time Won't Let Me	Outsiders	(7/10)
10	Soul & Inspiration	**Righteous Bros.**	(8/12)
11	Paint It Back	**Rolling Stones**	(-/3)
12	Backstage	Gene Pitney	(16/7)
13	I Am A Rock	Simon & Garfunkel	(-/6)
14	Rhapsody In Rain	Lou Christie	(14/10)
15	EVOL-Not Love	Five Americans	(17/7)
16	Batman & Grandma	Dickie Goodman	(-/1)
17	Sloop John B.	Beach Boys	(10/9)
18	Green Grass	Gary Lewis	(-/3)
19	How Does That Grab You	**Nancy Sinatra**	(18/6)
20	Sun Ain't Gonna Shine	Walker Bros.	(19/5)

(Number in parentheses indicates last week's position / Weeks heard on WCFL)

S0UND 10 STAIRWAY

Barefootin'	Robt. Parker
Dirty Water	Standells
Don't Bring Me Down	Animals
Girl In Love	Outsiders
Let's Start All Over	Ronnie Dove
Message To Michael	Dionne Warwick
Twinkle Toes	Roy Orbison
When Man Loves Woman	Percy Sledge
You Don't Love Me	Gary Walker
Love Is Like Itching	Supremes

CHICAGO PREMIERS

		Projection
Crying	Jay & Americans	71%
Race With The Wind	The Robbs	73%

WEEK ENDING MAY 26, 1966

	Title	Artist			Title	Artist
1	Monday, Monday	(1/9) **Mamas & Papas**		11	Soul & Inspiration	(10/13) **Righteous Bros.**
2	Kicks	(2/12) **Paul Revere**		12	Eight Miles High	(8/9) The Byrds
3	A Groovy Kind Of Love	(3/8) Mindbenders		13	Green Grass	(18/4) Gary Lewis
4	Paint It Black	(11/4) Rolling Stones		14	Sun Ain't Gonna Shine	(20/6) Walker Bros.
5	Rainy Day Woman	(4/7) **Bob Dylan**		15	How Does That Grab You	(19/7) Nancy Sinatra
6	Make Up Your Mind	(6/6) **Lovin' Spoonful**		16	Don't Bring Me Down	(-/2) The Animals
7	I Am A Rock	(13/7) **Simon & Garfunkel**		17	Message To Michael	(-/7) Dionne Warwick
8	Good Lovin'	(5/13) Young Rascals		18	Strangers In The Night	(-/3) Frank Sinatra
9	EVOL-Not Love	(15/8) Five Americans		19	Rhapsody In Rain	(14/11) **Lou Christie**
10	Batman & Grandma	(16/2) **Dickie Goodman**		20	When A Man Loves A Woman	(-/7) Percy Sledge

(Number in parentheses indicates last week's position / Weeks heard on WCFL)

		SOUND 10 STAIRWAY		
Barefootin'	Robert Parker		Hold On	Sam & Dave
Come On, Let's Go	McCoys		Man's Man's World	James Brown
Cool Jerk	Capitols		Oh Yeah	Shadows Of Knight
Dirty Water	Standells		Sweet Talkin' Guy	Chiffons
Girl In Love	Outsiders		S.Y.S.J.F.M.	Joe Tex

CHICAGO PREMIERS

		Projection
Greatest Moments In A Girl's Life	Tokens	68%
Popsicle	Jan & Dean	72%

WEEK ENDING JUNE 2, 1966

	Title	(Last/Weeks)	Artist
1	Paint It Black	(4/5)	**Rolling Stones**
2	Monday, Monday	(1/10)	**Mamas & Papas**
3	A Groovy Kind Of Love	(3/9)	Mindbenders
4	I Am A Rock	(7/8)	Simon & Garfunkel
5	Make Up Your Mind	(6/7)	**Lovin' Spoonful**
6	Kicks	(2/13)	**Paul Revere**
7	EVOL-Not Love	(9/9)	**Five Americans**
8	Rainy Day Woman	(5/8)	Bob Dylan
9	Good Lovin'	(8/14)	Young Rascals
10	Green Grass	(13/5)	**Gary Lewis**
11	Don't Bring Me Down	(16/3)	**Animals**
12	When A Man Loves A Woman	(20/8)	Percy Sledge
13	Girl In Love	(-/6)	Outsiders
14	Batman & Grandma	(10/3)	Dickie Goodman
15	Strangers In The Night	(18/4)	Frank Sinatra
16	Dirty Water	(-/4)	Standells
17	Cool Jerk	(-/3)	Capitols
18	Message To Michael	(17/8)	Dionne Warwick
19	Come ON Let's Go	(-/4)	**McCoys**
20	Sweet Talkin' Guy	(-/5)	Chiffons

(Number in parentheses indicates last week's position / Weeks heard on WCFL)

Title	Artist			Title	Artist
Barefootin'	Robert Parker	S	S T	Man's Man's World	James Brown
Day For Decision	Johnny Sea	0	A I	Oh Yeah	Shadows Of Knight
Hold On	Sam & Dave	U 10	R W	Opus #17	Four Seasons
Little Red Book	Love	N	A Y	The More I See You	Chris Montez
Mama	B.J. Thomas	D		You Wouldn't Listen To Me	Ides Of March

CHICAGO PREMIERS

		Projection
Hanky Panky	The Shondells	75%
It's That Time Of Year	Len Barry	72%
Paperback Writer/Rain	Beatles	100%

WEEK ENDING JUNE 9, 1966

	Title	Artist	(Last/Weeks)
1	Paint It Black	**Rolling Stones**	(1/6)
2	I Am A Rock	**Simon & Garfunkel**	(4/9)
3	A Groovy Kind Of Love	Mindbenders	(3/10)
4	Monday, Monday	Mamas & Papas	(2/11)
5	Kicks	**Paul Revere**	(6/14)
6	Make Up Your Mind	**Lovin' Spoonful**	(5/8)
7	Strangers In The Night	**Frank Sinatra**	(15/5)
8	Green Grass	Gary Lewis	(10/6)
9	Dirty Water	Standells	(16/5)
10	EVOL-Not Love	Five Americans	(7/10)
11	Don't Bring Me Down	**Animals**	(11/4)
12	Batman & Grandma	Dickie Goodman	(14/4)
13	When A Man Loves A Woman	Percy Sledge	(12/9)
14	Girl In Love	Outsiders	(13/7)
15	Hanky Panky	Shondells	(-/3)
16	Come On Let's Go	McCoys	(19/5)
17	Oh Yeah	Shadows Of Knight	(-/3)
18	Cool Jerk	Capitols	(17/4)
19	Day For Decision	**Johnny Sea**	(-/2)
20	Paperback Writer/Rain	The Beatles	(-/2)

(Number in parentheses indicates last week's position / Weeks heard on WCFL)

S0UND 10 STAIRWAY

Title	Artist
Barefootin'	Robert Parker
Hey Joe	The Leaves
Land Of Milk & Honey	The Vogues
Mama	B. J. Thomas
Oh How Happy	Shades Of Blue
Red Rubber Ball	The Cyrcle
The More I See You	Chris Montez
You Don't Have To Say You Love Me	Dusty Springfield
You Wouldn't Listen To Me	Ides Of March
Race With The Wind	The Robbs

CHICAGO PREMIERS

Title	Artist	***Projection***
High On Love	Knickerbockers	73%
You'd Better Run	Young Rascals	71%

WEEK ENDING JUNE 16, 1966

#	Title	Artist	(Last/Weeks)
1	Paint It Black	**Rolling Stones**	(1/7)
2	Hanky Panky	**Shondells**	(15/4)
3	Strangers In The Night	Frank Sinatra	(7/6)
4	I Am A Rock	Simon & Garfunkel	(2/10)
5	Make Up Your Mind	**Lovin' Spoonful**	(6/9)
6	Dirty Water	**Standells**	(9/6)
7	Paperback Writer/Rain	**Beatles**	(20/3)
8	A Groovy Kind Of Love	Mindbenders	(3/11)
9	When Man Loves A Woman	Percy Sledge	(13/10)
10	Cool Jerk	**Capitols**	(18/5)
11	Don't Bring Me Down	**Animals**	(11/5)
12	Girl In Love	Outsiders	(14/8)
13	Green Grass	Gary Lewis	(8/7)
14	You Wouldn't Listen To Me	Ides Of March	(-/6)
15	Oh Yeah	Shadows Of Knight	(17/4)
16	Day For Decision	Johnny Sea	(19/3)
17	Come On Let's Go	McCoys	(16/6)
18	Sweet Talkin' Guy	Chiffons	(-/7)
19	Hey Joe	**The Leaves**	(-/4)
20	You Don't Have To Say You Love Me	Dusty Springfield	(-/4)

(Number in parentheses indicates last week's position / Weeks heard on WCFL)

SOUND 10 — STAIRWAY

Title	Artist
Ain't Too Proud To Beg	Temptations
I Call Your Name	Buckinghams
Land Of Milk & Honey	Vogues
Mama	B. J. Thomas
Oh How Happy	Shades Of Blue
Pied Piper	Crispian St. Peter
Please Tell Me Why	Dave Clark 5
Race With The Wind	Robbs
Red Rubber Ball	The Cyrkle
The More I See You	Chris Montez

CHICAGO PREMIERS

Title	Artist
Hungry	Paul Revere
Sunny Afternoon	Kinks
Train Of Live	Roger Miller

WEEK ENDING JUNE 23, 1966

1	Hanky Panky	**Shondells**	(2/5)
2	Paint It Back	**Rolling Stones**	(1/8)
3	Paperback Writer/Rain	Beatles	(7/4)
4	Strangers In The Night	Frank Sinatra	(3/7)
5	I Am A Rock	**Simon & Garfunkel**	(4/11)
6	Dirty Water	**Standells**	(6/7)
7	Cool Jerk	**Capitols**	(10/6)
8	When Man Loves A Woman	Percy Sledge	(9/11)
9	Hey Joe	Leaves	(19/5)
10	You Wouldn't Listen To Me	**Ides Of March**	(14/7)
11	A Groovy Kind Of Love	**Mindbenders**	(8/12)
12	Don't Bring Me Down	Animals	(11/6)
13	Pied Piper	Crispian St. Peter	(-/3)
14	Girl In Love	Outsiders	(12/9)
15	You Don't Have To Say You Love Me	Dusty Springfield	(20/5)
16	Oh Yeah	Shadows Of Knight	(15/5)
17	The More I See You	Chris Montez	(-/9)
18	Land Of Milk & Honey	Vogues	(-/5)
19	Sweet Talkin' Guy	**Chiffons**	(18/8)
20	Red Rubber Ball	Cyrkle	(-/4)

(Number in parentheses indicates last week's position / Weeks heard on WCFL)

SOUND 10 STAIRWAY

Along Comes Mary	Association
Hungry	Paul Revere
I Call Your Name	Buckinghams
Little Girl	Syndicate Of Sound
My Little Red Book	Love
Oh How Happy	Shades Of Blue
Please Tell Me Why	Dave Clark 5
Race With The Wind	Robbs
Wild Thing	Troggs
You Better Run	Young Rascals

CHICAGO PREMIERS

Cadillac	New Colony Six
I Saw Her Again	Mamas & Papas
Work Song	Herb Alpert

WEEK ENDING JUNE 30, 1966

	Title	Artist	(Last/Weeks)
1	Hanky Panky	**Shondells**	(1/6)
2	Paperback Writer/Rain	**Beatles**	(3/5)
3	Dirty Water	Standells	(6/8)
4	Strangers In The Night	Frank Sinatra	(4/8)
5	Paint It Black	**Rolling Stones**	(2/9)
6	Pied Piper	**Crispian St. Peter**	(13/4)
7	I Am A Rock	**Simon & Garfunkel**	(5/12)
8	Wild Thing	Troggs	(-/3)
9	You Don't Have To Say	Dusty Springfield	(15/6)
10	You Wouldn't Listen To Me	**Ides Of March**	(10/8)
11	Red Rubber Ball	**Cyrkle**	(20/5)
12	Hey Joe	Leaves	(9/6)
13	The More I See You	Chris Montez	(17/10)
14	Cool Jerk	Capitols	(7/7)
15	Hungry	*Paul Revere*	(-/3)
16	Oh Yeah	Shadows Of Knight	(16/6)
17	Land Of Milk & Honey	Vogues	(18/6)
18	Oh How Happy	Shades Of Blue	(-/7)
19	Sweet Talkin' Guy	**Chiffons**	(19/9)
20	I Call Your Name	Buckinghams	(-/6)

(Number in parentheses indicates last week's position / Weeks heard on WCFL)

S0UND 10 STAIRWAY

Title	Artist
Ain't Too Proud To Beg	Temptations
Along Comes Mary	The Association
I Saw Her Again	Mamas & Papas
Li'l Red Riding Hood	Sam The Sham
Little Girl	Syndicate Of Sound
Please Tell Me Why	Dave Clark 5
Popsicle	Jan & Dean
Race With The Wind	Robbs
Sweet Pea	Tommy Roe
You Better Run	Young Rascals

CHICAGO PREMIERS

Title	Artist
Can I Trust You	Bachelors
Diddy Wah Diddy	Captain Beefheart
This Door Swings Both Ways	Herman's Hermits

WEEK ENDING JULY 7, 1966

1	Hanky Panky	(1/7)	**Shondells**
2	Wild Thing	(8/4)	**Troggs**
3	Paperback Writer/Rain	(2/6)	Beatles
4	Dirty Water	(3/9)	Standells
5	Strangers In The Night	(4/9)	**Frank Sinatra**
6	Pied Piper	(6/5)	**Crispian St. Peter**
7	Paint It Black	(5/10)	**Rolling Stones**
8	You Don't Have To Say	(9/7)	Dusty Springfield
9	Hungry	(15/4)	Paul Revere & Raiders
10	I Am A Rock	(7/13)	**Simon & Garfunkel**
11	You Wouldn't Listen	(10/9)	**Ides Of March**
12	Red Rubber Ball	(11/6)	Cyrkle
13	The More I See You	(13/11)	Chris Montez
14	I Saw Her Again	(-/3)	Mamas & Papas
15	Along Comes Mary	(-/4)	The Association
16	Land Of Milk & Honey	(17/7)	The Vogues
17	I Call Your Name	(20/7)	Buckinghams
18	Oh How Happy	(18/8)	Shades Of Blue
19	Sweet Pea	(-/4)	**Tommy Roe**
20	Li'l Red Riding Hood	(-/2)	Sam & Pharaohs

(Number in parentheses indicates last week's position / Weeks heard on WCFL)

SOUND 10 STAIRWAY

Happy Summer Days	Ronnie Dove
I Want You	Bob Dylan
Little Girl	Syndicate of Sound
Please Tell Me Why	Dave Clark 5
Popsicle	Jan & Dean
Pretty Flamingo	Manfred Mann
Sugar & Spice	Cryan Shames
This Door Swings Both Ways	Herman's Hermits
Work Song	Herb Alpert
You Better Run	Young Rascals

CHICAGO PREMIERS

Come On Home	Wayne Fontana
I Couldn't Live Without Your Love	Petula Clark
Summer In The City	Lovin' Spoonful

WEEK ENDING JULY 14, 1966

#	Title	Artist	(Last/Weeks)
1	Wild Thing	**Troggs**	(2/5)
2	Hanky Panky	**Shondells**	(1/8)
3	Pied Piper	Crispian St. Peter	(6/6)
4	Paperback Writer/Rain	Beatles	(3/7)
5	Hungry	**Paul Revere & Raiders**	(9/5)
6	I Saw Her Again	**Mamas & Papas**	(14/4)
7	Red Rubber Ball	**Cyrkle**	(12/7)
8	Dirty Water	Standells	(4/10)
9	You Don't Have To Say	Dusty Springfield	(8/8)
10	Along Comes Mary	**The Association**	(15/5)
11	Strangers In The Night	**Frank Sinatra**	(5/10)
12	Sugar & Spice	Cryan Shames	(-/2)
13	You Wouldn't Listen	Ides Of March	(11/10)
14	The More I See You	Chris Montez	(13/12)
15	Pretty Flamingo	Manfred Mann	(-/5)
16	This Door Swings Both Ways	Herman's Hermits	(-/3)
17	Land Of Milk & Honey	The Vogues	(16/8)
18	Li'l Red Riding Hood	Sam & Pharoahs	(20/3)
19	Sweet Pea	**Tommy Roe**	(19/5)
20	Little Girl	Syndicate of Sound	(-/5)

(Number in parentheses indicates last week's position / Weeks heard on WCFL)

SOUND 10 STAIRWAY

Cadillac	New Colony Six
I Want You	Bob Dylan
Last Time Around	Del-Vetts
Mother's Little Helper	Rolling Stones
Muddy Water	Johnny Rivers
Over, Under, Sideways, Down	Yardbirds
Popsicle	Jan & Dean
Summer In The City	Lovin' Spoonful
Work Song	Herb Alpert
You Better Run	Young Rascals

CHICAGO PREMIERS

Outside Chance	Turtles
Say I Am	Shondells
You Make Me Feel So Good	McCoys

WEEK ENDING JULY 21, 1966

1	Wild Thing	(1/6)	**Troggs**
2	Hanky Panky	(2/9)	**Shondells**
3	Pied Piper	(3/7)	Crispian St. Peter
4	Hungry	(5/6)	Paul Revere & Raiders
5	I Saw Her Again	(6/5)	**Mamas & Papas**
6	Red Rubber Ball	(7/8)	**Cyrkle**
7	Paperback Writer/Rain	(4/8)	**Beatles**
8	Coming To Take Me Away	(-/2)	Napoleon XIV
9	Summer In The City	(-/3)	Lovin' Spoonful
10	Sugar & Spice	(12/3)	**Cryan Shames**
11	Sweet Pea`	(19/6)	**Tommy Roe**
12	You Don't Have To Say	(9/9)	Dusty Springfield
13	Li'l Red Riding Hood	(18/4)	Sam & Pharaohs
14	Along Comes Mary	(10/6)	The Association
15	Pretty Flamingo	(15/6)	Manfred Mann
16	This Door Swings Both Ways	(16/4)	Herman's Hermits
17	Mother's Little Helper	(-/5)	Rolling Stones
18	Over, Under, Sideways, Down	(-/6)	Yardbirds
19	Little Girl	(20/6)	**Syndicate of Sound**
20	Popsicle	(-/9)	Jan & Dean

(Number in parentheses indicates last week's position / Weeks heard on WCFL)

SOUND 10 STAIRWAY

Title	Artist
Born A Woman	Sandy Posey
Bus Stop	Hollies
Friday's Child	Nancy Sinatra
I Couldn't Live Without Your Love	Petula Clark
I Want You	Bob Dylan
Love Letters	Elvis Presley
Sweet Dreams	Tommy McLain
Trains, Boats, Planes	Dianne Warwick
Work Song	Herb Alpert
You Better Run	Young Rascals

CHICAGO PREMIERS

Title	Artist
God Only Knows	Beach Boys
Respectable	Outsiders
With A Girl Like You	Troggs

WEEK ENDING JULY 28, 1966

1	Wild Thing	**Troggs**	(1/7)
2	Coming To Take Me Away	**Napoleon XIV**	(8/3)
3	Pied Piper	Crispian St. Peter	(3/8)
4	I Saw Her Again	Mamas & Papas	(5/6)
5	Hungry	**Paul Revere & Raiders**	(4/7)
6	Summer In The City	**Lovin' Spoonful**	(9/4)
7	Sugar & Spice	**Cryan Shames**	(10/4)
8	Red Rubber Ball	Cyrkle	(6/9)
9	Hanky Panky	Shondells	(2/10)
10	Li'l Red Riding Hood	**Sam & Pharaohs**	(13/5)
11	Along Comes Mary	**The Association**	(14/7)
12	This Door Swings Both Ways	Herman's Hermits	(16/5)
13	Sweet Pea	Tommy Roe	(11/7)
14	Mother's Helper/Lady Jane	Rolling Stones	(17/6)
15	You Don't Have To Say	Dusty Springfield	(12/10)
16	Bus Stop	The Hollies	(-/3)
17	Pretty Flamingo	Manfred Mann	(15/7)
18	Over, Under, Sideways, Down	Yardbirds	(18/7)
19	Couldn't Live Without Your Love	**Petula Clark**	(-/4)
20	Sweet Dreams	Tommy McLain	(-/9)

(Number in parentheses indicates last week's position / Weeks heard on WCFL)

SOUND 10

Cadillac	New Colony Six
Come On And See Me	Tammi Terrell
I Love Onions	Susan Christie
I Want You	Bob Dylan
Love Letters	Elvis Presley

STAIRWAY

Open The Door To Your Heart	Darrell Banks
Somewhere My Love	Roy Coniff Singers
Sunny	Bobby Hebb
Trains, Boats, Planes	D. Warwick
Work Song	Herb Alpert

CHICAGO PREMIERS

Goin' Back	Dusty Springfield
My Heart's Symphony	Gary Lewis
The Dangling Conversation	Simon & Garfunkel

WEEK ENDING AUGUST 4, 1966

1	Summer In The City	**Lovin' Spoonful**	(6/5)
2	Wild Thing	**Troggs**	(1/8)
3	Pied Piper	Crispian St. Peter	(3/9)
4	Coming To Take Me	Napoleon XIV	(2/4)
5	Hungry	**Paul Revere & Raiders**	(5/8)
6	I Saw Her Again	**Mamas & Papas**	(4/7)
7	Bus Stop	**The Hollies**	(16/4)
8	Sugar & Spice	Cryan Shames	(7/5)
9	Li'l Red Riding Hood	Sam & Pharaohs	(10/6)
10	Sweet Pea	**Tommy Roe**	(13/8)
11	Mother's Helper/Lady Jane	**Rolling Stones**	(14/7)
12	Without Your Love	*Petula Clark*	(19/5)
13	This Door Swings Both Ways	Herman's Hermits	(12/6)
14	Hanky Panky	Shondells	(9/11)
15	Over, Under, Sideways	*Yardbirds*	(18/8)
16	Sweet Dreams	Tommy McLain	(20/10)
17	Sunny	Bobby Hebb	(-/6)
18	Say I Am	Shondells	(-/4)
19	Love Letters	**Elvis Presley**	(-/6)
20	See You In September	The Happenings	(-/3)

(Number in parentheses indicates last week's position / Weeks heard on WCFL)

S0UND 10 STAIRWAY

Alfie	Cilla Black
Born A Woman	Sandy Posey
Cadillac	New Colony Six
Distant Shores	Chad & Jeremy
Feel So Good	McCoys
Misty	Groove Holmes
Respectable	Outsiders
Summertime	Billy Stewart
Trains, Boats, & Planes	D. Warwick
Wade In The Water	Ramsey Lewis

CHICAGO PREMIERS

Cherish	Association
Satisfied With You	Dave Clark Five
You Can't Hurry Love	Supremes

WEEK ENDING AUGUST 11, 1966

1	Summer In The City	**Lovin' Spoonful**	(1/6)
2	Bus Stop	***The Hollies***	(7/5)
3	Li'l Red Riding Hood	Sam & Pharaohs	(9/7)
4	I Saw Her Again	*Mamas & Papas*	(6/8)
5	Hungry	**Paul Revere & Raiders**	(5/5)
6	Wild Thing	**Troggs**	(2/9)
7	Without Your Love	**Petula Clark**	(12/6)
8	Sugar & Spice	*Cryan Shames*	(8/6)
9	Sunny	Bobby Hebb	(17/7)
10	Pied Piper	**Crispian St. Peter**	(3/10)
11	Sweet Pea	**Tommy Roe**	(10/9)
12	Door Swings Both Ways	*Herman's Hermits*	(13/7)
13	Mother's Helper/Lady Jane	Rolling Stones	(11/8)
14	Say I Am	Shondells	(18/5)
15	Over, Under, Sideways	Yardbirds	(15/9)
16	See You In September	The Happenings	(20/4)
17	Sweet Dreams	Tommy McLain	(16/11)
18	Sunshine Superman	Donovan	(-/3)
19	Born A Woman	**Sandy Posey**	(-/2)
20	Love Letters	Elvis Presley	(19/7)

(Number in parentheses indicates last week's position / Weeks heard on WCFL)

S0UND 10 STAIRWAY

Alfie	Cilla Black
Distant Shores	Chad & Jeremy
God Only Knows	Beach Boys
Misty	Groove Holmes
Respectable	Outsiders
Summertime	Billy Stewart
Wade In The Water	Ramsey Lewis
Where Were You When I Needed You	Grass Roots
With A Girl Like You	The Troggs
Wouldn't It Be Nice	Beach Boys

CHICAGO PREMIERS

Black Is Black	Los Bravos
Eleanor Rigby/Yellow Submarine	The Beatles

WEEK ENDING AUGUST 18, 1966

1	Summer In The City	(1/7)	**Lovin' Spoonful**
2	Bus Stop	(2/6)	**The Hollies**
3	Say I Am	(14/6)	Shondells
4	Sunny	(9/8)	Bobby Hebb
5	Without Your Love	(7/7)	**Petula Clark**
6	Li'l Red Riding Hood	(3/8)	**Sam & Pharoahs**
7	See You In September	(16/5)	**The Happenings**
8	Sunshine Superman	(18/4)	Donovan
9	Wild Thing	(6/10)	Troggs
10	Hungry	(5/6)	***Paul Revere & Raiders***
11	Sugar & Spice	(8/7)	**Cryan Shames**
12	I Saw Her Again	(4/9)	Mamas & Papas
13	Respectable	(-/5)	Outsiders
14	Wouldn't It Be Nice/God Only Knows	(-/5)	Beach Boys
15	Pied Piper	(10/11)	Crispian St. Peter
16	Sweet Pea	(11/10)	Tommy Roe
17	Over, Under, Sideways	(15/10)	Yardbirds
18	Summertime	(-/4)	Billy Stewart
19	Yellow Submarine/Eleanor Rigby	(-/2)	**Beatles**
20	Alfie	(-/7)	Cilla Black

(Number in parentheses indicates last week's position / Weeks heard on WCFL)

S0UND 10 STAIRWAY

Distant Shores	Chad & Jeremy
Girl Like You	The Troggs
Indication	The Zombies
Land of 1000 Dances	Wilson Pickett
My Heart's Symphony	Gary Lewis

Sunny Afternoon	The Kinks
Wade In The Water	Ramsey Lewis
Where Were You When I Needed You	Grass Roots
You Can't Hurry Love	Supremes
Searching For Your Love	B. Moore

CHICAGO PREMIERS

Bad Little Woman	Shadows Of Knight
Kissin' My Life Away	The Handells
Roller Coaster	Ides Of March

WEEK ENDING AUGUST 25, 1966

	Title	Artist	(Last/Weeks)
1	Summer In The City	Lovin' Spoonful	(1/8)
2	Bus Stop	Hollies	(2/7)
3	Sunshine Superman	Donovan	(8/5)
4	Wouldn't It Be Nice/God Only Knows	Beach Boys	(14/6)
5	See You In September	Happenings	(7/6)
6	Say I Am	Shondells	(3/7)
7	Sunny	Bobby Hebb	(4/9)
8	Li'l Red Riding Hood	Sam & Pharaohs	(6/9)
9	Without Your Love	Petula Clark	(5/8)
10	Yellow Submarine/Eleanor Rigby	Beatles	(19/3)
11	Respectable	Outsiders	(13/6)
12	Hungry	Paul Revere & Raiders	(10/7)
13	I Saw Her Again	Mamas & Papas	(12/10)
14	Sugar & Spice	Cryan Shames	(11/8)
15	Summertime	Billy Stewart	(18/5)
16	Cherish	Association	(-/4)
17	Over, Under, Sideways	Yardbirds	(17/11)
18	You Can't Hurry Love	Supremes	(-/4)
19	Sunny Afternoon	Kinks	(-/6)
20	With A Girl Like You	Troggs	(-/6)

(Number in parentheses indicates last week's position / Weeks heard on WCFL)

SOUND 10 STAIRWAY

Title	Artist
Beauty Is Only Skin Deep	Temptations
Dangling Conversation	Simon & Garfunkel
Guantanamera	Sandpipers
Indication	Zombies
Land of 1000 Dances	W. Pickett
My Heart's Symphony	Gary Lewis
Satisfied With You	Dave Clark 5
There Will Never Be Another You	Chris Montez
Wade In The Water	Ramsey Lewis
Wipeout	Surfaris

CHICAGO PREMIERS

Title	Artist
Flamingo	Herb Alpert
I've Got You Under My Skin	Four Seasons
Raining In My Heart	Legend

WEEK ENDING SEPTEMBER 1, 1966

1	Bus Stop	(2/8)	**Hollies**
2	Sunshine Superman	(3/6)	**Donovan**
3	Say I Am	(6/8)	Shondells
4	Summer In The City	(1/9)	**Lovin' Spoonful**
5	Yellow Submarine/Eleanor Rigby	(10/4)	**Beatles**
6	Sunny	(7/10)	**Bobby Hebb**
7	See You In September	(5/7)	**Happenings**
8	Cherish	(16/5)	Association
9	Wouldn't It Be Nice/God only Knows	(4/7)	Beach Boys
10	I Couldn't Live Without Your Love	(9/9)	**Petula Clark**
11	Respectable	(11/7)	**Outsiders**
12	Li'l Red Riding Hood	(8/10)	Sam & Pharoahs
13	Hungry	(12/8)	Paul Revere & Raiders
14	You Can't Hurry Love	(18/5)	Supremes
15	With A Girl Like You	(20/7)	Troggs
16	Summertime	(15/6)	Billy Stewart
17	Born A Woman	(-/5)	Sandy Posey
18	Sunny Afternoon	(19/7)	Kinks
19	Land of 1,000 Dances	(-/6)	**Wilson Pickett**
20	Guantanamera	(-/8)	Sandpipers

(Number in parentheses indicates last week's position / Weeks heard on WCFL)

S0UND 10

Wade In The Water	Ramsey Lewis
I've Been Wrong	Buckinghams
Searching For My Love	Bobby Moore
Indication	Zombies
What Becomes of Broken Hearted	Jimmy Ruffin

STAIRWAY

Black Is Black	Los Bravos
Satisfied With You	Dave Clark 5
Joker Went Wild	Brian Hyland
Working In The Coal Mine	Lee Dorsey
Beauty Is Only Skin Deep	Temptations

CHICAGO PREMIERS

I Can Make It With You	Jackie DeShannon
Girl On A Swing	Gerry & Pacemakers
Hooray For Hazel	Tommy Roe

WEEK ENDING SEPTEMBER 8, 1966

#	Title	Artist	(Last Week/Weeks)
1	Sunshine Superman	**Donovan**	(2/7)
2	Cherish	***Association***	(8/6)
3	Yellow Submarine/Eleanor Rigby	Beatles	(5/5)
4	See You In September	Happenings	(7/8)
5	Summer In The City	**Lovin' Spoonful**	(4/10)
6	Bus Stop	**Hollies**	(1/9)
7	You Can't Hurry Love	**Supremes**	(14/6)
8	Say I Am	Shondells	(3/9)
9	Sunny	Bobby Hebb	(6/11)
10	Wouldn't It Be Nice/God Only Knows	**Beach Boys**	(9/8)
11	Wipe Out	**Surfaris**	(-/4)
12	Respectable	*Outsiders*	(11/8)
13	Li'l Red Riding Hood	Sam & Pharoahs	(12/11)
14	Guantanamera	Sandpipers	(20/9)
15	With A Girl Like You	Troggs	(15/8)
16	I Couldn't Live Without Your Love	*Petula Clark*	(10/10)
17	Sunny Afternoon	Kinks	(18/8)
18	Black Is Black	Los Bravos	(-/5)
19	Land of 1,000 Dances	**Wilson Pickett**	(19/7)
20	Born A Woman	*Sandy Posey*	(17/6)

(Number in parentheses indicates last week's position / Weeks heard on WCFL)

S0UND 10 STAIRWAY

Title	Artist
Beauty Is Only Skin Deep	Temptations
Dangling Conversation	Simon & Garfunkel
Every Day, Every Night	Trolls
I've Been Wrong	Buckinghams
Joker Went Wild	Brian Hyland
Jug Band Music	Mugwumps
Another You	Chris Montez
Turn Down Day	Cyrkle
What Becomes Of Broken Hearted	Jimmy Ruffin
Working In The Cool Mine	Lee Dorsey

CHICAGO PREMIERS

Title	Artist
Changes	Crispian St. Peter
Last Train To Clarksville	Monkees
Poor Side Of Town	Johnny Rivers

WEEK ENDING SEPTEMBER 15, 1966

#	Title	Artist	(Last/Weeks)
1	Cherish	**Association**	(2/7)
2	Sunshine Superman	**Donovan**	(1/8)
3	Yellow Submarine/Eleanor Rigby	Beatles	(3/6)
4	You Can't Hurry Love	***Supremes***	(7/7)
5	See You In September	**Happenings**	(4/9)
6	Wipe Out	**Surfaris**	(11/5)
7	Wouldn't It Be Nice/God Only Knows	***Beach Boys***	(10/9)
8	Bus Stop	*Hollies*	(6/10)
9	Summer In The City	Lovin' Spoonful	(5/11)
10	Say I Am	**Shondells**	(8/10)
11	Sunny	**Bobby Hebb**	(9/12)
12	Black Is Black	Los Bravos	(18/6)
13	Guantanamera	*Sandpipers*	(14/10)
14	Joker Went Wild	Brian Hyland	(-/6)
15	With A Girl Like You	Troggs	(15/9)
16	Sunny Afternoon	Kinks	(17/9)
17	Turn Down Day	*Cyrkle*	(-/7)
18	Beauty Is Only Skin Deep	*Temptations*	(-/4)
19	Every Day, Every Night	**Trolls**	(-/3)
20	I've Been Wrong	Buckinghams	(-/7)

(Number in parentheses indicates last week's position / Weeks heard on WCFL)

SOUND 10 STAIRWAY

Another You	Chris Montez
Bad Little Woman	Shadows Of Knight
Cherry Cherry	Neil Diamond
Jug Band Music	Mugwumps
Let's Call It A Day	Razors Edge
Mr. Diengly Sad	Critters
Reach Out, I'll Be There	4 Tops
Roller Coaster	Ides Of March
Working In Coal Mine	Lee Dorsey
96 Tears	Question Mark

CHICAGO PREMIERS

Next Time You See Me	Robbs
No Salt On Her Tail	Mamas & Papas
Stop, Look & Listen	Chiffons

WEEK ENDING SEPTEMBER 22, 1966

1	Cherish	**Association**	(1/8)	11	Sunny	***Bobby Hebb***	(11/31)
2	You Can't Hurry Love	**Supremes**	(4/8)	12	Bus Stop	Hollies	(8/11)
3	Sunshine Superman	Donovan	(2/9)	13	Turn Down Day	Cyrkle	(17/8)
4	Wipe Out	Surfaris	(6/6)	14	Say I Am	Shondells	(10/11)
5	See You In September	**Happenings**	(5/10)	15	I've Been Wrong	Buckinghams	(20/8)
6	Yellow Submarine/Eleanor Rigby	**Beatles**	(3/7)	16	Mr. Dieingly Sad	Critters	(-/4)
7	Wouldn't It Be Nice/God Only Knows	**Beach Boys**	(12/7)	17	Sunny Afternoon	*Kinks*	(16/10)
8	Black Is Black	Los Bravos	(12/7)	18	Every Day, Every Night	*Trolls*	(19/4)
9	Guantanamera	Sandpipers	(13/11)	19	Beauty Is Only Skin Deep	**Temptations**	(18/5)
10	Joker Went Wild	**Brian Hyland**	(14/7)	20	Cherry Cherry	*Neil Diamond*	(-/4)

(Number in parentheses indicates last week's position / Weeks heard on WCFL)

Another You	Chris Montez	S		S T	Last Train To Clarksville	Monkees
Bad Little Women	Shadows Of Knight	0		A I	Let's Call It A Day	Razors Edge
Flamingo	Herb Alpert	U	10	R W	Reach Out, I'll Be There	Four Tops
Girl On A Swing	Gerry & Pacemakers	N		A Y	Roller Coaster	Ides Of March
Summer Samba	Walter Wonderly	D			96 Tears	Question Mark

CHICAGO PREMIERS

Go Away Little Girl	Happenings
Hair On My Chinny Chin Chin	Sam & Pharaohs
Little Man	Sonny & Cher

WEEK ENDING SEPTEMBER 29, 1966

1	Cherish	**Association**	(1/9)
2	You Can't Hurry Love	**Supremes**	(2/9)
3	Wipe Out	Surfaris	(4/7)
4	Black Is Black	Los Bravos	(8/8)
5	Sunshine Superman	**Donovan**	(3/10)
6	Joker Went Wild	**Brian Hyland**	(10/8)
7	Yellow Submarine/Eleanor Rigby	**Beatles**	(6/8)
8	Guantanamera	Sandpipers	(9/12)
9	See You In September	**Happenings**	(5/11)
10	96 Tears	**Question Mark**	(-/3)
11	Cherry Cherry	**Neil Diamond**	(20/5)
12	Mr. Dieingly Sad	Critters	(16/5)
13	Every Day, Every Night	Trolls	(18/5)
14	Turn Down Day	Cyrkle	(13/9)
15	Reach Out, I'll Be There	Four Tops	(-/3)
16	Beauty Is Only Skin Deep	Temptations	(19/6)
17	Last Train To Clarksville	Monkees	(-/4)
18	Girl On A Swing	Gerry & Pacemakers	(-/5)
19	Bad Little Woman	**Shadows Of Knight**	(-/8)
20	Summer Samba	Walter Wonderly	(-/3)

(Number in parentheses indicates last week's position / Weeks heard on WCFL)

S0UND 10 STAIRWAY

Ain't Gonna Lie	Keith
All I See Is You	Dusty Springfifeld
Flamingo	Herb Alpert
Get Away	Georgie Fame
Hooray For Hazel	Tommy Roe
Just Like A Woman	Bob Dylan
Poor Side Of Town	Johnny Rivers
Psychotic Reaction	Count Five
Summer Wind	Frank Sinatra
What Becomes Broken Hearted	Jimmy Ruffin

CHICAGO PREMIERS

I Can't Control Myself	Troggs
Paint Me A Picture	Gary Lewis
Secret Love	Billy Stewart

WEEK ENDING OCTOBER 6, 1966

#	Title	Artist	(Last/Weeks)
1	Cherish	**Association**	(1/10)
2	You Can't Hurry Love	**Supremes**	(2/10)
3	Black Is Black	Los Bravos	(4/9)
4	96 Tears	Question Mark	(10/4)
5	Wipe Out	**Surfaris**	(3/8)
6	Yellow Submarine/Eleanor Rigby	**Beatles**	(7/9)
7	Sunshine Superman	**Donovan**	(5/11)
8	Last Train To Clarksville	*Monkees*	(17/5)
9	Joker Went Wild	Brian Hyland	(6/9)
10	Cherry Cherry	**Neil Diamond**	(11/6)
11	Reach Out, I'll Be There	**Four Tops**	(15/4)
12	Every Day, Every Night	*Trolls*	(13/6)
13	Mr. Dieingly Sad	*Critters*	(12/6)
14	Guantanamera	Sandpipers	(8/13)
15	Beauty Is Only Skin Deep	Temptations	(16/7)
16	Girl On A Swing	Gerry & Pacemakers	(18/6)
17	Hooray For Hazel	Tommy Roe	(-/6)
18	Psychotic Reaction	Count Five	(-/3)
19	Ain't Gonna Lie	**Keith**	(-/8)
20	Summer Samba	Walter Wonderly	(20-4)

(Number in parentheses indicates last week's position / Weeks heard on WCFL)

Title	Artist			Title	Artist
Changes	Crispian St. Peter	S	S T	I've Got You Under My Skin	Four Seasons
Dandy	Herman's Hermits	0	A I	Poor Side Of Town	Johnny Rivers
Flamingo	Herb Alpert	U 10	R W	Summer Wind	Frank Sinatra
Get Away	Georgie Fame	N	A Y	What Becomes Broken Hearted	Jimmy Ruffin
Great Airplane Strike	Paul Revere	D		I Really Don't Want To Know	Ronnie Dove

CHICAGO PREMIERS

Title	Artist
Devil With Blue Dress On	Mitch Ryder
I've Got To Do A Little Bit Better	Joe Tex
Lady Godiva	Peter & Gordon

WEEK ENDING OCTOBER 13, 1966

#	Title	Artist	(Last/Weeks)
1	Cherish	**Association**	(1/11)
2	96 Tears	**Question Mark**	(4/5)
3	You Can't Hurry Love	Supremes	(2/11)
4	Reach Out, I'll Be There	Four Tops	(11/5)
5	Black Is Black	**Los Bravos**	(3/10)
6	Cherry Cherry	**Neil Diamond**	(10/7)
7	Last Train To Clarksville	**Monkees**	(8/6)
8	Wipe Out	Surfaris	(5/9)
9	Joker Went Wild	Brian Hyland	(9/10)
10	Hooray For Hazel	**Tommy Roe**	(17/7)
11	Sunshine Superman	**Donovan**	(7/12)
12	Mr. Dieingly Sad	Critters	(13/7)
13	Eleanor Rigby/Yellow Submarine	Beatles	(7-10)
14	Guantanamera	Sandpipers	(14/14)
15	Girl On A Swing	Gerry & Pacemakers	
16	Every Day, Every Night	Trolls	(12/7)
17	Beauty Is Only Skin Deep	Temptations	(15/8)
18	Great Airplane Strike	Paul Revere	(-/4)
19	Ain't Gonna Lie	**Keith**	(19/9)
20	Psychotic Reaction	Count Five	(18/4)

(Number in parentheses indicates last week's position / Weeks heard on WCFL)

S
0
U 10
N
D

S
T
A
I
R
W
A
Y

Title	Artist
All I See Is You	Dusty Springfield
Born Free	Roger Williams
Changes	Crispian St. Peter
Dandy	Herman's Hermits
Have You Seen Your Mother	Rolling Stones
I'm Your Puppet	J.&B. Purify
I've Got You Under My Skin	4 Seasons
Love Is A Hurtin' Thing	Lou rawls
Poor Side Of Town	Johnny Rivers
What Becomes Broken Hearted	Jimmy Ruffin

CHICAGO PREMIERS

Title	Artist
Look Through My Window	Mamas & Papas
Louie Louie	Sandpipers
Who Am I	Petula Clark

WEEK ENDING OCTOBER 20, 1966

#	Title	Artist	(Last week/Weeks)
1	96 Tears	**Question Mark**	(2/6)
2	Reach Out, I'll Be There	**Four Tops**	(4/6)
3	Last Train To Clarksville	Monkees	(7/7)
4	Cherish	Association	(1/12)
5	Cherry Cherry	**Neil Diamond**	(6/8)
6	Hooray For Hazel	**Tommy Roe**	(10/8)
7	You Can't Hurry Love	***Supremes***	(3/12)
8	Wipe Out	Surfaris	(8/10)
9	Black Is Black	Los Bravos	(5/11)
10	Dandy	**Herman's Hermits**	(-/4)
11	Girl On A Swing	**Gerry & Pacemakers**	(15/8)
12	Mr. Dieingly Sad	Critters	(12/8)
13	Psychotic Reaction	*Count Five*	(20/5)
14	Have You Seen Your Mother	Rolling Stones	(-/5)
15	Great Airplane Strike	Paul Revere	(18/5)
16	Poor Side Of Town	Johnny Rivers	(-/7)
17	Beauty Is Only Skin Deep	Temptations	(17/9)
18	Ain't Gonna Lie	Keith	(19/10)
19	What Becomes Of Broken Hearted	**Jimmy Ruffin**	(-/6)
20	I'm Your Puppet	J.&B. Purify	(-/2)

(Number in parentheses indicates last week's position / Weeks heard on WCFL)

SOUND 10 STAIRWAY

All I See Is You	Dusty Springfield
Changes	Crispian St. Peter
Go Away Little Girl	Happenings
I Can Make It With You	Pozo Secon Singers
I Can't Control Myself	Troggs
I've Got You Under My Skin	4 Seasons
Little Man	Sonny & Cher
Love Is A Hurtin' Thing	Lou Rawls
Paint Me A Picture	Gary Lewis
See See Rider	Eric Burdon & Animals

CHICAGO PREMIERS

Title	Artist
Help Me Girl	Outsiders
It's Only Love	Shondells
You Keep Me Hangin' On	Supremes

WEEK ENDING OCTOBER 27, 1966

#	Title	Artist	(Last/Weeks)
1	96 Tears	**Question Mark**	(1/7)
2	Reach Out, I'll Be There	**Four Tops**	(2/7)
3	Last Train To Clarksville	Monkees	(3/8)
4	Cherry Cherry	Neil Diamond	(5/9)
5	Hooray For Hazel	**Tommy Roe**	(6/9)
6	Cherish	**Association**	(4/13)
7	Poor Side Of Town	**Johnny Rivers**	(16/8)
8	Dandy	Herman's Hermits	(10/5)
9	Psychotic Reaction	Count Five	(13/6)
10	Have You Seen Your Mother	**Rolling Stones**	(14/6)
11	Wipe Out	**Surfaris**	(8/11)
12	Great Airplane Strike	Paul Revere	(15/6)
13	What Becomes Of Broken Hearted	*Jimmy Ruffin*	(19/7)
14	Girl On A Swing	Gerry & Pacemakers	(11/9)
15	Ain't Gonna Lie	Keith	(18/11)
16	Go Away Little Girl	Happenings	(-/6)
17	Love Is A Hurtin' Thing	Lou Rawls	(-/3)
18	See See Rider	Eric Burdon & Animals	(-/6)
19	Rain On The Roof	**Lovin' Spoonful**	(-/4)
20	I've Got You Under My Skin	Four Seasons	(-/10)

(Number in parentheses indicates last week's position / Weeks heard on WCFL)

SOUND 10 STAIRWAY

Good Vibrations — Beach Boys
I Can Make It With You — Pozo Seco Singers
I Can't Control Myself — Troggs
I Wanna Meet You — Cryan Shames
If I Were A Carpenter — Bobby Darin
Little Man — Sonny & Cher
Look Through My Window — Mamas & Papas
Paint Me A Picture — Gary Lewis
Spinout — Elvis Presley
Winchester Cathedral — New Vaudeville Band

CHICAGO PREMIERS

Behind The Door — Cher
Hazy Shade Of Winter — Simon & Garfunkel
I Got The Feelin' — Neil Diamond

WEEK ENDING NOVEMBER 3, 1966

#	Title	Artist	(Last week/Weeks)
1	96 Tears	**Question Mark**	(1/8)
2	Reach Out, I'll Be There	**Four Tops**	(2/8)
3	Last Train To Clarksville	Monkees	(3/9)
4	Hooray For Hazel	Tommy Roe	(5/10)
5	Cherry Cherry	**Neil Diamond**	(4/10)
6	Poor Side Of Town	**Johnny Rivers**	(7/9)
7	Good Vibrations	**Beach Boys**	(-/4)
8	Dandy	Herman's Hermits	(8/6)
9	Psychotic Reaction	Count Five	(9/7)
10	Cherish	**Association**	(6/14)
11	Have You Seen Your Mother	**Rolling Stones**	(10/7)
12	Great Airplane Strike	Paul Revere	(12/7)
13	Rain On The Roof	*Lovin' Spoonful*	(19/5)
14	What Becomes Of Broken Hearted	Jimmy Ruffin	(13/8)
15	I Wanna Meet You	Cryan Shames	(-/3)
16	See See Rider	Eric Burdon & Animals	(18/7)
17	If I Were A Carpenter	Bobby Darin	(-/6)
18	Winchester Cathedral	New Vaudeville Band	(-/3)
19	Ain't Gonna Lie	**Keith**	(15/12)
20	Go Away Little Girl	Happenings	(16/7)

(Number in parentheses indicates last week's position / Weeks heard on WCFL)

SOUND 10

Title	Artist
I Can Make It With You	Pozo Seco Singers
I'm Your Puppet	J.&B. Purify
Little Man	Sonny & Cher
Look Through My Window	Mamas & Papas
Louie Louie	Sandpipers

STAIRWAY

Paint Me A Picture	Gary Lewis
Pipeline	Chantays
Stop Stop Stop	Hollies
Wish You Were Here, Buddy	Pat Boone
You Keep Me Hangin' On	Supremes

CHICAGO PREMIERS

Title	Artist
Hi Hi Hazel	Gary & Hornets
I'm Gonna Make You Mine	Shadows of Knight
Mellow Yellow	Donovan

WEEK ENDING NOVEMBER 10, 1966

#	Title	Artist	(Last/Weeks)
1	Reach Out, I'll Be There	**Four Tops**	(2/9)
2	Good Vibrations	**Beach Boys**	(7/5)
3	Hooray For Hazel	Tommy Roe	(4/11)
4	96 Tears	Question Mark	(1/9)
5	Poor Side Of Town	**Johnny Rivers**	(6/10)
6	Last Train To Clarksville	**Monkees**	(3/10)
7	Winchester Cathedral	**New Vaudeville Band**	(18/4)
8	You Keep Me Hangin' On	Supremes	(-/4)
9	Psychotic Reaction	Count Five	(9/8)
10	I Wanna Meet You	**Cryan Shames**	(15/4)
11	Dandy	**Herman's Hermits**	(8/7)
12	Stop Stop Stop	Hollies	(-/5)
13	What Becomes Of Broken Hearted	Jimmie Ruffin	(13/9)
14	Great Airplane Strike	Paul Revere	(12/8)
15	Rain On The Roof	Lovin' Spoonful	(13/6)
16	If I Were A Carpenter	Bobby Darin	(17/7)
17	Devil With Blue Dress On	Mitch Ryder	(-/6)
18	Look Through My Window	Mamas & Papas	(-/5)
19	See See Rider	**Eric Burdon & Animals**	(16/8)
20	Paint Me A Picture	Gary Lewis	(-/7)

(Number in parentheses indicates last week's position / Weeks heard on WCFL)

SOUND 10		STAIRWAY	
Hair On My Chinny Chin Chin	Sam Sham	Lady Godiva	Peter & Gordon
I'm Ready For Love	Martha & Vandellas	Louie Louie	Sandpipers
I'm Your Puppet	J.& B. Purify	Pipeline	Chantays
It's Only Love	Shondells	Secret Love	Billy Stewart
Spinout	Elvis Presley	Who Am I	Petula Clark

CHICAGO PREMIERS

A Fool Am I	Cilla Black
I'm The One You Need	Miracles
Mame	Herb Alpert

WEEK ENDING NOVEMBER 17, 1966

1	Good Vibrations	**Beach Boys**	(2/6)
2	Reach Out, I'll Be There	**Four Tops**	(1/10)
3	Winchester Cathedral	New Vaudeville Band	(7/5)
4	Poor Side Of Town	Johnny Rivers	(5/11)
5	Last Train To Clarksville	**Monkees**	(6/11)
6	You Keep Me Hangin' On	**Supremes**	(8/5)
7	Hooray For Hazel	**Tommy Roe**	(3/12)
8	96 Tears	Question Mark	(4/10)
9	I Wanna Meet You	Cryan Chames	(10/5)
10	Stop Stop Stop	**Hollies**	(12/6)
11	What Becomes Of Broken Hearted	**Jimmie Ruffin**	(13/10)
12	Dandy	Herman's Hermits	(11/8)
13	Devil With Blue Dress On	Mitch Ryder	(17/7)
14	Rain On The Roof	Lovin' Spoonful	(15/7)
15	If I Were A Carpenter	Bobby Darin	(16/8)
16	See See Rider	Eric Burdon & Animals	(19/9)
17	Lady Godiva	Peter & Gordon	(-/7)
18	Paint Me A Picture	Gary Lewis	(20-8)
19	Pipeline	**Chantays**	(-/4)
20	Secret Love	Billy Stewart	(-/8)

(Number in parentheses indicates last week's position / Weeks heard on WCFL)

S0UND 10 STAIRWAY

19 Days	Dave Clark 5
Hair On Chinny Chin	Sam The Sham
Hazy Shade Of Winter	Simon & Garfunkel
Help Me Girl	Outsiders
Time After Time	Chris Montez
I'm Ready For Love	Martha & Vandelias
I'm Your Puppet	J. & B. Purify
It's Only Love	Shondells
Mellow Yellow	Donovan
Who Am I	Petula Clark

CHICAGO PREMIERS

A Place In The Sun	Stevie Wonder
Dusty	Floyd & Jerry
Pandor's Golden Heebie Jeebies	The Association

WEEK ENDING NOVEMBER 24, 1966

1	Good Vibratons	**Beach Boys**	(1/7)
2	Winchester Cathedral	**New Vaudeville Band**	(3/6)
3	You Keep Me Hangin' On	Supremes	(6/6)
4	Reach Out, I'll Be There	Four Tops	(2/11)
5	Poor Side Of Town	**Johnny Rivers**	(4/12)
6	Last Train To Clarksville	**Monkees**	(5/12)
7	Devil With Blue Dress On	**Mitch Ryder**	(13/8)
8	Lady Godiva	Peter & Gordon	(17/8)
9	Stop Stop Stop	Hollies	(10/7)
10	What Becomes Of Broken Hearted	**Jimmie Ruffin**	(11/11)
11	Hooray For Hazel	**Tommy Roe**	(7/13)
12	I Wanna Meet You	Cryan Shames	(9/6)
13	Mellow Yellow	*Donovan*	(-/4)
14	Rain On The Roof	Lovin' Spoonful	(14/8)
15	Dandy	Herman's Hermits	(12/9)
16	See See Rider	Eric Burdon & Animals	(16/10)
17	It's Only Love	Shondells	(-/6)
18	Hair On My Chinny Chin	Sam The Sham	(-/10)
19	Talk Talk	**Music Machine**	(-/5)
20	Who Am I	Petula Clark	(-/7)

(Number in parentheses indicates last week's position / Weeks heard on WCFL)

S0UND 10 STAIRWAY

19 Days	Dave Clark 5
Coming On Strong	Brenda Lee
Hazy Shade Of Winter	Simon & Garfunkel
Help Me Girl	Outsiders
I Got The Feeling	Neil Diamond
I'm Ready For Love	Martha & Vandellas
Kind Of A Drag	Buckinghams
Louie Louie	Sandpipers
Sugar Town	Nancy Sinatra
Time After Time	Chris Montez

CHICAGO PREMIERS

Communications Breakdown	Roy Orbison
Going Nowhere	Los Bravos
The Girl That Stood Beside Me	Bobby Darin

WEEK ENDING DECEMBER 1, 1966

1	Winchester Cathedral	**New Vaudeville Band**	(2/7)
2	Good Vibrations	**Beach Boys**	(1/8)
3	You Keep Me Hangin' On	Supremes	(3/7)
4	Poor Side Of Town	*Johnny Rivers*	(5/13)
5	Lady Godiva	**Peter & Gordon**	(8/9)
6	Mellow Yellow	***Donovan***	(13/5)
7	Reach Out, I'll Be There	**Four Tops**	(4/12)
8	Devil With Blue Dress On	Mitch Ryder	(7/9)
9	Stop Stop Stop	Hollies	(9/8)
10	What Becomes Of Broken Hearted	**Jimmie Ruffin**	(10/12)
11	I Wanna Meet You	**Cryan Shames**	(12/7)
12	Sugar Town	Nancy Sinatra	(-/5)
13	It's Only Love	Shondells	(17/7)
14	Rain On The Roof	Lovin' Spoonful	(14/9)
15	Kind Of A Drag	Buckinghams	(-/4)
16	Help Me Girl	Outsiders	(-/7)
17	Talk Talk Talk	*Music Machine*	(19/6)
18	Time After Time	Chris Montez	(-/5)
19	I'm Ready For Love	**Martha & Vandellas**	(-/7)
20	Louie Louie	Sandpipers	(-/9)

(Number in parentheses indicates last week's position / Weeks heard on WCFL)

S0UND 10 STAIRWAY

19 Days	David Clark 5
Coming On Strong	Brenda Lee
Hazy Shade Of Winter	Simon & Garfunkel
I Got The Feeling	Neil Diamond
I Know I'm Losing You	Temptations
I'm Gonna Make You Mine	Shadows Of Night
Mame	Herb Alpert
Please Don't Ever Leave Me	Cyrkle
She Comes To Me	Chicago Loop
That's Life	Frank Sinatra

CHICAGO PREMIERS

Baby What I Mean	Drifters
Good Thing	Paul Revere & Raiders
I'm A Believer	Monkees

WEEK ENDING DECEMBER 8, 1966

1	Winchester Cathedral	**New Vaudeville Band**	(1/8)
2	Good Vibrations	**Beach Boys**	(2/9)
3	Mellow Yellow	Donovan	(6/6)
4	You Keep Me Hangin' On	Supremes	(3/8)
5	Lady Godiva	**Peter & Gordon**	(5/10)
6	Devil With Blue Dress On	**Mitch Ryder**	(8/10)
7	Kind Of A Drag	**Buckinghams**	(15/5)
8	Poor Side Of Town	Johnny Rivers	(4/14)
9	Talk Talk Talk	Music Machine	(17/7)
10	Stop Stop Stop	**Hollies**	(9/9)
11	I Wanna Meet You	***Cryan Shames***	(11/8)
12	Sugar Town	Nancy Sinatra	(12/6)
13	What Becomes Of Broken Hearted	Jimmie Ruffin	(10/13)
14	It's Only Love	Shondells	(13/8)
15	Hazy Shade Of Winter	*Simon & Garfunkel*	(-/7)
16	Good Thing	Paul Revere & Raiders	(-/2)
17	Time After Time	Chris Montez	(18/6)
18	I'm Ready For Love	Martha & Vandellas	(19/8)
19	Mame	***Herb Alpert***	(-/5)
20	Cry	Ronnie Dove	(-/4)

(Number in parentheses indicates last week's position / Weeks heard on WCFL)

SOUND 10 STAIRWAY

Coming On Strong	Brenda Lee
Dancing In Street/Words of Love	Mamas & Papas
East West	Herman's Hermits
I Got The Feeling	Neil Diamond
I Know I'm Losing You	Temptations
I'm Gonna Make You Mine	Shadows Of Knight
Just One Smile	Gene Pitney
Please Don't Ever Leave Me	Cyrkle
Single Girl	Sandy Posey
That's Life	Frank Sinatra

CHICAGO PREMIERS

I've Passed This Way Before	Jimmy Ruffin
Standing In The Shadows Of Love	Four Tops
Tell It To The Rain	Four Seasons

WEEK ENDING DECEMBER 15, 1966

1	Winchester Cathedral	(1/9)	**New Vaudeville Band**
2	Good Vibrations	(2/10)	**Beach Boys**
3	Mellow Yellow	(3/7)	Donovan
4	You Keep Me Hangin' On	(4/9)	Supremes
5	Devil With Blue Dress On	(6/11)	**Mitch Ryder**
6	Kind Of A Drag	(7/6)	**Buckinghams**
7	Lady Godiva	(5/11)	**Peter & Gordon**
8	Sugar Town	(12/7)	Nancy Sinatra
9	Talk Talk Talk	(9/8)	Music Machine
10	It's Only Love	(14/9)	**Shondells**
11	Good Thing	(16/3)	**Paul Revere & Raiders**
12	I Wanna Meet You	(11/9)	Cryan Shames
13	Stop Stop Stop	(10/10)	Hollies
14	Dancing In Streets/Words of Love	(-/4)	Mamas & Papas
15	Snoopy Vs. Red Baron	(-/2)	Royal Guardsmen
16	East West	(-/3)	Herman's Hermits
17	I'm A Believer/Steppin' Stone	(-/3)	Monkees
18	Cry	(20/5)	Ronnie Dove
19	Mame	(19/6)	**Herb Alpert**
20	Time After Time	(17/7)	Chris Montez

(Number in parentheses indicates last week's position / Weeks heard on WCFL)

SOUND 10 STAIRWAY

Grizzly Bear	Youngbloods
Happenings 10 Years Ago	Yardbirds
Help Me Girl	Eric Burdon & Animals
I Got The Feeling	Neil Diamond
Single Girl	Sandy Posey
I Know I'm Losing You	Temptations
Just One Smile	Gene Pitney
Please Don't Ever Leave Me	Cyrkle
Shadows Of Your Love	Four Tops
That's Life	Frank Sinatra

CHICAGO PREMIERS

Color My World	Petula Clark
How Do You Catch A Girl	Sam The Sham
Knight In Rusty Armor	Peter & Gordon

WEEK ENDING DECEMBER 22, 1966

1	Winchester Cathedral	***New Vaudeville Band***	(1/10)
2	Kind Of A Drag	***Buckinghams***	(6/7)
3	Snoopy Vs. Red Baron	*Royal Guardsmen*	(15/3)
4	I'm A Believer/Steppin' Stone	*Monkees*	(17/4)
5	Mellow Yellow	***Donovan***	(3/8)
6	Devil With Blue Dress On	***Mitch Ryder***	(5/12)
7	Good Vibrations	***Beach Boys***	(2/11)
8	Lady Godiva	*Peter & Gordon*	(7/12)
9	You Keep Me Hangin' On	*Supremes*	(4/10)
10	Good Thing	***Paul Revere & Raiders***	(11/4)
11	Talk Talk Talk	***Music Machine***	(9/9)
12	Sugar Town	*Nancy Sinatra*	(8/8)
13	Dancing In Streets/Words of Love	*Mamas & Papas*	(14/5)
14	I Wanna Meet You	*Cryan Shames*	(12/10)
15	It's Only Love	*Shondells*	(10/10)
16	East West	*Herman's Hermits*	(16/4)
17	Just One Smile	*Gene Pitney*	(-/6)
18	That's Life	*Frank Sinatra*	(-/7)
19	Tell It Like It Is	***Aaron Neville***	(-/2)
20	Cry	*Ronnie Dove*	(18/6)

(Number in parentheses indicates last week's position / Weeks heard on WCFL)

S0UND 10

98.6	Keith
Coming Home Soldier	Bobby Vinton
Grizzly Bear	Youngbloods
Help Me Girl	Outsiders/Eric Burdon & Animals
I Got The Feeling	Neil Diamond

STAIRWAY

I Need Somebody	? & Mysterians
Nashville Cats	Lovin' Spoonful
Please Don't Ever Leave Me	Cyrkle
Single Girl	Sandy Posey
Shadows Of Your Love	Four Tops

CHRISTMAS PREMIERS

If Every Day Was Christmas	Elvis Presley
Silent Night	Barbra Streisand
Winters Children	Capes of Good Hope

1967

WEEK ENDING JANUARY 5, 1967

1	Snoopy Vs. Red Baron	***Royal Guardsmen***	(1/5)
2	Kind Of A Drag	***Buckinghams***	(7/9)
3	I'm A Believer/Steppin' Stone	*Monkees*	(2/6)
4	Good Thing	*Paul Revere & Raiders*	5/6)
5	Winchester Cathedral	***New Vaudeville Band***	(3/12)
6	Words Of Love/Dancing In Streets	***Mamas & Papas***	(8/7)
7	Mellow Yellow	***Donovan***	(4/10)
8	Just One Smile	*Gene Pitney*	(13/8)
9	Sugar Town	*Nancy Sinatra*	(9/10)
10	Devil With Blue Dress On	***Mitch Ryder***	(6/14)
11	East West	***Herman's Hermits***	(11/6)
12	Shadows Of Your Love	*Four Tops*	(17/5)
13	Tell It Like It Is	*Aaron Neville*	(16/4)
14	I Love You So Much	*New Colony Six*	(-/3)
15	That's Life	*Frank Sinatra*	(18/9)
16	Nashville Cats	*Lovin' Spoonful*	(-/5)
17	Grizzly Bear	*Youngbloods*	(20/6)
18	98.6	*Keith*	(-/6)
19	Georgy Girl	***Seekers***	(-3)
20	Ain't Got Nothin' Yet	*Blues Magoos*	(-/5)

(Number in parentheses indicates last week's position / Weeks heard on WCFL)

S0UND 10 STAIRWAY

All	James Darren
Coming Home Soldier	Bobby Vinton
Communication Breakdown	Roy Orbison
Galiant Men	Senator Everett McKinley Dirksen
I Had Too Much To Dream	Electric Prunes
I Need Somebody	? & Mysterians
It's Now Winters Day	Tommy Roe
Knight In Rusty Armour	Gordon
There's Got To Be A Word	Innocence
Where Will The Words Come From	Gary Lewis

HAPPY NEW YEAR TO ALL OF YOU
FROM THE WCFL VIP'S.

WEEK ENDING DECEMBER 29, 1966

1	Snoopy Vs. Red Baron	***Royal Guardsmen***	(3/4)
2	I'm A Believer/Steppin' Stone	***Monkees***	(4/5)
3	Winchester Cathedral	*New Vaudeville Band*	(1/11)
4	Mellow Yellow	*Donovan*	(5/9)
5	Good Thing	***Paul Revere Raiders***	(10/5)
6	Devil With Blue Dress On	***Mitch Ryder***	(6/13)
7	Kind Of A Drag	***Buckinghams***	(2/8)
8	Words Of Love/Dancing In Streets	*Mamas & Papas*	(13/6)
9	Sugar Town	*Nancy Sinatra*	(12/9)
10	Talk Talk	***Music Machine***	(11/10)
11	East West	***Herman's Hermits***	(16/5)
12	Good Vibrations	*Beach Boys*	(7/12)
13	Just One Smile	*Gene Pitney*	(17/7)
14	Lady Godiva	*Peter & Gordon*	(8/13)
15	You Keep Me Hangin' On	*Supremes*	(9/11)
16	Tell It Like It Is	*Aaron Neville*	(19/3)
17	Shadows Of Your Love	*Four Tops*	(-/4)
18	That's Life	*Frank Sinatra*	(18/8)
19	Cry	***Ronnie Dove***	(20/7)
20	Grizzly Bear	*Youngbloods*	(-/5)

(Number in parentheses indicates last week's position / Weeks heard on WCFL)

SOUND 10 STAIRWAY

98.6	Keith
Ain't Got Nothin' Yet	Blues Magoos
Communications Breakdown	Roy Orbison
Gallant Men	Senator Everett McKinley Dirksen
Nashville Cats	Lovin' Spoonful

I Got The Feeling	Neil Diamond
I Know I'm Losing You	Temptations
I Need Somebody	? & Mysterians
Love You So Much	New Colony Six
Single Girl	Sandy Posey

CHICAGO PREMIERS

To Where You Wanna Go	5th Dimension
I've Got To Have A Reason	Dave Clark 5

WEEK ENDING JANUARY 12, 1967

	Title	Artist	(Last/Weeks)
1	Snoopy Vs. Red Baron	***Royal Guardsmen***	(1/6)
2	I'm A Believer/Steppin' Stone	***Monkees***	(3/7)
3	Good Thing	*Paul Revere & Raiders*	(4/7)
4	Kind Of A Drag	*Buckinghams*	(2/10)
5	Winchester Cathedral	***New Vaudeville Band***	(5/13)
6	Words Of Love/Dancing In Streets	***Mamas & Papas***	(6/8)
7	Mellow Yellow	***Donovan***	(7/11)
8	Just One Smile	*Gene Pitney*	(8/9)
9	Devil With Blue Dress On	*Mitch Ryder*	(10/15)
10	Sugar Town	***Nancy Sinatra***	(9/11)
11	Tell It Like It Is	***Aaron Neville***	(13/5)
12	Shadows Of Your Love	*Four Tops*	(12/6)
13	I Love You So Much	*New Colony Six*	(14/4)
14	Georgy Girl	*Seekers*	(19/4)
15	Ain't Got Nothin' Yet	*Blue Magoos*	(20/6)
16	98.6	*Keith*	(18/7)
17	Nashville Cats	*Lovin' Spoonful*	(16/6)
18	Grizzly Bear	*Youngbloods*	(17/7)
19	Coming Home Soldier	***Bobby Vinton***	(-/5)
20	East West	*Herman's Hermits*	(11/7)

(Number in parentheses indicates last week's position / Weeks heard on WCFL)

SOUND 10 STAIRWAY

All	James Darren
Gallant Men	Senator Everett
I Had Too Much To Dream	Electric Prunes
It's Now Winters Day	Tommy Roe
Knight In Rusty Armour	Peter Gordon
Pushin' Too Hard	Seeds
Tell It To The Rain	Four Seasons
There's Got To Be A Word	Innocence
Where Will The Words Come From	Gary Lewis
Wild Thing	Hardly Worth It Players

CHICAGO PREMIERS

Lovin' You	Bobby Darin
Ride Ride Ride	Brenda Lee

WEEK ENDING JANUARY 19, 1967

#	Title	Artist	(Last/Weeks)
1	I'm A Believer/Steppin' Stone	***Monkees***	(2/8)
2	Snoopy Vs. Red Baron	***Royal Guardsmen***	(1/7)
3	Good Thing	*Paul Revere & Raiders*	(3/8)
4	Kind Of A Drag	*Buckinghams*	(4/11)
5	Words Of Love/Dancing In Streets	***Mamas & Papas***	(6/9)
6	Georgy Girl	***Seekers***	(14/5)
7	I Love You So Much	***New Colony Six***	(13/5)
8	Ain't Got Nothin' Yet	*Blues Magoos*	(15/7)
9	Tell It Like It Is	*Aaron Neville*	(11/6)
10	Sugar Town	***Nancy Sinatra***	(10/12)
11	Just One Smile	***Gene Pitney***	(8/10)
12	Shadows Of Your Love	*Four Tops*	(12/7)
13	Pushin' Too Hard	*Seeds*	(-/6)
14	98.6	*Keith*	(16/8)
15	It's Now Winters Days	*Tommy Roe*	(-/6)
16	Nashville Cats	*Lovin' Spoonful*	(17-7)
17	I Had Too Much To Dream	*Electric Prunes*	(-/5)
18	All	*James Dareen*	(-/5)
19	Knight In Rusty Armour	***Peter & Gordon***	(-/6)
20	Wild Thing	*Sen. Bobby & Hardly Worth It Players*	(-/3)

(Number in parentheses indicates last week's position / Weeks heard on WCFL)

S 0 U N D 10 S T A I R W A Y

Title	Artist
Can't You See	The Flock
Color My World	Petula Clark
Gimme Some Lovin'	Spencer Davis Group
I've Got To Have A Reason	Dave Clark 5
I've Passed This Way Before	Jimmy Ruffin
Music To Watch Girls By	Bob Crewe Generation
Ruby Tuesday	Rolling Stones
Tell It To The Rain	Four Seasons
Tramp	Lowell Fursom
Where Will The Words Come From	Gary Lewis

CHICAGO PREMIERS

Title	Artist
Love Is Here And Now You're Gone	Supremes
So You Want To Be A Rock 'N Roll Star	Byrds
The Beat Goes on	Sonny & Cher

WEEK ENDING JANUARY 26, 1967

#	Title	Artist	(Last/Weeks)
1	I'm A Believer	***Monkees***	(1/9)
2	Georgy Girl	*Seekers*	(6/6)
3	Snoopy Vs. Red Baron	*Royal Guardsmen*	(2/8)
4	I Love You So Much	*New Colony Six*	(7/6)
5	Ain't Got Nothin' Yet	***Blues Magoos***	(8/8)
6	Good Thing	*Paul Revere & Raiders*	(3/9)
7	Kind Of A Drag	*Buckinghams*	(4/12)
8	Pushin' Too Hard	*Seeds*	(13/7)
9	Words Of Love	*Mamas & Papas*	(5/10)
10	Shadows Of Love	***Four Tops***	(12/8)
11	I Had Too Much To Dream	***Electric Prunes***	(17/6)
12	Tell It Like It Is	*Aaron Neville*	(9/7)
13	It's Now Winters Day	*Tommy Roe*	(15/7)
14	98.6	*Keith*	(14/9)
15	Gimme Some Lovin'	*Spencer Davis Group*	(-/3)
16	Knight In Rusty Armour	*Peter & Gordon*	(19/7)
17	Ruby Tuesday	*Rolling Stones*	(-/2)
18	All	*James Darren*	(18/6)
19	Where Will Words Come From	*Gary Lewis*	(-/9)
20	Music To Watch Girls By	*Bob Crewe Generation*	(-/4)

(Number in parentheses indicates last week's position / Weeks heard on WCFL)

SOUND 10 STAIRWAY

Title	Artist
Can't You See	The Flock
Color My World	Petula Clark
Green Green Grass Of Home	Tom Jones
Hey Leroy!	Jimmy Castor
I've Got To Have A Reason	Dave Clark 5
I've Passed This Way Before	Jimmy Ruffin
Stand By Me	Spyder Turner
Tell It To The Rain	Four Seasons
Tramp	Lowell Fulsom
Wish Me A Rainbow	Gunter Kallman Chorus

CHICAGO PREMIERS

Title	Artist
Baby I Need Your Loving	Johnny Rivers
Happy Together	Turtles
I Think We're Alone Now	Tommy James & Shondells

WEEK ENDING FEBRUARY 2, 1967

1.	**Georgy Girl**	***Seekers***	2/7
2.	**I'm A Believer**	***Monkees***	1/10
3.	**Pushin' Too Hard**	***Seeds***	8/8
4.	**I Love You So Much**	***New Colony Six***	4/7
5.	**Ain't Got Nothin' Yet**	***Blues Magoos***	5/9
6.	**Snoopy Vs. Red Baron**	***Royal Guardsmen***	3/9
7.	**I Had Too Much To Dream**	***Electric Prunes***	11/7
8.	**Ruby Tuesday**	***Rolling Stones***	17/3
9.	**Kind Of A Drag**	***Buckinghams***	7/13
10.	**Tell It Like It Is**	***Aaron Neville***	12/8
11.	Gimme Some Lovin'	*Spencer Davis Group*	15/4
12.	It's Now Winters Day	*Tommy Roe*	13/8
13.	Shadows Of Love	*Four Tops*	10/9
14.	98.6	*Keith*	14/10
15.	All	*James Darren*	18/7
16.	Knight in Rusty Armour	*Peter & Gordon*	16/8
17.	Music To Watch Girls By	*Bob Crewe Generation*	20/5
18.	Green Green Grass Of Home	*Tom Jones*	-/4
19.	Where Will The Words Come From	*Gary Lewis*	19/10
20.	Tell It To The Rain	*Four Seasons*	-/9

(First number indicates last week's position/Second number is the weeks heard on WCFL)

ACTION TEN

21.	ARE YOU LONELY FOR ME	FREDDY SCOTT
22.	CAN'T YOU SEE I STILL LOVE YOU	THE FLOCK
23.	HELLO HELLO	SOPWITH CAMEL
24.	I'VE GOT TO HAVE A REASON	DAVE CLARK 5
25.	LOVE IS HERE AND NOW YOU'RE GONE	SUPREMES
26.	LOVIN' YOU	BOBBY DARIN
27.	OOH BABY	BO DIDDLEY
28.	THE BEAT GOES ON	SONNY & CHER
29.	THEN YOU CAN TELL ME GOODBYE	CASINOS
30.	TRAMP	LOWELL FULSOM

CHICAGO PREMIERES

DON'T DO IT	MICKY DOLENZ
MR. FARMER	THE SEEDS
SOCK IT TO BE BABY	MITCH RYDER

WEEK ENDING FEBRUARY 9, 1967

1.	**Georgy Girl**	**Seekers**	1/8
2.	**Pushin' Too Hard**	***Seeds***	3/9
3.	**Ruby Tuesday**	***Rolling Stones***	8/4
4.	**I'm A Believer**	***Monkees***	2/11
5.	**I Had Too Much To Dream**	***Electric Prunese***	7/8
6.	**I Love You So Much**	***New Colony Six***	4/8
7.	**Ain't Got Nothin' Yet**	***Blues Magoos***	5/10
8.	**Snoopy Vs. Red Baron**	***Royal Guardsmen***	6/10
9.	**Gimme Some Lovin'**	***Spencer Davis Group***	11/5
10.	**Tell It Like It Is**	***Aaron Neville***	10/9
11.	It's Now Winters Day	*Tommy Roe*	12/9
12.	Shadows Of Love	*Four Tops*	13/10
13.	Green Green Grass Of Home	*Tom Jones*	18/5
14.	Love Is Here & Now You're Gone	*Supremes*	-/4
15.	The Beat Goes On	*Sonny & Cher*	-/4
16.	Music To Watch Girls By	*Bob Crewe Generation*	17/6
17.	Where Will The Words Come From	*Gary Lewis*	19/11
18.	Then You Can Tell Me Goodbye	*Casinos*	-/7
19.	Hello Hello	*Sopwith Camel*	-/5
20.	All	*James Darren*	15/8

(First number indicates last week's position/Second number is the weeks heard on WCFL)

ACTION TEN

21.	I THINK WE'RE ALONE NOW	TOOMMY JAMES & SHONDELLS
22.	PRETTY BALLERINA	LEFT BANKE
23.	I'VE PASSED THIS WAY BEFORE	JIMMY RUFFIN

24. LOVIN' YOU	BOBBY DARIN
25. MECHANICAL MAN	BENT BOLT
26. SOCK IT TO ME BABY	MITCH RYDER
27. MERCY MERCY	CANNONBALL ADDERLEY
28. HAPPY TOGETHER	TURTLES
29. YOU GOT TO ME	NEIL DIAMOND
30. NO FAIR AT ALL	ASSOCIATION

CHICAGO PREMIERES

DARLING BE HOME SOON	LOVIN' SPOONFUL
IN THE MIDNIGHT HOUR	DICK WHITTINGTON'S CATS
SAVE ME	DAVE, DEE, DOZY, BEAKY, MICK & TICH

WEEK ENDING FEBRUARY 16, 1967

1.	**Ruby Tuesday**	**Rolling Stones**	3/5
2.	**Georgy Girl**	***Seekers***	1/9
3.	**Pushin' Too Hard**	***Seeds***	2/10
4.	**I Had Too Much To Dream**	***Electric Prunes***	5/9
5.	**I'm A Believer**	***Monkees***	4/12
6.	**Gimme Some Lovin'**	***Spencer Davis Group***	9/6
7.	**I Think We're Alone Now**	***Tommy James & Shondells***	-/4
8.	**Ain't Got Nothin' Yet**	***Blues Magoos***	7/11
9.	**I Love You So Much**	***New Colony Six***	6/9
10.	**Love Is Here & Now You're Gone**	***Supremes***	14/5
11.	The Beat Goes On	*Sonny & Cher*	15/5
12.	Green Green Grass Of Home	*Tom Jones*	13/6
13.	Music To Watch Girls By	*Bob Crewe Generation*	16/7
14.	It's Now Winters Day	*Tommy Roe*	11/10
15.	Then You Can Tell Me Goodbye	*Casinos*	18/8
16.	There's A Kind Of A Hush	*Herman's Hermits*	-/3
17.	Pretty Ballerina	*Left Banke*	-/7
18.	Where Will The Words Come From	*Gary Lewis*	17/12
19.	You Got To Me	*Neil Diamond*	-/5
20.	Hello Hello	*Sopwith Camel*	19/6

(First number indicates last week's position/Second number is the weeks heard on WCFL)

ACTION TEN

21.	BABY I NEED YOUR LOVIN'	JOHNNY RIVERS
22.	NO FAIR AT ALL	THE ASSOCIATION
23.	HAPPY TOGETHER	THE TURTLES
24.	MERCY MERCY	CANNONBALL ADDERLEY
25.	SOCK IT TO ME BABY	MITCH RYDER
26.	THE HUNTER GETS CAPTURED	MARVELETTES
27.	MECHANICAL MAN	BENT BOLT
28.	SO YOU WANT TO BES A ROCK & ROLL STAR	THE BYRDS
29.	FOR WHAT IT'S WORTH	BUFFALO SPRINGFIELD
30.	INDESCRIBABLY BLUE	ELVIS PRESLEY

CHICAGO PREMIERES

DEDICATED TO THE ONE I LOVE	MAMAS & PAPAS
PENNY LANE	THE BEATLES
STRAWBERRY FIELDS FOREVER	THE BEATLES

WEEK ENDING FEBRUARY 23, 1967

1.	**Ruby Tuesday**	**Rolling Stones**	1/6
2.	**I Think We're Alone Now**	***Tommy James & Shondells***	7/5
3.	**Georgy Girl**	***Seekers***	2/10
4.	**I Had Too Much To Dream**	***Electric Prunes***	4/10
5.	**Pushin' Too Hard**	***Seeds***	3/11
6.	**Gimme Some Lovin'**	***Spencer Davis Group***	6/7
7.	**Love Is Here & Now You're Gone**	***Supremes***	10/6
8.	**The Beat Goes On**	***Sonny & Cher***	11/6
9.	**I'm A Believer**	***Monkees***	5/13
10.	**There's a Kind Of Hush**	***Herman's Hermits***	16/4
11.	I Love You So Much	*New Colony Six*	9/10
12.	Ain't Got Nothin' Yet	*Blues Magoos*	8/12
13.	Then You Can Tell Me Goodbye	*Casinos*	15/9
14.	Happy Together	*The Turtles*	-/5
15.	Sock It To Me Baby	*Mitch Ryder*	-/4
16.	Pretty Ballerina	*Left Banke*	17/8
17.	Green Green Grass Of Home	*Tom Jones*	12/7
18.	You Got To Me	*Neil Diamond*	19/6
19.	No Fair At all	*The Association*	-/7
20.	Baby I Need Your Lovin'	*Johnny Rivers*	-/5

(First number indicates last week's position/Second number is the weeks heard on WCFL)

ACTION TEN

21.	MERCY MERCY	CANNONBALL ADDERLY
22.	SO YOU WANT TO BE A ROCK AND ROLL STAR	THE BYRDS
23.	THE HUNTER GETS CAPTURED	MARVELETTES
24.	MECHANICAL MAN	BENT BOLT
25.	I'VE BEEN LONELY TOO LONG	YOUNG RASCALS
26.	UPS AND DOWNS	PAUL REVERE & RAIDERS
27.	MY CUP RUNNETH OVER	ED AMES
28.	INDESCRIBABLY BLUE	ELVIS PRESLEY
29.	EPISTLE TO DIPPY	DONOVAN
30.	LOVIN' YOU	BOBBY DARIN

CHICAGO PREMIERES

BECAUSE OF YOU	CHRIS MONTEZ
BEGGIN'	FOUR SEASONS
THE LOSER	GARY LEWIS

WEEK ENDING MARCH 2, 1967

1.	**I Think We're Alone Now**	**Tommy James & Shondells**	2/6
2.	**Ruby Tuesday**	***Rolling Stones***	1/7
3.	**Pushin' Too Hard**	***Seeds***	5/12
4.	**There's A Kind Of Hush**	***Herman's Hermits***	10/5
5.	**Georgy Girl**	***Seekers***	3/11
6.	**Happy Together**	***The Turtles***	14/5
7.	**The Beat Goes On**	***Sonny & Cher***	8/7
8.	**Love Is Here & Now You're Gone**	***Supremes***	7/7
9.	**I'm A Believer**	***Monkeees***	9/14
10.	**I Had Too Much To Dream**	***Electric Prunes***	4/11
11.	For What It's Worth	*Buffalo Springfield*	-/4
12.	Sock It To Me Baby	*Mitch Ryder*	15/5
13.	Then You Can Tell Me Goodbye	*Casinos*	13/10
14.	Pretty Ballerina	*Left Banke*	16/9
15.	Ups And Downs	*Paul Revere & Raiders*	-/4
16.	Baby I Need Your Lovin'	*Johnny Rivers*	20/6
17.	You Got To Me	*Neil Diamond*	18/7
18.	No Fair At All	*The Association*	19/8
19.	Dedicated To The One I Love	*Mamas & Papas*	-/3
20.	Mercy Mercy Mercy	*Cannonball Adderly*	-/2

(First number indicates last week's position/Second number is the weeks heard on WCFL)

ACTION TEN

21.	GO WHERE YOU WANNA GO	5TH DIMENSION
22.	MY CUP RUNNETH OVER	ED AMES
23.	STRAWBERRY FIELDS FOREVER/PENNY LANE	BEATLES
24.	RETURN OF THE RED BARON	ROYAL GUARDSMEN
25.	THE HUNTER GETS CAPTURED BY THE GAME	MARVELETTES
26.	I'VE BEEN LONELY TOO LONG	YOUNG RASCALS
27.	DARLING BE HOME SOON	LOVIN' SPOONFUL
28.	WALK TALL	TWO OF CLUBS
29.	EPISTLE TO DIPPY	DONOVAN
30.	WESTERN UNION	FIVE AMERICANS

CHICAGO PREMIERES

BERNADETTE	FOUR TOPS
DETROIT CITY	TOM JONES
GET ME TO THE WORLD ON TIME	ELECTRIC PRUNES

WEEK ENDING MARCH 9, 1967

1.	**I Think We're Alone Now**	**Tommy James & Shondells**	1/7
2.	**Happy Together**	***The Trutles***	6/6
3.	**There's A Kind Of Hush**	***Herman's Hermits***	4/6
4.	**Ruby Tuesday**	***Rolling Stones***	2/8
5.	**For What It's Worth**	***Buffalo Springfield***	11/5
6.	**The Beat Goes On**	***Sonny & Cher***	7/8
7.	**Georgy Girl**	***Seekers***	5/12
8.	**Pushin' Too Hard**	***The Seeds***	2/13
9.	**Sock It To Me Baby**	***Mitch Ryder***	12/6
10.	**Love Is Here & Now You're Gone**	***Supremes***	8/8
11.	Penny Lane/Strawberry Fields Forever	*Beatles*	-/4
12.	Ups And Downs	*Paul Revere & Raiders*	15/5
13.	Then You Can Tell Me Goodbye	*Casinos*	13/11
14.	Dedicated To The One I Love	*Mamas & Papas*	19/4
15.	Pretty Ballerina	*Left Banke*	14/10
16.	You Got To Me	*Neil Diamond*	17/8
17.	Baby I Need Your Lovin'	*Johnny Rivers*	16/7
18.	My Cup Runneth Over	*Ed Ames*	-/5
19.	No Fair At All	*The Association*	18/9
20.	Western Union	*Five Americans*	-/5

(First number indicates last week's position/Second number is the weeks heard on WCFL)

ACTION TEN

21.	GO WHERE YOU WANNA GO	FIFTH DIMENSION
22.	WALK TALL	TWO OF CLUBS
23.	I'VE BEEN LONELY TOO LONG	YOUNG RASCALS
24.	EPISTLE TO DIPPY	DONOVAN
25.	RETURN OF THE RED BARON	ROYAL GUARDSMEN
26.	I'LL TAKE CARE OF YOUR CARES	FRANKIE LAINE
27.	DARLING BE HOME SOON	LOVIN' SPOONFUL
28.	DADDY'S LITTLE GIRL	AL MARTINO
29.	LAUDY MISS CLAUDY	BUCKINGHAMS
30.	MAIRZY DOATS	THE INNOCENCE

CHICAGO PREMIERES

I'LL TRY ANYTHING	DUSTY SPRINGFIELD
LOVE EYES	NANCY SINATRA
SUNDAY FOR TEA	PETER & GORDON

WEEK ENDING MARCH 16, 1967

1.	**Happy Together**	***The Turtles***	2/7
2.	**I Think We're Alone Now**	***Tommy James & Shondells***	1/8
3.	**There's a Kind Of Hush**	***Herman's Hermits***	3/7
4.	**For What It's Worth**	***Buffalo Springfield***	5/6
5.	**Penny Lane/Strawberry Fields Forever**	***Beatles***	11/5
6.	**Dedicated To The One I Love**	***Mamas & Papas***	14/5
7.	**Ruby Tuesday**	***Rolling Stones***	4/9
8.	**The Beat Goes On**	***Sonny & Cher***	6/8
9.	**Sock It To Me Baby**	***Mitch Ryder***	9/7
10.	**Baby I Need Your Lovin'**	***Johnny Rivers***	17/8
11.	Love Is Here & Now You're Gone	*Supremes*	10/9
12.	Western Union	*Five Americans*	20/6
13.	Ups And Downs	*Paul Revere & Raiders*	12/6
14.	Then You Can Tell Me Goodbye	*Casinos*	13/12
15.	Walk Tall	*2 Of Clubs*	-/2
16.	My Cup Runneth Over	*Ed Ames*	18/6
17.	Go Where You Wanna Go	*Fifth Dimension*	-/12
18.	This Is My Song	*Petula Clark*	-/4
19.	No Fair At all	*The Association*	19/10
20.	Mr. Unreliable	*Cryan Shames*	-/4

(First number indicates last week's position/Second number is the weeks heard on WCFL)

ACTION TEN

21.	ON A CAROUSEL	THE HOLLIES
22.	59TH STREET BRIDGE	HARPERS BIZARRE
23.	RETURN OF THE RED BARON	ROYAL GUARDSMEN
24.	LAUDY MISS CLAUDY	BUCKINGHAMS
25.	I'VE BEEN LONELY TOO LONG	YOUNG RASCALS
26.	DARLING BE HOME SOON	LOVIN' SPOONFUL
27.	I'LL TAKE CARE OF YOUR CARES	FRANKIE LAINE
28.	SWEETS FOR MY SWEET	RIDDLES
29.	JIMMY MACK	MARTHA & VANDELLAS
30.	DON'T YOU CARE	BURKINGHAMS

CHICAGO PREMIERES

AT THE ZOO	SIMON & GARFUNKEL
OH THAT'S GOOD, NO THAT'S BAD	SAM THE SHAM & PHARAOHS
YOU'RE GONNA BE MINE	NEW COLONY SIX

WEEK ENDING MARCH 23, 1967

1.	**I Think We're Alone Now**	**Tommy James & Shondells**	2/9
2.	**Happy Together**	***The Turtles***	1/8
3.	**For What It's Worth**	***Buffalo Springfield***	4/7
4.	**Penny Lane/Strawberry Fields Forever**	***Beatles***	5/6
5.	**There's A Kind Of Hush**	***Herman's Hermits***	3/8
6.	**Dedicated To The One I Love**	***Mamas & Papas***	6/6
7.	**Ups And Downs**	***Paul Revere & Raiders***	13/7
8.	**Baby I Need Your Lovin'**	***Johnny Rivers***	10/9
9.	**Western Union**	***Five Americans***	12/7
10.	**The Beat Goes On**	***Sonny & Cher***	8/9
11.	Sock It To Me Baby	*Mitch Ryder*	9/8
12.	Love Is Here & now You're Gone	*Supremes*	11/10
13.	Ruby Tuesday	*Rolling Stones*	7/10
14.	My Cup Runneth Over	*Ed Ames*	16/7
15.	Walk Tall	*2 Of Clubs*	15/3
16.	This Is My Song	*Petula Clark*	18/5
17.	Then You Can Tell Me Goodbye	*The Casinos*	14/13
18.	Mr. Unreliable	*Cryan Shames*	20/5
19.	On A Carousel	*The Hollies*	-/5
20.	Go Where You Wanna Go	*Fifth Dimension*	17/13

(First number indicates last week's position/Second number is the weeks heard on WCFL)

ACTION TEN

21.	RETURN OF THE RED BARON	ROYAL GUARDSMEN
22.	59TH STREET BRIDGE	HARPERS BIZARRE
23.	A LITTLE BIT ME, A LITTLE BIT YOU	THE MONKEES
24.	DARLING BE HOME SOON	LOVIN' SPOONFUL
25.	I'VE BEEN LOVELY TOO LONG	YOUNG RASCALS
26.	SOMETHIN' STUPID	NANCY & FRANK SINATRA
27.	SWEETS FOR MY SWEET	THE RIDDLES
28.	THE LOVE I SAW IN YOU	THE MIRACLES
29.	I'LL TAKE CARE OF YOUR CARES	FRANKIE LAINE
30.	MAIRZY DOATS	THE INNOCENCE

CHICAGO PREMIERES

MY BACK PAGES	THE BYRDS
YELLOW BALLOON	YELLOW BALLOONS
YOU GOT WHAT IT TAKES	DAVE CLARK 5

WEEK ENDING MARCH 30, 1967

1.	**Happy Together**	**The Turtles**	2/9
2.	**I Think We're Alone Now**	***Tommy James & Shondells***	1/10
3.	**For What It's Worth**	***Buffalo Springfield***	3/8
4.	**Penny Lane/Strawberry Fields Forever**	***Beatles***	4/7
5.	**Dedicated To The One I Love**	***Mamas & Papaas***	6/7
6.	**There's a Kind Of Hush**	***Herman's Hermits***	5/9
7.	**Western Union**	***Five Americans***	9/8
8.	**Baby I Need Your Lovin'**	***Johnny Rivers***	8/10
9.	**This Is My Song**	***Petula Clark***	16/6
10.	**Ups & Downs**	***Paul Revere & Raiders***	7/8
11.	On A Carousel	*Hollies*	19/6
12.	A Little Bit Me, A Little Bit You	*The Monkees*	-/3
13.	Mr. Unreliable	*Cryan Shames*	18/6
14.	Walk Tall	*2 Of Clubs*	15/4
15.	The Beat Goes On	*Sonny & Cher*	10/10
16.	Somethin' Stupid	*Nancy & Frank Sinatra*	-/4
17.	The 59th Street Bridge Song	*Harpers Bizarre*	-/5
18.	Bernadette	*Four Tops*	-/5
19.	California Nights	*Lesley Gore*	-/8
20.	Jimmy Mack	*Martha & Vandellas*	-/5

(First number indicates last week's position/Second number is the weeks heard on WCFL)

ACTION TEN

21.	HIPPY DIPPY WEATHER MAN	GEORGE CARLIN
22.	DON'T YOU CARE	BUCKINGHAMS
23.	AT THE ZOO	SIMON & GARFUNKEL
24.	SWEETS FOR MY SWEET	THE RIDDLES
25.	MORNINGTOWN RIDE	THE SEEKERS
26.	WILLIE JEAN	SHADOWS OF KNIGHT
27.	DETROIT CITY	TOM JONES
28.	YOU'RE GONNA BE MINE	NEW COLONY SIX
29.	BEGGIN'	FOUR SEASONS
30.	I NEVER LOVED A MAN THE WAY I LOVE YOU	ARETHA FRANKLIN

CHICAGO PREMIERES

31.	GIRL YOU'LL BE A WOMAN SOON	NEIL DIAMOND
32.	THE HAPPENING	SUPREMES
33.	WHEN I WAS YOUNG	ERIC BURDON

WEEK ENDING APRIL 6, 1967

1.	**Happy Together**	**The Turtles**	1/10
2.	**I Think We're Alone Now**	***Tommy James & Shondells***	2/11
3.	**A Little Bit Me, A Little Bit You**	***The Monkees***	12/4
4.	**For What It's Worth**	***Buffalo Springfield***	3/9
5.	**Dedicated To The One I Love**	***mamas & Papas***	5/8
6.	**Penny Lane/Strawberry Fields Forever**	***Beatles***	4/8
7.	**Somethin' Stupid**	***Nancy & Frank Sinatra***	16/5
8.	**Western Union**	***Five Americans***	7/9
9.	**On A Carousel**	***Hollies***	11/7
10.	**This Is My Song**	***Petula Clark***	9/7
11.	There's A Kind Of Hush	*Herman's Hermits*	6/10
12.	Mr. Unreliable	*Cryan Shames*	13/7
13.	Walk Tall	*2 Of Clubs*	14/5
14.	Jimmy Mack	*Martha & Vandellas*	20/6
15.	Bernadette	*Four Tops*	18/6
16.	California Nights	*Leslie Gore*	19/9
17.	The 59th Street Bridge Song	*Harpers Bizarre*	17/6
18.	Don't You Care	*Buckinghams*	-/5
19.	At The Zoo	*Simon & Garfunkel*	-/5
20.	Hippy Dippy Weather Man	*George Carlin*	-/3

(First number indicates last week's position/Second number is the weeks heard on WCFL)

ACTION TEN

21.	YOU'RE GONNE BE MINE	NEW COLONY SIX
22.	SWEETS FOR MY SWEET	THE RIDDLES
23.	YOU GOT WHAT IT TAKES	DAVE CLARK FIVE
24.	BEGGIN'	FOUR SEASONS
25.	I'M A MAN	SPENCER DAVID GROUP
26.	THE LOSER	GARY LEWIS
27.	DETROIT CITY	TOM JONES
28.	LOVE EYES	NANCY SINATRA
29.	I'LL TRY ANYTHING	DUSTY SPRINGFIELD
30.	HERE COMES MY BABY	TREMELOS

CHICAGO PREMIERES

CASINO ROYALE	HERB ALPERT
LITTLE GAMES	YARDBIRDS
THERE WAS A TIME	THE TROLLS

WEEK ENDING APRIL 13, 1967

1.	**Happy Together**	**The Turtles**	1/11
2.	**A Little Bit Me, A Little Bit You**	***The Monkees***	3/5
3.	**Somethin' Stupid**	***Nancy & Frank Sinatra***	7/6
4.	**For What It's Worth**	***Buffalo Springfield***	4/10
5.	**I Think We're Alone Now**	***Tommy James & Shondells***	2/12
6.	**On A Carousel**	***Hollies***	9/8
7.	**Dedicated To The One I Love**	***Mamas & Papas***	5/9
8.	**Western Union**	***Five Americans***	8/10
9.	**Penny Lane/Strawberry Fields Forever**	***Beatles***	6/9
10.	**Walk Tall**	***2 Of Clubs***	13/6
11.	This Is My Song	*Petula Clark*	10/8
12.	Jimmy Mack	*Martha & Vandellas*	14/7
13.	Mr. Unreliable	*Cryan Shames*	12/8
14.	California Nights	*Leslie Gore*	16/10
15.	Don't You Care	*Buckinghams*	18/6
16.	The 59th Street Bridge Song	*Harpers Bizarre*	17/7
17.	Bernadette	*Four Tops*	15/7
18.	At The Zoo	*Simon & Garfunkel*	19/6
19.	You're Gonna Be Mine	*New Colony Six*	-/5
20.	You Got What It Takes	*Dave Clark Five*	-/4

(First number indicates last week's position/Second number is the weeks heard on WCFL)

ACTION TEN

21.	HERE COMES MY BABY	TREMELOES
22.	BEGGIN'	FOUR SEASONS
23.	SWEETS FOR MY SWEET	THE RIDDLES

24.	THE HAPPENING	SUPREMES
25.	I'M A MAN	SPENCER DAVIS GROUP
26.	GIRL YOU'LL BE A WOMAN SOON	NEIL DIAMOND
27.	SWEET SOUL MUSIC	ARTHUR CONLEY
28.	LOVE EYES	NANCY SINATRA
29.	DETROIT CITY	TOM JONES
30.	SUNDAY FOR TEA	PETER & GORDON

CHICAGO PREMIERES

ANOTHER DAY, ANOTHER HEARTACHE	FIFTH DIMENSION
MIRAGE	TOMMY JAMES & SHONDELLS
SHAKE A TAIL FEATHER	JAMES & BOBBY PURIFY

WEEK ENDING APRIL 20, 1967

1.	**Happy Together**	**The Turtles**	1/13
2.	**A Little Bit Me, A Little Bit You**	***The Monkees***	2/6
3.	**Somethin' Stupid**	***Nancy & Frank Sinatra***	3/7
4.	**On A Carousel**	***Hollies***	6/9
5.	**For What It's Worth**	***Buffalo Springfields***	4/11
6.	**I Think We're Alone Now**	***Tommy James & Shondells***	5/13
7.	**Penny Lane/Strawberry Fields Forever**	***Beatles***	9/10
8.	**Don't You Care**	***Buckinghams***	15/7
9.	**Western Union**	***Five Americans***	8/11
10.	**Jimmy Mack**	***Martha & Vandellas***	12/8
11.	California Nights	*Leslie Gore*	14/11
12.	At The Zoo	*Simon & Garfunkel*	18/7
13.	This Is My Song	*Petula Clark*	11/9
14.	Mr. Unreliable	*Cryan Shames*	13/9
15.	You Got What It Takes	*Dave Clark Five*	20/5
16.	The 59th Street Bridge Song	*Harpers Bizarre*	16/8
17.	Here Comes My Baby	*Tremeloes*	-/5
18.	Bernadette	*Four Tops*	17/8
19.	You'e Gonna Be Mine	*New Colony Six*	19/6
20.	The Happening	*Supremes*	-/4

(First number indicates last week's position/Second number is the weeks heard on WCFL)

ACTION TEN

21.	SWEET SOUL MUSIC	ARTHUR CONLEY
22.	I'M A MAN	SPENCER DAVIS GROUP
23.	BEGGIN'	FOUR SEASONS
24.	GIRL YOU'LL BE A WOMAN SOON	NEIL DIAMOND
25.	LOVE EYES	NANCY SINATRA
26.	MUSIC TO WATCH GIRLS BY	ANDY WILLIAMS
27.	SUNDAY FOR TEA	PETER & GORDON
28.	DEAD END STREET	LOU RAWLS
29.	WHEN I WAS YOUNG	ERIC BURDON & ANIMALS
30.	WILLIE JEAN	SHADOWS OF KNIGHT

CHICAGO PREMIERES

A BEAUTIFUL STORY	SONNY & CHER
HIM OR ME, WHAT'S IT GONNA BE	PAUL REVERE & RAIDERS
SIX O'CLOCK	LOVIN' SPOONFUL

WEEK ENDING APRIL 27, 1967

1.	**A Little Bit Me, A Little Bit You**	**The Monkees**	2/7
2.	**Somethin' Stupid**	***Nancy & Frank Sinatra***	3/8
3.	**Happy Together**	***The Turtles***	1/14
4.	**On A Carousel**	***Hollies***	4/10
5.	**Don't You Care**	***Buckinghams***	8/8
6.	**Jimmy Mack**	***Martha & Vandellas***	10/9
7.	**Here Comes My Baby**	***Tremeloes***	17/6
8.	**California Nights**	***Leslie Gore***	11/12
9.	**The Happening**	***Supremes***	20/5
10.	**You Got What It Takes**	***Dave Clark Five***	15/6
11.	For What It's Worth	*Buffalo Springfifeld*	5/12
12.	You're Gonna Be Mine	*New Colony Six*	19/7
13.	At The Zoo	*Simon & Garfunkel*	12/8
14.	Sweet Soul Music	*Arthur Conley*	-/5
15.	Mr. Unreliable	*Cryan Shames*	14/10
16.	This Is My Song	*Petula Clark*	13/10
17.	Girl You'll Be A Woman Soon	*Neil Diamond*	-/5
18.	The 59th Street Bridge Song	*Harpers Bizarre*	16/9
19.	Bernadette	*Four Tops*	18/9
20.	Beggin'	*Four Seasons*	-/9

(First number indicates last week's position/Second number is the weeks heard on WCFL)

ACTION TEN

21.	MIDNIGHT HOUR	THE MESSENGERS
22.	I'M A MAN	SPENCER DAVID GROUP
23.	MUSIC TO WATCH GIRLS BY	ANDY WILLIAMS
24.	DEAD END STREET	LOU RAWLS
25.	LOVE EYES	NANCY SINATRA
26.	SOMEBODY TO LOVE	JEFFERSON AIRPLANE
27.	WHEN I WAS YOUNG	ERIC BURDON & ANIMALS
28.	CAN'T SEEM TO MAKE YOU MINE	THE SEEDS
29.	SHAKE HANDS	LOU CHRISTIE
30.	CLOSE YOUR EYES	PEACHES & HERB

CHICAGO PREMIERES

ALL I NEED	THE TEMPTATIONS
BOWLING GREEN	EVERLY BROTHERS
CREEQUE ALLEY	MAMAS & PAPAS

WEEK ENDING MAY 4, 1967

	Title	Artist	
1.	**A Little Bit Me, A Little Bit You**	**Monkees**	1/8
2.	**Somethin' Stupid**	***Nancy & Frank Sinatra***	2/9
3.	**Don't You Care**	***Buckinghams***	5/9
4.	**On A Carousel**	***Hollies***	4/11
5.	**Here Comes My Baby**	***Tremeloes***	7/7
6.	**You Got What It Takes**	***David Clark Five***	10/7
7.	**Jimmy Mack**	***Martha & Vandellas***	6/10
8.	**Sweet Soul Music**	***Arthur Conley***	14/6
9.	**Happy Together**	***The Turtles***	3/15
10.	**California Nights**	***Leslie Gore***	8/13
11.	Midnight Hour	*The Messengers*	-/1
12.	The Happening	*Supremes*	9/6
13.	Girl You'll Be A Woman Soon	*Neil Diamond*	17/6
14.	You're Gonna Be Mine	*New Colony Six*	12/8
15.	Mr. Unreliable	*Cryan Shames*	15/11
16.	Mirage	*Tommy James & Shondells*	-/4
17.	Him Or Me, What's It Gonna Be	*Paul Revere & Raiders*	-/3
18.	Somebody To Love	*Jefferson Airplane*	-/2
19.	Respect	*Aretha Franklin*	-/1
20.	Beggin'	*Four Seasons*	20/10

(First number indicates last week's position/Second number is the weeks heard on WCFL)

ACTION TEN

21.	DEAD END STREET	LOU RAWLS
22.	MUSIC TO WATCH GIRLS BY	ANDY WILLIAMS
23.	I'M A MAN	SPENCER DAVIS GROUP
24.	I WAS KAISER BILL'S BATMAN	WHISTLING JACK SMITH
25.	WHEN I WAS YOUNG	ERIC BURDON & ANIMALS
26.	CLOSE YOUR EYES	PEACHES & HERB
27.	CAN'T SEEM TO MAKE YOU MINE	THE SEEDS
28.	SHAKE HANDS	LOU CHRISTIE
29.	I GOT RHYTHM	THE HAPPENINGS
30.	SWEETS FOR MY SWEET	RIDDLES

CHICAGO PREMIERES

GIRLS IN LOVE	GARY LEWIS & PLAYBOYS
ROUND ROUND	JONATHON KING
SOUNDS OF LOVE	FIVE AMERICANS

WEEK ENDING MAY 11, 1967

1.	**Somethin' Stupid**	**Nancy & Frank Sinatra**	2/10
2.	**A Little Bit Me, A Little Bit You**	***Monkees***	1/9
3.	**Don't You Care**	***Buckinghams***	3/10
4.	**Sweet Soul Music**	***Arthur Conley***	8/7
5.	**Here Comes My Baby**	***Tremeloes***	5/8
6.	**You Got What It Takes**	***Dave Clark Five***	6/8
7.	**Mirage**	***Tommy James & Shondells***	16/5
8.	**Girl You'll Be A Woman Soon**	***Neil Diamond***	13/7
9.	**Midnight Hour**	***The Messengers***	11/2
10.	**On A Carousel**	***Hollies***	4/12
11.	The Happening	*Supremes*	12/7
12.	Somebody To Love	*Jefferson Airplane*	18/3
13.	I Got Rhythm	*The Happenings*	-/6
14.	Jimmy Mack	*Martha & Vandellas*	7/11
15.	Him Or Me, What's It Gonna Be	*Paul revere & Raiders*	17/4
16.	Respect	*Aretha Franklin*	19/2
17.	Music To Watch Girls By	*Andy Williams*	-/8
18.	Can't Seem To Make You Mine	*The Seeds*	-/2
19.	I Was Kaiser Bill's Batman	*Whistling Jack Smith*	-/4
20.	Dead End Street	*Lou Rawls*	-/5

(First number indicates last week's position/Second number is the weeks heard on WCFL)

ACTION TEN

21.	WHEN I WAS YOUNG	ERIC BURDON & ANIMALS
22.	SHAKE HANDS	LOU CHRISTIE
23.	RELEASE ME	ENGELBERT HUMPERDINCK
24.	CLOSE YOUR EYES	PEACHES & HERB
25.	GROOVIN'	YOUNG RASCALS
26.	CASINO ROYALE	HERB ALPERT
27.	MY BABE	RONNIE DOVE
28.	BUY FOR ME THE RAIN	NITTY GRITTY DIRT BAND
29.	SIX O'CLOCK	LOVIN' SPOONFUL
30.	CREEQUE ALLEY	MAMAS & PAPAS

CHICAGO PREMIERES

FUNNY FAMILIAR FORGOTTEN FEELINGS	TOM JONES
SHE'D RATHER BE WITH ME	THE TURTLES
TWO IN THE AFTERNOON	DINO, DESI & BILLY

WEEK ENDING MAY 18, 1967

1.	**Mirage**	**Tommy James & Shondells**	7/6
2.	**Don't You Care**	***Buckinghams***	3/11
3.	**Somebody To Love**	***Jefferson Airplane***	12/4
4.	**Here Comes My Baby**	***Tremeloes***	5/9
5.	**Somethin' Stupid**	***Nancy & Frank Sinatra***	1/11
6.	**I Got Rhythm**	***The Happenings***	13/7
7.	**A Little Bit Me, A Little Bit You**	***Monkees***	2/10
8.	**Him Or Me, What's It Gonna Be**	***Paul Revere & Raiders***	15/5
9.	**Sweet Soul Music**	***Arthur Conley***	4/8
10.	**Midnight Hour**	***The Messengers***	9/3
11.	The Happening	*Supremes*	11/8
12.	You Got What It Takes	*Dave Clark Five*	6/9
13.	Girl You'll Be A Woman Soon	*Neil Diamond*	8/8
14.	Respect	*Aretha Franklin*	16/3
15.	Can't Seem To Make You Mine	*The Seeds*	18/3
16.	Release Me	*Engelbert Humperdinck*	-/5
17.	Creeque Alley	*Mamas & Papas*	-/4
18.	Groovin'	*Young Rascals*	-/4
19.	Dead End Street	*Lou Rawls*	20/6
20.	I Was Kaiser Bill's Batman	*Whistling Jack Smith*	19/5

(First number indicates last week's position/Second number is the weeks heard on WCFL)

ACTION TEN

21.	SIX O'CLOCK	LOVIN' SPOONFUL
22.	CASINO ROYALE	HERB ALPERT
23.	LITTLE BIT O'SOUL	MUSIC EXPLOSION
24.	OOOGUM BOOGUM SONG	BRENTON WOOD
25.	BUY FOR ME THE RAIN	NITTY GRITTY DIRT BAND
26.	SHAKE HANDS	LOU CHRISTIE
27.	MY GIRL JOSEPHINE	JERRY JAYE
28.	YELLOW BALLOON	YELLOW BALLOONS
29.	TOO MANY FISH IN THE SEA	MITCH RYDER
30.	MAKING MEMORIES	FRANKIE LAINE

CHICAGO PREMIERES

7 ROOMS OF GLOOM	FOUR TOPS
PAY YOU BACK WITH INTEREST	THE HOLLIES
WINDY	THE ASSOCIATION

WEEK ENDING MAY 25, 1967

1.	**Somebody To Love**	**Jefferson Airplane**	3/5
2.	**I Got Rhythm**	***The Happenings***	6/8
3.	**Mirage**	***Tommy James & Shondells***	1/7
4.	**Him Or Me, What's It Gonna Be**	***Paul Revere & Raiders***	8/6
5.	**Here Comes My Baby**	***Tremeloes***	4/10
6.	**Don't You Care**	***Buckinghamns***	2/12
7.	**Midnight Hour**	***The Messengers***	10/4
8.	**A Little Bit O' Soul**	***Music Explosion***	-/4
9.	**Sweet Soul Music**	***Arthur Conley***	9/9
10.	**Respect**	***Aretha Franklin***	14/4
11.	Groovin'	*Young Rascals*	18/5
12.	Somethin' Stupid	*Nancy & Frank Sinatra*	5/12
13.	Release Me	*Engelbert Humperdinck*	16/6
14.	Windy	*The Association*	-/3
15.	The Happening	*Supremes*	11/9
16.	A Little Bit Me, A Little Bit You	*Monkees*	7/11
17.	Can't Seem To Make You Mine	*The Seeds*	15/4
18.	Six O'Clock	*Lovin' Spoonful*	-/6
19.	Yellow Balloon	*Yellow Balloons*	-/10
20.	Casino Royale	*Herb Alpert*	-/8

(First number indicates last week's position/Second number is the weeks heard on WCFL)

ACTION TEN

21.	SHE'D RATHER BE WITH ME	THE TURTLES
22.	CREEQUE ALLEY	MAMAS & PAPAS
23.	OOGUM BOOGUM SONG	BRENTON WOOD
24.	MY GIRL JOSEPHINE	JERRY JAYE
25.	PAY YOU BACK WITH INTEREST	THE HOLLIES
26.	TOO MANY FISH IN THE SEA	MITCH RYDER
27.	A BEAUTIFUL STORY	SONNY & CHER
28.	GIRL	THE PRIDE & JOY
29.	LET'S LIVE FOR TODAY	GRASS ROOTS
30.	MAKING MEMORIES	FRANKIE LAINE

CHICAGO PREMIERES

DON'T SLEEP IN THE SUBWAY	PETULA CLARK
PLASTIC MAN	SONNY & CHER
THE TRACKS OF MY TEARS	JOHNNY RIVERS

WEEK ENDING JUNE 1, 1967

1.	**Somebody To Love**	**Jefferson Airplane**	1/6
2.	**I Got Rhythm**	***The Happenings***	2/9
3.	**A Little Bit O' Soul**	***Music Explosion***	8/5
4.	**Windy**	***The Association***	14/4
5.	**Midnight Hour**	***The Messengers***	7/5
6.	**Mirage**	***Tommy James & Shondells***	3/8
7.	**Groovin'**	***Young Rascals***	11/6
8.	**Him Or Me, What's It Gonna Be**	***Paul Revere & Raiders***	4/7
9.	**Don't You Care**	***Buckinghams***	6/13
10.	**Respect**	***Aretha Franklin***	10/5
11.	Sweet Soul Music	*Arthur Conley*	9/10
12.	She'd Rather Be With Me	*The Turtles*	-/4
13.	Release Me	*Engelbert Humperdinck*	13/7
14.	Yellow Balloon	*Yellow Balloons*	19/11
15.	Can't Seem To Make You Mine	*The Seeds*	17/5
16.	Creeque Alley	*Mamas & Papas*	-/6
17.	Six O'Clock	*Lovin' Spoonful*	18/7
18.	Casino Royale	*Herb Alpert*	20/9
19.	Oogum Boogum Song	*Brenton Wood*	-/4
20.	Pay You Back With Interest	*The Hollies*	-/3

(First number indicates last week's position/Second number is the weeks heard on WCFL)

ACTION TEN

21.	LET'S LIVE FOR TODAY	GRASS ROOTS
22.	MY GIRL JOSEPHINE	JERRY JAYE
23.	DING DONG THE WITCH IS DEAD	THE FIFTH ESTATE
24.	CAN'T TAKE MY EYES OFF YOU	FRANKIE VALLI
25.	UP UP AND AWAY	5TH DIMENSION
26.	GIRL	THE PRIDE & JOY
27.	SOUNDS OF LOVE	FIVE AMERICANS
28.	GIRLS IN LOVE	GARY LEWIS
29.	SUNSHINE GIRL	THE PARADE
30.	NEW YORK MINING DISASTER	THE BEE GEES

CHICAGO PREMIERES

C'MON MARIANNE	FOUR SEASONS
DAYLIGHT SAVING TIME	KEITH
YOU MUST HAVE BEEN A BEAUTIFUL BABY	DAVE CLARK 5

WEEK ENDING JUNE 8, 1967

1.	**Windy**	**The Association**	4/5
2.	**A Little Bit O' Soul**	***Music Explosion***	3/6
3.	**Somebody To Love**	***Jefferson Airplane***	1/7
4.	**Groovin'**	***Young Rascals***	7/7
5.	**I Got Rhythm**	***The Happenings***	2/10
6.	**She'd Rather Be With Me**	***The Turtles***	12/5
7.	**Mirage**	***Tommy James & Shondells***	6/9
8.	**Midnight Hour**	***The Messengers***	5/6
9.	**Respect**	***Aretha Franklin***	10/6
10.	**Him Or Me, What's It Gonna Be**	***Paul Revere & Raiders***	8/8
11.	Sweet Soul Music	*Arthur Conley*	11/11
12.	Yellow Balloon	*Yellow Balloons*	14/12
13.	Release Me	*Engelbert Humperdinck*	13/8
14.	Let's Live For Today	*Grass Roots*	-/6
15.	Can't Seem To Make You Mine	*The Seeds*	15/6
16.	Up Up And Away	*5th Dimension*	-/3
17.	Casino Royale	*Herb Alpert*	18/10
18.	Can't Take My Eyes Off You	*Frankie Valli*	-/5
19.	Oogum Boogum Song	*Brenton Wood*	19/5
20.	Pay You Back With Interest	*The Hollies*	20/4

(First number indicates last week's position/Second number is the weeks heard on WCFL)

ACTION TEN

21.	DING DONG THE WITCH IS DEAD	THE FIFTH ESTATE
22.	SOUNDS OF LOVE	FIVE AMERICANS
23.	NEW YORK MINING DISASTER 1941	BEE GEES
24.	7 ROOMS OF GLOOM	FOUR TOPS
25.	COME ON DOWN TO MY BOAT	EVERY MOTHERS' SON
26.	STEP OUT OF YOUR MIND	THE AMERICAN BREED
27.	SUNSHINE GIRL	THE PARADE
28.	SUNDAY WILL NEVER BE THE SAME	SPANKY AND OUR GANG
29.	GIRLS IN LOVE	GARY LEWIS
30.	DON'T SLEEP IN THE SUBWAY	PETULA CLARK

CHICAGO PREMIERES

CARRIE ANNE	THE HOLLIES
I'M JUST WAITING ANTICIPATING FOR HER TO SHOW UP	THE NEW COLONY SIX
WHITE RABBIT	JEFFERSON AIRPLANE

WEEK ENDING JUNE 15, 1967

1.	**Windy**	**The Association**	1/6
2.	**A Little Bit O' Soul**	***Music Explosion***	2/7
3.	**Somebody To Love**	***Jefferson Airplane***	3/8
4.	**I Got Rhythm**	***The Happenings***	5/11
5.	**Groovin'**	***Young Rascals***	4/8
6.	**She'd Rather Be With Me**	***The Turtles***	6/6
7.	**Respect**	***Aretha Franklin***	9/7
8.	**Up Up And Away**	***5th Dimension***	16/4
9.	**Let's Live For Today**	***Grass Roots***	14/7
10.	**Can't Take My Eyes Off You**	***Frankie Valli***	18/6
11.	Midnight Hour	*The Messengers*	8/7
12.	Yellow Balloon	*Yellow Balloons*	12/13
13.	Release Me	*Engelbert Humperdinck*	13/9
14.	Him Or Me, What's It Gonna Be	*Paul Revere & Raiders*	10/9
15.	Pay You Back With Interest	*The Hollies*	20/5
16.	Ding Dong The Witch Is Dead	*The Fifth Estate*	-/6
17.	Sounds Of Love	*Five Americans*	-/7
18.	Come On Down To My Boat	*Every Mothers' Son*	-/3
19.	Casino Royale	*Herb Alpert*	17/11
20.	San Francisco	*Scott McKenzie*	-/4

(First number indicates last week's position/Second number is the weeks heard on WCFL)

ACTION TEN

21.	STEP OUT OF YOUR MIND	THE AMERICAN BREED
22.	SUNDAY WILL NEVER BE THE SAME	SPANKY & OUR GANG
23.	NEW YORK MINING DISASTER 1941	BEE GEES
24.	DON'T SLEEP IN THE SUBWAY	PETULA CLARK
25.	DO IT AGAIN	JON & ROBIN
26.	7 ROOM OF GLOOM	FOUR TOPS
27.	THE TRACKS OF MY TEARS	JOHNNY RIVERS
28.	YOU MUST HAVE BEEN A BEAUTIFUL BABY	DAVE CLARK 5
29.	C'MON MARIANNE	FOUR SEASONS
30.	MERCY MERCY	THE BUCKINGHAMS

CHICAGO PREMIERES

DON'T GO OUT INTO THE RAIN	HERMANS' HERMITS
I LIKE THE WAY	TOMMY JAMES & SHONDELLS
ONLY ALL THE TIME	BARBARA LEWIS

WEEK ENDING JUNE 22, 1967

1.	**Windy**	**The Association**	1/7
2.	**A Little Bit O' Soul**	***Music Explosion***	2/8
3.	**Groovin'**	***Young Rascals***	5/9
4.	**She'd Rather Be With Me**	***The Turtles***	6/7
5.	**Up Up And Away**	***5th Dimension***	8/5
6.	**Somebody To Love**	***Jefferson Airplane***	3/9
7.	**Can't Take My Eyes Off You**	***Frankie Valli***	10/7
8.	**Let's Live For Today**	***Grass Roots***	9/8
9.	**Respect**	***Aretha Franklin***	7/8
10.	**I Got Rhythm**	***The Happenings***	4/12
11.	Come On Down To My Boat	*Every Mothers' Son*	18/4
12.	San Francisco	*Scott McKenzie*	20/5
13.	Pay You Back With Interest	*The Hollies*	15/6
14.	Yellow Balloon	*Yellow Balloons*	12/14
15.	Do It Again	*Jon & Robin*	-/6
16.	Step Out Of Your Mind	*The American Breed*	-/6
17.	Ding Dong The Witch Is Dead	*The Fifth Estate*	16/7
18.	The Tracks Of My Tears	*Johnny Rivers*	-/5
19.	Sounds Of Love	*Five Americans*	17/8
20.	Sunday Will Never Be The Same	*Spanky & Our Gang*	-/7

(First number indicates last week's position/Second number is the weeks heard on WCFL)

ACTION TEN

21.	DON'T SLEEP IN THE SUBWAY	PETULA CLARK
22.	HOLD ON	THE MAUDS
23.	NEW YORK MINING DISASTER 1941	BEE GEES
24.	7 ROOMS OF GLOOM	FOUR TOPS
25.	MERCY MERCY	THE BUCKINGHAMS
26.	C'MON MARIANNE	FOUR SEASONS
27.	YOU MUST HAVE BEEN A BEAUTIFUL BABY	DAVE CLARK 5
28.	HERE WE GO AGAIN	RAY CHARLES
29.	DON'T GO OUT INTO THE RAIN	HERMAN'S HERMITS
30.	CARRIE ANN	THE HOLLIES

CHICAGO PREMIERES

GIVE ME TIME	DUSTY SPRINGFIELD
LONELY DRIFTER	PIECES OF EIGHT
SUNRISE HIGHWAY	PETER ANDERS

WEEK ENDING JUNE 29, 1967

1.	**Windy**	**The Association**	1/8
2.	**A Little Bit O' Soul**	***Music Explosion***	2/9
3.	**Groovin'**	***Young Rascals***	3/10
4.	**Let's Live For Today**	***Grass Roots***	8/9
5.	**Can't Take My Eyes Off You**	***Frankie Valli***	7/8
6.	**Up Up And Away**	***5th Dimension***	5/6
7.	**Come On Down To My Boat**	***Every Mothers' Son***	11/5
8.	**She'd Rather Be With Me**	***The Turtles***	4/8
9.	**San Francisco**	***Scott McKenzie***	12/6
10.	**Respect**	***Aretha Franklin***	9/9
11.	Step Out Of Your Mind	*The American Breed*	16/7
12.	Do It Again	*Jon & Robin*	15/7
13.	C'Mon Marianne	*Four Seasons*	-/5
14.	The Tracks Of My Tears	*Johnny Rivers*	18/6
15.	Don't Sleep In The Subway	*Petula Clark*	-/6
16.	Mercy Mercy	*Buckinghams*	-/4
17.	Hold On	*The Mauds*	-/2
18.	Sunday Will Never Be The Same	*Spanky & Our Gang*	20/8
19.	Ding Dong The Witch Is Dead	*The Fifth Estate*	17/8
20.	Sounds Of Love	*Five Americans*	19/9

(First number indicates last week's position/Second number is the weeks heard on WCFL)

ACTION TEN

21.	NEW YORK MINING DISASTER 1941	BEE GEES
22.	HERE WE GO AGAIN	RAY CHARLES
23.	YOU MUST HAVE BEEN A BEAUTIFUL BABY	DAVE CLARK 5
24.	WHITE RABBIT	JEFFERSON AIRPLANE
25.	I LIKE THE WAY	TOMMY JAMES & SHONDELLS
26.	TWO IN THE AFTERNOON	DINO, DESI & BILLY
27.	DON'T GO OUT INTO THE RAIN	HERMAN'S HERMITS
28.	MARY IN THE MORNING	AL MARTINO
29.	I'M JUST WAITING ANTICIPATING FOR HER TO SHOW UP	NEW COLONY SIX
30.	CARRIE ANN	THE HOLLIES

CHICAGO PREMIERES

BLUEBIRD	BUFFALO SPRINGFIELD
I'LL NEVER NEED MORE THAN THIS	IKE & TINA TURNER
IT COULD BE WE'RE IN LOVE	CRYAN SHAMES

WEEK ENDING JULY 6, 1967

1.	**Windy**	***The Association***	1/9
2.	**Can't Take My Eyes Off You**	***Frankie Valli***	5/9
3.	**Little Bit O' Soul**	***Music Explosion***	2/10
4.	**Let's Live For Today -**	***Grass Roots***	4/10
5.	**Come On Down To My Boat – Every Mother's Son**	***Every Mother's Son***	7/6
6.	**San Francisco – Scott McKenzie**	***Scott McKenzie***	9/7
7.	**Up Up And Away – 5th Dimension**	***5th Dimension***	6/7
8.	**Groovin' – Young Rascals**	***Young Rascals***	3/11
9.	**Mercy Mercy Mercy – Buckinghams**	***Buckinghams***	16/5
10.	**Do It Again – Jon & Robin**	***Jon & Robin***	12/8
11.	Step Out Of Your Mind – The American Breed	*The American Breed*	11/8
12.	C'Mon Marianne – Four Seasons	*Four Seasons*	13/6
13.	The Tracks Of My Tears – Johnny Rivers	*Johnny Rivers*	14/7
14.	Hold On – The Mauds	*The Mauds*	17/3
15.	Don't Sleep In The Subway – Petula Clark	*Petula Clark*	15/7
16.	Sunday Will Never Be The Same – Spanky & Our Gang	*Spanky & Our Gang*	18/9
17.	Sounds Of Love – Five Americans	*Five Americans*	20/10
18.	White Rabbit	*Jefferson Airplane*	-/5
19.	I Like The Way	*Tommy James & Shondells*	-/4
20.	I Was Made To Love Her	*Stevie Wonder*	-/1

(First number indicates last week's position/Second number is the weeks heard on WCFL)

ACTION TEN

21.	HERE WE GO AGAIN	RAY CHARLES
22.	I'M JUST WAITING ANTICIPATING FOR HER TO SHOW UP	NEW COLONY SIX
23.	YOU MUST HAVE BEEN A BEAUTIFUL BABY	DAVE CLARK 5
24.	DON'T GO OUT INTO THE RAIN	HERMAN'S HERMITS
25.	ROMEO & JULIET	MICHAEL & MESSENGERS
26.	LIGHT MY FIRE	THE DOORS
27.	IT COULD BE WE'RE IN LOVE	CRYAN SHAMES
28.	TWO IN THE AFTERNOON	DINO, DESI & BILLY
29.	WHITER SHADE OF PALE	PROCAL HARUM
30.	CARRIE ANN	THE HOLLIES

CHICAGO PREMIERES

A GIRL LIKE YOU	YOUNG RASCALS
MY MAMMY	THE HAPPENINGS
THE HAPPENING	HERB ALPERT

WEEK ENDING JULY 13, 1967

1.	**Windy**	***The Association***	1/10
2.	**Can't Take My Eyes Off You**	***Frankie Valli***	2/10
3.	**Little Bit O' Soul**	***Music Explosion***	3/11
4.	**San Francisco**	***Scott McKenzie***	6/8
5.	**Come On Down To My Boat**	***Eery Mothers' Son***	5/8
6.	**Let's Live For Today**	***Grass Roots***	4/11
7.	**Up Up And Away**	***5th Dimension***	7/8
8.	**Do It Again**	***Jon & Robin***	10/9
9.	**Mercy Mercy Mercy**	***Buckinghams***	9/6
10.	**Step Out Of Your Mind**	***American Breed***	11/9
11.	White Rabbit	*Jefferson Airplane*	18/6
12.	Groovin'	*Young Rascals*	8/12
13.	C'Mon Marianne	*Four Seasons*	12/7
14.	Don't Sleep In The Subway	*Petula Clark*	15/8
15.	Hold On	*The Mauds*	14/4
16.	I'm Just Waiting Anticipating	*New Colony Six*	-/6
17.	The Tracks Of My Tears	*Johnny Rivers*	13/8
18.	Sunday Will Never Be The Same	*Spanky & Our Gang*	16/10
19.	I Like The Way	*Tommy James & Shondells*	19/5
20.	I Was Made To Love Her	*Stevie Wonder*	20/2

(First number indicates last week's position/Second number is the weeks heard on WCFL)

ACTION TEN

21.	HERE WE GO AGAIN	RAY CHARLES
22.	ROMEO & JULIET	MICHAEL & MESSENGERS
23.	IT COULD BE WE'RE IN LOVE	CRYAN SHAMES
24.	BLUES THEME	THE ARROWS
25.	DON'T GO OUT INTO THE RAIN	HERMAN'S HERMITS
26.	LIGHT MY FIRE	THE DOORS
27.	MY WORLD FELL DOWN	SAGITTARIUS
28.	CARRIE ANN	THE HOLLIES
29.	A WHITER SHADE OF PALE	PROCAL HARUM
30.	JACKSON	NANCY SINATRA/LEE HAZELWOOD

CHICAGO PREMIERES

ALL YOU NEED IS LOVE/BABY YOU'RE A RICH MAN	THE BEATLES
MY GIRL	ROLLING STONES
PLEASANT VALLEY SUNDAY	MONKEES

WEEK ENDING JULY 20, 1967

1.	**Come On Down To My Boat**	**Every Mothers' Son**	5/9
2.	**Can't Take My Eyes Off You**	***Frankie Valli***	2/11
3.	**Windy**	***The Association***	1/11
4.	**Let's Live For Today**	***Grass Roots***	6/12
5.	**Little Bit O' Soul**	***Music Explosion***	3/12
6.	**San Francisco**	***Scott McKenzie***	4/9
7.	**White Rabbit**	***Jefferson Airplane***	11/7
8.	**Mercy Mercy Mercy**	***The Buckinghams***	9/7
9.	**Do It Again**	***Jon & Robin***	8/10
10.	**C'Mon Marianne**	***Four Seasons***	13/8
11.	Hold On	*The Mauds*	15/5
12.	I Was Made To Love Her	*Stevie Wonder*	20/3
13.	Blues Theme	*The Arrows*	-/2
14.	Step Out Of Your Mind	*American Breed*	10/10
15.	It Could Be We're In Love	*Cryan Shames*	-/3
16.	Light My Fire	*The Doors*	-/5
17.	I'm Just Waiting Anticipating	*New Colony Six*	16/7
18.	I Like The Way	*Tommy James & Shondells*	19/6
19.	Sunday Will Never Be The Same	*Spanky & Our Gang*	18/11
20.	My World Fell Down	*Sagittarius*	-/1

(First number indicates last week's position/Second number is the weeks heard on WCFL)

ACTION TEN

21.	WORDS/PLEASANT VALLEY SUNDAY	THE MONKEES
22.	ROMEO & JULIET	MICHAEL & MESSENGERS
23.	SOCIETY'S CHILD	JANIS IAN
24.	A WHITER SHADE OF PALE	PROCAL HARUM
25.	CARRIE ANN	THE HOLLIES
26.	SILENCE IS GOLDEN	THE TREMELOES
27.	JACKSON	NANCY SINATRA/LEE HAZELWOOD
28.	LONELY DRIFTER	PIECES OF EIGHT
29.	MY MAMMY	THE HAPPENINGS
30.	SOMEBODY HELP ME	SPENCER DAVIS GROUP

CHICAGO PREMIERES

BABY I LOVE YOU	ARETHA FRANKLIN
FAKIN' IT	SIMON & GARFUNKEL
IT'S THE LITTLE THINGS	SONNY & CHER

WEEK ENDING JULY 27, 1967

1.	**Can't Take My Eyes Off You**	**Frankie Valli**	2/12
2.	**Come On Down To My Boat**	***Every Mothers' Son***	1/10
3.	**Little Bit O' Soul**	***Music Explosion***	5/13
4.	**White Rabbit**	***Jefferson Airplane***	7/8
5.	**Blues Theme**	***The Arrows***	13/3
6.	**San Francisco**	***Scott McKenzie***	6/10
7.	**Mercy Mercy Mercy**	***The Buckinghams***	8/8
8.	**It Could Be We're In Love**	***Cryan Shames***	15/4
9.	**Windy**	***The Association***	3/12
10.	**I Was Made To Love Her**	***Stevie Wonder***	12/4
11.	Words/Pleasant Valley Sunday	*The Monkees*	-/2
12.	C'Mon Marianne	*Four Seasons*	10/9
13.	Light My Fire	*The Doors*	16/6
14.	I Like The Way	*Tommy James & Shondells*	18/7
15.	Romeo & Juliet	*Michael & Messengers*	-/8
16.	I'm Just Waiting Anticipating	*New Colony Six*	17/8
17.	My Mammy	*The Happenings*	-/4
18.	Carrie Ann	*The Hollies*	-/8
19.	My World Fell Down	*Sagittarius*	20/2
20.	Society's Child	*Janis Ian*	-/7

(First number indicates last week's position/Second number is the weeks heard on WCFL)

ACTION TEN

21.	ALL YOU NEED IS LOVE/BABY YOU'RE A RICH MAN	THE BEATLES
22.	A WHITER SHADE OF PALE	PROCAL HARUM
23.	SILENCE IS GOLDEN	THE TREMELOES
24.	SOMEBODY HELP ME	SPENCER DAVID GROUP
25.	LONELY DRIFTER	PIECES OF EIGHT
26.	DON'T GO OUT INTO THE RAIN	HERMAN'S HERMITS
27.	BABY I LOVE YOU	ARETHA FRANKLIN
28.	THERE GOES MY EVERYTHING	ENGELBERT HUMPERDINCK
29.	JACKSON	NANCY SINATRA/LEE HAZELWOOD
30.	THANK THE LORD FOR THE NIGHT TIME	NEIL DIAMOND

CHICAGO PREMIERES

HA HA SAID THE CLOWN	THE YARDBIRDS
HEROES & VILLAINS	BEACH BOYS
THE WORLD WE KNEW	FRANK SINATRA

WEEK ENDING AUGUST 3, 1967

1.	**Can't Take My Eyes Off You**	**Frankie Valli**	1/13
2.	**Come On Down To My Boat**	***Every Mothers' Son***	2/11
3.	**Blues Theme**	***The Arrows***	5/4
4.	**It Could Be We're In Love**	***Cryan Shames***	8/5
5.	**White Rabbit**	***Jefferson Airplane***	4/9
6.	**Light My Fire**	***The Doors***	13/7
7.	**Words/Pleasant Valley Sunday**	***The Monkees***	11/3
8.	**Mercy Mercy Mercy**	***The Buckinghams***	7/9
9.	**San Francisco**	***Scott McKenzie***	6/11
10.	**Windy**	***The Association***	9/13
11.	I Was Made To Love Her	*Stevie Wonder*	10/5
12.	C'Mon Marianne	*Four Seasons*	12/10
13.	All You Need Is Love/Baby You're A Rich Man	*Beatles*	21/2
14.	Romeo & Juliet	*Michael & Messengers*	15/9
15.	Society's Child	*Janis Ian*	20/8
16.	Carrie Ann	*The Hollies*	18/9
17.	Silence Is Golden	*The Tremeloes*	23/2
18.	A Whiter Shade Of Pale	*Procol Harum*	-/-
19.	I Like The Way	*Tommy James & Shondells*	14/8
20.	My Mammy	*The Happenings*	17/5

(First number indicates last week's position/Second number is the weeks heard on WCFL)

ACTION TEN

21.	MY WORLD FELL DOWN	SAGITTARIUS
22.	SOMEBODY HELP ME	SPENCER DAVIS GROUP
23.	BABY I LOVE YOU	ARETHA FRANKLIN
24.	LONELY DRIFTER	PIECES OF EIGHT
25.	THANK THE LORD FOR THE NIGHT TIME	NEIL DIAMOND
26.	THERE GOES MY EVERYTHING	ENGELBERT HUMPERDINCK
27.	TAKE ME BACK	FLOCK
28.	DON'T GO OUT INTO THE RAIN	HERMAN'S HERMITS
29.	A GIRL LIKE YOU	YOUNG RASCALS
30.	THE HAPPENING	HERB ALPERT

CHICAGO PREMIERES

REFLECTIONS	SUPREMES
A LITTLE BIT NOW	DAVE CLARK 5
LADY FRIEND	BYRDS

WEEK ENDING AUGUST 10, 1967

1.	**Can't Take My Eyes Off You**	***Frankie Valli***	1/14
2.	**Light My Fire**	***The Doors***	6/8
3.	**It Could Be We're In Love**	***Cryan Shames***	4/6
4.	**Blues Theme**	***The Arrows***	3/5
5.	**Words/Pleasant Valley Sunday**	***The Monkees***	7/4
6.	**Come On Down To My Boat**	***Every Mothers' Son***	2/12
7.	**I Was Made To Love Her**	***Stevie Wonder***	11/6
8.	**All You Need Is Love/Baby You're A Rich Man**	***Beatles***	13/3
9.	**White Rabbit**	***Jefferson Airplane***	5/10
10.	**Mercy Mercy Mercy**	***The Buckinghams***	8/10
11.	San Francisco	*Scott McKenzie*	9/12
12.	C'Mon Marianne	*Four Seasons*	12/11
13.	Society's Child	*Janis Ian*	15/9
14.	Carrie Ann	*The Hollies*	16/10
15.	A Whiter Shade Of Pale	*Procol Harum*	18/2
16.	Silence Is Golden	*The Tremeloes*	17/3
17.	My Mammy	*The Happenings*	20/6
18.	Romeo & Juliet	*Michael & Messengers*	14/10
19.	Thank The Lord For The Night Time	*Neil Diamond*	25-3
20.	Somebody Help Me	*Spencer Davis Group*	22-3

(First number indicates last week's position/Second number is the weeks heard on WCFL)

ACTION TEN

21.	TAKE ME BACK	FLOCK
22.	A GIRL LIKE YOU	YOUNG RASCALS
23.	BABY I LOVE YOU	ARETHA FRANKLIN
24.	HEROES & VILLAINS	BEACH BOYS
25.	AIN'T NO MOUNTAIN HIGH ENOUGH	MARVIN GAYE
26.	HOLD ON	THE MAUDS
27.	JACKSON	NANCY SINATRA/LEE HAZELWOOD
28.	COME BACK WHEN YOU GROW UP	BOBBY VEE
29.	REFLECTIONS	SUPREMES
30.	OUT & ABOUT	TOMMY BOYCE/BOBBY HART

CHICAGO PREMIERES

LITTLE OLD WINE DRINKER	DEAN MARTIN
BROWN EYED GAL	DAN MORRISON
I HAD A DREAM	PAUL REVERE & THE RAIDERS

WEEK ENDING AUGUST 17, 1967

1.	**It Could Be We're In Love**	**Cryan Shames**	4/7
2.	**Words/Pleasant Valley Sunday**	***The Monkees***	5/5
3.	**Light My Fire**	***The Doors***	2/9
4.	**Can't Take My Eyes Off You**	***Frankie Valli***	1/15
5.	**Blues Theme**	***The Arrows***	4/6
6.	**I Was Made To Love Her**	***Stevie Wonder***	7/7
7.	**Come On Down To My Boat**	***Every Mothers' Son***	6/13
8.	**All You Need Is Love/Baby You're A Rich Man**	***Beatles***	8/4
9.	**White Rabbit**	***Jefferson Airplane***	9/11
10.	**Carrie Ann**	***The Hollies***	14/11
11.	Thank The Lord For The Night Time	*Neil Diamond*	19/4
12.	Society's Child	*Janis Ian*	13/10
13.	C'Mon Marianne	*Four Seasons*	12/12
14.	Silence Is Golden	*The Tremeloes*	16/4
15.	Ode To Billy Joe	*Bobby Gentry*	-/2
16.	A Whiter Shade Of Pale	*Procol Harum*	15/10
17.	My Mammy	*The Happenings*	17/7
18.	Come Back When You Grow Up	*Bobby Vee*	-/6
19.	A Girl Like You	*Young Rascals*	-/7
20.	Heroes & Villains	*Beach Boys*	-/4

(First number indicates last week's position/Second number is the weeks heard on WCFL)

ACTION TEN

21.	BABY I LOVE YOU	ARETHA FRANKLIN
22.	REFLECTIONS	DIANA ROSS & SUPREMES
23.	ZIP CODE	FIVE AMERICANS
24.	YOU'RE MY EVERYTHING	THE TEMPTATIONS
25.	FAKIN' IT	SIMON & GARFUNKEL
26.	FUNKY BROADWAY	WILSON PICKETT
27.	I HAD A DREAM	PAUL REVERE & RAIDERS
28.	RUN RUN RUN	THIRD RAIL
29.	OUT AND ABOUT	TOMMY BOYCE/BOBBY HART
30.	YOU KNOW WHAT I MEAN	THE TURTLES

CHICAGO PREMIERES

GETTING TOGETHER	TOMMY JAMES & SHONDELLS
MAKING EVERY MINUTE COUNT	SPANKY & OUR GANG
TWELVE THIRTY	MAMAS & PAPAS

WEEK ENDING AUGUST 24, 1967

1.	**It Could Be We're In Love**	**Cryan Shames**	1/8
2.	**Light My Fire**	***The Doors***	3/10
3.	**Blues Theme**	***The Arrows***	5/7
4.	**Pleasant Valley Sunday/Words**	***The Monkees***	2/6
5.	**I Was Made To Love Her**	***Stevie Wonder***	6/8
6.	**Ode To Billy Joe**	***Bobbie Gentry***	15/3
7.	**All You Need Is Love/Baby You're A Rich Man**	***Beatles***	8/5
8.	**Can't Take My Eyes Off You**	***Frankie Valli***	4/16
9.	**Thank The Lord For The Night Time**	***Neil Diamond***	11/5
10.	**Carrie Ann**	***The Hollies***	10/12
11.	Silence Is Golden	*The Tremeloes*	14/5
12.	Reflections	*Diana Ross & Supremes*	-/3
13.	Come Back When You Grown Up	*Bobby Vee*	18/7
14.	Society's Child	*Janis Ian*	12/11
15.	A Girl Like You	*Young Rascals*	19/8
16.	Heroes & Villains	*Beach Boys*	20/5
17.	Take Me Back	*The Flock*	-/3
18.	Baby I Love You	*Aretha Franklin*	-/6
19.	Zip Code	*Five Americans*	-/5
20.	Fakin' It	*Simon & Garfunkel*	-/6

(First number indicates last week's position/Second number is the weeks heard on WCFL)

ACTION TEN

21.	YOU KNOW WHAT I MEAN	THE TURTLES
22.	I HAD A DREAM	PAUL REVERE & RAIDERS
23.	YOU'RE MY EVERYTHING	THE TEMPTATIONS

24.	RUN RUN RUN	THIRD RAIL
25.	FUNKY BROADWAY	WILSON PICKETT
26.	DON'T LET THE RAIN FALL DOWN ON ME	THE CRITTERS
27.	THE WORLD WE KNEW	FRANK SINATRA
28.	A LITTLE BIT NOW	DAVE CLARK 5
29.	OUT AND ABOUT	TOMMY BOYCE/BOBBY HART
30.	THINGS I SHOULD HAVE SAID	GRASS ROOTS

CHICAGO PREMIERES

I MADE A FOOL OF MYSELF	FRANKIE VALLI
THE BALLAD OF YOU & ME & POONEIL	JEFFERSON AIRPLANE
THE CAT IN THE WINDOW	PETULA CLARK

WEEK ENDING AUGUST 31, 1967

1.	**It Could Be We're In Love**	**Cryan Shames**	1/9
2.	**Ode To Billy Joe**	***Bobbie Gentry***	6/4
3.	**Light My Fire**	***The Doors***	2/11
4.	**All You Need Is Love/Baby You're A Rich Man**	***Beatles***	7/6
5.	**I Was Made To Love Her**	***Stevie Wonder***	5/9
6.	**Blues Theme**	***The Arrorws***	3/8
7.	**Pleasant Valley Sunday/Words**	***The Monkees***	4/8
8.	**Reflections**	***Diana Ross Supremes***	12/6
9.	**Come Back When You Grow Up**	***Bobby Vee***	13/8
10.	**Carrie Ann**	***The Hollies***	10/13
11.	Thank The Lord For The Night Time	*Neil Diamond*	9/6
12.	Silence Is Golden	*The Tremeloes*	11/6
13.	Society's Child	*Janis Ian*	14/12
14.	Heroes & Villains	*Beach Boys*	16/6
15.	A Girl Like You	*Young Rascals*	15/9
16.	I Had A Dream	*Paul Revere & Raiders*	-/3
17.	Zip Code	*Five Americans*	19/6
18.	Baby I Love You	*Aretha Franklin*	18/7
19.	You Know What I Mean	*The Turtles*	-/5
20.	Never My Love	*The Association*	-/6

(First number indicates last week's position/Second number is the weeks heard on WCFL)

ACTION TEN

21.	YOU'RE MY EVERYTHING	THE TEMPTATIONS
22.	FUNKY BROADWAY	WILSON PICKETT
23.	RUN RUN RUN	THIRD RAIL
24.	BROWN EYED GIRL	VAN MORRISON
25.	GETTIN' TOGETHER	TOMMY JAMES & SHONDELLS
26.	THE LETTER	THE BOX TOPS
27.	I DIG ROCK & ROLL MUSIC	PETER, PAUL & MARY
28.	HA HA SAID THE CLOWN	YARDBIRDS
29.	DON'T LET THE RAIN FALL DOWN ON ME	THE CRITTERS
30.	I MAKE A FOOL OF MYSELF	FRANKIE VALLI

CHICAGO PREMIERES

A BANDA	HERB ALPERT & TIJUANA BRASS
DANDELIONS	ROLLING STONES
HOW CAN I BE SURE	YOUNG RASCALS

WEEK ENDING SEPTEMBER 7, 1967

1.	**Ode To Billy Joe**	**Bobbie Gentry**	2/5
2.	**It Could Be We're In Love**	***Cryan Shames***	1/10
3.	**Light My Fire**	***The Doors***	3/12
4.	**Come Back When You Grow Up**	***Bobby Vee***	9/9
5.	**Reflections**	***Diana Ross & Supremes***	9/7
6.	**Never My Love**	***The Association***	20/7
7.	**The Letter**	***The Box Tops***	-/2
8.	**All You Need Is Love/Baby You're A Rich Man**	***Beatles***	4/7
9.	**Thank The Lord For The Night Time**	***Neil Diamond***	11/7
10.	**Pleasant Valley Sunday/Words**	***The Monkees***	7/9
11.	I Had A Dream	*Paul Revere & Raiders*	16/4
12.	Funky Broadway	*Wilson Pickett*	-/5
13.	Brown Eyed Girl	*Van Morrison*	-/4
14.	I Dig Rock & Roll Music	*Peter, Paul & Mary*	-/4
15.	Baby I Love You	*Aretha Franklin*	18/8
16.	You Know What I Mean	*The Turtles*	19/6
17.	Zip Code	*Five Americans*	17/7
18.	Heroes & Villains	*Beach Boys*	14/7
19.	You're My Everything	*The Temptations*	-/5
20.	Gettin' Together	*Tommy James & Shondells*	-/4

(First number indicates last week's position/Second number is the weeks heard on WCFL)

ACTION TEN

21.	I MAKE A FOOL OF MYSELF	FRANKIE VALLI
22.	GIMME LITTLE SIGN	BRENTON WOOD
23.	LITTLE OLE MAN	BILL COSBY
24.	APPLES, PEACHES & PUMPKIN PIE	JAY & TECHNIQUES
25.	TWELVE THIRTY	MAMAS & PAPAS
26.	BALLAD OF YOU & ME & POONEIL	JEFFERSON AIRPLANE
27.	THERE IS A MOUNTAIN	DONOVAN
28.	MUSEUM	HERMAN'S HERMITS
29.	HEY BABY (THEY'RE PLAYING OUR SONG)	BUCKINGHAMS
30.	ANYTHING GOES	HARPERS BIZARRE

CHICAGO PREMIERES

THERE'S ALWAYS ME	ELVIS PRESLEY
YOU KEEP RUNNING AWAY	FOUR TOPS

WEEK ENDING SEPTEMBER 14, 1967

1.	**Ode To Billy Joe**	**Bobbie Gentry**	1/6
2.	**Come Back When You Grow Up**	***Bobby Vee***	4/10
3.	**Light My Fire**	***The Doors***	3/13
4.	**The Letter**	***The Box Tops***	7/3
5.	**It Could Be We're In Love**	***Cryan Shames***	2/11
6.	**Never My Love**	***The Association***	6/8
7.	**Reflections**	***Diana Ross & Supremes***	5/8
8.	**Brown Eyed Girl**	***Van Morrison***	13/5
9.	**Thank The Lord For the Night Time**	***Neil Diamond***	9/10
10.	**Funky Broadway**	***Wilson Pickett***	12/6
11.	I Had A Dream	*Paul Revere & Raiders*	11/5
12.	Gettin' Together	*Tommy James & Shondells*	20/5
13.	All You Need Is Love	*Beatles*	8/8
14.	I Dig Rock & Roll Music	*Peter, Paul & Mary*	14/5
15.	You Know What I Mean	*The Turtles*	16/7
16.	I Make A Fool Of Myself	*Frankie Valli*	-/4
17.	You're My Everything	*The Temptations*	19/6
18.	Little Ole Man	*Bill Cosby*	-/3
19.	Gimme Little Sign	*Brenton Wood*	-/5
20.	Hey Baby (They're Playing Our Song)	*Buckinghams*	-/2

(First number indicates last week's position/Second number is the weeks heard on WCFL)

ACTION TEN

21.	TWELVE THIRTY	MAMAS & PAPAS
22.	APPLES, PEACHES & PUMPKIN PIE	JAY & TECHNIQUES
23.	THERE IS A MOUNTAIN	DONOVAN
24.	ANYTHING GOES	HARPERS BIZARRE
25.	GET ON UP	THE ESQUIRES
26.	DANDELION	ROLLING STONES
27.	HOW CAN I BE SURE	YOUNG RASCALS
28.	HIGHER & HIGHER	JACKIE WILSON
29.	THE CAT IN THE WINDOW	PETULA CLARK
30.	MUSEUM	HERMAN'S HERMITS

CHICAGO PREMIERES

LIGHTNING'S GIRL	NANCY SINATRA
WHAT NOW MY LOVE	MITCH RYDER
YOUNGER GENERATION BLUES	JANIS IAN

WEEK ENDING SEPTEMBER 21, 1967

1.	**The Letter**	**The Box Tops**	4/4
2.	**Ode To Billy Joe**	***Bobbie Gentry***	1/7
3.	**Never My Love**	***The Association***	6/9
4.	**Come Back When You Grow Up**	***Bobby Vee***	2/11
5.	**Brown Eyed Girl**	***Van Morrison***	8/6
6.	**Funky Broadway**	***Wilson Pickett***	10/7
7.	**Reflections**	***Diana Ross & Supremes***	7/7
8.	**Little Ole Man**	***Bill Cosby***	18/4
9.	**Light My Fire**	***The Doors***	3/14
10.	**I Had A Dream**	***Paul Revere & Raiders***	11/6
11.	It Could Be We're In Love	*Cryan Shames*	5/12
12.	Gettin' Together	*Tommy James & Shondells*	12/6
13.	Thank The Lord For The Night Time	*Neil Diamond*	9/11
14.	I Dig Rock & Roll Music	*Peter, Paul & Mary*	14/6
15.	You Know What I Mean	*The Turtles*	15/8
16.	I Make A Fool Of Myself	*Frankie Valli*	16/5
17.	Gimme Little Sign	*Brenton Wood*	19/6
18.	Hey Baby (They're Playing Our Song)	*Buckinghams*	20/3
19.	Apples, Peaches & Pumpkin Pie	*Jay & Techniques*	-/8
20.	Dandelion	*Rolling Stones*	-/4

(First number indicates last week's position/Second number is the weeks heard on WCFL)

ACTION TEN

21.	TWELVE THIRTY	MAMAS & PAPAS
22.	THERE IS A MOUNTAIN	DONAVON
23.	GET ON UP	ESQUIRES
24.	ANYTHING GOES	HARPERS BIZARRE
25.	HIGHER & HIGHER	JACKIE WILSON
26.	GROOVIN'	BOOKER T. & MG'S
27.	HOW CAN I BE SURE	YOUNG RASCALS
28.	BALLAD OF YOU & ME & POONEIL	JEFFERSON AIRPLANE
29.	THE CAT IN THE WINDOW	PETULA CLARK
30.	IT MUST BE HIM	VIKKI CARR

CHICAGO PREMIERES

I CAN'T HAPPEN WITHOUT YOU	POWER PLANT
PEOPLE ARE STRANGE	THE DOORS
TREAT HER GROOVY	NEW COLONY SIX

WEEK ENDING SEPTEMBER 28, 1967

1.	**The Letter**	**The Box Tops**	1/5
2.	**Ode To Billy Joe**	***Bobbie Gentry***	2/8
3.	**Never My Love**	***The Association***	3/10
4.	**Brown Eyed Girl**	***Van Morrison***	5/7
5.	**Come Back When You Grow Up**	***Bobby Vee***	4/12
6.	**Reflections**	***Diana Ross & Supremes***	7/8
7.	**Funky Broadway**	***Wilson Pickett***	6/8
8.	**Little Ole Man**	***Bill Cosby***	8/5
9.	**Gimme Little Sign**	***Brenton Wood***	17/7
10.	**Apples, Peaches & Pumpkin Pie**	***Jay & Techniques***	19/9
11.	Hey Baby (They're Playing Our Song)	*Buckinghams*	18/4
12.	I Make A Fool Of Myself	*Frankie Valli*	16/6
13.	Gettin' Together	*Tommy James & Shondells*	12/7
14.	I Dig Rock & Roll Music	*Peter, Paul & Mary*	14/7
15.	You Know What I Mean	*The Turtles*	15/9
16.	How Can I Be Sure	*Young Rascals*	-/5
17.	Dandelion	*Rolling Stones*	20/5
18.	Higher & Higher	*Jackie Wilson*	-/3
19.	Anything Goes	*Harpers Bizarre*	-/7
20.	Twelve Thirty	*Mamas & Papas*	-/7

(First number indicates last week's position/Second number is the weeks heard on WCFL)

ACTION TEN

21.	TO SIR WITH LOVE	LULU
22.	IT MUST BE HIM	VIKKI CARR
23.	GET ON UP	THE ESQUIRES
24.	YOUR PRECIOUS LOVE	MARVIN GAYE/TAMMY TERRELL
25.	LET IT OUT	HOMBRES
26.	YOU'VE GOT TO PAY THE PRICE	AL KENT
27.	SOUL MAN	SAM & DAVE
28.	GROOVIN'	BOOKER T. & MG'S
29.	THE CAT IN THE WINDOW	PETULA CLARK
30.	WINDOWS OF THE WORLD	DIONNE WARWICK

CHICAGO PREMIERES

A NATURAL WOMAN	ARETHA FRANKLIN
KING MIDAS IN REVERSE	THE HOLLIES

WEEK ENDING OCTOBER 5, 1967

1.	**The Letter**	**The Box Tops**	1/6
2.	**Never My Love**	***The Association***	3/11
3.	**Ode To Billy Joe**	***Bobbie Gentry***	2/9
4.	**Brown Eyed Girl**	***Van Morrison***	4/8
5.	**Come Back When You Grow Up**	***Bobby Vee***	5/13
6.	**Funky Broadway**	***Wilson Pickett***	7/9
7.	**Little Ole Man**	***Bill Cosby***	8/6
8.	**Gimme Little Sign**	***Brenton Wood***	9/8
9.	**Reflections**	***Diana Ross & Supremes***	6/9
10.	**Apples, Peaches & Pumpkin Pie**	***Jay & Techniques***	10/10
11.	Hey Baby (They're Playing Our Song)	*Buckinghams*	11/5
12.	Let It Out	*Hombres*	-/2
13.	How Can I Be Sure	*Young Rascals*	16/6
14.	I Make A Fool Of Myself	*Frankie Valli*	12/7
15.	You Know What I Mean	*The Turtles*	15/10
16.	To Sir With Love	*Lulu*	-/5
17.	Higher & Higher	*Jackie Wilson*	18/4
18.	It Must Be Him	*Vikki Carr*	-/7
19.	I Dig Rock & Roll Music	*Peter, Paul & Mary*	14/8
20.	Anything Goes	*Harpers Bizarre*	19/8

(First number indicates last week's position/Second number is the weeks heard on WCFL)

ACTION TEN

21.	DANDELION	ROLLING STONES
22.	SOUL MAN	SAM & DAVE
23.	GET ON UP	THE ESQUIRES
24.	YOUR PRECIOUS LOVE	MARVIN GAYE/TAMMY TERRELL
25.	PURPLE HAZE	JIMI HENDRIX EXPERIENCE
26.	PEOPLE ARE STRANGE	THE DOORS
27.	YOU GOTTA PAY THE PRICE	AL KENT
28.	EXPRESSWAY TO YOUR HEART	SOUL SURVIVORS
29.	YOU KEEP RUNNING AWAY	FOUR TOPS
30.	WINDOWS OF THE WORLD	DIONNE WARWICK

CHICAGO PREMIERES

HARD TO GET THING CALLED LOVE	LOU RAWLS
I'M WONDERING	STEVIE WONDER
KENTUCKY WOMAN	NEIL DIAMOND

WEEK ENDING OCTOBER 12, 1967

1.	**The Letter**	**The Box Tops**	1/7
2.	**Little Ole Man**	***Bill Cosby***	7/7
3.	**Never My Love**	***The Association***	2/12
4.	**Apples, Peaches & Pumpkin Pie**	***Jay & Techniques***	10/11
5.	**Gimme Little Sign**	***Brenton Wood***	8/9
6.	**Brown Eyed Girl**	***Van Morrison***	4/9
7.	**Let It Out**	***Hombres***	12/3
8.	**Ode To Billy Joe**	***Bobbie Gentry***	3/10
9.	**How Can I Be Sure**	***Young Rascals***	13/7
10.	**Reflections**	***Diana Ross & Supremes***	9/10
11.	Hey Baby (They're Playing Our Song)	*Buckinghams*	11/6
12.	Funky Broadway	*Wilson Pickett*	6/10
13.	To Sir With Love	*Lulu*	16/6
14.	It Must Be Him	*Vikki Carr*	18/8
15.	I Make A Fool Of Myself	*Frankie Valli*	14/8
16.	You Know What I Mean	*The Turtles*	15/11
17.	Soul Man	*Sam & Dave*	-/2
18.	Higher & Higher	*Jackie Wilson*	17/5
19.	Purple Haze	*Jimi Hendrix Experience*	-/2
20.	Expressway To Your Heart	*Soul Survivors*	-/1

(First number indicates last week's position/Second number is the weeks heard on WCFL)

ACTION TEN

21.	YOUR PRECIOUS LOVE	MARVIN GAYE/TAMMY TERRELL
22.	HUSH	BILLY JOE ROYAL
23.	GET ON UP	THE ESQUIRES
24.	YOU KEEP RUNNING AWAY	FOUR TOPS
25.	PEOPLE ARE STRANGE	THE DOORS
26.	THE RAIN, THE PARK & OTHER THINGS	THE COWSILLS
27.	VISIT TO A SAD PLANET	LEONARD NEMOY
28.	A NATURAL WOMAN	ARETHA FRANKLIN
29.	CHILD OF CLAY	JIMMIE RODGERS
30.	TREAT HER GROOVY	NEW COLONY SIX

CHICAGO PREMIERES

EXCERPT FROM "A TEENAGE OPERA"	KEITH WEST
LIKE AN OLD TIME MOVIE	SCOTT McKENZIE
WAKE UP, WAKE UP	GRASSROOTS

WEEK ENDING OCTOBER 19, 1967

1.	**The Letter**	**The Box Tops**	1/8
2.	**Never My Love**	***The Association***	3/13
3.	**Brown Eyed Girl**	***Van Morrison***	6/10
4.	**How Can I Be Sure**	***Young Rascals***	9/8
5.	**To Sir With Love**	***Lulu***	13/7
6.	**Hey Baby (They're Playing Our Song)**	***Buckinghams***	11/7
7.	**Apples, Peaches & Pumpkin Pie**	***Jay & Techniques***	4/12
8.	**Little Ole Man**	***Bill Cosby***	2/8
9.	**Let It Out**	***Hombres***	7/4
10.	**Gimme Little Sign**	***Brenton Wood***	5/10
11.	It Must Be Him	*Vikki Carr*	14/9
12.	Funky Broadway	*Wilson Pickett*	12/11
13.	Purple Haze	*Jimi Hendrix Experience*	19/3
14.	Soul Man	*Sam & Dave*	17/3
15.	Expressway To Your Heart	*Soul Survivors*	20/2
16.	Higher & Higher	*Jackie Wilson*	18/6
17.	Hush	*Billy Joe Royal*	-/6
18.	Your Precious Love	*Marvin Gaye/Tammy Terrell*	-/4
19.	The Rain, The Park & Other Things	*The Cowsills*	-/3
20.	Incense & Peppermints	*Strawberry Alarm Clock*	-/3

(First number indicates last week's position/Second number is the weeks heard on WCFL)

ACTION TEN

21.	GET ON UP	THE ESQUIRES
22.	PEOPLE ARE STRANGE	THE DOORS
23.	CHILD OF CLAY	JIMMIE RODGERS
24.	TREAT HER GROOVY	NEW COLONY SIX
25.	A NATURAL WOMAN	ARETHA FRANKLIN
26.	LOOK OF LOVE	DUSTY SPRINGFIELD
27.	THE LAST WALTZ	ENGELBERT HUMPERDINCK
28.	IT'S YOU THAT I NEED	THE TEMPTATIONS
29.	LIGHTNING'S GIRL	NANCY SINATRA
30.	KENTUCKY WOMAN	NEIL DIAMOND

CHICAGO PREMIERES

DIFFERENT DRUM	STONE PONEYS
GLAD TO BE UNHAPPY	MAMAS & PAPAS
I WANT ACTION	RUBY WINTERS

wned and operated by The Chicago Federation of Labor

WEEK ENDING OCTOBER 26, 1967

1.	**The Letter**	**The Box Tops**	1/9
2.	**Never My Love**	***The Association***	2/14
3.	**Let It Out**	***Hombres***	9/5
4.	**To Sir With Love**	***Lulu***	5/8
5.	**How Can I Be Sure**	***Young Rascals***	4/9
6.	**Little Ole Man**	***Bill Cosby***	8/9
7.	**Brown Eyed Girl**	***Van Morrison***	3/11
8.	**Hey Baby (They're Playing Our Song)**	***Buckinghams***	6/8
9.	**Incense & Peppermints**	***Strawberry Alarm Clock***	20/4
10.	**It Must Be Him**	***Vikki Carr***	11/10
11.	The Rain, The Park & Other Things	*The Cowsills*	19/4
12.	Purple Haze	*Jimi Hendrix Experience*	13/4
13.	Soul Man	*Sam & Dave*	14/4
14.	Hush	*Billy Joe Royal*	17/7
15.	Expressway To Your Heart	*Soul Survivors*	15/3
16.	Look Of Love	*Dusty Springfield*	-/6
17.	Your Precious Love	*Marvin Gaye/Tammi Terrell*	18/5
18.	Treat Her Groovy	*New Colony Six*	-/6
19.	Child Of Clay	*Jimmie Rodgers*	-/3
20.	People Are Strangers	*The Doors*	-/6

(First number indicates last week's position/Second number is the weeks heard on WCFL)

ACTION TEN

21.	KENTUCKY WOMAN	NEIL DIAMOND
22.	A NATURAL WOMAN	ARETHA FRANKLIN
23.	IT'S YOU THAT I NEED	TEMPTATIONS
24.	THE LAST WALTZ	ENGELBERT HUMPERDINCK
25.	LOVE IS STRANGE	PEACHES & HERB
26.	I'M WONDERING	STEVIE WONDER
27.	LIGHTNING'S GIRL	NANCY SINATRA
28.	PLEASE LOVE ME FOREVER	BOBBY VINTON
29.	I CAN'T HAPPEN WITHOUT YOU	POWER PLANT
30.	EVERLASTING LOVE	ROBERT KNIGHT

CHICAGO PREMIERES

LITTLE BITTY PRETTY ONE	JEWEL AKENS
SHE'S MY GIRL	THE TURTLES
STAG-O-LEE	WILSON PICKETT

wned and operated by The Chicago Federation of Labor

WEEK ENDING NOVEMBER 2, 1967

1.	**To Sir With Love**	**Lulu**	4/9
2.	**The Letter**	***The Box Tops***	1/10
3.	**Let It Out**	***The Hombres***	3/6
4.	**Incense & Peppermints**	***Strawberry Alarm Clock***	9/5
5.	**The Rain, The Park & Other Things**	***The Cowsills***	11/5
6.	**How Can I Be Sure**	***Young Rascals***	5/10
7.	**Purple Haze**	***Jimi Hendrix Experience***	12/5
8.	**Expressway To Your Heart**	***Soul Survivors***	15/4
9.	**It Must Be Him**	***Vikki Carr***	10/11
10.	**Never My Love**	***The Association***	2/15
11.	Look Of Love	*Dusty Springfield*	16/7
12.	Soul Man	*Sam & Dave*	13/5
13.	Hush	*Billy Joe Royal*	14/8
14.	Your Precious Love	*Marvin Gaye/Tammi Terrell*	17/6
15.	Treat Her Groovy	*New Colony Six*	18/7
16.	Kentucky Woman	*Neil Diamond*	-/5
17.	Child Of Clay	*Jimmie Rodgers*	19/4
18.	A Natural Woman	*Aretha Franklin*	-/6
19.	Love Is Strange	*Peaches & Herb*	-/5
20.	Please Love Me Forever	*Bobby Vinton*	-/4

(First number indicates last week's position/Second number is the weeks heard on WCFL)

ACTION TEN

21.	IT'S YOU THAT I NEED	TEMPTATIONS
22.	LIGHTNING'S GIRL	NANCY SINATRA
23.	I'M WONDERING	STEVIE WONDER
24.	EVERLASTING LOVE	ROBERT KNIGHT
25.	GLAD TO BE UNHAPPY	MAMAS & PAPAS
26.	WATCH THE FLOWERS GROW	FOUR SEASONS
27.	STAG-O-LEE	WILSON PICKETT
28.	FACE THE AUTUMN	THE FAMILY
29.	LAZY DAY	SPANKY & OUR GANG
30.	SUZANNE	NOEL HARRISON

CHICAGO PREMIERES

DAYDREAM BELIEVER	THE MONKEES
ITCHYCOO PARK	SMALL FACES
YOU BETTER SIT DOWN, KIDS	CHER

wned and operated by The Chicago Federation of Labor

WEEK ENDING NOVEMBER 9, 1967

1.	**To Sir With Love**	**Lulu**	1/10
2.	**Incense & Peppermints**	***Strawberry Alarm Clock***	4/6
3.	**The Rain, The Park & Other Things**	***The Cowsills***	5/6
4.	**Let It Out**	***The Hombres***	3/7
5.	**The Letter**	***The Box Tops***	2/11
6.	**How Can I Be Sure**	***Young Rascals***	6/11
7.	**Purple Haze**	***Jimi Hendrix Experience***	7/6
8.	**It Must Be Him**	***Vikki Carr***	9/12
9.	**Look Of Love**	***Dusty Springfield***	11/8
10.	**Hush**	***Billy Joe Royal***	13/9
11.	Expressway To Your Heart	*Soul Survivors*	8/5
12.	Soul Man	*Sam & Dave*	12/6
13.	Please Love Me Forever	*Bobby Vinton*	20/5
14.	Treat Her Groovy	*New Colony Six*	15/8
15.	Kentucky Woman	*Neil Diamond*	16/6
16.	Everlasting Love	*Robert Knight*	-/8
17.	Love Is Strange	*Peaches & Herb*	19/6
18.	Child Of Clay	*Jimmie Rodgers*	17/5
19.	Glad To Be Unhappy	*Mamas & Papas*	-/4
20.	Watch The Flowers Grow	*Four Seasons*	-/3

(First number indicates last week's position/Second number is the weeks heard on WCFL)

ACTION TEN

21.	IT'S YOU THAT I NEED	TEMPTATIONS
22.	I HEARD IT THROUGH THE GRAPEVINE	GLADYS KNIGHT & PIPS
23.	THE LAST WALTZ	ENGELBERT HUMPERDINCK
24.	LIGHTNING'S GIRL	NANCY SINATRA
25.	I'M WONDERING	STEVIE WONDER
26.	SKINNY LEGS & ALL	JOE TEX
27.	STAG-O-LEE	WILSON PICKETT
28.	OUT OF THE BLUE	TOMMY JAMES & SHONDELLS
29.	FACE THE AUTUMN	THE FAMILY
30.	AN OPEN LETTER TO MY TEENAGE SON	VICTOR LUNDBERT

CHICAGO PREMIERES

IN AND OUT OF LOVE	DIANA ROSS & SUPREMES
PEACE OF MIND	PAUL REVERE & RAIDERS

wned and operated by The Chicago Federation of Labor

WEEK ENDING NOVEMBER 16, 1967

1.	**The Rain, The Park & Other Things**	**The Cowsills**	3/7
2.	**Incense & Peppermints**	***Strawberry Alarm Clark***	2/7
3.	**To Sir With Love**	***Lulu***	1/11
4.	**Let It Out**	***The Hombres***	4/8
5.	**Purple Haze**	***Jimi Hendrix Experience***	7/7
6.	**The Letter**	***The Box Tops***	5/12
7.	**Look Of Love**	***Dusty Springfield***	9/9
8.	**How Can I Be Sure**	***Young Rascals***	6/12
9.	**Daydream Believer**	***The Monkees***	-/3
10.	**Expressway To Your Heart**	***Soul Survivors***	11/6
11.	Please Love Me Forever	***Bobby Vinton***	13/6
12.	It Must Be Him	*Vikki Carr*	8/13
13.	Hush	*Billy Joe Royal*	10/10
14.	Everlasting Love	*Robert Knight*	16/9
15.	Soul Man	*Sam & Dave*	12/7
16.	Treat Her Groovy	*New Colony Six*	14/9
17.	Kentucky Woman	*Neil Diamond*	15/7
18.	Glad To Be Unhappy	*Mamas & Papas*	19/5
19.	An Open Letter To My Teenage Son	*Victor Lundberg*	-/2
20.	Watch The Flowers Grow	*Four Seasons*	20/4

(First number indicates last week's position/Second number is the weeks heard on WCFL)

ACTION TEN

21.	FACE THE AUTUMN	THE FAMILY
22.	SKINNY LEGS & ALL	JOE TEX
23.	I HEARD IT THROUGH THE GRAPEVINE	GLADYS KNIGHT & PIPS
24.	STAG-O-LEE	WILSON PICKETT
25.	I SAY A LITTLE PRAYER	DIONNE WARWICK
26.	OUT OF THE BLUE	TOMMY JAMES SHONDELLS
27.	SHE'S MY GIRL	THE TURTLES
28.	SHE IS STILL A MYSTERY	LOVIN' SPOONFUL
29.	YOU BETTER SIT DOWN KIDS	CHER
30.	PAPER CUP	FIFTH DIMENSION

CHICAGO PREMIERES

BABY YOU'VE GOT IT	BRENTON WOOD
DEAR ELOISE	THE HOLLIES
MYSTERY TOUR	THE BEATLES

wned and operated by The Chicago Federation of Labor

WEEK ENDING NOVEMBER 23, 1967

1.	**The Rain, The Park & Other Things**	***The Cowsills***	1/8
2.	**Incense & Peppermints**	***Strawberry Alarm Clock***	2/8
3.	**To Sir With Love**	***Lulu***	3/12
4.	**Daydream Believer**	***The Monkees***	9/4
5.	**Purple Haze**	***Jimi Hendrix Experience***	5/8
6.	**Please Love Me Forever**	***Bobby Vinton***	11/7
7.	**Let It Out**	***The Hombres***	4/9
8.	**Everlasting Love**	***Robert Knight***	14/10
9.	**Look Of Love**	***Dusty Springfield***	7/10
10.	**Hush**	***Billy Joe Royal***	13/11
11.	Expressway To Your Heart	*Soul Survivors*	10/7
12.	Treat Her Groovy	*New Colony Six*	16/10
13.	Face The Autumn	*The Family*	-/3
14.	Skinny Legs & All	*Joe Tex*	-/4
15.	Watch The Flowers Grow	*Four Seasons*	20/5
16.	Glad To Be Unhappy	*Mamas & Papas*	18/6
17.	An Open Letter To My Teenage Son	*Victor Lundberg*	19/3
18.	You Better Sit Down Kids	*Cher*	-/4
19.	I Say A Little Prayer	*Dionne Warwick*	-/6
20.	Out Of The Blue	*Tommy James & Shondells*	-/5

(First number indicates last week's position/Second number is the weeks heard on WCFL)

ACTION TEN

21.	I CAN SEE FOR MILES	THE WHO
22.	SHE IS STILL A MYSTERY	LOVIN' SPOONFUL
23.	STAG-O-LEE	WILSON PICKETT
24.	SHE'S MY GIRL	THE TURTLES
25.	BEAUTIFUL PEOPLE	KENNY O'DELL
26.	I SECOND THAT EMOTION	SMOKEY ROBINSON & MIRACLES
27.	IN AND OUT OF LOVE	DIANA ROSS SUPREMES
28.	NEON RAINBOW	THE BOX TOPS
29.	FELICIDAD	SALLY FIELD
30.	PAPER CUP	FIFTH DIMENSION

CHICAGO PREMIERES

HELLO GOODBYE/I AM THE WALRUS	THE BEATLES
OKOLONA RIVER BOTTOM BAND	BOBBIE GENTRY
SUSAN	THE BUCKINGHAMS

WEEK ENDING NOVEMBER 30, 1967

1.	**The Rain, The Park & Other Things**	**The Cowsills**	1/9
2.	**Daydream Believer**	***The Monkees***	4/5
3.	**To Sir With Love**	***Lulu***	3/13
4.	**Incense & Peppermints**	***Strawberry Alarm Clock***	2/9
5.	**Everlasting Love**	***Robert Knight***	8/11
6.	**Please Love Me Forever**	***Bobby Vinton***	6/8
7.	**Expressway To Your Heart**	***Soul Survivors***	11/8
8.	**You Better Sit Down Kids**	***Cher***	18/5
9.	**An Open Letter To My Teenage Son**	***Victor Lundberg***	17/4
10.	**I Say A Little Prayer**	***Dionne Warwick***	19/7
11.	Look Of Love	*Dusty Springfield*	9/11
12.	Watch The Flowers Grow	*Four Seasons*	15/6
13.	Skinny Legs & All	*Joe Tex*	14/5
14.	Face The Autumn	*The Family*	13/4
15.	She's My Girl	*The Turtles*	-/6
16.	Glad To Be Unhappy	*Mamas & Papas*	16/7
17.	Beautiful People	*Kenny O'Dell*	-/5
18.	I Can See For Miles	*The Who*	-/6
19.	Out Of The Blue	*Tommy James & Shondells*	20/6
20.	I Second That Emotion	*Smokey Robinson & Miracles*	-/3

(First number indicates last week's position/Second number is the weeks heard on WCFL)

ACTION TEN

21.	I HEARD IT FROM THE GRAPEVINE	GLADYS KNIGHT & PIPS
22.	FELICIDAD	SALLY FIELD
23.	BOOGALOO DOWN BROADWAY	THE FANTASTIC JOHNNY C
24.	NEON RAINBOW	THE BOX TOPS
25.	BEND ME, SHAPE ME	THE AMERICAN BREED
26.	IN & OUT OF LOVE	DIANA ROSS & SUPREMES
27.	PAPER CUP	FIFTH DIMENSION
28.	PEACE OF MIND	PAUL REVERE & RAIDERS
29.	HELLO GOODBYE/I AM THE WALRUS	THE BEATLES
30.	BEG, BORROW & STEAL	OHIO EXPRESS

CHICAGO PREMIERES

CHAIN OF FOOLS	ARETHA FRANKLIN
FOXEY LADY	JIMI HENDRIX
IT'S WONDERFUL	YOUNG RASCALS

WEEK ENDING DECEMBER 7, 1967

1.	**Daydream Believer**	**Monkees**	2/6
2.	**The Rain, The Park & Other Things**	***The Cowsills***	1/10
3.	**Incense & Peppermints**	***Strawberry Alarm Clock***	4/10
4.	**You Better Sit Down Kids**	***Cher***	8/6
5.	**Everlasting Love**	***Robert Knight***	5/12
6.	**Hello Goodbye/I Am The Walrus**	***The Beatles***	-/3
7.	**I Say A Little Prayer**	***Dionne Warwick***	10/8
8.	**To Sir With Love**	***Lulu***	3/14
9.	**Please Love Me Forever**	***Bobby Vinton***	6/9
10.	**Skinny Legs & All**	***Joe Tex***	13/6
11.	Bend Me, Shape Me	*The American Breed*	-/4
12.	She's My Girl	*The Turtles*	15/7
13.	I Can See For Miles	*The Who*	18/7
14.	Boogaloo Down Broadway	*The Fantastic Johnny C*	-/1
15.	Neon Rainbow	*The Box Tops*	-/5
16.	I Second That Emotion	*Smokey Robinson & Miracles*	20/4
17.	Judy In Disguise	*John Fred*	-/3
18.	In & Out Of Love	*Diana Ross & Supremes*	-/5
19.	I Heard It From The Grapevine	*Gladys Knight & Pips*	-/5
20.	Beautiful People	*Kenny O'Dell*	17/6

(First number indicates last week's position/Second number is the weeks heard on WCFL)

ACTION TEN

21.	SNOOPY'S CHRISTMAS	THE ROYAL GUARDSMEN
22.	SUSAN	THE BUCKINGHAMS
23.	WOMAN, WOMAN	UNION GAP
24.	BABY YOU GOT IT	BRENTON WOOD
25.	STAG-O-LEE	WILSON PICKETT
26.	BEG, BORROW & STEAL	OHIO EXPRESS
27.	PEACE OF MIND	PAUL REVERE & RAIDERS
28.	PAPER CUP	FIFTH DIMENSION
29.	MASSACHUSETTS	THE BEE GEES
30.	YESTERDAY	RAY CHARLES

CHICAGO PREMIERES

DARLING	THE BEACH BOYS
THEIR SATANIC MAJESTIES REQUEST	ROLLING STONES ALBUM
WHO WILL ANSWER	ED AMES

WEEK ENDING DECEMBER 14, 1967

1.	**Daydream Believer**	**The Monkees**	1/7
2.	**Hello Goodbye/I Am The Walrus**	***The Beatles***	6/4
3.	**You Better Sit Down Kids**	***Cher***	4/7
4.	**The Rain, The Park & Other Things**	***The Cowsills***	2/11
5.	**Incense & Peppermints**	***Strawberry Alarm Clock***	3/11
6.	**Bend Me, Shape Me**	***The American Breed***	11/5
7.	**I Say A Little Prayer**	***Dionne Warwick***	7/9
8.	**Judy In Disguise**	***John Fred***	17/4
9.	**I Can See For Miles**	***The Who***	13/8
10.	**Skinny Legs & All**	***Joe Tex***	10/7
11.	Boogaloo Down Broadway	*The Fantastic Johnny C*	14/2
12.	She's My Girl	*The Turtles*	12/8
13.	Neon Rainbow	*The Box Tops*	15/6
14.	Woman, Woman	*Union Gap*	-/6
15.	I Heard It Through The Grapevine	*Gladys Knight & Pips*	19/6
16.	Snoopy's Christmas	*Royal Guardsmen*	-/3
17.	I Second That Emotion	*Smokey Robinson & Miracles*	16/5
18.	Beautiful People	*Kenny O'Dell*	20/7
19.	Susan	*The Buckinghams*	-/4
20.	In & Out Of Love	*Diana Ross & Supremes*	18/6

(First number indicates last week's position/Second number is the weeks heard on WCFL)

ACTION TEN

21.	CHAIN OF FOOLS	ARETHA FRANKLIN
22.	BABY YOU GOT IT	BRENTON WOOD
23.	PEACE OF MIND	PAUL REVERE & RAIDERS
24.	BEG, BORROW & STEAL	OHIO EXPRESS
25.	GREEN TAMBOURINE	LEMON PIPERS
26.	MASSACHUSETTS	THE BEE GEES
27.	IF I COULD BUILD MY WORLD AROUND YOU	M. GAYE/T. TERRELL
28.	I'M IN LOVE	WILSON PICKETT
29.	KEEP THE BALL ROLLING	JAY & TECHNIQUES
30.	YESTERDAY	RAY CHARLES

CHICAGO PREMIERES

31.	GOOD COMBINATION	SONNY & CHER
32.	JUST AS MUCH AS EVEVR	BOBBY VINTON
33.	MAGICAL MYSTERY TOUR ALBUM	THE BEATLES

WEEK ENDING DECEMBER 21, 1967

1.	**Hello Goodbye/I Am The Walrus**	**The Beatles**	2/5
2.	**Daydream Believer**	***The Monkees***	1/8
3.	**Judy In Disguise**	***John Fred***	8/5
4.	**Bend Me, Shape Me**	***The American Breed***	6/6
5.	**You Better Sit Down Kids**	***Cher***	3/8
6.	**Snoopy's Christmas**	***Royal Guardsmen***	16/4
7.	**Woman, Woman**	***Union Gap***	14/7
8.	**I Can See For Miles**	***The Who***	9/9
9.	**Boogaloo Down Broadway**	***The Fantastic Johnny C***	11/3
10.	**Neon Rainbow**	***The Box Tops***	13/7
11.	She's My Girl	*The Turtles*	12/9
12.	Skinny Legs & All	*Joe Tex*	10/8
13.	Chain Of Fools	*Aretha Franklin*	-/4
14.	I Second That Emotion	*Smokey Robinson & Miracles*	17/6
15.	I Heard It Through The Grapevine	*Gladys Knight & Pips*	15/7
16.	Susan	*The Buckinghams*	19/5
17.	Green Tambourine	*Lemon Pipers*	-/2
18.	In & Out Of Love	*Diana Ross & Supremes*	20/7
19.	It's Wonderful	*Young Rascals*	-/4
20.	Beautiful People	*Kenny O'Dell*	18/8

(First number indicates last week's position/Second number is the weeks heard on WCFL)

ACTION TEN

21.	SUMMER RAIN	JOHNNY RIVERS
22.	IF I COULD BUILD MY WORLD AROUND YOU	M. GAYE/T. TERRELL
23.	BABY YOU GOT IT	BRENTON WOOD
24.	BEG, BORROW & STEAL	OHIO EXPRESS
25.	WEAR YOUR LOVE LIKE HEAVEN	DONOVAN
26.	HONEY CHILE	MARTHA & VANDELLAS
27.	KEEP THE BALL ROLLING	JAY & TECHNIQUES
28.	TELL MAMA	ETTA JAMES
29.	BY THE TIME I GET TO PHOENIX	GLEN CAMPBELL
30.	GOIN' OUT OF MY HEAD/CAN'T TAKE MY EYE OFF YOU	LETTERMEN

CHICAGO PREMIERES

I'M COMING HOME	TOM JONES
TO GIVE	FRANKIE VALLI
YOU HAVEN'T SEEN MY LOVE	THE ONES

WEEK ENDING DECEMBER 28, 1967

1.	**Hello Goodbye/I Am The Walrus**	**The Beatles**	1/6
2.	**Judy In Disguise**	***John Fred***	3/6
3.	**Bend Me, Shape Me**	***The American Breed***	4/7
4.	**Woman, Woman**	***The Union Gap***	7/8
5.	**Daydream Believer**	***The Monkees***	2/9
6.	**Snoopy's Christmas**	***Royal Guardsmen***	6/5
7.	**Chain of Fools**	***Aretha Franlin***	13/5
8.	**Neon Rainbow**	***The Box Tops***	10/8
9.	**I Can See For Miles**	***The Who***	8/10
10.	**I Second That Emotion**	***Smokey Robinson & Miracles***	14/7
11.	Boogaloo Down Broadway	*The Fantastic Johnny C*	9/4
12.	Green Tambourine	*Lemon Pipers*	17/3
13.	Susan	*The Buckinghams*	16/6
14.	I Heard It Through The Grapevine	*Gladys Knight & Pips*	15/8
15.	Summer Rain	*Johnny Rivers*	-/5
16.	In & Out Of Love	*Diana Ross & Supremes*	18/8
17.	It's Wonderful	*Young Rascals*	19/5
18.	By The Time I Get To Phoenix	*Glen Campbell*	-/4
19.	I'm In Love	*Wilson Pickett*	-/3
20.	Beautiful People	*Kenny O'Dell*	20/9

(First number indicates last week's position/Second number is the weeks heard on WCFL)

ACTION TEN

21.	TELL MAMA	ETTA JAMES
22.	WHO WILL ANSWER	ED AMES
23.	WEAR YOUR LOVE LIKE HEAVEN	DONOVAN
24.	COME SEE ABOUT ME	JR. WALKER & ALL STARS
25.	GOIN' OUT OF MY HEAD/CAN'T TAKE MY EYES OFF YOU	LETTERMEN
26.	HONEY CHILE	MARTHA & VANDELLAS
27.	BOTTLE OF WINE	THE FIREBALLS
28.	THE OTHER MAN'S GRASS IS ALWAYS GREENER	PETULA CLARK
29.	SKIP A ROPE	HENSON CARGILL
30.	EXPLOSION IN YOUR SOUL	SOUL SURVIVORS

CHICAGO PREMIERES

BLESSED ARE THE LONELY	ROBERT KNIGHT
NEW ORLEANS	NEIL DIAMOND

1968

WEEK ENDING JANUARY 4, 1968

1.	**Judy In Disguise**	**John Fred**	2/7
2.	**Hello Goodbye/I Am The Walrus**	***The Beatles***	1/7
3.	**Bend Me, Shape Me**	***The American Breed***	3/8
4.	**Daydream Believer**	***The Monkees***	5/10
5.	**Woman, Woman**	***The Union Gap***	4/9
6.	**Chain Of Fools**	***Aretha Franklin***	7/6
7.	**Green Tambourine**	***Lemon Pipers***	12/4
8.	**I Heard It Through The Grapevine**	***Gladys Knight & Pips***	14/9
9.	**Susan**	***The Buckinghams***	13/7
10.	**Snoopy's Christmas**	***Royal Guardsmen***	6/6
11.	Boogaloo Down Broadway	*The Fantastic Johnny C*	11/5
12.	I Second That Emotion	*Smokey Robinson & Miracles*	10/8
13.	Neon Rainbow	*The Box Tops*	8/9
14.	She's My Girl	*The Turtles*	-/10
15.	Summer Rain	*Johnny Rivers*	15/6
16.	It's Wonderful	*Young Rascals*	17/6
17.	Bottle of Wine	*The Fireballs*	-/5
18.	If I Could Build My Life Around You	*M. Gaye/T. Terrell*	-/5
19.	I'm In Love	*Wilson Pickett*	19/4
20.	Who Will Answer	*Ed Ames*	-/5

(First number indicates last week's position/Second number is the weeks heard on WCFL)

ACTION TEN

21.	TELL MAMA	ETTA JAMES
22.	SKIP A ROPE	HENSON CARGILL
23.	GOIN' OUT OF MY HEAD/CAN'T TAKE MY EYES OFF YOU	LETTERMEN
24.	I WONDER WHAT SHE'S DOING TONIGHT	T. BOYCE/B. HART
25.	THE OTHER MAN'S GRASS IS ALWAYS GREENER	PETULA CLARK
26.	TO GIVE	FRANKIE VALLI
27.	EXPLOSION IN YOUR SOUL	SOUL SURVIVORS
28.	EVERYBODY KNOWS	DAVE CLARK 5
29.	AM I THAT EASY TO FORGET	ENGELBERT HUMPERDINCK
30.	JUST AS MUCH AS EVER	BOBBY VINTON

CHICAGO PREMIERES

31.	CARMEN	HERB ALPERT
32.	RED, GREEN, YELLOW & BLUE	DICKEY LEE
33.	WE CAN FLY	THE COWSILLS

WEEK ENDING JANUARY 11, 1968

1.	**Judy In Disguise**	**John Fred**	1/8
2.	**Bend Me, Shape Me**	***The American Breed***	3/9
3.	**Green Tambourine**	***Lemon Pipers***	7/5
4.	**Chain Of Fools**	***Aretha Franklin***	6/7
5.	**Woman, Woman**	***The Union Gap***	5/10
6.	**Hello Goodbye/I am The Walrus**	***The Beatles***	2/8
7.	**Daydream Believer**	***The Monkees***	4/11
8.	**Susan**	***The Buckinghams***	9/8
9.	**I Heart It Through the Grapevine**	***Gladys Knight & Pips***	8/10
10.	**Boogaloo Down Broadway**	***The Fantastic Johnny C***	11/6
11.	It's Wonderful	*Young Rascals*	16/7
12.	I Second That Emotion	*Smokey Robinson & Miracles*	12/9
13.	She's My Girl	*The Turtles*	14/11
14.	Summer Rain	*Johnny Rivers*	15/7
15.	I'm In Love	*Wilson Pickett*	19/5
16.	To Give	*Frankie Valli*	-/4
17.	I Wonder What She's Doing Tonight	*T. Boyce/B. Hart*	-/2
18.	Tell Mama	*Etta James*	-/7
19.	Out Of My Head/Can't Take Eyes Off You	*Lettermen*	-/7
20.	Who Will Answer	*Ed Ames*	20/6

(First number indicates last week's position/Second number is the weeks heard on WCFL)

ACTION TEN

21.	IF I COULD BUILD MY WHOLE WORLD AROUND YOU	M. GAYE/T. TERRELL
22.	SKIP A ROPE	HENSON CARGILL
23.	BOTTLE OF WINE	FIREBALLS
24.	HONEY CHILE	MARTHA & VANDELLAS
25.	EXPLOSION IN YOUR SOUL	SOUL SURVIVORS
26.	SUNSHINE OF YOUR LOVE	THE CREAM
27.	COME SEE ABOUT ME	JR. WALKER & ALL STARS
28.	AM I THAT EASY TO FORGET	ENGELBERT HUMPERDINCK
29.	DIFFERENT DRUM	THE STONE PONEYS
30.	JUST AS MUCH AS EVER	BOBBY VINTON

CHICAGO PREMIERES

I WISH IT WOULD RAIN	THE TEMPTATIONS
PICK UP THE PIECES	CARLA THOMAS
SPOOKY	CLASSIC IV

WEEK ENDING JANUARY 18, 1968

1.	**Bend Me, Shape Me**	**The American Breed**	2/10
2.	**Judy In Disguise**	***John Fred***	1/9
3.	**Green Tambourine**	***Lemon Pipers***	3/6
4.	**Chain Of Fools**	***Aretha Franklin***	4/8
5.	**Woman, Woman**	***The Union Gap***	5/11
6.	**Susan**	***The Buckinghams***	8/9
7.	**Daydream Believer**	***The Monkees***	7/12
8.	**Hello Goodbye**	***The Beatles***	6/9
9.	**Summer Rain**	***Johnny Rivers***	14/8
10.	**It's Wonderful**	***Young Rascals***	11/8
11.	Out Of My Head/Can't Take Eyes Off You	*Lettermen*	19/8
12.	To Give	*Frankie Valli*	16/5
13.	I Wonder What She's Doing Tonight	*T. Boyce/B. Hart*	17/3
14.	I'm In Love	*Wilson Pickett*	15/6
15.	Who Will Answer	*Ed Ames*	20/7
16.	Skip A Rope	*Henson Cargill*	-/6
17.	Different Drum	*The Stone Poneys*	-/7
18.	Sunshine Of Your Love	*The Cream*	-/2
19.	Bottle Of Wine	*The Fireballs*	-/7
20.	There Was A Time	*James Brown*	-/1

(First number indicates last week's position/Second number is the weeks heard on WCFL)

ACTION TEN

21.	LOVE IS BLUE	PAUL MAURIAT
22.	WEAR YOUR LOVE LIKE HEAVEN	DONOVAN
23.	JUST AS MUCH AS EVER	BOBBY VINTON
24.	TOMORROW	STRAWBERRY ALARM CLOCK
25.	EXPLOSION IN MY SOUL	SOUL SURVIVORS
26.	I WISH IT WOULD RAIN	THE TEMPTATIONS
27.	WE'RE A WINNER	THE IMPRESSIONS
28.	EVERYBODY KNOWS	DAVE CLARK FIVE
29.	THERE IS	THE DILLS
30.	NOBODY BUT ME	THE HUMAN BEINZ

CHICAGO PREMIERES

GET OUT NOW	TOMMY JAMES & SHONDELLS
WALK AWAY RENEE	THE FOUR TOPS
WORDS	THE BEE GEES

WEEK ENDING JANUARY 25, 1968

1.	**Bend Me, Shape Me**	***The American Breed***	1/11
2.	**Green Tambourine**	***Lemon Pipers***	3/7
3.	**Judy In Disguise**	***John Fred***	2/10
4.	**Love Is Blue**	***Paul Mauriat***	-/8
5.	**Chain Of Fools**	***Aretha Franklin***	4/9
6.	**I Wonder What She's Doing Tonight**	***T. Boyce/B. Hart***	13/4
7.	**Susan**	***The Buckinghams***	6/10
8.	**Out Of My Head/Can't Take Eyes Off You**	***Lettermen***	11/9
9.	**Woman, Woman**	***The Union Gap***	5/12
10.	**Summer Rain**	***Johnny Rivers***	9/9
11.	To Give	*Frankie Valli*	12/6
12.	Different Drum	*The Stone Poneys*	17/8
13.	Sunshine Of Your Love	*The Cream*	18/3
14.	Skip A Rope	*Henson Cargill*	16/7
15.	It's Wonderful	*Young Rascals*	10/9
16.	Nobody But Me	*The Human Beinz*	-/4
17.	Tomorrow	*Strawberry Alarm Clock*	-/3
18.	I Wish It Would Rain	*The Temptations*	-/3
19.	There Is	*The Dells*	-/5
20.	There Was A Time	*James Brown*	20/2

(First number indicates last week's position/Second number is the weeks heard on WCFL)

ACTION TEN

21.	IF I COULD BUILD MY WHOLE WORLD	M. GAYE/T. TERRELL
22.	COME SEE ABOUT ME	JR. WALKER & ALL STARS
23.	SPOOKY	CLASSICS IV
24.	JUST AS MUCH AS EVER	BOBBY VINTON
25.	WE CAN FLY	THE COWSILLS
26.	WE'RE A WINNER	THE IMPRESSIONS
27.	AM I THAT EASY TO FORGET	ENGELBERT HUMPERDINCK
28.	EVERYBODY KNOWS	DAVE CLARK FIVE
29.	BURNING SPEAR	SOULFUL STRINGS
30.	I CAN TAKE OR LEAVE YOUR LOVING	HERMAN'S HERMITS

CHICAGO PREMIERES

EVERYTHING THAT TOUCHES YOU	THE ASSOCIATION
GREENLIGHT	THE AMERICAN BREED
I WILL ALWAYS THINK ABOUT YOU	NEW COLONY SIX

WEEK ENDING FEBRUARY 1, 1968

1.	**Bend Me, Shape Me**	***The American Breed***	1/12
2.	**Love Is Blue**	***Paul Mauriat***	4/9
3.	**I Wonder What She's Doing Tonight**	***T. Boyce/B. Hart***	6/5
4.	**Green Tambourine**	***Lemon Pipers***	2/8
5.	**Judy In Disguise**	***John Fred***	3/11
6.	**Out Of My Head/Can't Take Eyes Off You**	***Lettermen***	8/10
7.	**Woman, Woman**	***The Union Gap***	9/13
8.	**Different Drum**	***The Stone Poneys***	12/9
9.	**Chain Of Fools**	***Aretha Franklin***	5/10
10.	**Summer Rain**	***Johnny Rivers***	10/10
11.	Nobody But Me	*The Human Beinz*	16/5
12.	Sunshine Of Your Love	*The Cream*	13/4
13.	To Give	*Frankie Valli*	11/7
14.	There Was A Time	*James Brown*	20/3
15.	Tomorrow	*Strawberry Alarm Clock*	17/4
16.	There Is	*The Dells*	19/6
17.	I Wish It Would Rain	*The Temptations*	18/4
18.	Spooky	*Classics IV*	-/4
19.	Skip A Rope	*Henson Cargill*	14/8
20.	Itchycoo Park	*The Small Faces*	-/6

(First number indicates last week's position/Second number is the weeks heard on WCFL)

ACTION TEN

21.	IF I COULD BUILD MY WHOLE WORLD	M. GAYE/T. TERRELL
22.	BABY, NOW THAT I FOUND YOU	THE FOUNDATIONS
23.	BURNING SPEAR	SOULFUL STRINGS
24.	WE CAN FLY	THE COWSILLS
25.	WE'RE A WINNER	THE IMPRESSIONS
26.	I CAN TAKE OR LEAVE YOUR LOVING	HERMAN'S HERMITS
27.	JUST AS MUCH AS EVER	BOBBY VINTON
28.	AM I THAT EASY TO FORGET	ENGELBERT HUMPERDINCK
29.	EVERYTHING THAT TOUCHES YOU	THE ASSOCIATION
30.	YOU	MARVIN GAYE

CHICAGO PREMIERES

DOTTIE I LIKE IT	TOMMY ROE
MEN ARE GETTING SCARCE	JOE TEX
TOO MUCH TALK	PAUL REVERE & RAIDERS

WEEK ENDING FEBRUARY 8, 1968

1.	**Love Is Blue**	**Paul Mauriat**	2/10
2.	**I Wonder What She's Doing Tonight**	***T. Boyce/B. Hart***	3/6
3.	**Bend Me, Shape Me**	***The American Breed***	1/13
4.	**Different Drum**	***The Stone Poneys***	8/10
5.	**Green Tambourine**	***Lemon Pipers***	4/9
6.	**Nobody But Me**	***The Human Beinz***	11/6
7.	**Out Of My Head/Can't Take Eyes Off You**	***Lettermen***	6/11
8.	**Spooky**	***Classics IV***	18/5
9.	**Woman, Woman**	***The Union Gap***	7/14
10.	**Judy In Disguise**	***John Fred***	5/12
11.	Sunshine Of Your Love	*The Cream*	12/5
12.	Itchycoo Park	*The Small Faces*	20/7
13.	To Give	*Frankie Valli*	13/8
14.	Baby, Now That I Found You	*The Foundations*	-/5
15.	Tomorrow	*Strawberry Alarm Clock*	15/5
16.	I Can Take Or Leave Your Loving	*Herman's Hermits*	-/4
17.	There Was A Time	*James Brown*	14/4
18.	There Is	*The Dells*	16/7
19.	(Theme From) Valley Of The Dolls	*Dionne Warwick*	-/3
20.	I Wish It Would Rain	*The Temptations*	17/5

(First number indicates last week's position/Second number is the weeks heard on WCFL)

ACTION TEN

21.	WE CAN FLY	THE COWSILLS
22.	BURNING SPEAR	SOULFUL STRINGS
23.	WE'RE A WINNER	THE IMPRESSIONS
24.	EVERYTHING THAT TOUCHES YOU	THE ASSOCIATION
25.	I THANK YOU	SAM & DAVE
26.	AM I THAT EASY TO FORGET	ENGELBERT HUMPERDINCK
27.	GET OUT NOW	TOMMY JAMES & SHONDELLS
28.	WALK AWAY RENEE	THE FOUR TOPS
29.	YOU	MARVIN GAYE
30.	DARLIN'	THE BEACH BOYS

CHICAGO PREMIERES

HANG UP CITY	BERKLEY KITES
JEALOUS LOVE	WILSON PICKETT
KISS ME GOODBYE	PETULA CLARK

WEEK ENDING FEBRUARY 15, 1968

1.	**Love Is Blue**	**Paul Mauriat**	1/11
2.	**I Wonder What She's Doing Tonight**	***T. Boyce/B. Hart***	2/7
3.	**Nobody But Me**	***The Human Beinz***	6/7
4.	**Itchycoo Park**	***The Small Faces***	12/8
5.	**Different Drum**	***The Stone Poneys***	4/11
6.	**Bend Me, Shape Me**	***The American Breed***	3/14
7.	**Baby, Now That I Found You**	***The Foundations***	14/6
8.	**Out Of My Head/Can't Take Eyes Off You**	***Lettermen***	7/12
9.	**Spooky**	***Classics IV***	8/16
10.	**Sunshine Of Your Love**	***The Cream***	11/6
11.	Green Tambourine	*Lemon Pipers*	5/10
12.	I Wish It Would Rain	*The Temptations*	20/6
13.	(Theme From) Valley Of The Dolls	*Dionne Warwick*	19/4
14.	Tomorrow	*Strawberry Alarm Clock*	15/6
15.	There Was A Time	*James Brown*	17/5
16.	I Can Take Or Leave Your Loving	*Herman's Hermits*	16/5
17.	Everything That Touches You	*The Association*	/-4
18.	There Is	*The Dells*	18/8
19.	Simon Says	*1910 Fruitgum Co.*	-/6
20.	Walk Away Renee	*The Four Tops*	-/5

(First number indicates last week's position/Second number is the weeks heard on WCFL)

ACTION TEN

21.	WE CAN FLY	THE COWSILLS
22.	(SITTIN' ON) THE DOCK OF THE BAY	OTIS REDDING
23.	BURNING SPEAR	SOULFUL STRINGS
24.	I THANK YOU	SAM & DAVE
25.	WE'RE A WINNER	THE IMPRESSIONS
26.	TOO MUCH TALK	PAUL REVERE & RAIDERS
27.	GET OUT NOW	TOMMY JAMES & SHONDELLS
28.	YOU	MARVIN GAYE
29.	A QUESTION OF TEMPERATURE	THE BALLOON FARM
30.	THIS IS THE THANKS I GET	BARBARA LYNN

CHICAGO PREMIERES

SCARBOROUGH FAIR CANTICLE	SIMON & GARFUNKEL
SINCE YOU'VE BEEN GONE	ARETHA FRANKLIN
WILL YOU LOVE ME TOMORROW	FOUR SEASONS

WEEK ENDING FEBRUARY 22, 1968

1.	**Love Is Blue**	**Paul Mauriat**	1/12
2.	**Spooky**	***Classics IV***	9/7
3.	**I Wonder What She's Doing Tonight**	***T. Boyce/B. Hart***	2/8
4.	**Itchycoo Park**	***The Small Faces***	4/9
5.	**Nobody But Me**	***The Human Beinz***	3/8
6.	**Baby, Now That I Found You**	***The Foundations***	7/7
7.	**(Theme From) Valley Of The Dolls**	***Dionne Warwick***	13/5
8.	**Different Drum**	***The Stone Poneys***	5/12
9.	**Simon Says**	***1910 Fruitgum Co.***	19/7
10.	**Out Of My Head/I Can't Take Eyes Off You**	***Lettermen***	8/13
11.	Walk Away Renee	*The Four Tops*	20/6
12.	I Wish It Would Rain	*The Temptations*	12/7
13.	Everything That Touches You	*The Association*	17/5
14.	I Will Always Think About You	*New Colony Six*	-/5
15.	Tomorrow	*Strawberry Alarm Clock*	14/7
16.	(Sittin' On) The Dock Of The Bay	*Otis Redding*	-/4
17.	Too Much Talk	*Paul Revere & Raiders*	-/4
18.	Just Dropped In	*The First Edition*	-/3
19.	I Thank You	*Sam & Dave*	-/4
20.	A Question Of Temperature	*The Balloon Farm*	-/1

(First number indicates last week's position/Second number is the weeks heard on WCFL)

ACTION TEN

21.	WE'RE A WINNER	THE IMPRESSIONS
22.	GET OUT NOW	TOMMY JAMES & SHONDELLS
23.	YOU'VE GOT TO BE LOVED	THE MONTANAS
24.	MISSION-IMPOSSIBLE	LALO SCHIFRIN
25.	KISS ME GOODBYE	PETULA CLARK
26.	DANCE TO THE MUSIC	SLY & THE FAMILY STONE
27.	UP ON THE ROOF	THE CRYAN SHAMES
28.	BALLAD OF BONNIE & CLYDE	GEORGIE FAME
29.	THE END OF OUR ROAD	GLADYS KNIGHT & PIPS
30.	MEN ARE GETTIN' SCARCE	JOE TEX

CHICAGO PREMIERES

BACK ON MY FEET AGAIN	THE FOUNDATIONS
TIN SOLDIER	THE SMALL FACES
YOUNG GIRL	UNION GAP

WEEK ENDING FEBRUARY 29, 1968

1.	**Love Is Blue**	**Paul Mauriat**	1/13
2.	**Spooky**	***Classics IV***	2/8
3.	**Nobody But Me**	***The Human Beinz***	5/9
4.	**Baby, Now That I Found You**	***The Foundations***	6/8
5.	**I Wonder What She's Doing Tonight**	***T. Boyce/B. Hart***	3/9
6.	**Itchycoo Park**	***The Small Faces***	4/10
7.	**(Sittin' On) The Dock Of The Bay**	***Otis Redding***	16/5
8.	**Simon Says**	***1910 Fruitgum Co.***	9/8
9.	**I Wish It Would Rain**	***The Temptations***	12/8
10.	**(Theme From) Valley Of The Dolls**	***Dionne Warwick***	7/6
11.	Everything That Touches You	*The Association*	13/6
12.	Just Dropped In	*The First Edition*	18/4
13.	I Will Always Think About You	*New Colony Six*	14/6
14.	Walk Away Renee	*The Four Tops*	11/7
15.	Kiss Me Goodbye	*Petula Clark*	-/4
16.	I Thank You	*Sam & Dave*	19/5
17.	A Question Of Temperature	*The Balloon Farm*	20/2
18.	Too Much Talk	*Paul Revere & Raiders*	17/5
19.	Up On The Roof	*The Cryan Shames*	-/5
20.	Dance To The Music	*Sly & The Family Stone*	-/3

(First number indicates last week's position/Second number is the weeks heard on WCFL)

ACTION TEN

21.	YOU'VE GOT TO BE LOVED	THE MONTANAS
22.	MISSION: IMPOSSIBLE	LALO SCHIFRIN
23.	WILL YOU LOVE ME TOMORROW	THE FOUR SEASONS
24.	GREENLIGHT	THE AMERICAN BREED
25.	SINCE YOU'VE BEEN GONE	ARETHA FRANKLIN
26.	THE END OF OUR ROAD	GLADYS KNIGHT & PIPS
27.	THIS IS THE THANKS I GET	BARBARA LYNN
28.	BALLAD OF BONNIE & CLYDE	GEORGIE FAME
29.	LA-LA-MEANS I LOVE YOU	THE DELFONICS
30.	LOVE IS ALL AROUND	THE TROGGS

CHICAGO PREMIERES

FUNKY NORTH PHILLY	BILL COSBY
JENNIFER ECCLES	THE HOLLIES
SOUND ASLEEP	THE TURTLES

WEEK ENDING MARCH 7, 1968

1.	**Love Is Blue**	**Paul Mauriat**	1/14
2.	**Spooky**	***Classics IV***	2/9
3.	**(Sittin' On) The Dock Of The Bay**	***Otis Redding***	7/6
4.	**I Will Always Think About You**	***New Colony Six***	13/7
5.	**Simon Says**	***1910 Fruitgum Co.***	8/9
6.	**Just Dropped IN**	***The First Edition***	12/5
7.	**Nobody But Me**	***The Human Beinz***	3/10
8.	**I Wonder What She's Doing Tonight**	***T. Boyce/B. Hart***	5/10
9.	**Up On The Roof**	***Cryan Shames***	19/6
10.	**(Theme From) Valley Of The Dolls**	***Dionne Warwick***	10/7
11.	Everything That Touches You	*The Association*	11/7
12.	Too Much Talk	*Paul Revere & Raiders*	18/6
13.	Valleri/Tapioca Tundra	*The Monkees*	-/2
14.	A Question Of Temperature	*The Balloon Farm*	17/3
15.	Walk Away Renee	*The Four Tops*	14/8
16.	I Thank You	*Sam & Dave*	16/6
17.	Dance To The Music	*Sly & The Family Stone*	20/4
18.	Kiss Me Goodbye	*Petula Clark*	15/5
19.	You've Got To Be Loved	*The Montanas*	-/4
20.	Since You've Been Gone	*Aretha Franklin*	-/3

(First number indicates last week's position/Second number is the weeks heard on WCFL)

ACTION TEN

21.	BALLAD OF BONNIE & CLYDE	GEORGIE FAME
22.	GREENLIGHT	THE AMERICAN BREED
23.	THE END OF OUR ROAD	GLADYS KNIGHT & PIPS
24.	I'M GONNA MAKE YOU LOVE ME	MADELINE BELL
25.	LA-LA-MEANS I LOVE YOU	THE DELFONICS
26.	WILL YOU LOVE ME TOMORROW	THE FOUR SEASONS
27.	MIGHTY QUINN	MANFRED MANN
28.	YOUNG GIRL	THE UNION GAP
29.	LOOKING FOR A FOX	CLARENCE CARTER
30.	LOVE IS ALL AROUND	THE TROGGS

CHICAGO PREMIERES

IN NEED OF A FRIEND	THE COWSILLS
SHERRY DON'T GO	THE LETTERMEN
SIT WITH THE GURU	STRAWBERRY ALARM CLOCK

WEEK ENDING MARCH 14, 1968

1.	**Love Is Blue**	**Paul Mauriat**	1/15
2.	**Simon Says**	***1910 Fruitgum Co.***	5/10
3.	**(Sittin' On) The Dock Of The Bay**	***Otis Redding***	3/7
4.	**Just Dropped In**	***The First Edition***	6/6
5.	**Valleri/Tapioca Tundra**	***The Monkees***	13/1
6.	**I Will Always Think About You**	***New Colony Six***	4/8
7.	**Up On The Roof**	***Cryan Shames***	9/7
8.	**Spooky**	***Classics IV***	2/10
9.	**(Theme From) Valley Of The Dolls**	***Dionne Warwick***	10/8
10.	**Kiss Me Goodbye**	***Petula Clark***	18/6
11.	Too Much Talk	*Paul Revere & Raiders*	12/7
12.	A Question Of Temperature	*The Balloon Farm*	14/4
13.	Everything That Touches You	*The Association*	11/8
14.	Dance To The Music	*Sly & The Family Stone*	17/7
15.	Since You've Been Gone	*Aretha Franklin*	20/4
16.	I Thank You	*Sam & Dave*	16/7
17.	La-La-Means I Love You	*The Delfonics*	-/6
18.	Ballad Of Bonnie & Clyde	*Georgie Fame*	-/7
19.	You've Got To Be Loved	*The Montanas*	19/5
20.	Mighty Quinn	*Manfred Mann*	-/6

(First number indicates last week's position/Second number is the weeks heard on WCFL)

ACTION TEN

21.	YOUNG GIRL	THE UNION GAP
22.	WILL YOU LOVE ME TOMORROW	THE FOUR SEASONS
23.	GREENLIGHT	THE AMERICAN BREED
24.	THE END OF OUR ROAD	GLADYS KNIGHT & PIPS
25.	LOVE IS ALL AROUND	THE TROGGS
26.	SUMMERTIME BLUES	BLUE CHEER
27.	IF YOU CAN WANT	S. ROBINSON/MIRACLES
28.	I'M GONNA MAKE YOU LOVE ME	MADELINE BELL
29.	CAB DRIVER	THE MILLS BROTHERS
30.	I GOT THE FEELIN'	JAMES BROWN

CHICAGO PREMIERES

FOREVER CAME TODAY	DIANA ROSS & SUPREMES
HONEY	BOBBY GOLDSBORO
LADY MADONNA/INNER LIFE	THE BEATLES

WEEK ENDING MARCH 21, 1968

1.	**Valleri/Tapioca Tundra**	**The Monkees**	5/4
2.	**Love Is Blue**	***Paul Mauriat***	1/16
3.	**I Will Always Think About You**	***New Colony Six***	6/9
4.	**Just Dropped In**	***The First Edition***	4/7
5.	**Simon Says**	***1910 Fruitgum Co.***	2/11
6.	**(Sittin' On) The Dock Of The Bay**	***Otis Redding***	3/8
7.	**Up On The Roof**	***Cryan Shames***	7/8
8.	**Young Girl**	***The Union Gap***	-/5
9.	**Since You've Been Gone**	***Aretha Franklin***	15/5
10.	**Kiss Me Goodbye**	***Petula Clark***	10/7
11.	A Question Of Temperature	*The Balloon Farm*	12/5
12.	Ballad Of Bonnie & Clyde	*Georgie Fame*	18/8
13.	Too Much Talk	*Paul Revere & Raiders*	11/8
14.	Dance To The Music	*Sly & The Family Stone*	14/8
15.	Mighty Quinn	*Manfred Mann*	20/7
16.	I Thank You	*Sam & Dave*	16/8
17.	La-La-Means I Love You	*The Delfonics*	17/7
18.	Green Light	*The American Breed*	-/9
19.	You've Got To Be Loved	*The Montanas*	19/6
20.	Lady Madonna/The Inner Light	*The Beatles*	-/2

(First number indicates last week's position/Second number is the weeks heard on WCFL)

ACTION TEN

21.	SUMMERTIME BLUES	BLUE CHEER
22.	WILL YOU LOVE ME TOMORROW	THE FOUR SEASONS
23.	LOVE IS ALL AROUND	THE TROGGS
24.	CRY LIKE A BABY	THE BOX TOPS
25.	THE GOOD, THE BAD AND THE UGLY	HUGO MONTENEGRO
26.	SWEET INSPIRATION	THE SWEET INSPIRATIONS
27.	IF YOU CAN WANT	S. ROBINSON/MIRACLES
28.	I GOT THE FEELIN'	JAMES BROWN
29.	CAB DRIVER	MILLS BROTHERS
30.	MEN ARE GETTING SCARCE	JOE TEX

CHICAGO PREMIERES

DELILAH	TOM JONES
MONY, MONY	T. JAMES/SHONDELLS

WEEK ENDING MARCH 28, 1968

1.	**Valeri/ Tapioca Tundra**	**The Monkees**	1/5
2.	**I Will Always Think About You**	***New Colony Six***	3/10
3.	**(Sittin' On) The Dock Of The Bay**	***Otis Redding***	6/9
4.	**Simon Says**	***1910 Fruitgum Co.***	5/12
5.	**Lady Madonna/The Inner Light**	***The Beatles***	20/3
6.	**Young Girl**	***The Union Gap***	8/6
7.	**Love Is Blue**	***Paul Mauriat***	2/17
8.	**Since You've Been Gone**	***Aretha Franklin***	9/6
9.	**La-La-Means I Love You**	***The Delfonics***	17/8
10.	**Just Dropped IN**	***The First Edition***	4/8
11.	Up On The Roof	*Cryan Shames*	7/9
12.	Kiss Me Goodbye	*Petula Clark*	10/8
13.	Dance To The Music	*Sly & The Family Stone*	14/9
14.	Cry Like A Baby	*The Box Tops*	-/3
15.	Mighty Quinn	*Manfred Mann*	15/8
16.	Summertime Blues	*Blue Cheer*	-/3
17.	Ballad Of Bonnie & Clyde	*Georgia Fame*	12/9
18.	Honey	*Bobby Goldsboro*	-/3
19.	Sweet Inspiration	*The Sweet Inspirations*	-/2
20.	The Good, The Bad & The Ugly	*Hugo Montenegro*	-/5

(First number indicates last week's position/Second number is the weeks heard on WCFL)

ACTION TEN

21.	LOVE IS ALL AROUND	THE TROGGS
22.	WILL YOU LOVE ME TOMORROW	THE FOUR SEASONS
23.	I GOT THE FEELIN'	JAMES BROWN
24.	IF YOU CAN WANT	S. ROBINSON/MIRACLES
25.	JENNIFER JUNIPER	DONOVAN
26.	SCARBOROUGH FAIR (CANTICLE)	SIMON & GARFUNKEL
27.	CAB DRIVER	MILLS BROTHERS
28.	SOUND ASLEEP	THE TURTLES
29.	BACK ON MY FEET AGAIN	FOUNDATIONS
30.	FOREVER CAME TODAY	DIANA ROSS & SUPREMES

CHICAGO PREMIERES

FOGGY MOUNTAIN BREAKDOWN	FLATT & SCRUGGS
MAY I TAKE A GIANT STEP	1910 FRUITGUM CO.
SHE'S LOOKIN' GOOD	WILSON PICKETT

WEEK ENDING APRIL 4, 1968

1.	**I Will Always Think About You**	**The New Colony Six**	2/11
2.	**Lady Madonna/Inner Light**	***The Beatles***	5/4
3.	**Valleri/Tapioca Tundra**	***The Monkees***	1/6
4.	**Young Girl**	***The Union Gap***	6/7
5.	**(Sittin' On) The Dock Of The Bay**	***Otis Redding***	3/10
6.	**Honey**	***Bobby Goldsboro***	18/4
7.	**La-La-Means I Love You**	***The Delfonics***	9/9
8.	**Cry Like A Baby**	***The Box Tops***	14/4
9.	**Since You've Been Gone**	***Aretha Franklin***	8/7
10.	**Summertime Blues**	***Blue Cheer***	16/4
11.	Up On The Roof	*Cryan Shames*	11/10
12.	The Good, The Bad and The Ugly	*Hugo Montenegro*	20/6
13.	Ballad Of Bonnie & Clyde	*Georgie Fame*	17/10
14.	Mighty Quinn	*Manfred Mann*	15/9
15.	Sweet Inspiration	*The Sweet Inspirations*	19/3
16.	I Got The Feelin'	*James Brown*	-/4
17.	Love Is All Around	***The Troggs***	-/6
18.	Tighten Up	*Archie Bell/Drells*	-/1
19.	If You Can Want	*S. Robinson/Miracles*	-/5
20.	Sound Asleep	*The Turtles*	-/7

(First number indicates last week's position/Second number is the weeks heard on WCFL)

ACTION TEN

21.	THE UNICORN	IRISH ROVERS
22.	SCARBOROUGH FAIR CANTICLE	SIMON & GARFUNKEL
23.	FOREVER CAME TODAY	D. ROSS & SUPREMES
24.	CAB DRIVER	MILLS BROTHERS
25.	JENNIFER JUNIPER	DONOVAN
26.	BACK ON MY FEET AGAIN	FOUNDATIONS
27.	MONY, MONY	T. JAMES/SHONDELLS
28.	FUNKY STREET	ARTHUR CONLEY
29.	SOUL SERENADE	WILLIE MITCHELL
30.	BIG BOY	JACKSON FIVE

CHICAGO PREMIERES

WEAR IT ON OUR FACE	THE DELLS
BEAUTIFUL MORNING	THE RASCALS
AIN'T NO WAY	ARETHA FRANKLIN

WEEK ENDING APRIL 11, 1968

1.	**Lady Madonna/Inner Light**	**The Beatles**	2/5
2.	**Honey**	***Bobby Goldsboro***	6/5
3.	**Young Girl**	***The Union Gap***	4/8
4.	**I Will Always Think About You**	***New Colony Six***	1/12
5.	**Cry Like A Baby**	***The Box Tops***	8/5
6.	**Valleri/Tapioca Tundra**	***The Monkees***	3/7
7.	**La-La-Means I Love You**	***The Delfonics***	7/10
8.	**Since You're Been Gone**	***Aretha Franklin***	9/8
9.	**The Good, The Bad & The Ugly**	***Hugo Montenegro***	12/6
10.	**Summertime Blues**	***Blue Cheer***	10/5
11.	Tighten Up	*Archie Bell & Drells*	18/2
12.	Mighty Quinn	*Manfred Mann*	14/10
13.	Ballad Of Bonnie & Clyde	*Georgia Fame*	13/11
14.	Up On The Roof	*Cryan Shames*	11/11
15.	Sweet Inspiration	*The Sweet Inspirations*	15/4
16.	Mony, Mony	*T. James & Shondells*	-/4
17.	I Got The Feelin'	*James Brown*	16/5
18.	The Unicorn	*Irish Rovers*	-/4
19.	Jennifer Juniper	*Donovan*	-/7
20.	Scarborough Fair (/Canticle)	*Simon & Garfunkel*	-/9

(First number indicates last week's position/Second number is the weeks heard on WCFL)

ACTION TEN

21.	IF YOU CAN WANT	S. ROBINSON & MIRACLES
22.	FUNKY STREET	ARTHUR CONLEY
23.	U.S. MALE	ELVIS PRESLEY
24.	FOREVER CAME TODAY	DIANA ROSS & SUPREMES
25.	SOUL SERENADE	WILLIE MITCHELL
26.	CAB DRIVER	MILLS BROTHERS
27.	MY GIRL, HEY GIRL	BOBBY VEE
28.	TAKE TIME TO KNOW HER	PERCY SLEDGE
29.	ME, THE PEACEFUL HEART	LULU
30.	A BEAUTIFUL MORNING	THE RASCALS

CHICAGO PREMIERES

FRIENDS	THE BEACH BOYS
IF I WERE A CARPENTER	FOUR TOPS
UNKNOWN SOLDIER	THE DOORS

WEEK ENDING APRIL 18, 1968

1.	**Honey**	**Bobby Goldsboro**	2/6
2.	**Young Girl**	***The Union Gap***	3/9
3.	**Cry Like A Baby**	***The Box Tops***	5/6
4.	**The Good, The Bad & The Ugly**	***Hugo Montenegro***	9/7
5.	**Lady Madonna/Inner Light**	***The Beatles***	1/6
6.	**Summertime Blues**	***Blue Cheer***	10/6
7.	**Tighten Up**	***Archie Bell & Drells***	11/3
8.	**I Will Always Think About You**	***New Colony Six***	4/13
9.	**Mony, Mony**	***T. James & Shondells***	16/5
10.	**Valleri/Tapioca Tundra**	***The Monkees***	6/8
11.	Sweet Inspiration	*The Sweet Inspirations*	15/5
12.	Mighty Quinn	*Manfred Mann*	12/11
13.	Ballad Of Bonnie & Clyde	*Georgie Fame*	13/12
14.	The Unicorn	*Irish Rovers*	18/5
15.	I Got The Feelin'	*James Brown*	17/6
16.	Jennifer Juniper	*Donovan*	19/8
17.	U.S. Male	*Elvis Presley*	-/3
18.	Funky Street	*Arthur Conley*	-/4
19.	Forever Came Today	*Diana Ross & Supremes*	-/6
20.	A Beautiful Morning	*The Rascals*	-/3

(First number indicates last week's position/Second number is the weeks heard on WCFL)

ACTION TEN

21.	MY GIRL, HEY GIRL	BOBBY VEE
22.	COWBOYS TO GIRLS	THE INTRUDERS
23.	ME, THE PEACEFUL HEART	LULU
24.	TAKE TIME TO KNOW HER	PERCY SLEDGE
25.	SHOO-BE-DOO-BE-DOO-DA-DAY	STEVIE WONDER
26.	CALL ME LIGHTNING	THE WHO
27.	DELILAH	TOM JONES
28.	TAKE GOOD CARE OF MY BABY	BOBBY VINTON
29.	SHE'S LOOKIN' GOOD	WILSON PICKETT
30.	JENNIFER ECCLES	THE HOLLIES

CHICAGO PREMIERES

MACARTHUR PARK	RICHARD HARRIS
MRS. ROBINSON	SIMON & GARFUNKEL
SLEEPY JOE	HERMAN'S HERMITS

WEEK ENDING APRIL 25, 1968

1.	**Honey**	**Bobby Goldsboro**	1/7
2.	**Cry Like A Baby**	***The Box Tops***	3/7
3.	**The Good, The Bad & The Ugly**	***Hugo Montenegro***	4/8
4.	**Mony, Mony**	***T. James Shondells***	9/6
5.	**Summertime Blues**	***The Blue Cheer***	6/7
6.	**Young Girl**	***The Union Gap***	2/10
7.	**Tighten Up**	***Archie Bell & Drells***	7/4
8.	**Lady Madonna/Inner Light**	***The Beatles***	5/7
9.	**I Will Always Think About You**	***The New Colony Six***	8/14
10.	**I Got The Feelin'**	***James Brown***	15/7
11.	A Beautiful Morning	*The Rascals*	20/4
12.	The Unicorn	*Irish Rovers*	14/6
13.	Mighty Quinn	*Manfred Mann*	12/12
14.	Funky Street	*Arthur Conley*	18/5
15.	My Girl, Hey Girl	*Bobby Vee*	-/4
16.	U.S. Male	*Elvis Presley*	17/4
17.	Sweet Inspiration	*The Sweet Inspirations*	11/6
18.	Jennifer Juniper	*Donovan*	16/9
19.	Cowboys To Girls	*The Intruders*	-/4
20.	Forever Came Today	*Diana Ross & Supremes*	19/7

(First number indicates last week's position/Second number is the weeks heard on WCFL)

ACTION TEN

21.	SHOO-BE-DOO-BE-DOO-DA-DAY	STEVIE WONDER
22.	CALL ME LIGHTNING	THE WHO
23.	AIN'T NO WAY	ARETHA FRANKLIN
24.	TAKE TIME TO KNOW HER	PERCY SLEDGE
25.	SHE'S LOOKIN' GOOD	WILSON PICKETT
26.	TAKE GOOD CARE OF MY BABY	BOBBY VINTON
27.	BIG BOY	JACKSON FIVE
28.	DELILAH	TOM JONES
29.	JENNIFER ECCLES	THE HOLLIES
30.	LOOK TO YOUR SOUL	JOHNNY RIVERS

CHICAGO PREMIERES

READY, WILLING AND ABLE	*THE AMERICAN BREED*
SURPRISE, SURPRISE I NEED YOU	THE TROGGS
THE HAPPY SONG	OTIS REDDING

WEEK ENDING MAY 2, 1968

1.	**Honey**	**Bobby Goldsboro**	1/8
2.	**Mony, Mony**	***T. James & Shondells***	4/7
3.	**The Good, The Bad & The Ugly**	***Hugo Montenegro***	3/9
4.	**Tighten Up**	***Archie Bell & Drells***	7/5
5.	**Cry Like A Baby**	***The Box Tops***	2/8
6.	**Young Girl**	***The Union Gap***	6/11
7.	**A Beautiful Morning**	***The Rascals***	11/5
8.	**Summertime Blues**	***The Blue Cheer***	5/8
9.	**The Unicorn**	***Irish Rovers***	12/7
10.	**Lady Madonna/Inner Light**	***The Beatles***	8/8
11.	Funky Street	*Arthur Conley*	14/6
12.	Jennifer Juniper	*Donovan*	18/10
13.	Cowboys To Girls	*The Intruders*	19/5
14.	I Got The Feelin'	*James Brown*	10/8
15.	My Girl, Hey Girl	*Bobby Vee*	15/5
16.	Shoo-Be-Doo-Be-Doo-Da-Day	*Stevie Wonder*	-/3
17.	Call Me Lightning	*The Who*	-/7
18.	U.S. Male	*Elvis Presley*	16/5
19.	Jennifer Eccles	*The Hollies*	-/10
20.	Take Good Care Of My Baby	*Bobby Vinton*	-/5

(First number indicates last week's position/Second number is the weeks heard on WCFL)

ACTION TEN

21.	AIN'T NO WAY	ARETHA FRANKLIN
22.	SHE'S LOOKIN' GOOD	WILSON PICKETT
23.	DELILAH	TOM JONES
24.	LOOK TO YOUR SOUL	JOHNNY RIVERS
25.	TAKE TIME TO KNOW HER	PERCY SLEDGE
26.	YUMMY YUMMY YUMMY	THE OHIO EXPRESS
27.	DO YOU KNOW THE WAY TO SAN JOSE	DIONNE WARWICK
28.	MRS. ROBINSON	SIMON & GARFUNKEL
29.	DOES YOUR MAMA KNOW ABOUT ME	B. TAYLOR & VANCOUVERS
30.	SHERRY, DON'T GO	THE LETTERMEN

CHICAGO PREMIERES

A MAN WITH OUT LOVE	ENGELBERT HUMPERDINCK
REACH OUT OF THE DARKNESS	FRIEND & LOVER
THE BIRDS, THE BEES & THE MONKEES (ALBUM)	THE MONKEES

WEEK ENDING MAY 9, 1968

1.	**Mony, Mony**	***T. James & Shondells***	2/8
2.	**Honey**	***Bobby Goldsboro***	1/9
3.	**A Beautiful Morning**	***The Rascals***	7/6
4.	**Tighten Up**	***Archie Bell & Drells***	4/6
5.	**The Good, The Bad & The Ugly**	***Hugo Montenegro***	3/10
6.	**Cry Like A Baby**	***The Box Tops***	5/9
7.	**My Girl, Hey Girl**	***Bobby Vee***	15/6
8.	**Yummy Yummy Yummy**	***The Ohio Express***	-/3
9.	**The Unicorn**	***Irish Rovers***	9/8
10.	**Funky Street**	***Arthur Conley***	11/7
11.	Shoo-Be-Doo-Be-Doo-Da-Day	*Stevie Wonder*	16/4
12.	Summertime Blues	*The Blue Cheer*	8/9
13.	I Got The Feelin'	*James Brown*	14/9
14.	Cowboys To Girls	*The Intruders*	13/6
15.	U.S. Male	*Elvis Presley*	18/6
16.	Jennifer Eccles	*The Hollies*	19/11
17.	Take Good Care Of My Baby	*Bobby Vinton*	20/6
18.	Mrs. Robinson	*Simon & Garfunkel*	-/4
19.	Ain't No Way	*Aretha Franklin*	-/6
20.	She's Lookin' Good	*Wilson Pickett*	-/7

(First number indicates last week's position/Second number is the weeks heard on WCFL)

ACTION TEN

21.	LOOK TO YOUR SOUL	JOHNNY RIVERS
22.	DO YOU KNOW THE WAY TO SAN JOSE	DIONNE WARWICK
23.	DELILAH	TOM JONES
24.	MASTER JACK	FOUR JACKS & JILL
25.	DOES YOUR MAMA KNOW ABOUT ME	B. TAYLOR & VANCOUVERS
26.	AIN'T NOTHING LIKE THE REAL THING	M. GAYE/T. TERRELL
27.	SHERRY, DON'T GO	THE LETTERMEN
28.	LIKE TO GET TO KNOW YOU	SPANKY & OUR GANG
29.	IF I WERE A CARPENTER	THE FOUR TOPS
30.	READY, WILLING & ABLE	THE AMERICAN BREED

CHICAGO PREMIERES

THE PAPAS & THE MAMAS (ALBUM)	MAMAS & PAPAS
THIS GUY'S IN LOVE WITH YOU	HERB ALPERT
TIME FOR LIVIN'	THE ASSOCIATION

WEEK ENDING MAY 16, 1968

1.	**Mony, Mony**	**T. James & Shondells**	1/9
2.	**Yummy Yummy Yummy**	***The Ohio Express***	8/4
3.	**Tighten Up**	***Archie Bell & Drells***	4/7
4.	**A Beautiful Morning**	***The Rascals***	3/7
5.	**Honey**	***Bobby Goldsboro***	2/10
6.	**The Good, The Bad & The Ugly**	***Hugh Montenegro***	5/11
7.	**The unicorn**	***Irish Rovers***	9/9
8.	**Mrs. Robinson**	***Simon & Garfunkel***	18/5
9.	**My Girl, Hey Girl**	***Bobby Vee***	7/7
10.	**Cry Like A Baby**	***The Box Tops***	6/10
11.	Shoo-Be-Doo-Be-Doo-Da-Day	*Stevie Wonder*	11/5
12.	Cowboys To Girls	*The Intruders*	14/7
13.	Funky Street	*Arthur Conley*	10/8
14.	Jennifer Eccles	*The Hollies*	16/12
15.	Master Jack	*Four Jacks & Jill*	-/5
16.	Do You Know The way To San Jose	*Dionne Warwick*	-/9
17.	She's Lookin' Good	*Wilson Pickett*	20/8
18.	Take Good Care Of My Baby	*Bobby Vinton*	17/7
19.	U.S. Male	*Elvis Presley*	15/7
20.	Reach Out Of The Darkness	*Friend & Lover*	-/3

(First number indicates last week's position/Second number is the weeks heard on WCFL)

ACTION TEN

21.	DELILAH	TOM JONES
22.	IF I WERE A CARPENTER	FOUR TOPS
23.	DOES YOUR MAMA KNOW ABOUT ME	B. TAYLOR & VANCOUVERS
24.	AIN'T NOTHING LIKE THE REAL THING	M. GAYE/T. TERRELL
25.	I COULD NEVER LOVE ANOTHER	THE TEMPTATIONS
26.	LIKE TO GET TO KNOW YOU	SPANKY & OUR GANG
27.	READY, WILLING & ABLE	THE AMERICAN BREED
28.	MACARTHUR PARK	RICHARD HARRIS
29.	HOW'D WE EVER GET THIS WAY	ANDY KIM
30.	THE HAPPY SONG	OTIS REDDING

CHICAGO PREMIERES

BRING A LITTLE LOVIN'	LOS BRAVOS
STONED SOUL PICNIC	THE 5TH DIMENSION
THINK	ARETHA FRANKLIN

WEEK ENDING MAY 23, 1968

1.	**Do you Know the Way To San Jose**	**T. James & Shondells**	1/10
2.	**Yummy Yummy Yummy**	***The Ohio Express***	2/5
3.	**A Beautiful Morning**	***The Rascals***	4/8
4.	**Mrs. Robinson**	***Simon & Garfunkel***	8/6
5.	**Tighten Up**	***Archie Bell & Drells***	3/8
6.	**My Girl, Hey Girl**	***Bobby Vee***	9/8
7.	**Honey**	***Bobby Goldsboro***	5/11
8.	**Reach Out Of The Darkness**	***Friend & Lover***	20/4
9.	**The Unicorn**	***The Irish Rovers***	7/10
10.	**The Good, The Bad & The Ugly**	***Hugo Montenegro***	6/12
11.	Cowboys To Girls	*The Intruders*	12/8
12.	Shoo-Be-Doo-Be-Doo-Da-Day	*Stevie Wonder*	11/6
13.	Do You Know The Way To San Jose	*Dionne Warwick*	16/10
14.	Master Jack	*Four Jacks & Jill*	15/6
15.	Funky Street	*Arthur Conley*	13/9
16.	She's Lookin' Good	*Wilson Pickett*	17/9
17.	This Guy's In Love With You	*Herb Alpert*	-/3
18.	If I Were A Carpenter	*The Four Tops*	-/7
19.	Like To Get To Know You	*Spanky & Our Gang*	-/6
20.	MacArthur Park	*Richard Harris*	-/5

(First number indicates last week's position/Second number is the weeks heard on WCFL)

ACTION TEN

21.	AIN'T NOTHING LIKE THE REAL THING	M. GAYE/T. TERRELL
22.	DOES YOUR MAMA KNOW ABOUT ME	B. TAYLOR & VANCOUVERS
23.	I COULD NEVER LOVE ANOTHER	THE TEMPTATIONS
24.	HOW'D WE EVER GET THIS WAY	ANDY KIM
25.	THINK	ARETHA FRANKLIN
26.	TIME FOR LIVIN'	THE ASSOCIATION
27.	BROOKLYN ROADS	NEIL DIAMOND
28.	THE HAPPY SONG	OTIS REDDING
29.	A MAN WITHOUT LOVE	ENGELBERT HUMPERDINCK
30.	READY, WILLING & ABLE	THE AMERICAN BREED

CHICAGO PREMIERES

BACK IN LOVE AGAIN	THE BUCKINGHAMS
INDIAN LAKE	THE COWSILLS
YOUNG BIRDS FLY	THE CRYAN SHAMES

WCFL

1000 WCFL-AM SOUND 10 SURVEY-CHICAGO

WEEK ENDING JUNE 6, 1968

This is the first WCFL survey chart that was not available to take home.

1.	YUMMY, YUMMY, YUMMY	THE OHIO EXPRESS
2.	MRS. ROBINSON	SIMON AND GARFUNKEL
3.	THIS GUY'S IN LOVE WITH YOU	HERB ALPERT
4.	REACH OUT OF THE DARKNESS	FRIEND AND LOVER
5.	MONY, MONY	TOMMY JAMES & THE SHONDELLS
6.	MASTER JACK	4 JACKS AND A JILL
7.	LIKE TO GET TO KNOW YOU	SPANKY AND OUR GANG
8.	SHOO-BE-DOO-BE-DOO-DA-DAY	STEVIE WONDER
9.	BEAUTIFUL MORNING	THE RASCALS
10.	TIGHTEN UP	ARCHIE BELL AND THE DRELLS
11.	HOW'D WE EVER GET THIS WAY	ANDY KIM
12.	MAC ARTHUR PARK	RICHARD HARRIS
13.	IF I WERE A CARPENTER	FOUR TOPS
14.	ANGEL IN THE MORNING	MERILEE RUSH
15.	DO YOU KNOW THE WAY TO SAN JOSE	DIONNE WARWICK
16.	I LOVE YOU	PEOPLE
17.	THINK	ARETHA FRANKLIN
18.	BROOKLYN ROADS	NEIL DIAMOND
19.	TIME FOR LIVIN'	THE ASSOCIATION
20.	YOUNG BIRDS FLY	THE CRYAN' SHAMES

ACTION 10

21.	I COULD NEVER LOVE ANOTHER	THE TEMPTATIONS
22.	AIN'T NOTHIN' LIKE THE REAL THING	MARVIN GAYE / TAMMI TERRELL
23.	LICKEN STICK –LICKEN STICK	JAMES BROWN
24.	A MAN WITHOUT LOVE	ENGLEBERT HUMPERDINCK
25.	JELLY JUNGLE	THE LEMON PIPERS
26.	THE LOOK OF LOVE	SERGIO MENDES
27.	FACE IT GIRL, IT'S OVER	NANCY WILSON
28.	SHE'S A HEARTBREAKER	GENE PITNEY
29.	TIP-TOE THRU THE TULIPS WITH ME	TINY TIM
30.	BACK IN LOVE AGAIN	THE BUCKINGHAMS

WCFL
1000 WCFL-AM SOUND 10 SURVEY-CHICAGO

WEEK ENDING JUNE 13, 1968

1.	THIS GUY'S IN LOVE WITH YOU	HERB ALPERT	3
2.	YUMMY, YUMMY, YUMMY	THE OHIO EXPRESS	1
3.	MRS. ROBINSON	SIMON AND GARFUNKEL	2
4.	REACHOUT OF THE DARKNESS	FRIEND AND LOVER	4
5.	I LOVE YOU	PEOPLE	16
6.	LIKE TO GET TO KNOW YOU	SPANKY AND OUR GANG	7
7.	ANGEL IN THE MORNING	MERILEE RUSH	14
8.	MASTER JACK	4 JACKS AND A JILL	6
9.	HOW'D WE EVER GET THIS WAY	ANDY KIM	11
10.	MAC ARTHUR PARK	RICHARD HARRIS	12
11.	IF I WERE A CARPENTER	FOUR TOPS	13
12.	THINK	ARETHA FRANKLIN	17
13.	THE LOOK OF LOVE	SERGIO MENDES	26
14.	TIME FOR LIVIN'	THE ASSOCIATION	19
15.	AIN'T NOTHIN' LIKE THE REAL THING	MARVIN GAYE / TAMMI TERRELL	22
16.	YOUNG BIRDS FLY	THE CRYAN' SHAMES	20
17.	I NEED LOVE	THE THIRD BOOTH	--
18.	I COULD NEVER LOVE ANOTHER	THE TEMPTATIONS	21
19.	A MAN WITHOUT LOVE	ENGLEBERT HUMPERDINCK	24
20.	LICKEN STICK –LICKEN STICK	JAMES BROWN	23

ACTION 10

21.	JELLY JUNGLE	THE LEMON PIPERS	25
22.	PICTURES OF MATCHSTICK MEN	STATUS QUO	--
23.	SHE'S A HEARTBREAKER	GENE PITNEY	28
24.	TIP-TOE THRU THE TULIPS WITH ME	TINY TIM	29
25.	FACE IT GIRL, IT'S OVER	NANCY WILSON	27
26.	HERE COMES THE JUDGE	SHORTY LONG	--
27.	BACK IN LOVE AGAIN	THE BUCKINGHAMS	30
28.	CHOO CHOO TRAIN	THE BOX TOPS	--
29.	THE HORSE	CLIFF NOBELS & CO.	--
30.	JUMPIN' JACK FLASH	THE ROLLING STONES	--

WCFL

1000 WCFL-AM SOUND 10 SURVEY-CHICAGO

WEEK ENDING JUNE 20, 1968

1.	THIS GUY'S IN LOVE WITH YOU	HERB ALPERT	1
2.	MRS. ROBINSON	SIMON AND GARFUNKEL	3
3.	YUMMY, YUMMY, YUMMY	THE OHIO EXPRESS	2
4.	I LOVE YOU	PEOPLE	5
5.	ANGEL IN THE MORNING	MERILEE RUSH	7
6.	REACHOUT OF THE DARKNESS	FRIEND AND LOVER	4
7.	MAC ARTHUR PARK	RICHARD HARRIS	10
8.	LIKE TO GET TO KNOW YOU	SPANKY AND OUR GANG	6
9.	HOW'D WE EVER GET THIS WAY	ANDY JIM	9
10.	MASTER JACK	4 JACKS AND A JILL	8
11.	THE LOOK OF LOVE	SERGIO MENDES	13
12.	PICTURES OF MATCHSTICK MEN	STATUS QUO	22
13.	THINK	ARETHA FRANKLIN	12
14.	IF I WERE A CARPENTER	FOUR TOPS	11
15.	I NEED LOVE	THE THIRD BOOTH	17
16.	YOUNG BIRDS FLY	THE CRYAN' SHAMES	16
17.	SHE'S A HEARTBREAKER	GENE PITNEY	23
18.	I COULD NEVER LOVE ANOTHER	THE TEMPTATIONS	18
19.	HERE COMES THE JUDGE	SHORTY LONG	26
20.	TIP-TOE THRU THE TULIPS WITH ME	TINY TIM	24

ACTION 10

21.	BACK IN LOVE AGAIN	THE BUCKINGHAMS	27
22.	LICKEN STICK –LICKEN STICK	JAMES BROWN	20
23.	NEVER GIVE YOU UP	JERRY BUTLER	--
24.	LADY WILLPOWER	GARY PUCKETT & THE UNION GAP	--
25.	THE HORSE	CLIFF NOBELS & CO.	29
26.	CHOO CHOO TRAIN	THE BOX TOPS	28
27.	CAN'T YOU SEE ME CRY	THE NEW COLONY SIX	--
28.	FACE IT GIRL, IT'S OVER	NANCY WILSON	25
29.	GRAZING IN THE GRASS	HUGH MASEKELA	--
30.	STONED SOUL PICNIC	THE 5TH DIMENSION	--

WCFL
1000 WCFL-AM SOUND 10 SURVEY-CHICAGO

WEEK ENDING JUNE 27, 1968

1.	THIS GUY'S IN LOVE WITH YOU	HERB ALPERT	1
2.	ANGEL IN THE MORNING	MERILEE RUSH	5
3.	I LOVE YOU	PEOPLE	4
4.	PICTURES OF MATCHSTICK MEN	STATUS QUO	12
5.	MRS. ROBINSON	SIMON AND GARFUNKEL	2
6.	HOW'D WE EVER GET THIS WAY	ANDY KIM	9
7.	SHE'S A HEARTBREAKER	GENE PITNEY	17
8.	THE LOOK OF LOVE	SERGIO MENDEZ	11
9.	I NEED LOVE	THE THIRD BOOTH	15
10.	YUMMY, YUMMY, YUMMY	THE OHIO EXPRESS	3
11.	REACH OUT OF THE DARKNESS	FRIEND AND LOVER	6
12.	YOUNG BIRDS FLY	THE CRYAN' SHAMES	16
13.	HERE COMES THE JUDGE	SHORTY LONG	19
14.	MAC ARTHUR PARK	RICHARD HARRIS	7
15.	THE HORSE	CLIFF NOBELS & CO.	25
16.	THINK	ARETHA FRANKLIN	13
17.	CAN'T YOU SEE ME CRY	THE NEW COLONY SIX	27
18.	GRAZING IN THE GRASS	HUGH MASEKELA	29
19.	LADY WILLPOWER	GARY PUCKETT & THE UNION GAP	24
20.	NEVER GIVE YOU UP	JERRY BUTLER	23

ACTION 10

21.	IT'S NICE TO BE WITH YOU	THE MONKEES	--
22.	FACE IT GIRL, IT'S OVER	NANCY WILSON	28
23.	LICKIN' STICK –LICKIN' STICK	JAMES BROWN	22
24.	JUMPIN' JACK FLASH	THE ROLLING STONES	--
25.	INDIAN LAKE	THE COWSILLS	--
26.	STONED SOUL PICNIC	THE 5TH DIMENSION	30
27.	CHOO CHOO TRAIN	THE BOX TOPS	26
28.	DON'T TAKE IT SO HARD	PAUL REVERE AND THE RAIDERS	--
29.	TURN AROUND LOOK AT ME	THE VOGUES	--
30.	LET YOURSELF GO	ELVIS PRESLEY	--

WCFL
1000 WCFL-AM SOUND 10 SURVEY-CHICAGO

WEEK ENDING JULY 4, 1968

1.	THIS GUY'S IN LOVE WITH YOU	HERB ALPERT	1
2.	PICTURES OF MATCHSTICK MEN	STATUS QUO	4
3.	I LOVE YOU	PEOPLE	3
4.	JUMPIN' JACK FLASH	THE ROLLING STONES	24
5.	LADY WILLPOWER	GARY PUCKETT & THE UNION GAP	19
6.	ANGEL IN THE MORNING	MERILEE RUSH	2
7.	SHE'S A HEARTBREAKER	GENE PITNEY	7
8.	HOW'D WE EVER GET THIS WAY	ANDY KIM	6
9.	STONED SOUL PICNIC	THE 5TH DIMENSION	26
10.	MRS. ROBINSON	SIMON AND GARFUNKEL	5
11.	I NEED LOVE	THE THIRD BOOTH	9
12.	THE LOOK OF LOVE	SERGIO MENDES	8
13.	THE HORSE	CLIFF NOBELS & CO.	15
14.	GRAZING IN THE GRASS	HUGH MASEKELA	18
15.	HERE COMES THE JUDGE	SHORTY LONG	13
16.	INDIAN LAKE	THE COWSILLS	25
17.	YOUNG BIRDS FLY	THE CRYAN SHAMES	12
18.	DON'T TAKE IT SO HARD	PAUL REVERE AND THE RAIDERS	28
19.	CAN'T YOU SEE ME CRY	THE NEW COLONY SIX	17
20.	FACE IT GIRL, IT'S OVER	NANCY WILSON	22

ACTION 10

21.	NEVER GIVE YOU UP	JERRY BUTLER	20
22.	TURN AROUND LOOK AT ME	THE VOGUES	29
23.	D.W. WASHBURN / IT'S NICE TO BE WITH YOU	THE MONKEES	21
24.	I'M MIDNIGHT MOVER	WILSON PICKETT	--
25.	LET YOURSELF GO	ELVIS PRESLEY	30
26.	SOME THINGS YOU NEVER GET USED TO	DIANA ROSS AND THE SUPREMES	--
27.	HURDY GURDY MAN	CONOVAN	--
28.	HERE I AM BABY		--
29.	THE STORY OF ROCK AND ROLL	THE TURTLES	--
30.	AUTUMN OF MY LIFE	BOBBY GOLDSBORO	--

WCFL
1000 WCFL-AM SOUND 10 SURVEY-CHICAGO

WEEK ENDING JULY 11, 1968

1.	JUMPIN' JACK FLASH	THE ROLLING STONES	4
2.	PICTURES OF MATCHSTICK MEN	STATUS QUO	2
3.	STONED SOUL PICNIC	THE 5TH DIMENSION	9
4.	I LOVE YOU	PEOPLE	3
5.	THIS GUY'S IN LOVE WITH YOU	HERB ALPERT	1
6.	GRAZING IN THE GRASS	HUGH MASEKELA	14
7.	LADY WILLPOWER	GARY PUCKETT & THE UNION GAP	5
8.	THE HORSE	CLIFF NOBELS & CO.	13
9.	TURN AROUND LOOK AT ME	THE VOGUES	22
10.	ANGEL IN THE MORNING	MERILEE RUSH	6
11.	HERE COMES THE JUDGE	SHORTY LONG	15
12.	I NEED LOVE	THE THIRD BOOTH	11
13.	SHE'S A HEARTBREAKER	GENE PITNEY	7
14.	INDIAN LAKE	THE COWSILLS	16
15.	CAN'T YOU SEE ME CRY	THE NEW COLONY SIX	19
16.	FACE IT GIRL, IT'S OVER	NANCY WILSON	20
17.	NEVER GIVE YOU UP	JERRY BUTLER	21
18.	DON'T TAKE IT SO HARD	PAUL REVERE AND THE RAIDERS	18
19.	HURDY GURDY MAN	DONOVAN	27
20.	SEALED WITH A KISS	GARY LEWIS AND THE PLAYBOYS	--

ACTION 10

21.	BORN TO BE WILD	STEPPENWOLF	--
22.	I'M A MIDNIGHT MOVER	WILSON PICKETT	24
23.	AUTUMN OF MY LIFE	BOBBY GOLDSBORO	30
24.	HELLO, I LOVE YOU (Won't You Tell Me Your Name)	THE DOORS	--
25.	STAY IN THE CORNER	THE DELLS	--
26.	CLASSICAL GAS	MASON WILLIAMS	--
27.	LOVER'S HOLIDAY	PEGGY SCOTT / JO JO BENSON	--
28.	PEOPLE SURE ACT FUNNY	ARTHUR CONLEY	--
29.	DREAMS OF THE EVERYDAY HOUSEWIFE	GLEN CAMPBELL	--
30.	SOMEBODY CARES	TOMMY JAMES & THE SHONDELLS	--

WCFL
1000 WCFL-AM SOUND 10 SURVEY-CHICAGO

WEEK ENDING JULY 18, 1968

1.	JUMPIN' JACK FLASH	THE ROLLING STONES	1
2.	STONED SOUL PICNIC	THE 5TH DIMENSION	3
3.	PICTURES OF MATCHSTICK MEN	STATUS QUO	2
4.	LADY WILLPOWER	GARY PUCKETT & THE UNION GAP	7
5.	TURN AROUND LOOK AT ME	THE VOGUES	9
6.	GRAZING IN THE GRASS	HUGH MASEKELA	6
7.	INDIAN LAKE	THE COWSILLS	14
8.	I LOVE YOU	PEOPLE	4
9.	THE HORSE	CLIFF NOBELS & CO.	8
10.	THIS GUY'S IN LOVE WITH YOU	HERB ALPERT	5
11.	SHE'S A HEARTBREAKER	GENE PITNEY	13
12.	I NEED LOVE	THE THIRD BOOTH	12
13.	CAN'T YOU SEE ME CRY	THE NEW COLONY SIX	15
14.	DON'T TAKE IT SO HARD	PAUL REVERE AND THE RAIDERS	18
15.	BORN TO BE WILD	STEPPENWOLF	21
16.	SEALED WITH A KISS	GARY LEWIS AND THE PLAYBOYS	20
17.	FACE IT GIRL, IT'S OVER	NANCY WILSON	16
18.	HURDY GURDY MAN	DONOVAN	19
19.	I'M A MIDNIGHT MOVER	WILSON PICKETT	22
20.	NEVER GIVE YOU UP	JERRY BUTLER	17

ACTION 10

21.	HELLO, I LOVE YOU (Won't You Tell Me Your Name)	THE DOORS	24
22.	CLASSICAL GAS	MASON WILLIAMS	26
23.	AUTUMN OF MY LIFE	BOBBY GOLDSBORO	23
24.	SKY PILOT	ERIC BURDON AND THE ANIMALS	--
25.	STAY IN MY CORNER	THE DELLS	25
26.	TWO BIT MANCHILD	NEIL DIAMOND	--
27.	LOVER'S HOLIDAY	PEGGY SCOTT / JO JO BENSON	27
28.	DREAMS OF THE EVERYDAY HOUSEWIFE	GLEN CAMPBELL	29
29.	PEOPLE GOT TO BE FREE	THE RASCALS	--
30.	WITH PEN IN HAND	BILLY VERA	--

WCFL
1000 WCFL-AM SOUND 10 SURVEY-CHICAGO

WEEK ENDING JULY 25, 1968

1.	JUMPIN' JACK FLASH	THE ROLLING STONES	1
2.	STONED SOUL PICNIC	THE 5TH DIMENSION	2
3.	PICTURES OF MATCHSTICK MEN	STATUS QUO	3
4.	TURN AROUND LOOK AT ME	THE VOGUES	5
5.	INDIAN LAKE	THE COWSILLS	7
6.	GRAZING IN THE GRASS	HUGH MASEKELA	6
7.	BORN TO BE WILD	STEPPENWOLF	15
8.	LADY WILLPOWER	GARY PUCKETT & THE UNION GAP	4
9.	HURDY GURDY MAN	DONOVAN	18
10.	SEALED WITH A KISS	GARY LEWIS AND THE PLAYBOYS	16
11.	THE HORSE	CLIFF NOBELS & CO.	9
12.	I NEED LOVE	THE THIRD BOOTH	12
13.	DON'T TAKE IT SO HARD	PAUL REVERE AND THE RAIDERS	14
14.	HELLO, I LOVE YOU (Won't You Tell Me Your Name)	THE DOORS	21
15.	CLASSICAL GAS	MASON WILLIAMS	22
16.	SHE'S A HEARTBREAKER	GENE PITNEY	11
17.	CAN'T YOU SEE ME CRY	THE NEW COLONY SIX	13
18.	I'M A MIDNIGHT MOVER	WILSON PICKETT	19
19.	AUTUMN OF MY LIFE	BOBBY GOLDSBORO	23
20.	PEOPLE GOT TO FREE	THE RASCALS	29

ACTION 10

21.	SKY PILOT	ERIC BURDON AND THE ANIMALS	24
22.	TWO BIT MANCHILD	NEIL DIAMOND	26
23.	DREAMS OF THE EVERYDAY HOUSEWIFE	GLEN CAMPBELL	28
24.	STAY IN MY CORNER	THE DELLS	25
25.	LOVER'S HOLIDAY	PEGGY SCOTT / JO JO BENSON	27
26.	JOURNEY TO THE CENTER OF THE MIND	AMBOY DUKES	--
27.	SOMEBODY CARES	TOMMY JAMES & THE SHONDELLS	--
28.	1, 2, 3, RED LIGHT	1910 FRUITGUM CO.	--
29.	(You're Still My Favorite Girlfriend) ALICE LONG	TOMMY BOYCE AND BOBBY HART	--
30.	DREAM A LITTLE DREAM OF ME	MAMA CASS	--

WCFL
1000 WCFL-AM SOUND 10 SURVEY-CHICAGO

WEEK ENDING AUGUST 1, 1968

1.	JUMPIN' JACK FLASH	THE ROLLING STONES	1
2.	STONED SOUL PICNIC	THE 5TH DIMENSION	2
3.	HELLO, I LOVE YOU (Won't You Tell Me Your Name)	THE DOORS	14
4.	BORN TO BE WILD	STEPPENWOLF	7
5.	TURN AROUND LOOK AT ME	THE VOGUES	4
6.	GRAZING IN THE GRASS	HUGH MASEKELA	6
7.	HURDY GURDY MAN	DONOVAN	9
8.	CLASSICAL GAS	MASON WILLIAMS	15
9.	PICTURES OF MATCHSTICK MEN	STATUS QUO	3
10.	LADY WILLPOWER	GARY PUCKETT & THE UNION GAP	8
11.	SEALED WITH A KISS	GARY LEWIS AND THE PLAYBOYS	10
12.	DON'T TAKE IT SO HARD	PAUL REVERE AND THE RAIDERS	13
13.	SKY PILOT	ERIC BURDON AND THE ANIMALS	21
14.	PEOPLE GOT TO BE FREE	THE RASCALS	20
15.	JOURNEY TO THE CENTER OF THE MIND	AMBOY DUKES	26
16.	1, 2, 3, RED LIGHT	1910 FRUITGUM CO.	--
17.	LIGHT MY FIRE	JOSE FELICIANO	--
18.	AUTUMN OF MY LIFE	BOBBY GOLDSBORO	19
19.	I'M A MIDNIGHT MOVER	WILSON PICKETT	18
20.	SLIP AWAY	CLARENCE CARTER	--

ACTION 10

21.	STAY IN MY CORNER	THE DELLS	24
22.	(You're Still My Favorite Girlfriend) ALICE LONG	TOMMY BOYCE AND BOBBY HART	29
23.	SOUL LIMBO	BOOKER T. AND THE MG'S	--
24.	TWO BIT MANCHILD	NEIL DIAMOND	22
25.	SOMEBODY CARES	TOMMY JAMES & THE SHONDELLS	27
26.	LOVER'S HOLIDAY	PEGGY SCOTT / JO JO BENSON	25
27.	DREAM A LITTLE DREAM OF ME	MAMA CASS	30
28.	YOU MET YOUR MATCH	STEVIE WONDER	--
29.	WORKIN' ON A GROOVY THING	PATTI DREW	--
30.	I CAN'T STOP DANCING	ARCHIE BELL AND THE DRELLS	--

WCFL
1000 WCFL-AM SOUND 10 SURVEY-CHICAGO

WEEK ENDING AUGUST 8, 1968

1.	HELLO, I LOVE YOU (Won't You Tell Me Your Name)	THE DOORS	3
2.	BORN TO BE WILD	STEPPENWOLF	4
3.	TURN AROUND LOOK AT ME	THE VOGUES	5
4.	STONED SOUL PICNIC	THE 5TH DIMENSION	2
5.	JUMPIN' JACK FLASH	THE ROLLING STONES	1
6.	CLASSICAL GAS	MASON WILLIAMS	8
7.	SEALED WITH A KISS	GARY LEWIS AND THE PLAYBOYS	11
8.	PEOPLE GOT TO BE FREE	THE RASCALS	14
9.	GRAZING IN THE GRASS	HUGH MASEKELA	6
10.	HURDY GURDY MAN	DONOVAN	7
11.	SKY PILOT	ERIC BURDON AND THE ANIMALS	13
12.	LIGHT MY FIRE	JOSE FELICIANO	17
13.	1, 2, 3, RED LIGHT	1910 FRUITGUM CO.	16
14.	JOURNEY TO THE CENTER OF THE MIND	AMBOY DUKES	15
15.	(You're Still My Favorite Girlfriend) ALICE LONG	TOMMY BOYCE AND BOBBY HART	22
16.	AUTUMN OF MY LIFE	BOBBY GOLDSBORO	18
17.	SLIP AWAY	CLARENCE CARTER	20
18.	STAY IN MY CORNER	THE DELLS	21
19.	DREAM A LITTLE DREAM OF ME	MAMA CASS	27
20.	SOMEBODY CARES	TOMMY JAMES & THE SHONDELLS	25

ACTION 10

21.	SOUL LIMBO	BOOKER T. & THE MG'S	23
22.	SUNSHINE OF YOUR LOVE	CREAM	--
23.	HALFWAY TO PARADISE	BOBBY VINTON	--
24.	YOU KEEP ME HANGIN' ON	THE VANILLA FUDGE	--
25.	DO IT AGAIN	THE BEACH BOYS	--
26.	YOU'RE ALL I NEED TO GET BY	MARVIN GAYE & TAMMI TERRELL	--
27.	I GET THE SWEETEST FEELING	JACKIE WILSON	--
28.	WORKIN' ON A GROOVY THING	PATTI DREW	29
29.	I CAN'T STOP DANCING	ARCHIE BELL AND THE DRELLS	30
30.	YESTERDAY'S DREAMS	FOUR TOPS	--

WCFL

1000 WCFL-AM SOUND 10 SURVEY-CHICAGO

WEEK ENDING AUGUST 15, 1968

1.	BORN TO BE WILD	STEPPENWOLF	2
2.	TURN AROUND LOOK AT ME	THE VOGUES	3
3.	HELLO, I LOVE YOU (Won't You Tell Me Your Name)	THE DOORS	1
4.	CLASSICAL GAS	MASON WILLIAMS	6
5.	PEOPLE GOT TO BE FREE	THE RASCALS	8
6.	LIGHT MY FIRE	JOSE FELICIANO	12
7.	1, 2, 3, RED LIGHT	1910 FRUITGUM CO.	13
8.	STONED SOUL PICNIC	THE 5TH DIMENSION	4
9.	SEALED WITH A KISS	GARY LEWIS AND THE PLAYBOYS	7
10.	HURDY GURDY MAN	DONOVAN	10
11.	JOURNEY TO THE CENTER OF THE MIND	AMBOY DUKES	14
12.	SKY PILOT	ERIC BURDON AND THE ANIMALS	11
13.	SUNSHINE OF YOUR LOVE	CREAM	22
14.	(You're Still My Favorite Girlfriend) ALICE LONG	TOMMY BOYCE AND BOBBY HART	15
15.	SLIP AWAY	CLARENCE CARTER	17
16.	YOU KEEP ME HANGIN' ON THE	VANILLA FUDGE	24
17.	STAY IN MY CORNER	THE DELLS	18
18.	SOUL LIMBO	BOOKER T. & THE MG'S	21
19.	DREAM A LITTLE DREAM OF ME	MAMA CASS	19
20.	DO IT AGAIN	THE BEACH BOYS	25

ACTION 10

21.	I CAN'T STOP DANCING	ARCHIE BELL AND THE DRELLS	29
22.	HALFWAY TO PARADISE	BOBBY VINTON	23
23.	PLEASE RETURN YOU LOVE TO ME	THE TEMPTATIONS	--
24.	YOU'RE ALL I NEED TO GET BY	MARVIN GAYE & TAMMI TERRELL	--
25.	LOVE MAKES A WOMAN	BARBARA ACKLIN	--
26.	I GET THE SWEETEST FEELING	JACKIE WILSON	27
27.	DOWN AT LULU'S	OHIO EXPRESS	--
28.	WORKIN' ON A GROOVY THING	PATTI DREW	28
29.	I SAY A LITTLE PRAYER / THIS IS THE HOUSE THAT JACK BUILT	ARETHA FRANKLIN	--
30.	TUESDAY AFTERNOON	THE MOODY BLUES	--

WCFL

1000 WCFL-AM SOUND 10 SURVEY-CHICAGO

WEEK ENDING AUGUST 22, 1968

1.	LIGHT MY FIRE	JOSE FELICIANO	6
2.	PEOPLE GOT TO BE FREE	THE RASCALS	5
3.	BORN TO BE WILD	STEPPENWOLF	1
4.	1, 2, 3, RED LIGHT	1910 FRUITGUM CO.	7
5.	CLASSICAL GAS	MASON WILLIAMS	4
6.	JOURNEY TO THE CENTER OF THE MIND	AMBOY DUKES	11
7.	SEALED WITH A KISS	GARY LEWIS AND THE PLAYBOYS	9
8.	(You're Still My Favorite Girlfriend) ALICE LONG	TOMMY BOYCE AND BOBBY HART	14
9.	HELLO, I LOVE YOU (Won't You Tell Me Your Name)	THE DOORS	3
10.	TURNAROUND LOOK AT ME	THE VOGUES	2
11.	SUNSHINE OF YOUR LOVE	CREAM	13
12.	DO IT AGAIN	THE BEACH BOYS	20
13.	SLIP AWAY	CLARENCE CARTER	15
14.	YOU KEEP ME HANGIN' ON	THE VANILLA FUDGE	16
15.	STAY IN MY CORNER	THE DELLS	17
16.	DREAM A LITTLE DREAM OF ME	MAMA CASS	19
17.	DOWN AT LULU'S	OHIO EXPRESS	27
18.	SOUL LIMBO	BOOKER T. & THE MG'S	18
19.	I CAN'T STOP DANCING	ARCHIE BELL AND THE DRELLS	29
20.	HALFWAY TO PARADISE	BOBBY VINTON	22

ACTION 10

21.	LOVES MAKES A WOMAN	BARBARA ACKLIN	25
22.	YOU'RE ALL I NEED TO GET BY	MARVIN GAYE & TAMMI TERRELL	24
23.	I SAY A LITTLE PRAYER / THIS IS THE HOUSE THAT JACK BUILT	ARETHA FRANKLIN	29
24.	PLEASE RETURN YOUR LOVE TO ME	THE TEMPTATIONS	23
25.	WORKIN' ON A GROOVY THING	PATTI DREW	28
26.	TUESDAY AFTERNOON	THE MOODY BLUES	30
27.	FOOL ON THE HILL	SERGIO MENDEZ	--
28.	EYES OF A NEW YORK WOMAN	B.J. THOMAS	--
29.	SOUL DRIPPIN'	THE MAUDS	--
30.	YOU MET YOUR MATCH	STEVIE WONDER	--

WCFL
1000 WCFL-AM SOUND 10 SURVEY-CHICAGO

WEEK ENDING AUGUST 29, 1968

1.	PEOPLE GOT TO BE FREE	THE RASCALS	2
2.	LIGHT MY FIRE	JOSE FELICIANO	1
3.	1, 2, 3, RED LIGHT	1910 FRUITGUM CO.	4
4.	BORN TO BE WILD	STEPPENWOLF	3
5.	DO IT AGAIN	THE BEACH BOYS	12
6.	CLASSICAL GAS	MASON WILLIAMS	5
7.	(You're Still My Favorite Girlfriend) ALICE LONG	TOMMY BOYCE AND BOBBY HART	8
8.	DOWN AT LULU'S	OHIO EXPRESS	17
9.	JOURNEY TO THE CENTER OF THE MIND	AMBOY DUKES	6
10.	SEALED WITH A KISS	GARY LEWIS AND THE PLAYBOYS	7
11.	DREAM A LITTLE DREAM OF ME	MAMA CASS	16
12.	SLIP AWAY	CLARENCE CARTER	13
13.	HELLO, I LOVE YOU (Won't You Tell Me Your Name)	THE DOORS	9
14.	HARPER VALLEY P.T.A.	JEANNIE C. RILEY	--
15.	I CAN'T STOP DANCING	ARCHIE BELL AND THE DRELLS	19
16.	YOU KEEP ME HANGIN' ON	THE VANILLA FUDGE	14
17.	SOUL LIMBO	BOOKER T. & THE MG'S	18
18.	I SAY A LITTLE PRAYER / THIS IS THE HOUSE THAT JACK BUILT	ARETHA FRANKLIN	23
19.	YOU'RE ALL I NEED TO GET BY	MARVIN GAYE / TAMMI TERRELL	22
20.	LOVE MAKES A WOMAN	BARBARA ACKLIN	21

ACTION 10

21.	EYES OF A NEW YORK WOMAN	B.J. THOMAS	28
22.	TUESDAY AFTERNOON	THE MOODY BLUES	26
23.	PLEASE RETURN YOUR LOVE TO ME	THE TEMPTATIONS	24
24.	FOOL ON THE HILL	SERGIO MENDEZ	27
25.	MAGIC BUS	THE WHO	--
26.	SOUL DRIPPIN'	THE MAUDS	29
27.	SHAPES OF THINGS TO COME	MAX FROST AND THE TROOPERS	--
28.	SIX MAN BAND	THE ASSOCIATION	--
29.	SALLY HAD A PARTY	FLAVOR	--
30.	SPECIAL OCCASION	SMOKEY ROBINSON & THE MIRACLES	--

WCFL
1000 WCFL-AM SOUND 10 SURVEY-CHICAGO

WEEK ENDING SEPTEMBER 5, 1968

1.	1, 2, 3, RED LIGHT	1910 FRUITGUM CO.	3
2.	PEOPLE GOT TO BE FREE	THE RASCALS	1
3.	LIGHT MY FIRE	JOSE FELICIANO	2
4.	BORN TO BE WILD	STEPPENWOLF	4
5.	HARPER VALLEY P.T.A.	JEANNIE C. RILEY	14
6.	YOU KEEP ME HANGIN' ON	VANILLA FUDGE	16
7.	DO IT AGAIN	THE BEACH BOYS	5
8.	(You're Still My Favorite Girlfriend) ALICE LONG	TOMMY BOYCE AND BOBBY HART	7
9.	DOWN AT LULU'S	OHIO EXPRESS	8
10.	I CAN'T STOP DANCING	ARCHIE BELL AND THE DRELLS	15
11.	JOURNEY TO THE CENTER OF THE MIND	AMBOY DUKES	9
12.	I SAY A LITTLE PRAYER / THIS IS THE HOUSE THAT JACK BUILT	ARETHA FRANKLIN	18
13.	TUESDAY AFTERNOON	THE MOODY BLUES	22
14.	EYES OF A NEW YORK WOMAN	B.J. THOMAS	21
15.	YOU'RE ALL I NEED TO GET BY	MARVIN GAYE / TAMMI TERRELL	19
16.	MAGIC BUS	THE WHO	25
17.	LOVE MAKES A WOMAN	BARBARA ACKLIN	20
18.	FOOL ON THE HILL	SERGIO MENDEZ	24
19.	PLEASE RETURN YOUR LOVE TO ME	THE TEMPTATIONS	23
20.	SHAPES OF THINGS TO COME	MAX FROST AND THE TROOPERS	27

ACTION 10

21.	INDIAN RESERVATION	DON FARDON	--
22.	SOUL DRIPPIN'	THE MAUDS	26
23.	ON THE ROAD AGAIN	CANNED HEAT	--
24.	HUSH	DEEP PURPLE	--
25.	SIX MAN BAND	THE ASSOCIATION	28
26.	LITTLE GREEN APPLES	O.C. SMITH	--
27.	SALLY HAD A PARTY	FLAVOR	29
28.	DOWN ON ME	BIG BROTHER & THE HOLDING CO.	--
29.	MY SPECIAL ANGEL	THE VOGUES	--
30.	HEY JUDE / REVOLUTION	THE BEATLES	--

WCFL
1000 WCFL-AM SOUND 10 SURVEY-CHICAGO

WEEK ENDING SEPTEMBER 12, 1968

1.	DO IT AGAIN	THE BEACH BOYS	7
2.	PEOPLE GOT TO BE FREE	THE RASCALS	2
3.	1, 2, 3, RED LIGHT	1910 FRUITGUM CO.	1
4.	HARPER VALLEY P.T.A.	JEANNIE C. RILEY	5
5.	DOWN AT LULU'S	OHIO EXPRESS	9
6.	I SAY A LITTLE PRAYER / THIS IS THE HOUSE THAT JACK BUILT	ARETHA FRANKLIN	12
7.	EYES OF A NEW YORK WOMAN	B.J. THOMAS	14
8.	YOU'RE ALL I NEED TO GET BY	MARVIN GAYE / TAMMI TERRELL	15
9.	LIGHT MY FIRE	JOSE FELICIANO	3
10.	YOU KEEP ME HANGIN' ON	THE VANILLA FUDGE	6
11.	TUESDAY AFTERNOON	THE MOODY BLUES	13
12.	INDIAN RESERVATION	DON FARDON	21
13.	ON THE ROAD AGAIN	CANNED HEAT	23
14.	BORN TO BE WILD	STEPPENWOLF	4
15.	FOOL ON THE HILL	SERGIO MENDEZ	18
16.	MAGIC BUS	THE WHO	16
17.	LOVE MAKES A WOMAN	BARBARA ACKLIN	17
18.	(You're Still My Favorite Girlfriend) ALICE LONG	TOMMY BOYCE AND BOBBY HART	8
19.	SHAPES OF THINGS TO COME	MAX FROST AND THE TROOPERS	20
20.	PLEASE RETURN YOUR LOVE TO ME	THE TEMPTATIONS	19

ACTION 10

21.	HEY JUDE / REVOLUTION	THE BEATLES	30
22.	HUSH	DEEP PURPLE	24
23.	FIRE	CRAZY WORLD OF ARTHUR BROWN	--
24.	LITTLE GREEN APPLE	O.C. SMITH	26
25.	MIDNIGHT CONFESSIONS	THE GRASS ROOTS	--
26.	MY SPECIAL ANGEL	THE VOGUES	29
27.	SPECIAL OCCASION	SMOKEY ROBINSON AND THE MIRACLES	--
28.	DOWN ON ME	BIG BROTHER & THE HOLDING CO.	28
29.	SOUL DRIPPIN'	THE MAUDS	22
30.	HIP CITY (Part II)	JR. WALKER AND THE ALL STARS	--

WCFL
1000 WCFL-AM SOUND 10 SURVEY-CHICAGO

WEEK ENDING SEPTEMBER 26, 1968

1.	FIRE	CRAZY WORLD OF ARTHUR BROWN	4
2.	HEY JUDE / REVOLUTION	THE BEATLES	3
3.	DO IT AGAIN	THE BEACH BOYS	2
4.	HARPER VALLEY P.T.A.	JEANNIE C. RILEY	1
5.	INDIAN RESERVATION	DON FARDON	11
6.	HUSH	DEEP PURPLE	15
7.	ON THE ROAD AGAIN	CANNED HEAT	12
8.	SHAPES OF THINGS TO COME	MAX FROST AND THE TROOPERS	17
9.	MY SPECIAL ANGEL	THE VOGUES	16
10.	TUESDAY AFTERNOON	THE MOODY BLUES	7
11.	FOOL ON THE HILL	SERGIO MENDEZ	8
12.	I SAY A LITTLE PRAYER / THIS IS THE HOUSE THAT JACK BUILT	ARETHA FRANKLIN	13
13.	EYES OF A NEW YORK WOMAN	B.J. THOMAS	14
14.	DOWN AT LULU'S	OHIO EXPRESS	6
15.	1, 2, 3, RED LIGHT	1910 FRUITGUM CO.	5
16.	MIDNIGHT CONFESSIONS	THE GRASS ROOTS	18
17.	OVER YOU	GARY PUCKETT & THE UNION GAP	23
18.	LITTLE GREEN APPLES	O.C. SMITH	19
19.	SOUL DRIPPIN'	THE MAUDS	20
20.	DOWN ON ME	BIG BROTHER & THE HOLDING CO.	21

ACTION 10

21.	SPECIAL OCCASION	SMOKEY ROBINSON AND THE MIRACLES	22
22.	SUSIE Q (Part I)	CREEDENCE CLEARWATER REVIVAL	24
23.	I'VE GOTTA GET A MESSAGE TO YOU	THE BEE GEES	26
24.	GREENBURG, GLICKSTEIN, CHARLES DAVID SMITH & JONES	THE CRYAN' SHAMES	--
25.	POOR BABY	THE COWSILLS	27
26.	I MET HER IN CHURCH	THE BOX TOPS	30
27.	SAN FRANCISCO GIRLS	FEVER TREE	29
28.	GIRL WATCHER	THE O'KAYSIONS	--
29.	TIME HAS COME TODAY	THE CHAMBERS BROTHERS	--
30.	BABY, COME BACK	THE EQUALS	--

WCFL
1000 WCFL-AM SOUND 10 SURVEY-CHICAGO

WEEK ENDING SEPTEMBER 19, 1968

1.	HARPER VALLEY P.T.A.	JEANNIE C. RILEY	4
2.	DO IT AGAIN	THE BEACH BOYS	1
3.	HEY JUDE / REVOLUTION	THE BEATLES	21
4.	FIRE	CRAZY WORLD OF ARTHUR BROWN	23
5.	1, 2, 3, RED LIGHT	1910 FRUITGUM CO.	3
6.	DOWN AT LULU'S	OHIO EXPRESS	5
7.	TUESDAY AFTERNOON	THE MOODY BLUES	11
8.	FOOL ON THE HILL	SERGIO MENDEZ	15
9.	MAGIC BUS	THE WHO	16
10.	PEOPLE GOT TO BE FREE	THE RASCALS	2
11.	INDIAN RESERVATION	DON FARDON	12
12.	ON THE ROAD AGAIN	CANNED HEAT	13
13.	I SAY A LITTLE PRAYER / THIS IS THE HOUSE THAT JACK BUILT	ARETHA FRANKLIN	6
14.	EYES OF A NEW YORK WOMAN	B.J. THOMAS	7
15.	HUSH	DEEP PURPLE	22
16.	MY SPECIAL ANGEL	THE VOGUES	26
17.	SHAPES OF THINGS TO COME	MAX FROST AND THE TROOPERS	19

18.	MIDNIGHT CONFESSIONS	THE GRASS ROOTS	25
19.	LITTLE GREEN APPLES	O.C. SMITH	24
20.	SOUL DRIPPIN'	THE MAUDS	29

ACTION 10

21.	DOWN ON ME	BIG BROTHER & THE HOLDING CO.	28
22.	SPECIAL OCCASION	SMOKEY ROBINSON AND THE MIRACLES	27
23.	OVER YOU	GARY PUCKETT & THE UNION GAP	--
24.	SUSIE Q (Part I)	CREEDENCE CLEARWATER REVIVAL	--
25.	HIP CITY (Part II)	JR. WALKER AND THE ALL STARS	30
26.	I'VE GOTTA GET A MESSAGE TO YOU	THE BEE GEES	--
27.	POOR BABY	THE COWSILLS	--
28.	TO WAIT FOR YOU	HERB ALPERT	--
29.	SAN FRANCISCO GIRLS	FEVER TREE	--
30.	I MET HER IN CHURCH	THE BOX TOPS	--

WCFL
1000 WCFL-AM SOUND 10 SURVEY-CHICAGO

WEEK ENDING OCTOBER 3, 1968

1.	FIRE	CRAZY WORLD OF ARTHUR BROWN	1
2.	HEY / REVOLUTION	THE BEATLES	2
3.	SHAPES OF THINGS TO COME	MAX FROST AND THE TROOPERS	8
4.	MIDNIGHT CONFESSIONS	THE GRASS ROOTS	16
5.	LITTLE GREEN APPLES	O.C. SMITH	18
6.	ON THE ROAD AGAIN	CANNED HEAT	7
7.	HUSH	DEEP PURPLE	6
8.	MY SPECIAL ANGEL	THE VOGUES	9
9.	DO IT AGAIN	THE BEACH BOYS	3
10.	GIRL WATCHER	THE O'KAYSIONS	28
11.	HARPER VALLEY P.T.A.	JEANNIE C. RILEY	4
12.	FOOL ON THE HILL	SERGIO MENDEZ	11
13.	INDIAN RESERVATION	DON FARDON	5
14.	OVER YOU	GARY PUCKETT & THE UNION GAP	17
15.	DOWN AT LULU'S	OHIO EXPRESS	14
16.	I'VE GOTTA GET A MESSAGE TO YOU	THE BEE GEES	23
17.	SOUL DRIPPIN'	THE MAUDS	19
18.	GREENBURG, GLICKSTEIN, CHARLES DAVID SMITH & JONES	THE CRYAN' SHAMES	24
19.	SUSIE Q (Part I)	CREEDENCE CLEARWATER REVIVAL	22
20.	POOR BABY	THE COWSILLS	25

ACTION 10

21.	IN-A-GADDA-DA-VIDA	IRON BUTTERFLY	--
22.	BABY, COME BACK	THE EQUALS	30
23.	TIME HAS COME TODAY	THE CHAMBERS BROTHERS	29
24.	THE SNAKE	AL WILSON	--
25.	WHITE ROOM	CREAM	--
26.	I MET HER IN CHURCH	THE BOX TOPS	26
27.	SAN FRANCISCO GIRLS	FEVER TREE	27
28.	THOSE WERE THE DAYS	MARY HOPKIN	--
29.	ELENORE	THE TURTLES	--
30.	CINNAMON	DEREK	--

WCFL
1000 WCFL-AM SOUND 10 SURVEY-CHICAGO

WEEK ENDING OCTOBER 10, 1968

1.	FIRE	CRAZY WORLD OF ARTHUR BROWN	1
2.	HEY JUDE / REVOLUTION	THE BEATLES	2
3.	SHAPES OF THINGS TO COME	MAX FROST AND THE TROOPERS	3
4.	LITTLE GREEN APPLES	O.C. SMITH	5
5.	MIDNIGHT CONFESSIONS	THE GRASS ROOTS	4
6.	GIRL WATCHER	THE O'KAYSIONS	10
7.	OVER YOU	GARY PUCKETT & THE UNION GAP	14
8.	I'VE GOTTA GET A MESSAGE TO YOU	THE BEE GEES	16
9.	GREENSBURG, GLICKSTEIN, CHARLES DAVID SMITH & JONES	THE CRYAN' SHAMES	18
10.	TIME HAS COME TODAY	THE CHAMBERS BROTHERS	23
11.	SUSIE Q (Part I)	CREEDENCE CLEARWATER REVIVAL	19
12.	HUSH	DEEP PURPLE	7
13.	ON THE ROAD AGAIN	CANNED HEAT	6
14.	MY SPECIAL ANGEL	THE VOGUES	8
15.	FOOL ON THE HILL	SERGIO MENDEZ	12
16.	SOUL DRIPPIN'	THE MAUDS	17
17.	BABY, COME BACK	THE EQUALS	22
18.	WHITE ROOM	CREAM	25
19.	IN-A-GADDA-DA-VIDA	IRON BUTTERFLY	21
20.	FIRE	FIVE BY FIVE	--

HIT BOUND

21.	ELENORE	THE TURTLES	29
22.	POOR BABY	THE COWSILLS	20
23.	THOSE WERE THE DAYS	MARY HOPKIN	28
24.	CINNAMON	DEREK	30
25.	SAY IT LOUD-I'M BLACK AND I'M PROUD (Part I)	JAMES BROWN	--
26.	SAN FRANCISCO GIRLS	FEVER TREE	27
27.	MAGIC CARPET RIDE	STEPPENWOLF	--
28.	CHAINED	MARVIN GAYE	--
29.	FOOL FOR YOU	THE IMPRESSIONS	--
30.	HEY, WESTERN UNION MAN	JERRY BUTLER	--

WCFL
1000 WCFL-AM SOUND 10 SURVEY-CHICAGO

WEEK ENDING OCTOBER 17, 1968

1.	FIRE	CRAZY WORLD OF ARTHUR BROWN	1
2.	MIDNIGHT CONFESSIONS	THE GRASS ROOTS	5
3.	LITTLE GREEN APPLES	O.C. SMITH	4
4.	HEY JUDE / REVOLUTION	THE BEATLES	2
5.	GIRL WATCHER	THE O'KAYSIONS	6
6.	OVER YOU	GARY PUCKETT & THE UNION GAP	7
7.	I'VE GOTTA GET A MESSAGE TO YOU	THE BEE GEES	8
8.	SHAPES OF THINGS TO COME	MAX FROST AND THE TROOPERS	3
9.	TIME HAS COME TODAY	THE CHAMBERS BROTHERS	10
10.	SUSIE Q (Part I)	CREEDENCE CLEARWATER REVIVAL	11
11.	ELENORE	THE TURTLES	21
12.	BABY, COME BACK	THE EQUALS	17
13.	THOSE WERE THE DAYS	MARY HOPKIN	23
14.	GREENBURG, GLICKSTEIN, CHARLES DAVID SMITH & JONES	THE CRYAN' SHAMES	9
15.	MY SPECIAL ANGEL	THE VOGUES	14
16.	WHITE ROOM	CREAM	18
17.	IN-A-GADDA-DA-VIDA	IRON BUTTERFLY	19
18.	FIRE	FIVE BY FIVE	20
19.	MAGIC CARPET RIDE	STEPPENWOLF	27
20.	FOOL FOR YOU	THE IMPRESSIONS	29

HIT BOUND

21.	HI HEEL SNEAKERS	JOSE FELICIANO	--
22.	CINNAMON	DEREK	24
23.	SAY IT LOUD – I'M BLACK AND I'M PROUD (Part I)	JAMES BROWN	25
24.	ALL ALONG THE WATCHTOWER	JIMI HENDRIX EXPERIENCE	--
25.	HEY, WESTERN UNION MAN	JERRY BUTLER	30
26.	CHAINED	MARVIN GAYE	28
27.	RUN TO ME	MONTANAS	--
28.	PIECE OF MY HEART	BIG BROTHER & THE HOLDING CO.	--
29.	ALFIE	STEVIE WONDER	--
30.	GENTLE ON MY MIND	GLEN CAMPBELL	--

WCFL
1000 WCFL-AM SOUND 10 SURVEY-CHICAGO

WEEK ENDING OCTOBER 24, 1968

1.	LITTLE GREEN APPLES	O.C. SMITH	3
2.	OVER YOU	GARY PUCKETT & THE UNION GAP	6
3.	I'VE GOTTA GET A MESSAGE TO YOU	THE BEE GEES	7
4.	MIDNIGHT CONFESSIONS	THE GRASS ROOTS	2
5.	FIRE	THE CRAZY WORLD OF ARTHUR BROWN	1
6.	THOSE WERE THE DAYS	MARY HOPKIN	13
7.	GIRL WATCHER	THE O'KAYSIONS	5
8.	HEY JUDE / REVOLUTION	THE BEATLES	4
9.	ELENORE	THE TURTLES	11
10.	MAGIC CARPET RIDE	STEPPENWOLF	19
11.	SHAPES OF THINGS TO COME	MAX FROST AND THE TROOPERS	8
12.	SUSIE Q (Part I)	CREEDENCE CLEARWATER REVIVAL	10
13.	GREENBURG, GLICKSTEIN, CHARLES DAVID SMITH & JONES	THE CRYAN' SHAMES	14
14.	TIME HAS COME TODAY	THE CHAMBERS BROTHERS	9
15.	WHITE ROOM	CREAM	16
16.	FIRE	FIVE BY FIVE	18
17.	LOVE CHILD	DIANA ROSS AND THE SUPREMES	--
18.	HI HEEL SNEAKERS	JOSE FELICIANO	21
19.	FOOL FOR YOU	THE IMPRESSIONS	20
20.	CINNAMON	DEREK	22

HIT BOUND

21.	SAY IT LOUD-I'M BLACK AND I'M PROUD (Part I)	JAMES BROWN	23
22.	HEY, WESTERN UNION MAN	JERRY BUTLER	25
23.	ALL ALONG THE WATCHTOWER	JIMI HENDRIX EXPERIENCE	24
24.	RUN TO ME	MONTANAS	27
25.	PIECE OF MY HEART	BIG BROTHER & THE HOLDING CO.	28
26.	ALFIE	STEVIE WONDER	29
27.	CHEWY, CHEWY	OHIO EXPRESS	--
28.	GENTLE ON MY MIND	GLEN CAMPBELL	30
29.	HOLD ME TIGHT	JOHNNY NASH	--
30.	I'M IN A DIFFERENT WORLD	FOUR TOPS	--

WCFL
1000 WCFL-AM SOUND 10 SURVEY-CHICAGO

WEEK ENDING OCTOBER 31, 1968

1.	THOSE WERE THE DAYS	MARY HOPKIN	6
2.	I'VE GOTTA GET A MESSAGE TO YOU	THE BEE GEES	3
3.	MAGIC CARPET RIDE	STEPPENWOLF	10
4.	MIDNIGHT CONFESSIONS	THE GRASS ROOTS	4
5.	LITTLE GREEN APPLES	O.C. SMITH	1
6.	ELENORE	THE TURTLES	9
7.	FIRE	FIVE BY FIVE	16
8.	OVER YOU	GARY PUCKETT & THE UNION GAP	2
9.	FIRE	THE CRAZY WORLD OF ARTHUR BROWN	5
10.	LOVE CHILD	DIANA ROSS AND THE SUPREMES	17
11.	WHITE ROOM	CREAM	15
12.	HEY JUDE / REVOLUTION	THE BEATLES	8
13.	GIRL WATCHER	THE O'KAYSIONS	7
14.	CHEWY, CHEWY	OHIO EXPRESS	27
15.	GREENBURG, GLICKSTEIN, CHARLES DAVID SMITH & JONES	THE CRYAN SHAMES	13
16.	RUN TO ME	MONTANAS	24
17.	HI HEEL SNEAKERS	JOSE FELICIANO	18
18.	FOOL FOR YOU	THE IMPRESSIONS	19
19.	CINNAMON	DEREK	20
20.	ALL ALONG THE WATCHTOWER	JIMI HENDRIX EXPERIENCE	23

HIT BOUND

21.	HEY, WESTERN UNION MAN	JERRY BUTLER	22
22.	HOLD ME TIGHT	JOHNNY NASH	29
23.	PIECE OF MY HEART	BIG BROTHER & THE HOLDING CO.	25
24.	I'M IN A DIFFERENT WORLD	FOUR TOPS	30
25.	ALFIE	STEVIE WONDER	26
26.	COURT OF LOVE	THE UNIFICS	--
27.	WHO'S MAKIN' LOVE	JOHNNY TAYLOR	--
28.	SHAME, SHAME	MAGIC LANTERNS	--
29.	SHAKE	THE SHADOWS OF KNIGHT	--
30.	SWEET BLINDNESS	THE 5TH DIMENSION	--

WCFL

1000 WCFL-AM SOUND 10 SURVEY-CHICAGO

WEEK ENDING NOVEMBER 7, 1968

1.	THOSE WERE THE DAYS	MARY HOPKIN	1
2.	MAGIC CARPET RIDE	STEEPENWOLF	3
3.	LOVE CHILD	DIANA ROSS AND THE SUPREMES	10
4.	ELENORE	THE TURTLES	6
5.	FIRE	FIVE BY FIVE	7
6.	I'VE GOTTA GET A MESSAGE TO YOU	THE BEE GEES	2
7.	MIDNIGHT CONFESSIONS	THE GRASS ROOTS	4
8.	LITTLE GREEN APPLES	O.C. SMITH	5
9.	WHITE ROOM	CREAM	11
10.	CHEWY, CHEWY	OHIO EXPRESS	14
11.	OVER YOU	GARY PUCKETT & THE UNION GAP	8
12.	HOLD ME TIGHT	JOHNNY NASH	22
13.	RUN TO ME	MONTANAS	16
14.	HEY, WESTERN UNION MAN	JERRY BUTLER	21
15.	HEY JUDE / REVOLUTION	THE BEATLES	12
16.	FOOL FOR YOU	THE IMPRESSIONS	18
17.	HI HEEL SNEAKERS	JOSE FELICIANO	17
18.	CINNAMON	DEREK	19
19.	ALL ALONG THE WATCHTOWER	JIMI HENDRIX EXPERIENCE	20
20.	WHO'S MAKIN' LOVE	JOHNNY TAYLOR	27

HIT BOUND

21.	SHAME, SHAME	MAGIC LANTERNS	28
22.	QUICK JOEY SMALL (Run Joey Run)	KASENETZ-KATZ SINGING ORCHESTRAL CIRCUS	--
23.	BANG-SHANG-A-LANG	THE ARCHIES	--
24.	I'M IN A DIFFERENT WORLD	FOUR TOPS	24
25.	COURT OF LOVE	THE UNIFICS	26
26.	SHAKE	SHADOWS OF KNIGHT	29
27.	SWEET BLINDNESS	THE 5TH DIMENSION	30
28.	GOODY, GOODY GUMDROPS	1910 FRUITGUM CO.	--
29.	ABRAHAM, MARTIN & JOHN	DION	--
30.	TAKE ME FOR A LITTLE WHILE	THE VANILLA FUDGE	--

WCFL
1000 WCFL-AM SOUND 10 SURVEY-CHICAGO

WEEK ENDING NOVEMBER 14, 1968

1.	THOSE WERE THE DAYS	MARY HOPKIN	1
2.	LOVE CHILD	DIANA ROSS AND THE SUPREMES	3
3.	MAGIC CARPET RIDE	STEPPENWOLF	2
4.	ELENORE	THE TURTLES	4
5.	WHITE ROOM	CREAM	9
6.	CHEWY, CHEWY	OHIO EXPRESS	10
7.	FIRE	FIVE BY FIVE	5
8.	HOLD ME TIGHT	JOHNNY NASH	12
9.	I'VE GOTTA GET A MESSAGE TO YOU	THE BEE GEES	6
10.	RUN TO ME	MONTANAS	13
11.	SHAME, SHAME	MAGIC LANTERNS	21
12.	WHO'S MAKIN' LOVE	JOHNNY TAYLOR	20
13.	CINNAMON	DEREK	18
14.	BANG-SHANG-A-LANG	THE ARCHIES	23
15.	HEY JUDE / REVOLUTION	THE BEATLES	15
16.	QUICK JOEY SMALL (Run Joey Run)	KASENETZ-KATZ SINGING ORCHESTRAL CIRCUS	22
17.	ABRAHAM, MARTIN & JOHN	DION	29
18.	SWEET BLINDNESS	THE 5TH DIMENSION	27
19.	FOR ONCE IN MY LIFE	STEVIE WONDER	--
20.	ALWAYS TOGETHER	THE DELLS	--

HIT BOUND

21.	GOODY, GOODY GUMDROPS	1910 FRUITGUM CO.	28
22.	CYCLES	FRANK SINATRA	--
23.	COURT OF LOVE	THE UNIFICS	25
24.	SHAKE	THE SHADOWS OF KNIGHT	26
25.	KEEP ON LOVIN' ME HONEY	MARVIN GAYE & TAMMI TERRELL	--
26.	TAKE ME FOR A LITTLE WHILE	THE VANILLA FUDGE	30
27.	I HEARD IT THROUGH THE GRAPEVINE	MARVIN GAYE	--
28.	BOTH SIDES NOW	JUDY COLLINS	--
29.	I LOVE HOW YOU LOVE ME	BOBBY VINTON	--
30.	STORMY	CLASSICS IV	--

WCFL
1000 WCFL-AM SOUND 10 SURVEY-CHICAGO

WEEK ENDING NOVEMBER 21, 1968

1.	LOVE CHILD	DIANA ROSS AND THE SUPREMES	2
2.	THOSE WERE THE DAYS	MARY HOPKIN	1
3.	CHEWY, CHEWY	OHIO EXPRESS	6
4.	MAGIC CARPET RIDE	STEPPENWOLF	3
5.	SHAME, SHAME	MAGIC LANTERNS	11
6.	ABRAHAM, MARTIN & JOHN	DION	17
7.	WHITE ROOM	CREAM	5
8.	ELENORE	THE TURTLES	4
9.	RUN TO ME	MONTANAS	10
10.	BANG-SHANG-A-LANG	THE ARCHIES	14
11.	HOLD ME TIGHT	JOHNNY NASH	8
12.	CINNAMON	DEREK	13
13.	WHO'S MAKIN' LOVE	JOHNNY TAYLOR	12
14.	QUICK JOEY SMALL (Run Joey Run)	KASENETZ-KATZ SINGING ORCHESTRAL CIRCUS	16
15.	FIRE	FIVE BY FIVE	7
16.	FOR ONCE IN MY LIFE	STEVIE WONDER	19
17.	STORMY	CLASSICS IV	30
18.	I LOVE HOW YOU LOVE ME	BOBBY VINTON	29
19.	SWEET BLINDNESS	THE 5TH DIMENSION	18
20.	GOODY, GOODY GUMDROPS	1910 FRUITGUM CO.	21

HIT BOUND

21.	CYCLES	FRANK SINATRA	22
22.	ALWAYS TOGETHER	THE DELLS	20
23.	SHAKE	THE SHADOWS OF KNIGHT	24
24.	I HEART IT THROUGH THE GRAPEVINE	MARVIN GAYE	27
25.	BOTH SIDES NOW	JUDY COLLINS	28
26.	THING'S I'D LIKE TO SAY	THE NEW COLONY SIX	--
27.	BRING IT ON HOME TO ME	EDDIE FLOYD	--
28.	WICHITA LINEMAN	GLEN CAMPBELL	--
29.	PROMISES, PROMISES	DIONNE WARWICK	--
30.	TILL	THE VOGUES	--

WCFL
1000 WCFL-AM SOUND 10 SURVEY-CHICAGO

WEEK ENDING NOVEMBER 28, 1968

1.	ABRAHAM, MARTIN & JOHN	DION	6
2.	LOVE CHILD	DIANA ROSS AND THE SUPREMES	1
3.	CHEWY, CHEWY	OHIO EXPRESS	3
4.	SHAME, SHAME	MAGIC LANTERNS	5
5.	MAGIC CARPET RIDE	STEPPENWOLF	4
6.	THOSE WERE THE DAYS	MARY HOPKIN	2
7.	CINNAMON	DEREK	12
8.	I HEART IT THROUGH THE GRAPEVINE	MARVIN GAYE	24
9.	FOR ONCE IN MY LIFE	STEVIE WONDER	16
10.	WHO'S MAKIN' LOVE	JOHNNY TAYLOR	13
11.	QUICK JOEY SMALL (Run Joey Run)	KASENETZ-KATZ SINGING ORCHESTRAL CIRCUS	14
12.	STORMY	CLASSICS IV	17
13.	BANG-SHANG-A-LANG	THE ARCHIES	10
14.	HOLD ME TIGHT	JOHNNY NASH	11
15.	GOODY, GOODY GUMDROPS	1910 FRUITGUM CO.	20
16.	THING'S I'D LIKE TO SAY	THE NEW COLONY SIX	26
17.	I LOVE HOW YOU LOVE ME	BOBBY VINTON	18
18.	SHAKE	THE SHADOWS OF NIGHT	23
19.	BOTH SIDES NOW	JUDY COLLINS	25
20.	WICHITA LINEMAN	GLEN CAMPBELL	28

HIT BOUND

21.	TILL	THE VOGUES	30
22.	PROMISES, PROMISES	DIONNE WARWICK	29
23.	BRING IT ON HOME TO ME	EDDIE FLOYD	27
24.	THE STRAIGHT LIFE	BOBBY GOLDSBORO	--
25.	SCARBOROUGH FAIR	SERGIO MENDEZ	--
26.	HANG 'EM HIGH	BOOKER T. & THE MG'S	--
27.	KENTUCKY WOMAN	DEEP PURPLE	--
28.	CLOUD NINE	THE TEMPTATIONS	--
29.	I CAN'T TURN YOU LOOSE	THE CHAMBERS BROTHERS	--
30.	SEE SAW	ARETHA FRANKLIN	--

WCFL
1000 WCFL-AM SOUND 10 SURVEY-CHICAGO

WEEK ENDING DECEMBER 5, 1968

1.	ABRAHAM, MARTIN & JOHN	DION	1
2.	I HEARD IT THROUGH THE GRAPEVINE	MARTIN GAYE	8
3.	LOVE CHILD	DIANA ROSS AND THE SUPREMES	2
4.	FOR ONCE IN MY LIFE	STEVIE WONDER	9
5.	CHEWY, CHEWY	OHIO EXPRESS	3
6.	SHAME, SHAME	MAGIC LANTERNS	4
7.	CINNAMON	DEREK	7
8.	MAGIC CARPET RIDE	STEPPENWOLF	5
9.	THOSE WERE THE DAYS	MARY HOPKIN	6
10.	QUICK JOEY SMALL (Run Joey Run)	KASENETA-KATZ SINGING ORCHESTRAL CIRCUS	11
11.	STORMY	CLASSICS IV	12
12.	WHO'S MAKIN' LOVE	JOHNNY TAYLOR	10
13.	GOODY, GOODY GUMDROPS	1910 FRUITGUM CO.	15
14.	SHAKE	THE SHADOWS OF KNIGHT	18
15.	I LOVE HOW YOU LOVE ME	BOBBY VINTON	17
16.	THING'S I'D LIKE TO SAY	THE NEW COLONY SIX	16
17.	BOTH SIDES NOW	JUDY COLLINS	19
18.	WICHITA LINEMAN	GLEN CAMPBELL	20
19.	TILL	THE VOGUES	21
20.	PROMISES, PROMISES	DIONNE WARWICK	22

HIT BOUND

21.	HANG 'EM HIGH	BOOKER T. & THE MG'S	26
22.	SCARBOROUGH FAIR	SERGIO MENDEZ	25
23.	I CAN'T TURN YOU LOOSE	THE CHAMBERS BROTHERS	29
24.	THE STRAIGHT LIFE	BOBBY GOLDSBORO	24
25.	SEE SAW	ARETHA FRANKLIN	30
26.	CLOUD NINE	THE TEMPTATIONS	28
27.	KENTUCKY WOMAN	DEEP PURPLE	27
28.	NOBODY	THREE DOG NIGHT	--
29.	BELLA LINDA	THE GRASS ROOTS	--
30.	I'M GONNA MAKE YOU LOVE ME	DIANA ROSS AND THE SUPREMES / THE TEMPTATIONS	--

WCFL
1000 WCFL-AM SOUND 10 SURVEY-CHICAGO

WEEK ENDING DECEMBER 12, 1968

1.	I HEARD IT THROUGH THE GRAPEVINE	MARVIN GAYE	2
2.	ABRAHAM, MARTIN & JOHN	DION	1
3.	FOR ONCE IN MY LIFE	STEVIE WONDER	4
4.	LOVE CHILD	DIANA ROSS AND THE SUPREMES	3
5.	STORMY	CLASSICS IV	11
6.	THING'S I'D LIKE TO SAY	THE NEW COLONY SIX	16
7.	CHEWY, CHEWY	OHIO EXPRESS	5
8.	I LOVE HOW YOU LOVE ME	BOBBY VINTON	15
9.	CINNAMON	DEREK	7
10.	WICHITA LINEMAN	GLEN CAMPBELL	18
11.	GOODY, GOODY GUMDROPS	1910 FRUITGUM CO.	13
12.	WHO'S MAKIN' LOVE	JOHNNY TAYLOR	12
13.	SHAKE	THE SHADOWS OF KNIGHT	14
14.	BOTH SIDES NOW	JUDY COLLINS	17
15.	QUICK JOEY SMALL (Run Joey Run)	KASENETZ-KATZ SINGING ORCHESTRAL CIRCUS	10
16.	I'M GONNA MAKE YOU LOVE ME	DIANA ROSS AND THE SUPREMES / THE TEMPTATIONS	30
17.	I CAN'T TURN YOU LOOSE	THE CHAMBERS BROTHERS	23
18.	TILL	THE VOGUES	19
19.	HANG 'EM HIGH	BOOKER T. THE MG'S	21
20.	CLOUD NINE	THE TEMPTATIONS	26

HIT BOUND

21.	SCARBOROUGH FAIR	SERGIO MENDEZ	22
22.	PROMISES, PROMISES	DIONNE WARWICK	20
23.	THE STRAIGHT LIFE	BOBBY GOLDSBORO	24
24.	SEE SAW	ARETHA FRANKLIN	25
25.	KENTUCKY WOMAN	DEEP PURPLE	27
26.	SON-OF-A PREACHER MAN	DUSTY SPRINGFIELD	--
27.	BELLA LINDA	THE GRASS ROOTS	29
28.	SOULFUL STRUT	YOUNG HOLT UNLIMITED	--
29.	HOOKED ON A FEELING	B.J. THOMAS	--
30.	A RAY OF HOPE	THE RASCALS	--

WCFL
1000 WCFL-AM SOUND 10 SURVEY-CHICAGO

WEEK ENDING DECEMBER 19, 1968

1.	I HEART IT THROUGH THE GRAPEVINE	MARVIN GAYE	1
2.	FOR ONCE IN MY LIFE	STEVIE WONDER	3
3.	THING'S I'D LIKE TO SAY	THE NEW COLONY SIX	6
4.	I LOVE HOW YOU LOVE ME	BOBBY VINTON	8
5.	I'M GONNA MAKE YOU LOVE ME	DIANA ROSS AND THE SUPREMES / THE TEMPTATIONS	16
6.	ABRAHAM, MARTIN & JOHN	DION	2
7.	STORMY	CLASSICS IV	5
8.	CINNAMON	DEREK	9
9.	LOVE CHILD	DIANA ROSS AND THE SUPREMES	4
10.	SHAKE	THE SHADOWS OF KNIGHT	13
11.	BOTH SIDES NOW	JUDY COLLINS	14
12.	WICHITA LINEMAN	GLEN CAMPBELL	10
13.	WHO'S MAKIN' LOVE	JOHNNY TAYLOR	12
14.	CLOUD NINE	THE TEMPTATIONS	20
15.	I CAN'T TURN YOU LOOSE	THE CHAMBERS BROTHERS	17
16.	SON-OF-A PREACHER MAN	DUSTY SPRINGFIELD	26
17.	HANG'EM HIGH	BOOKER T. & THE MG'S	19
18.	TILL	THE VOGUES	18
19.	SOULFUL STRUT	YOUNG HOLT UNLIMITED	28
20.	PROMISES, PROMISES	DIONNE WARWICK	22

HIT BOUND

21.	SCARBOROUGH FAIR	SERGIO MENDEZ	21
22.	KENTUCKY WOMAN	DEEP PURPLE	25
23.	BELLA LINDA	THE GRASS ROOTS	27
24.	HOOKED ON A FEELING	B.J. THOMAS	29
25.	A RAY OF HOPE	THE RASCALS	30
26.	LO MUCHO QUE TE QUIERO	RENE / RENE	--
27.	GOING UP THE COUNTRY	CANNED HEAT	--
28.	TOO WEAK TO FIGHT	CLARENCE CARTER	--
29.	IF I CAN DREAM	ELVIS PRESLEY	--
30.	WORST THAT CAN HAPPEN	BROOKLYN BRIDGE	--

WCFL
1000 WCFL-AM SOUND 10 SURVEY-CHICAGO

WEEK ENDING DECEMBER 26, 1968

1.	I HEARD IT THROUGH THE GRAPEVINE	MARVIN GAYE	1
2.	FOR ONCE IN MY LIFE	STEVIE WONDER	2
3.	I LOVE HOW YOU LOVE ME	BOBBY VINTON	4
4.	I'M GONNA MAKE YOU LOVE ME	DIANA ROSS AND THE SUPREMES / THE TEMPTATIONS	5
5.	THING'S I'D LIKE TO SAY	THE NEW COLONY SIX	3
6.	STORMY	CLASSICS IV	7
7.	CLOUD NINE	THE TEMPTATIONS	14
8.	ABRAHAM, MARTIN & JOHN	DION	6
9.	CINNAMON	DEREK	8
10.	LOVE CHILD	DIANA ROSS AND THE SUPREMES	9
11.	WICHITA LINEMAN	GLEN CAMPBELL	12
12.	SON-OF-A PREACHER MAN	DUSTY SPRINGFIELD	16
13.	BOTH SIDES NOW	JUDY COLLINS	11
14.	SOULFUL STRUT	YOUNG HOLT UNLIMITED	19
15.	HANG 'EM HIGH	BOOKER T. & THE MG'S	17
16.	CRIMSON AND CLOVER	TOMMY JAMES & THE SHONDELLS	--
17.	WHO'S MAKIN' LOVE	JOHNNY TAYLOR	13
18.	SHAKE	THE SHADOWS OF KNIGHT	10
19.	I CAN'T TURN YOU LOOSE	THE CHAMBERS BROTHERS	15
20.	TILL	THE VOGUES	18

HIT BOUND

21.	SCARBOROUGH FAIR	SERGIO MENDEZ	21
22.	BELLA LINDA	THE GRASS ROOTS	23
23.	GOING UP THE COUNTRY	CANNED HEAT	27
24.	SEE SAW	ARETHA FRANKLIN	--
25.	KENTUCKY WOMAN	DEEP PURPLE	22
26.	TOO WEAK TO FIGHT	CLARENCE CARTER	28
27.	LO MUCHO QUE TE QUIERO	RENE / RENE	26
28.	HOOKED ON A FEELING	B.J. THOMAS	24
29.	A RAY OF HOPE	THE RASCALS	25
30.	IF I CAN DREAM	ELVIS PRESLEY	29

1969

WCFL
1000 WCFL-AM SOUND 10 SURVEY-CHICAGO

WEEK ENDING JANUARY 1, 1969
THIS IS THE YEAR END CHART FOR 1968 MONTH BY NUMBER

CHART #31

1.	LOVE IS BLUE	PAUL MAURIAT AND HIS ORCHESTRA	2
2.	THIS GUY'S IN LOVE WITH YOU	HERB ALPERT	6
3.	HONEY	BOBBY GOLDSBORO	4
4.	HEY JUDE / REVOLUTION	THE BEATLES	9
5.	HARPER VALLEY P.T.A.	JEANIE C. RILEY	9
6.	THOSE WERE THE DAYS	MARY HOPKIN	10
7.	I HEARD IT THROUGH THE GRAPEVINE	MARVIN GAYE	12
8.	FIRE	CRAZY WORLD OF ARTHUR BROWN	9
9.	JUMPIN' JACK FLASH	THE ROLLING STONES	7
10.	ABRAHAM, MARTIN & JOHN	DION	12
11.	YUMMY, YUMMY, YUMMY	OHIO EXPRESS	5
12.	MONY, MONY	TOMMY JAMES AND THE SHONDELLS	5
13.	VALERIE / TAPIOCA TUNDRA	THE MONKEES	3
14.	LOVE CHILD	DIANA ROSS AND THE SUPREMES	11
15.	MRS. ROBINSON	SIMON AND GARFUNKEL	5
16.	LIGHT MY FIRE	JOSE FELICIANO	8
17.	SPOOKY	CLASSICS IV	2
18.	(Sittin" On) THE DOCK OF THE BAY	OTIS REDDING	3
19.	I WONDER WHAT SHE'S DOING TONIGHT	TOMMY BOYCE AND BOBBY HART	2
20.	PEOPLE GOT TO BE FREE	THE RASCALS	8
21.	YOUNG GIRL	GARY PUCKETT & THE UNION GAP	4
22.	THE GOOD, THE BAD & THE UGLY	HUGO MONTENGRO	5
23.	GREEN TAMBOURINE	THE LEMON PIPERS	1
24.	GRAZING IN THE GRASS	HUGH MASEKELA	7
25.	HELLO, I LOVE YOU (Won't You Tell Me Your Name)	THE DOORS	8
26.	LITTLE GREEN APPLES	O.C. SMITH	10
27.	BORN TO BE WILD	STEPPENWOLF	8
28.	LADY WILLPOWER	GARY PUCKETT & THE UNION GAP	7
29.	LADY MADONNA / THE INNER LIGHT	THE BEATLES	4
30.	SIMON SAYS	1910 FRUITGUM CO.	3

WCFL
1000 WCFL-AM SOUND 10 SURVEY-CHICAGO

WEEK ENDING JANUARY 8, 1969

1.	I HEARD IT THROUGH THE GRAPEVINE	MARVIN GAYE	1
2.	I'M GONNA MAKE YOU LOVE ME	DIANA ROSS AND THE SUPREMES / THE TEMPTATIONS	4
3.			
4.	STORMY	CLASSICS IV	6
5.	I LOVE HOW YOU LOVE ME	BOBBY VINTON	3
6.	CLOUD NINE	THE TEMPTATIONS	7
7.	THING'S I'D LIKE TO SAY	THE NEW COLONY SIX	5
8.	SOULFUL STRUT	YOUNG HOLT UNLIMITED	14
9.	CRIMSON AND CLOVER	TOMMY JAMES & THE SHONDELLS	16
10.	HANG'EM HIGH	BOOKER T. & THE MG'S	15
11.	SON-OF-A PREACHER MAN	DUSTY SPRINGFIELD	12
12.			
13.	HOOKED ON A FEELING	B.J. THOMAS	28
14.	IF I CAN DREAM	ELVIS PRESLEY	30
15.			
16.	BELLA LINDA	THE GRASS ROOTS	22
17.	WORST THAT COULD HAPPEN	BROOKLYN BRIDGE	--
18.	LO MUCHO QUE TE QUIERO	RENE / RENE	27
19.	TOO WEAK TO FIGHT	CLARENCE CARTER	26
20.	TOUCH ME	THE DOORS	--

CHART 32

HIT BOUND

21.	CONDITION RED	THE GOODEES	--
22.	EVERYDAY PEOPLE	SLY AND THE FAMILY STONE	--
23.	GOODNIGHT MY LOVE	PAUL ANKA	--
24.	HEY JUDE	WILSON PICKETT	--
25.	I STARTED A JOKE	THE BEE GEES	--
26.			
27.	PAPA'S GOT A BRAND NEW BAG	OTIS REDDING	--
28.			
29.	BABY, LET'S WAIT	THE ROYAL GUARDSMEN	--
30.	RAMBLIN' GAMBLIN' MAN	BOB SEGER SYSTEM	

BECAUSE OF THE Y EAR-END, THIS WEEK NEVER APPEARED IN THE STORES. HERE ARE THE HITS THAT FELL OFF THE CHARTS BETWEEN 12/26 & 1/15.

2.	FOR ONCE IN MY LIFE	STEVIE WONDER
8.	ABRAHAM, MARTIN & JOHN	DION
9.	CINNAMON	DEREK
10.	LOVE CHILD	DIANA ROSS AND THE SUPREMES
11.	WICHITA LINEMAN	GLEN CAMPBELL
13.	BOTH SIDES NOW	JUDY COLLINS
17.	WHO'S MAKIN' LOVE	JOHNNY TAYLOR
18.	SHAKE	THE SHADOWS OF KNIGHT
19.	I CAN'T TURN YOU LOOSE	THE CHAMBERS BROTHERS
20.	TILL	THE VOGUES
21.	SCARBOROUGH FAIR	SERGIO MENDEZ
24.	SEE SAW	ARETHA FRANKLIN
25.	KENTUCKY WOMAN	DEEP PURPLE
29.	A RAY OF HOPE	

WCFL
1000 WCFL-AM SOUND 10 SURVEY-CHICAGO

WEEK ENDING JANUARY 15, 1969

1.	I'M GONNA MAKE YOU LOVE ME	DIANA ROSS AND THE SUPREMES / THE TEMPTATIONS	2
2.	I HEARD IT THROUGH THE GRAPEVINE	MARVIN GAYE	1
3.	CRIMSON AND CLOVER	TOMMY JAMES & THE SHONDELL	9
*4.	SOULFUL STRUT	YOUNG HOLT UNLIMITED	8
*5.	TOUCH ME	THE DOORS	20
*6.	HANG 'EM HIGH	BOOKER T. & THE MG'S	10
7.	I LOVE HOW YOU LOVE ME	BOBBY VINTON	5
8.	CLOUD NINE	THE TEMPTATIONS	6
9.	THING'S I'D LIKE TO SAY	THE NEW COLONY SIX	7
*10.	SON-OF-A PREACHERS MAN	DUSTY SPRINGFIELD	11
11.	HOOKED ON A FEELING	B.J. THOMAS	13
*12.	GOING UP THE COUNTRY	CANNED HEAT	--
13.	IF I CAN DREAM	ELVIS PRESLEY	14
14.	WORST THAT CAN HAPPEN	BROOKLYN BRIDGE	17
15.	BELLA LINDA	THE GRASS ROOTS	16
*16.	EVERYDAY PEOPLE	SLY AND THE FAMILY STONE	22
17.	STORMY	CLASSICS IV	4
18.	LO MUCHO QUE TE QUIERO	RENE / RENE	18
19.	CONDITION RED	THE GOODESS	21
20.	TOO WEAK TO FIGHT	CLARENCE CARTER	19

CHART #33

HIT BOUND

21.	GOODNIGHT MY LOVE	PAUL ANKA	23
22.	I STARTED A JOKE	THE BEE GEES	25
23.	HEY JUDE	WILSON PICKETT	24
*24.	BABY, LET'S WAIT	THE ROYAL GUARDSMEN	29
*25.	CAN I CHANGE MY MIND	TYRONE DAVIS	--
26.	YOU SHOWED ME	THE TURTLES	--
27.	PAPA'S GOT A BRAND NEW BAG	OTIS REDDING	27
28.	CALIFORNIA SOUL	THE 5TH DIMENSION	--
29.	RAMBLIN' GAMBLIN' MAN	BOB SEGER SYSTEM	30
30.	BABY, BABY DON'T CRY	SMOKEY ROBINSON AND THE MIRACLES	--

WCFL

1000 WCFL-AM SOUND 10 SURVEY-CHICAGO

WEEK ENDING JANUARY 22, 1969

1.	I'M GONNA MAKE YOU LOVE ME	DIANA ROSS AND THE SUPREMES / THE TEMPTATIONS	1
*2.	TOUCH ME	THE DOORS	5
*3.	SOULFUL STRUT	YOUNG HOLT UNLIMITED	4
*4.	EVERYDAY PEOPLE	SLY AND THE FAMILY STONE	16
5.	CRIMSON AND CLOVER	TOMMY JAMES & THE SHONDELLS	3
*6.	WORST THAT COULD HAPPEN	BROOKLYN BRIDGE	14
*7.	HOOKED ON A FEELING	B.J. THOMAS	11
*8.	I HEARD IT THROUGH THE GRAPEVINE	MARVIN GAYE	2
*9.	SON-OF-A PREACHER MAN	DUSTY SPRINGFIELD	10
*10.	GOING UP THE COUNTRY	CANNED HEAT	12
11.	HANG 'EM HIGH	BOOKER T. & THE MG'S	6
12.	IF I CAN DREAM	ELVIS PRESLEY	13
13.	CLOUD NINE	THE TEMPTATIONS	8
*14.	CONDITION RED	THE GOODEES	19
*15.	I STARTED A JOKE	THE BEE GEES	22
*16.	YOU SHOWED ME	THE TURTLES	26
17.	GOODNIGHT MY LOVE	PAUL ANKA	21
*18.	BABY, LET'S WAIT	THE ROYAL GUARDSMEN	24
19.	TOO WEAK TO FIGHT	CLARENCE CARTER	20
*20.	CAN I CHANGE MY MIND	TYRONE DAVIS	25

HIT BOUND

*21.	CALIFORNIA SOUL	THE 5TH DIMENSION	28
*22.	RAMBLIN' GAMBLIN' MAN	BOB SEGER SYSTEM	29
*23.	THERE'S GONNA BE A SHOWDOWN	ARCHIE BELL AND THE DRELLS	--
*24.	BABY, BABY DON'T CRY	SMOKEY ROBINSON AND THE MIRACLES	30
*25.	THIS MAGIC MOMENT	JAY AND THE AMERICANS	--
26.	OB-LA-DI, OB-LA-DA	ARTHUR CONLEY	--
27.	SOUL SISTER, BROWN SUGAR	SAM AND DAVE	--
28.	I GOT A LINE ON YOU	SPIRIT	--
29.	I'M LIVING IN SHAME	DIANA ROSS AND THE SUPREMES	--
30.	BUILD ME UP BUTTERCUP	THE FOUNDATIONS	--

WCFL
1000 WCFL-AM SOUND 10 SURVEY-CHICAGO

WEEK ENDING JANUARY 29, 1969

1.	EVERYDAY PEOPLE	SLY AND THE FAMILY STONE	4
2.	TOUCH ME	THE DOORS	2
3.	I'M GONNA MAKE YOU LOVE ME	DIANA ROSS AND THE SUPREMES / THE TEMPTATIONS	1
*4.	WORST THAT CAN HAPPEN	BROOKLYN BRIDGE	6
5.	CRIMSON AND CLOVER	TOMMY JAMES & THE SHONDELLS	5
*6.	HOOKED ON A FEELING	B.J. THOMAS	7
*7.	GOING UP THE COUNTRY	CANNED HEAT	10
8.	SOULFUL STRUT	YOUNG HOLT UNLIMITED	3
9.	I HEARD IT THROUGH THE GRAPEVINE	MARVIN GAYE	8
*10.	IF I CAN DREAM	ELVIS PRESLEY	12
*11.	YOU SHOWED ME	THE TURTLES	16
12.	HANG 'EM HIGH	BOOKER T. & THE MG'S	11
13.	I STARTED A JOKE	THE BEE GEES	15
14.	SON-OF-A PREACHER MAN	DUSTY SPRINGFIELD	9
15.	BABY, LET'S WAIT	THE ROYAL GUARDSMEN	18
*16.	RAMBLIN' GAMBLIN' MAN	BOB SEGER SYSTEM	22
17.	CAN I CHANGE MY MIND	TYRONE DAVIS	20
*18.	THIS MAGIC MOMENT	JAY AND THE AMERICANS	25
19.	CLOUD NINE	THE TEMPTATIONS	13
20.	CALIFORNIA SOUL	THE 5TH DIMENSION	21

HIT BOUND

21.	BABY, BABY DON'T CRY	SMOKEY ROBINSON AND THE MIRACLES	24
22.	THERE'S GONNA BE A SHOWDOWN	ARCHIE BELL AND THE DRELLS	23
*23.	I'M LIVING IN SHAME	DIANA ROSS AND THE SUPREMES	29
24.	I GOT A LINE ON YOU	SPIRIT	28
25.	OB-LA-DI, OB-LA-DA	ARTHUR CONLEY	26
26.	BUILD ME UP BUTTERCUP	THE FOUNDATIONS	30
27.	SHE'S NOT THERE	THE ROAD	--
*28.	GAMES PEOPLE PLAY	JOE SOUTH	--
29.	I'VE GOTTA BE ME	SAMMY DAVID JR.	--
30.	NOTHIN' BUT A HEARTACHE	THE FLIRTATIONS	--

WCFL

1000 WCFL-AM SOUND 10 SURVEY-CHICAGO

WEEK ENDING FEBRUARY 5, 1969

1.	EVERYDAY PEOPLE	SLY AND THE FAMILY STONE	1
2.	TOUCH ME	THE DOORS	2
*3.	WORST THAT COULD HAPPEN	BROOKLYN BRIDGE	4
4.	YOU SHOWED ME	THE TURTLES	11
*5.	I STARTED A JOKE	THE BEE GEES	13
6.	I'M GONNA MAKE YOU LOVE ME	DIANA ROSS AND THE SUPREMES / THE TEMPTATIONS	3
7.	CRIMSON AND CLOVER	TOMMY JAMES & THE SHONDELLS	5
*8.	THIS MAGIC MOMENT	JAY AND THE AMERICANS	18
*9.	RAMBLIN' GAMBLIN' MAN	BOB SEGER SYSTEM	16
10.	HOOKED ON A FEELING	B.J. THOMAS	6
*11.	I GOT A LINE ON YOU	SPIRIT	24
12.	IF I CAN DREAM	ELVIS PRESLEY	10
*13.	BUILD ME UP BUTTERCUP	THE FOUNDATIONS	26
*14.	THERE'S GONNA BE A SHOWDOWN	ARCHIE BELL AND THE DRELLS	22
*15.	I'M LIVING IN SHAME	DIANA ROSS AND THE SUPREMES	23
16.	CAN I CHANGE YOUR MIND	TYRONE DAVIS	17
17.	BABY, BABY DON'T CRY	SMOKEY ROBINSON AND THE MIRACLES	21
18.	BABY, LET'S WAIT	THE ROYAL GUARDSMEN	15
19.	CALIFORNIA SOUL	THE 5TH DIMENSION	20
20.	SOULFUL STRUT	YOUNG HOLT UNLIMITED	8

HIT BOUND

*21.	I'VE GOTTA BE ME	SAMMY DAVIS JR.	29
*22.	SHE'S NOT THERE	THE ROAD	27
*23.	GAMES PEOPLE PLAY	JOE SOUTH	28
*24.	BUT YOU KNOW I LOVE YOU	THE FIRST EDITION	--
*25.	HELLO, IT'S ME	NAZZ	--
26.	INDIAN GIVER	1910 FRUITGUM CO.	--
27.	NOTHIN' BUT A HEARTACHE	THE FLIRTATIONS	30
28.	TIME OF THE SEASON	THE ZOMBIES	--
29.	SOPHISTICATED CISSY	THE METERS	--
30.	PROUD MARY	CREEDENCE CLEARWATER REVIVAL	--

WCFL
1000 WCFL-AM SOUND 10 SURVEY-CHICAGO

WEEK ENDING FEBRUARY 12, 1969

1.	TOUCH ME	THE DOORS	2
*2.	WORST THAT COULD HAPPEN	BROOKLYN BRIDGE	3
*3.	YOU SHOWED ME	THE TURTLES	4
*4.	RAMBLIN' GAMBLIN' MAN	BOB SEGER SYSTEM	9
5.	EVERYDAY PEOPLE	SLY AND THE FAMILY STONE	1
*6.	BUILD ME UP BUTTERCUP	THE FOUNDATIONS	13
*7.	THIS MAGIC MOMENT	JAY AND THE AMERICAN	8
*8.	BABY, BABY DON'T CRY	SMOKEY ROBINSON AND THE MIRACLES	17
*9.	I GOT A LINE ON YOU	SPIRIT	11
10.	CRIMSON AND CLOVER	TOMMY JAMES & THE SHONDELL	7
11.	I STARTED A JOKE	THE BEE GEES	5
12.	THERE'S GONNA BE A SHOWDOWN	ARCHIE BELL AND THE DRELL	14
13.	I'M LIVING IN SHAME	DIANA ROSS AND THE SUPREMES	15
14.	HOOKED ON A FEELING	B.J. THOMAS	10
15.	IF I CAN DREAM	ELVIS PRESLEY	12
*16.	I'VE GOTTA BE ME	SAMMY DAVIS JR.	21
17.	BABY, LET'S WAIT	THE ROYAL GUARDSMEN	18
*18.	GAMES PEOPLE PLAY	JOE SOUTH	23
19.	I'M GONNA MAKE YOU LOVE ME	THE TEMPTATIONS	6
20.	BUT YOU KNOW I LOVE YOU	THE FIRST EDITION	24

HIT BOUND

*21.	INDIAN GIVER	1910 GRUITGUM CO.	26
*22.	TIME OF THE SEASON	THE ZOMBIES	28
*23.	PROUD MARY	CREEDENCE CLEARWATER REVIVAL	30
24.	HELLO, IT'S ME	NAZZ	25
*25.	CROSSROADS	CREAM	--
26.	NOTHIN' BUT A HEARTACHE	THE FLIRTATIONS	27
27.	SOUL SHAKE	PEGGY SCOTT / JO JO BENSON	--
28.	SOPHISTICATED CISSY	THE METERS	29
29.	DIZZY	TOMMY ROE	--
30.	WTTCHI TAI TO	JIM PEPPER	--

WCFL
1000 WCFL-AM SOUND 10 SURVEY-CHICAGO

WEEK ENDING FEBRUARY 19, 1969

*1.	RAMBLIN' GAMBLIN' MAN	BOB SEGER SYSTEM	4
2.	TOUCH ME	THE DOORS	1
*3.	BABY, BABY DON'T CRY	SMOKEY ROBINSON AND THE MIRACLES	8
*4.	BUILD ME UP BUTTERCUP	THE FOUNDATIONS	6
5.	WORST THAT CAN HAPPEN	BROOKLYN BRIDGE	2
*6.	THIS MAGIC MOMENT	JAY AND THE AMERICANS	7
7.	EVERYDAY PEOPLE	SLY AND THE FAMILY STONE	5
8.	YOU SHOWED ME	THE TURTLES	3
*9.	I'VE GOTTA BE ME	SAMMY DAVIS JR.	16
10.	I GOT A LINE ON YOU	SPIRIT	9
*11.	GAMES PEOPLE PLAY	JOE SOUTH	18
12.	CRIMSON AND CLOVER	TOMMY JAMES & THE SHONDELLS	10
13.	I'M LIVING IN SHAME	DIANA ROSS AND THE SUPREMES	13
*14.	INDIAN GIVER	1910 FRUITGUM CO.	21
*15.	TIME OF THE SEASON	THE ZOMBIES	22
16.	I STARTED A JOKE	THE BEE GEES	11
*17.	PROUD MARY	CREEDENCE CLEARWATER REVIVAL	23
*18.	DIZZY	TOMMY ROE	29
19.	BUT YOU KNOW I LOVE YOU	THE FIRST EDITION	20
20.	THERE'S GONNA BE A SHOWDOWN	ARCHIE BELL AND THE DRELLS	12

HIT BOUND

21.	CROSSROADS	CREAM	25
*22.	GIVE IT UP OR TURNIT LOOSE	JAMES BROWN	--
*23.	THIS GIRL'S IN LOVE WITH YOU	DIONNE WARWICK	--
24.	NOTHIN' BUT A HEARTACHE	THE FLIRTATIONS	26
25.	SOUL SHAKE	PEGGY SCOTT / JO JO BENSON	27
*26.	LOVIN' THINGS	THE GRASS ROOTS	--
27.	SOPHISTICATED CISSY	THE METERS	28
28.	HEAVEN	THE RASCALS	--
29.	MAY I	BILL DEAL AND THE RHONDELLS	--
30.	LONG GREEN	THE FIREBALLS	--

WCFL
1000 WCFL-AM SOUND 10 SURVEY-CHICAGO

WEEK ENDING FEBRUARY 26, 1969

*1.	RAMBLIN' GAMBLIN' MAN	BOB SEGER SYSTEM	1
*2.	BUILD ME UP BUTTERCUP	THE FOUNDATIONS	4
*3.	THIS MAGIC MOMENT	JAY AND THE AMERICANS	6
4.	BABY, BABY DON'T CRY	SMOKEY ROBINSON AND THE MIRACLES	3
*5.	I'VE GOTTA BE ME	SAMMY DAVIS JR.	9
6.	TOUCH ME	THE DOORS	2
*7.	I GOT A LINE ON YOU	SPIRIT	10
8.	EVERYDAY PEOPLE	SLY AND THE FAMILY STONE	7
9.	YOU SHOWED ME	THE TURTLES	8
*10.	GAMES PEOPLE PLAY	JOE SOUTH	11
11.	INDIAN GIVER	1910 FRUITGUM CO.	14
12.	TIME OF THE SEASON	THE ZOMBIES	15
13.	I'M LIVING IN SHAME	DIANA ROSS AND THE SUPREMES	13
14.	DIZZY	TOMMY ROE	18
15.	WORST THAT COULD HAPPEN	BROOKLYN BRIDGE	5
16.	PROUD MARY	CREEDENCE CLEARWATER REVIVAL	17
17.	CRIMSON AND CLOVER	TOMMY JAMES & THE SHONDELLS	12
18.	BUT YOU KNOW I LOVE YOU	FIRST EDITION	19
19.	CROSSROADS	CREAM	21
20.	THIS GIRL'S IN LOVE WITH YOU	DIONNE WARWICK	23

HIT BOUND

21.	GIVE IT UP OR TURNIT LOOSE	JAMES BROWN	22
*22.	MAY I	BILL DEAL AND THE RHONDELLS	29
*23.	HOT SMOKE & SASAFRASS	BUBBLE PUPPY	--
24.	LOVIN' THINGS	THE GRASS ROOTS	26
*25.	TRACES	CLASSICS IV	--
26.	HEAVEN	THE RASCALS	28
27.	LONG GREEN	THE FIREBALLS	30
28.	MY WHOLE WORLD ENDED (The Moment You Left Me)	DAVID RUFFIN	--
29.	GOOD LOVIN' AIN'T EASY TO COME BY	MARVIN GAYE / TAMMI TERRELL	--
30.	FIRST TRAIN TO CALIFORNIA	THE CRYAN' SHAMES	--

WCFL

1000 WCFL-AM SOUND 10 SURVEY-CHICAGO

WEEK ENDING MARCH 5, 1969

*1.	RAMBLIN' GAMBLIN' MAN	BOB SEGER SYSTEM	1
*2.	DIZZY	TOMMY ROE	14
*3.	I'VE GOTTA BE ME	SAMMY DAVIS JR.	5
4.	BABY, BABY DON'T CRY	SMOKEY ROBINSON AND THE MIRACLES	4
5.	BUILD ME UP BUTTERCUP	THE FOUNDATIONS	2
*6.	INDIAN GIVER	1910 FRUITGUM CO.	11
*7.	TIME OF THE SEASON	THE ZOMBIES	12
8.	I GOT A LINE ON YOU	SPIRIT	7
*9.	PROUD MARY	CREEDENCE CLEARWATER REVIVAL	16
10.	THIS MAGIC MOMENT	JAY AND THE AMERICANS	3
11.	EVERYDAY PEOPLE	SLY AND THE FAMILY STONE	8
12.	YOU SHOWED ME	THE TURTLES	9
*13.	MY WHOLE WORLD ENDED (The Moment You Left Me)	DAVID RUFFIN	28
14.	GAMES PEOPLE PLAY	JOE SOUTH	10
15.	I'M LIVIN' IN SHAME	DIANA ROSS AND THE SUPREMES	13
16.	THIS GIRLS' IN LOVE WITH YOU	DIONNE WARWICK	20
17.	TOUCH ME	THE DOORS	6
18.	CROSSROADS	CREAM	19
19.	HOT SMOEK & SASAFRASS	BUBBLE PUPPY	23
20.	MAY I	BILL DEAL AND THE RHONDELLS	22

HIT BOUND

21.	TRACES	CLASSICS IV	25
22.	LOVIN' THINGS	THE GRASS ROOTS	24
23.	LONG GREEN	THE FIREBALLS	27
*24.	NO, NOT MUCH	THE VOGUES	--
*25.	NOTHIN' BUT A HEARTACHE	THE FLIRTATIONS	--
26.	FIRST TRAIN TO CALIFORNIA	THE CRYAN' SHAMES	30
27.	DO YOUR THING	WATTS 103RD ST. BAND	--
28.	MENDOCINO	SIR DOUGLAS QUINTET	--
29.	MR. SUN, MR. MOON	PAUL REVERE AND THE RAIDERS	--
*30.	L.U.V.	TOMMY BOYCE AND BOBBY HART	--

WCFL
1000 WCFL-AM SOUND 10 SURVEY-CHICAGO

WEEK ENDING MARCH 12, 1969

*1.	DIZZY	TOMMY ROE	2
2.	RAMBLIN' GAMBLIN' MAN	BOB SEGER SYSTEM	1
*3.	TIME OF THE SEASON	THE ZOMBIES	7
*4.	INDIAN GIVER	1910 FRUITGUM CO.	6
5.	I'VE GOTTA BE ME	SAMMY DAVIS JR.	3
*6.	PROUD MARY	CREEDENCE CLEARWATER REVIVAL	9
7.	BABY, BABY DON'T CRY	SMOKEY ROBINSON AND THE MIRACLES	4
*8.	THIS GIRL'S IN LOVE WITH YOU	DIONNE WARWICK	16
*9.	TRACES	CLASSICS IV	21
10.	BUILD ME UP BUTTERCUP	THE FOUNDATIONS	5
*11.	NO, NOT MUCH	THE VOGUES	24
12.	MY WHOLE WORLD ENDED (The Moment You Left Me)	DAVID RUFFIN	13
*13.	HOT SMOKE & SASAFRASS	BUBBLE PUPPY	19
*14.	MAY I	BILL DEAL AND THE RHONDELLS	20
15.	EVERYDAY PEOPLE	SLY AND THE FAMILY STONE	11
16.	YOU SHOWED ME	THE TURTLES	12
17.	I GOT A LINE ON YOU	SPIRIT	8
18.	CROSSROADS	CREAM	18
*19.	DO YOUR THING	WATTS 103RD ST. BAND	27
20.	LONG GREEN	THE FIREBALLS	23

HIT BOUND

*21.	RUNAWAY CHILD, RUNNING WILD	THE TEMPTATIONS	--
*22.	MENDOCINO	SIR DOUGLAS QUINTET	28
*23.	ROCK ME	STEPPENWOLF	--
*24.	MR. SUN, MR. MOON	PAUL REVERE AND THE RAIDERS	29
*25.	IT'S YOUR THING	THE ISLEY BROTHERS	--
*26.	ONLY THE STRONG SURVIVE	JERRY BUTLER	--
27.	L.U.V.	TOMMY BOYCE AND BOBBY HART	30
28.	YOU GAVE ME A MOUNTAIN	FRANKIE LANE	--
29.	BACK DOOR MAN	DEREK	--
30.	THE LETTER	THE ARBORS	--

WCFL
1000 WCFL-AM SOUND 10 SURVEY-CHICAGO

WEEK ENDING MARCH 19, 1969

*1.	TIME OF THE SEASON	THE ZOMBIES	3
2.	DIZZY	TOMMY ROE	1
*3.	INDIAN GIVER	1910 FRUITGUM CO.	4
*4.	PROUD MARY	CREEDENCE CLEARWATER REVIVAL	6
5.	RAMBLIN' GAMBLIN' MAN	BOB SEGER SYSTEM	2
*6.	TRACES	CLASSICS IV	9
*7.	THIS GIRL'S IN LOVE WITH YOU	DIONNE WARWICK	8
8.	I'VE GOTTA BE ME	SAMMY DAVIS JR.	5
*9.	MY WHOLE WORLD ENDED (The Moment You Left Me)	DAVID RUFFIN	12
*10.	HOT SMOKE & SASAFRASS	BUBBLE PUPPY	13
11.	BABY, BABY DON'T CRY	SMOKEY ROBINSON AND THE MIRACLES	7
12.	NO, NOT MUCH	THE VOGUES	11
13.	MAY I	BILL DEAL AND THE RHONDELLS	14
*14.	DO YOUR THING	WATTS 103RD ST. BAND	19
*15.	IT'S YOUR THING	THE ISLEY BROTHERS	25
*16.	MENDOCINO	SIR DOUGLAS QUINTET	22
17.	RUNAWAY CHILD, RUNNING WILD	THE TEMPTATIONS	21
18.	LONG GREEN	THE FIREBALLS	20
*19.	HAIR	THE COWSILLS	--
20.	ROCK ME	STEPPENWOLF	23

HIT BOUND

21.	MR. SUN, MR. MOON	PAUL REVERE AND THE RAIDERS	24
*22.	AQUARIUS / LET THE SUNSHINE IN	THE 5TH DIMENSION	--
*23.	TWENTY-FIVE MILES	EDWIN STARR	--
24.	ONLY THE STRONG SURVIVE	JERRY BUTLER	26
25.	L.U.V.	TOMMY BOYCE AND BOBBY HART	27
*26.	YOU'VE MADE ME SO VERY HAPPY	BLOOD, SWEAT & TEARS	--
27.	YOU GAVE ME A MOUNTAIN	FRANKIE LANE	28
28.	BACK DOOR MAN	DEREK	29
29.	THE LETTER	THE ARBORS	30
30.	GALVESTON	GLEN CAMPBELL	--

WCFL
1000 WCFL-AM SOUND 10 SURVEY-CHICAGO

WEEK ENDING MARCH 26, 1969

*1.	TIME OF THE SEASON	THE ZOMBIES	1
*2.	INDIAN GIVER	1910 FRUITGUM CO.	3
*3.	TRACES	CLASSICS IV	6
4.	DIZZY	TOMMY ROE	2
*5.	HAIR	THE COWSILLS	19
*6.	IT'S YOUR THING	THE ISLEY BROTHERS	15
*7.	DO YOUR THING	WATTS 103RD ST. BAND	14
*8.	HOT SMOKE & SASAFRASS	BUBBLE PUPPY	10
9.	THIS GIRL'S IN LOVE WITH YOU	DIONNE WARWICK	7
*10.	RUNAWAY CHILD, RUNNING WILD	THE TEMPTATIONS	17
11.	MAY I	BILL DEAL AND THE RHONDELLS	13
12.	MY WHOLE WORLD ENDED (The Moment You Left Me)	DAVID RUFFIN	9
*13.	AQUARIUS / LET THE SUNSHINE IN	THE 5TH DIMENSION	22
14.	PROUD MARY	CREEDENCE CLEARWATER REVIVAL	4
*15.	ONLY THE STRONG SURVIVE	JERRY BUTLER	24
*16.	YOU'VE MADE ME SO VERY HAPPY	BLOOD, SWEAT & TEARS	26
17.	MR. SUN, MR. MOON	PAUL REVERE AND THE RAIDERS	21
18.	MENDOCINO	SIR DOUGLAS QUINTET	16
19.	ROCK ME	STEPPENWOLF	20
*20.	DON'T GIVE INTO HIM	GARY PUCKETT & THE UNION GAP	--

HIT BOUND

21.	TWENTY-FIVE MILES	EDWIN STARR	23
*22.	GALVESTON	GLEN CAMPBELL	30
*23.	BROTHER LOVES TRAVELING SALVATION SHOW	NEIL DIAMOND	--
24.	L.U.V.	TOMMY BOYCE AND BOBBY HART	25
25.	THE LETTER	THE ARBORS	29
26.	YOU GAVE ME A MOUNTAIN	FRANKIE LANE	27
27.	I'LL TRY SOMETHING NEW	DIANA ROSS AND THE SUPREMES / THE TEMPTATIONS	--
28.	I CAN HEAR MUSIC	THE BEACH BOYS	--
29.	GIMME GIMME GOOD LOVIN'	CRAZY ELEPHANT	--
30.	TIME IS TIGHT	BOOKER T. & THE MG'S	--

WCFL

1000 WCFL-AM SOUND 10 SURVEY-CHICAGO

WEEK ENDING APRIL 2, 1969

*1.	HAIR	THE COWSILLS	5
*2.	AQUARIUS / LET THE SUNSHINE IN	THE 5TH DIMENSION	13
*3.	ITS YOUR THING	THE ISLEY BROTHERS	6
4.	TRACES	CLASSICS IV	3
5.	DIZZY	TOMMY ROE	4
*6.	DO YOUR THING	WATTS 103 RD ST. BAND	7
7.	TIME OF THE SEASON	THE ZOMBIES	1
*8.	YOU'VE MADE ME SO VERY HAPPY	BLOOD, SWEAT & TEARS	16
9.	INDIAN GIVER	1910 FRUITGUM CO.	2
10.	HOT SMOKE & SASAFRASS	BUBBLE PUPPY	8
*11.	ROCK ME	STEPPENWOLF	19
12.	RUNAWAY CHILD, RUNNING WILD	THE TEMPTATIONS	10
13.	ONLY THE STRONG SURVIVE	JERRY BUTLER	15
*14.	TWENTY-FIVE MILES	EDWIN STARR	21
*15.	GALVESTON	GLEN CAMPBELL	22
16.	DON'T GIVE INTO HIM	GARY PUCKETT & THE UNION GAP	20
17.	PROUD MARY	CREEDENCE CLEARWATER REVIVAL	14
18.	MR. SUN, MR. MOON	PAUL REVERE AND THE RAIDERS	17
*19.	GIMME GIMME GOOD LOVIN'	CRAZY ELEPHANT	29
*20.	WILL YOU BE STAYING AFTER SUNDAY	PEPPERMINT RAINBOW	--

HIT BOUND

*21.	TIME IS TIGHT	BOOKER T. & THE MG'S	30
22.	THE LETTER	THE ARBORS	25
23.	L.U.V.	TOMMY BOYCE AND BOBBY HART	24
24.	I CAN HEAR MUSIC	THE BEACH BOYS	28
25.	BROTHER LOVES TRAVELING SALVATION	NEIL DIAMOND	23
26.	I'LL TRY ANYTHING	DIANA ROSS AND THE SUPREMES / THE TEMPTATIONS	27
27.	MERCY	OHIO EXPRESS	--
28.	MEMORIES	ELVIS PRESLEY	--
29.	THE BOXER	SIMON AND GARFUNKEL	--
30.	PINBALL WIZARD	THE WHO	--

WCFL
1000 WCFL-AM SOUND 10 SURVEY-CHICAGO

WEEK ENDING APRIL 9, 1969

*1.	HAIR	THE COWSILLS	1
*2.	AQUARIUS / LET THE SUNSHINE IN	THE 5TH DIMENSION	2
*3.	YOU'VE MADE ME SO VERY HAPPY	BLOOD, SWEAT & TEARS	8
4.	IT'S YOUR THING	THE ISLEY BROTHERS	3
*5.	GALVESTON	GLEN CAMPBELL	15
*6.	ROCK ME	STEPPENWOLF	11
*7.	MR. SUN, MR. MOON	PAUL REVERE AND THE RAIDERS	18
8.	DIZZY	TOMMY ROE	5
*9.	I CAN HEAR MUSIC	THE BEACH BOYS	24
*10.	ONLY THE STRONG SURVIVE	JERRY BUTLER	13
*11.	DON'T GIVE INTO HIM	GARY PUCKETT & THE UNION GAP	16
12.	TWENTY-FIVE MILES	EDWIN STARR	14
13.	TRACES	CLASSICS IV	4
*14.	BROTHER LOVES TRAVELING SALVATION SHOW	NEIL DIAMOND	25
15.	TIME OF THE SEASON	THE ZOMBIES	7
16.	GIMME GIMME GOOD LOVIN'	CRAZY ELEPHANT	19
17.	DO YOUR THING	WATTS 103RD ST. BAND	6
18.	WILL YOU BE STAYING AFTER SUNDAY	PEPPERMINT RAINBOW	20
19.	TIME IS TIGHT	BOOKER T. & THE MG'S	21
20.	THE LETTER	THE ARBORS	22

HIT BOUND

*21.	MERCY	OHIO EXPRESS	27
*22.	THE BOXER	SIMON AND GARFUNKEL	29
*23.	MEMORIES	ELVIS PRESLEY	28
*24.	PINBALL WIZARD	THE WHO	30
*25.	SWEET CHERRY WINE	TOMMY JAMES & THE SHONDELLS	--
*26.	TO KNOW YOU IS TO LOVE YOU	BOBBY VINTON	--
27.	HAWAII FIVE-O	THE VENTURES	--
28.	THE CHOKIN' KIND	JOE SIMON	--
29.	THESE EYES	THE GUESS WHO	--
30.	GITARZAN	RAY STEVENS	--

WCFL

1000 WCFL-AM SOUND 10 SURVEY-CHICAGO

WEEK ENDING APRIL 16, 1969

*1.	YOU'VE MADE ME SO VERY HAPPY	BLOOD, SWEAT & TEARS	3
2.	HAIR	THE COWSILLS	1
3.	AQUARIUS / LET THE SUNSHINE IN	THE 5TH DIMENSION	2
*4.	GALVESTON	GLEN CAMPBELL	5
*5.	MR. SUN, MR. MOON	PAUL REVERE AND THE RAIDERS	7
6.	IT'S YOUR THING	THE ISLEY BROTHERS	4
*7.	I CAN HEAR MUSIC	THE BEACH BOYS	9
*8.	ONLY THE STRONG SURVIVE	JERRY BUTLER	10
9.	ROCK ME	STEPPENWOLF	6
*10.	DON'T GIVE INTO HIM	GARY PUCKETT & THE UNION GAP	11
*11.	GIMME GIMME GOOD LOVIN'	CRAZY ELEPHANT	16
12.	BROTHER LOVES TRAVELING SALVATION	NEIL DIAMOND	14
*13.	TIME IS TIGHT	BOOKER T. & THE MG'S	19
14.	DIZZY	TOMMY ROE	8
15.	WILL YOU BE STAYING AFTER SUNDAY	PEPPERMINT RAINBOW	18
*16.	THE BOXER	SIMON AND GARFUNKEL	22
17.	TWENTY-FIVE MILES	EDWIN STARR	12
18.	MERCY	OHIO EXPRESS	21
*19.	GITARZAN	RAY STEVENS	30
20.	THE LETTER	THE ARBORS	20

HIT BOUND

21.	PINBALL WIZARD	THE WHO	24
*22.	THE CHOKIN' KIND	JOE SIMON	28
23.	MEMORIES	ELVIS PRESLEY	23
24.	TO KNOW YOU IS TO LOVE YOU	BOBBY VINTON	26
25.	HAWAII FIVE-O	THE VENTURES	27
26.	THESE EYES	THE GUESS WHO	29
27.	LOVE (Can Make You Happy)	MERCY	--
28.	THE COMPOSER	DIANA ROSS AND THE SUPREMES	--
29.	IT'S ONLY LOVE	B.J. THOMAS	--
30.	I'VE BEEN HURT	BILL DEAL AND THE RHONDELLS	--

WCFL
1000 WCFL-AM SOUND 10 SURVEY-CHICAGO

WEEK ENDING APRIL 23, 1969

*1.	AQUARIUS / LET THE SUNSHINE IN	THE 5TH DIMENSION	3
2.	HAIR	THE COWSILLS	2
*3.	MR. SUN, MR. MOON	PAUL REVERE AND THE RAIDERS	5
4.	YOU'VE MADE ME SO VERY HAPPY	BLOOD, SWEAT & TEARS	1
*5.	DON'T GIVE INTO HIM	GARY PUCKETT & THE UNION GAP	10
6.	IT'S YOUR THING	THE ISLEY BROTHERS	6
7.	I CAN HEAR MUSIC	THE BEACH BOYS	7
*8.	GIMME GIMME GOOD LOVIN'	CRAZY ELEPHANT	11
*9.	THE BOXER	SIMON AND GARFUNKEL	16
*10.	GITARZAN	RAY STEVENS	19
11.	ONLY THE STRONG SURVIVE	JERRY BUTLER	8
12.	BROTHER LOVES TRAVELING SALVATION	NEIL DIAMOND	12
*13.	PINBALL WIZARD	THE WHO	21
14.	GALVESTON	GLEN CAMPBELL	4
15.	MERCY	OHIO EXPRESS	18
*16.	THE CHOKIN' KIND	JOE SIMON	22
17.	TIME IS TIGHT	BOOKER T. & THE MG'S	13
18.	WILL YOU BE STAY AFTER SUNDAY	PEPPERMINT RAINBOW	15
*19.	THESE EYES	THE GUESS WHO	26
*20.	LOVE (Can Make You Happy)	MERCY	27

HIT BOUND

21.	HAWAII FIVE-O	THE VENTURES	25
22.	MEMORIES	ELVIS PRESLEY	23
23.	TO KNOW YOU IS TO LOVE YOU	BOBBY VINTON	24
24.	THE COMPOSER	DIANA ROSS AND THE SUPREMES	28
*25.	I'VE BEEN HURT	BILL DEAL AND THE RHONDELLS	30
26.	IT'S ONLY LOVE	B.J. THOMAS	29
27.	OH, HAPPY DAY	EDWIN HAWKINS SINGERS	--
28.	SWEET CHERRY WINE	TOMMY JAMES & THE SHONDELLS	--
29.	ATLANTIS	DONOVAN	--
30.	SORRY SUZANNE	THE HOLLIES	--

WCFL

1000 WCFL-AM SOUND 10 SURVEY-CHICAGO

WEEK ENDING APRIL 30, 1969

*1.	GITARZAN	RAY STEVENS	10
*2.	THE BOXER	SIMON AND GARFUNKEL	9
3.	AQUARIUS / LET THE SUNSHINE IN	THE 5TH DIMENSION	1
*4.	GIMME GIMME GOOD LOVIN'	CRAZY ELEPHANT	8
*5.	MERCY	OHIO EXPRESS	15
6.	HAIR	THE COWSILLS	2
*7.	PINBALL WIZARD	THE WHO	13
8.	IT'S YOUR THING	THE ISLEY BROTHERS	6
*9.	THESE EYES	THE GUESS WHO	19
10.	YOU'VE MADE ME SO VERY HAPPY	BLOOD, SWEAT & TEARS	4
11.	I CAN HEAR MUSIC	THE BEACH BOYS	7
12.	MR. SUN, MR. MOON	PAUL REVERE AND THE RAIDERS	3
13.	THE CHOKIN' KIND	JOE SIMON	16
*14.	LOVE (Can Make You Happy)	MERCY	20
15.	BROTHER LOVES TRAVELING SALVATION SHOW	NEIL DIAMOND	12
16.	DON'T GIVE INTO HIM	GARY PUCKETT & THE UNION GAP	5
17.	WILL YOU BE STAYING AFTER SUNDAY	PEPPERMINT RAINBOW	18
*18.	OH, HAPPY DAY	EDWIN HAWKINS SINGERS	27
*19.	THE COMPOSER	DIANA ROSS AND THE SUPREMES	24
*20.	I'VE BEEN HURT	BILL DEAL AND THE RHONDELLS	25

HIT BOUND

21.	HAWAII FIVE-O	THE VENTURES	21
22.	TO KNOW YOU IS TO LOVE YOU	BOBBY VINTON	23
23.	IT'S ONLY LOVE	B.J. THOMAS	26
24.	SWEET CHERRY WINE	TOMMY JAMES & THE SHONDELLS	28
*25.	SEATTLE	PERRY COMO	--
*26.	MORNING GIRL	NEON PHILHARMONIC	--
27.	ATLANTIS	DONOVAN	29
28.	SORRY SUZANNE	THE HOLLIES	30
29.	GET BACK / DON'T LET ME DOWN	THE BEATLES	--
30.	MEDICINE MAN (Part 1)	THE BUCHANAN BROTHERS	--

WCFL
1000 WCFL-AM SOUND 10 SURVEY-CHICAGO

WEEK ENDING MAY 7, 1969

*1.	GITARZAN	RAY STEVENS	1
2.	THE BOXER	SIMON AND GARFUNKEL	2
*3.	GIMME GIMME GOOD LOVIN'	CRAZY ELEPHANT	4
*4.	THESE EYES	THE GUESS WHO	9
5.	MERCY	OHIO EXPRESS	5
*6.	PINBALL WIZARD	THE WHO	7
7.	AQUARIUS / LET THE SUNSHINE IN	THE 5TH DIMENSION	3
*8.	LOVE (Can Make You Happy)	MERCY	14
9.	HAIR	THE COWSILLS	6
*10.	HAWAII FIVE-O	THE VENTURES	21
11.	I CAN HERE MUSIC	THE BEACH BOYS	11
12.	THE CHOKIN' KIND	JOE SIMON	13
13.	IT'S YOUR THING	THE ISLEY BROTHERS	8
14.	OH, HAPPY DAY	EDWIN HAWKINS SINGERS	18
15.	YOU'VE MADE ME SO VERY HAPPY	BLOOD, SWEAT & TEARS	10
16.	MR. SUN, MR. MOON	PAUL REVERE AND THE RAIDERS	12
17.	DON'T GIVE INTO HIM	GARY PUCKETT & THE UNION GAP	16
18.	THE COMPOSER	DIANA ROSS AND THE SUPREMES	19
19.	I'VE BEEN HURT	BILL DEAL & THE RHONDELLS	20
*20.	ATLANTIS	DONOVAN	27

HIT BOUND

21.	TO KNOW YOU IS TO LOVE YOU	BOBBY VINTON	22
*22.	GET BACK / DON'T LET ME DOWN	THE BEATLES	29
23.	SWEET CHERRY WINE	TOMMY JAMES & THE SHONDELLS	24
24.	IT'S ONLY LOVE	B.J. THOMAS	23
25.	SEATTLE	PERRY COMO	25
26.	MORNING GIRL	NEON PHILHARMONIC	26
27.	TOO BUSY THINKING ABOUT MY BABY	MARVIN GAYE	--
28.	SORRY SUZANNE	THE HOLLIES	28
29.	MEDICINE MAN (Part 1)	THE BUCHANAN BROTHERS	30
30.	GRAZING IN THE GRASS	THE FRIENDS OF DISTINCTION	--

WCFL
1000 WCFL-AM SOUND 10 SURVEY-CHICAGO

WEEK ENDING MAY 14, 1969

*1.	LOVE (Can Make You Happy)	MERCY	8
*2.	THESE EYES	THE GUESS WHO	4
*3.	GET BACK / DON'T LET ME DOWN	THE BEATLES	22
4.	THE BOXER	SIMON AND GARFUNKEL	2
*5.	PINBALL WIZARD	THE WHO	6
6.	GITARZAN	RAY STEVENS	1
7.	GIMME GIMME GOOD LOVIN'	CRAZY ELEPHANT	3
*8.	ATLANTIS	DONOVAN	20
*9.	MORNING GIRL	NEON PHILHARMONIC	26
10.	AQUARIUS / LET THE SUNSHINE IN	THE 5TH DIMENSION	7
11.	OH, HAPPY DAY	EDWIN HAWKINS SINGERS	14
12.	HAIR	THE COWSILLS	9
13.	MERCY	OHIO EXPRESS	5
14.	THE CHOKIN' KIND	JOE SIMON	12
15.	I'VE BEEN HURT	BILL DEAL & THE RHONDELLS	19
16.	HAWAII FIVE-O	THE VENTURES	10
*17.	TOO BUSY THINKING ABOUT MY BABY	MARVIN GAYE	27
18.	IT'S YOUR THING	THE ISLEY BROTHERS	13
*19.	I COULD NEVER LIE TO YOU	THE NEW COLONY SIX	--
*20.	MEDICINE MAN (Part 1)	THE BUCHANAN BROTHERS	29

HIT BOUND

21.	SWEET CHERRY WINE	TOMMY JAMES & THE SHONDELLS	23
22.	SEATTLE	PERRY COMO	25
*23.	GRAZING IN THE GRASS	THE FRIENDS OF DISTINCTION	30
*24.	LET ME	PAUL REVERE AND THE RAIDERS	--
*25.	SPECIAL DELIVERY	1910 FRUITGUM CO.	--
*26.	GOODBYE	MARY HOPKIN	--
27.	NOTHING BUT A HEARTACHE	THE FLIRTATIONS	--
28.	SORRY SUZANNE	THE HOLLIES	28
29.	DAY IS DONE	PETER PAUL AND MARY	--
30.	IN THE GHETTO	ELVIS PRESLEY	--

WCFL
1000 WCFL-AM SOUND 10 SURVEY-CHICAGO

WEEK ENDING MAY 21, 1969

*1.	GET BACK / DON'T LET ME DOWN	THE BEATLES	3
2.	LOVE (Can Make You Happy)	MERCY	1
3.	THESE EYES	THE GUESS WHO	2
*4.	OH, HAPPY DAY	EDWIN HAWKINS SINGERS	11
*5.	MORNING GIRL	NEON PHILHARMONIC	9
*6.	ATLANTIS	DONOVAN	8
7.	THE BOXER	SIMON AND GARFUNKEL	4
8.	PINBALL WIZARD	THE WHO	5
*9.	TOO BUSY THINKING ABOUT MY BABY	MARVIN GAYE	17
10.	GITARZAN	RAY STEVENS	6
11.	I'VE BEEN HURT	BILL DEAL & THE RHONDELLS	15
*12.	GRAZING IN THE GRASS	THE FRIENDS OF DISTINCTION	23
13.	MERCY	OHIO EXPRESS	13
*14.	MEDICINE MAN (Part 1)	THE BUCHANAN BROTHERS	20
15.	AQUARIUS / LET THE SUNSHINE IN	THE 5TH DIMENSION	10
16.	I COULD NEVER LIE TO YOU	THE NEW COLONY SIX	19
17.	HAIR	THE COWSILLS	12
18.	GIMME GIMME GOOD LOVIN'	CRAZY ELEPHANT	7
19.	THE CHOKIN' KIND	JOE SIMON	14
20.	LET ME	PAUL REVERE AND THE RAIDERS	24

HIT BOUND

21.	SEATTLE	PERRY COMO	22
*22.	IN THE GHETTO	ELVIS PRESLEY	30
23.	SPECIAL DELIVERY	1910 FRUITGUM CO.	25
24.	GOODBYE	MARY HOPKIN	26
25.	NOTHING BUT A HEARTACHE	THE FLIRTATIONS	27
26.	DAY IS DONE	PETER PAUL AND MARY	29
27.	SO I CAN LOVE YOU	THE EMOTIONS	--
28.	BAD MOON RISING / LODI	CREEDENCE CLEARWATER REVIVAL	--
29.	(Love Theme) FROM ROMEO & JULIET	HENRY MANCINI	--
30.	MORE TODAY THAN YESTERDAY	SPIRAL STAIRCASE	--

WCFL
1000 WCFL-AM SOUND 10 SURVEY-CHICAGO

WEEK ENDING MAY 28, 1969

*1.	GET BACK / DON'T LET ME DOWN	THE BEATLES	1
2.	LOVE (Can Make You Happy)	MERCY	2
3.	THESE EYES	THE GUESS WHO	3
*4.	MORNING GIRL	NEON PHILHARMONIC	5
5.	MEDICINE MAN (Part 1)	THE BUCHANAN BROTHERS	14
6.	OH, HAPPY DAY	EDWIN HAWKINS SINGER	4
*7.	BAD MOON RISING / LODI	CREEDENCE CLEARWATER REVIVAL	28
*8.	IN THE GHETTO	ELVIS PRESLEY	22
9.	TOO BUSY THINKING ABOUT MY BABY	MARVIN GAYE	9
*10.	LET ME	PAUL REVERE AND THE RAIDERS	20
11.	ATLANTIS	DONOVAN	6
12.	GRAZING IN THE GRASS	THE FRIENDS OF DISTINCTION	12
13.	I COULD NEVER LIE TO YOU	THE NEW COLONY SIX	16
*14.	SPECIAL DELIVERY	1910 FRUITGUM CO.	23
15.	THE BOXER	SIMON AND GARFUNKEL	7
16.	PINBALL WIZARD	THE WHO	8
17.	GITARZAN	RAY STEVENS	10
*18.	(Love Theme) FROM ROMEO & JULIET	HENRY MANCINI	29
*19.	DAY IS DONE	PETER PAUL AND MARY	26
*20.	GOOD MORNING STARSHINE	OLIVER	--

HIT BOUND

*21.	LOVE ME TONIGHT	TOM JONES	--
22.	NOTHING BUT A HEARTACHE	THE FLIRTATIONS	25
23.	GOODBYE	MARY HOPKIN	24
24.	SEATTLE	PERRY COMO	21
*25.	EVERYDAY WITH YOU GIRL	CLASSICS IV	--
*26.	BLACK PEARL	SONNY CHARLES AND THE CHECKMATES LTD.	--
27.	ONE	THREE DOG NIGHT	--
28.	SEE	THE RASCALS	--
29.	ISRAELITES	DESMOND DEKKER & THE ACES	--
30.	BUT IT'S ALRIGHT	J.J. JACKSON	--

WCFL
1000 WCFL-AM SOUND 10 SURVEY-CHICAGO

WEEK ENDING JUNE 4, 1969

*1.	GRAZING IN THE GRASS	THE FRIENDS OF DISTINCTION	12
*2.	TOO BUSY THINKING ABOUT MY BABY	MARVIN GAYE	9
3.	GET BACK / DON'T LET ME DOWN	THE BEATLES	1
*4.	BAD MOON RISING / LODI	CREEDENCE CLEARWATER REVIVAL	7
5.	MEDICINE MAN (Part 1)	THE BUCHANAN BROTHERS	5
*6.	IN THE GHETTO	ELVIS PRESLEY	8
7.	LOVE (Can Make You Happy)	MERCY	2
*8.	(Love Theme) FROM ROMEO & JULIET	HENRY MANCINI	18
9.	THESE EYES	THE GUESS WHO	3
10.	MORNING GIRL	NEON PHILHARMONIC	4
11.	SPECIAL DELIVERY	1910 FRUITGUM CO.	14
12.	LET ME	PAUL REVERE AND THE RAIDERS	10
13.	I COULD NEVER LIE TO YOU	THE NEW COLONY SIX	13
14.	OH, HAPPY DAY	EDWIN HAWKINS SINGERS	6
15.	DAY IS DONE	PETER PAUL AND MARY	19
*16.	LOVE ME TONIGHT	TOM JONES	21
*17.	EVERYDAY WITH YOU GIRL	CLASSICS IV	25
18.	GOOD MORNING STARSHINE	OLIVER	20
*19.	SEE	THE RASCALS	28
*20.	ISREALITES	DESMOND DEKKER & THE ACES	29

HIT BOUND

21.	NOTHING BUT A HEARTACHE	THE FLIRTATIONS	22
22.	GOODBYE	MARY HOPKIN	23
23.	ONE	THREE DOG NIGHT	27
*24.	BUT IT'S ALRIGHT	J.J. JACKSON	30
25.	BLACK PEARL	SONNY CHARLES AND THE CHECKMATES LTD.	26
*26.	NO MATTER WHAT SIGN YOU ARE	DIANA ROSS AND THE SUPREMES	--
27.	MOODY WOMAN	JERRY BUTLER	--
28.	COLOR HIM FATHER	THE WINSTONS	--
29.	CRYSTAL BLUE PERSUASION	TOMMY JAMES & THE SHONDELLS	--
30.	THE BALLAD OF JOHN AND YOKO / OLD BROWN SHOE	THE BEATLES	--

WCFL

1000 WCFL-AM SOUND 10 SURVEY-CHICAGO

WEEK ENDING JUNE 11, 1969

*1.	(Love Theme) FROM ROMEO & JULIET	HENRY MANCINI	8
*2.	BAD MOON RISING / LODI	CREEDENCE CLEARWATER REVIVAL	4
3.	GRAZING IN THE GRASS	THE FRIENDS OF DISTINCTION	1
4.	TOO BUSY THINKING ABOUT MY BABY	MARVIN GAYE	2
5.	GET BACK	THE BEATLES	3
*6.	GOOD MORNING STARSHINE	OLIVER	18
7.	MEDICINE MAN (Part 1)	THE BUCHANAN BROTHERS	5
*8.	LET ME	PAUL REVERE AND THE RAIDERS	12
9.	LOVE (Can Make You Happy)	MERCY	7
10.	IN THE GHETTO	ELVIS PRESLEY	6
*11.	ONE	THREE DOG NIGHT	23
12.	SPECIAL DELIVERY	1910 FRUITGUM CO.	11
13.	MORNING GIRL	NEON PHILHARMONIC	10
14.	EVERYDAY WITH YOU GIRL	CLASSICS IV	17
15.	THESE EYES	THE GUESS WHO	9
16.	DAY IS DONE	PETER PAUL AND MARY	15
17.	LOVE ME TONIGHT	TOM JONES	16
18.	I COULD NEVER LIE TO YOU	THE NEW COLONY SIX	13
19.	ISREALITES	DEMOND DEKKER & THE ACES	20
20.	SEE	THE RASCALS	19

HIT BOUND

*21.	NO MATTER WHAT SIGN YOU ARE	DIANA ROSS AND THE SUPREMES	26
*22.	WHAT DOES IT TAKE (To Win Your Love)	JR. WALKER & THE ALL STARS	--
23.	MOODY WOMAN	JERRY BUTLER	27
24.	BUT IT'S ALRIGHT	J.J. JACKSON	24
25.	BLACK PEARL	SONNY CHARLES AND THE CHECKMATES LTD.	25
26.	THE BALLAD OF JOHN AND YOKO / OLD BROWN SHOE	THE BEATLES	30
27.	COLOR HIM FATHER	THE WINSTONS	28
28.	CRYSTAL BLUE PERSUASION	TOMMY JAMES & THE SHONDELLS	29
29.	MY CHERIE AMOUR	STEVIE WONDER	--
30.	WITH PEN IN HAND	VIKKI CARR	--

WCFL
1000 WCFL-AM SOUND 10 SURVEY-CHICAGO

WEEK ENDING JUNE 18, 1969

*1.	(Love Theme) FROM ROMEO & JULIET	HENRY MANCINI	1
*2.	BAD MOON RISING / LODI	CREEDENCE CLEARWATER REVIVAL	2
*3.	GOOD MORNING STARSHINE	OLIVER	6
*4.	ONE	THREE DOG NIGHT	11
5.	TOO BUSY THINKING ABOUT MY BABY	MARVIN GAYE	4
*6.	ISREALITES	DESMOND DEKKER & THE ACES	19
7.	GET BACK	THE BEATLES	5
8.	MEDICINE MAN (Part 1)	THE BUCHANAN BROTHERS	7
9.	LET ME	PAUL REVERE AND THE RAIDERS	8
10.	GRAZING IN THE GRASS	THE FRIENDS OF DISTINCTION	3
11.	IN THE GHETTO	ELVIS PRESLEY	10
12.	SPECIAL DELIVERY	1910 FRUITGUM CO.	12
13.	LOVE ME TONIGHT	TOM JONES	17
*14.	WHAT DOES IT TAKE (To Win Your Love)	JR. WALKER AND THE ALL STARS	22
15.	LOVE (Can Make You Happy)	MERCY	9
16.	MORNING GIRL	NEON PHILHARMONIC	13
17.	EVERYDAY WITH YOU GIRL	CLASSICS IV	14
*18.	MOODY WOMAN J	JERRY BUTLER	23
*19.	THE BALLAD OF JOHN AND YOKO / OLD BROWN SHOE	THE BEATLES	26
20.	SEE	THE RASCALS	20

HIT BOUND

*21.	MY CHERIE AMOUR	STEVIE WONDER	29
*22.	COLOR HIM FATHER	THE WINSTONS	27
*23.	CRYSTAL BLUE PERSUASION	TOMMY JAMES & THE SHONDELLS	28
24.	BUT IT'S ALRIGHT	J.J. JACKSON	24
25.	BLACK PEARL	SONNY CHARLES AND THE CHECKMATES LTD.	25
26.	NO MATTER WHAT SIGN YOU ARE	DIANA ROSS AND THE SUPREMES	21
27.	RUBY, DON'T TAKE YOUR LOVE TO TOWN	KENNY ROGERS & THE FIRST EDITION	--
28.	BABY, I LOVE YOU	ANDY KIM	--
29.	WITH PEN IN HAND	VIKKI CARR	30
30.	MRS. ROBINSON	BOOKER T. & THE MG'S	--

WCFL
1000 WCFL-AM SOUND 10 SURVEY-CHICAGO

WEEK ENDING JUNE 25, 1969

*1.	(Love Theme) FROM ROMEO & JULIET	HENRY MANCINI	1
*2.	ONE	THREE DOG NIGHT	4
3.	BAD MOON RISING	CREEDENCE CLEARWATER REVIVAL	2
4.	GOOD MORNING STARSHINE	OLIVER	3
*5.	ISREALITES	DESMOND DEKKER & THE ACES	6
*6.	CRYSTAL BLUE PERSUASION	TOMMY JAMES & THE SHONDELLS	23
7.	TOO BUSY THINKING ABOUT MY BABY	MARVIN GAYE	5
8.	MEDICINE MAN (Part I)	THE BUCHANAN BROTHERS	8
9.	LET ME	PAUL REVERE AND THE RAIDERS	9
*10.	COLOR HIM FATHER	THE WINSTONS	22
11.	WHAT DOES IT TAKE (To Win Your Love)	JR. WALKER AND THE ALL STARS	14
12.	LOVE ME TONIGHT	TOM JONES	13
13.	EVERYDAY WITH YOU GIRL	CLASSICS IV	17
*14.	THE BALLAD OF JOHN AND YOKO / OLD BROWN SHOE	THE BEATLES	19
15.	IN THE GHETTO	ELVIS PRESLEY	11
*16.	BLACK PEARL	SONNY CHARLES AND THE CHECKMATES LTD.	25
17.	GET BACK	THE BEATLES	7
18.	SPECIAL DELIVERY	1910 FRUITGUM CO.	12
19.	MY CHERIE AMOUR	STEVIE WONDER	21
*20.	BABY, I LOVE YOU	ANDY KIM	28

HIT BOUND

21.	BUT IT'S ALRIGHT	J.J. JACKSON	24
*22.	RUBY, DON'T TAKE YOUR LOVE TO TOWN	KENNY ROGERS AND THE FIRST EDITION	27
23.	NO MATTER WHAT SIGN YOU ARE	DIANA ROSS AND THE SUPREMES	26
*24.	SPINNING WHEEL / MORE & MORE	BLOOD, SWEAT AND TEARS	--
*25.	MY PLEDGE OF LOVE	JOE JEFFREY GROUP	--
26.	WITH PEN IN HAND	VIKKI CARR	29
27.	MRS. ROBINSON	BOOKER T. & THE MG'S	30
28.	IN THE YEAR 2525 (Exordium & Terminus)	ZAGER & EVANS	--
29.	MOTHER POPCORN (Part 1)	JAMES BROWN	--
30.	DOGGONE RIGHT	SMOKEY ROBINSON AND THE MIRACLES	--

WCFL
1000 WCFL-AM SOUND 10 SURVEY-CHICAGO

WEEK ENDING JULY 2, 1969

*1.	ONE	THREE DOG NIGHT	2
2.	(Love Theme) FROM ROMEO & JULIET	HENRY MANCINI	1
*3.	CRYSTAL BLUE PERSUASION	TOMMY JAMES & THE SHONDELLS	6
4.	GOOD MORNING STARSHINE	OLIVER	4
5.	ISREALITES	DEMOND DEKKER & THE ACES	5
*6.	COLOR HIM FATHER	THE WINSTONS	10
*7.	IN THE YEAR 2525 (Exordium and Terminus)	ZAGER & EVANS	28
*8.	BLACK PEARL	SONNY CHARLES AND THE CHECKMATES LTD.	16
*9.	WHAT DOES IT TAKE (To Win Your Love)	JR. WALKER AND THE ALL STARS	11
10.	BALLAD OF JOHN AND YOKO / OLD BROWN SHOE	THE BEATLES	14
11.	LOVE ME TONIGHT	TOM JONES	12
12.	TOO BUSY THINKING ABOUT MY BABY	MARVIN GAYE	7
*13.	MY CHERIE AMOUR	STEVIE WONDER	19
*14.	BABY, I LOVE YOU	ANDY KIM	20
15.	BAD MOON RISING	CREEDENCE CLEARWATER REVIVAL	3
16.	LET ME	PAUL REVERE AND THE RAIDERS	9
*17.	RUBY, DON'T TAKE YOUR LOVE TO TOWN	THE FIRST EDITION	22
18.	MEDICINE MAN (Part 1)	THE BUCHANAN BROTHERS	8
*19.	SPINNING WHEEL / MORE & MORE	BLOOD SWEAT AND TEARS	24
*20.	MOTHER POPCORN (Part 1)	JAMES BROWN	29

HIT BOUND

21.	MY PLEDGE OF LOVE	JOE JEFFREY GROUP	25
*22.	MRS. ROBINSON	BOOKER T. & THE MG'S	27
*23.	QUENTIN'S THEME	CHARLES RANDOLPH GREAN SOUNDE	--
*24.	DOGGONE RIGHT	SMOKEY ROBINSON AND THE MIRACLES	--
*25.	PUT A LITTLE LOVE IN YOUR HEART	JACKIE DeSHANNON	--
*26.	GOOD OLD ROCK 'N ROLL (Medley)	MOTHER & THE ALL NIGHT NEWS BOYS	--
27.	RECONSIDER ME	JOHNNY ADAMS	--
28.	POLK SALAD ANNIE	TONY JOE WHITE	--
29.	YOU DON'T HAVE TO WALK IN THE RAIN	THE TURTLES	--
30.	I'LL NEVER FALL IN LOVE AGAIN	TOM JONES	--

WCFL

1000 WCFL-AM SOUND 10 SURVEY-CHICAGO

WEEK ENDING JULY 9, 1969

*1.	ONE	THREE DOG NIGHT	2
*2.	CRYSTAL BLUE PERSUASION	TOMMY JAMES & THE SHONDELLS	3
3.	IN THE YEAR 2525 (Exordium and Terminus)	ZAGER & EVANS	7
4.	(Love Theme) FROM ROMEO & JULIET	HENRY MANCINI	2
*5.	COLOR HIM FATHER	THE WINSTONS	6
6.	GOOD MORNING STARSHINE	OLIVER	4
*7.	BABY, I LOVE YOU	ANDY KIM	14
*8.	WHAT DOES IT TAKE (To Win Your Love)	JR. WALKER AND THE ALL STARS	9
*9.	THE BALLAD OF JOHN AND YOKO / OLD BROWN SHOE	THE BEATLES	10
*10.	RUBY, DON'T TAKE YOUR LOVE TO TOWN	KENNY ROGERS AND THE FIRST EDITION	17
11.	ISREALITES	DESMOND DEKKER & THE ACES	5
12.	MY CHERIE AMOUR	STEVIE WONDER	13
13.	BLACK PEARL	SONNY CHARLES AND THE CHECKMATES LTD.	8
*14.	SPINNING WHEEL / MORE & MORE	BLOOD SWEAT AND TEARS	19
15.	TOO BUSY THINKING ABOUT MY BABY	MARVIN GAYE	12
16.	MOTHER POPCORN (Part 1)	JAMES BROWN	20
17.	LOVE ME TONIGHT	TOM JONES	11
18.	MY PLEDGE OF LOVE	JOE JEFFREY GROUP	21
19.	LET ME	PAUL REVERE AND THE RAIDERS	16
20.	BAD MOON RISING	CREEDENCE CLEARWATER REVIVAL	15

HIT BOUND

21.	QUENTIN'S THEME	CHARLES RANDOLPH GREAN SOUNDE	23
22.	PUT A LITTLE LOVE IN YOUR HEART	JACKIE DeSHANNON	25
23.	DOGGONE RIGHT	SMOKEY ROBINSON AND THE MIRACLES	24
24.	POLK SALAD ANNIE	TONY JOE WHITE	28
25.	GOOD OLD ROCK 'N ROLL (Medley)	CAT MOTHER & THE ALL NIGHT NEWS BOYS	26
26.	RECONSIDER ME	JOHNNY ADAMS	27
27.	I'LL NEVER FALL IN LOVE AGAIN	TOM JONES	30
28.	MRS. ROBINSON	BOOKER T. & THE MG'S	22
29.	MARRAKESH EXPRESS	CROSBY STILLS & NASH	--
30.	SWEET CAROLINE	NEIL DIAMOND	--

WCFL
1000 WCFL-AM SOUND 10 SURVEY-CHICAGO

WEEK ENDING JULY 16, 1969

*1.	IN THE 2525 (Exordium and Terminus)	ZAGER & EVANS	3
2.	ONE	THREE DOG NIGHT	1
3.	CRYSTAL BLUE PERSUASION	TOMMY JAMES & THE SHONDELLS	2
*4.	RUBY, DON'T TAKE YOUR LOVE TO TOWN	KENNY ROGERS AND THE FIRST EDITION	10
*5.	BABY, I LOVE YOU	ANDY KIM	
*6.	WHAT DOES IT TAKE (To Win Your Love)	JR. WALKER AND THE ALL STARS	8
7.	COLOR HIM FATHER	THE WINSTONS	5
*8.	THE BALLAD OF JOHN AND ONO / OLD BROWN SHOE	THE BEATLES	9
*9.	MOTHER POPCORN (Part 1)	JAMES BROWN	16
*10.	MY CHERIE AMOUR	STEVIE WONDER	12
11.	(Love Theme) FROM ROMEO & JULIET	HENRY MANCINI	4
12.	GOOD MORNING STARSHINE	OLIVER	6
13.	SPINNING WHEEL / MORE & MORE	BLOOD SWEAT AND TEARS	14
*14.	QUENTIN'S THEME	CHARLES RANDOLPH GREAN SOUNDE	21
15.	MY PLEDGE OF LOVE	JOE JEFFREY GROUP	18
*16.	POLK SALAD ANNIE	TONY JOE WHITE	24
*17.	PUT A LITTLE LOVE IN YOUR HEART	JACKIE DeSHANNON	22
18.	ISRAELITES	DESMOND DEKKER & THE ACES	11
19.	BLACK PEARL	SONNY CHARLES & THE CHECKMATES LTD.	13
20.	LOVE ME TONIGHT	TOM JONES	17

HIT BOUND

*21.	SWEET CAROLINE	NEIL DIAMOND	30
22.	DOGGONE RIGHT	SMOKEY ROBINSON AND THE MIRACLES	23
23.	GOOD TIME ROCK 'N ROLL (Medley)	CAT MOTHER & THE ALL NIGHT NEWS BOYS	25
24.	I'LL NEVER FALL IN LOVE AGAIN	TOM JONES	27
25.	RECONSIDER ME	JOHNNY ADAMS	26
26.	MARRAKESH EXPRESS	CROSBY STILLS & NASH	29
27.	WHY I SING THE BLUES	B.B. KING	--
*28.	HONKY TONK WOMAN / YOU CAN'T ALWAYS GET WHAT YOU WANT	THE ROLLING STONES	--
*29.	SOUL DEEP	THE BOX TOPS	--
30.	YESTERDAY, WHEN I WAS YOUNG	ROY CLARK	--

WCFL

1000 WCFL-AM SOUND 10 SURVEY-CHICAGO

WEEK ENDING JULY 23, 1969

*1.	IN THE YEAR 2525 (Exordium and Terminus)	ZAGER & EVANS	1
*2.	RUBY, DON'T TAKE YOUR LOVE TO TOWN	KENNY ROGERS AND THE FIRST EDITION	4
*3.	BABY, I LOVE YOU	ANDY KIM	5
4.	CRYSTAL BLUE PERSUASION	TOMMY JAMES & THE SHONDELLS	3
5.	WHAT DOES IT TAKE (To Win Your Love)	JR. WALKER AND THE ALL STARS	6
6.	ONE	THREE DOG NIGHT	2
*7.	HONKY TONK WOMAN / YOU CAN'T ALWAYS GET WHAT YOU WANT	THE ROLLING STONES	28
*8.	QUENTIN'S THEME	CHARLES RANDOLPH GREEN SOUNDE	14
*9.	SWEET CAROLINE	NEIL DIAMOND	21
*10.	POLK SALAD ANNIE	TONY JOE WHITE	16
11.	MOTHER POPCORN (Part 1)	JAMES BROWN	9
12.	MY CHERIE AMOUR	STEVIE WONDER	10
13.	SPINNING WHEEL	BLOOD,SWEAT AND TEARS	13
14.	THE BALLAD OF JOHN AND YOKO	THE BEATLES	8
15.	PUT A LITTLE LOVE IN YOUR HEART	JACKIE DESHANNON	17
16.	GOOD MORNING STARSHINE	OLIVER	12
17.	MY PLEDGE OF LOVE	JOE JEFFERY GROUP	15
18.	(Love Theme) FROM ROMEO & JULIET	HENRY MANCINI	11
19.	COLOR HIM FATHER	THE WINSTONS	7
20.	GOOD OLD ROCK 'N ROLL (Medley)	CAT MOTHER AND THE ALL-NIGHT NEWS BOYS	23

HIT BOUND

*21.	BIRTHDAY	UNDERGROUND SUNSHINE	--
*22.	HURT SO BAD	THE LETTERMEN	--
*23.	SOUL DEEP	THE BOX TOPS	29
*24.	ALONG CAME JONES	RAY STEVENS	--
*25.	LAUGHING	THE GUESS WHO	--
26.	RECONSIDER ME	JOHNNY ADAMS	25
27.	WHY I SING THE BLUES	B.B. KING	27
28.	MARRAKESH EXPRESS	CROSBY, STILLS & NASH	26
29.	I'LL NEVER FALL IN LOVE AGAIN	TOM JONES	24
*30.	A BOY NAMED SUE	JOHNNY CASH	--

WCFL
1000 WCFL-AM SOUND 10 SURVEY-CHICAGO

WEEK ENDING JULY 30, 1969

*1.	HONKY TONK WOMEN / YOU CAN'T ALWAYS GET WHAT YOU WANT	THE ROLLING STONES	7
2.	IN THE YEAR 2525 (Exordium and Terminus)	ZAGER & EVANS	1
3.	BABY, I LOVE YOU	ANDY JIM	3
*4.	POLK SALAD ANNIE	TONY JOE WHITE	10
5.	CRYSTAL BLUE PERSUASION	TOMMY JAMES & THE SHONDELLS	4
*6.	SWEET CAROLINE	NEIL DIAMOND	9
7.	WHAT DOES IT TAKE (To Win Your Love)	JR. WALKER AND THE ALL STARS	5
8.	QUENTIN'S THEME	CHARLES RANDOLPH GREEN SOUNDS	8
*9.	BIRTHDAY	UNDERGROUND SUNSHINE	21
*10.	PUT A LITTLE LOVE IN YOUR HEART	JACKIE DeSHANNON	15
*11.	HURT SO BAD	THE LETTERMEN	22
12.	MOTHER POPCORN (Part 1)	JAMES BROWN	11
13.	MY CHERIE AMOUR	STEVIE WONDER	12
*14.	SOUL DEEP	THE BOX TOPS	23
15.	RUBY, DON'T TAKE YOUR LOVE TO TOWN	KENNY ROGERS & FIRST EDITION	2
*16.	A BOY NAMED SUE	JOHNNY CASH	30
*17.	LAUGHING	THE GUESS WHO	25
18.	ONE	THREE DOG NIGHT	6
19.	SPINNING WHEEL	BLOOD SWEAT AND TEARS	13
20.	GOOD MORNING STARSHINE	OLIVER	16

HIT BOUND

*21.	I'LL NEVER FALL IN LOVE AGAIN	TOM JONES	29
*22.	WORKIN' ON A GROOVY THING	THE 5TH DIMENSION	--
23.	RECONSIDER ME	JOHNNY ADAMS	26
*24.	GIVE PEACE A CHANCE	PLASTIC ONO BAND	--
25.	MARRAKESH EXPRESS	CROSBY STILLS & NASH	--
*26.	GREEN RIVER / COMMOTION	CREEDENCE CLEARWATER REVIVAL	--
27.	ABERGAVENNY	SHANNON	--
28.	LAY LADY LAY	BOB DYLAN	--
29.	YOUR GOOD THING	LOU RAWLS	--
30.	EVERYBODY KNOWS MATILDA	DUKE BAXTER	--

WCFL
1000 WCFL-AM SOUND 10 SURVEY-CHICAGO

WEEK ENDING AUGUST 6, 1969

*1.	HONKY TONK WOMAN / YOU CAN'T ALWAYS GET WHAT YOU WANT	THE ROLLING STONES	1
2.	HURT SO BAD	THE LETTERMEN	11
3.	IN THE YEAR 2525 (Exordium and Terminus)	ZAGER AND EVANS	2
*4.	SWEET CAROLINE	NEIL DIAMOND	6
*5.	MY CHERIE AMOUR	STEVIE WONDER	13
*6.	BIRTHDAY	UNDERGROUND SUNSHINE	9
7.	BABY, I LOVE YOU	ANDY KIM	3
8.	POLK SALAD ANNIE	TONY JOE WHITE	4
*9.	PUT A LITTLE LOVE IN YOUR HEART	JACKIE DeSHANNON	10
*10.	SOUL DEEP	THE BOX TOPS	14
*11.	A BOY NAMED SUE	JOHNNY CASH	16
12.	CRYSTAL BLUE PERSUASION	TOMMY JAMES & THE SHONDELLS	5
13.	LAUGHING	THE GUESS WHO	17
*14.	I'LL NEVER FALL IN LOVE AGAIN	TOM JONES	21
15.	QUENTIN'S THEME	CHARLES RANDOLPH GREEN SOUNDS	8
*16.	WORKIN' ON A GROOVY THING	THE 5TH DIMENSION	22
17.	MOTHER POPCORN (Part 1)	JAMES BROWN	12
*18.	MARRAKESH EXPRESS	CROSBY STILLS & NASH	25
19.	WHAT DOES IT TAKE (To Win Your Love)	JR. WALKER AND THE ALL STARS	7
20.	GIVE PEACE A CHANCE	PLASTIC ONO BAND	24

HIT BOUND

*21.	GREEN RIVER / COMMOTION	CREEDENCE CLEARWATER REVIVAL	26
*22.	ABERGAVENNY	SHANNON	27
*23.	LAY LADY LAY	BOB DYLAN	28
*24.	YOUR GOOD THING (Is About To End)	LOU RAWLS	29
*25.	GET TOGETHER	THE YOUNGBLOODS	--
*26.	I'D WAIT A MILLION YEARS	THE GRASS ROOTS	--
27.	EVERYBODY KNOWS MITILDA	DUKE BAXTER	30
28.	DID YOU SEE HER EYES	THE ILLUSION	--
29.	SUGAR, SUGAR	THE ARCHIES	--
30.	NITTY GRITTY	GLADYS KNIGHT AND THE PIPS	--

WCFL
1000 WCFL-AM SOUND 10 SURVEY-CHICAGO

WEEK ENDING AUGUST 13, 1969

*1.	HONKY TONK WOMAN / YOU CAN'T ALWAYS GET WHAT YOU WANT	THE ROLLING STONES	1
*2.	SWEET CAROLINE	NEIL DIAMOND	4
3.	HURT SO BAD	THE LETTERMEN	2
*4.	BIRTHDAY	UNDERGROUND SUNSHINE	6
5.	IN THE YEAR 2525 (Exordium and Terminus)	ZAGER & EVANS	3
*6.	A BOY NAMED SUE	JOHNNY CASH	11
*7.	PUT A LITTLE LOVE IN YOUR HEART	JACKIE DeSHANNON	9
*8.	SOUL DEEP	THE BOX TOPS	10
9.	POLK SALAD ANNIE	TONY JOE WHITE	8
*10.	LAUGHING	THE GUESS WHO	13
*11.	I'LL NEVER FALL IN LOVE AGAIN	TOM JONES	14
*12.	GREEN RIVER / COMMOTION	CREEDENCE CLEARWATER REVIVAL	21
13.	CRYSTAL BLUE PERSUASION	TOMMY JAMES & THE SHONDELLS	12
14.	WORKIN' ON A GROOVY THING	THE 5TH DIMENSION	16
15.	BABY, I LOVE YOU	ANDY KIM	7
16.	GIVE PEACE A CHANCE	PLASTIC ONO BAND	20
17.	MARRAKESH EXPRESS	CROSBY STILLS & NASH	18
*18.	LAY LADY LAY	BOB DYLAN	23
19.	QUENTIN'S THEME	CHARLES RANDOLPH GREEN SOUNDE	15
20.	MY CHERIE AMOUR	STEVIE WONDER	5

HIT BOUND

21.	GET TOGETHER	THE YOUNGBLOODS	25
*22.	EASY TO BE HARD	THREE DOG NIGHT	--
23.	YOUR GOOD THING (Is About To End)	LOU RAWLS	24
*24.	NITTY GRITTY	GLADYS KNIGHT AND THE PIPS	30
25.	I'D WAIT A MILLION YEARS	THE GRASS ROOTS	26
26.	ABERGAVENNY	SHANNON	22
27.	EVERYBODY KNOWS MATILDA	DUKE BAXTER	27
28.	DID YOU SEEN HER EYES	THE ILLUSION	28
29.	SUGAR, SUGAR	THE ARCHIES	29
30.	TRUE GRIT	GLEN CAMPBELL	--

WCFL

1000 WCFL-AM SOUND 10 SURVEY-CHICAGO

WEEK ENDING AUGUST 27, 1969

*1.	LAY LADY LAY	BOB DYLAN
2.	HONKY TONK WOMAN / YOU CAN'T ALWAYS GET WHAT YOU WANT	THE ROLLING STONES
*3.	SUGAR, SUGAR	THE ARCHIES
*4.	HURT SO BAD	THE LETTERMEN
*5.	SOUL DEEP	THE BOX TOPS
6.	GREEN RIVER / COMMOTION	CREEDENCE CLEARWATER REVIVAL
*7.	GET TOGETHER	THE YOUNGBLOODS
*8.	EASY TO BE HARD	THREE DOG NIGHT
*9.	I'D WAIT A MILLION YEARS	THE GRASS ROOTS
10.	A BOY NAMED SUE	JOHNNY CASH
11.	LAUGHING	THE GUESS WHO
12.	BIRTHDAY	UNDERGROUND SUNSHINE
13.	SWEET CAROLINE	NEIL DIAMOND
14.	I'LL NEVER FALL IN LOVE AGAIN	TOM JONES
15.	DID YOU SEE HER EYES	THE ILLUSION
*16.	THE TRAIN	1910 FRUITGUM CO.
17.	WORKIN' ON A GROOVY THING	THE 5TH DIMENSION
18.	KEEM-O-SABE	ELECTRIC INDIAN
19.	THIS GIRL'S A WOMAN NOW	GARY PUCKETT & THE UNION GAP
20.	PUT A LITTLE LOVE IN YOUR HEART	JACKIE DeSHANNON

HIT BOUND

21.	NITTY GRITTY	GLADYS KNIGHT AND THE PIPS
*22.	I WANT YOU TO KNOW	THE NEW COLONY SIX
*23.	YOU, I	THE RUGBYS
24.	WHAT KIND OF FOOL DO YOU THINK I AM	BILL DEAL AND THE RHONDELLS
25.	YOUR GOOD THING (Is About To End)	LOU RAWLS
*26.	I CAN'T GET NEXT TO YOU	THE TEMPTATIONS
27.	DADDY'S LITTLE MAN	O.C. SMITH
28.	WHEN I DIE	MOTHERLODE
29.	BREAK MY MIND	PAT BOONE ?
30.	CARRY ME BACK	THE RASCALS

WCFL
1000 WCFL-AM SOUND 10 SURVEY-CHICAGO

WEEK ENDING SEPTEMBER 3, 1969

*1.	SUGAR, SUGAR	THE ARCHIES	3
2.	HONKY TONK WOMAN	THE ROLLING STONES	2
3.	LAY LADY LAY	BOB DYLAN	1
*4.	SOUL DEEP	THE BOX TOPS	5
5.	HURT SO BAD	THE LETTERMEN	4
*6.	GET TOGETHER	THE YOUNGBLOODS	7
*7.	EASY TO BE HARD	THREE DOG NIGHT	8
*8.	I'D WAIT A MILLION YEARS	THE GRASS ROOTS	9
9.	GREEN RIVER	CREEDENCE CLEARWATER REVIVAL	6
*10.	THE TRAIN	1910 FRUITGUM CO.	16
11.	DID YOU SEE HER EYES	THE ILLUSION	15
12.	A BOY NAMED SUE	JOHNNY CASH	10
*13.	THIS GIRL'S A WOMAN NOW	GARY PUCKETT & THE UNION GAP	19
14.	I'LL NEVER FALL IN LOVE AGAIN	TOM JONES	14
15.	BIRTHDAY	UNDERGROUND SUNSHINE	12
16.	KEEM-O-SABE	ELECTRIC INDIAN	18
17.	SWEET CAROLINE	NEIL DIAMOND	13
18.	LAUGHING	THE GUESS WHO	11
*19.	YOU, I	THE RUGBYS	23
20.	WORKIN' ON A GROOVY THING	THE 5TH DIMENSION	17

HIT BOUND

21.	I WANT YOU TO KNOW	THE NEW COLONY SIX	22
22.	I CAN'T GET NEXT TO YOU	THE TEMPTATIONS	26
23.	WHAT KIND OF FOOL DO YOU THINK I AM	BILL DEAL AND THE RHONDELLS	24
24.	YOUR GOOD THING (Is About To End)	LOU RAWLS	25
25.	DADDY'S LITTLE MAN	O.C. SMITH	27
26.	WHEN I DIE	MOTHERLODE	28
27.	I'M GONNA MAKE YOU MINE	LOU CHRISTIE	--
28.	OH, WHAT A NIGHT	THE DELLS	--
29.	JEAN	OLIVER	--
30.	CARRY ME BACK	THE RASCALS	--

WCFL
1000 WCFL-AM SOUND 10 SURVEY-CHICAGO

WEEK ENDING SEPTEMBER 10, 1969

*1.	SUGAR, SUGAR	THE ARCHIES	1
*2.	THIS GIRL'S A WOMAN NOW	GARY PUCKETT & THE UNION GAP	13
*3.	THE TRAIN	1910 FRUITGUM CO.	10
4.	HONKY TONK WOMAN	THE ROLLING STONES	2
*5.	EASY TO BE HARD	THREE DOG NIGHT	7
*6.	HURT SO BAD	THE LETTERMEN	5
7.	I'D WAIT A MILLION YEARS	THE GRASS ROOTS	8
8.	LAY LADY LAY	BOB DYLAN	3
9.	SOUL DEEP	THE BOX TOPS	4
10.	GREEN RIVER	CREEDENCE CLEARWATER REVIVAL	9
11.	GET TOGETHER	THE YOUNGBLOODS	6
*12.	YOU, I	THE RUGBYS	19
13.	DID YOU SEE HER EYES	THE ILLUSION	11
*14.	I CAN'T GET NEXT TO YOU	THE TEMPTATIONS	22
*15.	I'M GONNA MAKE YOU MINE	LOU CHRISTIE	27
*16.	WHAT KIND OF FOOL DO YOU THINK I AM	BILL DEAL AND THE RHONDELLS	23
17.	I'LL NEVER FALL IN LOVE AGAIN	TOM JONES	14
*18.	LITTLE WOMAN	BOBBY SHERMAN	--
19.	A BOY NAMED SUE	JOHNNY CASH	12
20.	KEEM-O-SABE	ELECTRIC INDIAN	16

HIT BOUND

21.	I WANT YOU TO KNOW	THE NEW COLONY SIX	21
22.	YOUR GOOD THING (Is About To End)	LOU RAWLS	24
23.	DADDY'S LITTLE MAN	O.C. SMITH	25
*24.	JEAN	OLIVER	29
25.	WHEN I DIE	MOTHERLODE	26
26.	CARRY ME BACK	THE RASCALS	30
27.	EVERYBODY'S TALKIN'	NILSSON	--
28.	THAT'S THE WAY LOVE IS	MARVIN GAYE	--
29.	OH, WHAT A NIGHT	THE DELLS	28
30.	DON'T IT MAKE YOU WANNA GO HOME	JOE SOUTH	--

WCFL

1000 WCFL-AM SOUND 10 SURVEY-CHICAGO

WEEK ENDING SEPTEMBER 17, 1969

*1.	SUGAR, SUGAR	THE ARCHIES	1
2.	THIS GIRL'S A WOMAN NOW	GARY PUCKETT & THE UNION GAP	2
*3.	I'D WAIT A MILLION YEARS	THE GRASS ROOTS	7
*4.	EASY TO BE HARD	THREE DOG NIGHT	5
*5.	I CAN'T GET NEXT TO YOU	THE TEMPTATIONS	15
*6.	I'M GONNA MAKE YOU MINE	LOU CHRISTIE	16
7.	THE TRAIN	1910 FRUITGUM CO.	3
*8.	LITTLE WOMAN	BOBBY SHERMAN	18
*9.	YOU, I	THE RUGBYS	12
10.	HURT SO BAD	THE LETTERMEN	6
11.	GET TOGETHER	THE YOUNGBLOODS	11
12.	HONKY TONK WOMAN	THE ROLLING STONES	4
13.	GREEN RIVER	CREEDENCE CLEARWATER REVIVAL	10
14.	LAY LADY LAY	BOB DYLAN	8
15.	SOUL DEEP	THE BOX TOPS	9
*16.	JEAN	OLIVER	24
*17.	THAT'S THE WAY LOVE IS	MARVIN GAYE	28
*18.	WHEN I DIE	MOTHERLODE	25
*19.	EVERYBODY'S TALKIN-	NILSSON	27
20.	I WANT YOU TO KNOW	THE NEW COLONY SIX	21

HIT BOUND

*21.	WHAT'S THE USE IN BREAKING UP	JERRY BUTLER	--
22.	CARRY ME BACK	THE RASCALS	26
23.	YOUR GOOD THING (Is About To End)	LOU RAWLS	22
24.	DADDY'S LITTLE GIRL	O.C. SMITH	23
*25.	DON'T IT MAKE YOU WANNA GO HOME	JOE SOUTH	30
*26.	MAKE BELIEVE	WIND	--
27.	SUGAR ON SUNDAY	CLIQUE	--
28.	TRACY	THE CUFFLINKS	--
29.	JEALOUS KIND OF FELLA	GARLAND GREEN	--
30.	OH, WHAT A NIGHT	THE DELLS	29

WCFL
1000 WCFL-AM SOUND 10 SURVEY-CHICAGO

WEEK ENDING SEPTEMBER 24, 1969

*1.	SUGAR, SUGAR	THE ARCHIES	1
2.	I'M GONNA MAKE YOU MINE	LOU CHRISTIE	6
*3.	LITTLE WOMAN	BOBBY SHERMAN	8
4.	THIS GIRL'S A WOMAN NOW	GARY PUCKETT & THE UNION GAP	2
5.	I CAN'T GET NEXT TO YOU	THE TEMPTATIONS	5
6.	EASY TO BE HARD	THREE DOG NIGHT	4
7.	I'D WAIT A MILLION YEARS	THE GRASS ROOTS	3
8.	THE TRAIN	1910 FRUITGUM CO.	7
9.	YOU, I	THE RUGBYS	9
*10.	JEAN	OLIVER	16
*11.	THAT'S THE WAY LOVE IS	MARVIN GAYE	17
12.	GET TOGETHER	THE YOUNGBLOODS	11
*13.	EVERYBODY'S TALKIN'	NILSSON	19
*14.	I WANT YOU TO KNOW	THE NEW COLONY SIX	20
15.	GREEN RIVER	CREEDENCE CLEARWATER REVIVAL	13
16.	HURT SO BAD	THE LETTERMEN	10
17.	WHEN I DIE	MOTHERLODE	18
18.	LAY LADY LAY	BOB DYLAN	14
*19.	TRACY	THE CUFF LINKS	28
20.	HONKY TONK WOMAN	THE ROLLING STONES	12

HIT BOUND

21.	WHAT'S THE USE IN BREAKING UP	JERRY BUTLER	21
22.	CARRY ME BACK	THE RASCALS	22
23.	DON'T IT MAKE YOU WANNA GO HOME	JOE SOUTH	25
24.	MAKE BELIEVE	WIND	26
25.	SUGAR ON SUNDAY	CLIQUE	27
26.	JEALOUS KIND OF FELLA	GARLAND GREEN	29
27.	YOUR GOOD THING (Is About To End)	LOU RAWLS	23
28.	DADDY'S LITTLE MAN	O.C. SMITH	24
29.	JACK & JILL	TOMMY ROE	--
30.	BABY, IT'S YOU	SMITH	--

WCFL
1000 WCFL-AM SOUND 10 SURVEY-CHICAGO

WEEK ENDING OCTOBER 1, 1969

*1.	LITTLE WOMAN	BOBBY SHERMAN	3
2.	I'M GONNA MAKE YOU MINE	LOU CHRISTIE	2
3.	SUGAR, SUGAR	THE ARCHIES	1
4.	JEAN	OLIVER	10
5.	I CAN'T GET NEXT TO YOU	THE TEMPTATIONS	5
6.	EASY TO BE HARD	THREE DOG NIGHT	6
7.	YOU, I	THE RUGBYS	9
8.	THIS GIRL'S A WOMAN NOW	GARY PUCKETT & THE UNION GAP	4
9.	THE TRAIN	1910 FRUITGUM CO.	8
10.	TRACY	THE CUFF LINKS	19
11.	EVERYBODY'S TALKIN'	NILSSON	13
12.	I'D WAIT A MILLION YEARS	THE GRASS ROOTS	7
13.	THAT'S THE WAY LOVE IS	MARVIN GAYE	11
14.	GET TOGETHER	THE YOUNGBLOODS	12
*15.	MAKE BELIEVE	WIND	24
16.	WHEN I DIE	MOTHERLODE	7
17.	GREEN RIVER	CREEDENCE CLEARWATER REVIVAL	15
18.	HURT SO BAD	THE LETTERMEN	16
19.	I WANT YOU TO KNOW	THE NEW COLONY SIX	14
*20.	BABY, IT'S YOU	SMITH	30

HIT BOUND

21.	WHAT'S THE USE IN BREAKING UP	JERRY BUTLER	21
22.	CARRY ME BACK	THE RASCALS	22
23.	DON'T IT MAKE YOU WANNA GO HOME	JOE SOUTH	23
24.	SUGAR ON SUNDAY	CLIQUE	25
*25.	RUBEN JANIES	KENNY ROGERS & THE FIRST EDITION	
*26.	WEDDING BELL BLUES	THE 5TH DIMENSION	--
27.	JACK & JILL	TOMMY ROE	29
*28.	SUSPICIOUS MINDS	ELVIS PRESLEY	--
29.	YOUR GOOD THING (Is About To End)	LOU RAWLS	27
30.	DADDY'S LITTLE MAN	O.C. SMITH	28

WCFL

1000 WCFL-AM SOUND 10 SURVEY-CHICAGO

WEEK ENDING OCTOBER 8, 1969

*1.	LITTLE WOMAN	BOBBY SHERMAN	1
2.	I'M GONNA MAKE YOU MINE	LOU CHRISTIE	2
*3.	I CAN'T GET NEXT TO YOU	THE TEMPTATIONS	5
4.	JEAN	OLIVER	4
*5.	TRACY	THE CUFF LINKS	10
6.	SUGAR, SUGAR	THE ARCHIES	3
7.	YOU, I	THE RUGBYS	7
*8.	EVERYBODY'S TALKIN'	NILSSON	11
*9.	THAT'S THE WAY LOVE IS	MARVIN GAYE	13
10.	THIS GIRL'S A WOMAN NOW	GARY PUCKETT & THE UNION GAP	8
11.	EASY TO BE HARD	THREE DOG NIGHT	6
*12.	SUSPICIOUS MINDS	ELVIS PRESLEY	28
13.	MAKE BELIEVE	WIND	15
*14.	BABY, IT'S YOU	SMITH	20
*15.	WEDDING BELL BLUES	THE 5TH DIMENSION	26
16.	THE TRAIN	1910 FRUITGUM CO.	9
17.	I'D WAIT A MILLION YEARS	THE GRASS ROOTS	12
*18.	DON'T IT MAKE YOU WANNA GO HOME	JOE SOUTH	23
*19.	RUBEN JAMES	ROGERS AND THE FIRST EDITION	25
20.	WHEN I DIE	MOTHERLODE	16
21.	SUGAR ON SUNDAY	CLIQUE	24
*22.	HOT FUN IN THE SUMMERTIME	SLY AND THE FAMILY STONE	--
*23.	JESUS IS A SOUL MAN	LAWRENCE REYNOLDS	--
24.	GET TOGETHER	THE YOUNGBLOODS	14
*25.	SUITE: JUDY BLUE EYES	CROSBY, STILLS & NASH	--
26.	JACK & JILL	TOMMY ROE	27
27.	WHAT'S THE USE OF BREAKING UP	JERRY BUTLER	21
28.	CARRY ME BACK	THE RASCALS	22
29.	BALL OF FIRE	TOMMY JAMES & THE SHONDELLS	--
30.	IS THAT ALL THERE IS	PEGGY LEE	--

1970

WC
FL

BIG 10 COUNT DOWN

FOR THE WEEK OF JANUARY 12, 1970

1.	"Venus"	Shocking Blue	Colossus
2.	Raindrops Keep Falling On My Head	B.J. Thomas	Scepter
3.	Jam Up, Jelly Tight	Tommy Roe	ABC
4.	Jingle Jangle	Archies	Kirshner
5.	Whole Lotta Love	Led Zeppelin	Atlantic
6.	La La La (If I Had You)	Bobby Sherman	Metromedia
7.	Don't Cry Daddy	Elvis Presley	RCA
8.	I Want You Back	Jackson Five	Motown
9.	No Time	Guess Who	RCA
10.	Someday We'll Be Together	D. Ross/Supremes	Motown
11.	Without Love (There Is Nothing)	Tom Jones	Parrot
12.	Midnight Cowboy	Ferrante & Teicher	United Artists
13.	I'll Never Fall In Love Again	Dionne Warwick	Scepter
14.	Going In Circles	Friends of Distinction	RCA
15.	Hey There Lonely Girl	Eddie Holman	ABC
16.	Wonderful World, Wonderful People	Jimmy Cliff	A & M
17.	Evil Woman	Crow	Amaret
18.	Winter World Of Love	Engel Humperdinck	Parrot
19.	Na Na Hey Hey	Steam	Fontana
20.	Thank You For Lettin' Me Be Myself	Sly/Family Stone	Epic
21.	If I Never Knew Your Name	Vic Dana	Liberty
22.	Down On The Corner	C.C. Revival	Fantasy
23.	Yester-Me, Yester-You, Yesterday	Stevie Wonder	Tamla
24.	Up On Cripple Creek	The Band	Capitol
25.	Early In The Morning	Vanity Fare	Page One
26.	The Thrill Is Gone	B.B. King	Bluesway
27.	Heaven Knows	Grass Roots	Dunhill
28.	Blowing Away	Fifth Dimension	Soul City
29.	Barbara, I Love You	New Colony Six	Mercury
30.	She	Tommy James/Shondells	Roulette
31.	Psychedelic Shack	Temptations	Gordy
32.	Arizona	Mark Lindsay	Columbia
33.	Ma Belle Amie	Tee Set	Colossus
34.	Cupid	Johnny Nash	Jade
35.	Ticket To Ride	Carpenters	A & M
36.	Walk A Mile In My Shoes	Joe South	Capitol
37.	Walkin' In The Rain	Jay/Americans	United Artists
38.	Cold Turkey	Plastic Ono Band	Apple
39.	Holly Holy	Neil Diamond	Uni
40.	Leaving On A Jet Plane	Peter, Paul & Mary	Warner Bros.

HOT BOUNDS

You Got Me Hummin'	Cold Blood	San Francisco
Breaking Up Is Hard To Do	Lenny Welch	Commonwealth United
She Come In Through The Bathroom Wind.	Jon Cocker	A & M

WC
FL

BIG 10 COUNT DOWN

FOR THE WEEK OF JANUARY 19, 1970

				Last Week
*1.	"Venus"	Shocking Blue	Colossus	(1)
*2.	Thank You For Lettin' Me Be Myself	Sly/Family Stone	Epic	(20)
3.	Raindrops Keep Falling On My Head	B.J. Thomas	Scepter	(2)
*4.	Hey There Lonely Girl	Eddie Holman	ABC	(15)
*5.	No Time	Guess Who	RCA	(9)
6.	Jam Up, Jelly Tight	Tommy Roe	ABC	(3)
*7.	Without Love (There Is Nothing)	Tom Jones	Parrot	(11)
8.	Jingle Jangle	Archies	Kirshner	(4)
*9.	I'll Never Fall In Love Again	Dionne Warwick	Scepter	(13)
10.	Whole Lotta Love	Led Zeppelin	Atlantic	(5)
11.	Don't Cry Daddy	Elvis Presley	RCA	(7)
12.	Someday We'll Be Together	D. Ross/Supremes	Motown	(10)
13.	I Want You Back	Jackson Five	Motown	(8)
14.	La La La (If I Had You)	Bobby Sherman	Metromedia	(6)
15.	Winter World Of Love	Engelbert Humperdinck	Parrot	(18)
16.	Wonderful World, Beautiful People	Jimmy Cliff	A & M	(16)
*17.	Blowing Away	Fifth Dimension	Soul City	(28)
18.	If I Never Knew Your Name	Vic Dana	Liberty	(21)
19.	Midnight Cowboy	Ferrante & Teicher	United Artists	(12)
20.	Going In Circles	Friends Of Distinction	RCA	(14)
*21.	Psychedelic Shack	Temptations	Gordy	(31)
*22.	The Thrill Is Gone	B.B. King	Bluesway	(26)
*23.	Walk A Mile In My Shoes	Joe South	Capitol	(36)
*24.	Arizona	Mark Lindsay	Columbia	(32)
*25.	Ma Belle Amie	Tee Set	Colossus	(33)
*26.	Ticket To Ride	Carpenters	A M	(35)
27.	Na Na Hey Hey	Steam	Fontana	(19)
28.	Evil Woman	Crow	Amaret	(17)
29.	Up ON Cripple Creek	The Band	Capitol	(24)
30.	Barbara, I Love You	New Colony Six	Mercury	(29)
31.	Down On The Corner	C.C. Revival	Fantasy	(22)
32.	Yester-Me, Yester-You, Yesterday	Stevie Wonder	Tamla	(23)
33.	Early In The Morning	Vanity Fare	Page One	(25)
*34.	Cold Turkey	Plastic Ono Band	Apple	(38)
35.	Walkin' In The Rain	Jay/Americans	United Artists	(37)
36.	She	Tommy James/Shondells	Roulette	(30)
37.	He Ain't Heavy, He's My Brother	Hollies	Epic	(N)
38.	She Came In Through The Bathroom Wind.	Joe Cocker	A & M	(N)
39.	You Got Me Hummin'	Cold Blood	San Francisco	(N)
40.	Rainy Night In Georgia	Brook Benton	Cotillion	(N)

HIT BOUND

Travelin' Band/Who'll Stop The Rain	C.C. Revival	Fantasy
The Rapper	The Jaggerz	Kama Sutra
Always Something There To Remind Me	R.B. Greaves	Atco

*Denotes Big 10 Climbers

WC
FL

BIG 10 COUNT DOWN

FOR THE WEEK OF JANUARY 26, 1970

				Last Week
*1.	Thank You For Lettin' Me Be Myself	Sly/Family Stone	Epic	(2)
*2.	No Time	Guess Who	RCA	(5)
3.	Hey There Lonely Girl	Eddie Holman	ABC	(4)
4.	"Venus"	Shocking Blue	Colossus	(1)
5.	Without Love (There Is Nothing)	Tom Jones	Parrot	(7)
6.	I'll Never Fall In Love Again	Dionne Warwick	Scepter	(9)
7.	Raindrops Keep Falling On My Head	B.J. Thomas	Scepter	(3)
8.	Jam Up, Jelly Tight	Tommy Roe	ABC	(6)
*9.	Psychedelic Shack	Temptations	Gordy	(21)
*10.	Arizona	Mark Lindsay	Columbia	(24)
*11.	Walk A Mile In My Shoes	Joe South	Capitol	(23)
12.	Whole Lotta Love	Led Zeppelin	Atlantic	(10)
13.	Jingle Jangle	Archies	Kirshner	(8)
*14.	Ma Bella Amie	Tee Set	Colossus	(25)
15.	Don't Cry Daddy	Elvis Presley	RCA	(11)
16.	Blowing Away	Fifth Dimension	Soul City	(17)
17.	Someday We'll Be Together	D. Ross/Supremes	Motown	(12)
18.	The Thrill Is Gone	B.B. King	Bluesway	(22)
*19.	Walkin' In The Rain	Jay/Americans	United Artists	(35)
20.	If I Never Knew Your Name	Vic Dana	Liberty	(18)
*21.	Ticket To Ride	Carpenters	A & M	(26)
*22.	Travelin' Band	C.C. Revival	Fantasy	(HB)
*23.	He Ain't Heavy, He's My Brother	Hollies	Epic	(37)
24.	I Want You Back	Jackson Five	Motown	(13)
25.	Winter World Of Love	Engelbert Humperdinck	Parrot	(15)
26.	Midnight Cowboy	Ferrante & Teicher	United Artists	(19)
27.	La La La (If I Had You)	Bobby Sherman	Metromedia	(14)
*28.	Cold turkey	Plastic Ono Band	Apple	(34)
29.	Wonderful World, Beautiful People	Jimmy Cliff	A & M	(16)
30.	Going In Circles	Friends Of Distinction	RCA	(20)
*31.	Rainy Night In Georgia	Brook Benton	Cotillion	(40)
*32.	She Came In Through The Bathroom Wind.	Joe Cocker	A & M	(38)
*33.	You Got Me Hummin'	Cold Blood	San Francisco	(39)
*34.	Shilo	Neil Diamond	Bang	(N)
35.	Evil Woman	Crow	Amaret	(28)
36.	Barbara, I Love You	New Colony Six	Mercury	(30)
37.	Always Something There To Remind Me	R.B. Greaves	Atco	(HB)
38.	Save The Country	Thelma Houston	Dunhill	(N)
39.	The Rapper	The Jaggerz	Kama Sutra	(HB)
40.	She Let Her Hair Down	Don Young	Bang	(N)

HIT BOUND

Rock Of All Ages	Badfinger	Apple
Give Me Just A Little More Time	The Chairman of the Board	Invictus
Baby Take Me In Your Arms	Jefferson	Janus

*Denotes Big 10 Climbers

WC
FL

BIG 10 COUNT DOWN

FOR THE WEEK OF FEBRUARY 2, 1970

				Last Week
*1.	No Time	Guess Who	RCA	(2)
2.	Thank You For Lettin' Me Be Myself	Sly/Family Stone	Epic	(1)
3.	Hey There Lonely Girl	Eddie Holman	ABC	(3)
*4.	Arizona	Mark Lindsay	Columbia	(10)
5.	Without Love (There Is Nothing)	Tom Jones	Parrot	(5)
6.	I'll Never Fall In Love Again	Dionne Warwick	Scepter	(6)
*7.	Psychedelic Shack	Temptations	Gordy	(9)
*8.	Walk A Mile In My Shoes	Joe South	Capitol	(11)
*9.	Ma Belle Amie	The Tee Set	Colossus	(14)
10.	"Venus"	Shocking Blue	Colossus	(4)
11.	Raindrops Keep Falling On My Head	B.J. Thomas	Scepter	(7)
*12.	He Ain't Heavy, He's My Brother	Hollies	Epic	(23)
*13.	Travelin' Band	C.C. Revival	Fantasy	(22)
*14.	The Thrill Is Gone	B.B. King	Bluesway	(18)
*15.	Ticket To Ride	Carpenters	A & M	(21)
*16.	The Rapper	The Jaggerz	Kama Sutra	(39)
*17.	Rainy Night In Georgia	Brook Benton	Cotillion	(31)
18.	Jam Up, Jelly Tight	Tommy Roe	ABC	(8)
19.	Whole Lotta Love	Led Zeppelin	Atlantic	(12)
20.	Walkin' In The Rain	Jay/Americans	United Artists	(19)
21.	Someday We'll Be Together	D. Ross/Supremes	Motown	(17)
22.	Don't Cry Daddy	Elvis Presley	RCA	(15)
23.	Jingle Jangle	Archies	Kirshner	(13)
24.	Blowing Away	Fifth Dimension	Soul City	(16)
25.	Winter World Of Love	Engelbert Humperdinck	Parrot	(25)
26.	Honey, Come Back	Glen Campbell	Capitol	(N)
27.	If I Never Knew Your Name	Vic Dana	Liberty	(20)
28.	Always Something There To Remind Me	R.B. Greaves	Atco	(37)
29.	Cold Turkey	Plastic Ono Band	Apple	(28)
30.	Breaking Up Is Hard To Do	Lenny Welch	Commonwealth United	(N)
31.	Who'll Stop The Rain	C.C.Revival	Fantasy	(N)
32.	Save The Country	Thelma Houston	Dunhill	(38)
33.	Bridge Over Troubled Water	Simon & Garfunkel	Columbia	(N)
34.	She Came In Through The Bathroom Wind.	Joe Cocker	A & M	(32)
35.	You Got Me Hummin'	Cold Blood	San Francisco	(33)
36.	Rock Of All Ages	Badfinger	Apple	(HB)
37.	She Let Her Hair Down	Don Young	Bang	(40)
38.	Baby Take Me In Your Arms	Jefferson	Janus	(HB)
39.	Give Me Just A Little More Time	Chairmen Of The Board	Invictus	(HB)
*40.	Until It's Time For You To Go	Neil Diamond	Uni	(N)

HIT BOUND

Kentucky Rain	Elvis Presley	RCA
Easy Come, Easy Go	Bobby Sherman	Metromedia

*Denotes Big 10 Climbers

WC
FL

BIG 10 COUNT DOWN

FOR THE WEEK OF FEBRUARY 9, 1970

				Last Week
*1.	No Time	Guess Who	RCA	(1)
2.	Hey There Lonely Girl	Eddie Holman	ABC	(3)
*3.	Psychedelic Shack	Temptations	Gordy	(7)
4.	Arizona	Mark Lindsay	Columbia	(4)
5.	Thank You For Lettin' Me Be Myself	Sly/Family Stone	Epic	(2)
6.	I'll Never Fall In Love Again	Dionne Warwick	Scepter	(6)
*7.	Ma Belle Amie	The Tee Set	Colossus	(9)
*8.	Bridge Over Troubled Water	Simon & Garfunkel	Columbia	(33)
*9.	He Ain't Heavy, He's My Brother	Hollies	Epic	(12)
*10.	Traveliln' Band/Who'll Stop The Rain	Creedence Clearwater Revival	Fantasy	(13)
*11.	Walkin' In The Rain	Jay/Americans	United Artists	(20)
*12.	The Rapper	The Jaggerz	Kama Sutra	(16)
13.	Without Love (There Is Nothing)	Tom Jones	Parrot	(5)
14.	Ticket To Ride	Carpenters	A & M	(15)
15.	Rainy Night In Georgia	Brook Benton	Cotillion	(17)
16.	Walk A Mile In My Shoes	Joe South	Capitol	(8)
17.	Raindrops Keep Falling On My Head	B.J. Thomas	Scepter	(11)
18.	"Venus"	Shocking Blue	Colossus	(10)
19.	The Thrill Is Gone	B.B. King	Bluesway	(14)
*20.	Always Something There To Remind Me	R.B. Greaves	Atco	(28)
*21.	Give Me Just A Little More Time	Chairmen Of The Board	Invictus	(39)
*22.	Baby Take Me In Your Arms	Jefferson	Janus	(38)
23.	Save The Country	Thelma Houston	Dunhill	(32)
24.	Cold Turkey	Plastic Ono Band	Apple	(29)
25.	Honey, Come Back	Glen Campbell	Capitol	(26)
26.	Someday We'll Be Together	D. Ross/Supremes	Motown	(21)
27.	Whole Lotta Love	Led Zeppelin	Atlantic	(19)
28.	Jam Up, Jelly Tight	Tommy Roe	ABC	(18)
29.	Breaking Up Is Hard To Do	Lenny Welch	Commonwealth United	(30)
*30.	She Let Her Hair Down	Don Young	Bang	(37)
31.	Blowing Away	Fifth Dimension	Soul City	(24)
32.	Winter World Of Love	Engelbert Humperdinck	Parrot	(25)
33.	You Got Me Hummin'	Cold Blood	San Francisco	(35)
34.	Rock Of All Ages	Badfinger	Apple	(36)
35.	Until It's Time For You To Go	Neil Diamond	Uni	(40)
36.	If I Never Knew Your Name	Vic Dana	Liberty	(27)
37.	Didn't I (Blow Your Mind This Time)	Delfonics	Philly Groove	(N)
38.	Oh Me, Oh My	Lulu	Atco	(N)
39.	Easy Come, Easy Go	Bobby Sherman	Metromedia	(HB)
*40.	Love Grows (Where My Rosemary Goes)	Edison Lighthouse	Bell	(N)

HIT BOUND

Gotta Get Back To You Baby	Tommy James/Shondells	Roulette
Celebrate	Three Dog Night	Dunhill

*Denotes Big 10 Climbers

WC
FL

BIG 10 COUNT DOWN

FOR THE WEEK OF FEBRUARY 16, 1970

				Last Week
*1.	Bridge Over Troubled Water	Simon & Garfunkel	Columbia	(8)
2.	Hey There Lonely Girl	Eddie Holman	ABC	(2)
*3.	The Rapper	The Jaggerz	Kama Sutra	(12)
4.	No Time	Guess Who	RCA	(1)
5.	Arizona	Mark Lindsay	Columbia	(4)
6.	Psychedelic Shack	Temptations	Gordy	(3)
7.	Thank You For Lettin' Me Be Myself	Sly/Family Stone	Epic	(5)
*8.	He Ain't Heavy, He's My Brother	Hollies	Epic	(9)
*9.	Travelin' Band/Who'll Stop The Rain	Creedence Clearwater Revival	Fantasy	(10)
*10.	Give Me Just A Little More Time	Chairmen Of The Board	Invictus	(21)
11.	Rainy Night In Georgia	Brook Benton	Cotillion	(15)
12.	Ma Belle Amie	The Tee Set	Colossus	(7)
13.	I'll Never Fall In Love Again	Dionne Warwick	Scepter	(6)
14.	Walkin' In The Rain	Jay/Americans	United Artists	(11)
*15.	Baby Take Me In Your Arms	Jefferson	Janus	(22)
16.	Without Love (There Is Nothing)	Tom Jones	Parrot	(13)
17.	Always Something There To Remind Me	R.B. Greaves	Atco	(20)
*18.	She Let Her Hair Down	Don Young	Bang	(30)
19.	Ticket To Ride	Carpenters	A & M	(14)
*20.	Didn't I (Blow Your Mind This Time)	Delfonics	Philly Groove	(37)
21.	Raindrops Keep Fallin' On My Head	B.J. Thomas	Scepter	(17)
*22.	Easy Come, Easy Go	Bobby Sherman	Metromedia	(39)
23.	Walk A Male In My Shoes	Joe South	Capitol	(16)
*24.	Love Grows (Where My Rosemary Goes)	Edison Lighthouse	Bell	(40)
25.	"Venus"	Shocking Blue	Colossus	(18)
26.	The Thrill Is Gone	B.B. King	Bluesway	(19)
27.	Save The Country	Thelma Houston	Dunhill	(23)
*28.	House Of The Rising Sun	Frijid Pink	Parrot	(N)
*29.	You Got Me Hummin'	Cold Blood	San Francisco	(33)
*30.	Kentucky Rain	Elvis Presley	RCA	(N)
31.	Oh Me, Oh My	Lulu	Atco	(38)
*32.	Why Should I Cry	The Gentrys	Sun	(N)
33.	Rock Of All Ages	Badfinger	Apple	(34)
34.	Until It's Time For You To Go	Neil Diamond	Uni	(35)
35.	Cold Turkey	Plastic Ono Band	Apple	(24)
*36.	Celebrate	Three Dog Night	Dunhill	(HB)
37.	Gotta Get Back To You	Tommy James/Shondells	Roulette	(HB)
38.	Blowing Away	Fifth Dimension	Soul City	(31)
39.	Evil Ways	Santana	Columbia	(N)
40.	Winter World Of Love	Engelbert Humperdinck	Parrot	(32)

HIT BOUND

Time To Get It Together	Country Coalition	Bluesway
Spirit In The Sky	Norman Greenbaum	Reprise

*Denotes Big 10 Climbers

BIG 10 COUNT DOWN

FOR THE WEEK OF FEBRUARY 23, 1970

				Last Week
*1.	Bridge Over Troubled Water	Simon & Garfunkel	Columbia	(1)
*2.	The Rapper	The Jaggerz	Kama Sutra	(3)
*3.	Give Me Just A Little More Time	Chairmen Of The Board	Invictus	(10)
4.	Hey There Lonely Girl	Eddie Holman	ABC	(2)
*5.	Rainy Night In Georgia	Brook Benton	Cotillion	(11)
6.	He Ain't Heavy, He's My Brother	Hollies	Epic	(8)
7.	Travelin' Band/Who'll Stop The Rain	Creedence Clearwater Revival	Fantasy	(9)
8.	Psychedelic Shack	Temptations	Gordy	(6)
9.	Arizona	Mark Lindsay	Columbia	(5)
*10.	Didn't I (Blow Your Mind This Time)	Delfonics	Philly Groove	(20)
11.	No Time	Guess Who	RCA	(4)
12.	Thank You For Lettin' Me Be Myself	Sly/Family Stone	Epic	(7)
13.	Ma Belle Amie	The Tee Set	Colossus	(12)
*14.	House Of The Rising Sun	Frijid Pink	Parrot	(28)
15.	Baby Take Me In Your Arms	Jefferson	Janus	(15)
*16.	Love Grows (Where My Rosemary Goes)	Edison Lighthouse	Bell	(24)
17.	She Let Her Hair Down	Don Young	Bang	(18)
*18.	Kentucky Rain	Elvis Presley	RCA	(30)
19.	Always Something There To Remind Me	R.B. Greaves	Atco	(17)
20.	Easy Come, Easy Go	Bobby Sherman	Meteromedia	(22)
*21.	Oh Me, Oh My	Lulu	Atco	(31)
*22.	Celebrate	Three Dog Night	Dunhill	(36)
23.	Walkin' In The Rain	Jay/Americans	United Artists	(14)
24.	I'll Never Fall In Love Again	Dionne Warwick	Scepter	(13)
25.	Raindrops Keep Fallin' On My Head	B.J. Thomas	Scepter	(21)
26.	Ticket To Ride	Carpenters	A & M	(19)
27.	Without Love (There Is Nothing)	Tom Jones	Parrot	(16)
*28.	Gotta Get Back To You	Tommy James/Shondells	Roulette	(37)
*29.	Evil Ways	Santana	Columbia	(39)
30.	Why Should I Cry	The Gentrys	Sun	(32)
31.	You Got Me Hummin'	Cold Blood	San Francisco	(29)
32.	Rock Of All Ages/Come And Get It	Badfinger	Apple	(33)
33.	Until It's Time For You To Go	Neil Diamond	Uni	(34)
34.	"Venus"	Shocking Blue	Colossus	(25)
35.	Time To Get It Together	Country Coalition	Bluesway	(HB)
36.	Spirit In The Sky	Normam Greenbaum	Reprise	(HB)
37.	Instant Karma	John Ono Lennon	Apple	(N)
38.	Take A Look Around	Smith	Dunhill	(N)
39.	Gotta Hold On To This Feeling	Jr. Walker/All Stars	Soul	(N)
40.	Hello It's Me	Nazz	SGC	(N)

HIT BOUND

Stir It Up And Serve It	Tommy Roe	ABC
Who's Your Baby?	The Archies	Kirshner

*Denotes Big 10 Climbers

WC
FL

BIG 10 COUNT DOWN

FOR THE WEEK OF MARCH 2, 1970

				Last Week
*1.	Bridge Over Troubled Water	Simon & Garfunkel	Columbia	(1)
*2.	The Rapper	The Jaggerz	Kama Sutra	(2)
*3.	Give Me Just A Little More Time	Chairmen Of The Board	Invictus	(3)
*4.	Didn't I (Blow Your Mind This Time)	Delfonics	Philly Grooves	(10)
5.	Rainy Night In Georgia	Brook Benton	Cotillion	(5)
*6.	Houses Of The Rising Sun	Frijid Pink	Parrot	(14)
*7.	Love Grows (Where My Rosemary Goes)	Edison Lighthouse	Bell	(16)
8.	He Ain't Heavy, He's My Brother	Hollies	Epic	(6)
9.	Hey There Lonely Girl	Eddie Holman	ABC	(4)
*10.	Kentucky Rain	Elvis Presley	RCA	(18)
11.	Travelin' Band/Who'll Stop The Rain	Creedence Clearwater Revival	Fantasy	(7)
12.	Psychedelic Shack	Temptations	Gordy	(8)
*13.	Celebrate	Three Dog Night	Dunhill	(22)
14.	Arizona	Mark Lindsay	Columbia	(9)
*15.	Oh Me, Oh My	Lulu	Atco	(21)
*16.	Easy Come, Easy Go	Bobby Sherman	Metromedia	(20)
17.	Baby Take Me In Your Arms	Jefferson	Janus	(15)
18.	Thank You For Lettin' Me Be Myself	Sly/Family Stone	Epic	(12)
*19.	Spirit In The Sky	Norman Greenbaum	Reprise	(36)
*20.	Instant Karma	John Ono Lennon	Apple	(37)
21.	She Let Her Hair Down	Dan Young	Bang	(17)
22.	No Time	Guess Who	RCA	(11)
23.	Ma Belle Amie	The Tee Set	Colossus	(13)
*24.	Evil Ways	Santana	Columbia	(29)
*25.	Take A Look Around	Smith	Dunhill	(38)
26.	Ticket To Ride	Carpenters	A & M	(26)
27.	Gotta Get Back To You	Tommy James/Shondells	Roulette	(28)
*28.	Come And Get It	Badfinger	Apple	(32)
29.	Always Something There To Remind Me	R.B. Greaves	Atco	(19)
30.	Until It's Time For You To go	Neil Diamond	Uni	(33)
31.	Gotta Hold On To This Feeling	Jr. Walker/All Stars	Soul	(39)
32.	Time To get It Together	Country Coalition	Bluesway	(35)
33.	Raindrops Keep Fallin' On My Head	B.J. Thomas	Scepter	(25)
34.	Why Should I Cry	The Gentrys	Sun	(30)
35.	Call Me	Aretha Franklin	Atlantic	(N)
36.	Stir It Up And Serve It	Tommy Roe	ABC	(HB)
37.	Who's Your Baby?	The Archies	Kirshner	(HB)
38.	Up The Ladder To The Roof	The Supremes	Motown	(N)
39.	Hello It's Me	Nazz	SGC	(40)
40.	Long Lonesome Highway	Michael Parks	MGM	(N)

HIT BOUND

Capture The Moment	Jay/Americans	United Artists
Love Or Let Me Be Lonely	The Friends of Distinction	RCA
Let It Be	The Beatles	Apple

*Denotes Big 10 Climbers

WC
FL

BIG 10 COUNT DOWN

FOR THE WEEK OF MARCH 9, 1970

				Last Week
1.	Bridge Over Troubled Water	Simon & Garfunkel	Columbia	(1)
*2.	House Of The Rising Sun	Frijid Pink	Parrot	(6)
*3.	Love Grows (Where My Rosemary Goes)	Edison Lighthouse	Bell	(7)
4.	Give Me Just A Little More Time	Chairmen Of The Board	Invictus	(3)
5.	Didn't I (Blow Your Mind This Time)	Delfonics	Philly Groove	(4)
6.	The Rapper	The Jaggerz	Kama Sutra	(2)
*7.	Celebrate	Three Dog Night	Dunhill	(13)
*8.	Spirit In The Sky	Norman Greenbaum	Reprise	(19)
*9.	Instant Karma	John Ono Lennon	Apple	(20)
10.	Kentucky Rain	Elvis Presley	RCA	(10)
11.	Easy Come, Easy Go	Bobby Sherman	Metromedia	(16)
*12.	Evil Ways	Santana	Columbia	(24)
13.	Rainy Night In Georgia	Brook Benton	Cotillion	(5)
14.	He Ain't Heavy, He's My Brother	Hollies	Epic	(8)
15.	Oh Me, Oh My	Lulu	Atco	(15)
16.	Psychedelic Shack	Temptations	Gordy	(12)
17.	Arizona	Mark Lindsay	Columbia	(14)
18.	Hey There Lonely Girl	Eddie Holman	ABC	(9)
19.	Who'll Stop The Rain/Travelin' Band	Creedence Clearwater Revival	Fantasy	(11)
*20.	Call Me	Aretha Franklin	Atlantic	(35)
21.	Come And Get It	Badfinger	Apple	(28)
*22.	Who's Your Baby?	The Archies	Kirshner	(37)
*23.	Stir It Up And Serve It	Tommy Roe	ABC	(36)
24.	Take A Look Around	Smith	Dunhill	(25)
25.	Gotta Hold On To This Feeling	Jr. Walker/All Stars	Soul	(31)
*26.	Up The Ladder To The Roof	The Supremes	Motown	(38)
*27.	Let It Be	The Beatles	Apple	(HB)
28.	Thank You For Lettin' Me Be Myself	Sly/Family Stone	Epic	(18)
29.	Gotta Get Back To You	Tommy James/Shondells	Roulette	(27)
30.	No Time	Guess Who	RCA	(22)
31.	Time To Get It Together	Country Coalition	Bluesway	(32)
32.	She Let Her Hair Down	Don Young	Bang	(21)
*33.	Long Lonesome Highway	Michael Parks	MGM	(40)
*34.	ABC	Jackson Five	Motown	(N)
35.	Ma Belle Amie	The Tee Set	Colossus	(23)
36.	Jennifer Tomkins	Street People	Musicor	(N)
37.	Capture The Moment	Jay/Americans	United Artists	(HB)
38.	Love Or Let Me Be Lonely	The Friends Of Distinction	RCA	(HB)
39.	Raindrops Keep Fallin' On My Head	B.J. Thomas	Scepter	(33)
40.	Baby Take Me In Your Arms	Jefferson	Janus	(17)

HIT BOUND

Vehicle	Ides Of March	Columbia

*Denotes Big 10 Climbers

WC
FL

BIG 10 COUNT DOWN

FOR THE WEEK OF MARCH 16, 1970

				Last Week
1.	Spirit In The Sky	Norman Greenbaum	Reprise	(8)
2.	House Of The Rising Sun	Frijid Pink	Parrot	(2)
3.	Love Grows (Where My Rosemary Goes)	Edison Lighthouse	Bell	(3)
*4.	Instant Karma	John Ono Lennon	Apple	(9)
5.	Bridge Over Troubled Water	Simon & Garfunkel	Columbia	(1)
*6.	Evil Ways	Santana	Columbia	(12)
7.	Celebrate	Three Dog Night	Dunhill	(7)
8.	Didn't I (Blow Your Mind This Time)	Delfonics	Philly Groove	(5)
*9.	Let It Be	The Beatles	Apple	(27)
10.	Give Me Just A Little More Time	Chairmen Of The Board	Invictus	(4)
11.	Easy Come, Easy Go	Bobby Sherman	Metromedia	(11)
12.	The Rapper	The Jaggerz	Kama Sutra	(6)
*13.	Come And Get It	Badfinger	Apple	(21)
*14.	ABC	Jackson Five	Motown	(34)
*15.	Who's Your Baby?	The Archies	Kirshner	(22)
*16.	Up The Ladder To The Roof	The Supremes	Motown	(26)
17.	Kentucky Rain	Elvis Presley	RCA	(10)
18.	Call Me	Aretha Franklin	Atlantic	(20)
*19.	Jennifer Tomkins	Street People	Musicor	(36)
*20.	Long Lonesome Highway	Michael Parks	MGM	(33)
21.	Oh Me, Oh My	Lulu	Atco	(15)
22.	Stir It Up And Serve It	Tommy Roe	ABC	(23)
23.	Gotta Hold On To This Feeling	Jr. Walker/All Stars	Soul	(25)
24.	Rainy Night In Georgia	Brook Benton	Cotillion	(13)
25.	He Ain't Heavy, He's My Brother	Hollies	Epic	(14)
26.	Psychedelic Shack	Temptations	Gordy	(16)
27.	Arizona	Mark Lindsay	Columbia	(17)
28.	Take A Look Around	Smith	Dunhill	(24)
29.	Hey There Lonely Girl	Eddie Holman	ABC	(18)
*30.	Capture The Moment	Jay/Americans	United Artists	(37)
31.	Time To Get It Together	Country Coalition	Bluesway	(31)
*32.	Vehicle	Ides Of March	Warner Bros.	(HB)
*33.	"Do The Funky Chicken"	Rufus Thomas	Stax	(N)
34.	Who'll Stop The Rain	Creedence Clearwater Revival	Fantasy	(19)
35.	Love Or Let Me Be Lonely	The Friends Of Distinction	RCA	(38)
*36.	Reflections Of My Life	The Marmalade	London	(N)
*37.	"Mighty Joe"	The Shocking Blue	Colossus	(N)
*38.	Turn Back The Hands Of Time	Tyrone Davis	Dakar	(N)
*39.	Shilo	Neil Diamond	Bang	(N)
40.	Thank You For Lettin' Me Be Myself	Sly/Family Stone	Epic	(28)

HIT BOUND

Cecilia	Simon & Garfunkel	Columbia
Angelica	Oliver	Crewe

*Denotes Big 10 Climbers

WC
FL

BIG 10 COUNT DOWN

FOR THE WEEK OF MARCH 23, 1970

				Last Week
*1.	Let It Be	The Beatles	Apple	(9)
2.	Spirit In The Sky	Norman Greenbaum	Reprise	(1)
3.	Instant Karma	John Ono Lennon	Apple	(4)
4.	Love Grows (Where My Rosemary Goes)	Edison Lighthouse	Bell	(3)
5.	House Of The Rising Sun	Frijid Pink	Parrot	(2)
6.	Evil Ways	Santana	Columbia	(6)
*7.	ABC	Jackson Five	Motown	(14)
*8.	Come And Get It	Badfinger	Apple	(13)
9.	Bridge Over Troubled Water	Simon & Garfunkel	Columbia	(5)
10.	Easy Come, Easy Go	Bobby Sherman	Metromedia	(11)
11.	Celebrate	Three Dog Night	Dunhill	(7)
*12.	Long Lonesome Highway	Michael Parks	MGM	(20)
*13.	Up The Ladder To The Roof	The Supremes	Motown	(16)
14.	Who's Your Baby?	The Archies	Kirshner	(15)
*15.	Jennifer Tomkins	Street People	Musicor	(19)
16.	Didn't I (Blow Your Mind This Time)	Delfonics	Philly Groove	(8)
*17.	Vehicle	Ides Of March	Warner Bros.	(32)
18.	The Rapper	The Jaggerz	Kama Sutra	(12)
19.	Give Me Just A Little More Time	Chairmen Of The Board	Invictus	(10)
20.	Call Me	Aretha Franklin	Atlantic	(18)
*21.	Shilo	Neil Diamond	Bang	(39)
22.	Stir It Up And Serve It	Tommy Roe	ABC	(22)
23.	Gotta Hold On To This Feeling	Jr. Walker/All Stars	Soul	(23)
24.	Oh Me, Oh My	Lulu	Atco	(21)
*25.	Reflections Of My Life	The Marmalade	London	(36)
*26.	Capture The Moment	Jay/Americans	United Artists	(30)
27.	"Do The Funky Chicken"	Rufus Thomas	Stax	q
28.	Kentucky Rain	Elvis Presley	RCA	(17)
29.	Love Or Let Me Be Lonely	Friends Of Distinction	RCA	(35)
30.	"Mighty Joe"	The Shocking Blue	Colossus	(37)
*31.	Turn Back The Hands Of Time	Tyrone Davis	Dakar	(38)
*32.	American Woman/No Sugar Tonight	The Guess Who	RCA	(N)
*33.	Tennessee Bird Walk	Jack Blanchard & Misty Morgan	Wayside	(N)
34.	He Ain't Heavy, He's My Brother	Hollies	Epic	(25)
35.	Arizona	Mark Lindsay	Columbia	(27)
*36.	Temma Harbour	Mary Hopkin	Apple	(N)
*37.	For The Love Of Him	Bobbi Martin	United Artists	(N)
38.	Psychedelic Shack	Temptations	Gordy	(26)
39.	Something's Burning	Kenny Rogers/First Edition	Reprise	(N)
40.	Rainy Night in Georgia	Brook Benton	Cotillion	(24)

HIT BOUND

Little Green Bag	George Baker Selection	Colossus
Which Way You Goin' Billy?	The Poppy Family	London

*Denotes Big 10 Climbers

WC
FL

BIG 10 COUNT DOWN

FOR THE WEEK OF MARCH 30, 1970

				Last Week
1.	Let It Be	The Beatles	Apple	(1)
*2.	ABC	Jackson Five	Motown	(7)
3.	Instant Karma	John Ono Lennon	Apple	(3)
*4.	Come And Get It	Badfinger	Apple	(8)
5.	Spirit In The Sky	Norman Greenbaum	Reprise	(2)
*6.	Up The Ladder To The Roof	The Supremes	Motown	(13)
7.	Love Grows (Where My Rosemary Goes)	Edison Lighthouse	Bell	(4)
8.	Easy Come, Easy Go	Bobby Sherman	Metromedia	(10)
9.	Long Lonesome Highway	Michael Parks	MGM	(12)
*10.	Vehicle	Ides Of March	Warner Bros.	(17)
11.	House Of The Rising Sun	Frijid Pink	Parrot	(5)
12.	Bridge Over Troubled Water	Simon & Garfunkel	Columbia	(9)
13.	Who's Your Baby?	The Archies	Kirshner	(14)
14.	Evil Ways	Santana	Columbia	(6)
15.	Jennifer Tomkins	Street People	Musicor	(15)
*16.	American Woman/No Sugar Tonight	The Guess Who	RCA	(32)
*17.	Reflections Of My Life	The Marmalade	London	(25)
18.	Celebrate	Three Dog Night	Dunhill	(11)
*19.	For The Love Of Him	Bobbi Martin	United Artists	(37)
*20.	Something's Burning	Kenny Rogers/First Edition	Reprise	(39)
21.	Shilo	Neil Diamond	Bang	(21)
22.	Didn't I (Blow Your Mind This Time)	Delfonics	Philly Groove	(16)
23.	Stir It Up And Serve It	Tommy Roe	ABC	(22)
*24.	Turn Back The Hands Of Time	Tyrone Davis	Dakar	(31)
25.	Love Or Let Me Be Lonely	Friends OF Distinction	RCA	(29)
26.	Capture The Moment	Jay/Americans	United Artists	(26)
27.	The Rapper	The Jaggerz	Kama Sutra	(18)
*28.	Tennessee Bird Walk	Jack Blanchard & Misty Morgan	Wayside	(33)
29.	"Mighty Joe"	The Shocking Blue	Colossus	(30)
30.	Temma Harbour	Mary Hopkin	Apple	(36)
*31.	Which Way You Goin' Billy?	The Poppy Family	London	(HB)
32.	Give Me Just A Little More Time	Chairmen Of The Board	Invictus	(19)
33.	Gotta Hold On To This Feeling	Jr. Walker/All Stars	Soul	(23)
34.	Call Me	Aretha Franklin	Atlantic	(20)
*35.	Hey Lawdy Mama	Steppenwolf	Dunhill	(N)
*36.	Little Green Bag	George Baker Selection	Colossus	(HB)
37.	Kentucky Rain	Elvis Presley	RCA	(28)
*38.	Fire And Rain	R.B. Greaves	Atlantic	(N)
39.	Oh Me, Oh My	Lulu	Atco	(24)
40.	He Ain't Heavy, He's My Brother	Hollies	Epic	(34)

HIT BOUND

What Is Truth	Johnny Cash	Columbia
Airport Love Theme	Vincent Bell	Decca

*Denotes Big 10 Climbers

WC
FL

BIG 10 COUNT DOWN

FOR THE WEEK OF APRIL 6, 1970

				Last Week
*1.	ABC	Jackson Five	Motown	(2)
2.	Let It Be	The Beatles	Apple	(1)
*3.	Up The Ladder To The Roof	The Supremes	Motown	(6)
4.	Come And Get It	Badfinger	Apple	(4)
5.	Instant Karma	John Ono Lennon	Apple	(3)
*6.	Long Lonesome Highway	Michael Parks	MGM	(9)
7.	Spirit In The Sky	Norman Greenbaum	Reprise	(5)
*8.	Vehicle	Ides Of March	Warner Bros.	(10)
9.	Easy Come, Easy Go	Bobby Sherman	Metromedia	(8)
*10.	American Woman/No Sugar Tonight	The Guess Who	RCA	(16)
*11.	Something's Burning	Kenny Rogers/First Edition	Reprise	(20)
*12.	Shilo	Neil Diamond	Bang	(21)
*13.	Reflections Of My Life	The Marmalade	London	(17)
14.	Love Grows (Where My Rosemary Goes)	Edison Lighthouse	Bell	(7)
15.	Bridge Over Troubled Water	Simon & Garfunkel	Columbia	(12)
*16.	Turn Back The Hands Of Time	Tyrone Davis	Dakar	(24)
*17.	For The Love Of Him	Bobbi Martin	United Artists	(19)
*18.	Which Way You Goin' Billy?	The Poppy Family	London	(31)
*19.	Hey Lawdy Mama	Steppenwolf	Dunhill	(35)
*20.	Little Green Bag	George Baker Selection	Colossus	(36)
21.	Tennessee Bird Walk	Jack Blanchard & Misty Morgan	Wayside	(28)
22.	House Of The Rising Sun	Frijid Pink	Parrot	(11)
23.	Who's Your Baby?	The Archies	Kirshner	(13)
24.	Love Or Let Me Be Lonely	Friends Of Distinction	RCA	(25)
25.	Jennifer Tomkins	Street People	Musicor	(15)
26.	Evil Ways	Santana	Columbia	(14)
27.	Celebrate	Three Dog Night	Dunhill	(18)
28.	"Mighty Joe"	The Shocking Blue	Colossus	(29)
29.	Didn't I (Blow Your Mind This Time)	Delfonics	Philly Groove	(22)
30.	Fire And Rain	R.B. Greaves	Atlantic	(38)
*31.	You're The One	Little Sister	Stone Flower	(N)
32.	Nothing Succeeds Like Success	Bill Deal/Rhondels	Heritage	(N)
33.	What Is Truth	Johnny Cash	Columbia	(HB)
34.	You Need Love Like I Do	Gladys Knight/Pips	Soul	(N)
35.	Airport Love Theme	Vincent Bell	Decca	(HB)
36.	The Rapper	The Jaggerz	Kama Sutra	(27)
37.	Temma Harbour	Mary Hopkin	Apple	(30)
38.	Give Me Just A Little More Time	Chairmen Of The Board	Invictus	(32)
39.	Capture The Moment	Jay/Americans	United Artists	(N)
40.	Everybody's Out Of Town	B.J. Thomas	Scepter	(N)

HIT BOUND

Get Ready	Rare Earth	Rare Earth

*Denotes Big 10 Climbers

WC
FL

BIG 10 COUNT DOWN

FOR THE WEEK OF APRIL 13, 1970

				Last Week
1.	ABC	Jackson Five	Motown	(1)
2.	Let It Be	The Beatles	Apple	(2)
*3.	American Woman/No Sugar Tonight	The Guess Who	RCA	(10)
*4.	Vehicle	Ides Of March	Warner Bros.	(8)
*5.	Something's Burning	Kenny Rogers/First Edition	Reprise	(11)
6.	Long Lonesome Highway	Michael Parks	MGM	(6)
7.	Up The Ladder To The Roof	The Supremes	Motown	(3)
8.	Come And Get It	Badfinger	Apple	(4)
*9.	Turn Back The Hands Of Time	Tyrone Davis	Dakar	(16)
10.	Reflections Of My Life	The Marmalade	London	(13)
11.	Instant Karma	John Ono Lennon	Apple	(5)
12.	Shilo	Neil Diamond	Bang	(12)
*13.	Which Way You Goin' Billy?	The Poppy Family	London	(18)
14.	For The Love Of Him	Bobbi Martin	United Artists	(17)
*15.	Hey Lawdy Mama	Steppenwolf	Dunhill	(19)
*16.	Little Green Bag	George Baker Selection	Colossus	(20)
17.	Easy Come, Easy Go	Bobby Sherman	Metromedia	(9)
*18.	Love Or Let Me Be Lonely	Friends Of Distinction	RCA	(24)
19.	Spirit In The Sky	Norman Greenbaum	Reprise	(7)
20.	Tennessee Bird Walk	Jack Blanchard & Misty Morgan	Wayside	(21)
*21.	You Need Love Like I Do	Gladys Knight/Pips	Soul	(34)
22.	Love Grows (Where My Rosemary Goes)	Edison Lighthouse	Bell	(14)
23.	Bridge Over Troubled Water	Simon & Garfunkel	Columbia	(15)
*24.	What Is Truth	Johnny Cash	Columbia	(33)
25.	Nothing Succeeds Like Success	Bill Deal/Rhondels	Heritage	(32)
*26.	Cecilia	Simon & Garfunkel	Columbia	(N)
27.	Airport Love Theme	Vincent Bell	Decca	(35)
28.	You're The One	Little Sister	Stone Flower	(31)
*29.	Hitchin' A Ride	Vanity Fare	Page One	(N)
30.	Fire And Rain	R.B. Greaves	Atlantic	(30)
31.	House Of The Rising Sun	Frijid Pink	Parrot	(22)
32.	Jennifer Tomkins	Street People	Musicor	(25)
*33.	Get Ready	Rare Earth	Rare Earth	(HB)
*34.	The Seeker	The Who	Decca	(N)
*35.	Up Around The Bend	Creedence Clearwater Revival	Fantasy	(N)
36.	Who's Your Baby?	The Archies	Kirshner	(23)
37.	Everybody's Out Of Town	B.J. Thomas	Scepter	(N)
38.	Didn't I (Blow Your Mind This Time)	Delfonics	Philly Groove	(29)
39.	Evil Ways	Santana	Columbia	(26)
40.	Celebrate	Three Dog Night	Dunhill	(27)

HIT BOUND

Puppet Man	Fifth Dimension	Bell
The Letter	Joe Cocker	A & M
Grover Henson Feels Forgotten	Bill Crosby	Uni

*Denotes Big 10 Climbers

WC
FL

BIG 10 COUNT DOWN

FOR THE WEEK OF APRIL 20, 1970

				Last Week
*1.	American Woman/No Sugar Tonight	The Guess Who	RCA	(3)
*2.	Something's Burning	Kenny Rogers/First Edition	Reprise	(5)
3.	Vehicle	Ides Of March	Warner Bros.	(4)
4.	ABC	Jackson Five	Motown	(1)
5.	Let It Be	The Beatles	Apple	(2)
*6.	For The Love Of Him	Bobbi Martin	United Artists	(14)
*7.	Reflections Of My Life	The Marmalade	London	(10)
8.	Turn Back The Hands Of Time	Tyrone Davis	Dakar	(9)
*9.	Which Way You Goin' Billy?	The Poppy Family	London	(13)
*10.	Little Green Bag	George Baker Selection	Colossus	(16)
11.	Up The Ladder To The Roof	The Supremes	Motown	(7)
*12.	Love Or Let Me Be Lonely	Friends Of Distinction	RCA	(18)
13.	Come And Get It	Badfinger	Apple	(8)
14.	Hey Lawdy Mama	Stepenwolf	Dunhill	(15)
15.	Long Lonesome Highway	Michael Parks	MGM	(6)
*16.	Cecilia	Simon & Garfunkel	Columbia	(26)
*17.	Hitchin' A Ride	Vanity Fare	Page One	(29)
*18.	Nothing Succeeds Like Success	Bill Deal/Rhondels	Heritage	(25)
*19.	You Need Love Like I Do	Gladys Knight/Pips	Soul	(21)
20.	Shilo	Neil Diamond	Bang	(12)
21.	What Is Truth	Johnny Cash	Columbia	(24)
22.	Instant Karma	John Ono Lennon	Apple	(11)
*23.	Up Around The Bend	Creedence Clearwater Revival	Fantasy	(35)
*24.	The Seeker	The Who	Decca	(34)
*25.	Get Ready	Rare Earth	Rare Earth	(33)
26.	Airport Love Theme	Vincent Bell	Decca	(27)
27.	Tennessee Bird Walk	Jack Blanchard & Misty Morgan	Wayside	(20)
28.	Spirit In The Sky	Norman Greenbaum	Reprise	(19)
29.	Fire And Rain	R.B. Greaves	Atlantic	(30)
30.	Easy Come, Easy Go	Bobby Sherman	Metromedia	(17)
31.	Bridge Over Troubled Water	Simon & Garfunkel	Columbia	(23)
32.	You're The One	Little Sister	Stone Flower	(28)
33.	Love Grows (Where My Rosemary Goes)	Edison Lighthouse	Bell	(22)
34.	Everybody's Out Of Town	B.J. Thomas	Scepter	(37)
35.	Puppet Man	Fifth Dimension	Bell	(HB)
36.	The Letter	Joe Cocker	A &M	(HB)
37.	Love On A Two-Way Street	The Moments	Stang	(N)
38.	Soolaimon	Neil Diamond	Uni	(N)
39.	Come Running	Van Morrison	Warner Bros.	(N)
40.	Make Me Smile	Chicago	Columbia	(N)

HIT BOUND

My Baby Loves Lovin'	White Plains	Deram

*Denotes Big 10 Climbers

WC
FL

BIG 10 COUNT DOWN

WCFL
SPECIAL MERIT AWARD
AFTER 20 WEEKS – STILL REPORTED "TOP TEN"
"RAINDROPS KEEP FALLIN' ON MY HEAD" – B. J. THOMAS

FOR THE WEEK OF APRIL 27, 1970

				Last Week
*1.	American Woman/No Sugar Tonight	The Guess Who	RCA	(1)
*2.	Something's Burning	Kenny Rogers/First Edition	Reprise	(2)
*3.	Little Green Bag	George Baker Selection	Colossus	(10)
4.	Vehicle	Ides Of March	Warner Bros.	(3)
5.	For The Love Of Him	Bobbi Martin	United Artists	(6)
6.	ABC	Jackson Five	Motown	(4)
7.	Let It Be	The Beatles	Apple	(5)
*8.	Love Or Let Me Be Lonely	Friends Of Distinction	RCA	(12)
*9.	Cecilia	Simon & Garfunkel	Columbia	(16)
*10.	Hitchin' A Ride	Vanity Fare	Page One	(17)
11.	Reflections Of My Life	The Marmalade	London	(7)
12.	Which Way You Goin' Billy?	The Poppy Family	London	(9)
13.	Turn Back The Hands Of Time	Tyrone Davis	Dakar	(8)
*14.	Up Around The Bend	Creedence Clearwater Revival	Fantasy	(23)
*15.	Get Ready	Rare Earth	Rare Earth	(25)
16.	Nothing ?Succeeds Like Success	Bill Deal/Rhondels	Heritage	(18)
17.	Hey Lawdy Mama	Steppenwolf	Dunhill	(14)
18.	Up The Ladder To The Roof	The Supremes	Motown	(11)
*19.	Puppet Man	Fifth Dimension	Bell	(35)
*20.	Soolaimon	Neil Diamond	Uni	(38)
21.	What Is Truth	Johnny Cash	Columbia	(21)
22.	The Seeker	The Who	Decca	(24)
23.	Come And Get It	Badfinger	Apple	(13)
24.	You Need Love Like I Do	Gladys Knight/Pips	Soul	(19)
25.	Long Lonesome Highway	Michael Parks	MGM	(15)
26.	Airport Love Theme	Vincent Bell	Decca	(26)
*27.	Love On A Two-Way Street	The Moments	Stang	(37)
*28.	The Letter	Joe Cocker	A & M	(36)
*29.	Come Running	Van Morrison	Warner Bros.	(39)
30.	Everybody's Out Of Town	B.J. Thomas	Scepter	(34)
*31.	Make Me Smile	Chicago	Columbia	(40)
32.	Shilo	Neil Diamond	Bang	(20)
*33.	Hum A Song (From Your Heart)	Lulu	Atco	(N)
34.	Spirit In The Sky	Norman Greenbaum	Reprise	(28)
35.	Tennessee Bird Walk	Jack Blanchard & Misty Morgan	Wayside	(27)
36.	Bridge Over Troubled Water	Simon & Garfunkel	Columbia	(31)
37.	Daughter Of Darkness	Tom Jones	Parrot	(N)
38.	Come Saturday Morning	The Sandpipers	A & M	(N)
39.	Reach Out And Touch (Somebody's Hand)	Diana Ross	Motown	(N)
40.	Instant Karma	John Ono Lennon	Apple	(22)

*Denotes Big 10 Climbers

WC
FL

BIG 10 COUNT DOWN

FOR THE WEEK OF MAY 4, 1970

				Last Week
*1.	Cecilia	Simon & Garfunkel	Columbia	(9)
2.	American Woman/So Sugar Tonight	The Guess Who	RCA	(1)
3.	Little Green Bag	George Baker Selection	Colossus	(3)
4.	Something's Burning	Kenny Rogers/First Edition	Reprise	(2)
*5.	Hitchin' A Ride	Vanity Fare	Page One	(10)
*6.	Up Around The Bend	Creedence Clearwater Revival	Fantasy	(14)
7.	Vehicle	Ides Of March	Warner Bros.	(4)
8.	Love Or Let Me Be Lonely	Friends Of Distinction	RCA	(8)
9.	For The Love Of Him	Bobbi Martin	United Artists	(5)
*10.	Get Ready	Rare Earth	Rare Earth	(15)
11.	ABC	Jackson Five	Motown	(6)
12.	Let It Be	The Beatles	Apple	(7)
*13.	Soolaimon	Neil Diamond	Uni	(20)
14.	Reflections Of My Life	The Marmalade	London	(11)
15.	Which Way You Goin' Billy?	The Poppy Family	London	(12)
*16.	Puppet Man	Fifth Dimension	Bell	(19)
*17.	The Seeker	The Who	Decca	(22)
*18.	The Letter	Joe Cocker	A & M	(28)
*19.	Love On A Two Way Street	The Moments	Stang	(27)
*20.	Make Me Smile	Chicago	Columbia	(31)
21.	What Is Truth	Johnny Cash	Columbia	(21)
22.	Turn Back The Hands Of Time	Tyrone Davis	Dakar	(13)
23.	Hey Lawdy Mama	Steppenwolf	Dunhill	(17)
24.	Nothing Succeeds Like Success	Bill Dean/Rhondels	Heritage	(16)
*25.	Hum A Song (From Your Heart)	Lulu	Atco	(33)
*26.	Daughter Of Darkness	Tom Jones	Parrot	(37)
27.	Up The Ladder To The Roof	The Supremes	Motown	(18)
*28.	Come Saturday Morning	The Sandpipers	A & M	(38)
29.	Everybody's Out Of Town	B.J. Thomas	Scepter	(30)
*30.	Ride Captain Ride	Blues Image	Atco	(N)
31.	Come And Get It	Badfinger	Apple	(23)
32.	Long Lonesome Highway	Michael Parks	MGM	(25)
33.	Reach Out And Touch (Somebody's Hand)	Diana Ross	Motown	(39)
34.	You Need Love Like I Do	Gladys Knight/Pips	Soul	(24)
35.	Airport Love Theme	Vincent Bell	Decca	(26)
36.	Lay Down (Candles In The Rain)	Melanie	Buddah	(N)
37.	My Baby Loves Lovin'	White Plains	Deram	(N)
38.	Easy To Be Free	Happy Day	Varpo	(N)
39.	The Wonder Of You	Elvis Presley	RCA	(N)
40.	Spirit In The Sky	Norman Greenbaum	Reprise	(34)

HIT BOUND

So Much Love	Faith Hope & Charity	Maxwell
Friends	Feather	White Whale

*Denotes Big 10 Climbers

WC
FL

BIG 10 COUNT DOWN

FOR THE WEEK OF MAY 11, 1970

					Last Week
1.	Cecilia	Simon & Garfunkel	Columbia	(1)
*2.	Hitchin' A Ride	Vanity Fare	Page One	(5)
*3.	Up Around The Bend	Creedence Clearwater Revival	Fantasy	(6)
4.	American Woman/No Sugar Tonight	The Guess Who	RCA	(2)
5.	Little Green Bag	George Baker Selection	Colossus	(3)
*6.	Get Ready	Rare Earth	Rare Earth	(10)
7.	Something's Burning	Kenny Rogers/First Edition	Reprise	(4)
8.	Vehicle	Ides Of March	Warner Bros.	(7)
9.	Love Or Let Me Be Lonely	Friends Of Distinction	RCA	(8)
*10.	The Letter	Joe Cocker	A & M	(18)
*11.	Soolaimon	Neil Diamond	Uni	(13)
12.	For The Love Of Him	Bobbi Martin	United Artists	(9)
*13.	Love On A Two Way Street	The Moments	Stang	(19)
14.	ABC	Jackson Five	Motown	(11)
15.	Puppet Man	Fifth Dimension	Bell	(16)
16.	The Seeker	The Who	Decca	(17)
*17.	Make Me Smile	Chicago	Columbia	(20)
*18.	Daughter Of Darkness	Tom Jones	Parrot	(26)
19.	Reflections Of My Life	The Marmalade	London	(14)
20.	Ride Captain Ride	Blues Image	Atco	(30)
*21.	Let It Be	The Beatles	Apple	(12)
22.	Which Way You Goin' Billy?	The Poppy Family	London	(15)
23.	Hum A Song (From Your Heart)	Lulu	Atco	(25)
24.	Come Saturday Morning	The Sandpipers	A & M	(28)
*25.	My Baby Loves Lovin'	White Plains	Deram	(37)
26.	What Is Truth	Johnny Cash	Columbia	(21)
27.	Turn Back The Hands Of Time	Tyrone Davis	Dakar	(22)
*28.	Lay Down (Candles In The Rain)	Melanie	Buddah	(36)
29.	Reach Out And Touch (Somebody's Hand)	Diana Ross	Motown	(33)
30.	The Wonder Of You	Elvis Presley	RCA	(39)
*31.	Easy To Be Free	Happy Day	Varpo	(38)
*32.	United We Stand	Brotherhood Of Man	Deram	(N)
33.	Question	The Moody Blues	Threshold	(N)
34.	So Much Love	Faith Hope & Charity	Maxwell	(HB)
35.	Friends	Feather	White Whale	(HB)
36.	Up The Ladder To The Roof	The Supremes	Motown	(27)
37.	Everything Is Beautiful	Ray Stevens	Barnaby	(N)
38.	Airport Love Theme	Vincent Bell	Decca	(35)
39.	Hey Lawdy Mama	Steppenwolf	Dunhill	(23)
40.	Everybody's Out Of Town	B.J. Thomas	Scepter	(29)

HIT BOUND

Hey, Mister Sun	Bobby Sherman	Metromedia

*Denotes Big 10 Climbers

BIG 10 COUNT DOWN

FOR THE WEEK OF MAY 18, 1970

				Last Week
*1.	Hitchin' A Ride	Vanity Fare	Page One	(2)
*2.	Get Ready	Rare Earth	Rare Earth	(6)
3.	Cecilia	Simon & Garfunkel	Columbia	(1)
4.	Up Around The Band	Creedence Clearwater Revival	Fantasy	(3)
5.	Little Green Bag	George Baker Selection	Colossus	(5)
6.	American Woman/No Sugar Tonight	The Guess Who	RCA	(4)
*7.	Love On A Two Way Street	The Moments	Stang	(13)
*8.	The Letter	Joe Cocker	A & M	(10)
*9.	Daughter Of Darkness	Tom Jones	Parrot	(18)
*10.	My Baby Loves Lovin'	White Plains	Deram	(25)
11.	Soolaimon	Neil Diamond	Uni	(11)
12.	Something's Burning	Kenny Rogers/First Edition	Reprise	(7)
13.	Love Or Let Me Be Lonely	Friends Of Distinction	RCA	(9)
14.	Vehicle	Ides Of March	Warner Bros.	(8)
*15.	Make Me Smile	Chicago	Columbia	(17)
*16.	Everything Is Beautiful	Ray Stevens	Barnaby	(37)
*17.	Ride Captain Ride	Blues Image	Atco	(20)
*18.	United We Stand	Brotherhood Of Man	Deram	(32)
19.	The Seeker	The Who	Decca	(16)
20.	Puppet Man	Fifth Dimension	Bell	(15)
21.	Reflections Of My Life	The Marmalade	London	(19)
22.	Which Way You Goin' Billy?	The Poppy Family	London	(22)
23.	Come Saturday Morning	The Sandpipers	A & M	(24)
24.	Hum A Song (From Your Heart)	Lulu	Atco	(23)
25.	For The Love Of Him	Bobbi Martin	United Artists	(12)
*26.	Lay Down (Candles In The Rain)	Melanie	Buddah	(28)
27.	The Wonder Of You	Elvis Presley	RCA	(30)
*28.	Easy To Be Free	Happy Day	Varpo	(31)
29.	Reach Out And Touch (Somebody's Hand)	Diana Ross	Motown	(29)
*30.	Question	The Moody Blues	Threshold	(33)
31.	ABC	Jackson Five	Motown	(14)
32.	So Much Love	Faith Hope & Charity	Maxwell	(34)
33.	Friends	Feather	White Whale	(35)
34.	Let It Be	The Beatles	Apple	(21)
35.	Turn Back The Hands Of Time	Tyrone Davis	Dakar	(27)
*36.	Hey Mister Sun	Bobby Sherman	Metromedia	(HB)
*37.	Love Land	Watts 103rd St. Rhythm Band	Warner Bros.	(N)
38.	Mama Told Me (Not To Come)	Three Dog Night	Dunhill	(N)
39.	The Long And Winding Road/For You Blue	The Beatles	Apple	(N)
40.	What Is Truth	Johnny Cash	Columbia	(26)

HIT BOUND

Good Morning Freedom	Daybreak	Uni
Ball Of Confusion	The Temptations	Gordy

*Denotes Big 10 Climbers

WC
FL

BIG 10 COUNT DOWN

FOR THE WEEK OF MAY 25, 1970

				Last Week
1.	Hitchin' A Ride	Vanity Fare	Page One	(1)
2.	Get Ready	Rare Earth	Rare Earth	(2)
*3.	My Baby Loves Lovin'	White Plains	Deram	(10)
*4.	Daughter Of Darkness	Tom Jones	Parrot	(9)
5.	Cecilia	Simon & Garfunkel	Columbia	(3)
6.	Love On A Two Way Street	The Moments	Stang	(7)
7.	Up Around The Bend	Creedence Clearwater Revival	Fantasy	(4)
8.	The Letter	Joe Cocker	A & M	(8)
*9.	Make Me Smile	Chicago	Columbia	(15)
*10.	Ride Captain Ride	Blues Image	Atco	(17)
*11.	Everything Is Beautiful	Ray Stevens	Barnaby	(16)
*12.	United We Stand	Brotherhood Of Man	Deram	(18)
13.	Little Green Bag	George Baker Selection	Colossus	(5)
14.	American Woman/No Sugar Tonight	The Guess Who	RCA	(6)
*15.	Come Saturday Morning	The Sandpipers	A & M	(23)
*16.	Lay Down (Candles In The Rain)	Melanie	Buddah	(26)
*17.	The Wonder Of You	Elvis Presley	RCA	(27)
*18.	Question	The Moody Blues	Threshold	(30)
19.	Soolaimon	Neil Diamond	Uni	(11)
*20.	Hey Mister Sun	Bobby Sherman	Metromedia	(36)
21.	Something's Burning	Kenny Rogers/First Edition	Reprise	(12)
22.	Love Or Let Me Be Lonely	Friends Of Distinction	RCA	(13)
23.	Vehicle	Ides Of March	Warner Bros.	(14)
*24.	Easy To Be Free	Happy Day	Varpo	(28)
*25.	Mama Told Me (Not To Come)	Three Dog Night	Dunhill	(38)
*26.	For You Blue/The Long And Winding Road	The Beatles	Apple	(39)
*27.	Love Land	Watts 103rd St. Rhythm Band	Warner Bros.	(37)
28.	Which Way You Goin' Billy?	The Poppy Family	London	(22)
29.	Reflections Of My Life	The Marmalade	London	(21)
30.	Friends	Feather	White Whale	(33)
31.	Reach Out And Touch (Somebody's Hand)	Diana Ross	Motown	(29)
32.	Ball Of Confusion	The Temptations	Gordy	(HB)
33.	Good Morning Freedom	Daybreak	Uni	(HB)
34.	It's All In The Game	Four Tops	Motown	(N)
35.	The Love You Save	The Jackson 5	Motown	(N)
36.	Gimme Dat Ding	The Pipkins	Capitol	(N)
37.	For The Love Of Him	Bobbi Martin	United Artists	(25)
38.	Let It Be	The Beatles	Apple	(34)
39.	Puppet Man	Fifth Dimension	Bell	(20)
40.	The Seeker	The Who	Decca	(19)

HIT BOUND

A Song Of Joy	Miguel Rios	A & M

*Denotes Big 10 Climbers

WC
FL

BIG 10 COUNT DOWN

FOR THE WEEK OF JUNE 1, 1970

				Last Week
*1.	Love On A Two Way Street	The Moments	Stang	(6)
2.	Hitchin' A Ride	Vanity Fare	Page One	(1)
*3.	My Baby Loves Lovin'	White Plains	Deram	(3)
4.	Get Ready	Rare Earth	Rare Earth	(2)
*5.	Make Me Smile	Chicago	Columbia	(9)
*6.	Ride Captain Ride	Blues Image	Atco	(10)
*7.	The Letter	Joe Cocker	A & M	(8)
8.	Daughter Of Darkness	Tom Jones	Parrot	(4)
*9.	Lay Down (Candles In The Rain)	Melanie	Buddah	(16)
*10.	United We Stand	Brotherhood Of Man	Deram	(12)
11.	Everything Is Beautiful	Ray Stevens	Barnaby	(11)
12.	Cecilia	Simon & Garfunkel	Columbia	(5)
13.	Up Around The Bend	Creedence Clearwater Revival	Fantasy	(7)
*14.	Love Land	Watts 103rd St. Rhythm Band	Warner Bros.	(27)
15.	Come Saturday Morning	The Sandpipers	A & M	(15)
16.	The Wonder Of You	Elvis Presley	RCA	(17)
*17.	The Question	The Moody Blues	Threshold	(18)
*18.	Hey Mister Sun	Bobby Sherman	Metromedia	(20)
*19.	Ball Of Confusion	The Temptations	Gordy	(32)
*20.	Mama Told Me (Not To Come)	Three Dog Night	Dunhill	(25)
21.	For You Blue/The Long And Winding Road	The Beatles	Apple	(26)
*22.	The Love You Save	The Jackson 5	Motown	(35)
23.	Little Green Bag	George Baker Selection	Colossus	(13)
24.	Easy To Be Free	Happy Day	Varpo	(24)
25.	Soolaimon	Neil Diamond	Uni	(19)
26.	Love Or Let Me Be Lonely	Friends Of Distinction	RCA	(22)
*27.	Good Morning Freedom	Daybreak	Uni	(33)
*28.	It's All In The Game	Four Tops	Motown	(34)
29.	Gimme Dat Ding	The Pipkins	Capitol	(36)
30.	Friends	Feather	White Whale	(30)
31.	Reach Out And Touch (Somebody's Hand)	Diana Ross	Motown	(31)
32.	American Woman/So Sugar Tonight	The Guess Who	RCA	(14)
33.	Something's Burning	Kenny Rogers/First Edition	Reprise	(21)
34.	A Song Of Joy	Miguel Rios	A & M	(HB)
35.	Band Of Gold	Freda Payne	Invictus	(N)
36.	Vehicle	Ides Of March	Warner Bros.	(23)
37.	Which Way You Goin' Billy?	The Poppy Family	London	(28)
38.	Reflections Of My Life	The Marmalade	London	(29)
39.	Mississippi	John Phillips	Dunhill	(N)
40.	The Witch	The Rattles	Probe	(N)

*Denotes Big 10 Climbers

WC
FL

BIG 10 COUNT DOWN

FOR THE WEEK OF JUNE 8, 1970

					Last Week
*1.	My Baby Loves Lovin'	White Plains	Deram	(3)
2.	Love On A Two Way Street	The Moments	Stang	(1)
3.	Hitchin' A Ride	Vanity Fare	Page One	(2)
*4.	Ride Captain Ride	Blues Image	Atco	(6)
*5.	Love Land	Watts 103rd St. Rhythm Band	Warner Bros.	(14)
6.	Make Me Smile	Chicago	Columbia	(5)
*7.	Lay Down (Candles In The Rain)	Melanie	Buddah	(9)
*8.	Ball Of Confusion	The Temptations	Gordy	(19)
*9.	United We Stand	Brotherhood Of Man	Deram	(10)
*10.	Question	The Moody Blues	Threshold	(17)
11.	Get Ready	Rare Earth	Rare Earth	(4)
12.	Daughter Of Darkness	Tom Jones	Parrot	(8)
13.	The Letter	Joe Cocker	A & M	(7)
*14.	Mama Told Me (Not To Come)	Three Dog Night	Dunhill	(20)
15.	The Wonder Of You	Elvis Presley	RCA	(16)
*16.	Hey Mister Sun	Bobby Sherman	Metromedia	(18)
*17.	For You Blue/The Long And Winding Road	The Beatles	Apple	(21)
*18.	The Love You Save	The Jackson 5	Motown	(22)
19.	Cecilia	Simon & Garfunkel	Columbia	(12)
20.	Everything Is Beautiful	Ray Stevens	Barnaby	(11)
21.	Come Saturday Morning	The Sandpipers	A & M	(15)
22.	Up Around The Bend	Creedence Clearwater Revival	Fantasy	(13)
*23.	A Song Of Joy	Miguel Rios	A & M	(34)
24.	Gimme Dat Ding	The Pipkins	Capitol	(29)
*25.	It's All In The Game	Four Tops	Motown	(28)
*26.	Good Morning Freedom	Daybreak	Uni	(27)
*27.	Band Of Gold	Freda Payne	Invictus	(35)
*28.	Mississippi	John Phillips	Dunhill	(39)
29.	Friends	Feather	White Whale	(30)
30.	Little Green Bag	George Baker Selection	Colossus	(23)
31.	Love Or Let Me Be Lonely	Friends Of Distinction	RCA	(26)
32.	Reach Out And Touch (Somebody's Hand)	Diana Ross	Motown	(31)
33.	Soolaimon	Neil Diamond	Uni	(25)
34.	Save The Country	Fifth Dimension	Bell	(N)
35.	Lay A Little Lovin' On Me	Robin McNamara	Steed	(N)
36.	The Witch	The Rattles	Probe	(40)
*37.	O-O-H Child	The 5 Stairsteps	Buddah	(N)
38.	American Woman/No Sugar Tonight	The Guess Who	RCA	(32)
39.	Which Way You Goin' Billy?	The Poppy Family	London	(37)
40.	Reflections Of My Life	The Marmalade	London	(38)

HIT BOUND

Pearl	Tommy Roe	ABC

*Denotes Big 10 Climbers

BIG 10 COUNT DOWN

FOR THE WEEK OF JUNE 15, 1970

				Last Week
1.	My Baby Loves Lovin'	White Plains	Deram	(1)
*2.	Love Land	Watts 103rd St. Rhythm Band	Warner Bros.	(5)
*3.	Ride Captain Ride	Blues Image	Atco	(4)
*4.	Ball Of Confusion	The Temptations	Gordy	(8)
*5.	Lay Down (Candles In The Rain)	Melanie	Buddah	(7)
6.	Love On A Two Way Street	The Moments	Stang	(2)
7.	Hitchin' A Ride	Vanity Fare	Page One	(3)
*8.	Question	The Moody Blues	Threshold	(10)
*9.	Mama Told Me (Not To Come)	Three Dog Night	Dunhill	(14)
*10.	The Love You Save	Jackson 5	Motown	(18)
*11.	For You Blue/The Long And Winding Road	The Beatles	Apple	(17)
12.	Make Me Smile	Chicago	Columbia	(6)
13.	United We Stand	Brotherhood Of Man	Deram	(9)
14.	The Wonder Of You	Elvis Presley	RCA	(15)
15.	Hey Mister Sun	Bobby Sherman	Metromedia	(16)
16.	Get Ready	Rare Earth	Rare Earth	(11)
*17.	Band Of Gold	Freda Payne	Invictus	(27)
18.	Gimme Dat Ding	The Pipkins	Capitol	(24)
19.	Daughter Of Darkness	Tom Jones	Parrot	(12)
*20.	A Song Of Joy	Miguel Rios	A & M	(23)
21.	The Letter	Joe Cocker	A & M	(13)
*22.	Mississippi	John Phillips	Dunhill	(28)
23.	It's All In The Game	Four Tops	Motown	(25)
24.	O-O-H Child	The 5 Stairsteps	Buddah	(37)
25.	Good Morning Freedom	Daybreak	Uni	(26)
*26.	Save The Country	Fifth Dimension	Bell	(34)
27.	Cecilia	Simon & Garfunkel	Columbia	(19)
28.	Everything Is Beautiful	Ray Stevens	Barnaby	(20)
*29.	Lay A Little Lovin' On Me	Robin McNamara	Steed	(35)
*30.	The Witch	The Rattles	Probe	(36)
31.	Come Saturday Morning	The Sandpipers	A & M	(21)
32.	Up Around The Bend	Creedence Clearwater Revival	Fantasy	(22)
33.	Pearl	Tommy Roe	ABC	(HB)
34.	Tighter, Tighter	Alive And Kicking	Roulette	(N)
35.	Heighty-Hi	Happy Day	Varpo	(N)
36.	Teach Your Children	Crosby, Stills, Nash & Young	Atlantic	(N)
37.	Check Out Your Mind	The Impressions	Curtom	(N)
38.	Reach Out And Touch (Somebody's Hand)	Diana Ross	Motown	(32)
39.	Little Green Bag	George Baker Selection	Colossus	(30)
40.	Love Or Let Me Be Lonely	Friends Of Distinction	RCA	(31)

*Denotes Big 10 Climbers

WC
FL

BIG 10 COUNT DOWN

FOR THE WEEK OF JUNE 22, 1970

				Last Week
*1.	Mama Told Me (Not To Come)	Three Dog Night	Dunhill	(9)
2.	Love Land	Watts 103rd St. Rhythm Band	Warner Bros.	(2)
*3.	Ball Of Confusion	The Temptations	Gordy	(4)
4.	Ride Captain Ride	Blues Image	Atco	(3)
*5.	The Love You Save	Jackson 5	Motown	(10)
6.	My Baby Loves Lovin'	White Plains	Deram	(1)
7.	Lay Down (Candles In The Rain)	Melanie	Buddah	(5)
*8.	Band Of Gold	Freda Payne	Invictus	(17)
9.	For You Blue/The Long And Winding Road	The Beatles	Apple	(11)
10.	Question	The Moody Blues	Threshold	(8)
*11.	Mississippi	John Phillips	Dunhill	(22)
12.	Hitchin' A Ride	Vanity Fare	Page One	(7)
13.	Love On A Two Way Street	The Moments	Stang	(6)
*14.	A Song Of Joy	Miguel Rios	A & M	(20)
15.	Gimme Dat Ding	The Pipkins	Capitol	(18)
16.	The Wonder Of You	Elvis Presley	RCA	(14)
*17.	O-O-H Child	The 5 Stairsteps	Buddah	(24)
18.	Make Me Smile	Chicago	Columbia	(12)
19.	Get Ready	Rare Earth	Rare Earth	(16)
20.	United We Stand	Brotherhood Of Man	Deram	(13)
*21.	Lay A Little Lovin' On Me	Robin McNamara	Steed	(29)
*22.	Tighter, Tighter	Alive And Kicking	Roulette	(34)
23.	It's All In The Game	Four Tops	Motown	(23)
24.	Save The Country	Fifth Dimension	Bell	(26)
*25.	The Witch	The Rattles	Probe	(30)
26.	Hey Mister Sun	Bobby Sherman	Metromedia	(15)
27.	Daughter Of Darkness	Tom Jones	Parrot	(19)
*28.	Pearl	Tommy Roe	ABC	(33)
29.	The Letter	Joe Cocker	A & M	(21)
*30.	Check Out Your Mind	The Impressions	Curtom	(37)
31.	Cecilia	Simon & Garfunkel	Columbia	(27)
32.	Everything Is Beautiful	Ray Stevens	Barnaby	(28)
33.	Heighty-Hi	Happy Day	Varpo	(35)
34.	Teach Your Children	Crosby, Stills, Nash & Young	Atlantic	(36)
35.	Come Saturday Morning	The Sandpipers	A & M	(31)
36.	Close To You	Carpenters	A & M	(N)
37.	My Marie	Engelbert Humperdinck	Parrot	(N)
38.	The Sly, Slick, And The Wicked	The Lost Generation	Brunswick	(N)
39.	The Boys In The Band	The Boys In The Band	Spring	(N)
40.	Are You Ready?	Pacific Gas & Electric	Columbia	(N)

HIT BOUND

Why Can't I Touch You?	Ronnie Dyson	Columbia

*Denotes Big 10 Climbers

WC
FL

BIG 10 COUNT DOWN

FOR THE WEEK OF JUNE 29, 1970

				Last Week
*1.	The Love You Save	Jackson 5	Motown	(5)
2.	Mama Told Me (Not To Come)	Three Dog Night	Dunhill	(1)
*3.	Band Of Gold	Freda Payne	Invictus	(8)
4.	Ball Of Confusion	The Temptations	Gordy	(3)
5.	Love Land	Watts 103rd St. Rhythm Band	Warner Bros.	(2)
6.	Ride Captain Ride	Blues Image	Atco	(4)
*7.	O-O-H Child	The 5 Stairsteps	Buddah	(17)
*8.	Mississippi	John Phillips	Dunhill	(11)
9.	For You Blue/The Long And Winding Road	The Beatles	Apple	(9)
10.	My Baby Loves Lovin'	White Plains	Deram	(6)
11.	Lay Down (Candles In The Rain)	Melanie	Buddah	(7)
*12.	A Song Of Joy	Miguel Rios	A & M	(14)
*13.	Tighter, Tighter	Alive And Kicking	Roulette	(22)
*14.	Lay A Little Lovin' On Me	Robin McNamara	Steed	(21)
15.	Gimme Dat Ding	The Pipkins	Capitol	(15)
*16.	Close To You	Carpenters	A & M	(36)
17.	Question	The Moody Blues	Threshold	(10)
18.	Hitchin' A Ride	Vanity Fare	Page One	(12)
*19.	Save The Country	Fifth Dimension	Bell	(24)
*20.	Check Out Your Mind	The Impressions	Curtom	(30)
21.	Love On A Two Way Street	The Moments	Stang	(13)
*22.	The Witch	The Rattles	Probe	(25)
*23.	Pearl	Tommy Roe	ABC	(28)
24.	The Wonder Of You	Elvis Presley	RCA	(16)
25.	Make Me Smile	Chicago	Columbia	(18)
*26.	The Sly, Slick, And The Wicked	The Lost Generation	Brunswick	(38)
27.	Get Ready	Rare Earth	Rare Earth	(19)
28.	Are You Ready?	Pacific Gas & Electric	Columbia	(40)
29.	It's All In The Game	Four Tops	Motown	(23)
30.	United We Stand	Brotherhood Of Man	Deram	(20)
*31.	Make It With You	Bread	Elektra	(N)
32.	Heighty-Hi	Happy Day	Varpo	(33)
33.	My Marie	Engelbert Humperdinck	Parrot	(37)
34.	The Boys In The Band	The Boys In The Band	Spring	(39)
*35.	In The Summertime	Mungo Jerry	Janus	(N)
36.	Why Can't I Touch You?	Ronnie Dyson	Columbia	(HB)
37.	Mississippi Queen	Mountain	Windfall	(N)
38.	Spill The Wine	Eric Burdon And War	MGM	(N)
39.	Hey Mister Sun	Bobby Sherman	Metromedia	(26)
40.	Daughter Of Darkness	Tom Jones	Parrot	(27)

*Denotes Big 10 Climbers

WC
FL

BIG 10 COUNT DOWN

FOR THE WEEK OF JULY 6, 1970

				Last Week
1.	The Love You Save	Jackson 5	Motown	(1)
2.	Mama Told Me (Not To Come)	Three Dog Night	Dunhill	(2)
3.	Band Of Gold	Freda Payne	Invictus	(3)
4.	Ball Of Confusion	The Temptations	Gordy	(4)
*5.	O-O-H Child	The 5 Stairsteps	Buddah	(7)
*6.	Close To You	Carpenters	A & M	(16)
*7.	Mississippi	John Phillips	Dunhill	(8)
*8.	Tighter, Tighter	Alive And Kicking	Roulette	(13)
9.	Ride Captain Ride	Blues Image	Atco	(6)
*10.	Lay A Little Lovin' On Me	Robin McNamara	Steed	(14)
11.	Love Land	Watts 103rd St. Rhythm Band	Warner Bros.	(5)
*12.	Make It With You	Bread	Elektra	(31)
13.	A Song Of Joy	Miguel Rios	A & M	(12)
*14.	Check Out Your Mind	The Impressions	Curtom	(20)
15.	My Baby Loves Lovin'	White Plains	Deram	(10)
16.	Lay Down (Candles In The Rain)	Melanie	Buddah	(11)
*17.	Are You Ready?	Pacific Gas & Electric	Columbia	(28)
18.	For You Blue/The Long And Winding Road	The Beatles	Apple	(9)
*19.	Pearl	Tommy Roe	ABC	(23)
*20.	The Witch	The Rattles	Probe	(22)
*21.	The Sly, Slick, And The Wicked	The Lost Generation	Brunswick	(26)
22.	Gimme Dat Ding	The Pipkins	Capitol	(15)
*23.	Mississippi Queen	Mountain	Windfall	(37)
*24.	Spill The Wine	Eric Burdon and War	MGM	(38)
*25.	Heighty-Hi	Happy Day	Varpo	(32)
*26.	In The Summertime	Mungo Jerry	Janus	(35)
27.	Question	The Moody Blues	Threshold	(17)
28.	Save The Country	Fifth Dimension	Bell	(19)
29.	The Boys In The Band	The Boys In The Band	Spring	(34)
30.	Why Can't I Touch You?	Ronnie Dyson	Columbia	(36)
31.	My Marie	Engelbert Humperdinck	Parrot	(33)
32.	Hitchin' A Ride	Vanity Fare	Page One	(18)
33.	The Wonder Of You	Elvis Presley	RCA	(24)
34.	Love On A Two Way Street	The Moments	Stang	(21)
35.	Get Ready	Rare Earth	Rare Earth	(27)
36.	Make Me Smile	Chicago	Columbia	(25)
37.	United We Stand	Brotherhood Of Man	Deram	(30)
38.	It's All In The Game	Four Tops	Motown	(29)
39.	Signed, Sealed, Delivered I'm Yours	Stevie Wonder	Tamla	(N)
40.	Crazy Love	The Happenings	Jubilee	(N)

HIT BOUND

It's Your Life	Andy Kim	Steed
Hand Me Down World	The Guess Who	RCA

*Denotes Big 10 Climbers

WC
FL

BIG 10 COUNT DOWN

FOR THE WEEK OF JULY 13, 1970

				Last Week
*1.	Close To You	Carpenters	A & M	(6)
*2.	O-O-H Child	The 5 Stairsteps	Buddah	(5)
3.	The Love You Save	Jackson 5	Motown	(1)
*4.	Tighter, Tighter	Alive And Kicking	Roulette	(8)
5.	Ball Of Confusion	The Temptations	Gordy	(4)
6.	Mama Told Me (Not To Come)	Three Dog Night	Dunhill	(2)
7.	Band Of Gold	Freda Payne	Invictus	(3)
*8.	Make It With You	Bread	Elektra	(12)
9.	Lay A Little Lovin' On Me	Robin McNamara	Steed	(10)
10.	Mississippi	John Phillips	Dunhill	(7)
*11.	Spill The Wine	Eric Burdon And War	MGM	(24)
12.	Ride Captain Ride	Blues Image	Atco	(9)
*13.	Mississippi Queen	Mountain	Windfall	(23)
*14.	Why Can't I Touch You?	Ronnie Dyson	Columbia	(30)
15.	Check Out Your Mind	The Impressions	Curtom	(14)
16.	Are You Ready?	Pacific Gas & Electric	Columbia	(17)
17.	Pearl	Tommy Roe	ABC	(19)
*18.	In The Summertime	Mungo Jerry	Janus	(26)
19.	A Song Of Joy	Miguel Rios	A & M	(13)
20.	The Sly, Slick, And The Wicked	The Lost Generation	Brunswick	(21)
21.	Love Land	Watts 103rd St. Rhythm Band	Warner Bros.	(11)
*22.	The Boys In The Band	The Boys In The Band	Spring	(29)
23.	My Baby Loves Lovin'	White Plains	Deram	(15)
24.	Heighty-Hi	Happy Day	Varpo	(25)
25.	Lay Down (Candles In The Rain)	Melanie	Buddah	(16)
26.	The Witch	The Rattles	Probe	(20)
27.	Gimme Dat Ding	The Pipkins	Capitol	(22)
28.	For You Blue/The Long And Winding Road	The Beatles	Apple	(18)
*29.	Signed, Sealed, Delivered I'm Yours	Stevie Wonder	Tamla	(39)
30.	Question	The Moody Blues	Threshold	(27)
31.	My Marie	Engelbert Humperdinck	Parrot	(31)
*32.	Crazy Love	The Happenings	Jubilee	(40)
33.	Hitchin' A Ride	Vanity Fare	Page One	(32)
34.	The Wonder Of You	Elvis Presley	RCA	(33)
35.	Silver Bird	Mark Lindsay	Columbia	(N)
36.	I Just Can't Help Believing	B.J. Thomas	Scepter	(N)
37.	Cottage Cheese	Crow	Amaret	(N)
38.	Get Ready	Rare Earth	Rare Earth	(35)
39.	Make Me Smile	Chicago	Columbia	(36)
40.	Save The Country	Fifth Dimension	Bell	(28)

HIT BOUND

Morning Much Better	Ten Wheel Drive With Genya Ravan	Polydor

*Denotes Big 10 Climbers

WC
FL

BIG 10 COUNT DOWN

FOR THE WEEK OF JULY 20, 1970

				Last Week
*1.	Close To You	Carpenters	A & M	(1)
*2.	O-O-H Child	The 5 Stairsteps	Buddah	(2)
*3.	Make It With You	Bread	Elektra	(8)
4.	Tighter, Tighter	Alive And Kicking	Roulette	(4)
*5.	Spill The Wine	Eric Burdon And War	MGM	(11)
6.	The Love You Save	Jackson 5	Motown	(3)
*7.	Why Can't I Touch You?	Ronnie Dyson	Columbia	(14)
8.	Ball Of Confusion	The Temptations	Gordy	(5)
9.	Mama Told Me (Not To Come)	Three Dog Night	Dunhill	(6)
10.	Band Of Gold	Freda Payne	Invictus	(7)
11.	Lay A Little Lovin' On Me	Robin McNamara	Steed	(9)
*12.	Mississippi Queen	Mountain	Windfall	(13)
13.	Mississippi	John Phillips	Dunhill	(10)
*14.	Are You ready?	Pacific Gas & Electric	Columbia	(16)
*15.	Signed, Sealed, Delivered I'm Yours	Stevie Wonder	Tamla	(29)
*16.	In The Summertime	Mungo Jerry	Janus	(18)
17.	Pearl	Tommy Roe	ABC	(17)
18.	Ride Captain Ride	Blues Image	Atco	(12)
19.	The Sly, Slick, And The Wicked	The Lost Generation	Brunswick	(20)
*20.	The Boys In The Band	The Boys In The Band	Spring	(22)
21.	Check Out Your Mind	The Impressions	Curtom	(15)
*22.	Silver Bird	Mark Lindsay	Columbia	(35)
23.	A Song Of Joy	Miguel Rios	A & M	(19)
24.	Love Lalnd	Watts 103rd St. Rhythm Band	Warner Bros.	(21)
25.	The Witch	The Rattles	Probe	(26)
*26.	Cottage Cheese	Crow	Amaret	(37)
*27.	Crazy Love	The Happenings	Jubilee	(32)
28.	My Baby Loves Lovin'	White Plains	Deram	(23)
29.	Lay Down (Candles In The Rain)	Melanie	Buddah	(25)
*30.	I Just Can't Help Believing	B.J. Thomas	Scepter	(36)
31.	Gimme Dat Ding	The Pipkins	Capitol	(27)
32.	For You Blue/The Long And Winding Road	The Beatles	Apple	(28)
33.	Overture From Tommy	The Assembled Multitude	Atlantic	(N)
34.	It's Your Life	Andy Kim	Steed	(N)
35.	Morning Much Better	Ten Wheel Drive With Genya Ravan	Polydor	(HB)
36.	Hitchin' A Ride	Vanity Fare	Page One	(33)
37.	Hi-De-Ho	Blood, Sweat & Tears	Columbia	(N)
38.	Question	The Moody Blues	Threshold	(30)
39.	The Wonder Of You	Elvis Presley	RCA	(34)
40.	Get Ready	Rare Earth	Rare Earth	(38)

HIT BOUND

Julie, Do Y Love Me	Bobby Sherman	Metromedia
25 Or 6 To 4	Chicago	Columbia
War	Edwin Starr	Gordy

*Denotes Big 10 Climbers

WC
FL

BIG 10 COUNT DOWN

FOR THE WEEK OF JULY 27, 1970

				Last Week
*1.	Make It With You	Bread	Elektra	(3)
2.	Close To You	Carpenters	A & M	(1)
*3.	Spill The Wine	Eric Burdon And War	MGM	(5)
4.	O-O-H Child	The 5 Stairsteps	Buddah	(2)
*5.	Why Can't I Touch You?	Ronnie Dyson	Columbia	(7)
6.	Tighter, Tighter	Alive And Kicking	Roulette	(4)
*7.	Signed, Sealed, Delivered I'm Yours	Stevie Wonder	Tamla	(15)
*8.	Are You Ready?	Pacific Gas & Electric	Columbia	(14)
9.	The Love You Save	Jackson 5	Motown	(6)
*10.	Mississippi Queen	Mountain	Windfall	(12)
11.	Lay A Little Lovin' On Me	Robin McNamara	Steed	(11)
*12.	In The Summertime	Mungo Jerry	Janus	(16)
13.	Ball Of Confusion	The Temptations	Gordy	(8)
14.	Mama Told Me (Not To Come)	Three Dog Night	Dunhill	(9)
15.	Band Of Gold	Freda Payne	Invictus	(10)
*16.	Cottage Cheese	Crow	Amaret	(26)
*17.	I Just Can't Help Believing	B.J. Thomas	Scepter	(30)
*18.	Overture From Tommy	The Assembled Multitude	Atlantic	(33)
19.	Mississippi	John Phillips	Dunhill	(13)
*20.	Silver Bird	Mark Lindsay	Columbia	(22)
21.	Pearl	Tommy Roe	ABC	(17)
*22.	Crazy Love	The Happenings	Jubilee	(27)
23.	The Sly, Sick, And The Wicked	The Lost Generation	Brunswick	(19)
24.	Ride Captain Ride	Blues Image	Atco	(18)
25.	The Boys In The Band	The Boys In The Band	Spring	(20)
*26.	Morning Much Better	Ten Wheel Drive With Genya Ravan	Polydor	(35)
*27.	Hi-De-Ho	Blood, Sweat & Tears	Columbia	(37)
*28.	It's Your Life	Andy Kim	Steed	(34)
29.	A Song Of Joy	Miguel Rios	A & M	(23)
*30.	Julie, Do Ya Love Me	Bobby Sherman	Metromedia	(HB)
31.	The Witch	The Rattles	Probe	(25)
32.	25 Or 6 To 4	Chicago	Columbia	(HB)
33.	War	Edwin Starr	Gordy	(HB)
34.	I Want To Take You Higher	Ike & Tina Turner & The Ikettes	Liberty	(N)
35.	Love Land	Watts 103rd St. Rhythm Band	Warner Bros.	(24)
36.	Check Out Your Mind	The Impressions	Curtom	(21)
37.	My Baby Loves Lovin'	White Plains	Deram	(28)
38.	Lay Down (Candles In The Rain)	Melanie	Buddah	(29)
39.	Hitchin' A Ride	Vanity Fare	Page One	(36)
40.	Question	The Moody Blues	Threshold	(38)

HIT BOUND

Patches	Clarence Carter	Atlantic
Ain't No Mountain High Enough	Diana Ross	Motown

*Denotes Big 10 Climbers

WC
FL

BIG 10 COUNT DOWN

FOR THE WEEK OF AUGUST 3, 1970

				Last Week
1.	Make It With You	Bread	Elektra	(1)
2.	Close To You	Carpenters	A & M	(2)
3.	Spill The Wine	Eric Burdon & War	MGM	(3)
*4.	Why Can't I Touch You?	Ronnie Dyson	Columbia	(5)
*5.	Signed, Sealed, Delivered I'm Yours	Stevie Wonder	Tamla	(7)
6.	O-O-H Child	The 5 Stairsteps	Buddah	(4)
*7.	Mississippi Queen	Mountain	Windfall	(10)
8.	Tighter, Tighter	Alive & Kicking	Roulette	(6)
*9.	In The Summertime	Mungo Jerry	Janus	(12)
*10.	Cottage Cheese	Crow	Amaret	(16)
*11.	I Just Can't Help Believing	B.J. Thomas	Scepter	(17)
*12.	War	Edwin Starr	Gordy	(33)
13.	Are You Ready?	Pacific Gas & Electric	Columbia	(8)
14.	The Love You Save	Jackson 5	Motown	(9)
*15.	Julie, Do Ya Love Me	Bobby Sherman	Metromedia	(30)
16.	Overture From Tommy	The Assembled Multitude	Atlantic	(18)
17.	Lay A Little Lovin' On Me	Robin McNamara	Steed	(11)
*18.	Morning Much Better	Ten Wheel Drive With Genya Ravan	Polydor	(26)
19.	Crazy Love	The Happenings	Jubilee	(22)
20.	Silver Bird	Mark Lindsay	Columbia	(20)
*21.	25 Or 6 To 4	Chicago	Columbia	(32)
22.	Ball Of Confusion	The Temptations	Gordy	(13)
23.	Band Of Gold	Freda Payne	Invictus	(15)
*24.	Hi-De-Ho	Blood, Sweat & Tears	Columbia	(27)
*25.	It's Your Life	Andy Kim	Steed	(28)
26.	Mama Told Me (Not To Come)	Three Dog Night	Dunhill	(14)
*27.	I Want To Take You Higher	Ike & Tina Turner & Ikettes	Liberty	(34)
*28.	Ain't No Mountain High Enough	Diana Ross	Motown	(HB)
*29.	Patches	Clarence Carter	Atlantic	(HB)
30.	The Sly, Slick, And The Wicked	The Lost Generation	Brunswick	(23)
31.	Mississippi	John Phillips	Dunhill	(19)
32.	Pearl	Tommy Roe	ABC	(21)
33.	Ride Captain Ride	Blues Image	Atco	(24)
34.	Solitary Man	Neil Diamond	Bang	(N)
35.	Do What You Wanna Do	Five Flights Up	T-A	(N)
36.	Lookin' Out My Back Door	Creedence Clearwater Revival	Fantasy	(N)
37.	Hand Me Down World	The Guess Who	RCA	(N)
38.	A Song Of Joy	Miguel Rios	A & M	(29)
39.	The Boys In The Band	The Boys In The Band	Spring	(25)
40.	My Baby Loves Lovin'	White Plains	Derah	(37)

HIT BOUND

Two Little Rooms	Trella Hart	Capitol

*Denotes Big 10 Climbers

WC
FL

BIG 10 COUNT DOWN

FOR THE WEEK OF AUGUST 10, 1970

				Last Week
*1.	War	Edwin Starr	Gordy	(12)
*2.	Spill The Wine	Eric Burdon & War	MGM	(3)
3.	Make It With You	Bread	Elektra	(1)
*4.	Signed, Sealed & Delivered	Stevie Wonder	Tamla	(5)
5.	Close To You	Carpenters	A & M	(2)
*6.	In The Summertime	Mungo Jerry	Janus	(9)
*7.	I Just Can't Help Believing	B.J. Thomas	Scepter	(11)
*8.	Cottage Cheese	Crow	Amaret	(10)
9.	Why Can't I Touch You?	Ronnie Dyson	Columbia	(4)
*10.	25 Or 6 To 4	Chicago	Columbia	(21)
11.	Tighter, Tighter	Alive & Kicking	Roulette	(8)
*12.	Ain't No Mountain High Enough	Diana Ross	Motown	(28)
13.	Mississippi Queen	Mountain	Windfall	(7)
*14.	Julie, Do Ya Love Me	Bobby Sherman	Metromedia	(15)
*15.	Overture From Tommy	Assembled Multitude	Atlantic	(16)
16.	O-O-H Child	5 Stairsteps	Buddah	(6)
*17.	Morning Much Better	Ten Wheel Drive With Genya Ravan	Poyndor	(18)
*18.	Patches	Clarence Carter	Atlantic	(29)
19.	Silver Bird	Mark Lindsay	Columbia	(20)
*20.	I Want To Take You Higher	Ike & Tina Turner & Ikettes	Liberty	(27)
21.	Are You Ready?	Pacific Gas & Electric	Columbia	(13)
22.	The Love You Save	Jackson 5	Motown	(14)
23.	Lay A Little Lovin' On Me	Robin McNamara	Steed	(17)
24.	Hi-De-Ho	Blood, Sweat & Tears	Columbia	(24)
25.	It's Your Life	Andy Kim	Steed	(25)
*26.	Lookin' Out My Back Door	Creedence Clearwater Revival	Fantasy	(36)
*27.	Solitary Man	Neil Diamond	Bang	(34)
28.	Ball Of Confusion	The Temptations	Gordy	(22)
29.	Band Of Gold	Freda Payne	Invictus	(23)
30.	Crazy Love	Happenings	Jubilee	(19)
31.	Hand Me Down World	The Guess Who	RCA	(37)
32.	Mama Told Me (Not To Come)	Three Dog Night	Dunhill	(26)
33.	Do What You Wanna Do	Five Flights Up	T-A	(35)
34.	Candida	Dawn	Bell	(N)
35.	Groovy Situation	Gene Chandler	Mercury	(N)
36.	Soul Shake	Delaney & Bonnie & Friends	Atco	(N)
37.	Cracklin' Rosie	Neil Diamond	Uni	(N)
38.	The Sly, Slick And The Wicked	The Lost Generation	Brunswick	(30)
39.	Pearl	Tommy Roe	ABC	(32)
40.	Two Little Rooms	Trella Hart	Capitol	(HB)

*Denotes Big 10 Climbers

WC
FL

BIG 10 COUNT DOWN

FOR THE WEEK OF AUGUST 17, 1970

				Last Week
1.	War	Edwin Starr	Gordy	(1)
*2.	Ain't No Mountain High Enough	Diana Ross	Motown	(12)
3.	Spill The Wine	Eric Burdon & War	MGM	(2)
*4.	In The Summertime	Mungo Jerry	Janus	(6)
*5.	Cottage Cheese	Crow	Amaret	(8)
6.	I Just Can't Help Believing	B.J. Thomas	Scepter	(7)
*7.	25 Or 6 To 4	Chicago	Columbia	(10)
8.	Make It With You	Bread	Elektra	(3)
*9.	Julie, Do Ya Love Me	Bobby Sherman	Metromedia	(14)
*10.	Overture From Tommy	Assembled Multitude	Atlantic	(15)
11.	Close To You	Carpenters	A & M	(5)
*12.	Patches	Clarence Carter	Atlantic	(18)
13.	Signed, Sealed & Delivered	Stevie Wonder	Tamla	(4)
*14.	I Want To Take You Higher	Ike & Tine Turner & Ikettes	Liberty	(20)
15.	Tighter Tighter	Alive & Kicking	Roulette	(11)
16.	Morning Much Better	Ten Wheel Drive With Genya Ravan	Polydor	(17)
17.	Why Can't I Touch You?	Ronnie Dyson	Columbia	(9)
*18.	Lookin' Out My Back Door	Creedence Clearwater Revival	Fantasy	(26)
*19.	Solitary Man	Neil Diamond	Bang	(27)
*20.	Groovy Situation	Gene Chandler	Mercury	(35)
21.	O-O-Child	5 Stairsteps	Buddah	(16)
22.	Hi-De-Ho	Blood, Sweat & Tears	Columbia	(24)
*23.	Cracklin' Rosie	Neil Diamond	Uni	(37)
24.	Mississippi Queen	Mountain	Windfall	(13)
25.	Silver Bird	Mark Lindsay	Columbia	(19)
*26.	Candida	Dawn	Bell	(34)
*27.	Neanderthal Man	Hotlegs	Capitol	(N)
28.	Are You Ready?	Pacific Gas & Electric	Columbia	(21)
29.	Lay A Little Lovin' On Me	Robin McNamara	Steed	(23)
30.	Hand Me Down World	The Guess Who	RCA	(31)
31.	Soul Shake	Delaney & Bonnie & Friends	Atco	(36)
32.	Do What You Wanna Do	Five Flights Up	T-A	(33)
33.	The Love You Save	Jackson 5	Motown	(22)
34.	It's Your Life	Andy Kim	Steed	(25)
35.	Go Back	Crabby Appleton	Elektra	(N)
36.	Two Little Rooms	Trella Hart	Capitol	(40)
37.	Ball Of Confusion	The Temptations	Gordy	(28)
38.	I (Who Have Nothing)	Tom Jones	Parrot	(N)
39.	Band Of Gold	Freda Payne	Invictus	(29)
40.	Mama Told Me (Not To Come)	Three Dog Night	Dunhill	(32)

*Denotes Big 10 Climbers

WC
FL

BIG 10 COUNT DOWN

FOR THE WEEK OF AUGUST 24, 1970

				Last Week
1.	War	Edwin Starr	Gordy	(1)
2.	Ain't No Mountain High Enough	Diana Ross	Motown	(2)
*3.	25 Or 6 To 4	Chicago	Columbia	(7)
*4.	Cottage Cheese	Crow	Amaret	(5)
5.	In The Summertime	Mungo Jerry	Janus	(4)
6.	Spill The Wine	Eric Burdon & War	MGM	(3)
*7.	Julie, Do Ya Love Me	Bobby Sherman	Metromedia	(9)
*8.	Overture From Tommy	Assembled Multitude	Atlantic	(10)
*9.	Patches	Clarence Carter	Atlantic	(12)
*10.	Lookin' Out My Back Door	Creedence Clearwater Revival	Fantasy	(18)
*11.	Solitary Man	Neil Diamond	Bang	(19)
*12.	Neanderthal Man	Hotlegs	Capitol	(27)
13.	I Want To Take You Higher	Ike & Tina Turner & Ikettes	Liberty	(14)
*14.	Groovy Situation	Gene Chandler	Mercury	(20)
15.	I Just Can't Help Believing	B.J. Thomas	Scepter	(6)
*16.	Cracklin' Rosie	Neil Diamond	Uni	(23)
17.	Make It With You	Bread	Elektra	(8)
18.	Close To You	Carpenters	A & M	(11)
19.	Morning Much Better	Ten Wheel Drive With Genya Ravan	Polydor	(16)
20.	Signed, Sealed, Delivered I'm Yours	Stevie Wonder	Tamla	(13)
*21.	Candida	Dawn	Bell	(26)
*22.	Go Back	Crabby Appleton	Elektra	(35)
*23.	I (Who Have Nothing)	Tom Jones	Parrot	(38)
24.	Hi-De-Ho	Blood, Sweat & Tears	Columbia	(22)
25.	Tighter, Tighter	Alive & Kicking	Roulette	(15)
*26.	Do What You Wanna Do	Five Flights Up	T-A	(32)
27.	Soul Shake	Delaney & Bonnie & Friends	Atco	(31)
28.	Why Can't I Touch You	Ronnie Dyson	Columbia	(17)
29.	O-O-H Child	5 Stairsteps	Buddah	(21)
*30.	(I Know) I'm Losing You	Rare Earth	Rare Earth	(N)
31.	Hand Me Down World	The Guess Who	RCA	(30)
32.	Mississippi Queen	Mountain	Windfall	(24)
33.	Silver Bird	Mark Lindsay	Columbia	(25)
34.	Lay A Little Lovin' On Me	Robin McNamara	Steed	(29)
35.	Two Little Rooms	Trella Hart	Capitol	(36)
36.	Are You Ready?	Pacific Gas & Electric	Columbia	(28)
37.	Rubber Duckie	Ernie (Jim Henson)	Columbia	(N)
38.	Out In The Country	Three Dog Night	Dunhill	(N)
39.	Indiana Wants Me	R. Dean Taylor	Rare Earth	(N)
40.	The Love You Save	Jackson 5	Motown	(33)

HIT BOUND

All Right Now	Free	A & M

*Denotes Big 10 Climbers

WC
FL

BIG 10 COUNT DOWN

FOR THE WEEK OF AUGUST 31, 1970

				Last Week
1.	War	Edwin Starr	Gordy	(1)
2.	Ain't No Mountain High Enough	Diana Ross	Motown	(2)
*3.	Solitary Man	Neil Diamond	Bang	(11)
4.	25 Or 6 To 4	Chicago	Columbia	(3)
*5.	Neanderthal Man	Hotlegs	Capitol	(12)
*6.	Patches	Clarence Carter	Atlantic	(9)
*7.	Lookin' Out My Back Door	Creedence Clearwater Revival	Fantasy	(10)
*8.	Groovy Situation	Gene Chandler	Mercury	(14)
9.	In The Summertime	Mungo Jerry	Janus	(5)
10.	Spill The Wine	Eric Burdon & War	MGM	(6)
11.	Julie, Do Ya Love Me	Bobby Sherman	Metromedia	(7)
12.	Cottage Cheese	Crow	Amaret	(4)
13.	I Want To Take You Higher	Ike & Tina Turner & Ikettes	Liberty	(13)
14.	Overture From Tommy	Assembled Multitude	Atlantic	(8)
*15.	Cracklin' Rosie	Neil Diamond	Uni	(16)
*16.	Candida	Dawn	Bell	(21)
*17.	I (Who Have Nothing)	Tom Jones	Parrot	(23)
*18.	Go Back	Crabby Appleton	Elektra	(22)
*19.	Rubber Duckie	Ernie (Jim Henson)	Columbia	(37)
*20.	(I Know) I'm Losing You	Rare Earth	Rare Earth	(30)
21.	I Just Can't Help Believing	B.J. Thomas	Scepter	(15)
22.	Close To You	Carpenters	A & M	(18)
*23.	Do What You Wanna Do	Five Flights Up	T-A	(26)
24.	Make It With You	Bread	Elektra	(17)
25.	Signed, Sealed, Delivered I'm Yours	Stevie Wonder	Tamla	(20)
26.	Hi-De-Ho	Blood, Sweat & Tears	Columbia	(24)
*27.	Indiana Wants Me	R. Dean Taylor	Rare Earth	(39)
*28.	Out In The Country	Three Dog Night	Dunhill	(38)
29.	Tighter, Tighter	Alive & Kicking	Roulette	(25)
30.	Morning Much Better	Ten Wheel Drive With Genya Ravan	Polydor	(19)
*31.	All Right Now	Free	A & M	(HB)
*32.	We Can Make Music	Tommy Roe	ABC	(N)
33.	Why Can't I Touch You?	Ronnie Dyson	Columbia	(28)
34.	Two Little Rooms	Trella Hart	Capitol	(35)
35.	Hand Me Down World	The Guess Who	RCA	(31)
36.	O-O-H-Child	5 Stairsteps	Buddah	(29)
37.	Soul Shake	Delaney & Bonnie & Friends	Atco	(27)
38.	Silver Bird	Mark Lindsay	Columbia	(33)
39.	Mississippi Queen	Mountain	Windfall	(32)
40.	Lay A Little Lovin' On Me	Robin McNamara	Steed	(34)

HIT BOUND

Our World	Blue Mink	Philips

*Denotes Big 10 Climbers

WC
FL

BIG 10 COUNT DOWN

FOR THE WEEK OF SEPTEMBER 7, 1970

				Last Week
1.	War	Edwin Starr	Gordy	(1)
*2.	Solitary Man	Neil Diamond	Bang	(3)
*3.	Groovy Situation	Gene Chandler	Mercury	(8)
*4.	Patches	Clarence Carter	Atlantic	(6)
5.	Neanderthal Man	Hotlegs	Capitol	(5)
6.	Ain't No Mountain High Enough	Diana Ross	Motown	(2)
7.	Lookin' Out My Back Door	Creedence Clearwater Revival	Fantasy	(7)
*8.	Candida	Dawn	Bell	(16)
*9.	Rubber Duckie	Ernie (Jim Henson)	Columbia	(19)
10.	25 Or 6 To 4	Chicago	Columbia	(4)
11.	Cracklin' Rosie	Neil Diamond	Uni	(15)
*12.	(I Know) I'm Losing You	Rare Earth	Rare Earth	(20)
*13.	Go Back	Crabby Appleton	Elektra	(18)
14.	I Want To Take You Higher	Ike & Tine Turner & Ikettes	Liberty	(13)
15.	I (Who Have Nothing)	Tom Jones	Parrot	(17)
16.	Julie, Do Ya Love Me	Bobby Sherman	Metromedia	(11)
*17.	All Right Now	Free	A & M	(31)
*18.	Indiana Wants Me	R. Dean Taylor	Rare Earth	(27)
19.	Spill The Wine	Eric Burdon & War	MGM	(10)
20.	Do What You Wanna Do	Five Flights Up	T-A	(23)
*21.	Out In The Country	Three Dog Night	Dunhill	(28)
22.	In The Summertime	Mungo Jerry	Janus	(9)
23.	Cottage Cheese	Crow	Amaret	(12)
24.	Overture From Tommy	Assembled Multitude	Atlantic	(14)
*25.	We Can Make Music	Tommy Roe	ABC	(32)
26.	Close To You	Carpenters	A & M	(22)
27.	I Just Can't Help Believing	B.J. Thomas	Scepter	(21)
28.	Hi-De-Ho	Blood, Sweat & Tears	Columbia	(26)
29.	Make It With You	Bread	Elektra	(24)
30.	Signed, Sealed, Delivered I'm Yours	Stevie Wonder	Tamla	(25)
31.	Lola	The Kinks	Reprise	(N)
32.	Our World	Blue Mink	Philips	(HB)
33.	Look What They've Done To My Song Ma	The New Seekers	Elektra	(N)
34.	We've Only Just Begun	Carpenters	A & M	(N)
35.	Tighter, Tighter	Alive & Kicking	Roulette	(29)
36.	Two Little Rooms	Trella Hart	Capitol	(34)
37.	Why Can't I Touch You?	Ronnie Dyson	Columbia	(33)
38.	-O-O-H Child	5 Stairsteps	Buddah	(36)
39.	I'll Be There	The Jackson 5	Motown	(N)
40.	Silver Bird	Mark Lindsay	Columbia	(38)

*Denotes Big 10 Climbers

WC
FL

BIG 10 COUNT DOWN

FOR THE WEEK OF SEPTEMBER 14, 1970

				Last Week
1.	War	Edwin Starr	Gordy	(1)
*2.	Groovy Situation	Gene Chandler	Mercury	(3)
3.	Solitary Man	Neil Diamond	Bang	(2)
4.	Patches	Clarence Carter	Atlantic	(4)
*5.	Lookin' Out My Back Door	Creedence Clearwater Revival	Fantasy	(7)
*6.	Candida	Dawn	Bell	(8)
*7.	Rubber Duckie	Ernie (Jim Henson)	Columbia	(9)
*8.	(I Know) I'm Losing You	Rare Earth	Rare Earth	(12)
*9.	Cracklin' Rosie	Neil Diamond	Uni	(11)
10.	Neanderthal Man	Hotlegs	Capitol	(5)
*11.	I (Who Have Nothing)	Tom Jones	Parrot	(15)
*12.	Go Back	Crabby Appleton	Elektra	(13)
*13.	Indiana Wants Me	R. Dean Taylor	Rare Earth	(18)
*14.	Out In The Country	Three Dog Night	Dunhill	(21)
*15.	All Right Now	Free	A & M	(17)
16.	Ain't No Mountain High Enough	Diana Ross	Momtown	(6)
*17.	Do What You Wanna Do	Five Flights Up	T-A	(20)
18.	25 Or 6 To 4	Chicago	Columbia	(10)
19.	Julie, Do Ya Love Me	Bobby Sherman	Metromedia	(16)
20.	I Want To Take You Higher	Ike & Tina Turner & Ikettes	Liberty	(14)
*21.	Lola	The Kinks	Reprise	(31)
*22.	Look What They've Done To My Song Ma	The New Seekers	Elektra	(33)
23.	Spill The Wine	Eric Burdon & War	MGM	(19)
24.	We Can Make Music	Tommy Roe	ABC	(25)
*25.	We've Only Just Begun	Carpenters	A & M	(34)
*26.	Our World	Blue Mink	Philips	(32)
27.	In The Summertime	Mungo Jerry	Janus	(22)
28.	Cottage Cheese	Crow	Amaret	(23)
29.	Close To You	Carpenters	A & M	(26)
*30.	I'll Be There	The Jackson 5	Motown	(39)
31.	Hi-De-Ho	Blood, Sweat & Tears	Columbia	(28)
32.	Overture From Tommy	Assembled Multitude	Atlantic	(24)
33.	I Just Can't Help Believing	B.J. Thomas	Scepter	(27)
34.	Green-Eyed Lady	Sugarloaf	Liberty	(N)
35.	Signed, Sealed, Delivered I'm Yours	Stevie Wonder	Tamla	(30)
36.	Joanne	Michael Nesmith & The First National Band	RCAQ	(N)
37.	Make It With You	Bread	Elektra	(29)
38.	Tighter, Tighter	Alive & Kicking	Roulette	(35)
39.	Two Little Rooms	Trella Hart	Capitol	(36)
40.	Why Can't I Touch You?	Ronnie Dyson	Columbia	(37)

*Denotes Big 10 Climbers

WC
FL

BIG 10 COUNT DOWN

FOR THE WEEK OF SEPTEMBER 21, 1970

				Last Week
*1.	Candida	Dawn	Bell	(6)
*2.	Lookin' Out My Back Door	Creedence Clearwater Revival	Fantasy	(5)
*3.	Cracklin' Rosie	Neil Diamond	Uni	(9)
4.	War	Edwin Starr	Gordy	(1)
5.	Patches	Clarence Carter	Atlantic	(4)
*6.	(I Know) I'm Losing You	Rare Earth	Rare Earth	(8)
*7.	Indiana Wants Me	R. Dean Taylor	Rare Earth	(13)
*8.	All Right Now	Free	A & M	(15)
9.	Groovy Situation	Gene Chandler	Mercury	(2)
*10.	Solitary Man	Neil Diamond	Bang	(3)
*11.	Out In The Country	Three Dog Night	Dunhill	(14)
*12.	Do What You Wanna Do	Five Flights Up	T-A	(17)
13.	Go Back	Crabby Appleton	Elektra	(12)
14.	I (Who Have Nothing)	Tom Jones	Parrot	(11)
*15.	Lola	The Kinks	Reprise	(21)
16.	Rubber Duckie	Ernie (Jim Henson)	Columbia	(7)
17.	Neanderthal Man	Hotlegs	Capitol	(10)
*18.	We Can Make Music	Tommy Roe	ABC	(24)
*19.	I'll Be There	Jackson 5	Motown	(30)
*20.	We've Only Just Begun	Carpenters	A & M	(25)
*21.	Look What They've Done To My Song Ma	The New Seekers	Elektra	(22)
*22.	Our World	Blue Mink	Philips	(26)
*23.	Green-Eyed Lady	Sugarloaf	Liberty	(34)
24.	Ain't No Mountain High Enough	Diana Ross	Motown	(16)
25.	Julie, Do Ya Love Me	Bobby Sherman	Metromedia	(19)
*26.	Joanne	Michael Nesmith & The First National Band	RCA	(36)
27.	25 Or 6 To 4	Chicago	Columbia	(18)
28.	Spill The Wine	Eric Burdon & War	MGM	(23)
29.	I Want To Take You Higher	Ike & Tine Turner & Ikettes	Liberty	(20)
*30.	Snowbird	Anne Murray	Capitol	(N)
31.	In The Summertime	Mungo Jerry	Janus	(27)
32.	It Don't Matter To Me	Bread	Elektra	(N)
33.	Close To You	Carpenters	A & M	(29)
34.	Cottage Cheese	Crow	Amaret	(28)
35.	That's Where I Went Wrong	The Poppy Family	London	(N)
36.	Hi-De-Ho	Blood, Sweat & Tears	Columbia	(31)
37.	I Just Can't Help Believing	B.J. Thomas	Scepter	(33)
38.	Signed, Sealed, Delivered I'm Yours	Stevie Wonder	Tamla	(35)
39.	You Better Think Twice	Poco	Epic	(N)
40.	Overture From Tommy	Assembled Multitude	Atlantic	(32)

HIT BOUND

Closer To Home	Grand Funk Railroad	Capitol
Somebody's Been Sleeping	100 Proof	Hot Wax
God, Love And Rock & Roll	Teegarden & Van Winkle	Westbound

*Denotes Big 10 Climbers

WC
FL

BIG 10 COUNT DOWN

FOR THE WEEK OF SEPTEMBER 28, 1970

				Last Week
*1.	Candida	Dawn	Bell	(1)
*2.	Cracklin' Rosie	Neil Diamond	Uni	(3)
*3.	Indiana Wants Me	R. Dean Taylor	Rare Earth	(7)
*4.	All Right Now	Free	A & M	(8)
5.	(I Know) I'm Losing You	Rare Earth	Rare Earth	(6)
6.	Lookin' Out My Back Door	Creedence Clearwater Revival	Fantasy	(2)
*7.	Lola	The Kinks	Reprise	(15)
*8.	I'll Be There	Jackson 5	Motown	(19)
*9.	Do What You Wanna Do	Five Flights Up	T-A	(12)
10.	Out In The Country	Three Dog Night	Dunhill	(11)
11.	Patches	Clarence Carter	Atlantic	(5)
*12.	Green-Eyed Lady	Sugar Loaf	Liberty	(23)
13.	War	Edwin Starr	Gordy	(4)
14.	Solitary Man	Neil Diamond	Bang	(10)
*15.	We've Only Just Begun	Carpenters	A & M	(20)
16.	Groovy Situation	Gene Chandler	Mercury	(9)
17.	Go Back	Crabby Appleton	Elektra	(13)
18.	We Can Make Music	Tommy Roe	ABC	(18)
19.	Our World	Blue Mink	Philips	(22)
20.	Look What They've Done To My Song Ma	The New Seekers	Elektra	(21)
*21.	Joanne	Michael Nesmith & The First National Band	RCA	(26)
22.	I (Who Have Nothing)	Tom Jones	Parrot	(14)
23.	Rubber Duckie	Ernie (Jim Henson)	Columbia	(16)
*24.	Snowbird	Anne Murray	Capitol	(30)
*25.	That's Where I Went Wrong	The Poppy Family	London	(35)
26.	Neanderthal Man	Hotlegs	Capitol	(17)
27.	Ain't No Mountain High Enough	Diana Ross	Motown	(24)
28.	It Don't Matter To Me	Bread	Elektra	(32)
29.	Julie, Do Ya Love Me	Bobby Sherman	Metromedia	(25)
*30.	Somebody's Been Sleeping	100 Proof	Hot Wax	(HB)
31.	Spill The Wine	Eric Burdon & War	MGM	(28)
32.	25 Or 6 To 4	Chicago	Columbia	(27)
33.	I Want To Take You Higher	Ike & Tine Turner & Ikettes	Liberty	(29)
34.	Closer To Home	Grand Funk Railroad	Capitol	(HB)
*35.	God, Love And Rock & Roll	Teegarden & Van Winkle	Westbound	(HB)
36.	You Better Think Twice	Poco	Epic	(39)
37.	In The Summertime	Mungo Jerry	Janus	(31)
38.	Close To You	Carpenters	A & M	(33)
39.	Cottage Cheese	Crow	Amaret	(34)
40.	Hi-De-Ho	Blood, Sweat & Tears	Columbia	(36)

HIT BOUND

Gypsy Woman	Brian Hyland	Uni

*Denotes Big 10 Climbers

WC
FL

BIG 10 COUNT DOWN

FOR THE WEEK OF OCTOBER 5, 1970

				Last Week
*1.	I'll Be There	Jackson 5	Motown	(8)
*2.	Indiana Wants Me	R. Dean Taylor	Rare Earth	(3)
*3.	All Right Now	Free	A & M	(4)
4.	Candida	Dawn	Bell	(1)
5.	(I Know) I'm Losing You	Rare Earth	Rare Earth	(5)
*6.	Lola	The Kinks	Reprise	(7)
7.	Cracklin' Rosie	Neil Diamond	Uni	(2)
*8.	Green-Eyed Lady	Sugarloaf	Liberty	(12)
9.	Do What You Wanna Do	Five Flights Up	T-A	(9)
*10.	We've Only Just Begun	Carpenters	A & M	(15)
11.	Out In The Country	Three Dog Night	Dunhill	(10)
12.	Lookin' Out My Back Door	Creedence Clearwater Revival	Fantasy	(6)
*13.	That's Where I Went Wrong	The Poppy Family	London	(25)
*14.	Joanne	Michael Nesmith & The First National Band	RCA	(21)
*15.	Look What They've Done To My Song Ma	The New Seekers	Elektra	(20)
16.	Patches	Clarence Carter	Atlantic	(11)
*17.	Snowbird	Anne Murray	Capitol	(24)
18.	War	Edwin Starr	Gordy	(13)
19.	Our World	Blue Mink	Philips	(19)
*20.	Closer To Home	Grand Funk Railroad	Capitol	(34)
*21.	It Don't Matter To Me	Bread	Elektra	(28)
*22.	Somebody's Been Sleeping	100 Proof	Hot Wax	(30)
23.	Solitary Man	Neil Diamond	Bang	(14)
24.	We Can Make Music	Tommy Roe	ABC	(18)
25.	Groovy Situation	Gene Chandler	Mercury	(16)
26.	I (Who Have Nothing)	Tom Jones	Parrot	(22)
27.	Go Back	Crabby Appleton	Elektra	(17)
*28.	God, Love And Rock And Roll	Teegarden & Van Winkle	Westbound	(35)
29.	Rubber Duckie	Ernie (Jim Henson)	Columbia	(23)
30.	Neanderthal Man	Hotlegs	Capitol	(26)
31.	Ain't No Mountain High Enough	Diana Ross	Motown	(27)
*32.	Fire And Rain	James Taylor	Warner Bros.	(N)
33.	Julie, Do Ya Love Me	Bobby Sherman	Metromedia	(29)
34.	Gypsy Woman	Brian Hyland	Uni	(HB)
35.	You Better Think Twice	Poco	Epic	(36)
36.	The Tears Of A Clown	Smokey Robinson And The Miracles	Tamla	(N)
37.	Spill The Wine	Eric Burdon & War	MGM	(31)
38.	I Want To Take You Higher	Ike & Tina Turner & Ikettes	Liberty	(33)
39.	25 Or 6 To 4	Chicago	Columbia	(32)
40.	It's A Shame	The Spinners	V.I.P.	(N)

*Denotes Big 10 Climbers

WC
FL

BIG 10 COUNT DOWN

FOR THE WEEK OF OCTOBER 12, 1970

				Last Week
1.	I'll Be There	Jackson 5	Motown	(1)
2.	Indiana Wants Me	R. Dean Taylor	Rare Earth	(2)
3.	All Right Now	Free	A & M	(3)
*4.	Lola	The Kinks	Reprise	(6)
*5.	Green-Eyed Lady	Sugarloaf	Liberty	(8)
*6.	We've Only Just Begun	Carpenters	A & M	(10)
7.	Candida	Dawn	Bell	(4)
8.	(I Know) I'm Losing You	Rare Earth	Rare Earth	(5)
*9.	That's Where I Went Wrong	The Poppy Family	London	(13)
*10.	Look What They've Done To My Song Ma	The New Seekers	Elektra	(15)
11.	Cracklin' Rosie	Neil Diamond	Uni	(7)
12.	Do What You Wanna Do	Five Flights Up	T-A	(9)
13.	Joanne	Michael Nesmith & The First National Band	RCA	(14)
*14.	Somebody's Been Sleeping	100 Proof	Hot Wax	(22)
15.	Out In The Country	Three Dog Night	Dunhill	(11)
*16.	It Don't Matter To Me	Bread	Elektra	(21)
*17.	Fire And Rain	James Taylor	Warner Bros.	(32)
18.	Closer To Home	Grand Funk Railroad	Capitol	(20)
19.	Lookin' Out My Back Door	Creedence Clearwater Revival	Fantasy	(12)
20.	Snowbird	Anne Murray	Capitol	(17)
21.	Gypsy Woman	Brian Hyland	Uni	(34)
22.	War	Edwin Starr	Gordy	(18)
23.	Patches	Clarence Carter	Atlantic	(16)
*24.	God, Love And Rock And Roll	Teegarden & Van Winkle	Westbound	(28)
*25.	The Tears Of A Clown	Smokey Robinson & The Miracles	Tamla	(36)
*26.	I Think I Love You	The Partridge Family	Bell	(N)
27.	Our World	Blue Mink	Philips	(19)
28.	Solitary Man	Neil Diamond	Bang	(23)
*29.	Mongoose	Elephant's Memory	Metromedia	(N)
*30.	King Of Rock & Roll	Crow	Amaret	(N)
31.	We Can Make Music	Tommy Roe	ABC	(24)
32.	You Better Think Twice	Poco	Epic	(35)
33.	I (Who Have Nothing)	Tom Jones	Parrot	(26)
34.	It's A Shame	The Spinners	V.I.P.	(40)
35.	Groovy Situation	Gene Chandler	Mercury	(25)
36.	Go Back	Crabby Appleton	Elektra	(27)
37.	Ain't No Mountain High Enough	Diana Ross	Motown	(31)
38.	Julie, Do Ya Love Me	Bobby Sherman	Metromedia	(33)
39.	Good Day	Bobby Trend	Ivanhoe	(N)
40.	Lucretia Mac Evil	Blood, Sweat & Tears	Columbia	(N)

*Denotes Big 10 Climbers

WC
FL

BIG 10 COUNT DOWN

FOR THE WEEK OF OCTOBER 19, 1970

				Last Week
*1.	Indiana Wants Me	R. Dean Taylor	Rare Earth	(2)
2.	I'll Be There	Jackson 5	Motown	(1)
*3.	Lola	The Kinks	Reprise	(4)
*4.	We've Only Just Begun	Carpenters	A & M	(6)
5.	Green-Eyed Lady	Sugarloaf	Liberty	(5)
6.	All Right Now	Free	A & M	(3)
*7.	It Don't Matter To Me	Bread	Elektra	(16)
*8.	Look What They've Done To My Song Ma	The New Seekers	Elektra	(10)
9.	That's Where I Went Wrong	The Poppy Family	London	(9)
*10.	Fire And Rain	James Taylor	Warner Bros.	(17)
*11.	Somebody's Been Sleeping	100 Proof	Hot Wax	(14)
12.	Candida	Dawn	Bell	(7)
13.	(I Know) I'm Losing You	Rare Earth	Rare Earth	(8)
14.	Cracklin' Rosie	Neil Diamond	Uni	(11)
*15.	I Think I Love You	The Partridge Family	Bell	(26)
16.	Closer To Home	Grand Funk Railroad	Capitol	(18)
*17.	Gypsy Woman	Brian Hyland	Uni	(21)
*18.	God, Love And Rock And Roll	Teegarden Van Winkle	Westbound	(24)
19.	Joanne	Michael Nesmith & The First National Band	RCA	(13)
*20.	Mongoose	Elephant's Memory	Metromedia	(29)
*21.	The Tears Of A Clown	Smokey Robinson & The Miracles	Tamla	(25)
*22.	King Of Rock And Roll	Crow	Amaret	(30)
23.	Do What You Wanna Do	Five Flights Up	T-A	(12)
24.	Out In The Country	Three Dog Night	Dunhill	(15)
25.	Lookin' Out My Back Door	Creedence Clearwater Revival	Fantasy	(19)
*26.	Good Day	Bobby Trend	Ivanhoe	(39)
27.	You Better Think Twice	Poco	Epic	(32)
*28.	Lucretia Mac Evil	Blood, Sweat & Tears	Columbia	(40)
*29.	Yellow River	Christie	Epic	(N)
30.	It's A Shame	The Spinners	V.I.P.	(34)
31.	War	Edwin Starr	Gordy	(22)
32.	Patches	Clarence Carter	Atlantic	(23)
33.	Snowbird	Anne Murray	Capitol	(20)
34.	Solitary Man	Neil Diamond	Bang	(28)
35.	El Condor Pasa	Simon & Garfunkel	Columbia	(N)
36.	Let's Work Together	Canned Heat	Liberty	(N)
37.	I (Who Have Nothing)	Tom Jones	Parrot	(33)
38.	We Can Make Music	Tommy Roe	ABC	(31)
39.	Ain't No Mountain High Enough	Diana Ross	Motown	(37)
40.	Still Water (Love)	Four Tops	Motown	(N)

HIT BOUND

Heaven Help Us All	Stevie Wonder	Tamla
Stand By Your Man	Candi Staton	Fame
Montego Bay	Bobby Bloom	L & R

*Denotes Big 10 Climbers

WC
FL

BIG 10 COUNT DOWN

FOR THE WEEK OF OCTOBER 26, 1970

				Last Week
*1.	We've Only Just Begun	Carpenters	A & M	(4)
*2.	Lola	The Kinks	Reprise	(3)
3.	Indiana Wants Me	R. Dean Taylor	Rare Earth	(1)
*4.	I Think I Love You	The Partridge Family	Bell	(15)
5.	I'll Be There	Jackson 5	Motown	(2)
6.	Green-Eyed Lady	Sugarloaf	Liberty	(5)
7.	It Don't Matter To Me	Bread	Elektra	(7)
8.	Look What They've Done To My Song Ma	The New Seekers	Elektra	(8)
*9.	Fire And Rain	James Taylor	Warner Bros.	(10)
*10.	Somebody's Been Sleeping	100 Proof	Hot Wax	(11)
*11.	Gypsy Woman	Brian Hyland	Uni	(17)
*12.	The Tears Of A Clown	Smokey Robinson & The Miracles	Tamla	(21)
13.	All Right Now	Free	A & M	(6)
14.	That's Where I Went Wrong	The Poppy Family	London	(9)
*15.	Mongoose	Elephant's Memory	Metromedia	(20)
*16.	Yellow River	Christie	Epic	(29)
17.	Candida	Dawn	Bell	(12)
18.	Cracklin' Rosie	Neil Diamond	Uni	(14)
*19.	Let's Work Together	Canned Heat	Liberty	(36)
*20.	King Of Rock And Roll	Crow	Amaret	(22)
*21.	Good Day	Bobby Trend	Ivanhoe	(26)
22.	(I Know) I'm Losing You	Rare Earth	Rare Earth	(13)
23.	God, Love And Rock And Roll	Teegarden Van Winkle	Westbound	(18)
24.	Closer To Home	Grand Funk Railroad	Capitol	(16)
25.	Joanne	Michael Nesmith & The First National Band	RCA	(19)
26.	You Better Think Twice	Poco	Epic	(27)
27.	Lucretia Mac Evil	Blood, Sweat & Tears	Columbia	(28)
*28.	Still Water (Love)	Four Tops	Motown	(40)
29.	Do What You Wanna Do	Five Flights Up	T-A	(23)
30.	It's A Shame	The Spinners	V.I.P.	(30)
31.	Out In The Country	Three Dog Night	Dunhill	(24)
32.	Lookin' Out My Back Door	Creedence Clearwater Revival	Fantasy	(25)
33.	Stand By Your Man	Candi Staton	Fame	(HB)
34.	Heaven Help Us All	Stevie Wonder	Tamla	(HB)
35.	Montego Bay	Bobby Bloom	L & R	(HB)
36.	El Condor Pasa	Simon & Garfunkel	Columbia	(35)
37.	Cry Me A River	Joe Cocker	A & M	(N)
38.	War	Edwin Starr	Gordy	(31)
39.	Patches	Clarence Carter	Atlantic	(32)
40.	Snowbird	Anne Murray	Capitol	(33)

*Denotes Big 10 Climbers

BIG 10 COUNT DOWN

FOR THE WEEK OF NOVEMBER 2, 1970

				Last Week
*1.	I Think I Love You	The Partridge Family	Bell	(4)
2.	We've Only Just Begun	Carpenters	A & M	(1)
*3.	Fire And Rain	James Taylor	Warner Bros.	(9)
4.	Lola	The Kinks	Reprise	(2)
*5.	Gypsy Woman	Brian Hyland	Uni	(11)
*6.	The Tears Of A Clown	Smokey Robinson & The Miracles	Tamla	(12)
*7.	Yellow River	Christie	Epic	(16)
8.	Indiana Wants Me	R. Dean Taylor	Rare Earth	(3)
9.	I'll Be There	Jackson 5	Motown	(5)
10.	Green-Eyed Lady	Sugarloaf	Liberty	(6)
11.	It Don't Matter To Me	Bread	Elektra	(7)
*12.	Let's Work Together	Canned Heat	Liberty	(19)
*13.	Mongoose	Elephant's Memory	Metromedia	(15)
14.	Somebody's Been Sleeping	100 Proof	Hot Wax	(10)
*15.	King Of Rock And Roll	Crow	Amaret	(20)
16.	Look What They've Done To My Song Ma	The New Seekers	Elektra	(8)
17.	All Right Now	Free	A & M	(13)
*18.	Montego Bay	Bobby Bloom	L & R	(35)
*19.	Heaven Help Us All	Stevie Wonder	Tamla	(34)
*20.	Good Day	Bobby Trend	Ivanhoe	(21)
21.	That's Where I Went Wrong	The Poppy Family	London	(14)
22.	Cracklin' Rosie	Neil Diamond	Uni	(18)
*23.	Still Water (Love)	Four Tops	Motown	(28)
24.	Candida	Dawn	Bell	(17)
*25.	Lucretia Mac Evil	Blood, Sweat & Tears	Columbia	(27)
*26.	Stand By Your Man	Candi Staton	Fame	(33)
27.	(I Know) I'm Losing You	Rare Earth	Rare Earth	(22)
28.	God, Love And Rock And Roll	Teegarden & Van Winkle	Westbound	(23)
29.	You Better Think Twice	Poco	Epic	(26)
*30.	Cry Me A River	Joe Cocker	A & M	(37)
31.	Closer To Home	Grand Funk Railroad	Capitol	(24)
32.	Be My Baby	Andy Kim	Steed	(N)
33.	Joanne	Michael Nesmith & The First National Band	RCA	(25)
34.	Do What You Wanna Do	Five Flights Up	T-A	(29)
35.	Does Anybody Really Know What Time It Is?	Chicago	Columbia	(N)
36.	El Condor Pasa	Simon & Garfunkel	Columbia	(36)
37.	It's A Shame	The Spinners	V.I.P.	(30)
38.	Lookin' Out My Back Door	Creedence Clearwater Revival	Fantasy	(32)
39.	Out In The Country	Three Dog Night	Dunhill	(31)
40.	War	Edwin Starr	Gordy	(38)

HIT BOUND

See Me, Feel Me	The Who	Decca
Everybody I Love You	Happy Day	Uni

*Denotes Big 10 Climbers

WC
FL

BIG 10 COUNT DOWN

FOR THE WEEK OF NOVEMBER 9, 1970

				Last Week
*1.	I Think I Love You	The Partridge Family	Bell	(1)
*2.	Gypsy Woman	Brian Hyland	Uni	(5)
3.	Fire And Rain	James Taylor	Warner Bros.	(3)
*4.	Yellow River	Christie	Epic	(7)
*5.	The Tears Of A Clown	Smokey Robinson & The Miracles	Tamla	(6)
6.	We've Only Just Begun	Carpenters	A & M	(2)
*7.	Montego Bay	Bobby Bloom	L & R	(18)
8.	Lola	The Kinks	Reprise	(4)
*9.	Let's Work Together	Canned Heat	Liberty	(12)
*10.	Mongoose	Elephant's Memory	Metromedia	(13)
11.	I'll Be There	Jackson 5	Motown	(9)
12.	Indiana Wants Me	R. Dean Taylor	Rare Earth	(8)
13.	Green-Eyed Lady	Sugarloaf	Liberty	(10)
*14.	Heaven Help Us All	Stevie Wonder	Tamla	(19)
15.	King Of Rock And Roll	Crow	Amaret	(15)
16.	It Don't Matter To Me	Bread	Elektra	(11)
*17.	Still Water (Love)	Four Tops	Motown	(23)
18.	Somebody's Been Sleeping	100 Proof	Hot Wax	(14)
19.	Good Day	Bobby Trend	Ivanhoe	(20)
*20.	Cry Me A River	Joe Cocker	A & M	(30)
*21.	See Me, Feel Me	The Who	Decca	(HB)
*22.	Stand By Your Man	Candi Staton	Fame	(26)
23.	Look What They've Done To My Song Ma	The New Seekers	Elektra	(16)
24.	All Right Now	Free	A & M	(17)
*25.	Does Anybody Really Know What Time It Is?	Chicago	Columbia	(35)
*26.	Be My Baby	Andy Kim	Steed	(32)
27.	Cracklin' Rosie	Neil Diamond	Uni	(22)
28.	That's Where I Went Wrong	The Poppy Family	London	(21)
29.	Candida	Dawn	Bell	(24)
*30.	As The Years Go By	Mashmakhan	Epic	(N)
*31.	After Midnight	Eric Clapton	Atco	(N)
32.	Lucretia Mac Evil	Blood, Sweat & Tears	Columbia	(25)
33.	No Matter What	Badfinger	Apple	(N)
34.	One Man Band	Three Dog Night	Dunhill	(N)
35.	(I Know) I'm Losing You	Rare Earth	Rare Earth	(27)
36.	God, Love And Rock And Roll	Teegarden Van Winkle	Westbound	(28)
37.	Closer To Home	Grand Funk Railroad	Capitol	(31)
38.	You Better Think Twice	Poco	Epic	(29)
39.	Joanne	Michael Nesmith & The First National Band	RCA	(33)
40.	El Condor Pasa	Simon & Garfunkel	Columbia	(36)

HIT BOUND

Knock Three Times	Dawn	Bell
When The Party Is Over	Robert John	A & M

*Denotes Big 10 Climbers

WC
FL

BIG 10 COUNT DOWN

FOR THE WEEK OF NOVEMBER 16, 1970

					Last Week
*1.	I Think I Love You	The Partridge Family	Bell	(1)
*2.	Gypsy Woman	Brian Hyland	Uni	(2)
*3.	The Tears Of A Clown	Smokey Robinson & The Miracles	Tamla	(5)
4.	Yellow River	Christie	Epic	(4)
*5.	Montego Bay	Bobby Bloom	L & R	(7)
*6.	Mongoose	Elephant's Memory	Metromedia	(10)
*7.	Let's Work Together	Canned Heat	Liberty	(9)
8.	Fire And Rain	James Taylor	Warner Bros.	(3)
9.	We've Only Just Begun	Carpenters	A & M	(6)
*10.	See Me, Feel Me	The Who	Decca	(21)
*11.	Heaven Help Us All	Stevie Wonder	Tamla	(14)
12.	Lola	The Kinks	Reprise	(8)
13.	King Of Rock And Roll	Crow	Amaret	(15)
*14.	Still Water (Love)	Four Tops	Motown	(17)
*15.	Does Anybody Really Know What Time It Is?	Chicago	Columbia	(25)
16.	I'll Be There	Jackson 5	Motown	(11)
17.	Cry Me A River	Joe Cocker	A & M	(20)
*18.	As The Years Go By	Mashmakhan	Epic	(30)
*19.	Be My Baby	Andy Kim	Steed	(26)
20.	Stand By Your Man	Candi Staton	Fame	(22)
21.	After Midnight	Eric Clapton	Atco	(31)
22.	No Matter What	Badfinger	Apple	(33)
23.	Indiana Wants Me	R. Dean Taylor	Rare Earth	(12)
*24.	One Man Band	Three Dog Night	Dunhill	(34)
25.	Green-Eyed Lady	Sugarloaf	Liberty	(13)
26.	It Don't Matter To Me	Bread	Elektra	(16)
27.	Somebody's Been Sleeping	100 Proof	Hot Wax	(18)
*28.	Knock Three Times	Dawn	Bell	(HB)
*29.	When The Party Is Over	Robert John	A & M	(HB)
30.	Good Day	Bobby Trend	Ivanhoe	(19)
31.	All Right Now	Free	A & M	(24)
32.	Look What They've Done To My Song Ma	The New Seekers	Elektra	(23)
33.	Everybody I Love You	Happy Day	Uni	(N)
34.	Domino	Van Morrison	Warner Bros.	(N)
35.	Cracklin' Rosie	Neil Diamond	Uni	(27)
36.	Candida	Dawn	Bell	(29)
37.	That's Where I Went Wrong	The Poppy Family	London	(28)
38.	5-10-15-20 (25-30 Years Of Love)	The Presidents	Sussex	(N)
39.	(I Know) I'm Losing You	Rare Earth	Rare Earth	(35)
40.	Closer To Home	Grand Funk Railroad	Capitol	(37)

HIT BOUND

I Dig Everything About You	The Mob	Colossus

*Denotes Big 10 Climbers

WC
FL

BIG 10 COUNT DOWN

FOR THE WEEK OF NOVEMBER 23, 1970

				Last Week
*1.	The Tears Of A Clown	Smokey Robinson & The Miracles	Tamla	(3)
2.	I Think I Love You	The Partridge Family	Bell	(1)
*3.	Montego Bay	Bobby Bloom	L & R	(5)
4.	Gypsy Woman	Brian Hyland	Uni	(2)
*5.	Mongoose	Elephant's Memory	Metromedia	(6)
6.	Let's Work Together	Canned Heat	Liberty	(7)
7.	Yellow River	Christie	Epic	(4)
*8.	See Me, Feel Me	The Who	Decca	(10)
*9.	Heaven Help Us All	Stevie Wonder	Tamla	(11)
*10.	Cry Me A River	Joe Cocker	A & M	(17)
*11.	As The Years Go By	Mashmakhan	Epic	(18)
*12.	Does Anybody Really Know What Time It Is?	Chicago	Columbia	(15)
13.	Still Water (Love)	Four Tops	Motown	(14)
*14.	After Midnight	Eric Clapton	Atco	(21)
*15.	Be My Baby	Andy Kim	Steed	(19)
16.	We've Only Just Begun	Carpenters	A & M	(9)
*17.	No Matter What	Badfinger	Apple	(22)
18.	Fire And Rain	James Taylor	Warner Bros.	(8)
*19.	Knock Three Times	Dawn	Bell	(28)
*20.	One Man Band	Three Dog Night	Dunhill	(24)
*21.	When The Party Is Over	Robert John	A & M	(29)
*22.	Share The Land	The Guess Who	RCA	(N)
*23.	One Less Bell To Answer	The 5th Dimension	Bell	(N)
24.	Lola	The Kinks	Reprise	(12)
25.	I'll Be There	Jackson 5	Motown	(16)
*26.	My Sweet Lord/Isn't It A Pity	George Harrison	Apple	(N)
27.	King Of Rock And Roll	Crow	Amaret	(13)
*28.	Domino	Van Morrison	Warner Bros.	(34)
*29.	You Don't Have To Say You Love Me	Elvis Presley	RCA	(N)
30.	Stand By Your Man	Candi Staton	Fame	(20)
*31.	5-10-15-20 (25-30 Years Of Love)	The Presidents	Sussex	(38)
32.	Everybody I Love You	Happy Day	Uni	(33)
33.	Green-Eyed Lady	Sugarloaf	Liberty	(25)
34.	Indiana Wants Me	R. Dean Taylor	Rare Earth	(23)
35.	Stoned Love	The Supremes	Motown	(N)
36.	I Dig Everything About You	The Mob	Colossus	(HB)
37.	It Don't Matter To Me	Bread	Elektra	(26)
38.	Somebody's Been Sleeping	100 Proof	Hot Wax	(27)
39.	All Right Now	Free	A & M	(31)
40.	Black Magic Woman	Santana	Columbia	(N)

*Denotes Big 10 Climbers

BIG 10 COUNT DOWN

FOR THE WEEK OF NOVEMBER 30, 1970

				Last Week
1.	The Tears Of A Clown	Smokey Robinson & The Miracles	Tamla	(1)
*2.	Montego Bay	Bobby Bloom	L & R	(3)
*3.	See Me, Feel Me	The Who	Decca	(8)
4.	I Think I Love You	The Partridge Family	Bell	(2)
5.	Gypsy Woman	Brian Hyland	Uni	(4)
*6.	Knock Three Times	Dawn	Bell	(19)
7.	Heaven Help Us All	Stevie Wonder	Tamla	(9)
*8.	Be My Baby	Andy Kim	Steed	(15)
*9.	After Midnight	Eric Clapton	Atco	(14)
10.	Does Anybody Really Know What Time It Is?	Chicago	Columbia	(12)
11.	As The Years Go By	Mashmakhan	Epic	(11)
12.	Cry Me A River	Joe Cocker	A & M	(10)
13.	Mongoose	Elephant's Memory	Metromedia	(5)
14.	Yellow River	Christie	Epic	(7)
15.	No Matter What	Badfinger	Apple	(17)
*16.	My Sweet Lord/Isn't It A Pity	George Harrison	Apple	(26)
*17.	Share The Land	The Guess Who	RCA	(22)
*18.	One Less Bell To Answer	The 5th Dimension	Bell	(23)
19.	When The Party Is Over	Robert John	A & M	(21)
20.	One Man Band	Three Dog Night	Dunhill	(20)
*21.	Domino	Van Morrison	Warner Bros.	(28)
22.	Let's Work Together	Canned Heat	Liberty	(6)
*23.	You Don't Have To Say You Love Me	Elvis Presley	RCA	(29)
24.	Still Water (Love)	Four Tops	Motown	(13)
25.	We've Only Just Begun	Carpenters	A & M	(16)
*26.	5-10-15-20 (25-30 Years Of Love)	The Presidents	Sussex	(31)
*27.	Stoned Love	The Supremes	Motown	(35)
*28.	Black Magic Woman	Santana	Columbia	(40)
*29.	I Dig Everything About You	The Mob	Colossus	(36)
30.	Everybody I Love You	Happy Day	Uni	(32)
31.	Fire And Rain	James Taylor	Warner Bros.	(18)
*32.	Lonely Days	Bee Gees	Atco	(N)
33.	I'll Be There	Jackson 5	Motown	(25)
34.	Lola	The Kinks	Reprise	(24)
35.	I'm Not My Brothers Keeper	The Flaming Ember	Hot Wax	(N)
36.	Stand By Your Man	Candi Staton	Fame	(30)
37.	Immigrant Song	Led Zeppelin	Atlantic	(N)
38.	Indiana Wants Me	R. Dean Taylor	Rare Earth	(34)
39.	Green-Eyed Lady	Sugarloaf	Liberty	(33)
40.	King Of Rock And Roll	Crow	Amaret	(27)

*Denotes Big 10 Climbers

WC
FL

BIG 10 COUNT DOWN

FOR THE WEEK OF DECEMBER 7, 1970

				Last Week
*1.	My Sweet Lord	George Harrison	Apple	(16)
2.	The Tears Of A Clown	Smokey Robinson & The Miracles	Tamla	(1)
3.	Montego Bay	Bobby Bloom	L & R	(2)
*4.	Knock Three Times	Dawn	Bell	(6)
5.	See Me, Feel Me	The Who	Decca	(3)
*6.	One Less Bell To Answer	The 5th Dimension	Bell	(18)
*7.	Be My Baby	Andy Kim	Steed	(8)
*8.	After Midnight	Eric Clapton	Atco	(9)
*9.	Does Anybody Really Know What Time It Is?	Chicago	Columbia	(10)
10.	I Think I Love You	The Partridge Family	Bell	(4)
11.	Gypsy Woman	Brian Hyland	Uni	(5)
12.	Heaven Help Us All	Stevie Wonder	Tamla	(7)
*13.	No Matter What	Badfinger	Apple	(15)
*14.	Share The Land	The Guess Who	RCA	(17)
15.	Cry Me A River	Joe Cocker	A & M	(12)
*16.	Stoned Love	The Supremes	Motown	(27)
*17.	Domino	Van Morrison	Warner Bros.	(21)
*18.	You Don't Have To Say You Love Me	Elvis Presley	RCA	(23)
19.	One Man Band	Three Dog Night	Dunhill	(20)
*20.	Black Magic Woman	Santana	Columbia	(28)
21.	When The Party Is Over	Robert John	A & M	(19)
*22.	5-10-15-20 (25-30 Years Of Love)	The Presidents	Sussex	(26)
*23.	I Dig Everything About You	The Mob	Colossus	(29)
*24.	Lonely Days	Bee Gees	Atco	(32)
25.	As The Years Go By	Mashmakhan	Epic	(11)
26.	Mongoose	Elephant's Memory	Metromedia	(13)
*27.	Immigrant Song	Led Zeppelin	Atlantic	(37)
28.	Yellow River	Christie	Epic	(14)
29.	Let's Work Together	Canned Heat	Liberty	(22)
30.	We've Only Just Begun	Carpenters	A & M	(25)
*31.	I'm Not My Brother's Keeper	The Flaming Ember	Hot Wax	(35)
32.	Still Water (Love)	Four Tops	Motown	(24)
33.	For The Good Times	Ray Price	Columbia	(N)
34.	River Deep-Mountain High	The Supremes & Four Tops	Motown	(N)
35.	Your Song	Elton John	Uni	(N)
36.	Stop The War Now	Edwin Starr	Gordy	(N)
37.	Everybody I Love You	Happy Day	Uni	(30)
38.	He Ain't Heavy...He's My Brother	Neil Diamond	Uni	(N)
39.	I'll Be There	Jackson 5	Motown	(33)
40.	Fire And Rain	James Taylor	Warner Bros.	(31)

HIT BOUND

Love The One You're With	Stephen Stills	Atlantic

*Denotes Big 10 Climbers

WC
FL

BIG 10 COUNT DOWN

FOR THE WEEK OF DECEMBER 14, 1970

					Last Week
*1.	My Sweet Lord	George Harrison	Apple	(1)
*2.	One Less Bell To Answer	The 5th Dimension	Bell	(6)
*3.	Knock Three Times	Dawn	Bell	(4)
4.	Montego Bay	Bobby Bloom	L & R	(3)
5.	The Tears Of A Clown	Smokey Robinson & The Miracles	Tamla	(2)
6.	See Me, Feel Me	The Who	Decca	(5)
7.	After Midnight	Eric Clapton	Atco	(8)
8.	Does Anybody Really Know What Time It Is?	Chicago	Columbia	(9)
*9.	Stoned Love	The Supremes	Motown	(16)
*10.	No Matter What	Badfinger	Apple	(13)
11.	Be My Baby	Andy Kim	Steed	(7)
*12.	Share The Land	The Guess Who	RCA	(14)
*13.	Domino	Van Morrison	Warner Bros.	(17)
14.	I Think I Love You	The Partridge Family	Bell	(10)
15.	You Don't Have To Say You Love Me	Elvis Presley	RCA	(18)
*16.	Black Magic Woman	Santana	Columbia	(20)
17.	Gypsy Woman	Brian Hyland	Uni	(11)
*18.	Lonely Days	Bee Gees	Atco	(24)
19.	Heaven Help Us All	Stevie Wonder	Tamla	(12)
*20.	Immigrant Song	Led Zeppelin	Atlantic	(27)
21.	I Dig Everything About You	The Mob	Colossus	(23)
22.	Cry Me A River	Joe Cocker	A & M	(15)
*23.	River-Deep Mountain High	The Supremes & Four Tops	Motown	(34)
24.	One Man Band	Three Dog Night	Dunhill	(19)
*25.	Stop The War Now	Edwin Starr	Gordy	(36)
*26.	For The Good Times	Ray Price	Columbia	(33)
*27.	He Ain't Heavy...He's My Brother	Neil Diamond	Uni	(38)
*28.	I Hear You Knocking	Dave Edmunds	MAM	(N)
29.	Your Song	Elton John	Uni	(35)
30.	I'm Not My Brother's Keeper	The Flaming Ember	Hot Wax	(31)
31.	When The Party Is Over	Robert John	A & M	(21)
32.	5-10-15-20 (25-30 Years of Love)	The Presidents	Sussex	(22)
33.	If I Were Your Woman	Gladys Knight & The Pips	Soul	(N)
34.	1900 Yesterday	Liz Damon's Orient Express	White Whale	(N)
35.	Mongoose	Elephant's Memory	Metromedia	(26)
36.	We've Only Just Begun	Carpenters	A & M	(30)
37.	It's Impossible	Perry Como	RCA	(N)
38.	Rose Garden	Lynn Anderson	Columbia	(N)
39.	Yellow River	Christie	Epic	(28)
40.	As The Years Go By	Mashmakhan	Epic	(25)

CHRISTMAS 1970

Goin' Home	Bobby Sherman	Metromedia
Merry Christmas Darling	Carpenters	A & M
Santa Claus Is Comin' To Town	The Jackson 5	Motown
Feliz Navidad	Jose Feliciano	RCA
The Marvelous Toy	Irish Rovers	Decca
Amazing Grace	Judy Collins	Elektra
The Chant	Jane Avenue Bus Stop	Buddah

*Denotes Big 10 Climbers

WC
FL

BIG 10 COUNT DOWN

FOR THE WEEK OF DECEMBER 21, 1970

	Title	Artist	Label	Last Week
*1.	My Sweet Lord	George Harrison	Apple	(1)
*2.	Knock Three Times	Dawn	Bell	(3)
3.	One Less Bell To Answer	The 5th Dimension	Bell	(2)
*4.	Stoned Love	The Supremes	Motown	(9)
*5.	Does Anybody Really Know What Time It Is?	Chicago	Columbia	(8)
*6.	After Midnight	Eric Clapton	Atco	(7)
7.	Montego Bay	Bobby Bloom	L & R	(4)
*8.	Lonely Days	Bee Gees	Atco	(18)
*9.	Immigrant Song	Led Zeppelin	Atlantic	(20)
10.	No Matter What	Badfinger	Apple	(10)
11.	The Tears Of A Clown	Smokey Robinson & The Miracles	Tamla	(5)
12.	Share The Land	The Guess Who	RCA	(12)
*13.	Black Magic Woman	Santana	Columbia	(16)
14.	See Me, Feel Me	The Who	Decca	(6)
15.	Be My Baby	Andy Kim	Steed	(11)
*16.	Domino	Van Morrison	Warner Bros.	(13)
17.	You Don't Have To Say You Love Me	Elvis Presley	RCA	(15)
*18.	I Hear You Knocking	Dave Edmunds	MAM	(28)
19.	I Think I Love You	The Partridge Family	Bell	(14)
20.	I Dig Everything About You	The Mob	Colossus	(21)
21.	River Deep-Mountain High	The Supremes & Four Tops	Motown	(23)
*22.	Stop The War Now	Edwin Starr	Gordy	(25)
*23.	1900 Yesterday	Liz Damon's Orient Express	White Whale	(34)
*24.	If I Were Your Woman	Gladys Knight & The Pips	Soul	(33)
25.	For The Good Times	Ray Price	Columbia	(26)
26.	He Ain't Heavy...He's My Brother	Neil Diamond	Uni	(27)
*27.	Your Song	Elton John	Uni	(29)
28.	Gypsy Woman	Brian Hyland	Uni	(17)
*29.	Love The One You're With	Stephen Stills	Atlantic	(N)
30.	Cry Me A River	Joe Cocker	A & M	(22)
31.	Heaven Help Us All	Stevie Wonder	Tamla	(19)
32.	It's Impossible	Perry Como	RCA	(37)
33.	Rose Garden	Lynn Anderson	Columbia	(38)
34.	One Man Band	Three Dog Night	Dunhill	(24)
35.	I'm Not My Brother's Keeper	The Flaming Ember	Hot Wax	(30)
36.	Remember Me	Diana Ross	Motown	(N)
37.	Games	Redeye	Pentagram	(N)
38.	It's Up To You-Petula	Edison Lighthouse	Bell	(N)
39.	We've Only Just Begun	Carpenters	A & M	(36)
40.	Mongoose	Elephant's Memory	Metromedia	(35)

CHRISTMAS 1970

Goin' Home	Bobby Sherman	Metromedia
Merry Christmas Darling	Carpenters	A & M
Santa Claus Is Comin' to Town	The Jackson 5	Motown
Feliz Navidad	Jose Feliciano	RCA
The Marvelous Toy	Irish Rovers	Decca
Amazing Grace	Judy Collins	Elektra
The Chant	Jane Avenue Bus Stop	Buddah

*Denotes Big 10 Climbers

WC
FL

BIG 10 COUNT DOWN

WCFL – TOP 40 - 1970

1.	Raindrops Keep Fallin' On My Head	B.J. Thomas	Scepter
2.	Bridge Over Troubled Water	Simon & Garfunkel	Columbia
3.	War	Edwin Starr	Gordy
4.	Hitchin' A Ride	Vanity Fare	Page One
5.	Indiana Wants Me	R. Dean Taylor	Rare Earth
6.	The Tears Of A Clown	Smokey Robinson & The Miracles	Tamla
7.	I Think I Love You	The Partridge Family	Bell
8.	We've Only Just Begun	Carpenters	A & M
9.	Someday We'll Be Together	Diana Ross & The Supremes	Motown
10.	I'll Be There	Jackson 5	Motown
11.	Candida	Dawn	Bell
12.	Close To You	Carpenters	A & M
13.	Spirit In The Sky	Norman Greenbaum	Reprise
14.	Spill The Wine	Eric Burdon & War	MGM
15.	Let It Be	The Beatles	Apple
16.	Lookin' Out My Back Door	Creedence Clearwater Revival	Fantasy
17.	The Love You Save	Jackson 5	Motown
18.	My Sweet Lord	George Harrison	Apple
19.	American Woman/No Sugar Tonight	The Guess Who	RCA
20.	Thank You	Sly & The Family Stone	Epic
21.	Lola	The Kinks	Reprise
22.	Mama Told Me (Not To Come)	Three Dog Night	Dunhill
23.	My Baby Loves Lovin'	White Plains	Deram
24.	Cracklin' Rosie	Neil Diamond	Uni
25.	Get Ready	Rare Earth	Rare Earth
26.	Cecilia	Simon & Garfunkel	Columbia
27.	Whole Lotta Love	Led Zepplin	Atlantic
28.	No Time	The Guess Who	RCA
29.	A.B.C.	Jackson 5	Motown
30.	Venus	Shocking Blue	Colossus
31.	Ride Captain Ride	Blues Image	Atco
32.	Ball Of Confusion	Temptations	Gordy
33.	Make It With You	Bread	Elektra
34.	Down On The Corner	Creedence Clearwater Revival	Fantasy
35.	Love On A Two Way Street	The Moments	Stang
36.	Montego Bay	Bobby Bloom	L & R
37.	The Rapper	The Jaggerz	Kama Sutra
38.	Up Around The Bend	Creedence Clearwater Revival	Fantasy
39.	Hey There Lonely Girl	Eddie Holman	ABC
40.	I Want You Back	Jackson 5	Motown

HAPPY NEW YEAR FROM WCFL AND THE CHICAGO FEDERATION OF LABOR

1971

WC
FL

BIG 10 COUNT DOWN

FOR THE WEEK, JANUARY 4, 1971

				Last Week
*1.	Knock Three Times	Dawn	Bell	(2)
*2.	Stoned Love	The Supremes	Motown	(4)
3.	One Less Bell To Answer	The 5th Dimension	Bell	(3)
*4.	Immigrant Song	Led Zeppelin	Atlantic	(9)
5.	My Sweet Lord	George Harrison	Apple	(1)
*6.	Lonely Days	Bee Gees	Atco	(8)
*7.	I Hear You Knocking	Dave Edmunds	MAM	(18)
*8.	Black Magic Woman	Santana	Columbia	(13)
*9.	Love The One You're With	Stephen Stills	Atlantic	(29)
*10.	If I Were Your Woman	Gladys Knight & The Pips	Soul	(24)
11.	Does Anybody Really Know What Time It Is?	Chicago	Columbia	(5)
*12.	River Deep—Mountain High	The Supremes & Four Tops	Motown	(21)
13.	No Matter What	Badfinger	Apple	(10)
14.	Montego Bay	Bobby Bloom	L & R	(7)
15.	After Midnight	Eric Clapton	Atco	(6)
*16.	1900 Yesterday	Liz Damon's Orient Express	White Whale	(23)
*17.	Games	Redeye	Pentagram	(37)
*18.	Stoney End	Barbra Streisand	Columbia	(N)
19.	Stop The War Now	Edwin Starr	Gordy	(22)
*20.	For The Good Times	Ray Price	Columbia	(25)
21.	The Tears Of A Clown	Smokey Robinson & The Miracles	Tamla	(11)
*22.	One Bad Apple	The Osmonds	MGM	(N)
23.	Share The Land	The Guess Who	RCA	(12)
*24.	It's Up To You, Petula	Edison Lighthouse	Bell	(38)
25.	Your Song	Elton John	Uni	(27)
26.	I Think I Love You	The Partridge Family	Bell	(19)
*27.	It's Impossible	Perry Como	RCA	(32)
*28.	Rose Garden	Lynn Anderson	Columbia	(33)
*29.	Remember Me	Diana Ross	Motown	(36)
30.	I Dig Everything About You	The Mob	Colossus	(20)
31.	If You Could Read My Mind	Gordon Lightfoot	Reprise	(N)
32.	Burning Bridges	The Mike Curb Congregation	MGM	(N)
33.	Theme From Love Story	Henry Mancini	RCA	(N)
34.	Silver Moon	Michael Nesmith & The First National Band	RCA	(N)
35.	See Me, Feel Me	The Who	Decca	(14)
36.	Be My Baby	Andy Kim	Steed	(15)
37.	You Don't Have To Say You Love Me	Elvis Presley	RCA	(17)
38.	Domino	Van Morrison	Warner Bros.	(16)
39.	Gypsy Woman	Brian Hyland	Uni	(28)
40.	Do It	Neil Diamond	Bang	(N)

*Denotes Big 10 Climbers

WC
FL

BIG 10 COUNT DOWN

FOR THE WEEK OF JANUARY 11, 1971

				Last Week
*1.	Knock Three Times	Dawn	Bell	(1)
*2.	Immigrant Song	Led Zeppelin	Atlantic	(4)
3.	Stoned Love	The Supremes	Motown	(2)
*4.	I Hear You Knocking	Dave Edmunds	MAM	(7)
*5.	Love The One You're With	Stephen Stills	Atlantic	(9)
*6.	Black Magic Woman	Santana	Columbia	(8)
7.	One Less Bell To Answer	The 5th Dimension	Bell	(3)
8.	Lonely Days	Bee Gees	Atco	(6)
9.	My Sweet Lord	George Harrison	Apple	(5)
10.	If I Were Your Woman	Gladys Knight & The Pips	Soul	(10)
11.	River Deep-Mountain High	The Supremes & Four Tops	Motown	(12)
*12.	Stoney End	Barbra Streisand	Columbia	(18)
*13.	Games	Redeye	Pentagram	(17)
*14.	1900 Yesterday	Liz Damon's Orient Express	White Whale	(16)
*15.	One Bad Apple	The Osmonds	MGM	(22)
16.	Does Anybody Really Know What Time It Is?	Chicago	Columbia	(11)
*17.	It's Up To You, Petula	Edison Lighthouse	Bell	(24)
*18.	Remember Me	Diana Ross	Motown	(29)
*19.	Your Song	Elton John	Uni	(25)
20.	For The Good Times	Ray Price	Columbia	(20)
21.	Montego Bay	Bobby Bloom	L & R	(14)
22.	After Midnight	Eric Clapton	Atco	(15)
23.	No Matter What	Badfinger	Apple	(13)
*24.	Theme From Love Story	Henry Mancini	RCA	(33)
*25.	If You Could Read My Mind	Gordon Lightfoot	Reprise	(31)
*26.	Burning Bridges	The Mike Curb Congregation	MGM	(32)
27.	It's Impossible	Perry Como	RCA	(27)
28.	Rose Garden	Lynn Anderson	Columbia	(28)
*29.	Do It	Neil Diamond	Bang	(40)
*30.	Born To Wander	Rare Earth	Rare Earth	(N)
31.	Silver Moon	Michael Nesmith & The First National Band	RCA	(34)
32.	The Tears Of A Clown	Smokey Robinson & The Miracles	Tamla	(21)
33.	Stop The War Now	Edwin Starr	Gordy	(19)
*34.	When I'm Dead And Gone	Bob Summers	MGM	(N)
*35.	Groove Me	King Floyd	Chimneyville	(N)
36.	We Gotta Get You A Woman	Runt	Ampex	(N)
37.	I Think I Love You	The Partridge Family	Bell	(26)
38.	Share The Land	The Guess Who	RCA	(23)
39.	Most Of All	B.J. Thomas	Scepter	(N)
40.	Watching Scotty Grow	Bobby Goldsboro	United Artists	(N)

HIT BOUND

Apeman	The Kinks	Reprise
Hang On To Your Life	The Guess Who	RCA
Timothy	The Buoys	Scepter
Mama's Pearl	The Jackson 5	Motown

*Denotes Big 10 Climbers

WC
FL

BIG 10 COUNT DOWN

FOR THE WEEK OF JANUARY 18, 1971

				Last Week
*1.	I Hear You Knocking	Dave Edmunds	MAM	(4)
2.	Knock Three Times	Dawn	Bell	(1)
3.	Love The One You're With	Stephen Stills	Atlantic	(5)
4.	Black Magic Woman	Santana	Columbia	(6)
5.	Immigrant Song	Led Zeppelin	Atlantic	(2)
*6.	Stoney End	Barbra Streisand	Columbia	(12)
7.	Stoned Love	The Supremes	Motown	(3)
*8.	Games	Redeye	Pentagram	(13)
*9.	1900 Yesterday	Liz Damon's Orient Express	White Whale	(14)
*10.	One Bad Apple	The Osmonds	MGM	(15)
11.	River Deep-Mountain High	The Supremes & Four Tops	Motown	(11)
12.	Lonely Days	Bee Bees	Atco	(8)
13.	One Less Bell To Answer	The 5th Dimension	Bell	(7)
14.	My Sweet Lord	George Harrison	Apple	(9)
*15.	It's Up To You, Petula	Edison Lighthouse	Bell	(17)
*16.	Remember Me	Diana Ross	Motown	(18)
*17.	Your Song	Elton John	Uni	(19)
18.	If I Were Your Woman	Gladys Knight & The Pips	Soul	(10)
*19.	Theme From Love Story	Henry Mancini	RCA	(24)
*20.	Burning Bridges	The Mike Curb Congregation	MGM	(26)
*21.	If You Could Read My Mind	Gordon Lightfoot	Reprise	(25)
22.	Rose Garden	Lynn Anderson	Columbia	(28)
*23.	Born To Wander	Rare Earth	Rare Earth	(30)
24.	Does Anybody Really Know What Time It Is?	Chicago	Columbia	(16)
25.	Do It	Neil Diamond	Bang	(29)
26.	Montego Bay	Bobby Bloom	L & R	(21)
27.	For The Good Times	Ray Price	Columbia	(20)
*28.	When I'm Dead And Gone	Bob Summers	MGM	(34)
*29.	Apeman	The Kinks	Reprise	(HB)
*30.	Mama's Pearl	The Jackson 5	Motown	(HB)
31.	Silver Moon	Michael Nesmith & The First National Band	RCA	(31)
32.	Groove Me	King Floyd	Chimneyville	(35)
33.	We Gotta Get You A Woman	Runt	Ampex	(36)
34.	Most Of All	B.J. Thomas	Scepter	(39)
35.	Watching Scotty Grow	Bobby Goldsboro	United Artists	(40)
36.	Amos Moses	Jerry Reed	RCA	(N)
37.	Hang On To Your Life	The Guess Who	RCA	(N)
38.	I Really Don't Want To Know	Elvis Presley	RCA	(N)
39.	Sweet Mary	Wadsworth Mansion	Sussex	(N)
40.	Angel Baby	Dusk	Bell	(N)

HIT BOUND

Have You Ever Seen The Rain/ Hey Tonight	Creedence Clearwater Revival	Fantasy
Amazing Grace	Judy Collins	Elektra
Heavy Makes You Happy (Sha-Na-Boom Boom)	The Staple Singers	Stax

*Denotes Big 10 Climbers

WC
FL

BIG 10 COUNT DOWN

FOR THE WEEK OF JANUARY 25, 1971

				Last Week
1.	I Hear You Knocking	Dave Edmunds	MAM	(1)
*2.	Stoney End	Barbra Streisand	Columbia	(6)
3.	Games	Redeye	Pentagram	(8)
*4.	One Bad Apple	The Osmonds	MGM	(10)
5.	Love The One You're With	Stephen Stills	Atlantic	(3)
6.	Knock Three Times	Dawn	Bell	(2)
7.	Black Magic Woman	Santana	Columbia	(4)
*8.	Your Song	Elton John	Uni	(17)
9.	1900 Yesterday	Liz Damon's Orient Express	White Whale	(9)
*10.	Remember Me	Diana Ross	Motown	(16)
11.	Immigrant Song	Led Zeppelin	Atlantic	(5)
12.	Stoned Love	The Supremes	Motown	(7)
*13.	If You Could Read My Mind	Gordon Lightfoot	Reprise	(21)
*14.	Born To Wander	Rare Earth	Rare Earth	(23)
15.	It's Up To You, Petula	Edison Lighthouse	Bell	(15)
*16.	Mama's Pearl	The Jackson 5	Motown	(30)
17.	Burning Bridges	The Mike Curb Congregation	MGM	(20)
*18.	Apeman	The Kinks	Reprise	(29)
19.	Theme From Love Story	Henry Mancini/Francis Lai	RCA/Paramount	(19)
*20.	When I'm Dead And Gone	Bob Summers	MGM	(28)
21.	Rose Garden	Lynn Anderson	Columbia	(22)
22.	Lonely Days	Bee Gees	Atco	(12)
23.	One Less Bell To Answer	The 5th Dimension	Bell	(13)
24.	My Sweet Lord	George Harrison	Apple	(14)
25.	Do It	Neil Diamond	Bang	(25)
26.	If I Were Your Woman	Gladys Knight The Pips	Soul	(18)
*27.	Amos Moses	Jerry Reed	RCA	(36)
*28.	Hang On To Your Life	The Guess Who	RCA	(37)
29.	River Dee-Mountain High	The Supremes & Four Tops	Motown	(11)
30.	Most Of All	B.J. Thomas	Scepter	(34)
31.	Groove Me	King Floyd	Chimneyville	(32)
32.	We Gotta Get You A Woman	Runt	Ampex	(33)
33.	Watching Scotty Grow	Bobby Goldsboro	United Artists	(35)
34.	Sweet Mary	Wadsworth Mansion	Sussex	(39)
35.	Angel Baby	Dusk	Bell	(40)
36.	I Really Don't Want To Know	Elvis Presley	RCA	(38)
37.	Have You Ever Seen The Rain/Hey Tonight	Creedence Clearwater Revival	Fantasy	(HB)
38.	Proud Mary	Ike & Tina Turner	Liberty	(N)
39.	Mother	John Lennon/Plastic Ono Band	Apple	(N)
40.	Amazing Grace	Judy Collins	Elektra	(HB)

HIT BOUND

Heavy Makes You Happy (Sha-Na-Boom Boom)	The Staple Singers	Stax
Blue Money	Van Morrison	Warner Bros.
Precious, Precious	Jackie Moore	Atlantic
Doesn't Somebody Want To Be Wanted	The Partridge Family	Bell
Country Road	James Taylor	Warner Bros.
Bridget The Midget	Ray Stevens	Barnaby

*Denotes Big 10 Climbers

WC
FL

BIG 10 COUNT DOWN

FOR THE WEEK OF FEBRUARY 1, 1971

				Last Week
*1.	One Bad Apple	The Osmonds	MGM	(4)
2.	I Hear You Knocking	Dave Edmunds	MAM	(1)
3.	Stoney End	Barbra Streisand	Columbia	(2)
4.	Games	Redeye	Pentagram	(3)
*5.	Remember Me	Diana Ross	Motown	(10)
*6.	Mama's Pearl	The Jackson 5	Motown	(16)
7.	Your Song	Elton John	Uni	(8)
*8.	If You Could Read My Mind	Gordon Lightfoot	Reprise	(13)
*9.	Burning Bridges	The Mike Club Congregation	MGM	(17)
10.	Knock Three Times	Dawn	Bell	(6)
11.	Love The One You're With	Stephen Stills	Atlantic	(5)
12.	Born To Wander	Rare Earth	Rare Earth	(14)
13.	1900 Yesterday	Liz Damon's Orient Express	White Whale	(9)
*14.	Apeman	The Kinks	Reprise	(18)
15.	Black Magic Woman	Santana	Columbia	(7)
16.	When I'm Dead And Gone	Bob Summers	MGM	(20)
*17.	Amos Moses	Jerry Reed	RCA	(27)
*18.	Hang On To Your Life	The Guess Who	RCA	(28)
*19.	Have You Ever Seen The Rain/Hey Tonight	Creedence Clearwater Revival	Fantasy	(37)
*20.	Proud Mary	Ike & Tina Turner	Liberty	(38)
*21.	Sweet Mary	Wadsworth Mansion	Sussex	(34)
*22.	Mother	John Lennon/Plastic Ono Band	Apple	(39)
*23.	Amazing Grace	Judy Collins	Elektra	(40)
24.	Immigrant Song	Led Zeppelin	Atlantic	(11)
25.	Theme From Love Story	Henry Mancini/Francis Lai	RCA/Paramount	(19)
*26.	Doesn't Somebody Want To Be Wanted	The Partridge Family	Bell	(HB)
27.	Groove Me	King Floyd	Chimneyville	(31)
*28.	Watching Scotty Grow	Bobby Goldsboro	United Artists	(33)
*29.	Blue Money	Van Morrison	Warner Bros.	(HB)
*30.	Country Road	James Taylor	Warner Bros.	(HB)
31.	Heavy Makes You Happy (Sha-Na-Boom Boom)	The Staple Singers	Stax	(HB)
32.	For All We Know	Carpenters	A & M	(N)
33.	Cried Like A Baby	Bobby Sherman	Metromedia	(N)
34.	Angel Baby	Dusk	Bell	(35)
35.	Precious, Precious	Jackie Moore	Atlantic	(HB)
36.	Just My Imagination	Temptations	Gordy	(N)
37.	Me And Bobby McGee	Janis Joplin	Columbia	(N)
38.	Bridget The Midget	Ray Stevens	Barnady	(HB)
39.	Sugar Mountain	Neil Young	Reprise	(N)
40.	I Really Don't Want To Know	Elvis Presley	RCA	(36)

*Denotes Big 10 Climbers

WC
FL

BIG 10 COUNT DOWN

FOR THE WEEK OF FEBRUARY 8, 1971

				Last Week
*1.	One Bad Apple	The Osmonds	MGM	(1)
*2.	Mama's Pearl	The Jackson 5	Motown	(6)
*3.	If You Could Read My Mind	Gordon Lightfoot	Reprise	(8)
4.	Remember Me	Diana Ross	Motown	(5)
*5.	Born To Wander	Rare Earth	Rare Earth	(12)
6.	I Hear You Knocking	Dave Edmunds	MAM	(2)
*7.	Sweet Mary	Wadsworth Mansion	Sussex	(21)
*8.	Amos Moses	Jerry Reed	RCA	(17)
9.	Burning Bridges	Mike Curb Congregation	MGM	(9)
*10.	Apeman	The Kinks	Reprise	(14)
11.	Your Song	Elton John	Uni	(7)
12.	Games	Redeye	Pentagram	(4)
*13.	Hang On To Your Life	The Guess Who	RCA	(18)
*14.	Proud Mary	Ike & Tina Turner	Liberty	(20)
15.	When I'm Dead And Gone	Bob Summers	MGM	(16)
16.	Have You Ever Seen The Rain	Creedence Clearwater Revival	Fantasy	(19)
*17.	Doesn't Somebody Want To Be Wanted	The Partridge Family	Bell	(26)
18.	Stoney End	Barbra Streisand	Columbia	(3)
*19.	Heavy Makes You Happy (Sha-Na-Boom Boom)	The Staple Singers	Stax	(31)
20.	Amazing Grace	Judy Collins	Elektra	(23)
*21.	Country Road	James Taylor	Warner Bros.	(30)
22.	Mother	John Lennon/Plastic Ono Band	Apple	(22)
23.	Knock Three Times	Dawn	Bell	(10)
24.	Love The One You're With	Stephen Stills	Atlantic	(11)
*25.	Watching Scotty Grow	Bobby Goldsboro	United Artists	(28)
26.	Groove Me	King Floyd	Chimneyville	(27)
*27.	Angel Baby	Dusk	Bell	(34)
*28.	Blue Money	Van Morrison	Warner Bros.	(29)
*29.	For All We Know	Carpenters	A & M	(32)
*30.	Cried Like A Baby	Bobby Sherman	Metromedia	(33)
*31.	Just My Imagination	Temptations	Gordy	(36)
*32.	Me & Bobby McGee	Janis Joplin	Columbia	(37)
*33.	What's Going On	Marvin Gaye	Tamla	(HB)
*34.	Sugar Mountain	Neil Young	Reprise	(39)
35	Bridget The Midget	Ray Stevens	Barnaby	(38)
36.	The Glory Of Love	The Dells	Cadet	(HB)
37.	Love Makes The World Go Round	Odds & Ends	Today	(HB)
38.	Don't Let The Green Grass Fool You	Wilson Pickett	Atlantic	(N)
39.	She's A Lady	Tom Jones	Parrot	(N)
40.	Cherish What Is Dear To You	Freda Payne	Invictus	(N)

*Denotes Big 10 Climbers

BIG 10 COUNT DOWN

FOR THE WEEK OF FEBRUARY 15, 1971

	Title	Artist	Label	Last Week
*1.	If You Could Read My Mind	Gordon Lightfoot	Reprise	(3)
*2.	Mama's Pearl	The Jackson 5	Motown	(2)
3.	One Bad Apple	The Osmonds	MGM	(1)
*4.	Hang On To Your Life	The Guess Who	RCA	(13)
*5.	Sweet Mary	Wadsworth Mansion	Sussex	(7)
*6.	Amos Moses	Jerry Reed	RCA	(8)
7.	Born To Wander	Rare Earth	Rare Earth	(5)
8.	Apeman	The Kinks	Reprise	(10)
*9.	Proud Mary	Ike & Tina Turner	Liberty	(14)
*10.	Have You Ever Seen The Rain	Creedence Clearwater Revival	Fantasy	(16)
*11.	Doesn't Somebody Want To Be Wanted	The Partridge Family	Bell	(17)
12.	Remember Me	Diana Ross	Motown	(4)
13.	I Hear You Knocking	Dave Edmunds	MAM	(6)
*14.	Heavy Makes You Happy (Sha-Na-Boom Boom	The Staple Singers	Stax	(19)
*15.	Country Road	James Taylor	Warner Bros.	(21)
*16.	Watching Scotty Grow	Bobby Goldsboro	United Artists	(25)
*17.	Blue Money	Van Morrison	Warner Bros.	(28)
18.	Burning Bridges	Mike Curb Congregation	MGM	(9)
*19.	For All We Know	Carpenters	A & M	(29)
*20.	Just My Imagination	Temptations	Gordy	(31)
21.	Amazing Grace	Judy Collins	Elektra	(20)
*22.	Me & Bobby McGee	Janis Joplin	Columbia	(32)
*23.	What's Going ON	Marvin Gaye	Tamla	(33)
24.	Cried Like A Baby	Bobby Sherman	Metromedia	(30)
25.	Your Song	Elton John	Uni	(11)
26.	Groove Me	King Floyd	Chimneyville	(26)
27.	Angel Baby	Dusk	Bell	(27)
28.	She's A Lady	Tom Jones	Parrot	(39)
29.	Games	Redeye	Pentagram	(12)
*30.	Loves Makes The World Go Round	Odds & Ends	Today	(37)

HIT BOUND

	Title	Artist	Label
31.	Don't Let The Green Grass Fool You	Wilson Pickett	Atlantic
32.	The Glory Of Love	The Dells	Cadet
33.	Cherish What Is Dear To You	Freda Payne	Invictus
34.	Sugar Mountain	Neil Young	Reprise
35.	Bridget The Midget	Ray Stevens	Barnaby
36.	Love's Lines, Angles & Rhymes	The 5th Dimension	Bell
37.	Jody's Got Your Girl And Gone	Johnnie Taylor	Stax
38.	No Love At All	B.J. Thomas	Scepter
39.	Eighteen	Alice Cooper	Warner Bros.
40.	Free	Chicago	Columbia

*Denotes Big 10 Climbers

WC
FL

BIG 10 COUNT DOWN

FOR THE WEEK OF FEBRUARY 22, 1971

				Last Week
*1.	Mama's Pearl	The Jackson 5	Motown	(2)
2.	If You Could Read My Mind	Gordon Lightfoot	Reprise	(1)
*3.	Sweet Mary	Wadsworth Mansion	Sussex	(5)
4.	Hang On To Your Life	The Guess Who	RCA	(4)
*5.	Amos Moses	Jerry Reed	RCA	(6)
*6.	Have You Ever Seen The Rain	Creedence Clearwater Revival	Fantasy	(10)
*7.	Proud Mary	Ike & Tina Turner	Liberty	(9)
*8.	Doesn't Somebody Want To Be Wanted	The Partridge Family	Bell	(11)
9.	One Bad Apple	The Osmonds	MGM	(3)
*10.	Heavy Makes You Happy (Sha-Na-Boom Boom)	The Staple Singers	Stax	(14)
*11.	Watching Scotty Grow	Bobby Goldsboro	United Artists	(16)
*12.	Country Road	James Taylor	Warner Bros.	(15)
13.	Blue Money	Van Morrison	Warner Bros.	(17)
*14.	For All We Know	Carpenters	A & M	(19)
15.	Born To Wander	Rare Earth	Rare Earth	(7)
*16.	Just My Imagination	Temptations	Gordy	(20)
*17.	What's Going On	Marvin Gaye	Gamla	(23)
18.	Apeman	The Kinks	Reprise	(8)
*19.	Me & Bobby McGee	Janis Joplin	Columbia	(22)
20.	Remember Me	Diana Ross	Motown	(12)
21.	I Hear You Knocking	Dave Edmunds	MAM	(13)
22.	Cried Like A Baby	Bobby Sherman	Metromedia	(24)
23.	Burning Bridges	Mike Curb Congregation	MGM	(18)
*24.	She's A Lady	Tom Jones	Parrot	(28)
*25.	Love Makes The World Go Round	Odds & Ends	Today	(30)
*26.	Love's Lines, Angles & Rhymes	The 5th Dimension	Bell	(36)
*27.	Don't Let The Green Grass Fool You	Wilson Pickett	Atlantic	(31)
*28.	Free	Chicago	Columbia	(40)
*29.	Jody's Got Your Girl And Gone	Johnnie Taylor	Stax	(37)
*30.	Cherish What Is Dear To You	Freda Payne	Invictus	(33)

HIT BOUND

31.	No Love At All	B.J. Thomas	Scepter
32.	What Is Life/Apple Scruffs	George Harrison	Apple
33.	You're All I Need To Get By	Aretha Franklin	Atlantic
34.	Eighteen	Alice Cooper	Warner Bros.
35.	Mr. Bojangles	Nitty Gritty Dirty Band	Liberty
36.	L.A. Goodbye	The Ides Of March	Warner Bros.
37.	Chick-A-Boom	Daddy Dewdrop	Sunflower
38.	Woodstock	Matthews Southern Comfort	Decca
39.	Anytime Sunshine	Crazy Paving	Kapp
40.	Timothy	The Buoys	Scepter

*Denotes Big 10 Climbers

WC
FL

FOR THE WEEK OF MARCH 1, 1971

					Last Week
1.	Mama's Pearl	The Jackson 5	Motown	(1)
*2.	Sweet Mary	Wadsworth Mansion	Sussex	(3)
*3.	Amos Moses	Jerry Reed	RCA	(5)
4.	Have You Ever Seen The Rain	Creedence Clearwater Revival	Fantasy	(6)
*5.	Doesn't Somebody Want To Be Wanted	The Partridge Family	Bell	(8)
6.	Proud Mary	Ike & Tina Turner	Liberty	(7)
7.	If You Could Read My Mind	Gordon Lightfoot	Reprise	(2)
8.	Heavy Makes You Happy (Sha-Na-Boom Boom)	The Staple Singers	Stax	(10)
*9.	Blue Money	Van Morrison	Warner Bros.	(13)
*10.	Just My Imagination	Temptations	Gordy	(16)
11.	Country Road	James Taylor	Warner Bros.	(12)
12.	For All We Know	Carpenters	A & M	(14)
*13.	What's Going On	Marvin Gaye	Tamla	(17)
*14.	Me & Bobby McGee	Janis Joplin	Columbia	(19)
*15.	She's A Lady	Tom Jones	Parrot	(24)
16.	Hang On To Your Life	The Guess Who	RCA	(4)
17.	One Bad Apple	The Osmonds	MGM	(9)
18.	Watching Scotty Grow	Bobby Goldsboro	United Artists	(11)
*19.	Love's Lines, Angles & Rhymes	The 5th Dimension	Bell	(26)
*20.	Cried Like A Baby	Bobby Sherman	Metromedia	(22)
21.	Born To Wander	Rare Earth	Rare Earth	(15)
*22	Eighteen	Alice Cooper	Warner Bros.	(34)
*23.	Free	Chicago	Columbia	(28)
*24.	What Is Life/Apple Scruffs	George Harrison	Apple	(32)
25.	Love Makes The World Go Round	Odds & Ends	Today	(25)
26.	Don't Let The Green Grass Fool You	Wilson Pickett	Atlantic	(27)
27.	Jody's Got Your Girl And Gone	Johnnie Taylor	Stax	(29)
*28.	No Love At All	B.J. Thomas	Scepter	(31)
*29.	L.A. Goodbye	The Ides Of March	Warner Bros.	(36)
30.	Cherish What Is Dear To You	Freda Payne	Invictus	(30)
31.	You're All I Need To Get By	Aretha Franklin	Atlantic	(33)
32.	Mr. Bojangles	Nitty Gritty Dirt Band	Liberty	(35)
33.	Chick-A-Boom	Daddy Dewdrop	Sunflower	(37)
34.	Woodstock	Matthews Southern Comfort	Decca	(38)
35.	Anytime Sunshine	Crazy Paving	Kapp	(39)
36.	Timothy	The Buoys	Scepter	(40)
37.	Adrienne	Tommy James	Roulette	(HB)
38.	Oye Como Va	Santana	Columbia	(HB)
39.	We Can Work It Out	Stevie Wonder	Tamla	(HB)
40.	Another Day/Oh Woman, Oh Why	Paul McCartney	Apple	(HB)

*Denotes Big 10 Climbers

WC
FL

ALL HIT MUSIC

IN THE
MIDWEST

WAS IS Week Ending March 11, 1971

WAS	IS	
(5)	*1.	Doesn't Somebody Want To Be Wanted-Partridge Family-Bell
(6)	*2.	Proud Mary-Ike & Tina Turner-Liberty
(3)	3.	Amos Moses-Jerry Reed-RCA
(2)	4.	Sweet Mary-Wadsworth Mansion-Sussex
(10)	*5.	Just My Imagination-Temptations-Gordy
(8)	6.	Heavy Makes You Happy (Sha-Na-Boom Boom)-Staple Singers-Stax
(1)	7.	Mama's Pearl-Jackson 5-Motown
(13)	*8.	What's Going On-Marvin Gaye-Tamla
(12)	9.	For All We Know-Carpenters-A & M
(14)	*10.	Me & Bobby McGee-Janis Joplin-Columbia
(11)	11.	Country Road-James Taylor-Warner Brothers
(15)	*12.	She's A Lady-Tom Jones-Parrot
(4)	13.	Have You Ever Seen The Rain-Creedance Clearwater Revival-Fantasy
(9)	14.	Blue Money-Van Morrison-Warner Brothers
(19)	*15.	Love's Lines, Angles & Rhymes-5th Dimension-Bell
(7)	16.	If You Could Read My Mind-Gordon Lightfoot-Reprise
(22)	*17.	Eighteen-Alice Cooper-Warner Brothers
(24)	*18.	What Is Life/Apple Scruffs-George Harrison-Apple
(23)	*19.	Free-Chicago-Columbia
(20)	20.	Cried Like A Baby-Bobby Sherman-Metromedia
(29)	*21.	L.A. Goodbye-Ides Of March-Warner Brothers
(40)	*22.	Another Day/Oh Woman Oh Why-Paul McCartney-Apple
(38)	*23.	Oye Como Va-Santana-Columbia
(26)	24.	Don't Let The Green Grass Fool You-Wilson Pickett-Atlantic
(27)	25.	Jody's Got Your Gal and Gone-Johnnie Taylor-Stax
(25)	26.	Love Makes The World Go Round-Odds & Ends-Today
(28)	27.	No Love At All-B.J. Thomas-Scepter
(32)	28.	Mr. Bojangles-Nitty Gritty Dirt Band-Liberty
(39)	*29.	We Can Work It Out-Stevie Wonder-Tamla
(31)	30.	You're All I Need To Get By-Aretha Franklin-Atlantic
(34)	31.	Woodstock-Matthews Southern Comfort-Decca

BIG 10 ALBUMS

1. Pearl-Janis Joplin-Columbia
2. Chicago III-Chicago-Columbia
3. Tumbleweed Connection-Elton John-Uni
4. All Things Must Pass-George Harrison-Apple
5. Jesus Christ Superstar-Decca
6. The Cry Of Love-Jimi Hendrix-Reprise
7. Tea For The Tillerman-Cat Stevens-A & M
8. Emerson, Lake & Palmer-Emerson, Lake & Palmer-Cotillion
9. If I Could Only Remember My Name-David Crosby-Atlantic
10. If You Could Read My Mind-Gordon Lightfoot-Reprise

Hear the best Albums in the Midwest ONLY on Big 10!

SMALL PRINT DEPARTMENT: The WCFL "ALL HIT MUSIC" represents the relative popularity of records in the Midwest based on Industry tabulations, local record sales and requests. WCFL The Voice of Labor, is o

BIG 10 SOLID GOLD

MOST REQUESTED GOLD

Light My Fire-Doors-(1967)

19th Nervous Breakdown-Rolling Stones-(1966)

Day Tripper-Beatles-(1967)

Turn, Turn, Turn-Byrds-(1965)

Louie, Louie-Kingsmen-(1963)

BIG 10-HOME OF CHICAGO'S ONLY SOLID GOLD WEEKEND

(33)	32.	Chick-A-Boom-Daddy Dewdrop-Sunflower
(35)	33.	Anytime Sunshine-Crazy Paving-Kapp
(36)	34.	Timothy-Buoys-Scepter
(37)	35.	Adrienne-Tommy James-Roulette
(HB)	36.	One Toke Over The Line-Brewer & Shipley-Kama Sutra
(HB)	37.	Temptation Eyes-Grass Roots Dunhill
(HB)	38.	Stay Awhile-Bells-Polydorr
(HB)	39.	Layla-Derek And The Dominos-Atco
(HB)	40.	I Play And Sing-Dawn-Bell

*Denotes Big 10 Climbers

BIG 10 WCFL "THE CHICAGO LISTENER" LISTENS TO YOU AT 222-1650

WC FL **ALL HIT MUSIC** **IN THE MIDWEST**

WAS IS Week Ending March 18, 1971

WAS	IS	
(1)	1.	Doesn't Somebody Want To Be Wanted-Partridge Family-Bell
(5)	*2.	Just My Imagination-Temptations-Gordy
(12)	*3.	She's A Lady-Tom Jones-Parrot
(8)	*4.	What's Going ON-Marvin Gaye-Tamla
(9)	*5.	For All We Know-Carpenters-A & M
(10)	*6.	Me & Bobby McGee-Janis Joplin-Columbia
(4)	7.	Sweet Mary-Wadsworth Mansion-Sussex
(2)	8.	Proud Mary-Ike & Tina Turner-Liberty
(3)	9.	Amos Moses-Jerry Reed-RCA
(6)	10.	Heavy Makes You Happy-Staple Singers-Stax
(17)	*11.	Eighteen-Alice Cooper-WB
(18)	12.	What Is Life-George Harrison-Apple
(15)	13.	Love's Lines, Angles & Rhymes-5th Dimension-Bell
(7)	14.	Mama's Pearl-Jackson 5-Motown
(19)	15.	Free-Chicago-Columbia
(21)	16.	L.A. Goodbye-Ides Of March-WB
(22)	17.	Another Day-Paul McCartney-Apple
(14)	18.	Blue Money-Van Morrison-WB
(23)	19.	Oye Como Va-Santana-Columbia
(13)	20.	Have You Ever Seen The Rain-C.C. Revival-Fantasy
(11)	21.	Country Road-James Taylor-WB
(29)	22.	We Can Work It Out-Stevie Wonder-Tamla
(16)	23.	If You Could Read My Mind-G. Lightfoot-Reprise
(24)	24.	Don't Let The Green Grass-W. Pickett-Atlantic
(20)	25.	Cried Like A Baby-Bobby Sherman-Metromedia
(27)	26.	No Love At All-B.J. Thomas-Scepter
(31)	27.	Woodstock-Matthews Southern Comfort-Decca
(28)	28.	Mr. Bojangles-Nitty Gritty Dirt Band-Liberty
(30)	29.	You're All I Need-Aretha Franklin-Atlantic
(33)	30.	Anytime Sunshine-Crazy Paving-Kapp
(34)	31.	Timothy-Buoys-Scepter
(32)	32.	Chick-A-Boom-Daddy Dewdrop-Sunflower
(39)	33.	Layla-Derek & The Dominos-Atco
(40)	34.	I Play & Sing-Dawn-Bell
(35)	35.	Adrienne-Tommy James-Roulette
(38)	36.	Stay Awhile-Bells-Polydor
(37)	37.	Temptation Eyes-Grass Roots-Dunhill
(HB)	38.	I Am...I Said-Neil Diamond-Uni

BIG 10 ALBUMS

1. Pearl-Janis Joplin-Columbia
2. Chicago III-Chicago-Columbia
3. Cry Of Love-Jimi Hendrix-Reprise
4. Emerson, Lake & Palmer-Emerson, Lake & Palmer-Cotillion
5. Jesus Christ Superstar-Various Artists-Decca
6. Tea For The Tillerman-Cat Stevens-A & M
7. If I Could Only Remember My Name-David Crosby-Atlantic
8. If You Could Read My Mind-Gordon Lightfoot-Reprise
9. Paranoid-Black Sabbath-WB
10. Manna-Bread-Elektra

Hear the best Albums in the Midwest ONLY on Big 10!

SMALL PRINT DEPARTMENT: The WCFL "ALL HIT MUSIC" represents the relative popularity of records in the Midwest based on Industry tabulations, local record sales and requests. WCFL The Voice of Labor, is o

BIG 10 SOLID GOLD

MOST REQUESTED GOLD

1. Purple Haze-Jimi Hendrix-(1967)
2. Sky Pilot-Eric Burdon & The Animals-(1968)
3. Honky Tonk Woman-Rolling Stones-(1969)
4. War-Edwin Starr-(1970)
5. Can't Buy Me Love-Beatles-(1964)

BIG 10-HOME OF CHICAGO'S ONLY GOLDEN WEEKEND

(HB)	39.	Can't Find The Time-Rose Colored Glass-Bang
(HB)	40.	Me & You & A Dog Named Boo-Lobo-Big Tree

*Denotes Big 10 Climbers

Erin Go Bragh and BIG TEN's MEN wish YOU a happy St. Pat's Day!

WC
FL

ALL HIT MUSIC

IN THE MIDWEST

BIG 10 ALBUMS

IN THE MIDWEST

1. Pearl-Janis Joplin-Columbia
2. If I Could Only Remember My Name-David Crosby-Atlantic
3. Jesus Christ, Superstar-Various Artists-Decca
4. Emerson, Lake & Palmer-Emerson, Lake & Palmer-Cotillion
5. Cry Of Love-Jimi Hendrix-Reprise
6. Chicago III-Chicago-Columbia
7. Paranoid-Black Sabbath-WB
8. Tea For The Tillerman-Cat Stevens-A & M
9. Moving Toward Happiness-Mason Proffit-Happy Tiger
10. Abraxas-Santana-Columbia

SMALL PRINT DEPARTMENT: The WCFL "ALL HIT MUSIC" represents the relative popularity of records in the Midwest based on Industry tabulations, local record sales and requests. WCFL The Voice of Labor, is o

Hear the best Albums in the Midwest ONLY on BIG 10!

BIG 10 SOLID GOLD

MOST REQUESTED GOLD

1. MacArthur Park-Richard Harris-(1968)
2. Sounds Of Silence-Simon & Garfunkel-(1965)
3. Help Me Rhonda-Beach Boys-(1965)
4. Hair-Cowsills-(1969)
5. The Wanderer-Dion-(1961)

BIG 10-HOME OF CHICAGO'S ONLY GOLDEN WEEKEND

WAS	IS	Week Ending March 25, 1971
(2)	1.	Just My Imagination-Temptations-Gordy
(3)	*2.	She's A Lady-Tom Jones-Parrot
(4)	*3.	What's Going On-Marvin Gaye-Tamla
(5)	*4.	For All We Know-Carpenters-A & M
(1)	5.	Doesn't Somebody Want To Be Wanted-Partridge Family-Bell
(6)	6.	Me & Bobby McGee-Janis Joplin-Columbia
(17)	*7.	Another Day-Paul McCartney-Apple
(11)	*8.	Eighteen-Alice Cooper-WB
(12)	*9.	What Is Life-George Harrison-Apple
(13)	*10.	Love's Lines, Angles & Rhymes-5th Dimension-Bell
(16)	*11.	L.A. Goodbye-Ides Of March-WB
(7)	12.	Sweet Mary-Wadsworth Mansion-Sussex
(8)	13.	Proud Mary-Ike & Tina Turner-Liberty
(9)	14.	Amos Moses-Jerry Reed-RCA
(10)	15.	Heavy Makes You Happy-Staple Singers-Stax
(19)	16.	Oye Como Va-Santana-Columbia
(22)	17.	We Can Work It Out-Stevie Wonder-Tamla
(26)	*18.	No Love At All-B.J. Thomas-Scepter
(27)	*19.	Woodstock-Matthews Southern Comfort-Decca
(33)	*20.	Layla-Derek & The Dominos-Atco
(34)	*21.	I Play & Sing-Dawn-Bell
(36)	*22.	Stay Awhile-Bells-Polydor
(32)	23.	Chick-A-Boom-Daddy Dewdrop-Sunflower
(15)	24.	Free-Chicago-Columbia
(14)	25.	Mama's Pearl-Jackson 5-Motown
(20)	26.	Have You Ever Seen The Rain-C.C. Revival-Fantasy
(29)	27.	You're All I Need-Aretha Franklin-Atlantic
(31)	28.	Timothy-Buoys-Scepter
(18)	29.	Blue Money-Van Morrison-WB
(25)	30.	Cried Like A Baby-Bobby Sherman-Metromedia
(37)	31.	Temptation Eyes-Grass Roots-Dunhill
(38)	32.	I Am...I Said-Neil Diamond-Uni
(39)	33.	Can't Find The Time-Rose Colored Glass-Bang
(40)	34.	Me & You & A Dog Named Boo-Lobo-Big Tree
(HB)	35.	Wild World-Cat Stevens-A & M

(HB)	36.	Put Your Hand In The Hand-Ocean-Kama Sutra
(HB)	37.	Toast & Marmalade For Tea-Tin Tin-Atco
(HB)	38.	13 Questions-Seatrain-Capitol
(HB)	39.	If-Bread-Elektra
(HB)	40.	Never Can Say Goodbye-Jackson 5-Motown

*Denotes Big 10 Climbers

WCFL...Chicago's *ONLY* "TOP 40" station AND Chicago's *ONLY* Hitline: 222-1650!

APRIL FOOLS EDITION!

WC — IN THE MIDWEST

FL — *SLIGHT HIT MUSIC* — & West 21st ST.

Week Ending April 1, 1972

1. Doesn't Somebody Want To Be Wanted-The FBI-Zimbalist
2. Tip Toe Through The Tulips-Gordon Lightfoot-Flowerbed
3. Instant Karma-Betty Crocker-Bitter/Batter
4. I Want You Back, You Sides And You Front-R.B. Greed-Anatomy
5. Blue Money-Internal Revenue Service-Bogus
6. Raindrops Keep Fallin' On My Head-John Coleman-Guess
7. Vehicle-Chicago Motor Club-Monoxide
8. Who'll Stop The Reign?-Harold Wilson-Dethroned
9. ABC-The Beaters-Big Ten
10. Something In The Air-The Chicago River-Phew!
11. My Cup Runneth Over-Raquel Welch-Abundant
12. Come On Down To My Boat-Aristotle Onassis-Dynasty
13. Lemon Tree-Ralph Nader-Fraud
14. 98.6-1.4 short-10
15. Me & Janis-Bobby McGee-Mirror
16. It's Up To You, Rhona-Elvis Presley-Groupie
17. Rare Earth-Nitty Gritty Dirt Band-Flip
18. Nitty Gritty Dirt-Rare Earth-Flop
19. Just My Imagination-Hugh Hefner-Fold-Out
20. I Hear You Knockin'-Larry O'Brien's Car-Shifter
21. Heavy Makes You Happy-Totie Field-Crunch
22. Don't Sleep In The Subway-Chicago Police Choir-26th & Calif.
23. Early In The Morning-Clark Weber-Zzzzzzz
24. Green, Green Grass Of Home-Sally Field-Monocot
25. I Got A Line On You-The CIA-Obscurity

25¼ . Gonna Build A Mountain-The Carpenters-Avalanche

25½. If You Could Read My Mind-Jeanne Dixon-Cerebrum

26+17. Shake, Rattle And Roll-Bob Dearborn's Car-Tailpipe

43-4. Oye Como Vamos-Jerry Kay & Jeanette MacDonald-Soundtrack

40. Theme from "I Am Curious Yellow"-Carol Crayola-Burnt Sienna

* Denotes Big 10 Climbers

†Denotes Big 10 Greasers

THIS HOUR'S

least requested gold

†*THE PIZZA SONG-DICK BIONDI

HAPPY APRIL 1 FROM THE BIG 10 MEN!

WC
FL

ALL HIT MUSIC IN THE MIDWEST

WAS IS Week Ending April 1, 1971

WAS	IS	
(2)	1.	She's A Lady-Tom Jones-Parrot
(1)	2.	Just My Imagination-Temptations-Gordy
(3)	3.	What's Going On-Marvin Gaye-Tamla
(4)	4.	For All We Know-Carpenters-A & M
(6)	*5.	Me & Bobby McGee-Janis Joplin-Columbia
(7)	*6.	Another Day-Paul McCartney-Apple
(8)	*7.	Eighteen-Alice Cooper-WB
(11)	*8.	L.A. Goodbye-Ides Of March-WB
(10)	*9.	Love's Lines, Angles & Rhymes-5th Dimension-Bell
(5)	10.	Doesn't Somebody Want To Be Wanted-Partridge Family-Bell
(9)	11.	What Is Life-George Harrison-Apple
(17)	*12.	We Can Work It Out-Stevie Wonder-Tamla
(16)	*13.	Oye Como Va-Santana-Columbia
(19)	*14.	Woodstock-Matthews Southern Comfort-Decca
(22)	*15.	Stay Awhile-Bells-Polydor
(21)	*16.	I Play & Sing-Dawn-Bell
(18)	17.	No Love At All-B.J. Thomas-Scepter
(15)	18.	Heavy Makes You Happy-Staple Singers-Stax
(20)	19.	Layla-Derek & The Dominos-Atco
(24)	20.	Free-Chicago-Columbia
(23)	21.	Chick-A-Boom-Daddy Dewdrop-Sunflower
(28)	*22.	Timothy-Buoys-Scepter
(36)	*23.	Put Your Hand In The Hand-Ocean-Kama Sutra
(40)	*24.	Never Can Say Goodbye-Jackson 5-Motown
(32)	*25.	I Am...I Said-Neil Diamond-Uni
(14)	26.	Amos Moses-Jerry Reed-RCA
(12)	27.	Sweet Mary-Wadsworth Mansion-Sussex
(13)	28.	Proud Mary-Ike & Tina Turner-Liberty
(39)	*29.	If-Bread-Elektra
(34)	30.	Me & You & A Dog Named Boo-Lobo-Big Tree
(30)	31.	Cried Like A Baby-Bobby Sherman-Metromedia
(33)	32.	Can't Find The Time-Rose Colored Glass-Bang
(35)	33.	Wild World-Cat Stevens-A & M
(26)	34.	Have You Ever Seen The Rain-C.C. Revival-Fantasy
(37)	35.	Toast & Marmalade For Tea-Tin

BIG 10 ALBUMS

IN THE MIDWEST

1. Jesus Christ, Superstar-Various Artists-Decca
2. If I Could Only Remember My Name-David Crosby-Atlantic
3. Pearl-Janis Joplin-Columbia
4. Paranoid-Black Sabbath-WB
5. The Cry Of Love-Jimi Hendrix-Reprise
6. Emerson, Lake & Palmer-Emerson, Lake & Palmer-Cotillion
7. Tea For The Tillerman-Cat Stevens-A & M
8. Chicago III-Chicago-Columbia
9. Movin' Toward Happiness-Mason Proffit-Happy Tiger
10. Abraxas-Santana-Columbia

Hear the best Albums in the Midwest ONLY on BIG 10!

SMALL PRINT DEPARTMENT: The WCFL "ALL HIT MUSIC" represents the relative popularity of records in the Midwest based on Industry tabulations, local record sales and requests. WCFL The Voice of Labor, is o

BIG 10 SOLID GOLD

MOST REQUESTED GOLD

1. Close To You-The Carpenters-(1970)
2. Cherish-The Association-(1966)
3. Somebody To Love-Jefferson Airplane-(1967)
4. Let It Be-The Beatles-(1970)
5. Lay Down (Candles In The Rain)-Melanie-(1970)

BIG 10-HOME OF CHICAGO'S ONLY GOLDEN WEEKEND

		Tin-Atco
(38)	36.	13 Questions-Seatrain-Capitol
(HB)	37.	Love Her Madly-Doors-Elektra
(HB)	38.	Joy To The World-Three Dog Night-Dunhill
(HB)	39.	I Don't Blame You At All-Smokey Robinson & The Miracles-Tamla
(HB)	40.	Do Me Right-Detroit Emeralds-Westbound

*Denotes Big 10 Climbers

Hold up to mirror
HAPPY APRIL 1st! !STIH rof ecalp *EHT* si NET GIB :ti ta kool uoy yaw *YNA*

WC
FL

ALL HIT MUSIC IN THE MIDWEST

WAS	IS	Week Ending April 8, 1971
(1)	1.	She's A Lady-Tom Jones-Parrot
(2)	2.	Just My Imagination-Temptations-Gordy
(4)	3.	For All We Know-Carpenters-A & M
(8)	*4.	L.A. Goodbye-Ides Of March-WB
(7)	*5.	Eighteen-Alice Cooper-WB
(6)	6.	Another Day-Paul McCartney-Apple
(3)	7.	What's Going On-Marvin Gaye-Tamla
(15)	*8.	Stay Awhile-Bells-Polydor
(9)	9.	Love's Lines, Angles & Rhymes-5th Dimension-Bell
(13)	*10.	Oye Como Va-Santana-Columbia
(12)	11.	We Can Work It Out-Stevie Wonder-Tamla
(14)	*12.	Woodstock-Matthews Southern Comfort-Decca
(5)	13.	Me & Bobby McGee-Janis Joplin-Columbia
(16)	14.	I Play & Sing-Dawn-Bell
(10)	15.	Doesn't Somebody-Partridge Family-Bell
(11)	16.	What Is Life-George Harrison-Apple
(17)	17.	No Love At All-B.J. Thomas-Scepter
(19)	18.	Layla-Derek & The Dominos-Atco
(24)	*19.	Never Can Say Goodbye-Jackson 5-Motown
(21)	20.	Chick-A-Boom-Daddy Dewdrop-Sunflower
(22)	21.	Timothy-Buoys-Scepter
(23)	22.	Put Your Hand In The Hand-Ocean-Kama Sutra
(25)	23.	I Am...I Said-Neil Diamond-Uni
(29)	24.	If-Bread-Elektra
(30)	*25.	Me & You & A Dog Named Boo-Lobo-Big Tree
(32)	*26.	Can't Find The Time-Rose Colored Glass-Bang
(33)	27.	Wild World-Cat Stevens-A & M
(35)	*28.	Toast & Marmalade For Tea-Tin Tin-Atco
(38)	29.	Joy To The World-Three Dog Night-Dunhill
(36)	*30.	13 Questions-Seatrain-Capitol
(37)	*31.	Love Her Madly-Doors-Elektra
(39)	32.	I Don't Blame You At All-The Miracles-Tamla
(40)	*33.	Do Me Right-Detroit Emeralds-Westbound
(20)	34.	Free-Chicago-Columbia
(18)	35.	Heavy Makes You Happy-Staple Singers-Stax

BIG 10 ALBUMS

IN THE MIDWEST

1. Jesus Christ Super Star-Various Artists-Decca
2. If I Could Only Remember My Name-David Crosby-Atlantic
3. Pearl-Janis Joplin-Columbia
4. Abraxas-Santana-Columbia
5. The Cry Of Love-Jimi Hendrix-Reprise
6. Emerson, Lake & Palmer-Emerson, Lake & Palmer-Cotillion
7. Tea For The Tillerman-Cat Stevens-A & M
8. Movin' Toward Happiness-Mason Proffit-Happy Tiger
9. Chicago III-Chicago-Columbia
10. Love Story-Andy Williams-Columbia

Hear the best Albums in the Midwest ONLY on BIG 10!

SMALL PRINT DEPARTMENT: The WCFL "ALL HIT MUSIC" represents the relative popularity of records in the Midwest based on Industry tabulations, local record sales and requests. WCFL The Voice of Labor, is o

---

BIG 10 SOLID GOLD

MOST REQUESTED GOLD

1. Help-The Beatles-(1965)
2. Sweet Caroline-Neil Diamond-(1969)
3. Eli's Coming-Three Dog Night-(1969)
4. Aquarius-5th Dimension-(1969)
5. Band of Gold-Freda Payne-(1970)

BIG 10-HOME OF CHICAGO'S ONLY GOLDEN WEEKEND

(HB)	36.	When There's No You-Engelbert Humperdinck-Parrot
(HB)	37.	Superstar-Murray Head-Decca
(HB)	38.	I Love You For All Seasons-The Fuzz-Calla
(HB)	39.	Sweet And Innocent-Donny Osmond-MGM
(HB)	40.	Dream Baby-Glen Campbell-Capitol

*Denotes Big 10 Climbers

WC FL — **ALL HIT MUSIC** — **IN THE MIDWEST**

WAS	IS	Week Ending April 15 1971
(8)	1.	Stay Awhile-Bells-Polydor
(4)	*2.	L.A. Goodbye-Ides Of March-WB
(2)	3.	Just My Imagination-Temptations-Gordy
(5)	4.	Eighteen-Alice Cooper-WB
(6)	5.	Another Day-Paul McCartney-Apple
(1)	6.	She's A Lady-Tom Jones-Parrot
(29)	*7.	Joy To The World-Three Dog Night-Dunhill
(19)	*8.	Never Can Say Goodbye-Jackson 5-Motown
(22)	*9.	Put Your Hand In The Hand-Ocean-Kama Sutra
(11)	10.	We Can Work It Out-Stevie Wonder-Tamla
(3)	11.	For All We Know-Carpenters-A & M
(12)	12.	Woodstock-Matthews Southern Comfort-Decca
(24)	*13.	If-Bread-Elektra
(14)	14.	I Play & Sing-Dawn-Bell
(20)	*15.	Chick-A-Boom-Daddy Dewdrop-Sunflower
(17)	16.	No Love At All-B.J. Thomas-Scepter
(21)	*17.	Timothy-Buoys-Scepter
(25)	*18.	Me & You & A Dog Named Boo-Lobo-Big Tree
(27)	*19.	Wild World-Cat Stevens-A & M
(9)	20.	Love's Lines, Angles & Rhymes-5th Dimension-Bell
(10)	21.	Oye Como Va-Santana-Columbia
(23)	22.	I Am...I Said-Neil Diamond-Uni
(7)	23.	What's Going On-Marvin Gaye-Tamla
(31)	24.	Love Her Madly-Doors-Elektra
(26)	25.	Can't Find The Time-Rose Colored Glass-Bang
(28)	26.	Toast & Marmalade For Tea-Tin Tin-Atco
(30)	27.	13 Questions-Seatrain-Capitol
(32)	28.	I Don't Blame You At All-The Miracles-Tamla
(16)	29.	What Is Life-George Harrison-Apple
(18)	30.	Layla-Derek & The Dominos-Atco
(37)	31.	Superstar-Murray Head-Decca
(33)	32.	Do Me Right-Detroit Emeralds-Westbound
(15)	33.	Doesn't Somebody-Partridge Family-Bell
(13)	34.	Me & Bobby McGee-Janis Joplin-Columbia
(36)	35.	When There's No You-E. Humperdinck-Parrot

BIG 10 ALBUMS

IN THE MIDWEST

1. Jesus Christ Superstar-Various Artists-Decca
2. Woodstock Two-Various Artists-Cotillion
3. Pearl-Janis Joplin-Columbia
4. Tea For The Tillerman-Cat Stevens-A & M
5. The Cry Of Love-Jimi Hendrix-Reprise
6. Emerson, Lake & Palmer-Emerson, Lake & Palmer-Cotillion
7. Movin' Toward Happiness-Mason Proffit-Happy Tiger
8. Up To Date-Partridge Family-Bell
9. Love Story-Andy Williams-Columbia
10. If I Could Only Remember My Name-David Crosby-Atlantic

The Original Album Survey from BIG TEN-the Station that PLAYS the Albums

SMALL PRINT DEPARTMENT: The WCFL "ALL HIT MUSIC" represents the relative popularity of records in the Midwest based on Industry tabulations, local record sales and requests. WCFL The Voice of Labor, is o

WCFL and Bob Dearborn
cordially request the honour of your presence during their presentation of the Rock opera
JESUS CHRIST, SUPERSTAR

Easter Sunday, the eleventh of
April, nineteen hundred and seventy-one
Eight-Fifteen to Ten in the evening
1,000 on your dial
R.S.V.P. not necessary

(38)	36.	I Love You For All Seasons-The Fuzz-Calla
(39)	37.	Sweet And Innocent-Donny Osmond-MGM
(40)	38.	Dream Baby-Glen Campbell-Capitol
(HB)	39.	I Don't Know How To Love Him-Yvonne Elliman-Decca
(HB)	40.	Cool Aid-Paul Humphrey-Lizard

*Denotes Big 10 Climbers

WC FL **ALL HIT MUSIC** **IN THE MIDWEST**

BIG 10 ALBUMS

1. Jesus Christ Superstar-Various Artists-Decca
2. Woodstock Two-Various Artists-Cotillion
3. Tea For The Tillerman-Cat Stevens-A & M
4. Pearl-Janis Joplin-Columbia
5. The Cry Of Love-Jimi Hendrix-Reprise
6. Movin' Toward Happiness-Mason Proffit-Happy Tiger
7. Emerson, Lake & Palmer-Emerson, Lake & Palmer-Cotillion
8. Love Story-Andy Williams-Columbia
9. Up To Date-Partridge Family-Bell
10. If I Could Only Remember My Name-David Crosby-Atlantic

The Original Album Survey from BIG TEN – the Station that PLAYS the Albums

SMALL PRINT DEPARTMENT: The WCFL "ALL HIT MUSIC" represents the relative popularity of records in the Midwest based on Industry tabulations, local record sales and requests. WCFL The Voice of Labor, is o

BIG 10 REQUEST GOLD

1. Magic Carpet Ride-Steppenwolf-(1968)
2. Don't You Care-The Buckinghams-(1967)
3. Cecelia-Simon & Garfunkel-(1970)
4. Close To You-The Carpenters-(1970)
5. My Girl-The Temptations-(1965)
6. Rock Around The Clock-Bill Haley-(1955)
7. Born To Be Wild-Steppenwolf-(1968)
8. Medicine Man-Buchannan Brothers-(1969)
9. I'd Wait A Million Years-Grass Roots-(1969)
10. Montego Bay-Bobby Bloom-(1970)

WAS	IS	Week Ending April 22, 1971
(1)	1.	Stay Awhile-Bells-Polydor
(7)	2.	Joy To The World-Three Dog Night-Dunhill
(8)	3.	Never Can Say Goodbye-Jackson 5-Motown
(5)	4.	Another Day-Paul McCartney-Apple
(9)	*5.	Put Your Hand In The Hand-Ocean-Kama Sutra
(2)	6.	L.A. Goodbye-Ides Of March-WB
(13)	*7.	If-Bread-Elektra
(6)	8.	She's A Lady-Tom Jones-Parrot
(4)	9.	Eighteen-Alice Cooper-WB
(10)	10.	We Can Work It Out-Stevie Wonder-Tamla
(3)	11.	Just My Imagination-Temptations-Gordy
(15)	*12.	Chick-A-Boom-Daddy Dewdrop-Sunflower
(18)	*13.	Me & You & A Dog Named Boo-Lobo-Big Tree
(19)	*14.	Wild World-Cat Stevens-A & M
(16)	15.	No Love At All-B.J. Thomas-Scepter
(17)	16.	Timothy-Buoys-Scepter
(22)	*17.	I Am...I Said-Neil Diamond-Uni
(11)	18.	For All We Know-Carpenters-A & M
(14)	19.	I Play & Sing-Dawn-Bell
(12)	20.	Woodstock-Matthews Southern Comfort-Decca
(24)	21.	Love Her Madly-Doors-Elektra
(28)	*22.	I Don't Blame You At All-The Miracles-Tamla
(25)	23.	Can't Find The Time-Rose Colored Glass-Bang
(26)	24.	Toast & Marmalade for Tea-Tin Tin-Atco
(27)	25.	13 Questions-Seatrain-Capitol
(31)	26.	Superstar-Murray Head-Decca
(20)	27.	Love's Lines, Angles & Rhymes-5th Dimension-Bell
(37)	*28.	Sweet and Innocent-Donny Osmond-MGM
(23)	29.	What's Going On-Marvin Gaye-Tamla
	30.	Oye Come Va-Santana-Columbia
(21)		
(35)	31.	When There's No You-E. Humperdinck-Parrot
(38)	32.	Dream Baby-Glen Campbell-Capitol
(36)	33.	I Love You For All Season-The Fuzz-Calla
(39)	34.	I Don't Know How To Love Him-

The Original Gold Survey from BIG TEN-home of Chicago's ORIGINAL GOLDEN WEEKEND

		Yvonne Elliman-Decca
(40)	35.	Cool Aid-Paul Humphrey-Lizard
(29)	36.	What Is Life-George Harrison-Apple
(HB)	37.	Brown Sugar-Rolling Stones-Atco
(HB)	38.	Reach Out I'll Be There-Diana Ross-Motown
(HB)	39.	Hot Love-T. Rex-Reprise
(HB)	40.	I'll Meet You Halfway-Partridge Family-Bell

*Denotes Big 10 Climbers

WC FL **ALL HIT MUSIC** **IN THE MIDWEST**

WAS	IS	Week Ending April 29, 1971
(2)	1.	Joy To The World-Three Dog Night-Dunhill
(1)	2.	Stay Awhile-Bells-Polydor
(3)	3.	Never Can Say Goodbye-Jackson 5-Motown
(5)	4.	Put Your Hand In The Hand-Ocean-Kama Sutra
(7)	*5.	If-Bread-Elektra
(13)	*6.	Me & You & A Dog Named Boo-Lobo-Big Tree
(12)	*7.	Chick-A-Boom-Daddy Dewdrop-Sunflower
(10)	8.	We Can Work It Out-Stevie Wonder-Tamla
(14)	*9.	Wild World-Cat Stevens-A & M
(4)	10.	Another Day-Paul McCartney-Apple
(6)	11.	L.A. Goodbye-Ides Of March-WB
(8)	12.	She's A Lady-Tom Jones-Parrot
(9)	13.	Eighteen-Alice Cooper-WB
(16)	14.	Timothy-Buoys-Scepter
(21)	*15.	Love Her Madly-Doors-Elektra
(17)	16.	I Am...I Said-Neil Diamond-Uni
(22)	*17.	I Don't Blame You At All-The Miracles-Tamla
(11)	18.	Just My Imagination-Temptations-Gordy
(20)	19.	Woodstock-Matthews Southern Comfort-Decca
(26)	*20.	Superstar-Murray Head-Decca
(25)	*21.	13 Questions-Seatrain-Capitol
(23)	22.	Can't Find The Time-Rose Colored Glass-Bang
(24)	23.	Toast & Marmalade For Tea-Tin Tin-Atco
(28)	24.	Sweet & Innocent-Donny Osmond-MGM
(18)	25.	For All We Know-Carpenters-A & M
(15)	26.	No Love At All-B.J. Thomas-Scepter
(19)	27.	I Play & Sing-Dawn-Bell
(31)	28.	When There's No You-E. Humperdinck-Parrot
(34)	29.	I Don't Know How To Love Him-Y. Elliman-Decca
(35)	30.	Cool Air-Paul Humphrey-Lizard
(32)	31.	Dream Baby-Glen Campbell-Capitol
(33)	32.	I Love You For All Seasons-The Fuzz-Calla
(37)	33.	Brown Sugar-Rolling Stones-Atco
(38)	34.	Reach Out I'll Be There-Diana Ross-Motown
(29)	35.	What's Going On-Marvin Gaye-Tamla
(39)	36.	Hot Love-T. Rex-Reprise
(40)	37.	I'll Meet You Halfway-Partridge Family-Bell
(HB)	38.	It Don't Come Easy-Ringo Starr-Bell

BIG 10 ALBUMS

1. Jesus Christ Superstar-Various Artists-Decca
2. 4 Way Street-Crosby, Stills, Nash & Young-Atlantic
3. Tea For The Tillerman-Cat Stevens-A & M
4. Emerson, Lake & Palmer-Emerson, Lake & Palmer-Cotillion
5. Movin' Toward Happiness-Mason Proffit-Happy Tiger
6. Tapestry-Carole King-Ode 70
7. Woodstock Two-Various Artists-Cotillion
8. Up To Date-Partridge Family-Bell
9. Golden Biscuits-Three Dog Night-Dunhill
10. Pearl-Janis Joplin-Columbia

SMALL PRINT DEPARTMENT: The WCFL "ALL HIT MUSIC" represents the relative popularity of records in the Midwest based on Industry tabulations, local record sales and requests. WCFL The Voice of Labor, is o

The Midwest's FIRST album survey from BIG TEN – the Station that PLAYS the Albums

BIG 10 REQUEST GOLD

1. Get Ready-Rare Earth-(1970)
2. Born To Be Wild-Steppenwolf-(1968)
3. In-A-Gadda-Da-Vida-Iron Butterfly-(1968)
4. You Can't Hurry Love-The Supremes-(1966)
5. Honky Tonk Woman-The Rolling Stones-(1969)
6. Purple Haze-Jimi Hendrix-(1967)
7. Let It Be-The Beatles-(1970)
8. Rock Around The Clock-Bill Haley & The Comets-(1955)
9. Lay Down (Candles In The Rain)-Melanie-(1970)
10. Apache-Jorgen Ingemann-(1961)

Chicago's ONLY Gold Survey – from BIG TEN – home of Chicago's ORIGINAL GOLDEN WEEKEND

(HB)	39.	Want Ads-The Honey Cone-Hot Wax
(HB)	40.	Right On The Tip Of My Tongue-Brenda/Tabulations-Top & Bottom

*Denotes Big 10 Climbers

WC FL **ALL HIT MUSIC** **IN THE MIDWEST**

WAS	IS	Week Ending May 6, 1971
(1)	1.	Joy To The World-Three Dog Night-Dunhill
(2)	2.	Stay Awhile-Bells-Polydor
(3)	3.	Never Can Say Goodbye-Jackson 5-Motown
(4)	4.	Put Your Hand In The Hand-Ocean-Kama Sutra
(5)	5.	If-Bread-Elektra
(6)	6.	Me & You & A Dog Named Boo-Lobo-Big Tree
(7)	7.	Chick-A-Boom-Daddy Dewdrop-Sunflower
(15)	*8.	Love Her Madly-Doors-Elektra
(9)	9.	Wild World-Cat Stevens-A & M
(14)	*10.	Timothy-Buoys-Scepter
(17)	*11.	I Don't Blame You At All-The Miracles-Tamla
(8)	12.	We Can Work It Out-Stevie Wonder-Tamla
(10)	13.	Another Day-Paul McCartney-Apple
(20)	*14.	Superstar-Murray Head-Decca
(24)	*15.	Sweet & Innocent-Donny Osmond-MGM
(21)	*16.	13 Questions-Seatrain-Capitol
(22)	*17.	Can't Find The Time-Rose Colored Glass-Bang
(11)	18.	L.A. Goodbye-Ides Of March-WB
(13)	19.	Eighteen-Alice Cooper-WB
(29)	*20.	I Don't Know How To Love Him-Y. Elliman-Decca
(12)	21.	She's A Lady-Tom Jones-Parrot
(33)	*22.	Brown Sugar-Rolling Stones-Atco
(30)	*23.	Cool Aid-Paul Humphrey-Lizard
(18)	24.	Just My Imagination-Temptations-Gordy
(16)	25.	I Am...I Said-Neil Diamond-Uni
(32)	26.	I Love You For All Seasons-The Fuzz-Calla
(34)	27.	Reach Out I'll Be There-Diana Ross-Motown
(19)	28.	Woodstock-Matthews Southern Comfort-Decca
(23)	29.	Toast & Marmalade For Tea-Tin Tin-Atco
(38)	30.	It Don't Come Easy-Ringo Starr-Apple
(37)	31.	I'll Meet You Halfway-Partridge Family-Bell
(36)	32.	Hot Love-T. Rex-Reprise
(39)	33.	Want Ads- The Honey Cone-Hot Wax

BIG 10 ALBUMS

1. Jesus Christ Superstar-Various Artists-Decca
2. 4 Way Street-Crosby, Stills, Nash, Young-Atlantic
3. Up To Date-Partridge Family-Bell
4. Tea For The Tillerman-Cat Stevens-A & M
5. Emerson, Lake, & Palmer-Emerson, Lake, & Palmer-Cotillion
6. Golden Biscuits-Three Dog Night-Dunhill
7. Mud Slide Slim-James Taylor-WB
8. Woodstock II-Various Artists-Cotillion
9. Tapestry-Carole King-Ode 70
10. If I Could Only Remember My Name-David Crosby-Atlantic

SMALL PRINT DEPARTMENT: The WCFL "ALL HIT MUSIC" represents the relative popularity of records in the Midwest based on Industry tabulations, local record sales and requests. WCFL The Voice of Labor, is o

The Midwest's FIRST album survey from BIG TEN – the Station that PLAYS the Albums

BIG 10 REQUEST GOLD

1. Mama Told Me Not To Come-Three Dog Night-(1970)
2. Born To Be Wild-Steppenwolf-(1968)
3. Suite: Judy Blue Eyes-Crosby, Stills & Nash-(1969)
4. Piece Of My Heart-Big Brother & The Holding Co.-(1968)
5. Brown Eyed Girl-Van Morrison-(1967)
6. Reach Out And I'll Be There-Four Tops-(1966)
7. He Ain't Heavy-The Hollies-(1970)
8. American Woman-Guess Who-(1970)
9. Bridge Over Troubled Water-Simon & Garfunkel-(1970)
10. 25 or 6 to 4-Chicago-(1970)

(27)	34.	I Play & Sing-Dawn-Bell
(40)	35.	Right On The Tip Of My Tongue-Brenda/Tabulations-Top & Bottom
(HB)	36.	Power To The People-John Lennon-Apple
(HB)	37.	That's The Way I've Heard-Carly Simon-Elektra
(HB)	38.	Here Comes That Rainy Day Feeling Again-The Fortunes-Capitol
(HB)	39.	Bridge Over Troubled Water-Aretha Franklin-Atlantic
(HB)	40.	Here Comes The Sun-Richie Havens-Stormy Forest

*Denotes Big 10 Climbers

Chicago's ONLY Gold Survey – from BIG TEN – home of Chicago's ORIGINAL GOLDEN WEEKEND

WC
FL

ALL HIT MUSIC

IN THE
MIDWEST

WAS IS Week Ending May 13, 1971

WAS	IS	
(1)	1.	Joy To The World-Three Dog Night-Dunhill
(3)	2.	Never Can Say Goodbye-Jackson 5-Motown
(4)	3.	Put Your Hand In The Hand-Ocean-Kama Sutra
(5)	4.	If-Bread-Elektra
(8)	*5.	Love Her Madly-Doors-Elektra
(11)	*6.	I Don't Blame You At All-The Miracles-Tamla
(14)	*7.	Superstar-Murray Head-Decca
(9)	8.	Wild World-Cat Stevens-A & M
(10)	9.	Timothy-Guoys-Scepter
(15)	*10.	Sweet And Innocent-Donny Osmond-MGM
(7)	11.	Chick-A-Boom-Daddy Dewdrop-Sunflower
(2)	12.	Stay Awhile-Bells-Polydor
(6)	13.	Me & You & A Dog Named Boo-Lobo-Big Tree
(16)	14.	13 Questions-Seatrain-Capitol
(17)	15.	Can't Find The Time-Rose Colored Glass-Bang
(20)	16.	I Don't Know How To Love Him-Y. Elliman-Decca
(22)	*17.	Brown Sugar-Rolling Stones-Atco
(13)	18.	Another Day-Paul McCartney-Apple
(23)	*19.	Cool Aid-Paul Humphrey-Lizard
(30)	*20.	It Don't Come Easy-Ringo Starr-Apple
(26)	*21.	I Love You For All Seasons-The Fuzz-Calla
(31)	*22.	I'll Meet You Halfway-Partridge Family-Bell
(27)	23.	Reach Out I'll Be There-Diana Ross-Motown
(12)	24.	We Can Work It Out-Stevie Wonder-Tamla
(33)	25.	Want Ads-The Honey Cone-Hot Wax
(32)	26.	Hot Love-T. Rex-Reprise
(36)	*27.	Power To The People-John Lennon-Apple
(18)	28.	L.A. Goodbye-Ides Of March-WB
(29)	29.	Toast & Marmalade For Tea-Tin Tin—Atco
(35)	30.	Right On The Tip Of My Tongue-Brenda/Tabulations-Top & Bottom
(37)	*31.	That's The Way I've Heard-Carly Simon-Elektra
(19)	32.	Eighteen-Alice Cooper-WB
(39)	33.	Bridge Over Troubled Water-Aretha Franklin-Atlantic

BIG 10 ALBUMS

1. Jesus Christ Superstar-Various Artists-Decca
2. 4 Way Street-Crosby, Stills, Nash, Young-Atlantic
3. Aqualung-Jethro Tull-Reprise
4. Mud Slide Slim-James Taylor-WB
5. Up To Date-Partridge Family-Bell
6. Tea For The Tillerman-Cat Stevens-A & M
7. Sticky Fingers-Rolling Stones-Atco
8. Golden Biscuits-Three Dog Night-Dunhill
9. Tapestry-Carole King-Ode 70
10. Emerson, Lake & Palmer-Emerson, Lake & Palmer-Cotillion

The Midwest's FIRST album survey from BIG TEN – the Station that PLAYS the Albums

SMALL PRINT DEPARTMENT: The WCFL "ALL HIT MUSIC" represents the relative popularity of records in the Midwest based on Industry tabulations, local record sales and requests. WCFL The Voice of Labor, is o

BIG 10 REQUEST GOLD

1. Purple Haze-Jimi Hendrix-(1967)
2. Raindrops Keep Falling On My Head-B.J. Thomas-(1970)
3. Vehicle-Ides Of March-(1970)
4. Daydreamer Believer-Monkees-(1968)
5. These Eyes-Guess Who-(1969)
6. Born To Be Wild-Steppenwolf-(1968)
7. Proud Mary-Credence Clearwater Revival-(1969)
8. Hey Jude-The Beatles-(1968)
9. Keem-O-Sabe-The Electric Indian-(1969)
10. Help-The Beatles-(1965)

Chicago's ONLY Gold Survey – from BIG TEN – home of Chicago's ORIGINAL GOLDEN WEEKEND

(40)	34.	Here Comes The Sun-Richie Havens-Stormy Forest
(38)	35.	Here Comes That Rainy Day Feeling-Fortunes-Capitol
(21)	36.	She's A Lady-Tom Jones-Parrot
(25)	37.	I Am...I Said-Neil Diamond-Uni
(24)	38.	Just My Imagination-Temptations-Gordy
(HB)	39.	Rainy Days And Mondays-Carpenters-A & M
(HB)	40.	If Not For You-Olivia Newton-John-Uni

*Denotes Big 10 Climbers

WC
FL

ALL HIT MUSIC

IN THE
MIDWEST

WAS	IS	Week Ending May 20, 1971
(1)	1.	Joy To The World-Three Dog Night-Dunhill
(10)	*2.	Sweet & Innocent-Donny Osmond-MGM
(6)	3.	I Don't Blame You At All-The Miracles-Tamla
(5)	4.	Love Her Madly-Doors-Elektra
(7)	5.	Superstar-Murray Head-Decca
(2)	6.	Never Can Say Goodbye-Jackson 5-Motown
(17)	7.	Brown Sugar-Rolling Stones-Atco
(9)	8.	Timothy-Buoys-Scepter
(16)	*9.	I Don't Know How To Love Him-Y. Elliman-Decca
(20)	*10.	It Don't Come Easy-Ringo Starr-Apple
(3)	11.	Put Your Hand In The Hand-Ocean-Kama Sutra
(8)	12.	Wild World-Cat Stevens-A & M
(4)	13.	If-Bread-Elektra
(22)	*14.	I'll Meet You Halfway-Partridge Family-Bell
(23)	*15.	Reach Out I'll Be There-Diana Ross-Motown
(25)	*16.	Want Ads-The Honey Cone-Hot Wax
(11)	17.	Chick-A-Boom-Daddy Dewdrop-Sunflower
(19)	18.	Cool Aid-Paul Humphrey-Lizard
(27)	*19.	Power To The People-John Lennon-Apple
(21)	*20.	I Love You For All Seasons-The Fuzz-Calla
(31)	*21.	That's The Way I've Heard-Carly Simon-Elektra
(12)	22.	Stay Awhile-Bells-Polydor
(26)	23.	Hot Love-T. Tex-Reprise
(33)	24.	Bridge Over Troubled Water-Aretha Franklin-Atlantic
(30)	25.	Right On The Tip Of My Tongue-Brenda/Tabulations-Top & Bottom
(34)	26.	Here Comes The Sun-Richie Havens-Stormy Forest
(13)	27.	Me & You & A Dog Named Boo-Lobo-Big Tree
(15)	28.	Can't Find The Time-Rose Colored Glass-Bang
(14)	29.	13 Questions-Seatrain-Capitol
(35)	30.	Here Comes That Rainy Day Feeling-Fortunes-Capitol
(39)	31.	Rainy Days And Mondays-Carpenters-A & M
(18)	32.	Another Day-Paul McCartney-Apple

BIG 10 ALBUMS

1. Sticky Fingers-Rolling Stones-Atco
2. Jesus Christ Superstar-Various Artists-Decca
3. Aqualung-Jethro Tull-Reprise
4. Mud Slide Slim-James Taylor-WB
5. Tapestry-Carole King-Ode 70
6. 4-Way Street-Crosby-Stills, Nash, Young-Atlantic
7. Tea For The Tillerman-Cat Stevens-A & M
8. L.A. Woman-Doors-Elektra
9. Golden Biscuits-Three Dog Night-Dunhill
10. Survival-Grand Funk Railroad-Capitol

SMALL PRINT DEPARTMENT: The WCFL "ALL HIT MUSIC" represents the relative popularity of records in the Midwest based on Industry tabulations, local record sales and requests. WCFL The Voice of Labor, is o

Chicago's FIRST and MOST ACCURATE album survey

BIG 10 REQUEST GOLD

*1 Hey Jude-The Beatles-(1968)
*2 Let It Be-The Beatles-(1970)
*3 Spill The Wine-Eric Burdon & War-(1970)
*4 Get Ready-Rare Earth-(1970)
*5 Pinball Wizard-The Who-(1969)
*6 Cecelia-Simon & Garfunkel-(1970)
*7 Get Back-The Beatles-(1969)
8. Indiana Wants Me-R.B. Greaves-(1970)
*9 Julie, Do Ya Love Me-Bobby Sherman-(1970)
*10 Born To Be Wild-Steppenwolf-(1968)

*denotes currently available as gold single or on an album.

(40)	33.	If Not For You-Olivia Newton-John-Uni
(24)	34.	We Can Work It Out-Stevie Wonder-Tamla
(HB)	35.	Double Lovin'-The Osmonds-MGM
(HB)	36.	Don't Knock My Love-Wilson Pickett-Atlantic
(HB)	37.	Nathan Jones-The Supremes-Motown
(HB)	38.	Puppet Man-Tom Jones-Parrot
(HB)	39.	It's Too Late-Carole King-Ode 70
(HB)	40.	Done Too Soon-Neil Diamond-Uni

*Denotes Big 10 Climbers

WC FL **ALL HIT MUSIC** **IN THE MIDWEST**

WAS	IS	Week Ending May 27, 1971
(2)	1.	Sweet & Innocent-Donny Osmond-MGM
(1)	2.	Joy To The World-Three Dog Night-Dunhill
(4)	*3.	Love Her Madly-Doors-Elektra
(7)	*4.	Brown Sugar-Rolling Stones-Atco
(9)	*5.	I Don't Know How To Love Him-Y. Elliman-Decca
(10)	*6.	It Don't Come Easy-Ringo Starr-Apple
(16)	*7.	Want Ads-Honey Cone-Hot Wax
(14)	*8.	I'll Meet You Halfway-Partridge Family-Bell
(15)	*9.	Reach Out I'll Be There-Diana Ross-Motown
(19)	*10.	Power To The People-John Lennon-Apple
(21)	*11.	That's The Way I've Heard-Carly Simon-Elektra
(3)	12.	I Don't Blame You AT All-The Miracles-Tamla
(18)	13.	Cool Aid-Paul Humphrey-Lizard
(25)	*14.	Right On The Tip Of My Tongue-Brenda/Tabulations-Top & Bottom
(23)	*15.	Hot Love-T. Rex-Reprise
(24)	*16.	Bridge Over Trouble Water-Aretha Franklin-Atlantic
(6)	17.	Never Can Say Goodbye-Jackson 5-Motown
(26)	*18.	Here Comes The Sun-Richie Havens-Stormy Forest
(5)	19.	Superstar-Murray Head-Decca
(31)	*20.	Rainy Days & Mondays-Carpenters-A & M
(13)	21.	If-Bread-Elektra
(8)	22.	Timothy-Buoys-Scepter
(30)	23.	Here Comes That Rainy Day Feeling-Fortunes-Capitol
(11)	24.	Put Your Hand In The Hand-Ocean-Kama Sutra
(33)	25.	If Not For You-Olivia Newton-John-Uni
(17)	26.	Chick-A-Boom-Daddy Dewdrop-Sunflower
(35)	27.	Double Lovin'-The Osmonds-MGM
(12)	28.	Wild World-Cat Stevens-A & M
(39)	29.	It's Too Late-Carole King-Ode 70
(37)	30.	Nathan Jones-The Supremes-Motown
(38)	31.	Puppet Man-Tom Jones-Parrot
(20)	32.	I Love You For All Seasons-The Fuzz-Calla
(36)	33.	Don't Knock My Love-Wilson Pickett-Atlantic
(40)	34.	Done Too Soon-Neil Diamond-Uni
(28)	35.	Can't Find The Time-Rose Colored Glass-Bang

BIG 10 ALBUMS

1. Sticky Fingers-Rolling Stones-Atco
2. Aqualung-Jethro Tull-Reprise
3. Jesus Christ Superstar-Various Artists-Decca
4. Tapestry-Carole King-Ode 70
5. 4 Way Street-Crosby-Stills-Nash, Young-Atlantic
6. Mud Slide Slim-James Taylor-WB
7. Golden Biscuits-Three Dog Night-Dunhill
8. Survival-Grand Funk Railroad-Capitol
9. Carly Simon-Carly Simon-Elektra
10. Up To Date-Partridge Family-Bell

SMALL PRINT DEPARTMENT: The WCFL "ALL HIT MUSIC" represents the relative popularity of records in the Midwest based on Industry tabulations, local record sales and requests. WCFL The Voice of Labor, is o

BIG 10 REQUEST GOLD

*1. Close To You-The Carpenters-(1970)
*2. Hey Jude-The Beatles-(1968)
*3. Lay, Lady, Lay-Bob Dylan-(1969)
4. Hitchin' A Ride-Vanity Fair-(1970)
*5. One-Three Dog Night-(1969)
*6. American Woman-Guess Who-(1970)
7. The Rapper-The Jaggerz-(1970)
*8. House Of The Rising Sun-The Animals-(1964)
9. Be My Baby-The Ronettes-(1963)
10. Honky Tonk-Bill Doggett-(1956)

*denotes currently available as gold single or on an album.

ALL HIT or SOLID GOLD – Chicago's full-time hitline is just for YOU! 222-1650

(HB)	36.	The Drum-Bobby Sherman-Metromedia
(HB)	37.	I Hear Those Church Bells Ringing-Duck-Bell
(HB)	38.	Try Some, Buy Some-Ronnie Spector-Apple
(HB)	39.	You've Got A Friend-James Taylor-WB
(HB)	40.	Treat Her Like A Lady-Cornelius Brothers-UA

*Denotes Big 10 Climbers

WC
FL

ALL HIT MUSIC

IN THE MIDWEST

BIG 10 ALBUMS

1. Sticky Fingers-Rolling Stones-Atco
2. Tapestry-Carole King-Ode 70
3. Jesus Christ Superstar-Various Artists-Decca
4. Mud Slide Slim-James Taylor-WB
5. Aqualung-Jethro Tull-Reprise
6. Ram-Paul McCartney-Apple
7. 4 Way Street-Crosby, Stills-Nash, Young-Atlantic
8. Survival-Grand Funk Railroad-Capitol
9. Golden Biscuits-Three Dog Night-Dunhill
10. Carly Simon-Carly Simon-Elektra

TED ANTHONY – *soon to be* –
Weekends on BIG TEN!

SMALL PRINT DEPARTMENT: The WCFL "ALL HIT MUSIC" represents the relative popularity of records in the Midwest based on Industry tabulations, local record sales and requests. WCFL The Voice of Labor, is o

BIG 10 REQUEST GOLD

*1. Cherish-The Association-(1966)
*2. War-Edwin Starr-(1970)
*3. My Girl-The Temptations-(1965)
*4. In The Summertime-Mungo Jerry-(1970)
*5. Your Song-Elton John-(1970)
6. Yellow River-Christie-(1970)
*7. Let It Be-The Beatles-(1970)
*8. Lay Down (Candles In The Rain)-Melanie-(1970)
9. This Diamond Ring-Gary Lewis-(1965)
*10. Woodstock-Crosby, Stills, Nash & Young-(1970)

*denotes currently available as gold single or on an album.

WAS	IS	Week Ending June 3, 1971
(1)	1.	Sweet & Innocent-Donny Osmond-MGM
(4)	2.	Brown Sugar-Rolling Stones-Atco
(7)	*3.	Want Ads-The Honey Cone-Hot Wax
(6)	*4.	It Don't Come Easy-Ringo Starr-Apple
(5)	5.	I Don't Know How To Love Him-Y. Elliman-Decca
(11)	*6.	That's The Way I've Heard-Carly Simon-Elektra
(8)	7.	I'll Meet You Halfway-Partridge Family-Bell
(10)	8.	Power To The People-John Lennon-Apple
(9)	9.	Reach Out I'll Be There-Diana Ross-Motown
(13)	*10.	Cool Aid-Paul Humphrey-Lizard
(2)	11.	Joy To The World-Three Dog Night-Dunhill
(20)	*12.	Rainy Days & Mondays-Carpenters-A & M
(14)	13.	Right On The Tip Of My Tongue-Brenda-Tabulations-Top & Bottom
(3)	14.	Love Her Madly-Doors-Elektra
(16)	15.	Bridge Over Troubled Water-Aretha Franklin-Atlantic
(18)	16.	Here Comes The Sun-Richie Havens-Stormy Forest
(12)	17.	I Don't Blame You At All-The Miracles-Tamla
(29)	*18.	It's Too Late-Carole King-Ode 70
(23)	*19.	Here Comes That Rainy Day Feeling-Fortunes-Capitol
(27)	*20.	Double Lovin'-The Osmonds-MGM
(25)	21.	If Not For You-Olivia Newton-John-Uni
(15)	22.	Hot Love-T. Rex-Reprise
(31)	*23.	Puppet Man-Tom Jones-Parrot
(17)	24.	Never Can Say Goodbye-Jackson 5-Motown
(30)	25.	Nathan Jones-The Supremes-Motown
(33)	26.	Don't Knock My Love-Wilson Pickett-Atlantic
(34)	27.	Done Too Soon-Neil Diamond-Uni
(19)	28.	Superstar-Murray Head-Decca
(21)	29.	If-Bread-Elektra
(36)	30.	The Drum-Bobby Sherman-Metromedia
(22)	31.	Timothy-Buoys-Scepter
(37)	32.	I Hear Those Church Bells Ringing-Dusk-Bell
(24)	33.	Put Your Hand In The Hand-Ocean-Kama Sustra
(38)	34.	Try Some, Buy Some-Ronnie Spector-Apple
(39)	35.	You've Got A Friend-James Taylor-

		WB
(26)	36.	Chick-A-Boom-Daddy Dewdrop-Sunflower
(40)	37.	Treat Her Like A Lady-Cornelius Brothers-UA
(HB)	38.	Draggin' The Line-Tommy James-Roulette
(HB)	39.	Bring The Boys Home-Freda Payne-Invictus
(HB)	40.	Never Ending Song Of Love-Delaney & Bonnie & Friends-Atco

*Denotes Big 10 Climbers

WC FL | **ALL HIT MUSIC** | **IN THE MIDWEST**

WAS IS Week Ending June 10, 1971

WAS	IS	Title
(3)	1.	Want Ads-The Honey Cone-Hot Wax
(4)	2.	It Don't Come Easy-Ringo Starr-Apple
(6)	3.	That's The Way I've Heard-Carly Simon-Elektra
(5)	4.	I Don't Know How To Love Him-Y. Elliman-Decca
(1)	5.	Sweet & Innocent-Donny Osmond-MGM
(7)	6.	I'll Meet You Halfway-Partridge Family-Bell
(12)	*7.	Rainy Days & Mondays-Carpenters-A & M
(9)	8.	Reach Out I'll Be There-Diana Ross-Motown
(18)	*9.	It's Too Late-Carole King-Ode 70
(10)	10.	Cool Aid-Paul Humphrey-Lizard
(2)	11.	Brown Sugar-Rolling Stones-Atco
(13)	12.	Right On The Tip Of My Tongue-Brenda/Tabulations-Top & Bottom
(16)	13.	Here Comes The Sun-Richie Havens-Stormy Forest
(20)	*14.	Double Lovin'-The Osmonds-MGM
(19)	15.	Here Comes That Rainy Day Feeling-Fortunes-Capitol
(21)	*16.	If Not For You-Olivia Newton-John-Uni
(23)	17.	Puppet Man-Tom Jones-Parrot
(8)	18.	Power To The People-John Lennon-Apple
(14)	19.	Love Her Madly-Doors-Elektra
(26)	*20.	Don't Knock My Love-Wilson Pickett-Atlantic
(11)	21.	Joy To The World-Three Dog Night-Dunhill
(25)	22.	Nathan Jones-Supremes-Motown
(30)	*23.	The Drum-Bobby Sherman-Metromedia
(27)	24.	Done Too Soon-Neil Diamond-Uni
(15)	25.	Bridge Over Troubled Water-Aretha Franklin-Atlantic
(35)	*26.	You've Got A Friend-James Taylor-WB
(17)	27.	I Don't Blame You At All-The Miracles-Tamla
(32)	28.	I Hear Those Church Bells Ringing-Dusk-Bell
(37)	*29.	Treat Her Like A Lady-Cornelius Brothers-UA
(22)	30.	Hot Love-T. Rex-Reprise
(34)	31.	Try Some, Buy Some-Ronnie Spector-Apple
(24)	32.	Never Can Say Goodbye-Jackson 5-Motown
(29)	33.	If-Bread-Elektra
(28)	34.	Superstar-Murray Head-Decca

BIG 10 ALBUMS

1. Tapestry-Carole King-Ode 70
2. Sticky Fingers-Rolling Stones-Atco
3. Ram-Paul & Linda McCartney-Apple
4. Jesus Christ Superstar-Various Artists-Decca
5. Mud Slide Slim-James Taylor-WB
6. Carly Simon-Carly Simon-Elektra
7. Aqualung-Jethro Tull-Reprise
8. 4 Way Street-Croscy-Stills-Nash, Young-Atlantic
9. Carpenters-Carpenters-A & M
10. Survival-Grand Funk Railroad-Capitol

SMALL PRINT DEPARTMENT: The WCFL "ALL HIT MUSIC" represents the relative popularity of records in the Midwest based on Industry tabulations, local record sales and requests. WCFL The Voice of Labor, is o

BIG 10 REQUEST GOLD

*1. War-Edwin Starr-(1970)
*2. Cecelia-Simon & Garfunkel-(1970)
*3. Come Together-The Beatles-(1969)
*4. The Letter-Joe Cocker-(1969)
*5. Purple Haze-Jimi Hendrix-(1967)
*6. Let It Be-The Beatles-(1970)
*7. Louie, Louie-The Kingsmen-(1963)
8. Bobby's Girl-Marcie Blane-(1962)
*9. Reach Out I'll Be There-Four tops-(1966)
*10. Holly Holy-Neil Diamond-(1970)

*denotes currently available as gold single or on an album.

YOURS for a song, Chicago's HITLINE – 222-1650

(38)	35.	Draggin' The Line-Tommy James-Roulette
(39)	36.	Bring The Boys Home-Freda Payne-Invictus
(40)	37.	Never Ending Song Of Love-Delaney & Bonnie & Freinds-Atco
(HB)	38.	Rainy Jane-Davy Jones-Bell
(HB)	39.	Summer Sand-Dawn-Bell
(HB)	40.	Don't Pull Your Love-Hamilton, Joe Frank & Reynolds-Dunhill

*Denotes Big 10 Climbers

WC
FL

ALL HIT MUSIC

IN THE
MIDWEST

WAS IS Week Ending June 17, 1971

WAS	IS	
(1)	1.	Want Ads-The Honey Cone-Hot Wax
(7)	2.	Rainy Days & Mondays-Carpenters-A & M
(3)	3.	That's The Way I've Heard-Carly Simon-Elektra
(9)	*4.	It's Too Late-Carole King-Ode 70
(6)	5.	I'll Meet You Halfway-Partridge Family-Bell
(16)	*6.	If Not For You-Olivia Newton-John-Uni
(14)	*7.	Double Lovin'-The Osmonds-MGM
(17)	*8.	Puppet Man-Tom Jones-Parrot
(13)	*9.	Here Comes The Sun-Richie Havens-Stormy Forest
(20)	*10.	Don't Knock My Love-Wilson Pickett-Atlantic
(2)	11.	It Don't Come Easy-Ringo Starr-Apple
(5)	12.	Sweet & Innocent-Donny Osmond-MGM
(15)	13.	Here Comes That Rainy Day Feeling-Fortunes-Capitol
(4)	14.	I Don't Know How To Love Him-Y. Elliman-Decca
(23)	*15.	The Drum-Bobby Sherman-Metromedia
(22)	*16.	Nathan Jones-Supremes-Motown
(10)	17.	Cool Aid-Paul Humphrey-Lizard
(8)	18.	Reach Out I'll Be There-Diana Ross-Motown
(26)	*19.	You've Got A Friend-James Taylor-WB
(29)	*20.	Treat Her Like A Lady-Cornelius Brothers-UA
(12)	21.	Right On The Tip Of My Tongue-Brenda/Tabulations-Top & Bottom
(24)	22.	Done Too Soon-Neil Diamond-Uni
(28)	23.	I Hear Those Church Bells Ringing-Dusk-Bell
(11)	24.	Brown Sugar-Rolling Stones-Atco
(35)	*25.	Draggin' The Line-Tommy James-Roulette
(18)	26.	Power To The People-John Lennon-Apple
(40)	*27.	Don't Pull Your Love-Hamilton, Joe Frank & Reynolds-Dunhill
(--)	28.	Do You Know What Time It Is-P-Nut Gallery-Buddah
(19)	29.	Love Her Madly-Doors-Elektra
(37)	30.	Never Ending Song Of Love-Delaney & Bonnie & Friends-Atco
(21)	31.	Joy To The World-Three Dog Night-Dunhill
(25)	32.	Bridge Over Troubled Water-Aretha Franklin-Atlantic
(36)	33.	Bring The Boys Home-Freda Payne-Invictus

BIG 10 ALBUMS

1. Tapestry-Carole King-Ode 70
2. Sticky Fingers-Rolling Stones-Atco
3. Ram-Paul & Linda McCartney-Apple
4. Mud Slide Slim-James Taylor-WB
5. Aqualung-Jethro Tull-Reprise
6. Carpenters-Carpenters-A & M
7. Songs For Beginners-Graham Nash-Atlantic
8. Live At Fillmore West-Aretha Franklin-Atlantic
9. Carly Simon-Carly Simon-Elektra
10. Jesus Christ Superstar-Various Artists-Decca

New to "10" Chicago's favorite number is 222-1650...call it!

SMALL PRINT DEPARTMENT: The WCFL "ALL HIT MUSIC" represents the relative popularity of records in the Midwest based on Industry tabulations, local record sales and requests. WCFL The Voice of Labor, is o

BIG 10 REQUEST GOLD

*1. Let It Be-The Beatles-(1970)
*2. ABC-Jackson 5-(1970)
*3. MacArthur Park-Richard Harris-(1968)
*4. Help-The Beatles-(1965)
*5. The Last Time-The Rolling Stones-(1965)
*6. People Got To Be Free-The Rascals-(1968)
*7. Your Song-Elton John-(1970)
*8. Gimmie Some Lovin'-Spencer Davis Group-(1967)
*9. The Letter-Joe Cocker-(1969)
10. He's Sure The Boy I Love-The Crystals-(1963)

*denotes currently available as gold single or on an album.

(38)	34.	Rainy Jane-Davy Jones-Bell
(39)	35.	Summer Sand-Dawn-Bell
(HB)	36.	Funky Nassau-The Beginning Of The End-Alston
(HB)	37.	High Time We Went-Joe Cocker-A & M
(HB)	38.	I Love You Lady Dawn-The Bells-Polydor
(HB)	39.	Moon Shadow-Cat Stevens-A & M
(HB)	40.	How Can You Mend A Broken Heart-Bee Gees-Atco

*Denotes Big 10 Climbers

WC
FL

ALL HIT MUSIC IN THE MIDWEST

BIG 10 ALBUMS

1. Tapestry-Carole King-Ode 70
2. Ram-Paul and Linda McCartney-Apple
3. Mud Slide Slim-James Taylor-WB
4. Carpenters-Carpenters-A & M
5. Sticky Fingers-Rolling Stones-Atco
6. Carly Simon-Carly Simon-Elektra
7. Songs For Beginners-Graham Nash-Atlantic
8. Live At Fillmore West-Aretha Franklin-Atlantic
9. Aqualung-Jethro Tull-Reprise
10. Jesus Christ Superstar-Various Artists-Decca

SMALL PRINT DEPARTMENT: The WCFL "ALL HIT MUSIC" represents the relative popularity of records in the Midwest based on Industry tabulations, local record sales and requests. WCFL The Voice of Labor, is o

BIG 10 REQUEST GOLD

*1. MacArthur Park-Richard Harris-(1968)
*2. Your Song-Elton John-(1970)
*3. Woodstock-Crosby, Stills, Nash & Young-(1970)
4. Don't You Care-The Buckinghams-(1967)
*5. Game Of Love-Wayne Fontana & The Mindbenders-(1965)
*6. I'm So Lonesome I Could Cry-B.J. Thomas-(1966)
*7. Cry Me A River-Joe Cocker-(1970)
8. Indiana Wants Me-R.B. Greaves-(1970)
*9. Suite: Judy Blue Eyes-Crosby, Stills & Nash-(1969)
10. Jimmy Mack-Martha & The Vandellas-(1967)

WAS	IS	Week Ending June 24, 1971
(4)	1.	It's Too Late-Carole King-Ode 70
(2)	2.	Rainy Days & Mondays-Carpenters-A & M
(6)	3.	If Not For You-Olivia Newton-John-Uni
(7)	*4.	Double Lovin'-The Osmonds-MGM
(10)	*5.	Don't Knock My Love-Wilson Pickett-Atlantic
(1)	6.	Want Ads-The Honey Cone-Hot Wax
(16)	*7.	Nathan Jones-Supremes-Motown
(9)	8.	Here Comes The Sun-Richie Havens-Stormy Forest
(13)	*9.	Here Comes That Rainy Day Feeling-Fortunes-Capitol
(19)	*10.	You've Got A Friend-James Taylor-WB
(5)	11.	I'll Meet You Halfway-Partridge Family-Bell
(15)	12.	The Drum-Bobby Sherman-Metromedia
(20)	*13.	Treat Her Like A Lady-Cornelius Brothers-UA
(3)	14.	That's The Way I've Heard-Carly Simon-Elektra
(8)	15.	Puppet Man-Tom Jones-Parrot
(11)	16.	It Don't Come Easy-Ringo Starr-Apple
(12)	17.	Sweet & Innocent-Donny Osmond-MGM
(25)	*18.	Draggin' The Line-Tommy James-Roulette
(23)	19.	I Hear Those Church Bells Ringing-Dusk-Bell
(27)	*20.	Don't Pull Your Love-Hamilton, Joe Frank, & Reynolds-Dunhill
(22)	21.	Done Too Soon-Neil Diamond-Uni
(14)	22.	I Don't Know How To Love Him-Y. Elliman-Decca
(30)	23.	Never Ending Song Of Love-Delaney & Bonnie-Atco
(17)	24.	Cool Aid-Paul Humphrey-Lizard
(28)	25.	Do You Know What Time It Is-P-Nut Galley-Buddah
(34)	*26.	Rainy Jane-Davy Jones-Bell
(18)	27.	Reach Out I'll Be There-Diana Ross-Motown
(35)	28.	Summer Sand-Dawn-Bell
(36)	29.	Funky Nassau-Beginning Of The End-Alston
(33)	30.	Bring The Boys Home-Freda Payne-Invictus
(21)	31.	Tip Of My Tongue-Brenda/ Tabulations-Top & Bottom
(37)	32.	High Time We Went-Joe Cocker-A & M

*denotes currently available as gold
single or on an album.

Have a GOLD HIT on us: 222-1650!

(39)	33.	Moon Shadow-Cat Stevens-A & M
(24)	34.	Brown Sugar-Rolling Stones-Atco
(40)	35.	How Can You Mend A Broken Heart-The Bee Gees-Atco
(38)	36.	I Love You Lady Dawn-The Bells-Polydor
(26)	37.	Power To The People-John Lennon-Apple
(29)	38.	Love Her Madly-Doors-Elektra
(HB)	39.	Mr. Big Stuff-Jean Knight-Stax
(HB)	40.	Mighty Clouds Of Joy-B.J. Thomas-Scepter

*Denotes Big 10 Climbers

WC
FL

ALL HIT MUSIC IN THE MIDWEST

WAS	IS	Week Ending July 1, 1971
(1)	1.	It's Too Late-Carole King-Ode 70
(2)	2.	Rainy Days & Mondays-Carpenters-A & M
(5)	3.	Don't Knock My Love-Wilson Pickett-Atlantic
(10)	*4.	You've Got A Friend-James Taylor-WB
(7)	5.	Nathan Jones-Supremes-Motown
(9)	6.	Here Comes That Rainy Day Feeling-Fortunes-Capitol
(8)	7.	Here Comes The Sun-Richie Havens-Stormy Forest
(12)	8.	The Drum-Bobby Sherman-Metromedia
(20)	*9.	Don't Pull Your Love-Hamilton, Joe Frank & Reynolds-Dunhill
(13)	10.	Treat Her Like A Lady-Cornelius Brothers-UA
(18)	*11.	Draggin' The Line-Tommy James-Roulette
(3)	12.	If Not For You-Olivia Newton-John-Uni
(19)	*13.	I Hear Those Church Bells Ringing-Dusk-Bell
(4)	14.	Double Lovin'-The Osmonds-MGM
(25)	*15.	Do You Know What Time It Is-P-Nut Gallery-Buddah
(6)	16.	Want Ads-The Honey Cone-Hot Wax
(23)	*17.	Never Ending Song Of Love-Delaney & Bonnie-Atco
(11)	18.	I'll Meet You Halfway-Partridge Family-Bell
(29)	*19.	Funky Nassau-Beginning Of The End-Alston
(14)	20.	That's The Way I've Heard-Carly Simon-Elektra
(26)	21.	Rainy Jane-Davy Jones-Bell
(28)	22.	Summer Sand-Dawn-Bell
(15)	23.	Puppet Man-Tom Jones-Parrot
(32)	*24.	High Time We Went-Joe Cocker-A & M
(33)	*25.	Moon Shadow-Cat Stevens-A & M
(16)	26.	It Don't Come Easy-Ringo Starr-Apple
(30)	27.	Bring The Boys Home-Freda Payne-Invictus
(35)	*28.	How Can You Mend A Broken Heart-Bee Gees-Atco
(17)	29.	Sweet & Innocent-Donny Osmond-MGM
(39)	*30.	Mr. Big Stuff-Jean Knight-Stax
(40)	31.	Mighty Clouds Of Joy-B.J. Thomas-Scepter
(36)	32.	I Love You Lady Dawn-The Bells-Polydor

BIG 10 ALBUMS

1. Tapestry-Carole King-Ode 70
2. Carpenters-Carpenters-A & M
3. Ram-Paul & Linda McCartney-Apple
4. Sticky Fingers-Rolling Stones-Atco
5. Mud Slide Slim-James Taylor-WB
6. Jesus Christ Superstar-Various Artists-Decca
7. Aqualung-Jethro Tull-Reprise
8. Live At Fillmore West-Aretha Franklin-Atlantic
9. What's Going ON-Marvin Gaye-Tamla
10. Every Picture Tells A Story-Rod Stewart-Mercury

The best of the newest albums-
24 hours a day on BIG TEN!

SMALL PRINT DEPARTMENT: The WCFL "ALL HIT MUSIC" represents the relative popularity of records in the Midwest based on Industry tabulations, local record sales and requests. WCFL The Voice of Labor, is o

BIG 10 REQUEST GOLD

*1. Those Were The Days-Mary Hopkin-(1968)
*2. Close To You-The Carpenters-(1970)
*3. 25 or 6 To 4-Chicago-(1970)
*4. Purple Haze-Jimi Hendrix-(1967)
*5. Venus-Shocking Blue-(1970)
*6. Suite: Judy Blue Eyes-Crosby-Stills & Nash-(1969)
*7. Help-The Beatles-(1965)
*8. Society's Child-Janis Ian-(1967)
9. La Bama-Ritchie Valens-(1967)
10. Little G.T.O.-Ronnie & The Daytonas-(1964)

*denotes currently available as gold single or on an album.

(22)	33. I Don't Know How To Love Him-Y. Elliman-Decca
(21)	34. Done Too Soon-Neil Diamond-Uni
(HB)	35. Walk Away-James Gang-ABC
(HB)	36. Beginnings-Chicago-Columbia
(HB)	37. Wild Horses-Rolling Stones-Atco
(HB)	38. Mercy Mercy Me-Marvin Gaye-Talma
(HB)	39. Riders On The Storm-The Doors-Elektra
(HB)	40. Take Me Home, Country Roads-John Denver-RCA

*Denotes Big 10 Climbers

WC
FL

ALL HIT MUSIC

IN THE
MIDWEST

WAS	IS	Week Ending July 8, 1971
(1)	1.	It's Too Late-Carole King-Ode 70
(4)	2.	You've Got A Friend-James Taylor-WB
(9)	*3.	Don't Pull Your Love-Hamilton, Joe Frank & Reynolds-Dunhill
(6)	4.	Here Comes That Rainy Day Feeling-Fortunes-Capitol
(5)	5.	Nathan Jones-Supremes-Motown
(11)	6.	Draggin' The Line-Tommy James-Roulette
(8)	7.	The Drum-Bobby Sherman-Metromedia
(2)	8.	Rainy Days & Mondays-Carpenters-A & M
(13)	9.	I Hear Those Church Bells Ringing-Dusk-Bell
(15)	*10.	Do You Know What Time It Is-P-Nut Gallery-Buddah
(17)	*11.	Never Ending Song Of Love-Delaney & Bonnie-Atco
(3)	12.	Don't Knock My Love-Wilson Pickett-Atlantic
(19)	*13.	Funky Nassau-Beginning Of The End-Alston
(21)	*14.	Rainy Jane-Davy Jones-Bell
(22)	*15.	Summer Sand-Dawn-Bell
(7)	16.	Here Comes The Sun-Richie Havens-Stormy Forest
(28)	*17.	How Can You Mend A Broken Heart-Bee Gees-Atco
(24)	*18.	High Time We Went-Joe Cocker-A & M
(10)	19.	Treat Her Like A Lady-Cornelius Brothers-UA
(25)	20.	Moon Shadow-Cat Stevens-A & M
(12)	21.	If Not For You-Olivia Newton-John-Uni
(--)	*22.	Indian Reservation-The Raiders-Columbia
(27)	23.	Bring The Boys Home-Freda Payne-Invictus
(30)	*24.	Mr. Big Stuff-Jean Knight-Stax
(14)	25.	Double Lovin'-The Osmonds-MGM
(36)	*26.	Beginnings-Chicago-Columbia
(31)	27.	Mighty Clouds Of Joy-B.J. Thomas-Scepter
(18)	28.	I'll Meet You Halfway-Partridge Family-Bell
(32)	29.	I Love You Lady Dawn-The Bells-Polydor
(16)	30.	Want Ads-The Honey Cone-Hot Wax
(35)	31.	Walk Away-James Gang-ABC
(38)	32.	Mercy Mercy Me-Marvin Gaye-Tamla
(37)	33.	Wild Horses-Rolling Stones-Atco
(40)	34.	Take Me Home, Country Roads-John Denver-RCA
(39)	35.	Riders On The Storm-The Doors-Elektra

BIG 10 ALBUMS

1. Tapestry-Carole King-Ode 70
2. Mud Slide Slim-James Taylor-WB
3. Sticky Fingers-Rolling Stones-Atco
4. Ram-Paul & Linda McCartney-Apple
5. Carpenters-Carpenters-A & M
6. Aqualung-Jethro Tull-Reprise
7. Live At Fillmore West-Aretha Franklin-Atlantic
8. Songs For Beginners-Graham Nash-Atlantic
9. Chase-Chase-Epic
10. Tarkus-Emerson, Lake & Palmer-Cotillion

SMALL PRINT DEPARTMENT: The WCFL "ALL HIT MUSIC" represents the relative popularity of records in the Midwest based on Industry tabulations, local record sales and requests. WCFL The Voice of Labor, is o

BIG 10 REQUEST GOLD

*1. Born To Be Wild-Steppenwolf-(1968)
*2. War-Edwin Starr-(1970)
*3. Help-The Beatles-(1965)
*4. Purple Haze-Jimi Hendrix-(1967)
*5. Downtown-Petula Clark-(1965)
*6. Sugar, Sugar-The Archies-(1970)
*7. The Boxer-Simon & Garfunkel-(1968)
8. Gloria-The Shadows Of Knight-(1966)
*9. Jailhouse Rock-Elvis Presley-(1958)
*10. Breaking Up Is Hard To Do-Neil Sedaka-(1962)

*denotes currently available as gold single or on an album.

Lay It On The Line! 222-1650

(23)	36.	Puppet Man-Tom Jones-Parrot
(20)	37.	That's The Way I've Heard-Carly Simon-Elektra
(HB)	38.	Sweet Hitch-Hiker-Creedence Clearwater Revival-Fantasy
(HB)	39.	Won't Get Fooled Again-The Who-Decca
(HB)	40.	Smiling Faces Sometimes-Undisputed Truth-Gordy

*Denotes Big 10 Climbers

WC
FL

ALL HIT MUSIC

IN THE
MIDWEST

WAS	IS	Week Ending July 15 1971
(2)	1.	You've Got A Friend-James Taylor-WB
(3)	2.	Don't Pull Your Love-Hamilton, Joe Frank & Reynolds-Dunhill
(6)	3.	Draggin' The Line-Tommy James-Roulette
(11)	*4.	Never Ending Song Of Love-Delaney & Bonnie-Atco
(1)	5.	It's Too Late-Carole King-Ode 70
(13)	*6.	Funky Nassau-Beginning Of The End-Alston
(4)	7.	Here Comes That Rainy Day Feeling-Fortunes-Capitol
(5)	8.	Nathan Jones-Supremes-Motown
(17)	*9.	How Can You Mend A Broken Heart-Bee Gees-Atco
(7)	10.	The Drum-Bobby Sherman-Metromedia
(8)	11.	Rainy Days & Mondays-Carpenters-A & M
(14)	12.	Rainy Jane-Davy Jones-Bell
(18)	*13.	High Time We Went-Joe Cocker-A & M
(9)	14.	I Hear Those Church Bells Ringing-Dusk-Bell
(15)	15.	Summer Sand-Dawn-Bell
(10)	16.	Do You Know What Time It Is-P-Nut Gallery-Buddah
(24)	*17.	Mr. Big Stuff-Jean Knight-Stax
(20)	18.	Moon Shadow-Cat Stevens-A & M
(22)	19.	Indian Reservation-The Raiders-Columbia
(26)	20.	Beginnings-Chicago-Columbia
(12)	21.	Don't Knock My Love-Wilson Pickett-Atlantic
(32)	*22.	Mercy Mercy Me-Marvin Gaye-Tamla
(35)	*23.	Riders On The Storm-The Doors-Elektra
(16)	24.	Here Comes The Sun-Richie Havens-Stormy Forest
(27)	25.	Mighty Clouds Of Joy-B.J. Thomas-Scepter
(31)	26.	Walk Away-James Gang-ABC
(29)	27.	I Love You Lady Dawn-The Bells-Polydor
(33)	28.	Wild Horses-Rolling Stones-Atco
(19)	29.	Treat Her Like A Lady-Cornelius Brothers-UA
(34)	30.	Take Me Home, Country Roads-John Denver-RCA
(23)	31.	Bring The Boys Home-Freda Payne-Invictus
(40)	32.	Smiling Faces Sometimes-Undisputed Truth-Gordy
(38)	33.	Sweet Hitch-Hiker-C.C. Revival-Fantasy
(25)	34.	Double Lovin'-The Osmonds-MGM
(39)	35.	Won't Get Fooled Again-The Who-

BIG 10 ALBUMS

1. Tapestry-Carole King-Ode 70
2. Mud Slide Slim-James Taylor-WB
3. Carpenters-Carpenters-A & M
4. Ram-Paul & Linda McCartney-Apple
5. Sticky Fingers-Rolling Stones-Atco
6. Live At Fillmore West-Aretha Franklin-Atlantic
7. Blood, Sweat 7 Tears #4-Blood, Sweat, & Tears-Columbia
8. Songs For Beginners-Graham Nash-Atlantic
9. What's Going On-Marvin Gaye-Tamla
10. Tarkus-Emerson, Lake, & Palmer-Cotillion

SMALL PRINT DEPARTMENT: The WCFL "ALL HIT MUSIC" represents the relative popularity of records in the Midwest based on Industry tabulations, local record sales and requests. WCFL The Voice of Labor, is o

BIG 10 REQUEST GOLD

* Hey Jude-The Beatles-(1968)

* Sounds Of Silence-Simon & Garfunkel-(1965)

* Turn, Turn, Turn-The Byrds-(1965)

* War-Edwin Starr-(1970)

* Lola-The Kinks-(1970)

* Question-The Moody Blues-(1970)

* Spinning Wheel-Blood, Sweat & Tears-(1969)

* Stoned Love-The Supremes-(1970)

* Sunny-Bobby Hebb-(1966)

Who Stole The Keeshaka?-The Matys Brothers-(1963)

*denotes currently available as gold single or on an album.

		Decca
(28)	36.	I'll Meet You Halfway-Partridge Familiy-Bell
(21)	37.	If Not For You-Olivia Newton-John-Uni
(HB)	38.	Mother Freedom-Bread-Elektra
(HB)	39.	Over And Over-The Delfonics-Philly Groove
(HB)	40.	Wedding Song (There Is Love)-Paul Stookey-WB

*Denotes Big 10 Climbers

WC
FL

ALL HIT MUSIC

IN THE
MIDWEST

WAS IS Week Ending July 22, 1971

WAS	IS	Title
(2)	1.	Don't Pull Your Love-Hamilton, Joe Frank & Reynolds-Dunhill
(3)	2.	Draggin' The Line-Tommy James-Roulette
(4)	3.	Never Ending Song Of Love-Delaney & Bonnie-Atco
(6)	4.	Funky Nassau-Beginning Of The End-Alston
(9)	5.	How Can You Mend A Broken Heart-Bee Gees-Atco
(17)	*6.	Mr. Big Stuff-Jean Knight-Stax
(12)	*7.	Rainy Jane-Davy Jones-Bell
(5)	8.	It's Too Late-Carole King-Ode- 70
(1)	9.	You've Got A Friend-James Taylor-WB
(13)	10.	High Time We Went-Joe Cocker-A & M
(19)	*11.	Indian Reservation-The Raiders-Columbia
(20)	*12.	Beginnings-Chicago-Columbia
(15)	13.	Summer Sand-Dawn-Bell
(22)	*14.	Mercy Mercy Me-Marvin Gaye-Tamla
(18)	15.	Moon Shadow-Cat Stevens-A & M
(23)	*16.	Riders On The Storm-The Doors-Elektra
(7)	17.	Here Comes That Rainy Day Feeling-Fortunes-Capitol
(25)	*18.	Mighty Clouds Of Joy-B.J. Thomas-Scepter
(8)	19.	Nathan Jones-Supremes-Motown
(11)	20.	Rainy Days & Mondays-Carpenters-A & M
(30)	*21.	Take Me Home, Country Roads-John Denver-RCA
(10)	22.	The Drum-Bobby Sherman-Metromedia
(28)	23.	Walk Away-James Gang-ABC
(32)	*24.	Smiling Faces Sometimes-Undisputed Truth-Gordy
(28)	25.	Wild Horses-Rolling Stones-Atco
(33)	*26.	Sweet Hitch-Hiker-C.C. Revival-Fantasy
(14)	27.	I Hear Those Church Bells Ringing-Dusk-Bell
(35)	*28.	Won't Get Fooled Again-The Who-Decca
(21)	29.	Don't Knock My Love-Wilson Pickett-Atlantic
(16)	30.	Do You Know What Time It Is?-P-Nut Gallery-Buddah
(39)	31.	Over And Over-The Delfonics-Philly Groove
(24)	32.	Here Comes The Sun-Richie Havens-Stormy Forest

BIG 10 ALBUMS

1. Tapestry-Carole King-Ode 70
2. Mud Slide Slim-James Taylor-WB
3. Ram-Paul & Linda McCartney-Apple
4. Carpenters-Carpenters-A & M
5. Stephen Stills 2-Stephen Stills-Atlantic
6. Sticky Fingers-Rolling Stones-Atco
7. Jesus Christ Superstar-Various Artists-Decca
8. Tarkus-Emerson, Lake and Palmer-Cotillion
9. What's Going On-Marvin Gaye-Tamla
10. Blood, Sweat & Tears #4-Blood, Sweat & Tears-Columbia

SPECIAL!

BIG 10 TOP REQUEST
What The World Needs Now -
Tom Clay - Mowest

BIG 10 REQUEST GOLD

- American Woman-The Guess Who-(1970)
- Lola-The Kinks-(1970)
- I Think I Love You-Partridge Family-(1970)
- Spirit In The Sky-Norman Greenbaum-(1970)
- Tears Of A Clown-Smokey Robinson & the Miracles-(1970)
- 26 Miles-Edwin Starr-(1970)
- My Sweet Lord-George Harrison-(1970)
- Could Be We're In Love-Cryan' Shames-(1967)
- Valleri-The Monkees-(1968)
- Wedding Bell Blues-Fifth Dimension-(1969)

*denotes currently available as gold single or on an album.

SMALL PRINT DEPARTMENT: The WCFL "ALL HIT MUSIC" represents the relative popularity of records in the Midwest based on Industry tabulations, local record sales and requests. WCFL The Voice of Labor, is o

(38)	33.	Mother Freedom-Bread-Elektra
(27)	34.	I Love You Lady Dawn-The Bells-Polydor
(40)	35.	Wedding Song (There Is Love)-Paul Stookey-WB
(29)	36.	Treat Her Like A Lady-Cornelius Brothers-UA
(HB)	37.	Liar-Three Dog Night-Dunhill
(HB)	38.	Rings-Cymarron-Entrance
(HB)	39.	Maybe Tomorrow-Jackson 5-Motown
(HB)	40.	Resurrection Shuffle-Ashton, Gardner & Dyke-Capitol

*Denotes Big 10 Climbers

WC FL **ALL HIT MUSIC** **IN THE MIDWEST**

BIG 10 ALBUMS

1. Tapestry-Carole King-Ode 70
2. Ram-Paul & Linda McCartney-Apple
3. Mud Slide Slim-James Taylor-WB
4. Carpenters-Carpenters-A & M
5. Stephen Stills 2-Stephen Stills-Atlantic
6. Blood, Sweat & Tears #4-Blood, Sweat & Tears-Columbia
7. Jesus Christ Superstar-Various Artists-Decca
8. Sticky Fingers-Rolling Stones-Atco
9. What's Going On-Marvin Gaye-Tamla
10. Blue-Joni Mitchell-Reprise

SPECIAL!

BIG 10 TOP REQUEST
What The World Needs Now – Tom Clay - Mowest.

BIG 10 REQUEST GOLD

* Light My Fire-The Doors-(1967)
* Vehicle-The Ides Of March-(1970)
* The Letter-Joe Cocker-(1970)
* Make Me Smile-Chicago-(1970)
* I Am A Rock-Simon & Garfunkel-(1966)
* Never My Love-The Association-(1968)
* 19th Nervous Breakdown-Rolling Stones-(1966)
* 98.6-Keith-(1967)
* Paperback Writer-The Beatles-(1966)
* Special Occasion-The Miracles-(1968)

*denotes currently available as gold single or on an album.

WAS	IS	Week Ending July 29, 1971
(5)	1.	How Can You Mend A Broken Heart-Bee Gees-Atco
(3)	2.	Never Ending Song Of Love-Delaney & Bonnie-Atco
(11)	*3.	Indian Reservation-The Raiders-Columbia
(6)	4.	Mr. Big Stuff-Jean Knight-Stax
(1)	5.	Don't Pull Your Love-Hamilton, Joe Frank & Reynolds-Dunhill
(12)	*6.	Beginnings-Chicago-Columbia
(2)	7.	Draggin' The Line-Tommy James-Roulette
(16)	*8.	Riders On The Storm-The Doors-Elektra
(4)	9.	Funky Nassau-Beginning Of The End-Alston
(14)	10.	Mercy Mercy Me-Marvin Gaye-Tamla
(7)	11.	Rainy Jane-Davy Jones-Bell
(15)	12.	Moon Shadow-Cat Stevens-A & M
(13)	13.	Summer Sand-Dawn-Bell
(21)	*14.	Take Me Home, Country Roads-John Denver-RCA
(8)	15.	It's Too Late-Carole King-Ode 70
(18)	16.	Mighty Clouds Of Joy-B.J. Thomas-Scepter
(23)	17.	Walk Away-James Gang-ABC
(9)	18.	You've Got A Friend-James Taylor-WB
(24)	*19.	Smiling Faces Sometimes-Undisputed Truth-Gordy
(26)	*20.	Sweet High-Hiker-C.C. Revival-Fantasy
(10)	21.	High Time We Went-Joe Cocker-A & M
(25)	22.	Wild Horses-Rolling Stones-Atco
(28)	23.	Won't Get Fooled Again-The Who-Decca
(38)	*24.	Rings-Cymarron-Entrance
(39)	*25.	Maybe Tomorrow-Jackson 5-Motown
(17)	26.	Here Comes That Rainy Day Feeling-Fortunes-Capitol
(35)	*27.	Wedding Song (There Is Love)-Paul Stookey-WB
(33)	*28.	Mother Freedom-Bread-Elektra
(37)	*29.	Liar-Three Dog Night-Dunhill
(31)	30.	Over And Over-The Delfonics-Philly Groove
(19)	31.	Nathan Jones-Supremes-Motown
(22)	32.	The Drum-Bobby Sherman-Metromedia
(20)	33.	Rainy Days & Mondays-Carpenters-A & M
(27)	34.	I Hear Those Church Bells Ringing-Dusk-Bell
(40)	35.	Resurrection Shuffle-Ashton, Gardner & Dyke-Capitol

SMALL PRINT DEPARTMENT: The WCFL "ALL HIT MUSIC" represents the relative popularity of records in the Midwest based on Industry tabulations, local record sales and requests. WCFL The Voice of Labor, is o

(29)	36.	Don't Knock My Love-Wilson Pickett-Atlantic
(HB)	37.	Uncle Albert/Admiral Halsey-Paul McCartney-Apple
(HB)	38.	Spanish Harlem-Aretha Franklin-Atlantic
(HB)	39.	Whatcha See Is Whatcha Get-Dramatics-Volt
(HB)	40.	Roll On-The New Colony Six-Sunlight

*Denotes Big 10 Climbers

WC FL **ALL HIT MUSIC** **IN THE MIDWEST**

WAS IS Week Ending August 5, 1971

WAS	IS	
(3)	1.	Indian Reservation-The Raiders-Columbia
(2)	2.	Never Ending Song Of Love-Delaney & Bonnie-Atco
(4)	3.	Mr. Big Stuff-Jean Knight-Stax
(1)	4.	How Can You Mend A Broken Heart-Bee Gees-Atco
(8)	5.	Riders On The Storm-The Doors-Elektra
(14)	*6.	Take Me Home, Country Roads-John Denver-RCA
(10)	7.	Mercy Mercy Me-Marvin Gaye-Tamla
(12)	8.	Moon Shadow-Cat Stevens-A & M
(19)	*9.	Smiling Faces Sometimes-Undisputed Truth-Gordy
(5)	10.	Don't Pull Your Love-Hamilton, Joe Frank Reynolds-Dunhill
(20)	*11.	Sweet Hitch-Hiker-C.C. Revival-Fantasy
(6)	12.	Beginnings-Chicago-Columbia
(24)	*13.	Rings-Cymarron-Entrance
(9)	14.	Funky Nassau-Beginning Of The End-Alston
(25)	*15.	Maybe Tomorrow-Jackson 5-Motown
(27)	*16.	Wedding Song (There Is Love)-Paul Stookey-WB
(7)	17.	Draggin' The Line-Tommy James-Roulette
(15)	18.	It's Too Late-Carole King-Ode 70
(23)	19.	Won't Get Fooled Again-The Who-Decca
(29)	*20.	Liar-Three Dog Night-Dunhill
(11)	21.	Rainy Jane-Davy Jones-Bell
(28)	22.	Mother Freedom-Bread-Elektra
(13)	23.	Summer Sand-Dawn-Bell
(16)	24.	Mighty Clouds Of Joy-B.J. Thomas-Scepter
(17)	25.	Walk Away-James Gang-ABC
(35)	*26.	Resurrection Shufle-Ashton, Gardner & Dyke-Capitol
(18)	27.	You've Got A Friend-James Taylor-WB
(37)	28.	Uncle Albert/Admiral Halsey-Paul McCartney-Apple
(21)	29.	High Time We Went-Joe Cocker-A & M
(38)	30.	Spanish Harlem-Aretha Franklin-Atlantic
(39)	31.	Whatcha See Is Whatcha Get-Dramatics-Volt
(40)	32.	Roll On-The New Colony Six-Sunlight
(33)	33.	Rainy Days & Mondays-Carpenters-A & M
(22)	34.	Wild Horses-Rolling Stones-Atco

BIG 10 ALBUMS

1. Ram-Paul & Linda McCartney-Apple
2. Stephen Stills 2-Stephens Stills-Atlantic
3. Carpenters-Carpenters-A & M
4. Tapestry-Carole King-Ode 70
5. Mud Slide Slim-James Taylor-WB
6. What's Going On-Marvin Gaye-Tamla
7. Blood, Sweat & Tears #4-Blood, Sweat & Tears-Columbia
8. Sticky Fingers-Rolling Stones-Atco
9. Blue-Joni Mitchell-Reprise
10. Jesus Christ Superstar-Various Artists-Decca

SPECIAL!

BIG 10 TOP REQUEST

What The World Needs Now – Tom Clay – Mowest

BIG 10 REQUEST GOLD

* People Got To Be Free-Rascals-(1968)
* Light My Fire-The Doors-(1967)
* Hello, Good-Bye-The Beatles-(1968)
* Gimmie Some Lovin'-Spencer Davis Group-(1967)
* A Beautiful Morning-Rascals-(1968)
* Do You Believe In Magic?-Lovin' Spoonful-(1965)
* Green-Eyed-Lady-Sugarloaf-(1970)
* I'd Wait A Million Years-Grassroots-(1969)
* It Hurts To Be In Love-Gene Pitney-(1964)
* Donna-Ritchie Valens-(1959)

*denotes currently available as gold single or on an album.

(30)	35.	Over And Over-The Defonics-Phillip Groove
(HB)	36.	I Just Want To Celebrate-Rare Earth-Rare Earth
(HB)	37.	Ain't No Sunshine-Bill Withers-Sussex
(HB)	38.	Go Away Little Girl-Donny Osmond-MGM
(HB)	39.	Bangla-Desh-George Harrison-Apple
(HB)	40.	Woke Up In Love This Morning-Partridge Family-Bell

*Denotes Big 10 Climbers

SMALL PRINT DEPARTMENT: The WCFL "ALL HIT MUSIC" represents the relative popularity of records in the Midwest based on Industry tabulations, local record sales and requests. WCFL The Voice of Labor, is o

WC
FL ALL HIT MUSIC IN THE MIDWEST

BIG 10 ALBUMS

1. Tapestry-Carole King-Ode 70
2. Ram-Paul & Linda McCartney-Apple
3. Carpenters-Carpenters-A & M
4. Mud Slide Slim-James Taylor-WB
5. Blood, Sweat & Tears #4-Blood, Sweat & Tears-Columbia
6. Stephen Stills #2-Stephen Stills-Atlanta
7. What's Going ON-Marvin Gaye-Tamla
8. Sticky Fingers-Rolling Stones-Atco
9. Jesus Christ Superstar-Various Artists-Decca
10. Aqualung-Jethro Tull-Reprise

What Gets Played Next??? You
Tell us: 222-1650!

SMALL PRINT DEPARTMENT: The WCFL "ALL HIT MUSIC" represents the relative popularity of records in the Midwest based on Industry tabulations, local record sales and requests. WCFL The Voice of Labor, is owned and operated by The Chicago Federation of Labor and Industrial Union Council.

BIG 10 REQUEST GOLD

- * My Girl-Temptations-(1965)
- * I'll Meet You Halfway-Partridge Family-(1971)
- * Make Me Smile-Chicago-(1970)
- * One Bad Apple-The Osmonds-(1971)
- * Love the One You're With-Stephen Stills-(1971)
- * Little Woman-Bobby Sherman-(19699)
- * Knock Three Times-Dawn-(1971)
- * One-Three Dog Night-(1969)
- * Help-Beatles-(1965)
- * Love Can Make You Happy-Mercy-(1969)

*denotes currently available as gold single or on an album.

WAS	IS	Week Ending August 12, 1971
(2)	1.	Never Ending Song Of Love-Delaney & Bonnie-Atco
(5)	2.	Riders On The Story-The Doors-Elektra
(3)	3.	Mr. Big Stuff-Jean Knight-Stax
(6)	4.	Take Me Home, Country Roads-John Denver-RCA
(9)	5.	Smiling Faces Sometimes-Undisputed Truth-Gordy
(7)	6.	Mercy Mercy Me-Marvin Gaye-Tamla
(8)	7.	Moon Shadow-Cal Stevens-A&M
(1)	8.	Indian Reservation-The Raiders-Columbia
(11)	9.	Sweet Hitch-Hiker-C.C. Revival-Fantasy
(15)	10.	Maybe Tomorrow-Jackson 5-Motown
(13)	11.	Rings-Cymarron-Entrance
(4)	12.	How Can You Mend A Broken Heart-Bea Gees-Atco
(18)	13.	Wedding Song (There Is Love)-Paul Stookey-WB
(19)	14.	Won't Get Fooled Again-The Who-Decca
(20)	15.	Liar-Three Dog Night-Dunhill
(10)	16.	Don't Pull Your Love-Hamilton, Joe Frank & Reynolds-Dunhill
(28)	*17.	Uncle Albert/Admiral Halsey-Paul McCartney-Apple
(12)	18.	Beginnings-Chicago-Columbia
(22)	19.	Mother Freedom-Bread-Elektra
(26)	*20.	Resurrection Shuffle-Ashten, Gardner & Dyke-Capitol
(30)	*21.	Spanish Harlem-Aretha Franklin-Atlantic
(14)	22.	Funky Nassau-Beginning Of The End-Alston
(17)	23.	Draggin The Line-Tommy James-Roulette
(18)	24.	It's Too Late-Carole King-Ode 70
(31)	*25.	Whatcha See Is Whatcha Get-Dramatics-Volt
(21)	26.	Rainy Jane-Davy Jones-Bell
(37)	*27.	Ain't No Sunshine-Bill Withers-Sussex
(32)	*28.	Roll On-The New Colony Six-Sunlight
(23)	29.	Summer Sand-Dawn-Bell
(24)	30.	Mighty Clouds Of Joy-B.J. Thomas-Scepter
(36)	31.	I Just Want To Celebrate-Rare Earth-Rare Earth
(38)	32.	Go Away Little Girl-Donny Osmond-MGM
(25)	33.	Walk Away-James Gang-ABC
(39)	34.	Bangla-Dash-George Harrison-Apple
(40)	35.	Woke Up In Love This Morning-Partridge Family-Bell
(27)	36.	You've Got A Friend-James Taylor-WB
(29)	37.	High Time We Went-Joe Cocker-A&M
(HB)	38.	Stagger Lee-Tommy Roe-ABC
(HB)	39.	I Ain't Got Time Anymore-The Glass Bottle-Avco
(HB)	40.	The Night They Drove Old Dixie Down-Joan Baez-Vanguard

*Denotes Big 10 Climbers

WC FL **ALL HIT MUSIC** **IN THE MIDWEST**

WAS	IS	Week Ending August 19, 1971
(5)	1.	Smiling Faces Sometimes-Undisputed Truth-Gordy
(4)	2.	Take Me Home-Country Roads-John Denver-RCA
(10)	*3.	Maybe Tomorrow-Jackson 5-Motown
(11)	*4.	Rings-Cymarron-Entrance
(1)	5.	Never Ending Song Of Love-Delaney & Bonnie-Atco
(6)	6.	Mercy Mercy Me-Marvin Gaye-Tamla
(13)	*7.	Wedding Song (There Is Love)-Paul Stookey-WB
(2)	8.	Riders On The Storm-The Doors-Elektra
(14)	9.	Won't Get Fooled Again-The Who-Decca
(17)	*10.	Uncle Albert/Admiral Halsey-Paul McCartney-Apple
(3)	11.	Mr. Big Stuff-Jean Knight-Stax
(20)	*12.	Resurrection Shuffle-Ashton, Gardner Dyke-Cap.
(7)	13.	Moon Shadow-Cat Stevens-A & M
(15)	14.	Liar-Three Dog Night-Dunhill
(21)	*15.	Spanish Harlem-Aretha Franklin-Atlantic
(9)	16.	Sweet Hitch-Hiker-C.C. Revival-Fantasy
(19)	17.	Mother Freedom-Bread-Elektra
(25)	*18.	Whatcha See Is Whatcha Get-Dramatics-Volt
(27)	*19.	Ain't No Sunshine-Bill Withers-Sussex
(8)	20.	Indian Reservation-The Raiders-Columbia
(12)	21.	How Can You Mend A Broken Heart-Bee Gees-Atco
(32)	*22.	Go Away Little Girl-Donny Osmond-MGM
(31)	*23.	I Just Want To Celebrate-Rare Earth-Rare Earth
(28)	24.	Roll On-The New Colony Six-Sunlight
(16)	25.	Don't Pull Your Love-H, J.F. & Reynolds-Dunhill
(40)	*26.	The Night They Drove Old Dixie Down-Joan Baez-Vanguard
(18)	27.	Beginnings-Chicago-Columbia
(35)	*28.	Woke Up In Love This Morning-Partridge Family-Bell
(22)	29.	Funky Nassau-Beginning Of The End-Alston
(34)	30.	Bangla-Desh-George Harrison-Apple
(23)	31.	Draggin' The Line-Tommy James-Roulette
(38)	32.	Stagger Lee-Tommy Roe-ABC
(39)	33.	I Ain't Got Time Anymore-The Glass Bottle-Avco
(24)	34.	It's Too Late-Carole King-Ode 70
(29)	35.	Summer Sand-Dawn-Bell
(26)	36.	Rainy Jane-Davy Jones-Bell
(30)	37.	Mighty Clouds of Joy-B.J. Thomas-Scepter

BIG 10 ALBUMS

1. Ram-Paul & Linda McCartney-Apple
2. Tapestry-Carole King-Ode 70
3. Stephen Stills #2-Stephen Stills-Atlantic
4. Carpenters-Carpenters-A & M
5. Mud Slide Slim-James Taylor-WB
6. Blood, Sweat & Tears #4-Blood, Sweat & Tears-Columbia
7. Sticky Fingers-Rolling Stones-Atco
8. What's Going On-Marvin Gaye-Tamla
9. Who's Next-The Who-Decca
10. Aqualung-Jethro Tull-Reprise

SMALL PRINT DEPARTMENT: The WCFL "ALL HIT MUSIC" represents the relative popularity of records in the Midwest based on Industry tabulations, local record sales and requests. WCFL The Voice of Labor, is o

BIG 10 REQUEST GOLD

* Groovin'-Young Rascals-(1967)
* A Different Drum-Stone Poneys-(1968)
* House Of The Rising Sun-Animals-(1964)
* My Pledge Of Love-Joe Jeffrey Group-(1969)
* Vehicle-Ides Of March-(1970)
* Mony, Mony-Tommy James & The Shondells-(1968)
* Spill The Wine-Eric Burdon & War-(1970)
* Purple Haze-Jimi Hendrix-(1967)
* Easy Come, Easy Go-Bobby Sherman-(1970)
* Come Together-Beatles-(1969)

*denotes currently available as gold single or on an album.

Take a Trip Back in Time with DICK BIONDI – "IN THE BEGINNING"-Tuesdays-1 AM!

(HB)	38.	Marianne-Stephen Stills-Atlantic	
(HB)	39.	I've Found Someone Of My Own- The Free Movement-Decca	
(HB)	40.	Take Me Girl, I'm Ready-Jr. Walker/ All Stars-Soul	

*Denotes Big 10 Climbers

WCFL **ALL HIT MUSIC** **IN THE MIDWEST**

Week Ending August 26, 1971

BIG 10 ALBUMS

1. Carpenters-Carpenters-A & M
2. Every Picture Tells A Story-Rod Stewart-Mercury
3. Tapestry-Carole King-Ode 70
4. Ram-Paul & Linda McCartney-Apple
5. Every Good Boy Deserves Favour-Moody Blues-London
6. Who's Next-The Who-Decca
7. Mud Slide Slim-James Taylor-WB
8. B, S & T #4-Blood, Sweat & Tears-Columbia
9. Stephen Stills #2-Stephen Stills-Atlantic
10. Blue-Joni Mitchell-Reprise

COMING NEXT WEEK:
A look inside the BIG TEN Plaything!

SMALL PRINT DEPARTMENT: The WCFL "ALL HIT MUSIC" represents the relative popularity of records in the Midwest based on Industry tabulations, local record sales and requests. WCFL The Voice of Labor, is o

BIG 10 REQUEST GOLD

* In The Summertime-Mungo Jerry-(1970)
* I Can't Get No Satisfaction-Rolling Stones-(1965)
* Love Grows Where My Rosemary Goes-Edison Lighthouse-(1970)
* Aquarius-5th Dimension-(1969)
* Sunshine Of Your Love-Cream-(1968)
* Superstar-Murray Head-(1970)
* Yellow Submarine-Beatles-(1968)
* I Started A Joke-Bee Gees-(1969)
* Goin' Up The Country-Canned Heat-(1969)

Deep Purple-Nino Temple/April Stevens-(1963)

WAS	IS	
(10)	1.	Uncle Albert/Admiral Halsey-Paul McCartney-Apple
(3)	2.	Maybe Tomorrow-Jackson 5-Motown
(4)	3.	Rings-Cymarron-Entrance
(7)	4.	Wedding Song (There Is Love)-Paul Stookey-WB
(1)	5.	Smiling Faces Sometimes-Undisputed Truth-Gordy
(9)	6.	Won't Get Fooled Again-The Who-Decca
(15)	*7.	Spanish Harlem-Aretha Franklin-Atlantic
(19)	*8.	Ain't No Sunshine-Bill Withers-Sussex
(2)	9.	Take Me Home, Country Roads-John Denver-RCA
(12)	10.	Resurrection Shuffle-Ashton, Gardner & Dyke-Cap.
(14)	11.	Liar-Three Dog Night-Dunhill
(17)	12.	Mother Freedom-Bread-Elektra
(6)	13.	Mercy Mercy Me-Marvin Gaye-Tamla
(22)	*14.	Go Away Little Girl-Donny Osmond-MGM
(23)	*15.	I Just Want To Celebrate-Rare Earth-Rare Earth
(5)	16.	Never Ending Song Of Love-Delaney & Bonnie-Atco
(18)	17.	Whatcha See Is Whatcha Get-Dramatics-Volt
(26)	*18.	Night They Drove Old Dixie-Joan Baez-Vanguard
(28)	*19.	Woke Up In Love This Morning-Partridge Family-Bell
(24)	20.	Roll On-The New Colony Six-Sunlight
(8)	21.	Riders On The Storm-The Doors-Elektra
(13)	22.	Moon Shadow-Cat Stevens-A & M
(16)	23.	Sweet Hitch-Hiker-C.C. Revival-Fantasy
(11)	24.	Mr. Big Stuff-Jean Knight-Stax
(30)	25.	Bangla-Desh-George Harrison-Apple
(20)	26.	Indian Reservation-The Raiders-Columbia
(33)	*27.	I Ain't Got Time Anymore-The Glass Bottle-Avco
(32)	28.	Stagger Lee-Tommy Roe-ABC
(21)	29.	How Can You Mend A Broken Heart-Bee Gees-Atco
(38)	*30.	Marianne-Stephen Stills-Atlantic
(25)	31.	Don't Pull Your Love-H., J.F. & Reynolds-Dunhill
(40)	32.	Take Me Girl, I'm Ready-Jr. Walker/All Stars-Soul

*denotes currently available as gold single or on an album.

(27)	33.	Beginnings-Chicago-Columbia
(39)	34.	I've Found Someone Of My Own-The Free Movement-Decca
(29)	35.	Funky Nassau-Beginning Of The End-Alston
(31)	36.	Draggin' The Line-Tommy James-Roulette
(HB)	37.	The Story In Your Eyes-The Moody Blues-Threshold
(HB)	38.	If You Really Love Me-Stevie Wonder-Tamla
(HB)	39.	He'd Rather Have The Rain-Heaven Bound-MGM
(HB)	40.	Sweet City Woman-Stampeders-Bell

*Denotes Big 10 Climbers

WC
FL

ALL HIT MUSIC

IN THE
MIDWEST

WAS	IS	Week End September 2, 1971
(1)	1.	Uncle Albert/Admiral Halsey-Paul McCartney-Apple
(2)	2.	Maybe Tomorrow-Jackson 5-Motown
(4)	3.	Wedding Song (There Is Love-Pau Stockey - WBl
(7)	4.	Spanish Harlem-Aretha Franklin-Atlantic
(6)	5.	Won't Get Fooled Again-The Who-Decca
(8)	6.	Ain't No Sunshine-Bill Withers-Sussex
(3)	7.	Rings-Cymarron-Entrance
(10)	8.	Resurrection Shuffle-Ashton, Gardner & Dyke-Cap.
(14)	*9.	Go Away Little Girl-Donny Osmond-MGM
(5)	10.	Smiling Faces Sometimes-Undisputed Truth-Gordy
(18)	*11.	Night They Drove Old Dixie-Joan Baez-Vanguard
(15)	12.	I Just Want To Celebrate-Rare Earth-Rare Earth
(17)	13.	Whatcha See Is Whatcha Get-Dramatics-Volt
(9)	14.	Take Me Home, Country Roads-John Denver-RCA
(19)	15.	Woke Up In Love This Morning-Partridge Family-Bell
(20)	16.	Roll On-New Colony Six-Sunlight
(11)	17.	Liar-Three Dog Night-Dunhill
(12)	18.	Mother Freedom-Bread-Elektra
(13)	19.	Mercy Mercy Me-Marvin Gaye-Tamla
(25)	20.	Bangla-Desh-George Harrison-Apple
(28)	*21.	Stagger Lee-Tommy Roe-ABC
(27)	22.	I Ain't Got Time Anymore-The Glass Bottle-Avco
(30)	*23.	Marianne-Stephen Stills-Atlantic
(16)	24.	Never Ending, Song Of Love-Delaney & Bonnie-Atco
(32)	*25.	Take Me Girl, I'm Ready-Jr. Walker/ All Stars-Soul
(40)	*26.	Sweet City Woman-Stampeders-Bell
(23)	27.	Sweet Hitch-Hiker-C.C. Revival-Fantasy
(21)	28.	Riders On The Storm-The Doors-Elektra
(34)	*29.	I've Found Someone Of My Own-The Free Movement-Decca
(22)	30.	Moon Shadow-Cat Stevens-A & M
(37)	31.	The Story In Your Eyes-The Moody Blues-Threshold
(38)	32.	If You Really Love Me-Stevie Wonder-Tamla
(26)	33.	Indian Reservation-The Raiders-Columbia
(31)	34.	Don't Pull Your Love-H., J.F. & Reynolds-Dunhill
(29)	35.	How Can You Mend A Broken Heart-Bee Gees-Atco

BIG 10 ALBUMS

1. Tapestry-Carole King-Ode 70
2. Every Picture Tells A Story-Rod Stewart-Mercury
3. Carpenters-Carpenters-A & M
4. Ram-Paul & Linda McCartney-Apple
5. Who's Next-The Who-Decca
6. Every Good Boy Deserves Favour-Moody Blues-London
7. What's Going On-Marvin Gaye-Tamla
8. Stephen Stills #2-Stephen Stills-Atlantic
9. Mud Slide Slim-James Taylor-WB
10. "Shaft" Sound Track-Isaac Hayes-Enterprise

SMALL PRINT DEPARTMENT: The WCFL "ALL HIT MUSIC" represents the relative popularity of records in the Midwest based on Industry tabulations, local record sales and requests. WCFL The Voice of Labor, is o

BIG 10 REQUEST GOLD

* Let It Be-The Beatles-(1970)
* The Last Time-The Rolling Stones-(1965)
* Sky Pilot-Eric Burdon & The Animals-(1968)
* Louie, Louie-The Kingsmen-(1963)
* Bad Moon Rising-Credence Clearwater Revival-(1969)
* To Sir With Love-Lulu-(1967)
* Mellow Yellow-Donovan-(1966)
* You Talk Too Much-Joe Jones-(1960)
* I Can See For Miles-The Who-(1967)
* Lay Down (Candles In The Rain)-Melanie-(1970)

*denotes currently available as gold single or on an album.

Chicago's ORIGINAL Golden Weekend begins every Friday on BIG 10!

(39)	36.	He'd Rather Have The Rain-Heaven Bound-MGM
(24)	37.	Mr. Big Stuff-Jean Knight-Stax
(HB)	38.	Do You Know What I Mean-Lee Michaels-A & M
(HB)	39.	MacArthur Park-Four Tops-Motown
(HB)	40.	Talk It Over In The Morning-Anne Murray-Capitol

*Denotes Big 10 Climbers

WC
FL

ALL HIT MUSIC

IN THE
MIDWEST

BIG 10 ALBUMS

1. Tapestry-Carole King-Ode 70
2. Every Good Boy Deserves Favour-Moody Blues-London
3. Every Picture Tells A Story-Rod Stewart-Mercury
4. Carpenters-Carpenters-A & M
5. Ram-Paul & Linda McCartney-Apple
6. Who's Next-The Who-Decca
7. Stephen Stills #2-Stephen Stills-Atlantic
8. Mud Slide Slim-James Taylor-WB
9. "Shalt" Sound Track-Isaac Hayes-Enterprise
10. Master of Reality-Black Sabbath-WB

SMALL PRINT DEPARTMENT: The WCFL "ALL HIT MUSIC" represents the relative popularity of records in the Midwest based on Industry tabulations, local record sales and requests. WCFL The Voice of Labor, is o

BIG 10 REQUEST GOLD

- * Make Me Smile-Chicago-(1970)
- * I Can See For Miles-The Who-(1967)
- * In The Summertime-Mungo Jerry-(1970)
- * Bridge Over Troubled Water-Simon & Garfunkel-(1970)
- * Lay Down (Candles In The Rain)-Melanie-(1970)
- * Cracklin' Rosie-Neil Diamond-(1970)
- * How Can I Be Sure-Young Rascals-(1967)
- * Windy-The Association-(1967)
- * Get A Job-The Silhouettes-(1958)
- * Easy Come, Easy Go-Bobby Sherman-(1970)

*denotes currently available as gold single or on an album.

WAS	IS	Week Ending September 9, 1971
(1)	1.	Uncle Albert/Admiral Halsey-Paul McCartney-Apple
(3)	2.	Wedding Song (There Is Love)-Paul Stookey-WB
(4)	3.	Spanish Harlem-Aretha Franklin-Atlantic
(5)	4.	Won't Get Fooled Again-The Who-Decca
(6)	5.	Ain't No Sunshine-Bill Withers-Sussex
(9)	6.	Go Away Little Girl-Donny Osmond-MGM
(11)	7.	Night They Drove Old Dixie-Joan Baez-Vanguard
(12)	8.	I Just Want To Celebrate-Rare Earth-Rare Earth
(2)	9.	Maybe Tomorrow-Jackson 5-Motown
(15)	10.	Woke Up In Love This Morning-Partridge Family-Bell
(17)	*11.	Liar-Three Dog Night-Dunhill
(13)	12.	Whatcha See Is Whatcha Get-Dramatics-Volt
(7)	13.	Rings-Cymarron-Entrance
(23)	*14.	Marianne-Stephen Stills-Atlantic
(16)	15.	Roll On-New Colony Six-Sunlight
(8)	16.	Resurrection Shuffle-Ashton, Gardner & Dyke-Cap.
(20)	17.	Bangla-Desh-George Harrison-Apple
(22)	18.	I Ain't Got Time Anymore-The Glass Bottle-Avco
(21)	19.	Stagger Lee-Tommy Roe-ABC
(10)	20.	Smiling Faces Sometimes-Undisputed Truth-Gordy
(26)	21.	Sweet City Woman-Stampeders-Bell
(25)	22.	Take Me Girl, I'm Ready-Jr. Walker/All Stars-Soul
(29)	*23.	I've Found Someone Of My Own-The Free Movement-Decca
(18)	24.	Mother Freedom-Bread-Elektra
(31)	*25.	The Story In Your Eyes-The Moody Blues-Threshold
(32)	*26.	If You Really Love Me-Stevie Wonder-Tamla
(14)	27.	Take Me Home, Country Roads-John Denver-RCA
(19)	28.	Mercy Mercy Me-Marvin Gaye-Tamla
(36)	*29.	He'd Rather Have The Rain-Heaven Bound-MGM
(24)	30.	Never Ending Song Of Love-Delaney & Bonnie-Atco
(27)	31.	Sweet Hitch-Hiker-C.C. Revival-Fantasy
(39)	32.	MacArthur Park-Four Tops-Motown
(28)	33.	Riders On The Storm-The Doors-Elektra
(38)	34.	Do You Know What I Mean-Lee Michaels-A & M

(40)	35.	Talk It Over In The Morning-Anne Murray-Capitol
(HB)	36.	Stick-Up-Honey Cone-Hot Wax
(HB)	37.	Maggie May-Rod Stewart-Mercury
(HB)	38.	Yo-Yo-The Osmonds-MGM
(HB)	39.	Superstar-Carpenters-A & M
(HB)	40.	Charity Ball-Fanny-Reprise

*Denotes Big 10 Climbers

WC
FL

ALL HIT MUSIC

IN THE
MIDWEST

BIG 10 ALBUMS

1. Every Good Boy Deserves Favour-Moody Blues-London
2. Every Picture Tells A Story-Rod Stewart-Mercury
3. Ram-Paul & Linda McCartney-Apple
4. Tapestry-Carole King-Ode 70
5. Carpenters-Carpenters-A & M
6. Master of Reality-Black Sabbath-WB
7. Stephen Stills #2-Stephen Stills-Atlantic
8. "Shaft" Sound Track-Isaac Hayes-Enterprise
9. Who's Next-The Who-Decca
10. Sound Magazine-The Partridge Family-Bell

SMALL PRINT DEPARTMENT: The WCFL "ALL HIT MUSIC" represents the relative popularity of records in the Midwest based on Industry tabulations, local record sales and requests. WCFL The Voice of Labor, is o

BIG 10 REQUEST GOLD

* Spinning Wheel-Blood, Sweat & Tears-(1969)
* Pinball Wizard-The Who-(1969)
* Candida-Dawn-(1970)
* Easy To Be Hard-3 Dog Night-(1969)
Rip It Up-Little Richard-(1956)
* Hey Jude-The Beatles-(1968)
* Magic Carpet Ride-Steppenwolf-(1968)
* One Bad Apple-The Osmonds-(1970)
* Light My Fire-The Doors-(1967)
Golden Days-Sally Field-(1966)

*denotes currently available as gold single or on an album.

HIT or NEW...It's For You...
222-1650!

WAS	IS	Week Ending September 16, 1971
(1)	1.	Uncle Albert/Admiral Halsey-Paul McCartney-Apple
(2)	2.	Wedding Song (There Is Love)-Paul Stookey-WB
(6)	3.	Go Away Little Girl-Donny Osmond-MGM
(7)	4.	Night They Drove Old Dixie-Joan Baez-Vanguard
(10)	*5.	Woke Up In Love This Morning-Partridge Family-Bell
(8)	6.	I Just Want To Celebrate-Rare Earth-Rare Earth
(14)	*7.	Marianne-Stephen Stills-Atlantic
(12)	8.	Whatcha See Is Whatcha Get-Dramatics-Volt
(21)	*9.	Sweet City Woman-Stampeders-Bell
(15)	*10.	Roll On-New Colony Six-Sunlight
(11)	11.	Liar-Three Dog Night-Dunhill
(3)	12.	Spanish Harlem-Aretha Franklin-Atlantic
(19)	13.	Stagger Lee-Tommy Ree-ABC
(4)	14.	Won't Get Fooled Again-The Who-Decca
(17)	15.	Bangla-Desh-George Harrison-Apple
(5)	16.	Ain't No Sunshine-Bill Withers-Sussex
(18)	17.	I Ain't Got Time Anymore-The Glass Bottle-Avco
(22)	18.	Take Me Girl, I'm Ready-Jr. Walker/All Stars-Soul
(9)	19.	Maybe Tomorrow-Jackson 5-Motown
(23)	20.	Found Someone Of My Own-Free Movement-Decca
(13)	21.	Rings-Cymarron-Entrance
(25)	22.	The Story In Your Eyes-The Moody Blues-Threshold
(26)	23.	If You Really Love Me-Stevie Wonder-Tamla
(16)	24.	Resurrection Shuffle-Ashton, Gardner & Dyke-Cap.
(29)	25.	He'd Rather Have The Rain-Heaven Bound-MGM
(37)	*26.	Maggie May-Rod Stewart-Mercury
(20)	27.	Smiling Faces Sometimes-Undisputed Truth-Gordy
(34)	*28.	Do You Know What I Mean-Lee Michaels-A & M
(36)	*29.	Stick-Up-Honey Cone-Hot Wax
(39)	*30.	Superstar-Carpenters-A & M
(32)	31.	MacArthur Park-Four Tops-Motown
(35)	32.	Talk It Over In The Morning-Anne Murray-Capitol
(38)	33.	Yo-Yo-The Osmonds-MGM
(24)	34.	Mother Freedom-Bread-Elektra
(40)	35.	Charity Ball-Fanny-Reprise
(27)	36.	Take Me Home, Country Roads-John Denver-RCA

(28)	37.	Mercy Mercy Me-Marvin Gaye-Tamla
(HB)	38.	I'm Comin' Home-Tommy James-Roulette
(HB)	39.	Birds Of A Feather-Raiders-Columbia
(HB)	40.	Only You Know And I Know-Delaney & Bonnie-Atco

*Denotes Big 10 Climbers

WC
FL

ALL HIT MUSIC

IN THE
MIDWEST

BIG 10 ALBUMS

1. Every Picture Tells A Story-Rod Stewart-Mercury
2. Every Good Boy Deserves Favour-Moody Blues-London
3. Ram-Paul & Linda McCartney-Apple
4. Tapestry-Carole King-Ode 70
5. "Shaft" Sound Track-Isaac Hayes-Enterprise
6. Who's Next-The Who-Decca
7. Carpenters-Carpenters-A & M
8. Master of Reality-Black Sabbath-WB
9. Bark-Jefferson Airplane-Grunt
10. Sound Magazine-Partridge Family-Bell

DICK BIONDI talks to the HITMAKERS every Sunday night at 10 on "DICK BIONDI AND FRIEND"

SMALL PRINT DEPARTMENT: The WCFL "ALL HIT MUSIC" represents the relative popularity of records in the Midwest based on Industry tabulations, local record sales and requests. WCFL The Voice of Labor, is o

BIG 10 REQUEST GOLD

* Another Day-Paul McCartney-(1971)
* Somebody Help Me-Spencer Davis Group-(1967)
* Come Together-The Beatles-(1965)
* Somebody to Love-Jefferson Airplane-(1967)
Overture From Tommy-Assembled Multitude-(1970)
* She's My Girl-The Turtles-(1967)
* Magic Bus-The Who-(1968)
Transfusion-Nervous Norvus-(1956)
* Georgy Girl-Seekers-(1967)
* Not Fade Away-Rolling Stones-(1964)

WAS	IS	Week Ending September 23, 1971
(3)	1.	Go Away Little Girl-Donny Osmond-MGM
(4)	2.	Night They Drove Old Dixie-Joan Baez-Vanguard
(5)	3.	Woke Up In Love This Morning-Partridge Family-Bell
(6)	4.	I Just Want To Celebrate-Rare Earth-Rare Earth
(9)	5.	Sweet City Woman-Stampeders-Bell
(1)	6.	Uncle Albert/Admiral Halsey-Paul McCartney-Apple
(7)	7.	Marianne-Stephen Stills-Atlantic
(2)	8.	Wedding Song-Paul Stookey-WB
(26)	*9.	Maggie May-Rod Stewart-Mercury
(10)	10.	Roll On-New Colony Six-Sunlight
(22)	*11.	The Story In Your Eyes-The Moody Blues-Threshold
(8)	12.	Whatcha See Is Whatcha Get-Dramatics-Volt
(13)	13.	Stagger Lee-Tommy Roe-ABC
(20)	*14.	Found Someone of My Own-Free Movement-Decca
(18)	15.	Take Me Girl, I'm Ready-Jr. Walker/All Stars-Soul
(28)	*16.	Do You Know What I Mean-Lee Michaels-A & M
(17)	17.	I Ain't Got Time Anymore-The Glass Bottle-Avco
(23)	*18.	If You Really Love Me-Stevie Wonder-Tamla
(11)	19.	Liar-Three Dog Night-Dunhill
(25)	*20.	He'd Rather Have The Rain-Heaven Bound-MGM
(30)	*21.	Superstar-Carpenters-A & M
(12)	22.	Spanish Harlem-Aretha Franklin-Atlantic
(29)	*23.	Stick-Up-Honey Cone-Hot Wax
(16)	24.	Ain't No Sunshine-Bill Withers-Sussex
(33)	*25.	Yo-Yo-The Osmonds-MGM
(14)	26.	Won't Get Fooled Again-The Who-Decca
(19)	27.	Maybe Tomorrow-Jackson 5-Motown
(31)	28.	MacArthur Park-Four Tops-Motown
(32)	29.	Talk It Over In The Morning-Anne Murray-Capitol
(15)	30.	Bangla-Desh-George Harrison-Apple
(35)	31.	Charity Ball-Fanny-Reprise
(38)	32.	I'm Comin' Home-Tommy James-

*denotes currently available as gold single or on an album.

		Roulette
(21)	33.	Rings-Cymarron-Entrance
(39)	34.	Birds Of A Feather-Raiders-Columbia
(40)	35.	Only You Know And I Know-Delaney & Bonnie-Atco
(24)	36.	Resurrection Shuffle-Ashton, Gardner, & Dyke-Cap.
(HB)	37.	Gypsys, Tramps, & Thieves-Cher-Kapp
(HB)	38.	Peace Train-Cat Stevens-A & M
(HB)	39.	What Are You Doing Sunday-Dawn-Bell
(HB)	40.	Tired Of Being Alone-Al Green-HI

*Denotes Big 10 Climbers

WC
FL

ALL HIT MUSIC IN THE MIDWEST

Week Ending September 30, 1971

WAS	IS	
(9)	1.	Maggie May-Rod Stewart-Mercury
(2)	2.	Night They Drove Old Dixie-Joan Baez-Vanguard
(5)	3.	Sweet City Woman-Stampeders-Bell
(7)	4.	Marianne-Stephen Stills-Atlantic
(1)	5.	Go Away Little Girl-Donny Osmond-MGM
(11)	6.	The Story In Your Eyes-The Moody Blue-Threshold
(16)	*7.	Do You Know What I Mean-Lee Michaels-A & M
(14)	*8.	Found Someone Of My Own-Free Movement-Decca
(11)	9.	The Story In Your Eyes-The Moody Blues-Threshold
(3)	10.	Woke Up In Love This Morning-Partridge Family-Bell
(21)	*11.	Superstar-Carpenters-A & M
(18)	*12.	If You Really Love Me-Stevie Wonder-Tamla
(10)	13.	Roll On-New Colony Six-Sunlight
(25)	*14.	Yo-Yo-The Osmonds-MGM
(4)	15.	I Just Want To Celebrate-Rare Earth-Rare Earth
(20)	16.	He'd Rather Have The Rain-Heaven Bound-MGM
(6)	17.	Uncle Albert/Admiral Halsey-Paul McCartney-Apple
(23)	*18.	Stick-Up-Honey Cone-Hot Wax
(8)	19.	Wedding Song-Paul Stookey-WB
(12)	20.	Whatcha See Is Whatcha Get-Dramatics-Volt
(31)	*21.	Charity Ball-Fanny-Reprise
(15)	22.	Take Me Girl, I'm Ready-Jr. Walker/All Stars-Soul
(17)	23.	I Ain't Got Time Anymore-The Glass Bottle-Avco
(32)	*24.	I'm Comin' Home-Tommy James-Roulette
(28)	25.	Mac Arthur Park-Four Tops-Motown
(35)	*26.	Only You Know And I Know-Delaney & Bonnie-Atco
(29)	27.	Talk It Over In The Morning-Anne Murray-Capitol
(34)	28.	Birds Of A Feather-Raiders-Columbia
(37)	29.	Gypsys, Tramps & Thieves-Cher-Kapp
(24)	30.	Ain't No Sunshine-Bill Withers-Sussex
(40)	31.	Tired Of Being Alone-Al Green-HI
(38)	32.	Peace Train-Cat Stevens-A & M
(39)	33.	What Are You Doing Sunday?-Dawn-Bell
(19)	34.	Liar-Three Dog Night-Dunhill
(22)	35.	Spanish Harlem-Aretha Franklin-Atlantic

BIG 10 ALBUMS

1. Every Picture Tells A Story-Rod Stewart-Mercury
2. Every Good Boy Deserves Favour-Moody Blues-London
3. "Shalt" Sound Track-Isaac Hayes-Enterprise
4. Tapestry-Carole King-Ode 70
5. Ram-Paul & Linda McCartney-Apple
6. Who's Next-The Who-Decca
7. Carpenters-Carpenters-A & M
8. Master of Reality-Black Sabbath-WB
9. Imagine-John Lennon-Apple
10. Bark-Jefferson Airplane-Grunt

SMALL PRINT DEPARTMENT: The WCFL "ALL HIT MUSIC" represents the relative popularity of records in the Midwest based on Industry tabulations, local record sales and requests. WCFL The Voice of Labor, is o

BIG 10 REQUEST GOLD

* Sweet Caroline-Neil Diamond-(1969)
* Bridge Over Troubled Water-Simon & Garfunkel-(1970)
* Lola-The Kinks-(1970)
* Somebody To Love-Jefferson Airplane-(1967)
* Sweet Soul Music-Arthur Conley-(1967)
* What's New Pussycat?-Tom Jones-(1965)
Liar, Liar-The Castaways-(1965)
* One Bad Apple-The Osmonds-(1970)
* Little Darlin'-The Diamonds-(1956)

*denotes currently available as gold single or on an album.

A Sure Cure for the Back-to-School Blues: LARRY O'BRIEN nightly 7-10

(HB)	36.	Love-The Lettermen-Capitol
(HB)	37.	Chirpy Chirpy Cheep Cheep-Mac & Katie Kissoon-ABC
(HB)	38.	Life Is A Carnival-The Band-Capitol
(HB)	39.	Inner City Blues-Marvin Gaye-Tamla
(HB)	40.	Your Move-Yes-Atlantic

*Denotes Big 10 Climbers

WC FL **ALL HIT MUSIC** **IN THE MIDWEST**

BIG 10 ALBUMS

1. Every Picture Tells A Story-Rod Stewart-Mercury
2. Tapestry-Carole King-Ode 70
3. Every Good Boy Deserves Favour-Moody Blues-London
4. Carpenters-Carpenters-A & M
5. Bark-Jefferson Airplane-Grunt
6. Imagine-John Lennon-Apple
7. Who's Next-The Who-Decca
8. Ram-Paul & Linda McCartney-Apple
9. Master Of Reality-Black Sabbath-WB
10. "Shaft" Sound Track-Isaac Hayes-Enterprise

Chicago's ORIGINAL Hitline...
222-1650...Yours For A Song

SMALL PRINT DEPARTMENT: The WCFL "ALL HIT MUSIC" represents the relative popularity of records in the Midwest based on Industry tabulations, local record sales and requests. WCFL The Voice of Labor, is o

BIG 10 REQUEST GOLD

- * I'm A Believer-The Monkees-(1967)
- Town Without Pity-Gene Pitney-(1962)
- * Help-The Beatles-(1965)
- * Get Off My Cloud-Rolling Stones-(1967)
- * Mellow Yellow-Donovan-(1968)
- * Close To You-The Carpenters-(1970)
- School Day-Chuck Berry-(1957)
- * She Loves You-The Beatles-(1964)
- Banana Boat Song-Stan Freberg-(1957)
- * Sweet And Innocent-Donny Osmond-(1971)

WAS	IS	Week Ending October 7, 1971
(1)	1.	Maggie May-Rod Stewart-Mercury
(3)	2.	Sweet City Woman-Stampeders-Bell
(6)	3.	The Story In Your Eyes-The Moody Blues-Threshold
(7)	4.	Do You Know What I Mean-Lee Michaels-A & M
(8)	5.	Found Someone Of My Own-Free Movement-Decca
(11)	6.	Superstar-Carpenters-A & M
(2)	7.	Night They Drove Old Dixie-Joan Baez-Vanguard
(12)	8.	If You Really Love Me-Stevie Wonder-Tamla
(14)	9.	Yo-Yo-The Osmonds-MGM
(4)	10.	Marianne-Stephen Stills-Atlantic
(5)	11.	Go Away Little Girl-Donny Osmond-MGM
(18)	*12.	Stick-Up-Honey Cone-Hot Wax
(21)	*13.	Charity Ball-Fanny-Reprise
(9)	14.	Stagger Lee-Tommy Roe-ABC
(10)	15.	Woke Up In Love This Morning-Partridge Family-Bell
(16)	16.	He'd Rather Have The Rain-Heaven Bound-MGM
(24)	*17.	I'm Comin' Home-Tommy James-Roulette
(29)	*18.	Gypsys, Tramps & Thieves-Cher-Kapp
(26)	*19.	Only You Know And I Know-Delaney & Bonnie-Atco
(28)	*20.	Birds Of A Feather-Raiders-Columbia
(13)	21.	Roll On-New Colony Six-Sunlight
(15)	22.	I Just Want To Celebrate-Rare Earth-Rare Earth
(32)	*23.	Peace Train-Cat Stevens-A & M
(25)	24.	Mac Arthur Park-Four Tops-Motown
(27)	25.	Talk It Over In The Morning-Anne Murray-Capitol
(31)	*26.	Tired Of Being Alone-Al Green-HI
(33)	*27.	What Are You Doing Sunday?-Dawn-Bell
(17)	28.	Uncle Albert/Admiral Halsey-Paul McCartney-Apple
(19)	29.	Wedding Song-Paul Stookey-WB
(22)	30.	Take Me Girl, I'm Ready-Jr. Walker/All Stars-Soul
(37)	31.	Chirpy Chirpy Cheep Cheep-Mac & Katie Kissoon-ABC
(39)	32.	Inner City Blues-Marvin Gaye-Tamla

*denotes currently available on a gold single or on an album.

(36)	33.	Love-The Lettermen-Capitol
(40)	34.	Your Move-Yes-Atlantic
(38)	35.	Life Is A Carnival-The Band-Capitol
(23)	36.	I Ain't Got Time Anymore-The Glass Bottle-Avco
(HB)	37.	Imagine-John Lennon-Apple
(HB)	38.	Wild Night-Van Morrison-WB
(HB)	39.	Absolutely Right-Five Man Electrical Band-Lionel
(HB)	40.	Bless You-Martha Reeves/ Vandellas-Gordy

*Denotes Big 10 Climbers

WC FL **ALL HIT MUSIC** **IN THE MIDWEST**

WAS IS Week Ending October 14, 1971

WAS	IS	
(1)	1.	Maggie May-Rod Stewart-Mercury
(4)	2.	Do You Know What I Mean-Lee Michaels-A & M
(6)	3.	Superstar-Carpenters-A & M
(5)	4.	Found Someone Of My Own-Free Movement-Decca
(9)	5.	Yo-Yo-The Osmonds-MGM
(8)	6.	If You Really Love Me-Stevie Wonder-Tamla
(2)	7.	Sweet City Woman-Stampeders-Bell
(13)	8.	Charity Ball-Fanny-Reprise
(12)	9.	Stick-Up-Honey Cone-Hot Wax
(18)	10.	Gypsys, Tramps & Thieves-Cher-Kapp
(3)	11.	The Story In Your Eyes-The Moody Blues-Threshold
(17)	*12.	I'm Comin' Home-Tommy James-Roulette
(7)	13.	Night They Drove Old Dixie-Joan Baez-Vanguard
(23)	*14.	Peace Train-Cat Stevens-A & M
(19)	*15.	Only You Know And I Know-Delaney & Bonnie-Atco
(20)	16.	Birds Of A Feather-Raiders-Columbia
(26)	*17.	Tired Of Being Alone-Al Green-HI
(27)	*18.	What Are You Doing Sunday?-Dawn-Bell
(31)	*19.	Chirpy Chirpy Cheep Cheep-Mac/ Katie Kissoon-ABC
(14)	20.	Stagger Lee-Tommy Roe-ABC
(10)	21.	Marianne-Stephen Stills-Atlantic
(32)	22.	Inner City Blues-Marvin Gaye-Tamla
(11)	23.	Go Away Little Girl-Donny Osmond-MGM
(33)	*24.	Love-The Lettermen-Capitol
(15)	25.	Woke Up In Love This Morning-Partridge Family-Bell
(34)	*26.	Your Move-Yes-Atlantic
(16)	27.	He'd Rather Have The Rain-Heaven Bound-MGM
(21)	28.	Roll On-New Colony Six-Sunlight
(35)	*29.	Life Is A Carnival-The Band-Capitol
(37)	*30.	Imagine-John Lennon-Apple
(38)	31.	Wild Night-Van Morrison-WB
(40)	32.	Bless You-Martha Reeves/ Vandellas-Gordy
(39)	33.	Absolutely Right-Five Man Electrical Band-Lionel
(24)	34.	Mac Arthur Park-Four Tops-Motown
(HB)	35.	One Fine Morning-Lighthouse-

BIG 10 ALBUMS

1. Every Picture Tells A Story-Rod Stewart-Mercury
2. Tapestry-Carole King-Ode 70
3. Imagine-John Lennon-Apple
4. Every Good Boy Deserves Favour-Moody Blues-London
5. Carpenters-Carpenters-A & M
6. Santana-Santana-Columbia
7. Teaser And The Firecat-Cat Stevens-A & M
8. Who's Next-The Who-Decca
9. Ram-Paul & Linda McCartney-Apple
10. "Shaft" Sound Track-Isaac Hayes-Enterprise

SMALL PRINT DEPARTMENT: The WCFL "ALL HIT MUSIC" represents the relative popularity of records in the Midwest based on Industry tabulations, local record sales and requests. WCFL The Voice of Labor, is o

BIG 10 REQUEST GOLD

* The Boxer-Simon & Garfunkel-(1969)
* Nowhere Man-The Beatles-(1966)
* People Got To Be Free-The Rascals-(1968)
* Did You Ever Have To Make Up Your Mind?-The Lovin' Spoonful-(1966)
* Love Is Here-The Supremes-(1967)
* Rainy Day Woman #12 & 35-Bob Dylan-(1966)
* Down In The Boondocks-Billy Joe Royal-(1965)
The Birds And The Bees-Jewel Akens-(1965)
Alley-Oop-Hollywood Argyles-(1960)
* Walk Right In-The Rooftop Singers-(1963)

*denotes currently available on a gold single or on an album.

Chicago's FIRST and ORIGINAL "ALL HIT MUSIC" survey...
available every Friday!

Evolution

(HB) 36. Easy Loving-Freddie Hart-Capitol

(HB) 37. Never My Love-The 5th Dimension-Bell

(HB) 38. An Old Fashioned Love Song-3 Dog Night-Dunhill

(HB) 39. Brand New Key-Melanie-Neighborhood

(HB) 40. Don't Wanna Live Inside Myself-Bee Bees-Atco

*Denotes Big 10 Climbers

WC
FL

ALL HIT MUSIC

IN THE
MIDWEST

WAS	IS	Week Ending October 21, 1971
(3)	1.	Superstar-Carpenters-A & M
(4)	2.	Found Someone Of My Own-Free Movement-Decca
(5)	3.	Yo-Yo-The Osmonds-MGM
(10)	*4.	Gypsys, Tramps & Thieves-Cher-Kapp
(8)	5.	Charity Ball-Fanny-Reprise
(1)	6.	Maggie May-Rod Stewart-Mercury
(2)	7.	Do You Know What I Mean-Lee Michaels-A & M
(12)	8.	I'm Comin' Home-Tommy James-Roulette
(14)	9.	Peace Train-Cat Stevens-A & M
(16)	10.	Birds Of A Feather-Raiders-Columbia
(6)	11.	If You Really Love Me-Stevie Wonder-Tamla
(19)	*12.	Chirpy Chirpy Cheep Cheep-Mac/ Katie Kissoon-ABC
(17)	13.	Tired Of Being Alone-Al Green-HI
(18)	14.	What Are You Doing Sunday?-Dawn-Bell
(15)	15.	Only You Know And I Know-Delaney & Bonnie-Atco
(7)	16.	Sweet City Woman-Stampeders-Bell
(22)	17.	Inner City Blues-Marvin Gaye-Tamla
(9)	18.	Stick-Up-Honey Cone-Hot Wax
(24)	*19.	Love-The Lettermen-Capitol
(11)	20.	The Story In Your Eyes-The Moody Blues-Threshold
(30)	*21.	Imagine-John Lennon-Apple
(26)	22.	Your Move-Yes-Atlantic
(13)	23.	Night They Drove Old Dixie-Joan Baez-Vanguard
(33)	*24.	Absolutely Right-Five Man Electrical Band-Lionel
(29)	25.	Life Is A Carnival-The Band-Capitol
(31)	*26.	Wild Night-Van Morrison-WB
(35)	*27.	One Fine Morning-Lighthouse-Evolution
(36)	*28.	Easy Loving-Freddie Hart-Capitol
(20)	29.	Stagger Lee-Tommy Roe-ABC
(32)	30.	Bless You-Martha Reeves/ Vandelias-Gordy
(37)	31.	Never My Love-The 5th Dimension-Bell
(21)	32.	Marianne-Stephen Stills-Atlantic
(39)	33.	Brand New Key-Melanie-Neighborhood
(23)	34.	Go Away Little Girl-Donny Osmond-MGM

BIG 10 ALBUMS

1. Imagine-John Lennon-Apple
2. Every Picture Tells A Story-Rod Stewart-Mercury
3. Tapestry-Carole King-Ode 70
4. Santana-Santana-Columbia
5. Teaser & The Firecat-Cat Stevens-A & M
6. "Shaft" Sound Track-Isaac Hayes-Enterprise
7. Every Good Boy Deserves Favour-Moody Blues-London
8. Carpenters-Carpenters-A & M
9. Who's Next-The Who-Decca
10. Ram-Paul & Linda McCartney-Apple

Solid HIT or Solid GOLD...
Lay it on the line...222-1650!

SMALL PRINT DEPARTMENT: The WCFL "ALL HIT MUSIC" represents the relative popularity of records in the Midwest based on Industry tabulations, local record sales and requests. WCFL The Voice of Labor, is o

BIG 10 REQUEST GOLD

* What The World Needs Now Is Love-Jackie DeShannon-(1965)
* Yesterday-The Beatles-(1965)
* My Sweet Lord-George Harrison-(1970)
* Everything That Touches You-Association-(1968)
* White Rabbit-Jefferson Airplane-(1967)
* Poor Side Of Town-Johnny Rivers-(1966)
* Wild Thing-The Troggs-(1966)
* Little Old Lady From Pasadena-Jan & Dean-(1964)
* ABC-Jackson 5-(1970)
Sugar Shack-Jimmy Gilmer-(1963)

*denotes currently available on a gold single or on an album.

(40)	35.	Don't Wanna Live Inside Myself-Bee Gees-Atco
(38)	36.	An Old Fashioned Love Song-3 Dog Night-Dunhill
(HB)	37.	Theme From "Shaft"-Isaac Hayes-Enterprise
(HB)	38.	If It's Alright With You-Rose Colored Glass-Bang
(HB)	39.	Baby I'm A-Want You-Bread-Elektra
(HB)	40.	Rock Steady-Aretha Franklin-Atlantic

*Denotes Big 10 Climbers

WC FL | **ALL HIT MUSIC** | **IN THE MIDWEST**

BIG 10 ALBUMS

1. Imagine-John Lennon-Apple
2. Every Picture Tells A Story-Rod Stewart-Mercury
3. Santana-Santana-Columbia
4. Tapestry-Carole King-Ode 70
5. "Shaft" Sound Track-Isaac Hayes-Enterprise
6. Teaser & The Firecat-Cat Stevens-A & M
7. Every Good Boy Deserves Favour-Moody Blues-London
8. Carpenters-Carpenters-A & M
9. Ram-Paul & Linda McCartney-Apple
10. Who's Next-The Who-Decca

SMALL PRINT DEPARTMENT: The WCFL "ALL HIT MUSIC" represents the relative popularity of records in the Midwest based on Industry tabulations, local record sales and requests. WCFL The Voice of Labor, is o

BIG 10 REQUEST GOLD

* California Nights-Leslie Gore-(1967)
* All Right Now-Free-(1970)
* Boy From New York City-The Ad-Libs-(1965)
* I Got You Babe-Sonny & Cher-(1965)
* I Am A Rock-Simon & Garfunkel-(1966)
Scratchy-Travis Wammack-(1964)
* 24 Hours From Tulsa-Gene Pitney-(1963)
* War-Edwin Starr-(1970)
Donna, The Prima Dona-Dion-(1963)
* Daydream-Lovin' Spoonful-(1966)

*denotes currently available on a gold single or on an album.

BOB DEARBORN features "long gold" nightly 10-1 AM!

WAS	IS	Week Ending October 28, 1971
(4)	1.	Gypsys, Tramps & Thieves-Cher-Kapp
(3)	2.	Yo-Yo-The Osmonds-MGM
(5)	3.	Charity Ball-Fanny-Reprise
(1)	4.	Superstar-Carpenters-A & M
(9)	5.	Peace Train-Cat Stevens-A & M
(2)	6.	Found Someone Of My Own-Free Movement-Decca
(8)	7.	I'm Comin' Home-Tommy James-Roulette
(10)	8.	Birds Of A Feather-Raiders-Columbia
(13)	9.	Tired Of Being Alone-Al Green-HI
(15)	10.	Only You Know And I Know-Delaney & Bonnie-Atco
(12)	11.	Chirpy Chirpy Cheep Cheep-Mac/ Katie Kissoon-ABC
(6)	12.	Maggie May-Rod Stewart-Mercury
(7)	13.	Do You Know What I Mean-Lee Michaels-A & M
(14)	14.	What Are You Doing Sunday?-Dawn-Bell
(21)	*15.	Imagine-John Lennon-Apple
(19)	16.	Love-The Lettermen-Capitol
(17)	17.	Inner City Blues-Marvin Gaye-Tamla
(22)	18.	Your Move-Yes-Atlantic
(11)	19.	If You Really Love Me-Stevie Wonder-Tamla
(25)	20.	Life Is A Carnival-The Band-Capitol
(24)	21.	Absolutely Right-Five Man Electrical Band-Lionel
(16)	22.	Sweet City Woman-Stampeders-Bell
(27)	23.	One Fine Morning-Lighthouse-Evolution
(26)	24.	Wild Night-Van Morrison-WB
(18)	25.	Stick-Up-Honey Cone-Hot Wax
(31)	*26.	Never My Love-The 5th Dimension-Bell
(28)	27.	Easy Loving-Freddie Hart-Capitol
(33)	*28.	Brand New Key-Melanie-Neighborhood
(30)	29.	Bless You-Martha Reeves/ Vandellas-Gordy
(37)	*30.	Theme From "Shaft"-Isaac Hayes-Enterprise
(20)	31.	The Story In Your Eyes-The Moody Blues-Threshold
(35)	32.	Don't Wanna Live Inside Myself-Bee Gees-Atco
(39)	33.	Baby I"m A-Want You-Bread-Elektra
(38)	34.	If It's Alright With You-Rose Colored Glass-Bang
(36)	35.	An Old Fashioned Love Song-3 Dog Night-Dunhill
(40)	36.	Rock Steady-Aretha Franklin-Atlantic

(32)	37.	Marianne-Stephen Stills-Atlantic
(HB)	38.	Got To Be There-Michael Jackson-Motown
(HB)	39.	All I Ever Need Is You-Sonny & Cher-Kapp
(HB)	40.	Do I Love You-Paul Anka-Buddah

*Denotes Big 10 Climbers

WC FL **ALL HIT MUSIC** **IN THE MIDWEST**

WAS IS Week Ending November 4, 1971

WAS	IS	Title
(1)	1.	Gypsys, Tramps & Thieves-Cher-Kapp
(2)	2.	Yo-Yo-The Osmonds-MGM
(5)	3.	Peace Train-Cat Stevens-A & M
(7)	4.	I'm Comin' Home-Tommy James-Roulette
(9)	5.	Tired Of Being Alone-Al Green-HI
(8)	6.	Birds Of A Feather-Raiders-Columbia
(3)	7.	Charity Ball-Fanny-Reprise
(15)	*8.	Imagine-John Lennon-Apple
(11)	9.	Chirpy Chirpy Cheep Cheep-Mac/Katie Kissoon-ABC
(10)	10.	Only You Know And I Know-Delaney & Bonnie-Atco
(4)	11.	Superstar-Carpenters-A & M
(16)	12.	Love-The Lettermen-Capitol
(14)	13.	What Are You Doing Sunday?-Dawn-Bell
(17)	14.	Inner City Blues-Marvin Gaye-Tamla
(18)	15.	Your Move-Yes-Atlantic
(6)	16.	Found Someone Of My Own-Free Movement-Decca
(21)	17.	Absolutely Right-Five Man Electrical Band-Lionel
(24)	*18.	Wild Night-Van Morrison-WB
(20)	19.	Life Is A Carnival-The Band-Capitol
(27)	*20.	Easy Loving-Freddie Hart-Capitol
(12)	21.	Maggie May-Rod Stewart-Mercury
(28)	*22.	Brand New Key-Melanie-Neighborhood
(30)	*23.	Theme From "Shaft"-Isaac Hayes-Enterprise
(13)	24.	Do You Know What I Mean-Lee Michaels-A & M
(23)	25.	One Fine Morning-Lighthouse-Evolution
(26)	26.	Never My Love-The 5th Dimension-Bell
(19)	27.	If You Really Love Me-Stevie Wonder-Tamla
(29)	28.	Bless You-Martha Reeves-Vandellas-Gordy
(33)	29.	Baby I'm A-Want You-Bread-Elektra
(32)	30.	Don't Wanna Live Inside Myself-Bee Gees-Atco
(35)	31.	An Old Fashioned Love Song-3 Dog Night-Dunhill
(34)	32.	If It's Alright With You-Rose Colored Glass-Bang
(36)	33.	Rock Steady-Aretha Franklin-Atlantic
(39)	34.	All I Ever Need Is You-Sonny & Cher-Kapp
(38)	35.	Got To Be There-Michael Jackson-Motown

BIG 10 ALBUMS

1. Santana-Santana-Columbia
2. Imagine-John Lennon-Apple
3. Every Picture Tells A Story-Rod Stewart-Mercury
4. Teaser & The Firecat-Cat Stevens-A & M
5. "Shaft" Sound Track-Isaac Hayes-Enterprise
6. Tapestry-Carole King-Ode 70
7. Carpenters-Carpenters-A & M
8. Every Good Boy Deserves Favour-Moody Blues-London
9. Who's Next-The Who-Decca
10. Bark-Jefferson Airplane-Grunt

SMALL PRINT DEPARTMENT: The WCFL "ALL HIT MUSIC" represents the relative popularity of records in the Midwest based on Industry tabulations, local record sales and requests. WCFL The Voice of Labor, is o

BIG 10 GHOUL GOLD

1. Monster Mash-Bobby (Boris) Pickett & The Crypt Kickers-(1958)
2. Haunted House-Johnny Fuller-(1965)
3. Witchcraft-Frank Sinatra-(1958)
4. Devil In Disguise-Elvis Presley-(1963)
5. Dinner With Dracula-John Zacherle-(1958)
6. Apples Peaches Pumpkin Pie-Jay & Techniques-(1967)
7. Spooky-Classics IV-(1968)
8. Dance By The Light Of The Moon-Olympics-(1961)
9. Ding, Dong The Witch Is Dead-Fifth Estate-(1965)
10. Monster-Steppenwolf-(1970)
11. The Cat In The Window-Petula Clark-(1967)
12. Purple People Eater-Sheb Wooley-(1958)
13. How'd We Ever Get This Way?-Andy Kim-(1968)

(40)	36.	Do I Love You-Paul Anka-Buddah
(22)	37.	Sweet City Woman-Stampeders-Bell
(HB)	38.	Stones-Neil Diamond-Uni
(HB)	39.	Have You Seen Her-Chi-Lites-Brunswick
(HB)	40.	Hallelujah-Sweathog-Columbia

*Denotes Big 10 Climbers

WC FL ALL HIT MUSIC **IN THE MIDWEST**

Week Ending November 11, 1971

WAS	IS	
(2)	1.	Yo-Yo-The Osmonds-MGM
(3)	2.	Peace Train-Cat Stevens-A & M
(8)	3.	Imagine-John Lennon-Apple
(9)	4.	Chirpy Chirpy Cheep Cheep-Mac/ Katie Kissoon-ABC
(1)	5.	Gypsys, Tramps & Thieves-Cher-Kapp
(10)	6.	Only You Know And I Know-Delaney & Bonnie-Atco
(13)	*7.	What Are You Doing Sunday?-Dawn-Bell
(14)	*8.	Inner City Blues-Marvin Gaye-Tamla
(17)	*9.	Absolutely Right-Five Man Electrical Band-Lionel
(15)	10.	Your Move-Yes-Atlantic
(4)	11.	I'm Comin' Home-Tommy James-Roulette
(12)	12.	Love-The Lettermen-Capitol
(18)	13.	Wild Night-Van Morrison-WB
(22)	*14.	Brand New Key-Melanie-Neighborhood
(23)	*15.	Theme From "Shaft"-Isaac Hayes-Enterprise
(5)	16.	Tired Of Being Alone-Al Green-HI
(19)	17.	Life Is A Carnival-The Band-Capitol
(6)	18.	Birds Of A Feather-Raiders-Columbia
(20)	19.	Easy Loving-Freddie Hart-Capitol
(7)	20.	Charity Ball-Fanny-Reprise
(29)	*21.	Baby I'm A-Want You-Bread-Elektra
(11)	22.	Superstar-Carpenters-A & M
(26)	23.	Never My Love-The 5th Dimension-Bell
(31)	*24.	An Old Fashioned Love Song-3 Dog Night-Dunhill
(33)	25.	Rock Steady-Aretha Franklin-Atlantic
(35)	*26.	Got To Be There-Michael Jackson-Motown
(30)	27.	Don't Wanna Live Inside Myself-Bee Gees-Atco
(28)	28.	Bless You-Martha Reeves/ Vandellas-Gordy
(25)	29.	One Fine Morning-Lighthouse-Evolution
(32)	30.	If It's Alright With You-Rose Colored Glass-Bang
(34)	31.	All I Ever Need Is You-Sonny & Cher-Kapp
(38)	32.	Stones-Neil Diamond-Uni
(39)	33.	Have You Seen Her-Chi-Lites-Brunswick
(36)	34.	Do I Love You-Paul Anka-Buddah
(16)	35.	Found Someone Of My Own-Free Movement-Decca
(40)	36.	Hallelujah-Sweathog-Columbia
(21)	37.	Maggie May-Rod Stewart-Mercury

BIG 10 ALBUMS

1. Imagine-John Lennon-Apple
2. Every Picture Tells A Story-Rod Stewart-Mercury
3. Santana-Santana-Columbia
4. "Shaft" Sound Track-Isaac Hayes-Enterprise
5. Teaser & The Firecat-Cat Stevens-A & M
6. Tapestry-Carole King- Ode 70
7. Carpenters-Carpenters-A & M
8. Ram-Paul & Linda McCartney-Apple
9. After All This Time-Bonnie Koloc-Ovation
10. Who's Next-The Who-Decca

SMALL PRINT DEPARTMENT: The WCFL "ALL HIT MUSIC" represents the relative popularity of records in the Midwest based on Industry tabulations, local record sales and requests. WCFL The Voice of Labor, is o

BIG 10 REQUEST GOLD

* War-Edwin Starr-(1970)

* Get Back-The Beatles-(1969)

* Close To You-The Carpenters-(1970)

* People Got To Be Free-The Rascals-(1968)

* Jean-Oliver-(1969)

* Remember Me-Diana Ross-(1971)

* Spirit In The Sky-Norman Greenbaum-(1970)

* Jailhouse Rock-Elvis Presley-(1957)

* Surfer Girl-Beach Boys-(1963)

Sandy-Ronny & The Daytonas-(1966)

*denotes currently available on a gold single or on an album.

BIG 10: Chicago's ONLY "Top 40" Station!

(HB) 38. Fool Me-Joe South-Capitol
(HB) 39. Cherish-David Cassidy-Bell
(HB) 40. Family Affair-Sly & The Family Stone-Epic

*Denotes Big 10 Climbers

WC FL — ALL HIT MUSIC — IN THE MIDWEST

WAS IS Week Ending November 18, 1971

WAS	IS	Title
(2)	1.	Peace Train-Cat Stevens-A & M
(3)	2.	Imagine-John Lennon-Apple
(9)	3.	Absolutely Right-Five Man Electrical Band-Lionel
(7)	4.	What Are You Doing Sunday?-Dawn-Bell
(10)	5.	Your Move-Yes-Atlantic
(15)	*6.	Theme From "Shaft"-Isaac Hayes-Enterprise
(1)	7.	Yo-Yo-The Osmonds-MGM
(8)	8.	Inner City Blues-Marvin Gaye-Tamla
(14)	9.	Brand New Key-Melanie-Neighborhood
(13)	10.	Wild Night-Van Morrison-WB
(21)	*11.	Baby I'm A-Want You-Bread-Elektra
(4)	12.	Chirpy Chirpy Cheep Cheep-Mac/Katie Kissoon-ABC
(19)	13.	Easy Loving-Freddie Hart-Capitol
(24)	*14.	An Old Fashioned Love Song-3 Dog Night-Dunhill
(5)	15.	Gypsys, Tramps & Thieves-Cher-Kapp
(6)	16.	Only You Know And I Know-Delaney & Bonnie-Atco
(26)	*17.	Got To Be There-Michael Jackson-Motown
(12)	18.	Love-The Lettermen-Capitol
(32)	*19.	Stones-Neil Diamond-Uni
(11)	20.	I'm Comin' Home-Tommy James-Roulette
(27)	21.	Don't Wanna Live Inside Myself-Bee Gees-Atco
(31)	*22.	All I Ever Need Is You-Sonny & Cher-Kapp
(30)	*23.	If It's Alright With You-Rose Colored Glass-Bang
(16)	24.	Tired Of Being Alone-Al Green-HI
(25)	25.	Rock Steady-Aretha Franklin-Atlantic
(33)	*26.	Have You Seen Her-Chi-Lites-Brunswick
(34)	*27.	Do I Love You-Paul Anka-Buddah
(18)	28.	Birds Of A Feather-Raiders-Columbia
(39)	*29.	Cherish-David Cassidy-Bell
(40)	*30.	Family Affair-Sly & The Family Stone-Epic
(17)	31.	Life Is A Carnival-The Band-Capitol
(20)	32.	Charity Ball-Fanny-Reprise
(36)	33.	Hallelujah-Sweathog-Columbia
(38)	34.	Fool Me-Joe South-Capitol

BIG 10 ALBUMS

1. Santana-Santana-Columbia
2. Imagine-John Lennon-Apple
3. Teaser & The Firecat-Cat Stevens-A & M
4. There's A Riot Goin' On-Sly & The Family Stone-Epic
5. Every Picture Tells A Story-Rod Stewart-Mercury
6. "Shaft" Sound Track-Isaac Hayes-Enterprise
7. Chicago At Carnegie Hall-Chicago-Columbia
8. After All This Time-Bonnie Koloc-Ovation
9. Carpenters-Carpenters-A & M
10. Tapestry-Carole King-Ode 70

SMALL PRINT DEPARTMENT: The WCFL "ALL HIT MUSIC" represents the relative popularity of records in the Midwest based on Industry tabulations, local record sales and requests. WCFL The Voice of Labor, is o

BIG 10 REQUEST GOLD

* Cherish-The Association-(1966)
* Kicks-Paul Revere & The Raiders-(1966)
* Rainy Days & Mondays-The Carpenters-(1971)
* Rag Doll-The Four Seasons-(1964)
Temma Harbour-Mary Hopkin-(1970)
Quick Joey Small-Kasenetz-Katz Singing Orchestral Circus-(1968)
* Hey Jude-The Beatles-(1968)
* MacArthur Park-Richard Harris-(1968)
* Lay Down (Candles In The Rain)-Melanie-(1970)
* Chick-A-Boom-Daddy Dewdrop-(1971)

*denotes currently available on a gold single or on an album.

(HB)	35.	Two Divided By Love-The Grass Roots-Dunhill
(HB)	36.	Sunshine-Jonathan Edwards-Capricorn
(HB)	37.	Long Time To Be Alone-The New Colony Six-Sunlight
(HB)	38.	Nothing To Hide-Tommy James-Roulette
(HB)	39.	Hey Girl-Donny Osmond-MGM
(HB)	40.	Respect Yourself-The Staple Singers-Stax

*Denotes Big 10 Climbers

WC
FL

ALL HIT MUSIC

IN THE
MIDWEST

Week Ending November 25, 1971

WAS	IS	
(6)	1.	Theme From "Shaft"-Isaac Hayes-Enterprise
(3)	2.	Absolutely Right-Five Man Electrical Band-Lionel
(4)	3.	What Are You Doing Sunday?-Dawn-Bell
(5)	4.	Your Move-Yes-Atlantic
(9)	5.	Brand New Key-Melanie-Neighborhood
(1)	6.	Peace Train-Cat Stevens-A & M
(11)	7.	Baby I'm A-Want You-Bread-Elektra
(10)	8.	Wild Night-Van Morrison-WB
(2)	9.	Imagine-John Lennon-Apple
(14)	10.	An Old Fashioned Love Song-3 Dog Night-Dunhill
(17)	*11.	Got To Be There-Michael Jackson-Motown
(19)	*12.	Stones-Neil Diamond-Uni
(7)	13.	Yo-Yo-The Osmonds-MGM
(22)	*14.	All I Ever Need Is You-Sonny & Cher-Kapp
(8)	15.	Inner City Blues-Marvin Gaye-Tamla
(23)	*16.	If It's Alright With You-Rose Colored Glass-Bang
(21)	17.	Don't Wanna Live Inside Myself-Bee Gees-Atco
(26)	*18.	Have You Seen Her-Chi-Lites-Brunswick
(29)	*19.	Cherish-David Cassidy-Bell
(30)	*20.	Family Affair-Sly & The Family Stone-Epic
(12)	21.	Chirpy Chirpy Cheep Cheep-Mac/Katie Kissoon-ABC
(25)	22.	Rock Steady-Aretha Franklin-Atlantic
(13)	23.	Easy Loving-Freddie Hart-Capitol
(27)	24.	Do I Love You-Paul Anka-Buddah
(18)	25.	Love-The Letterman-Capitol
(35)	*26.	Two Divided By Love-The Grass Roots-Dunhill
(16)	27.	Only You Know And I Know-Delaney & Bonnie-Atco
(33)	28.	Hallelujah-Sweathog-Columbia
(36)	29.	Sunshine-Jonathan Edwards-Capricorn
(15)	30.	Gypsys, Tramps & Thieves-Cher-Kapp
(34)	31.	Fool Me-Joe South-Capitol
(20)	32.	I'm Comin' Home-Tommy James-Roulette
(39)	33.	Hey Girl-Donny Osmond-MGM
(38)	34.	Nothing To Hide-Tommy James-Roulette

BIG 10 ALBUMS

1. Imagine-John Lennon-Apple
2. "Shaft" Sound Track-Isaac Hayes-Enterprise
3. Santana-Santana-Columbia
4. Teaser & The Firecat-Cat Stevens-A & M
5. There's A Riot Goin' On-Sly & The Family Stone-Epic
6. Every Picture Tells A Story-Rod Stewart-Mercury
7. Tapestry-Carole King-Ode 70
8. Chicago At Carnegie Hall-Chicago-Columbia
9. After All This Time-Bonnie Koloc-Ovation
10. Carpenters-Carpenters-A & M

SMALL PRINT DEPARTMENT: The WCFL "ALL HIT MUSIC" represents the relative popularity of records in the Midwest based on Industry tabulations, local record sales and requests. WCFL The Voice Of Labor, is o

BIG 10 REQUEST GOLD

- * La La La-Bobby Sherman-(1970)
- * (I Can't Get No) Satisfaction-Rolling Stones-(1970)
- * Woodstock-Crosby, Stills, Nash & Young-(1970)
- * Eli's Coming-Three Dog Night-(1969)
- * Love Child-Diana Ross & The Supremes-(1968)
- * My Pledge Of Love-Joe Jeffrey Group-(1969)
- * Peace Will Come-Melanie-(1970)
- * Atlantis-Donovan-(1969)
- * Ferry Cross The Mersey-Gerry & The Pacemakers-(1965)
- 5 O'Clock World-The Vogues-(1965)

*denotes currently available on a gold single or on an album.

50,000 Watts of all hit music,
24 hours a day

(37)	35.	Long Time To Be Alone-The New Colony Six-Sunlight
(40)	36.	Respect Yourself-The Staple Singers-Stax
(24)	37.	Tired Of Being Alone-Al Green-HI
(HB)	38.	Superstar-The Temptations-Gordy
(HB)	39.	Day After Day-Badfinger-Apple
(HB)	40.	White Lies, Blue Eyes-Bullet-Big Tree

*Denotes Big 10 Climbers

WC
FL

ALL HIT MUSIC

IN THE
MIDWEST

WAS	IS	Week Ending December 2, 1971
(1)	1.	Theme From "Shaft"-Isaac Hayes-Enterprise
(5)	2.	Brand New Key-Melanie-Neighborhood
(7)	3.	Baby I'm A-Want You-Bread-Elektra
(10)	*4.	An Old Fashioned Love Song-3 Dog Night-Dunhill
(11)	*5.	Got To Be There-Michael Jackson-Motown
(2)	6.	Absolutely Right-Five Man Electrical Band-Lionel
(8)	7.	Wild Night-Van Morrison-WB
(3)	8.	What Are You Doing Sunday?-Dawn-Bell
(14)	9.	All I Ever Need Is You-Sonny & Cher-Kapp
(12)	10.	Stones-Neil Diamond-Uni
(18)	*11.	Have You Seen Her-Chi-Lites-Brunswick
(20)	*12.	Family Affair-Sly & The Family Stone-Epic
(19)	*13.	Cherish-David Cassidy-Bell
(4)	14.	Your Move-Yes-Atlantic
(6)	15.	Peace Train-Cat Stevens-A & M
(26)	*16.	Two Divided By Love-The Grass Roots-Dunhill
(22)	17.	Rock Steady-Aretha Franklin-Atlantic
(9)	18.	Imagine-John Lennon-Apple
(24)	19.	Do I Love You-Paul Anka-Buddah
(16)	20.	If It's Alright With You-Rose Colored Glass-Bang
(29)	*21.	Sunshine-Jonathan Edwards-Capricorn
(34)	*22.	Nothing To Hide-Tommy James-Roulette
(33)	*23.	Hey Girl-Donny Osmond-MGM
(15)	24.	Inner City Blues-Marvin Gaye-Tamla
(28)	25.	Hallelujah-Sweathog-Columbia
(13)	26.	Yo-Yo-The Osmonds-MGM
(36)	*27.	Respect Yourself-The Staple Singers-Stax
(31)	28.	Fool Me-Joe South-Capitol
(35)	29.	Long Time To Be Alone-The New Colony Six-Sunlight
(38)	30.	Superstar-The Temptations-Gordy
(17)	31.	Don't Wanna Live Inside Myself-Bee Bees-Atco
(21)	32.	Chirpy Chirpy Cheep Cheep-Mac/Katie Kissoon-ABC
(25)	33.	Love-The Lettermen-Capitol
(39)	34.	Day After Day-Badfinger-Apple
(23)	35.	Easy Loving-Freddie Hart-Capitol

BIG 10 ALBUMS

1. Teaser & The Firecat-Cat Stevens-A & M
2. Santana-Santana-Columbia
3. Chicago At Carnegie Hall-Chicago-Columbia
4. There's A Riot Goin' On-Sly & The Family Stone-Epic
5. "Shaft" Sound Track-Isaac Hayes-Enterprise
6. Imagine-John Lennon-Apple
7. Every Picture Tells A Story-Rod Stewart-Mercury
8. Tapestry-Carole King-Ode 70
9. Led Zeppelin-Led Zeppelin-Atlantic
10. After All This Time-Bonnie Koloc-Ovation

Chicago's ORIGINAL album survey

SMALL PRINT DEPARTMENT: The WCFL "ALL HIT MUSIC" represents the relative popularity of records in the Midwest based on Industry tabulations, local record sales and requests. WCFL The Voice of Labor, is o

BIG 10 REQUEST GOLD

- * Let It Be-The Beatles-(1970)
- * These Eyes-The Guess Who-(1969)
- * Jumping Jack Flash-Rolling Stones-(1968)
- * Venus-Shocking Blue-(1970)
- * All Day And All Of The Night-The Kinks-(1965)
- * Dance To The Music-Sly & The Family Stone-(1968)
- * We've Only Just Begun-The Carpenters-(1970)
- * To Sir With Love-Lulu-(1967)
- * Roll Over Beethoven-Chuck Berry-(1956)
- 634-5789-Wilson Pickett-(1966)

*denotes currently available on a gold single or on an album.

(30)	36.	Gypsys, Tramps & Thieves-Cher-Kapp
(40)	37.	White Lies, Blue Eyes-Bullet-Big Tree
(HB)	38.	Where Did Our Love Go-Donnie Elbert-All Platinum
(HB)	39.	American Pie-Don McLean-United Artists
(HB)	40.	Scorpio-Dennis Coffey-Sussex

*Denotes Big 10 Climbers

WC
FL

ALL HIT MUSIC

IN THE
MIDWEST

BIG 10 ALBUMS

1. Chicago At Carnegie Hall-Chicago-Columbia
2. Teaser & The Firecat-Cat Stevens-A & M
3. Santana-Santana-Columbia
4. "Shaft" Sound Track-Isaac Hayes-Enterprise
5. There's A Riot Goin' On-Sly & The Family Stone-Epic
6. Led Zeppelin-Led Zeppelin-Atlantic
7. Imagine-John Lennon-Apple
8. Every Picture Tells A Story-Rod Stewart-Mercury
9. E Pluribus Funk-Grand Funk Railroad-Capitol
10. Stones-Neil Diamond-Uni

SMALL PRINT DEPARTMENT: The WCFL "ALL HIT MUSIC" represents the relative popularity of records in the Midwest based on Industry tabulations, local record sales and requests. WCFL The Voice of Labor, is o

BIG 10 CHRISTMAS GOLD

* The Man With All The Toys-Beach Boys-(1965)
Santa Claus Is Watching You-Ray Stevens-(1962)
* If Everyday Was Like Christmas-Elvis Presley-(1966)
* Santa Claus Is Comin' To Town-Jackson 5-(1970)
* Feliz Navidad-Jose Feliciano-(1970)
* Christmas All Alone-The Lettermen-(1969)
Merry Christmas-Melanie-(1970)
* Last Call For Peace-Rotary Connection-(1968)
Here Comes Santa Claus-David Seville & The Chipmunks-(1959)
* Love's What You're Gettin' For Christmas-Bobby Sherman-(1970)

WAS	IS	Week Ending December 9, 1971
(2)	1.	Brand New Key-Melanie-Neighborhood
(3)	2.	Baby I'm A-Want You-Bread-Elektra
(4)	3.	An Old Fashioned Love Song-3 Dog Night-Dunhill
(5)	4.	Got To Be There-Michael Jackson-Motown
(11)	*5.	Have You Seen Her-Chi-Lites-Brunswick
(9)	6.	All I Ever Need Is You-Sonny & Cher-Kapp
(10)	7.	Stones-Neil Diamond-Uni
(13)	8.	Cherish-David Cassidy-Bell
(12)	9.	Family Affair-Sly & The Family Stone-Epic
(16)	*10.	Two Divided By Love-The Grass Roots-Dunhill
(1)	11.	Theme From "Shaft"-Isaac Hayes-Enterprise
(6)	12.	Absolutely Right-Five Man Electrical Band-Lionel
(17)	13.	Rock Steady-Aretha Franklin-Atlantic
(21)	*14.	Sunshine-Jonathan Edwards-Capricorn
(7)	15.	Wild Night-Van Morrison-WB
(23)	*16.	Hey Girl-Donny Osmond-MGM
(19)	17.	Do I Love You-Paul Anka-Buddah
(8)	18.	What Are You Doing Sunday?-Dawn-Bell
(22)	19.	Nothing To Hide-Tommy James-Roulette
(27)	*20.	Respect Yourself-The Staple Singers-Stax
(25)	21.	Hallelujah-Sweathog-Columbia
(14)	22.	Your Move-Yes-Atlantic
(30)	*23.	Superstar-The Temptations-Gordy
(18)	24.	Imagine-John Lennon-Apple
(15)	25.	Peace Train-Cat Stevens-A & M
(29)	26.	Long Time To Be Alone-New Colony Six-Sunlight
(28)	27.	Fool Me-Joe South-Capitol
(39)	*28.	American Pie-Don McLean-United Artists
(20)	29.	If It's Alright With You-Rose Colored Glass-Bang
(34)	30.	Day After Day-Badfinger-Apple
(37)	31.	White Lies, Blue Eyes-Bullet-Big Tree
(38)	32.	Where Did Our Love Go-Donnie Elbert-All Platinum

*denotes currently available on a gold single or on an album.

(40)	33.	Scorpio-Dennis Coffey-Sussex
(24)	34.	Inner City Blues-Marvin Gaye-Tamla
(26)	35.	Yo-Yo-The Osmonds-MGM
(HB)	36.	Five Hundred Miles-Heaven Bound-MGM
(HB)	37.	Me And Bobby McGee-Jerry Lee Lewis-Mercury
(HB)	38.	Hey Big Brother-Rare Earth-Rare Earth
(HB)	39.	It's One Of Those Nights-The Partridge Family-Bell
(HB)	40.	Sugar Daddy-The Jackson 5-Motown

*Denotes Big 10 Climbers

WC
FL

ALL HIT MUSIC IN THE MIDWEST

BIG 10 ALBUMS

1. Chicago At Carnegie Hall-Chicago-Columbia
2. Teaser & The Firecat-Cat Stevens-A & M
3. There's A Riot Goin' On-Sly & The Family Stone-Epic
4. Santana-Santana-Columbia
5. Led Zeppelin-Led Zeppelin-Atlantic
6. Carole King Music-Carole King-Ode
7. "Shaft" Sound Track-Isaac Hayes-Enterprise
8. Imagine-John Lennon-Apple
9. American Pie-Don McLean-United Artists
10. Every Picture Tells A Story-Rod Stewart-Mercury

CHICAGO
Sunday-9:00 PM
DICK BIONDI and FRIEND

BIG 10 CHRISTMAS '71

1. Merry Christmas Baby-Elvis Presley-RCA
2. If Santa Were My Daddy-Little Jimmy Osmond-MGM
3. The Greatest Man Who Ever Lived-Dave Prince-Rare Earth
4. This Christmas-Donny Hathaway-Atco
5. So My Sheep May Safely Graze-Rod McKuen-WB
6. Happy Christmas-John & Yoko-Apple
7. What Christmas Means To Me-Stevie Wonder-Tamla
8. Merry Christmas Darling-The Carpenters-A & M
9. Goin' Home-Bobby Sherman-Metromedia
10. Amazing Grace-Judy Collins-Elektra

WAS	IS	Week Ending December 16, 1971
(1)	1.	Brand New Key-Melanie-Neighborhood
(5)	2.	Have You Seen Her-Chi-Lites-Brunswick
(9)	3.	Family Affair-Sly & The Family Stone-Epic
(4)	4.	Got To Be There-Michael Jackson-Motown
(6)	5.	All I Ever Need Is You-Sonny & Cher-Kapp
(8)	6.	Cherish-David Cassidy-Bell
(10)	7.	Two Divided By Love-The Grass Roots-Dunhill
(3)	8.	An Old Fashioned Love Song-3 Dog Night-Dunhill
(14)	9.	Sunshine-Jonathan Edwards-Capricorn
(16)	*10.	Hey Girl-Donny Osmond-MGM
(20)	*11.	Respect Yourself-The Staple Singers-Stax
(2)	12.	Baby I'm A-Want You-Bread-Elektra
(19)	*13.	Nothing To Hide-Tommy James-Roulette
(7)	14.	Stones-Neil Diamond-Uni
(17)	15.	Do I Love You-Paul Anka-Buddah
(28)	*16.	American Pie-Don McLean-United Artists
(23)	*17.	Superstar-The Temptations-Gordy
(11)	18.	Theme From "Shaft"-Isaac Hayes-Enterprise
(13)	19.	Rock Steady-Aretha Franklin-Atlantic
(26)	*20.	Long Time To Be Alone-New Colony Six-Sunlight
(12)	21.	Absolutely Right-Five Man Electrical Band-Lionel
(32)	*22.	Where Did Our Love Go-Donnie Elbert-All Platinum
(15)	23.	Wild Night-Van Morrison-WB
(33)	*24.	Scorpio-Dennis Coffey-Sussex
(31)	*25.	White Lies, Blue Eyes-Bullet-Big Tree
(30)	26.	Day After Day-Badfinger-Apple
(40)	*27.	Sugar Daddy-The Jackson 5-Motown
(18)	28.	What Are You Doing Sunday?-Dawn-Bell
(21)	29.	Hallelujah-Sweathog-Columbia
(36)	*30.	Five Hundred Miles-Heaven Bound-MGM
(38)	31.	Hey Big Brother-Rare Earth-Rare Earth
(22)	32.	Your Move-Yes-Atlantic
(27)	33.	Fool Me-Joe South-Capitol
(37)	34.	Me And Bobby McGee-Jerry Lee Lewis-Mercury
(39)	35.	It's One Of Those Nights-The Partridge Family-Bell

(24)	36.	Imagine-John Lennon-Apple
(25)	37.	Peace Train-Cat Stevens-A & M
(HB)	38.	Let's Stay Together-Al Green-HI
(HB)	39.	Looking For A Love-J. Geils Band-Atlantic
(HB)	40.	Black Dog-Led Zeppelin-Atlantic

*Denotes Big 10 Climbers

WC
FL

ALL HIT MUSIC IN THE MIDWEST

BIG 10 ALBUMS

1. Chicago At Carnegie Hall-Chicago-Columbia
2. Led Zeppelin-Led Zeppelin-Atlantic
3. Santana-Santana-Columbia
4. There's A Riot Goin' On-Sly & The Family Stone-Epic
5. Teaser & The Firecat-Cat Stevens-A & M
6. Carole King Music-Carole King-Ode
7. "Shalt" Sound Track-Isaac Hayes-Enterprise
8. E Pluribus Funk-Grand Funk Railroad-Capitol
9. American Pie-Don McLean-United Artists
10. Wings Wild Life-Paul & Linda McCartney-Apple

CHICAGO – PART TWO
Sunday-9:00 PM
DICK BIONDI and FRIEND

BIG 10 CHRISTMAS '71

1. The Marvelous Toy-Peter, Paul & Mary-WB
2. I Saw Mommy Kissing Santa Claus-Four Seasons-Vee-Jay
3. Christmas Holiday-Andy Williams-Columbia
4. Jingle Bell Rock-Bobby Helms-Decca
5. We Need A Little Christmas-Percy Faith-Columbia
6. My Favorite Things-The Supremes-Motown
7. Little Drummer Boy-Lou Rawls-Capitol
8. Merry Christmas-Darling-The Carpenters-A & M
9. The Bell That Couldn't Jingle-Burt Bacharach-A & M
10. Snoopy's Christmas-Royal Guardsmen-Laurie

WAS	IS	Week Ending December 23, 1971
(1)	1.	Brand New Key-Melanie-Neighborhood
(5)	2.	All I Ever Need Is You-Sonny & Cher-Kapp
(6)	3.	Cherish-David Cassidy-Bell
(9)	4.	Sunshine-Jonathan Edwards-Capricorn
(7)	5.	Two Divided By Love-The Grass Roots-Dunhill
(10)	6.	Hey Girl-Donny Osmond-MGM
(2)	7.	Have You Seen Her-Chi-Lites-Brunswick
(13)	8.	Nothing To Hide-Tommy James-Roulette
(3)	9.	Family Affair-Sly & The Family Stone-Epic
(11)	10.	Respect Yourself-The Staple Singers-Stax
(15)	11.	Do I Love You-Paul Anka-Buddah
(16)	12.	American Pie-Don McLean-United Artists
(4)	13.	Got To Be There-Michael Jackson-Motown
(17)	14.	Superstar-The Temptations-Gordy
(8)	15.	An Old Fashioned Love Song-3 Dog Night-Dunhill
(24)	*16.	Scorpio-Dennis Coffey-Sussex
(20)	17.	Long Time To Be Alone-New Colony Six-Sunlight
(12)	18.	Baby I'm A-Want You-Bread-Elektra
(27)	*19.	Sugar Daddy-The Jackson 5-Motown
(25)	20.	White Lies, Blue Eyes-Bullet-Big Tree
(26)	21.	Day After Day-Badfinger-Apple
(22)	22.	Where Did Our Love Go-Donnie Elbert-All Platinum
(14)	23.	Stones-Neil Diamond-Uni
(35)	*24.	It's One Of Those Nights-The Partridge Family-Bell
(18)	25.	Theme From "Shaft"-Isaac Hayes-Enterprise
(19)	26.	Rock Steady-Aretha Franklin-Atlantic
(30)	27.	Five Hundred Miles-Heaven Bound-MGM
(31)	28.	Hey Big Brother-Rare Earth-Rare Earth
(34)	*29.	Me And Bobby McGee-Jerry Lee Lewis-Mercury
(23)	30.	Wild Night-Van Morrison-WB
(21)	31.	Absolutely Right-Five Man Electrical Band-Lionel
(40)	32.	Black Dog-Led Zeppelin-Atlantic
(28)	33.	What Are You Dong Sunday?-Dawn-Bell

(38)	34.	Let's Stay Together-Al Green-HI
(39)	35.	Looking For A Love-J. Geils Band-Atlantic
(HB)	36.	I'd Like To Teach The World To Sing-The Hillside Singers-Metromedia
(HB)	37.	Stay With Me-Faces-WB
(HB)	38.	Kiss An Angel Good Mornin'-Charley Pride-RCA
(HB)	39.	Happy Xmas-John & Yoko/Plastic One Band-Apple
(HB)	40.	Don't Say You Don't Remember-Beverly Bremers-Scepter

*Denotes Big 10 Climbers

WC
FL

ALL HIT MUSIC

IN THE MIDWEST

BIG 10 ALBUMS

1. Chicago At Carnegie Hall-Chicago-Columbia
2. Carole King Music-Carole King-Ode
3. Led Zeppelin-Led Zeppelin-Atlantic
4. American Pie-Don McLean-United Artists
5. Santana-Santana-Columbia
6. Wings Wild Life-Paul & Linda McCartney-Apple
7. "Shaft" Sound Track-Isaac Hayes-Enterprise
8. Teaser & The Firecat-Cat Stevens-A & M
9. There's A Riot Goin' On-Sly & The Family Stone-Epic
10. Stones-Neil Diamond-Uni

SMALL PRINT DEPARTMENT: The WCFL "ALL HIT MUSIC" represents the relative popularity of records in the Midwest based on Industry tabulations, local record sales and requests. WCFL The Voice of Labor, is o

BIG 10 CHRISTMAS '71

1. Santa Claus Is Comin' To Town-Jackson 5-Motown
2. If Santa Were My Daddy-Little Jimmy Osmond-MGM
3. The Mistletoe And Me-Isaac Hayes-Enterprise
4. Twinkle Twinkle Little Me-The Supremes-Motown
5. Blue Christmas-Elvis Presley-RCA
6. A Christmas Song-Shawn Phillips-A & M
7. The Chant-Jane Avenue Bus Stop-Buddah
8. Jingle Bell Rock-Bobby Sherman-Metromedia
9. Feliz Navidad-Jose Feliciano-RCA
10. Santa Claus Is Watching You-Ray Stevens-Mercury

WAS	IS	Week Ending December 30, 1971
(4)	1.	Sunshine-Jonathan Edwards-Capricorn
(6)	2.	Hey Girl-Donny Osmond-MGM
(12)	*3.	American Pie-Don McLean-United Artists
(1)	4.	Brand New Key-Melanie-Neighborhood
(8)	5.	Nothing To Hide-Tommy James-Roulette
(10)	6.	Respect Yourself-The Staple Singers-Stax
(2)	7.	All I Ever Need Is You-Sonny & Cher-Kapp
(11)	8.	Do I Love You-Paul Anka-Buddah
(3)	9.	Cherish-David Cassidy-Bell
(16)	*10.	Scorpio-Dennis Coffey-Sussex
(19)	*11.	Sugar Daddy-The Jackson 5-Motown
(5)	12.	Two Divided By Love-The Grass Roots-Dunhill
(9)	13.	Family Affair-Sly & The Family Stone-Epic
(17)	14.	Long Time To Be Alone-New Colony Six-Sunlight
(7)	15.	Have You Seen Her-Chi-Lites-Brunswick
(21)	16.	Day After Day-Badfinger-Apple
(20)	17.	White Lies, Blue Eyes-Bullet-Big Tree
(24)	*18.	It's One Of Those Nights-The Partridge Family-Bell
(22)	19.	Where Did Our Love Go-Donnie Elbert-All Platinum
(13)	20.	Got To Be There-Michael Jackson-Motown
(14)	21.	Superstar-The Temptations-Gordy
(15)	22.	An Old Fashioned Love Song-3 Dog Night-Dunhill
(27)	23.	Five Hundred Miles-Heaven Bound-MGM
(34)	*24.	Let's Stay Together-Al Green-HI
(18)	25.	Baby I'm A-Want You-Bread-Elektra
(29)	26.	Me And Bobby McGee-Jerry Lee Lewis-Mercury
(28)	27.	Hey Big Brother-Rare Earth-Rare Earth
(36)	*28.	I'd Like To Teach The World To Sing-The Hillside Singers-Metromedia
(35)	*29.	Looking For A Love-J. Geils Band-Atlantic
(32)	30.	Black Dog-Led Zeppelin-Atlantic
(37)	31.	Stay With Me-Faces-WB
(39)	32.	Happy Xmas-John & Yoke/Plastic One Band-Apple

(25)	33.	Theme From "Shaft"-Isaac Hayes-Enterprise
(38)	34.	Kiss An Angel Good Mornin'-Charley Pride-RCA
(40)	35.	Don't Say You Don't Remember-Beverly Bremers-Scepter
(23)	36.	Stones-Neil Diamond-Uni
(26)	37.	Rock Steady-Aretha Franklin-Atlantic
(HB)	38.	The Witch Queen Of New Orleans-Redbone-Epic
(HB)	39.	Softly Whispering I Love You-The English Congregation-Atco
(HB)	40.	Precious And Few-Climax-Carousel

*Denotes Big 10 Climbers

SOLID GOLD '71

WCFL TOP 40 of 1971

BIG 10 ALBUMS '71

1. Tapestry-Carole King-Ode '70
2. Jesus Christ, Superstar-Various Artists-Decca
3. Carpenters-Carpenters-A & M
4. Ram-Paul & Linda McCartney-Apple
5. Every Picture Tells A Story-Rod Stewart-Mercury
6. Sticky Fingers-Rolling Stones-Atco
7. Every Good Boy Deserves Favour-Moody Blues-Threshold
8. Pearl-Janis Joplin-Columbia
9. Santana-Santana-Columbia
10. Imagine-John Lennon-Apple

SMALL PRINT DEPARTMENT: WCFL's "ALL HIT MUSIC" represents relative popularity of records based on Industry tabulations, record sales and requests. WCFL The Voice of Labor, is o

1. Let It Be-The Beatles-(1970)
2. Purple Haze-Jimi Hendrix-(1967)
3. Help-The Beatles-(1965)
4. War-Edwin Starr-(1970)
5. Close To You-The Carpenters-(1970)
6. Suite: Judy Blue Eyes-Crosby, Stills & Nash-(1969)
7. Lay Down (Candles In The Rain)-Melanie-(1970)
8. Hey Jude-The Beatles-(1968)
9. Born To Be Wild-Steppenwolf-(1968)
10. Bridge Over Troubled Water-Simon & Garfunkel-(1970)

1. Joy To The World-3 Dog Night-Dunhill
2. Uncle Albert/Admiral Halsey-Paul McCartney-Apple
3. Knock Three Times-Dawn-Bell
4. It's Too Late-Carole King-Ode 70
5. Maggie May-Rod Stewart-Mercury
6. Brand New Key-Melanie-Neighborhood
7. My Sweet Lord-George Harrison-Apple
8. Superstar-The Carpenters-A & M
9. Just My Imagination-The Temptations-Gordy
10. She's A Lady-Tom Jones-Parrot
11. Doesn't Somebody Want To Be Wanted-Partridge Family-Bell
12. I Hear You Knockin'-Dave Edmunds-MAM
13. Mama's Pearl-Jackson 5-Motown
14. Stay Awhile-The Bells-Polydor
15. One Bad Apple-The Osmonds-MGM
16. Theme From "Shaft"-Isaac Hayes-Enterprise
17. Want Ads-Honey Cone-Hot Wax
18. Never Ending Song Of Love-Delaney & Bonnie-Atco
19. Don't Pull Your Love-Hamilton, Joe Frank & Reynolds-Dunhill
20. If You Could Read My Mind-Gordon Lightfoot-Reprise
21. Go Away Little Girl-Donny Osmond-MGM
22. You've Got A Friend-James Taylor-WB
23. How Can You Mend A Broken Heart-Bee Gees-Atco
24. Smiling Faces Sometimes-Undisputed Truth-Gordy
25. Gypsys, Tramps & Thieves-Cher-Kapp
26. Indian Reservation-Raiders-Columbia
27. Wedding Song-Paul Stookey-WB
28. Rainy Days & Mondays-The Carpenters-A & M
28. Rainy Days & Mondays-The Carpenters-A & M
29. Never Can Say Goodbye-Jackson 5-Motown
30. Night They Drove Old Dixie Down-Joan Baez-Vanguard
31. L.A. Goodbye-Ides Of March-WB
32. Sweet Mary-Wadsworth Mansion-Sussex
33. Draggin' The Line-Tommy James-Roulette
34. Stoned Love-The Supremes-Motown
35. It Don't Come Easy-Ringo Starr-Apple
36. Riders On The Storm-The Doors-Elektra
37. Brown Sugar-Rolling Stones-Atco
38. Stoney End-Barbra Steisand-Columbia
39. For All We Know-The Carpenters-

A & M

40. Love Her Madly-The Doors-Elektra

1972

WC
FL

ALL HIT MUSIC **IN THE MIDWEST**

WAS	IS	Week Ending January 13, 1972
(3)	1.	American Pie-Don McLean-United Artists
(5)	2.	Nothing To Hide-Tommy James-Roulette
(1)	3.	Sunshine-Jonathan Edwards-Capricorn
(10)	*4.	Scorpio-Dennis Coffey-Sussex
(6)	5.	Respect Yourself-The Staple Singers-Stax
(2)	6.	Hey Girl-Donny Osmond-MGM
(16)	*7.	Day After Day-Badfinger-Apple
(11)	8.	Sugar Daddy-The Jackson 5-Motown
(17)	*9.	White Lies, Blue Eyes-Bullet-Big Tree
(18)	*10.	It's One Of Those Nights-The Partridge Family-Bell
(4)	11.	Brand New Key-Melanie-Neighborhood
(19)	*12.	Where Did Our Love Go-Donnie Elbert-All Platinum
(14)	13.	Long Time To Be Alone-New Colony Six-Sunlight
(23)	*14.	Five Hundred Miles-Heaven Bound-MGM
(7)	15.	All I Ever Need Is You-Sonny & Cher-Kapp
(30)	*16.	Black Dog-Led Zeppelin-Atlantic
(24)	*17.	Let's Stay Together-Al Green-HI
(9)	18.	Cherish-David Cassidy-Bell
(31)	*19.	Stay With Me-Faces-WB
(28)	*20.	I'd Like To Teach The World To Sing-The Hillside Singers-Metromedia
(12)	21.	Two Divided By Love-The Grass Roots-Dunhill
(29)	*22.	Looking For A Love-J. Geils Band-Atlantic
(13)	23.	Family Affair-Sly & The Family Stone-Epic
(15)	24.	Have You Seen Her-Chi-Lites-Brunswick
(26)	25.	Me And Bobby McGee-Jerry Lee Lewis-Mercury
(34)	*26.	Kiss An Angel Good Mornin'-Charley Pride-RCA
(35)	*27.	Don't Say You Don't Remember-Beverly Bremers-Scepter
(39)	*28.	Softly Whispering I Love you-The English Congregation-Atco
(38)	*29.	The Witch Queen Of New Orleans-Redbone-Epic
(40)	30.	Precious And Few-Climax-Carousel
(21)	31.	Superstar-The Temptations-Gordy
(HB)	32.	Hurting Each Other-Carpenters-A & M

BIG 10 ALBUMS

1. Chicago At Carnegie Hall-Chicago-Columbia
2. Carole King Music-Carole King-Ode
3. American Pie-Don McLean-United Artists
4. The Concert For Bangla Desh-Apple
5. Led Zeppelin-Led Zeppellin-Atlantic
6. Teaser & The Firecat-Cat Stevens-A & M
7. Santana-Santana-Columbia
8. Wings Wild Life-Paul & Linda McCartney-Apple
9. There's A Riot Goin' On-Sly & The Family Stone-Epic
10. Hot Rocks-The Rolling Stones-London

THE SHIRELLES
Sunday-9:00 PM
DICK BIONDI and FRIENDS

ENGLAND'S BIG 10

1. Ernie (The Fastest Milkman In The West)-Benny Hill-Columbia

*2. Softly Whispering I Love You-Congregation-Columbia

3. Something Tells Me-Cilla Black-Parlophone

*4. Jeepster-T. Tex-Fly

5. Coz I Love You-Slade-Polydor

6. No Matter How I Try-Gilbert O'Sullivan-MAM

7. Soley Soley-Middle Of The Road-RCA

*8. Fireball-Deep Purple-Harvest

*9. Gypsys, Tramps & Thieves-Cher-MCA

*10. Banks Of The Ohio-Olivia Newton-John-Pye

*denotes currently available in the U.S.

(HB)	33.	The Lion Sleeps Tonight-Robert John-Atlantic
(HB)	34.	Bang A Gong (Get It On)-T. Tex-Reprise
(HB)	35.	Without You-Nilsson-RCA
(HB)	36.	Anticipation-Carly Simon-Elektra
(HB)	37.	Drowning In The Sea Of Love-Joe Simon-Spring
(HB)	38.	Down By The Lazy River-The Osmonds-MGM
(HB)	39.	Joy-Apollo 100-Mega
(HB)	40.	Rock 'N Roll-Detroit/Featuring Mitch Ryder-Paramount

*Denotes Big 10 Climbers

WC
FL

ALL HIT MUSIC

IN THE
MIDWEST

WAS IS Week Ending January 20, 1972

WAS	IS	Title
(1)	1.	American Pie-Don McLean-United Artists
(8)	*2.	Sugar Daddy-The Jackson 5-Motown
(7)	3.	Day After Day-Badfinger-Apple
(4)	4.	Scorpio-Dennis Coffey-Sussex
(9)	5.	White Lies, Blue Eyes-Bullet-Big Tree
(10)	6.	It's One Of Those Nights-The Partridge Family-Bell
(12)	7.	Where Did Our Love Go-Donnie Elbert-All Platinum
(14)	*8.	Five Hundred Miles-Heaven Bound-MGM
(2)	9.	Nothing To Hide-Tommy James-Rouletete
(16)	*10.	Black Dog-Led Zeppelin-Atlantic
(17)	*11.	Let's Stay Together-Al Green-HI
(3)	12.	Sunshine-Jonathan Edwards-Capricorn
(19)	*13.	Stay With Me-Faces-WB
(5)	14.	Respect Yourself-The Staple Singers-Stax
(6)	15.	Hey Girl-Donny Osmond-MGM
(22)	*16.	Looking For A Love-J. Geils Band-Atlantic
(27)	*17.	Don't Say You Don't Remember-Beverly Bremers-Scepter
(20)	18.	I'd Like To Teach The World To Sing-The Hillside Singers-Metromedia
(11)	19.	Brand New Key-Melanie-Neighborhood
(26)	*20.	Kiss An Angel Good Mornin'-Charley Pride-RCA
(13)	21.	Long Time To Be Alone-New Colony Six-Sunlight
(29)	*22.	The Witch Queen Of New Orleans-Redbone-Epic
(30)	*23.	Precious And Few-Climax-Carousel
(28)	24.	Softly Whispering I Love You-The English Congregation-Atco
(32)	*25.	Hurting Each Other-Carpenters-A & M
(35)	*26.	Without You-Nilsson-RCA
(36)	*27.	Anticipation-Carly Simon-Elektra
(15)	28.	All I Ever Need Is You-Sonny & Cher-Kapp
(33)	29.	The Lion Sleeps Tonight-Robert John-Atlantic
(18)	30.	Cherish-David Cassidy-Bell
(34)	31.	Bang A Gong (Get It On)-T. Tex-Reprise
(39)	32.	Joy-Apollo 100-Mega
(38)	33.	Down By The Lazy River-The Osmonds-MGM

BIG 10 ALBUMS

1. The Concert For Bangla Desh-Apple
2. American Pie-Don McLean-United Artists
3. Carole King Music-Carole King-Ode
4. Chicago At Carnegie Hall-Chicago-Columbia
5. Led Zeppelin-Led Zeppelin-Atlantic
6. Wings Wild Life-Paul & Linda McCartney-Apple
7. Santana-Santana-Columbia
8. Teaser & The Firecat-Cat Stevens-A & M
9. The Low Spark Of High Heeled Boys-Traffic-Island
10. Hot Rocks-The Rolling Stones-London

THE SHIRELLES – Part Two
Sunday-9:00 PM
DICK BIONDI and FRIENDS

BIG 10 WINTER GOLD

*1. Our Winter Love-Bill Pursell-(1963)
*2. Riders On The Storm-The Doors-(1971)
3. When The Winds Blow in Chicago-Geo. McGern-(1963)
4. White On White-Danny Williams-(1964)
5. Snowman, Snowman-The Jaynettes-(1963)
*6. Windy-The Association-(1967)
*7. Popsicles And Icicles-The Murmaids-(1963)
*8. Winterlude-Bob Dylan-(1970)
9. Ice-Spirit-(1969)
*10. Snowqueen Of Texas-Mamas & Papas-(1971)

*denotes currently available on a gold single or on an album

(37)	34.	Drowning In The Sea Of Love-Joe Simon-Spring
(40)	35.	Rock 'N Roll-Detroit/Featuring Mitch Ryder-Paramount
(21)	36.	Two Divided By Love-The Grass Roots-Dunhill
(25)	37.	Me And Bobby McGee-Jerry Lee Lewis-Mercury
(HB)	38.	Everything I Own-Bread-Elektra
(HB)	39.	My World-Bee Gees-Atco
(HB)	40.	(Love Me) Love The Life I Lead-The Fantastics-Bell

*Denotes Big 10 Climbers

WC
FL

ALL HIT MUSIC

IN THE
MIDWEST

WAS IS Week Ending January 27, 1972

WAS	IS	
(1)	1.	American Pie-Don McLean-United Artists
(3)	2.	Day After Day-Badfinger-Apple
(7)	3.	Where Did Our Love Go-Donnie Elbert-All Platinum
(5)	4.	White Lies, Blue Eyes-Bullet-Big Tree
(8)	5.	Five Hundred Miles-Heaven Bound-MGM
(6)	6.	It's One Of Those Nights-The Partridge Family-Bell
(10)	7.	Black Dog-Led Zeppelin-Atlantic
(11)	8.	Let's Stay Together-Al Green-HI
(13)	9.	Stay With Me-Faces-WB
(2)	10.	Sugar Daddy-The Jackson 5-Motown
(17)	*11.	Don't Say You Don't Remember-Beverly Bremers-Scepter
(4)	12.	Scorpio-Dennis Coffey-Sussex
(25)	*13.	Hurting Each Other-Carpenters-A & M
(20)	*14.	Kiss An Angel Good Mornin'-Charley Pride-RCA
(23)	*15.	Precious And Few-Climax-Carousel
(16)	16.	Looking For A Love-J. Geils Band-Atlantic
(22)	17.	The Witch Queen Of New Orleans-Redbone-Epic
(18)	18.	I'd Like To Teach The World To Sing-The Hillside Singers-Metromedia
(24)	19.	Softly Whispering I Love You-The English Congregation-Atco
(33)	*20.	Down By The Lazy River-The Osmonds-MGM
(32)	*21.	Joy-Apollo 100-Mega
(9)	22.	Nothing To Hide-Tommy James-Roulette
(14)	23.	Respect Yourself-The Staple Singers-Stax
(26)	24.	Without You-Nilsson-RCA
(27)	25.	Anticipation-Carly Simon-Elektra
(12)	26.	Sunshine-Jonathan Edwards-Capricorn
(29)	27.	The Lion Sleeps Tonight-Robert John-Atlantic
(15)	28.	Hey Girl-Donny Osmond-MGM
(31)	29.	Bang A Gong (Get It On)-T. Tex-Reprise
(34)	30.	Drowning In The Sea Of Love-Joe Simon-Spring
(19)	31.	Brand New Key-Melanie-Neighborhood
(35)	32.	Rock 'N Roll-Detroit/Featuring Mitch Ryder-Paramount

BIG 10 ALBUMS

1. American Pie-Don McLean-United Artists
2. Led Zeppellin-Led Zeppelin-Atlantic
3. Carole King Music-Carole King-Ode
4. Chicago At Carnegie Hall-Chicago-Columbia
5. The Concert For Bangla Desh-Apple
6. Hot Rocks-Rolling Stones-London
7. Santana-Santana-Columbia
8. Teaser & The Firecat-Cat Stevens-A & M
9. Pictures At An Exhibition-Emerson, Lake & Palmer-Cotillion
10. Wings Wild Life-Paul & Linda McCartney-Apple

STEVE GOODMAN
Sunday-9:00 PM
DICK BIONDI and FRIEND

HAWAII'S BIG 10

*1. Have You Seen Her-Chi-Lites-Brunswick
*2. Respect Yourself-Staple Singers-Stax
*3. Superstar-The Temptations-Gordy
*4. You Are Everything-The Stylistics-Avco
*5. Cherish-David Cassidy-Bell
*6. Long Time To Be Alone-New Colony Six-Sunlight
*7. Got To Be There-Michael Jackson-Motown
*8. I Know I'm Losing You-Rod Stewart-Mercury
*9. Brand New Key-Melanie-Neighborhood
10. Come Run With Me-Coppernickle-Flair

*denotes BIG 10 Hit currently available on the mainland.

(38)	33.	Everything I Own-Bread-Elektra
(39)	34.	My World-Bee Gees-Atco
(40)	35.	(Love Me) Love The Life I Lead-The Fantastics-Bell
(21)	36.	Long Time To Be Alone-New Colony Six-Sunlight
(28)	37.	All I Ever Need Is You-Sonny & Cher-Kapp
(HB)	38.	Country Wine-The Raiders-Columbia
(HB)	39.	Heart Of Gold-Neil Young-Reprise
(HB)	40.	Floy Joy-The Supremes-Motown

*Denotes Big 10 Climbers

WC
FL

ALL HIT MUSIC IN THE MIDWEST

WAS	IS	Week Ending February 3, 1972
(9)	1.	Stay With Me-Faces-WB
(7)	2.	Black Dog-Led Zeppelin-Atlantic
(1)	3.	American Pie-Don McLean-United Artists
(15)	*4.	Precious And Few-Climax-Carousel
(2)	5.	Day After Day-Badfinger-Apple
(13)	*6.	Hurting Each Other-Carpenters-A & M
(8)	7.	Let's Stay Together-Al Green-HI
(11)	8.	Don't Say You Don't Remember-B. Bremers-Scepter
(4)	9.	White Lies, Blue Eyes-Bullet-Big Tree
(17)	*10.	The Witch Queen Of New Orleans-Redbone-Epic
(3)	11.	Where Did Our Love Go-Donnie Elbert-All Platinum
(19)	*12.	Softly Whispering-English Congregation-Atco
(14)	13.	Kiss An Angel Good Mornin'-Charley Pride-RCA
(21)	*14.	Joy-Apollo 100-Mega
(20)	15.	Down By The Lazy River-The Osmonds-MGM
(5)	16.	Five Hundred Miles-Heaven Bound-MGM
(24)	*17.	Without You-Nilsson-RCA
(6)	18.	It's One Of Those Nights-The Partridge Family-Bell
(10)	19.	Sugar Daddy-The Jackson 5-Motown
(27)	*20.	The Lion Sleeps Tonight-Robert John-Atlantic
(12)	21.	Scorpio-Dennis Coffey-Sussex
(25)	22.	Anticipation-Carly Simon-Elektra
(16)	23.	Looking For A Love-J. Geils Band-Atlantic
(33)	*24.	Everything I Own-Bread-Elektra
(18)	25.	Teach The World-Hillside Singers-Metromedia
(29)	26.	Bang A Gong (Get It On)-T. Tex-Reprise
(30)	27.	Drowning In The Sea Of Love-Joe Simon-Spring
(34)	*28.	My World-Bee Gees-Atco
(32)	29.	Rock 'N Roll-Detroit-Paramount
(35)	30.	(Love Me) Love The Life I Lead-The Fantastics-Bell
(23)	31.	Respect Yourself-The Staple Singers-Stax
(40)	32.	Floy Joy-The Supremes-Motown
(39)	33.	Heart Of God-Neil Young-Reprise
(38)	34.	Country Wine-The Raiders-

BIG 10 ALBUMS

1. American Pie-Don McLean-United Artists
2. The Concert For Bangla-Desh-Apple
3. Carole King Music-Carole King-Ode
4. Led Zeppelin-Led Zeppelin-Atlantic
5. Chicago At Carnegie Hall-Chicago-Columbia
6. Hot Rocks-Rolling Stones-London
7. Pictures At An Exhibition-Emerson, Lake & Palmer-Cotillion
8. Teaser & The Firecat-Cat Stevens-A & M
9. Wings Wild Life-Paul & Linda McCartney-Apple
10. The Low Spark Of High Heeled Boys-Traffic-Island

WAR
Sunday-9:00 PM
DICK BIONDI and FRIENDS

---

10 YEARS AGO TODAY

*1. The Twist-Chubby Checker-Parkway
*2. Town Without Pity-Gene Pitney-Musicor
3. Peppermint Twist-Joey Dee-Roulette
*4. The Lion Sleeps Tonight-The Tokens-RCA
5. Let There Be Drums-Sandy Nelson-Imperial
*6. Can't Help Falling In Love-Elvis Presley-RCA
*7. Run To Him-Bobby Vee-Liberty
*8. Moon River-Henry Mancini-RCA
*9. Happy Birthday Sweet Sixteen-Neil Sedaka-RCA
10. She's Everything-The Corsairs-Tuff

*denotes currently available on a gold single or on an album.

		Columbia
(28)	35.	Hey Girl-Donny Osmond-MGM
(26)	36.	Sunshine-Jonathan Edwards-Capricorn
(HB)	37.	The Way Of Love-Cher-Kapp
(HB)	38.	Roundabout-Yes-Atlantic
(HB)	39.	Rock And Roll Lullaby-B.J. Thomas-Scepter
(HB)	40.	Goin' Down (On The Road To L.A.)-Terry Black/Laurel Ward-Kama Sutra

*Denotes Big 10 Climbers

WC FL **ALL HIT MUSIC** **IN THE MIDWEST**

WAS IS Week Ending February 10, 1972

WAS	IS	
(1)	1.	Stay With Me-Faces-WB
(2)	2.	Black Dog-Led Zeppelin-Atlantic
(6)	3.	Hurting Each Other-Carpenters-A & M
(8)	4.	Don't Say You Don't Remember-B. Bremers-Scepter
(10)	5.	The Witch Queen Of New Orleans-Redbone-Epic
(12)	*6.	Softly Whispering-English Congregation-Atco
(14)	*7.	Joy-Apollo 100-Mega
(17)	*8.	Without You-Nilsson-RCA
(13)	9.	Kiss An Angel Good Mornin'-Charley Pride-RCA
(15)	10.	Down By The Lazy River-The Osmonds-MGM
(3)	11.	American Pie-Don McLean-United Artists
(4)	12.	Precious And Few-Climax-Carousel
(7)	13.	Let's Stay Together-Al Green-HI
(5)	14.	Day After Day-Badfinger-Apple
(20)	15.	The Lion Sleeps Tonight-Robert John-Atlantic
(11)	16.	Where Did Our Love Go-Donnie Elbert-All Platinum
(9)	17.	White Lies, Blue Eyes-Bullet-Big Tree
(22)	18.	Anticipation-Carly Simon-Elektra
(24)	*19.	Everything I Own-Bread-Elektra
(33)	*20.	Heart Of Gold-Neil Young-Reprise
(16)	21.	Five Hundred Miles-Heaven Bound-MGM
(18)	22.	It's One Of Those Nights-The Partridge Family-Bell
(26)	23.	Bang A Gong (Get It On)-T. Rex-Reprise
(23)	24.	Looking For A Love-J. Geils Band-Atlantic
(32)	*25.	Floy Joy-The Supremes-Motown
(19)	26.	Sugar Daddy-The Jackson 5-Motown
(28)	27.	My World-Bee Gees-Atco
(21)	28.	Scorpio-Dennis Coffey-Sussex
(30)	29.	(Love Me) Love The Life I Lead-The Fantastics-Bell
(34)	30.	Country Wine-The Raiders-Columbia
(37)	31.	The Way Of Love-Cher-Kapp
(38)	32.	Roundabout-Yes-Atlantic
(39)	33.	Rock And Roll Lullaby-B.J. Thomas-Scepter
(40)	34.	Goin' Down (On The Road to L.A.)-Terry Black/Laurel Ward-Kama Sutra
(HB)	35.	Ring The Living Bell-Melanie-

BIG 10 ALBUMS

1. American Pie-Don McLean-United Artists
2. The Concert For Bangla Desh-Apple
3. Hot Rocks-Rolling Stones-London
4. Carole King Music-Carole King-Ode
5. Led Zeppelin-Led Zeppelin-Atlantic
6. Pictures At An Exhibition-Emerson, Lake And Palmer-Cotillion
7. Chicago At Carnegie Hall-Chicago-Columbia
8. Baby I'm-A Want You-Bread-Elektra
9. The Low Spark Of High Heeled Boys-Traffic-Island
10. Teaser & The Firecat-Cat Stevens-A & M

WAR – Part Two
Sunday-9:00 PM
DICK BIONDI and FRIENDS

LOS ANGELES' BIG 10

*1. Brand New Key-Melanie-Neighborhood
*2. Let's Stay Together-Al Green-HI
*3. Scorpio-Dennis Coffey-Sussex
*4. You Are Everything-Stylistics-Avco
5. Never Been To Spain-3 Dog Night-Dunhill
*6. American Pie-Don McLean-United Artists
7. Levon-Elton John-Uni
*8. Sunshine-Jonathan Edwards-Capricorn
*9. Day After Day-Badfinger-Apple
*10. The Lion Sleeps Tonight-Robert John-Atlantic

*denotes former and current BIG 10 hit in Chicago.

		Neighborhood
(HB)	36.	Mother And Child Reunion-Paul Simon-Columbia
(HB)	37.	Willie Boy's A'Comin'-Tommy James-Roulette
(HB)	38.	Mercedes Benz-Goose Creek Symphony-Capitol
(HB)	39.	Could It Be Forever-David Cassidy-Bell
(HB)	40.	Glory Bound-The Grass Roots-Dunhill

*Denotes Big 10 Climbers

WC
FL

ALL HIT MUSIC

IN THE
MIDWEST

WAS IS Week Ending February 17, 1972

WAS	IS	
(3)	1.	Hurting Each Other-Carpenters-A & M
(8)	*2.	Without You-Nilsson-RCA
(5)	3.	The Witch Queen Of New Orleans-Redbone-Epic
(6)	4.	Softly Whispering-English Congregation-Atco
(7)	5.	Joy-Apollo 100-Mega
(1)	6.	Stay With Me-Faces-WB
(2)	7.	Black Dog-Led Zeppelin-Atlantic
(10)	8.	Down By The Lazy River-The Osmonds-MGM
(9)	9.	Kiss An Angel Good Mornin'-Charley Pride-RCA
(15)	10.	The Lion Sleeps Tonight-Robert John-Atlantic
(18)	*11.	Anticipation-Carly Simon-Elektra
(19)	*12.	Everything I Own-Bread-Elektra
(4)	13.	Don't Say You Don't Remember-B. Bremers-Scepter
(20)	*14.	Heart Of Gold-Neil Young-Reprise
(25)	*15.	Floy Joy-The Supremes-Motown
(12)	16.	Precious And Few-Climax-Carousel
(27)	*17.	My World-Bee Gees-Atco
(11)	18.	American Pie-Don McLean-United Artists
(23)	19.	Bang A Gong (Get It On)-T. Rex-Reprise
(13)	20.	Let's Stay Together-Al Green-HI
(30)	*21.	Country Wine-The Raiders-Columbia
(32)	*22.	Roundabout-Yes-Atlantic
(16)	23.	Where Did Our Love Go-Donnie Elbert-All Platinum
(31)	*24.	The Way Of Love-Cher-Kapp
(14)	25.	Day After Day-Badfinger-Apple
(29)	26.	(Love Me) Love The Life I Lead-The Fantastics-Bell
(17)	27.	White Lies, Blue Eyes-Bullet-Big Tree
(35)	*28.	Ring The Living Bell-Melanie-Neighborhood
(33)	29.	Rock And Roll Lullaby-B.J. Thomas-Scepter
(36)	30.	Mother And Child Reunion-Paul Simon-Columbia
(34)	31.	Goin' Down (On The Road To L.A.)-Terry Black/Laurel Ward-Kama Sutra
(37)	32.	Willie Boy's A'Comin'-Tommy James-Roulette
(39)	33.	Could It Be Forever-David Cassidy-Bell
(40)	34.	Glory Bound-The Grass Roots-

BIG 10 ALBUMS

1. American Pie-Don McLean-United Artists
2. Hot Rocks-Rolling Stones-London
3. The Concert For Bangla Desh-Apple
4. Carole King Music-Carole King-Ode
5. Baby I'm-A Want You-Bread-Elektra
6. Led Zeppelin-Led Zeppelin-Atlantic
7. Pictures At An Exhibition-Emerson, Lake & Palmer-Cotillion
8. Chicago At Carnegie Hall-Chicago-Columbia
9. Paul Simon-Paul Simon-Columbia
10. Fragile-Yes-Atlantic

CARLY SIMON
Sunday-9:00 PM
DICK BIONDI and FRIEND

BIG 10 VALENTINE GOLD

*1. Love-The Lettermen-(1971)

*2. 10 Commandments Of Love-Harvey & The Moonglows-(1958)

*3. And I Love Her-The Beatles-(1964)

4. Love Is Once In A Lifetime Thing-Dick & DeeDee-(1963)

*5. I Think I Love You-The Partridge Family-(1970)

6. True Love Never Runs Smooth-Gene Pitney-(1963)

7. Everybody Loves A Lover-The Shirelles-(1962)

*8. Love Is Blue-Paul Mauriat-(1968)

*9. Cupid-Sam Cooke-(1960)

*10. Love Her Madly-The Doors-(1971)

*denotes currently available on a gold single or on an album.

		Dunhill
(38)	35.	Mercedes Benz-Goose Creek Symphony-Capitol
(HB)	36.	Every Day Of My Life-Bobby Vinton-Epic
(HB)	37.	Crazy Mama-J.J. Cale-Shelter
(HB)	38.	A Horse With No Name-America-WB
(HB)	39.	Puppy Love-Donny Osmond-MGM
(HB)	40.	You Are The One-Sugar Bears-Big Tree

*Denotes Big 10 Climbers

WAS	IS	Week Ending February 24, 1972
(2)	1.	Without You-Nilsson-RCA
(5)	2.	Joy-Apollo 100-Mega
(4)	3.	Softly Whispering-English Congregation-Atco
(8)	4.	Down By The Lazy River-The Osmonds-MGM
(1)	5.	Hurting Each Other-Carpenters-A & M
(11)	*6.	Anticipation-Carly Simon-Elektra
(10)	7.	The Lion Sleeps Tonight-Robert John-Atlantic
(3)	8.	The Witch Queen Of New Orleans-Redbone-Epic
(12)	9.	Everything I Own-Bread-Elektra
(14)	10.	Heart Of Gold-Neil Young-Reprise
(17)	*11.	My World-Bee Gees-Atco
(15)	12.	Floy Joy-The Supremes-Motown
(6)	13.	Stay With Me-Faces-WB
(21)	*14.	Country Wine-The Raiders-Columbia
(7)	15.	Black Dog-Led Zeppelin-Atlantic
(24)	*16.	The Way Of Love-Cher-Kapp
(9)	17.	Kiss An Angel Good Mornin'-Charley Pride-RCA
(22)	18.	Roundabout-Yes-Atlantic
(28)	*19.	Ring The Living Bell-Melanie-Neighborhood
(13)	20.	Don't Say You Don't Remember-B. Bremers-Scepter
(16)	21.	Precious And Few-Climax-Carousel
(18)	22.	American Pie-Don McLean-United Artists
(26)	23.	(Love Me) Love The Life I Lead-The Fantastics-Bell
(33)	*24.	Could It Be Forever-David Cassidy-Bell
(34)	*25.	Glory Bound-The Grass Roots-Dunhill
(30)	26.	Mother And Child Reunion-Paul Simon-Columbia
(29)	27.	Rock And Roll Lullaby-B.J. Thomas-Scepter
(19)	28.	Bang A Gong (Get It On)-T. Rex-Reprise
(31)	29.	Goin' Down (On The Road To L.A.)-Terry Black/Laurel Ward-Kama Sutra
(32)	30.	Willie Boy's A'Comin'-Tommy James-Roulette
(35)	31.	Mercedes Benz-Goose Creek Symphony-Capitol
(39)	32.	Puppy Love-Donny Osmond-MGM
(36)	33.	Every Day Of My Life-Bobby Vinton-Epic
(37)	34.	Crazy Mama-J.J. Cale-Shelter
(38)	35.	A Horse With No Name-America-WB

BIG 10 ALBUMS

1. American Pie-Don McLean-United Artists
2. The Concert For Bangla Desh-Apple
3. Carole King Music-Carole King-Ode
4. Baby I'm-A Want You-Bread-Elektra
5. Hot Rocks-Rolling Stones-London
6. Led Zeppelin-Led Zeppelin-Atlantic
7. Fragile-Yes-Atlantic
8. The Low Spark Of High Heeled Boys-Traffic-Island
9. Pictures At An Exhibition-Emerson, Lake And Palmer-Cotillion
10. A Nod Is As Good As A Wink-Faces-WB

CARLY SIMON – Part Two
Sunday- 9:00 PM
DICK BIONDI and FRIEND

BIG 10 WINTER GOLD

1. American Pie-Don McLean-MCA
2. Down By The Lazy River-The Osmonds-Polydor
*3. Cotton Jenny-Anne Murray-Capitol
4. Let's Stay Together-Al Green-London
5. Without You-Nilsson-RCA
*6. Heart Of Gold-Neil Young-Kinney
7. Joy-Apollo 100-Quality
*8. Love Me, Love Me Love-Frank Mills-Polydor
*9. Good Friends-The Poppy Family-London
10. Hurting Each Other-The Carpenters-A & M

*denotes Canadian talent

(40)	36.	You Are The One-Sugar Bears-Big Tree
(20)	37.	Let's Stay Together-Al Green-HI
(HB)	38.	Son Of My Father-Chicory-Epic
(HB)	39.	Rock Me On The Water-Linda Ronstadt-Capitol
(HB)	40.	A Cowboy's Work Is Never Done-Sonny & Cher-Kapp

*Denotes Big 10 Climbers

WC FL — **ALL HIT MUSIC** — **IN THE MIDWEST**

WAS	IS	Week Ending March 2, 1972
(2)	1.	Joy-Apollo 100-Mega
(4)	2.	Down By The Lazy River-The Osmonds-MGM
(7)	3.	The Lion Sleeps Tonight-Robert John-Atlantic
(9)	4.	Everything I Own-Bread-Elektra
(1)	5.	Without You-Nilsson-RCA
(6)	6.	Anticipation-Carly Simon-Elektra
(10)	7.	Heart Of Gold-Neil Young-Reprise
(11)	8.	My World-Bee Gees-Atco
(3)	9.	Softly Whispering-English Congregation-Atco
(12)	10.	Floy Joy-The Supremes-Motown
(5)	11.	Hurting Each Other-Carpenters-A & M
(16)	12.	The Way Of Love-Cher-Kapp
(14)	13.	Country Wine-The Raiders-Columbia
(18)	14.	Roundabout-Yes-Atlantic
(19)	15.	Ring The Living Bell-Melanie-Neighborhood
(8)	16.	The Witch Queen Of New Orleans-Redbone-Epic
(24)	*17.	Could It Be Forever-David Cassidy-Bell
(26)	*18.	Mother And Child Reunion-Paul Simon-Columbia
(32)	*19.	Puppy Love-Donny Osmond-MGM
(27)	*20.	Rock And Roll Lullaby-B.J. Thomas-Scepter
(13)	21.	Stay With Me-Faces-WB
(25)	22.	Glory Bound-The Grass Roots-Dunhill
(35)	*23.	A Horse With No Name-America-WB
(15)	24.	Black Dog-Led Zeppelin-Atlantic
(33)	*25.	Every Day Of My Life-Bobby Vinton-Epic
(29)	26.	Goin' Down (On The Road to L.A.)-Terry Black/Laurel Ward-Kama Sutra
(30)	27.	Willie Boy's A'Comin'-Tommy James-Routelle
(17)	28.	Kiss An Angel Good Mornin'-Charley Pride-RCA
(31)	29.	Mercedes Benz-Goose Creek Symphony-Capitol
(36)	30.	You Are The One-Sugar Bears-Big Tree
(20)	31.	Don't Say You Don't Remember-B. Bremers-Specter
(34)	32.	Crazy Mama-J.J. Cate-Shelter
(22)	33.	American Pie-Don McLean-United Artists
(40)	34.	A Cowboy's Work Is Never Done-Sonny & Cher-Kapp

BIG 10 ALBUMS

1. America Pie-Don McLean-United Artists
2. Fragile-Yes-Atlantic
3. Carole King Music-Carole King-Ode
4. The Concert For Gangla Desh-Apple
5. Baby I'm-A Want You-Bread-Elektra
6. Harvest-Neil Young-Reprise
7. Hot Rocks-Rolling Stones-London
8. Led Zeppelin-Led Zeppelin-Atlantic
9. The Low Spark Of High Heeled Boys-Traffic-Island
10. Paul Simon-Paul Simon-Columbia

CAROL CHANNING
Sunday-9:00 PM
DICK BIOND and FRIEND

BIG 10 WINTER GOLD

*1. Winter Skies-Judy Collins-(1969)
*2. Winter World Of Love-Engelbert Humperdinck-(1969)
*3. Ice Melts In The Sun-Gary Lewis-(1967)
*4. Snowbird-Anne Murray-(1970)
*5. White Room-Cream-(1968)
6. Winter Warm-Bob Crewe-(1967)
7. Snow-Harper's Bizarre-(1966)
*8. Snowbound-Ferrante & Teicher-(1970)
9. A Winter's Song-The Trolls-(1967)
10. Winter Won't Come This Year-Bill Medley-(1969)

*denotes currently available on a gold single or on an album

(23)	35.	(Love Me) Love The Life I Lead-The Fantastics-Bell
(38)	36.	Son Of My Father-Chicory-Epic
(39)	37.	Rock Me On The Water-Linda Ronstadt-Capitol
(HB)	38.	Rockin' Robin-Michael Jackson-Motown
(HB)	39.	Doctor My Eyes-Jackson Browne-Asylum
(HB)	40.	The First Time Ever I Saw Your Face-Roberta Flack-Atlantic

*Denotes Big 10 Climbers

WC FL — **ALL HIT MUSIC** — **IN THE MIDWEST**

WAS IS Week Ending March 9, 1972

WAS	IS	
(2)	1.	Down By The Lazy River-The Osmonds-MGM
(3)	2.	The Lion Sleeps Tonight-Robert John-Atlantic
(7)	3.	Heart Of Gold-Neil Young-Reprise
(4)	4.	Everything I Own-Bread-Elektra
(12)	*5.	The Way Of Love-Cher-Kapp
(1)	6.	Joy-Apollo 100-Mega
(8)	7.	My World-Bee Gees-Atco
(10)	8.	Floy Joy-The Supremes-Motown
(14)	9.	Roundabout-Yes-Atlantic
(15)	10.	Ring The Living Bell-Melanie-Neighborhood
(5)	11.	Without You-Nilsson-RCA
(17)	12.	Could It Be Forever-David Cassidy-Bell
(6)	13.	Anticipation-Carly Simon-Elektra
(19)	14.	Puppy Love-Donny Osmond-MGM
(23)	*15.	A Horse With No Name-America-WB
(18)	16.	Mother And Child Reunion-Paul Simon-Columbia
(20)	17.	Rock And Roll Lullaby-B.J. Thomas-Scepter
(22)	18.	Glory Bound-The Grass Roots-Dunhill
(9)	19.	Softly Whispering-English Congregation-Atco
(11)	20.	Hurting Each Other-Carpenters-A & M
(25)	21.	Every Day Of My Life-Bobby Vinton-Epic
(16)	22.	The Witch Queen Of New Orleans-Redbone-Epic
(26)	23.	Goin' Down-Terry Black/Laurel Ward-Kama Sutra
(30)	*24.	You Are The One-Sugar Bears-Big Tree
(27)	25.	Willie Boy's A'Comin'-Tommy James-Roulette
(34)	*26.	A Cowboy's Work Is Never Done-Sonny & Cher-Kapp
(13)	27.	Country Wine-The Raiders-Columbia
(29)	28.	Mercedes Benz-Goose Creek Symphony-Capitol
(32)	29.	Crazy Mama-J.J. Cale-Shelter
(38)	*30.	Rockin' Robin-Michael Jackson-Motown
(36)	31.	Son Of My Father-Chicory-Epic
(40)	32.	The First Time Ever I Saw Your Face-Roberts Flack-Atlantic
(37)	33.	Rock Me On The Water-Linda Ronstadt-Capitol
(21)	34.	Stay With Me-Faces-WB
(39)	35.	Doctor My Eyes-Jackson Browne-Asylum

BIG 10 ALBUMS

1. American Pie-Don McLean-United Artists
2. Fragile-Yes-Atlantic
3. Harvest-Neil Young-Reprise
4. Baby I'm-A Want You-Bread-Elektra
5. The Low Spark Of High Heeled Boys-Traffic-Island
6. The Concert For Bangla-Desh-Apple
7. Hot Rocks-Rolling Stones-London
8. Nilsson Schmilsson-Nilisson-RCA
9. Carole King Music-Carole King-Ode
10. Paul Simon-Paul Simon-Columbia

REDBONE
Sunday-9:00 PM
DICK BIONDI and FRIEND

PHILADELPHIA'S BIG 10

*1. Precious And Few-Climax-Carousel
*2. Without You-Nilsson-RCA
*3. Hurting Each Other-The Carpenters-A & M
*4. Joy-Apollo 100-Mega
*5. Let's Stay Together-Al Green-HI
*6. Down By The Lazy River-The Osmonds-MGM
*7. The Lion Sleeps Tonight-Robert John-Atlantic
*8. The Way Of Love-Cher-Kapp
*9. Ring The Living Bell-Melanie Neighborhood
*10. Sweet Seasons-Carole King-Ode

*denotes former or current BIG 10 hit in Chicago.

(24)	36.	Black Dog-Led Zeppelin-Atlantic
(28)	37.	Kiss An Angel Good Mornin'-Charley Pride-RCA
(HB)	38.	Jungle Fever-The Chakachas-Polydor
(HB)	39.	Run Run Run-Jo Jo Gunne-Asylum
(HB)	40.	Someone, Sometime-The New Colony Six-Sunlight

*Denotes Big 10 Climbers

WC
FL

ALL HIT MUSIC

IN THE
MIDWEST

WAS IS Week Ending March 16, 1972

WAS	IS	Title
(1)	1.	Down By The Lazy River-The Osmonds-MGM
(4)	2.	Everything I Owen-Bread-Elektra
(3)	3.	Heart Of Gold-Neil Young-Reprise
(8)	4.	Floy Joy-The Supremes-Motown
(9)	5.	Roundabout-Yes-Atlantic
(14)	*6.	Puppy Love-Donny Osmond-MGM
(16)	*7.	Mother And Child Reunion-Paul Simon-Columbia
(10)	8.	Ring The Living Bell-Melanie-Neighborhood
(12)	9.	Could It Be Forever-David Cassidy-Bell
(15)	*10.	A Horse With No Name-America-WB
(2)	11.	The Lion Sleeps Tonight-Robert John-Atlantic
(5)	12.	The Way Of Love-Cher-Kapp
(18)	13.	Glory Bound-The Grass Roots-Dunhill
(17)	14.	Rock And Roll Lullaby-B.J. Thomas-Scepter
(21)	15.	Every Day Of My Life-Bobby Vinton-Epic
(6)	16.	Joy-Apollo 100-Mega
(30)	*17.	Rockin' Robin-Michael Jackson-Motown
(24)	18.	You Are The One-Sugar Bears-Big Tree
(7)	19.	My World-Bee Gees-Atco
(23)	20.	Goin' Down-Terry Black/Laurel Ward-Kama Sutra
(32)	*21.	The First Time Ever I Saw Your Face-Roberta Flack-Atlantic
(11)	22.	Without You-Nilsson-RCA
(26)	23.	A Cowboy's Work Is Never Done-Sonny & Cher-Kapp
(20)	24.	Hurting Each Other-Carpenters-A & M
(13)	25.	Anticipation-Carly Simon-Elektra
(31)	26.	Son Of My Father-Chicory-Epic
(35)	*27.	Doctor My Eyes-Jackson Browne-Asylum
(19)	28.	Softly Whispering-English Congregation—Atco
(33)	29.	Rock Me On The Water-Linda Ronstadt-Capitol
(38)	*30.	Jungle Fever-The Chakachas-Polydor
(25)	31.	Willie Boy's A'Comin'-Tommy James-Roulette
(39)	32.	Run Run Run-Jo Jo Gunne-Asylum
(28)	33.	Mercedes Benz-Goose Creek Symphony-Capitol
(40)	34.	Someone, Sometime-The New Colony Six-Sunlight
(29)	35.	Crazy Mama-J.J. Cale-Shelter

BIG 10 ALBUMS

1. Harvest-Neil Young-Reprise
2. Fragile-Yes-Atlantic
3. Baby I'm-A Want You-Bread-Elektra
4. American Pie-Don McLean-United Artists
5. Paul Simon-Paul Simon-Columbia
6. Hot Rocks-Rolling Stones-London
7. Nilsson Schmilsson-Nilsson-RCA
8. The Low Spark Of High Heeled Boys-Traffic-Island
9. America-America-WB
10. Carole King Music-Carole King-Ode

THANK YOU CHICAGO
for your acceptance of BIG 10
as *the* ALL HIT MUSIC authority

BIG 10 GREEN GOLD

*1. Green Green-New Christy Minstrels-(1963)
*2. Green Eyed Lady-Sugarloaf-(1970)
*3. Green Green Grass Of Home-Tom Jones-(1967)
*4. Green Onions-Booker T. & The M.G's-(1966)
*5. Green River-C.C. Revival-(1968)
*6. Green Tambourine-Lemon Pipers-(1966)
*7. Green Door-Jim Lowe-(1956)
*8. Green Grass-Gary Lewis-(1966)
*9. Greenback Dollar-Kingston Trio-(1960)
*10. Green Power-Little Richard-(1971)

*denotes currently available on a gold single or on an album

(22)	36.	The Witch Queen Of New Orleans-Redbone-Epic
(HB)	37.	Hot Rod Lincoln-Commander Cody-Paramount
(HB)	38.	Cheer-Potliquor-Janus
(HB)	39.	Day Dreaming-Aretha Franklin-Atlantic
(HB)	40.	Baby Blue-Badfinger-Apple

*Denotes Big 10 Climbers

WC FL **ALL HIT MUSIC** **IN THE MIDWEST**

Week Ending March 23, 1972

BIG 10 ALBUMS

1. Harvest-Neil Young-Reprise
2. America-America-WB
3. Fragile-Yes-Atlantic
4. Nilsson Schmilsson-Nilsson-RCA
5. Baby I'm-A Want You-Bread-Elektera
6. Paul Simon-Paul Simon-Columbia
7. American Pie-Don McLean-United Artists
8. Carole King Music-Carole King-Ode
9. Hot Rocks-Rolling Stones-London
10. The Low Spark Of High Heeled Boys-Traffic-Island

FRANKIE LAINE
Sunday-9:00 PM
DICK BIONDI and FRIEND

AUSTRALIA'S BIG 10

1. Brand New Key-Melanie-Buddah
2. Ernie (The Fastest Milkman In The West)-Benny Hill-Columbia
*3. Ranger's Waltz-Moms & Dads-Crescendo
4. Cherish-David Cassidy-Bell
5. American Pie-Don McLean-UA
*6. Day By Day-Colleen Hewitt-Festival
*7. Captain Zero-The Mixtures-Fable
8. Theme From "Shaft"-Isaac Hayes-Stax
9. Desiderata-Les Crane-WB
10. Imagine-John Lennon-Apple

*denotes Australian talent

WAS	IS	
(3)	1.	Heart Of Gold-Neil Young-Reprise
(4)	2.	Floy Joy-The Supremes-Motown
(6)	3.	Puppy Love-Donny Osmond-MGM
(5)	4.	Roundabout-Yes-Atlantic
(7)	5.	Mother And Child Reunion-Paul Simon-Columbia
(9)	6.	Could It Be Forever-David Cassidy-Bell
(10)	7.	A Horse With No Name-America-WB
(8)	8.	Ring The Living Bell-Melanie-Neighborhood
(1)	9.	Down By The Lazy River-The Osmonds-MGM
(13)	10.	Glory Bound-The Grass Roots-Dunhill
(2)	11.	Everything I Own-Bread-Elektra
(17)	12.	Rockin' Robin-Michael Jackson-Motown
(15)	13.	Every Day Of My Life-Bobby Vinton-Epic
(14)	14.	Rock And Roll Lullaby-B.J. Thomas-Scepter
(21)	*15.	The First Time Ever I Saw Your Face-Roberta Flack-Atlantic
(18)	16.	You Are The One-Sugar Bears-Big Tree
(12)	17.	The Way Of Love-Cher-Kapp
(23)	18.	A Cowboy's Work Is Never Done-Sonny & Cher-Kapp
(26)	*19.	Son Of My Father-Chicory-Epic
(11)	20.	The Lion Sleeps Tonight-Robert John-Atlantic
(30)	*21.	Jungle Fever-The Chakachas-Polydor
(27)	22.	Doctor My Eyes-Jackson Browne-Asylum
(16)	23.	Joy-Apollo 100-Mega
(20)	24.	Goin' Down-Terry Black/Laurel Ward-Kama Sutra
(32)	*25.	Run Run Run-Jo Jo Gunne-Asylum
(29)	26.	Rock Me On The Water-Linda Ronstadt-Capitol
(39)	*27.	Day Dreaming-Aretha Franklin-Atlantic
(19)	28.	My World-Bee Gees-Atco
(22)	29.	Without You-Nilsson-RCA
(24)	30.	Hurting Each Ohter-Carpenters-A & M
(34)	31.	Someone, Sometime-The New Colony Six-Sunlight
(38)	*32.	Cheer-Potliquor-Janus
(37)	33.	Hot Rod Lincoln-Commander Cody-Paramount
(40)	*34.	Baby Blue-Badfinger-Apple
(25)	35.	Anticipation-Carly Simon-Elektra

(31)	36.	Willie Boy's A'Comin'-Tommy James-Roulette
(HB)	37.	Simple Song Of Freedom-Buckwheat-London
(HB)	38.	Eve-Jim Capaldi-Island
(HB)	39.	Vahevella-Ken Loggins/Jim Messina-Columbia
(HB)	40.	Mister Can't You See-Buffy Sainte-Marie-Vanguard

*Denotes Big 10 Climbers

WC | | NEAR
FL | SMALL HIT MUSIC | PAULINA ST.

Week Ending April 1, 1973

*1. Let's Stay Together-Spiro Agnew-Vice
2. Everything I Own-Howard Hughes-$$$
3. Book Of Love-Kenny Rogers & The First Edition-Galley
4. Whatcha See Is Whatcha Get-U.S. Weather Bur.-Possibility
5. The Great Pretender-Clifford Irving-Imposter
6. We Belong Together-Steve & Edie Bono-Lunes
7. Only You Snow & I Know-R. Kissinger & Xuan Thuy-Stalemate
8. Won't Get Fooled Again-John Q. Public-Election
9. What Are You Dong Sunday?-Billy Graham-Pious
10 Have You Scene Her-Otto Preminger-Take One
11. Absolutely Right-William F. Buckley-Firing Line
12. An Old Fashioned Love Song-Rudy Vallee-Bygone
13. How Can You Mend A Broken Heart-Staple Singers-Aorta
14. It's One Of Those Nights-Count Dracula-Type O
15. Brand New Key-Ben Franklin-Com Edison
16. Stay With Me-Richard Nixon-Spiro
17. Heart Of Gold-Dr Christian Barnard-Pump
18. Drowning In The Sea Of Love-Ethel Waters-Waterbed
19. Nothing To Hide-Twiggy-Inflation

coming soon:
THE PIQUA, OHIO DEPT. OF SANITATION CHOIR
on DICK BIONDI and FRIEND

20. Five Hundred Miles-Rand McNally-Pedometer
21. Hurting Each Other-Marquis De Sade-Ouch
22. Down By The Lazy River-Mrs. Johnny Rivers-Siesta
23. Precious And Few-Dollars And Cents-Today
24. L.A. Goodbye-Mayor Yorty-Sabbatical
25. Two Divided By Love-Montagu And Capulet-Veronatown
26. Lady Godiva-Fred Astare-X
27. Love Means Always Having To Say You're Sorry-Don Juan-Byron
28. Puppy Love-American Kennel Club-Scooper
29. Canadian Sunset-Pierre Trudeau-Ballot
30. Mercedes Benz-R. Biondi & J. Kay-Push 'n' Shove

*denotes BIG 10 greasers.

MYOPIA DEPARTMENT: WCFL's "SMALL MUSIC" represents the relative unpopularity of the humor of Tom Konard, who is owned and operated by the WCFL Music Department.

WC | | **IN THE**
FL | **ALL HIT MUSIC** | **MIDWEST**

WAS IS Week Ending March 30, 1972

BIG 10 ALBUMS

1. Harvest-Neil Young-Reprise
2. America-America-WB
3. Fragile-Yes-Atlantic
4. Nilsson Schmilsson-Nilsson-RCA
5. Baby I'm-A Want You-Bread-Elektra
6. Paul Simon-Paul Simon-Columbia
7. American Pie-"Don McLean-United Artists
8. Carole King Music-Carole King-Ode

WAS	IS	
(3)	1.	Puppy Love-Donny Osmond-MGM
(5)	2.	Mother And Child Reunion-Paul Simon-Columbia
(8)	3.	Ring The Living Bell-Melanie-Neighborhood
(7)	4.	A Horse With No Name-America-WB
(6)	5.	Could It Be Forever-David Cassidy-Bell
(1)	6.	Heart Of Gold-Neil Young-Reprise
(10)	7.	Glory Bound-The Grass Roots-Dunhill
(15)	*8.	The First Time Ever I Saw Your Face-Roberta Flack-Atlantic
(12)	9.	Rockin' Robin-Michael Jackson-Motown
(4)	10.	Roundabout-Yes-Atlantic

9. Hot Rocks-Rolling Stones-London
10. First Take-Roberta Flack-Atlantic

PETER YARROW
Sunday-9:00 PM
DICK BIONDI and FRIEND

---

10 YEARS AGO TODAY

*1. She Cried-Jay & The Americans-UA
*2. Mashed Potato Time-Dee Dee Sharp-Cameo
*3. Soldier Boy-The Shirelles-Scepter
*4. Good Luck Charm-Elvis Presley-RCA
*5. Twistin' The Night Away-Sam Cooke-RCA
*6. Hey Baby-Bruce Channel-Smash
*7. Nut Rocker-B. Bumble & The Stingers-Rendezvous
8. Midnight In Moscow-Kenny Ball-Kapp
9. Slow Twistin'-Chubby Checker-Parkway
*10. Shout (Part I)-Joey Dee-Roulette

*denotes currently available on a gold single or on an album

(16) *11. You Are The One-Sugar Bears-Big Tree
(14) 12. Rock And Roll Lullaby-B.J. Thomas-Scepter
(2) 13. Floy Joy-The Supremes-Motown
(18) 14. A Cowboy's Work Is Never Done-Sonny & Cher-Kapp
(21) *15. Jungle Fever-The Chakachas-Polydor
(19) 16. Son Of My Father-Chicory-Epic
(22) 17. Doctor My Eyes-Jackson Browne-Asylum
(9) 18. Down By The Lazy River-The Osmonds-MGM
(11) 19. Everything I Own-Bread-Elektra
(25) 20. Run Run Run-Jo Jo Gunne-Asylum
(26) 21. Rock Me On The Water-Linda Ronstadt-Capitol
(33) *22. Hot Rod Lincoln-Commander Cody-Paramount
(27) 23. Day Dreaming-Aretha Franklin-Atlantic
(32) *24. Cheer-Potliquor-Janus
(13) 25. Every Day Of My Life-Bobby Vinton-Epic
(17) 26. The Way Of Love-Cher-Kapp
(34) *27. Baby Blue-Badfinger-Apple
(31) 28. Someone, Sometime-The New Colony Six-Sunlight
(20) 29. The Lion Sleeps Tonight-Robert John-Atlantic
(37) *30. Simple Song Of Freedom-Buckwheat-London
(40) *31. Mister Can't You See-Buffy Sainte Marie-Vanguard
(23) 32. Joy-Apollo 100-Mega
(38) 33. Eve-Jim Capaldi-Island
(24) 34. Goin' Down-Terry Black/Laurel Ward-Kama Sutra
(39) 35. Vahevella-Ken Loggins/Jim Messina-Columbia
(28) 36. My World-Bee Gees-Atco
(HB) 37. The Family Of Man-Three Dog Night-Dunhill
(HB) 38. Back Off Boogaloo-Ringo Starr-Apple
(HB) 39. You Could Have Been A Lady-April Wine-Big Tree
(HB) 40. Make You Break You-Gerry Rafferty-Signpost

*Denotes Big 10 Climbers

WC
FL

ALL HIT MUSIC

IN THE
MIDWEST

WAS	IS	Week Ending April 6, 1972
(4)	1.	A Horse With No Name-America-WB
(8)	2.	The First Time Ever I Saw Your Face-Roberta Flack-Atlantic
(9)	3.	Rockin' Robin-Michael Jackson-Motown
(1)	4.	Puppy Love-Donny Osmond-MGM
(2)	5.	Mother And Child Reunion-Paul Simon-Columbia
(12)	*6.	Rock And Roll Lullaby-B.J. Thomas-Scepter
(7)	7.	Glory Bound-The Grass Roots-Dunhill
(14)	*8.	A Cowboy's Work Is Never Done-Sonny & Cher-Kapp
(6)	9.	Heart Of Gold-Neil Young-Reprise
(11)	10.	You Are The One-Sugar Bears-Big Tree
(15)	11.	Jungle Fever-The Chakachas-Polydor
(17)	12.	Doctor My Eyes-Jackson Browne-Asylum
(16)	13.	Son Of My Father-Chicory-Epic
(3)	14.	Ring The Living Bell-Melanie-Neighborhood
(22)	*15.	Hot Rod Lincoln-Commander Cody-Paramount
(5)	16.	Could It Be Forever-David Cassidy-Bell
(20)	17.	Run Run Run-Jo Jo Gunne-Asylum
(24)	*18.	Cheer-Potliquor-Janus
(23)	19.	Day Dreaming-Aretha Franklin-Atlantic
(10)	20.	Roundabout-Yes-Atlantic
(27)	*21.	Baby Blue-Badfinger-Apple
(13)	22.	Floy Joy-The Supremes-Motown
(30)	*23.	Simple Song Of Freedom-Buckwheat-London
(28)	24.	Someone, Sometime-The New Colony Six-Sunlight
(18)	25.	Down By The Lazy River-The Osmonds-MGM
(31)	26.	Mister Can't You See-Buffy Sainte Marie-Vanguard
(19)	27.	Everything I Own-Bread-Elektra
(33)	28.	Eve-Jim Capaldi-Island
(38)	*29.	Back Off Boogaloo-Ringo Starr-Apple
(21)	30.	Rock Me On The Water-Linda Ronstadt-Capitol
(35)	31.	Vahevella-Ken Loggins/Jim Messina-Columbia
(37)	32.	The Family Of Man-Three Dog Night-Dunhill
(26)	33.	The Way Of Love-Cher-Kapp
(40)	34.	Make You, Break You-Gerry Rafferty-Signpost
(39)	35.	You Could Have Been A Lady-April

BIG 10 ALBUMS

1. America-America-WB
2. Harvest-Neil Young-Reprise
3. Fragile-Yes-Atlantic
4. Paul Simon-Paul Simon-Columbia
5. Nilsson Schmilsson-Nilsson-RCA
6. First Take-Roberta Flack-Atlantic
7. Baby I'm-A Want You-Bread-Elektra
8. American Pie-Don McLean-United Artists
9. The Low Spark Of High Heeled Boys-Traffic-Island
10. Carole King Music-Carole King-Ode

SMALL PRINT DEPARTMENT: WCFL's "ALL HIT MUSIC" represents relative popularity of records based on Industry tabulations, record sales and requests. WCFL The Voice of Labor, is o

BIG 10 FOOL'S GOLD

*1. Fool Me-The Youngbloods-(1967)
2. April Fool-Matt Monro-(1961)
*3. The Fool On The Hill-The Beatles-(1967)
*4. A Fool Such As I-Elvis Presley-(1959)
*5. A Fool In Love-Ike & Tina Turner-(1968)
*6. Foolish Little Girl-The Shirelles-(1962)
7. Fools Fall In Love-The Drifters-(1958)
*8. Foolin' Around-The Youngbloods-(1967)
9. Fools Rush In-Rick Nelson-(1961)
10. Fool Number One-Brenda Lee-(1960)

*denotes currently available in English Language releases.

		Wine-Big Tree
(HB)	36.	Suavecito-Malo-WB
(HB)	37.	In The Rain-Dramatics-Volt
(HB)	38.	Iko Iko-Dr. John-Atco
(HB)	39.	Taurus-Dennis Coffey-Sussex
(HB)	40.	Walk In The Night-Jr. Walker/All Stars-Soul

*Denotes Big 10 Climbers

WC
FL

ALL HIT MUSIC IN THE MIDWEST

WAS	IS	Week Ending April 13, 1972
(2)	1.	The First Time Ever I Saw Your Face-Roberta Flack-Atlantic
(3)	2.	Rockin' Robin-Michael Jackson-Motown
(1)	3.	A Horse With No Name-America-WB
(8)	4.	A Cowboy's Work Is Never Done-Sonny & Cher-Kapp
(11)	*5.	Jungle Fever-The Chakachas-Polydor
(10)	6.	You Are The One-Sugar Bears-Big Tree
(19)	*7.	Day Dreaming-Aretha Franklin-Atlantic
(4)	8.	Puppy Love-Donny Osmond-MGM
(15)	*9.	Hot Rod Lincoln-Commander Cody-Paramount
(13)	10.	Son Of My Father-Chicory-Epic
(12)	11.	Doctor My Eyes-Jackson Browne-Asylum
(17)	12.	Run Run Run-Jo Jo Gunne-Asylum
(9)	13.	Heart Of Gold-Neil Young-Reprise
(18)	14.	Cheer-Potliquor-Janus
(21)	15.	Baby Blue-Badfinger-Apple
(5)	16.	Mother And Child Reunion-Paul Simon-Columbia
(23)	*17.	Simple Song Of Freedom-Buckwheat-London
(6)	18.	Rock And Roll Lullaby-B.J. Thomas-Scepter
(26)	*19.	Mister Can't You See-Buffy Sainte-Marie-Vanguard
(29)	*20.	Back Off Boogaloo-Ringo Starr-Apple
(7)	21.	Glory Bound-The Grass Roots-Dunhill
(24)	22.	Someone, Sometime-The New Colony Six-Sunlight
(32)	*23.	The Family Of Man-Three Dog Night-Dunhill
(28)	24.	Eve-Jim Capaldi-Island
(14)	25.	Ring The Living Bell-Melanie-Neighborhood
(20)	26.	Roundabout-Yes-Atlantic
(37)	*27.	In The Rain-Dramatics-Volt
(31)	28.	Vahevella-Ken Loggins/Jim Messina-Columbia
(34)	29.	Make You, Break You-Gerry Rafferty-Signpost
(16)	30.	Could It Be Forever-David Cassidy-Bell
(39)	31.	Taurus-Dennis Coffey-Sussex
(35)	32.	You Could Have Been A Lady-April Wine-Big Tree
(40)	33.	Walk In The Night-Jr. Walker/All Stars-Soul
(36)	34.	Suavecito-Malo-WB
(38)	35.	Iko Iko-Dr. John-Atco

BIG 10 ALBUMS

1. America-America-WB
2. Harvest-Neil Young-Reprise
3. First Take-Roberta Flack-Atlantic
4. Paul Simon-Paul Simon-Columbia
5. Fragile-Yes-Atlantic
6. Nilsson Schmilsson-Nilsson-RCA
7. Baby I'm-A Want You-Bread-Elektra
8. Eat A Peach-Allman Brothers Band-Capricorn
9. American Pie-Don McLean-United Artists
10. The Concert For Bangla Desh-Apple

PETER YARROW – Part Two
Sunday-9:00 PM
DICK BIONDI and FRIEND

SOUTH AFRICA'S BIG 10

*1. Brand New Key-Melanie-Buddah
*2. Softly Whispering I Love You-Congregation-Columbia
3. Mother-Barbra Streisand-CBS
4. I Will Return-Springwater-Polydor
5. Mother-John Lennon-Parlophone
6. I Just Can't Help Believing-Elvis Presley-RCA
*7. Son Of My Father-Chicory Tip-CBS
8. Soley Soley-Middle Of The Road-RCA
*9. Banks Of The Ohio-Olivia Newton-John-Polydor
*10. Mother And Child Reunion-Paul Simon-CBS

*denotes former or current BIG 10 hit in Chicago.

(25)	36.	Down By The Lazy River-The Osmonds-MGM
(HB)	37.	Morning Has Broken-Cat Stevens-W & M
(HB)	38.	Sylvia's Mother-Dr. Hook & Medicine Show-Columbia
(HB)	39.	Little Bitty Pretty One-The Jackson 5-Motown
(HB)	40.	I Saw The Light-Todd Rundgren-Bearsville

*Denotes Big 10 Climbers

WC
FL

ALL HIT MUSIC

IN THE
MIDWEST

WAS	IS	Week Ending April 20, 1972
(1)	1.	The First Time Ever I Saw Your Face-Roberta Flack-Atlantic
(2)	2.	Rockin' Robin-Michael Jackson-Motown
(4)	3.	A Cowboy's Work Is Never Done-Sonny & Cher-Kapp
(9)	*4.	Hot Rod Lincoln-Commander Cody-Paramount
(3)	5.	A Horse With No Name-America-WB
(6)	6.	You Are The One-Sugar Bears-Big Tree
(7)	7.	Day Dreaming-Aretha Franklin-Atlantic
(11)	8.	Doctor My Eyes-Jackson Browne-Asylum
(10)	9.	Sun Of My Father-Chicory-Epic
(12)	10.	Run Run Run-JO Jo Gunne-Asylum
(14)	11.	Cheer-Potliquor-Janus
(8)	12.	Puppy Love-Donny Osmond-MGM
(15)	13.	Baby Blue-Badfinger-Apple
(17)	14.	Simple Song Of Freedom-Buckwheat-London
(20)	15.	Back Off Boogaloo-Ringo Starr-Apple
(5)	16.	Jungle Fever-The Chakachas-Polydor
(27)	*17.	In The Rain-Dramatics-Volt
(19)	18.	Mister Can't You See-Buffy Sainte Marie-Vanguard
(22)	19.	Someone, Sometime-The New Colony Six-Sunlight
(23)	20.	The Family Of Man-Three Dog Night-Dunhill
(13)	21.	Heart Of Gold-Neil Young-Reprise
(24)	22.	Eve-Jim Capaldi-Island
(28)	*23.	Vahevella-Ken Loggins/Jim Messina-Columbia
(31)	*24.	Taurus-Dennis Coffey-Sussex
(34)	*25.	Suavecito-Malo-WB
(32)	*26.	You Could Have Been A Lady-April Wine-Big Tree
(29)	27.	Make You, Break You-Gerry Rafferty-Signpost
(16)	28.	Mother And Child Reunion-Paul Simon-Columbia
(33)	29.	Walk In The Night-Jr. Walker/All Stars-Soul
(35)	30.	Iko Iko-Dr. John-Atco
(37)	*31.	Morning Has Broken-Cat Stevens-A & M
(38)	*32.	Sylvia's Mother-Dr. Hook & Medicine Show-Columbia
(18)	33.	Rock And Roll Lullaby-B.J. Thomas-Scepter
(21)	34.	Glory Bound-The Grass Roots-Dunhill
(39)	35.	Little Bitty Pretty One-The Jackson 5-Motown

BIG 10 ALBUMS

1. America-America-WB
2. Harvest-Neil Young-Reprise
3. First Take-Roberta Flack-Atlantic
4. Fragile-Yes-Atlantic
5. Paul Simon-Paul Simon-Columbia
6. Baby I'm-A Want You-Bread-Elektra
7. Nilsson Schmilsson-Nilsson-RCA
8. The Concert For Bangla Desh-Apple
9. Eat A Peach-Allman Brothers Band-Capricorn
10. Carole King Music-Carole King-Ode

GEORGE KIRBY
Sunday-9:00 PM
DICK BIONDI and FRIEND

ELVIS' BIG 10

1. Hound Dog/Don't Be Cruel-(1956)
2. Heartbreak Hotel-(1956)
3. It's Nor Or Never-(1960)
4. In The Ghetto-(1969)
5. All Shook Up-(1957)
6. Are You Lonesome Tonight-(1961)
7. Love Me Tender-(1956)
8. Stuck On You-(1960)
9. Return To Sender-(1963)
10. Kentucky Rain-(1970)

(40)	36.	I Saw The Light-Todd Rundgren-Bearsville
(25)	37.	Ring The Living Bell-Melanie-Neighborhood
(HB)	38.	Look What You Done For Me-Al Green-HI
(HB)	39.	Nice To Be With You-The Gallery-Sussex
(HB)	40.	Tumbling Dice-Rolling Stones-Rolling Stones

*Denotes Big 10 Climbers

WC
FL

ALL HIT MUSIC

IN THE
MIDWEST

WAS IS Week Ending April 27, 1972

WAS	IS	
(3)	1.	A Cowboy's Work Is Never Done-Sonny & Cher-Kapp
(4)	2.	Hot Rod Lincoln-Commander Cody-Paramount
(7)	3.	Day Dreaming-Aretha Franklin-Atlantic
(8)	4.	Doctor My Eyes-Jackson Browne-Asylum
(10)	5.	Run Run Run-Jo Jo-Gunne-Asylum
(2)	6.	Rockin' Robin-Michael Jackson-Motown
(13)	*7.	Baby Blue-Badfinger-Apple
(15)	*8.	Back Off Boogaloo-Ringo Starr-Apple
(11)	9.	Cheer-Potliquor-Janus
(1)	10.	The First Time Ever I Saw Your Face-Roberta Flack-Atlantic
(14)	11.	Simple Song Of Freedom-Buckwheat-London
(17)	12.	In The Rain-Dramatics-Volt
(5)	13.	A Horse With No Name-America-WB
(18)	14.	Mister Can't You See-Buffy Sainte Marie-Vanguard
(25)	*15.	Suavecito-Malo-WB
(20)	16.	The Family Of Man-Three Dog Night-Dunhill
(22)	*17.	Eve-Jim Capaldi-Island
(6)	18.	You Are The One-Sugar Bears-Big Tree
(23)	19.	Vahevella-Ken Loggins/Jim Messina-Columbia
(24)	20.	Taurus-Dennis Coffey-Sussex
(31)	*21.	Morning Has Broken-Cat Stevens-A & M
(19)	22.	Someone, Sometime-The New Colony Six-Sunlight
(26)	23.	You Could Have Been A Lady-April Wine-Big Tree
(29)	24.	Walk In The Night-Jr. Walker/All Star-Soul
(27)	25.	Make You, Break You-Gerry Rafferty-Signpost
(32)	*26.	Sylvia's Mother-Dr. Hook & Medicine Show-Columbia
(12)	27.	Puppy Love-Donny Osmond-MGM
(30)	28.	Iko Iko-Dr. John-Atco
(35)	*29.	Little Bitty Pretty One-The Jackson 5-Motown
(9)	30.	Son Of My Father-Chicory-Epic
(38)	*31.	Look What You Done For Me-Al Green-HI
(36)	32.	I Saw The Light-Todd Rundgren-Bearsville
(39)	*33.	Nice To Be With You-The Gallery-Sussex
(16)	34.	Jungle Fever-The Chakachas-Polydor
(40)	35.	Tumbling Dice-Rolling Stones-Rolling Stones

BIG 10 ALBUMS

1. First Take-Roberta Flack-Atlantic
2. America-America-WB
3. Harvest-Neil Young-Reprise
4. Fragile-Yes-Atlantic
5. Paul Simon-Paul Simon-Columbia
6. Nilsson Schmilsson-Nilsson-RCA
7. Eat A Peach-Allman Brothers Band-Capricorn
8. Baby I'm-A Want You-Bread-Elektra
9. The Concert For Bangla Desh-Apple
10. Smokin'-Humble Pie-A & M

JACKSON BROWNE
Sunday-9:00 PM
DICK BIONDI and FRIEND

BOSTON'S BIG 10

*1. A Horse With No Name-America-WB
2. Taxi-Harry Chapin-Elektra
3. Living Without You-Manfred Mann-Polydor
*4. A Cowboy's Work Is Never Done-Sonny & Cher-Kapp
5. Changes-David Bowie-RCA
*6. The Lion Sleeps Tonight-Robert John-Atlantic
*7. Glory Bound-Grass Roots-Dunhill
*8. Doctor My Eyes-Jackson Browne-Asylum
*9. The Way Of Love-Cher-Kapp
10. Vincent-Don McLean-UA

*denotes former or current BIG 10 hit in Chicago

(21)	36.	Heart Of Gold-Neil Young-Reprise
(28)	37.	Mother And Child Reunion-Paul Simon-Columbia
(HB)	38.	Oh Girl-Chi-Lites-Brunswick
(HB)	39.	Isn't Life Strange-The Moody Blues-Threshold
(HB)	40.	Song Sung Blue-Neil Diamond-Uni

*Denotes Big 10 Climbers

WC FL **ALL HIT MUSIC** **IN THE MIDWEST**

WAS IS Week Ending May 4, 1972

WAS	IS	
(1)	1.	A Cowboy's Work Is Never Done-Sonny & Cher-Kapp
(3)	2.	Day Dreaming-Aretha Franklin-Atlantic
(4)	3.	Doctor My Eyes-Jackson Browne-Asylum
(7)	4.	Baby Blue-Badfinger-Apple
(5)	5.	Run Run Run-Jo Jo Gunne-Asylum
(8)	6.	Back Off Boogaloo-Ringo Starr-Apple
(2)	7.	Hot Red Lincoln-Commander Cody-Paramount
(12)	8.	In The Rain-Dramatics-Volt
(15)	*9.	Suavecito-Malo-WB
(14)	10.	Mister Can't You See-Buffy Sainte Marie-Vanguard
(11)	11.	Simple Song Of Freedom-Buckwheat-London
(6)	12.	Rockin' Robin-Michael Jackson-Motown
(19)	*13.	Vahevella-Ken Loggins/Jim Messina-Columbia
(10)	14.	The First Time Ever I Saw Your Face-Roberta Flack-Atlantic
(16)	15.	The Family Of Man-Three Dog Night-Dunhill
(20)	16.	Taurus-Dennis Coffey-Sussex
(17)	17.	Eve-Jim Capaldi-Island
(21)	18.	Morning Has Broken-Cat Stevens-A & M
(9)	19.	Cheer-Potliquor-Janus
(23)	20.	You Could Have Been A Lady-April Wine-Big Tree
(24)	21.	Walk In The Night-Jr. Walker/All Stars-Soul
(26)	22.	Sylvia's Mother-Dr. Hook-Columbia
(35)	*23.	Tumbling Dice-Rolling Stones-Rolling Stones
(29)	*24.	Little Bitty Pretty One-The Jackson 5-Motown
(25)	25.	Make You, Break You-Gerry Rafferty-Signpost
(28)	26.	Iko Iko-Dr. John-Atco
(13)	27.	A Horse With No Name-America-WB
(33)	*28.	Nice To Be With You-The Gallery-Sussex
(31)	29.	Look What You Done For Me-Al Green-HI
(32)	30.	I Saw The Light-Todd Rundgren-Bearsville
(18)	31.	You Are The One-Sugar Bears-Big Tree
(22)	32.	Someone, Sometime-The New Colony Six-Sunlight
(38)	33.	Oh Girl-Chi-Lites-Brunswick
(39)	34.	Isn't Life Strange-The Moody Blues-

BIG 10 ALBUMS

1. First Take-Roberta Flack-Atlantic
2. America-America-WB
3. Harvest-Neil Young-Reprise
4. Fragile-Yes-Atlantic
5. Nilsson Schmilsson-Nilsson-RCA
6. Graham Nash-David Crosby-Atlantic
7. Eat A Peach-Allman Brothers Band-Capricorn
8. Baby I'm-A Want You-Bread-Elektra
9. Smokin' Humble Pie-A & M
10. Paul Simon-Paul Simon-Columbia

JACKSON BROWNE-Part Two
Sunday-9:00 PM
DICK BIONDI and FRIEND

BIG 10 MAYDAY GOLD

*1. Rose Garden-Lynn Anderson-(1970)
*2. Tiptoe Through The Tulips-Tiny Tim-(1968)
*3. Poison Ivy-The Coasters-(1959)
*4. Fresh As A Daisy-Emmitt Rhodes-(1970)
*5. Sweet Pea-Tommy Roe-(1968)
*6. Build Me Up Buttercup-The Foundations-(1968)
7. Lilacs And Violets-Leslie Gore-(1966)
*8. Sunflower-Mason Williams-(1968)
9. Dandelion-Rolling Stones-(1968)
*10. Which Way You Goin' Billy?-Poppy Family-(1970)

*denotes currently available on a gold single or on an album.

		Threshold
(27)	35.	Puppy Love-Donny Osmond-MGM
(40)	36.	Song Sung Blue-Neil Diamond-Uni
(30)	37.	Son Of My Father-Chicory-Epic
(HB)	38.	I'll Take You There-The Staple Singers-Stax
(HB)	39.	It's Going To Take Some Time-Carpenters-A & M
(HB)	40.	Automatically Sunshine-The Supremes-Motown

*Denotes Big 10 Climbers

WC
FL

ALL HIT MUSIC

IN THE
MIDWEST

Week Ending May 11, 1972

WAS	IS	
(3)	1.	Doctor My Eyes-Jackson Browne-Asylum
(5)	2.	Run Run Run-Jo Jo Gunne-Asylum
(9)	3.	Suavecito-Malo-WB
(1)	4.	A Cowboy's Work Is Never Done-Sonny & Cher-Kapp
(6)	5.	Back Off Boogaloo-Ringo Starr-Apple
(2)	6.	Day Dreaming-Aretha Franklin-Atlantic
(4)	7.	Baby Blue-Badfinger-Apple
(8)	8.	In The Rain-Dramatics-Volt
(10)	9.	Mister Can't You See-Buffy Sainte Marie-Vanguard
(13)	10.	Vahevella-Ken Loggins/Jim Messina-Columbia
(18)	*11.	Morning Has Broken-Cat Stevens-A & M
(7)	12.	Hot Rod Lincoln-Commander Cody-Paramount
(20)	*13.	You Could Have Been A Lady-April Wine-Big Tree
(15)	14.	The Family Of Man-Three Dog Night-Dunhill
(16)	15.	Taurus-Dennis Coffey-Sussex
(22)	*16.	Sylvia's Mother-Dr. Hook-Columbia
(21)	17.	Walk In The Night-Jr. Walker/All Stars-Soul
(11)	18.	Simple Song Of Freedom-Buckwheat-London
(24)	19.	Little Bitty Pretty One-The Jackson 5-Motown
(23)	20.	Tumbling Dice-Rolling Stones-Rolling Stones
(12)	21.	Rockin' Robin-Michael Jackson-Motown
(33)	*22.	Oh Girl-Chi-Lites-Brunswick
(28)	23.	Nice To Be With You-The Gallery-Sussex
(14)	24.	The First Time Ever-Roberta Flack-Atlantic
(26)	25.	Iko Iko-Dr. John-Atco
(29)	26.	Look What You Done For Me-Al Green-HI
(34)	*27.	Isn't Life Strange-The Moody Blues-Threshold
(30)	28.	I Saw The Light-Todd Rundgren-Bearsville
(36)	*29.	Song Sung Blue-Neil Diamond-Uni
(17)	30.	Eve-Jim Capaldi-Island
(19)	31.	Cheer-Potliquor-Janus
(38)	32.	I'll Take You There-The Staple Singers-Stax
(39)	33.	It's Going To Take Some Time-Carpenters-A & M
(25)	34.	Make You, Break You-Gerry Rafferty-Signpost
(40)	35.	Automatically Sunshine-The Supremes-Motown

BIG 10 ALBUMS

1. First Take-Roberta Flack-Atlantic
2. America-America-WB
3. Harvest-Neil Young-Reprise
4. Fragile-Yes-Atlantic
5. Eat A Peach-Allman Brothers Band-Capricorn
6. Nilsson Schmilsson-Nilsson-RCA
7. Paul Simon-Paul Simon-Columbia
8. Smokin'-Humble Pie-A & M
9. Graham Nash-David Crosby-Atlantic
10. Baby I'm-A Want You-Bread-Elektra

KEN LOGGINS/JIM MESSINA
Sunday-9:00 PM
DICK BIONDI and FRIENDS

ENGLAND'S BIG 10

1. Without You-Nilsson-RCA
2. Beg Steal Or Borrow-New Seekers-Polydor
*3. Alone Again-Gilbert O'Sullivan-MAM
4. American Pie-Don McLean-UA
*5. Hold Your Head Up-Argent-CBS
*6. Meet Me On The Corner-Lindisfarne-Charisma
7. Desiderata-Les Crane-WB
8. The Young New Mexican Puppeteer-Tom Jones-Decca
*9. Back Off Boogaloo-Ringo Starr-Apple
*10. Amazing Grace-Royal Scots Dragoon Guards Band-RCA

*denotes British talent.

(27)	36.	A Horse With No Name-America-WB
(HB)	37.	Walkin' In The Rain With The One I Love-Love Unlimited-Uni
(HB)	38.	(Last Night) I Didn't Get To Sleep At All-The 5th Dimension-Bell
(HB)	39.	How Can I Be Sure-David Cassidy-Bell
(HB)	40.	Dairy-Bread-Elektra

*Denotes Big 10 Climbers

WC FL — **ALL HIT MUSIC** — **IN THE MIDWEST**

WAS IS Week Ending May 18, 1972

WAS	IS	
(3)	1.	Suavecito-Malo-WB
(5)	2.	Back Off Boogaloo-Ringo Starr-Apple
(10)	*3.	Vahevella-Ken Loggins/Jim Messina-Columbia
(1)	4.	Doctor My Eyes-Jackson Browne-Asylum
(11)	*5.	Morning Has Broken-Cat Stevens-A & M
(8)	6.	In The Rain-Dramatics-Volt
(13)	*7.	You Could Have Been A Lady-April Wine-Big Tree
(2)	8.	Run Run Run-Jo Jo Gunne-Asylum
(7)	9.	Baby Blue-Badfinger-Apple
(16)	*10.	Sylvia's Mother-Dr. Hook-Columbia
(6)	11.	Day Dreaming-Aretha Franklin-Atlantic
(15)	12.	Taurus-Dennis Coffey-Sussex
(4)	13.	A Cowboy's Work Is Never Done-Sonny & Cher-Kapp
(14)	14.	The Family Of Man-Three Dog Night-Dunhill
(17)	15.	Walk In The Night-Jr. Walker/All Stars-Soul
(20)	16.	Tumbling Dice-Rolling Stones-Rolling Stones
(19)	17.	Little Bitty Pretty One-The Jackson 5-Motown
(22)	18.	Oh Girl-Chi-Lites-Brunswick
(23)	19.	Nice To Be With You-The Gallery-Sussex
(9)	20.	Mister Can't You See-Buffy Sainte-Marie-Vanguard
(27)	*21.	Isn't Life Strange-The Moody Blues-Threshold
(26)	22.	Look What You Done For Me-Al Green-HI
(12)	23.	Hot Rod Lincoln-Commander Cody-Paramount
(29)	24.	Song Sung Blue-Neil Diamond-Uni
(33)	*25.	It's Going To Take Some Time-Carpenters-A & M
(32)	*26.	I'll Take You There-The Staple Singers-Stax
(28)	27.	I Saw The Light-Todd Rundgren-Bearsville
(18)	28.	Simple Song Of Freedom-Buckwheat-London
(35)	*29.	Automatically Sunshine-The Supremes-Motown
(24)	30.	The First Time Ever-Roberta Flack-Atlantic
(21)	31.	Rockin' Robin-Michael Jackson-Motown
(37)	*32.	Walkin' In The Rain With The One I Love-Love Unlimited-Uni

BIG 10 ALBUMS

1. First Take-Roberta Flack-Atlantic
2. Harvest-Neil Young-Reprise
3. Thick As A Brick-Jethro Tull-Reprise
4. America-America-WB
5. Fragile-Yes-Atlantic
6. Graham Nash-David Crosby-Atlantic
7. Eat A Peach-Allman Brothers Band-Capricorn
8. Smokin'-Humble Pie-A & M
9. Nilsson Schmilsson-Nilsson-RCA
10. Paul Simon-Paul Simon-Columbia

KEN LOGGINS/JIM MESSINA-Part Two
Sunday-9:00 PM
DICK BIONDI and FRIENDS

BIG 10 SUMMER GOLD

*1. Sunshine-Jonathan Edwards-(1971)
*2. Sunny-Bobby Hebb-(1966)
*3. Sunshine Superman-Donovan-(1966)
*4. Her Comes The Sun-Richie Havens-(1971)
*5. Sun King-The Beatles-(1969)
*6. Sunlight-The Youngbloods-(1969)
*7. Ain't No Sunshine-Bill Withers-(1971)
*8. Sunset Strip-Ray Stevens-(1970)
9. Sunshine, Lollipops And Rainbows-Lesley Gore-(1965)
*10. The Sun Keeps On Rising-Kenny Rogers And The First Edition-(1968)

*denotes currently available on a gold single or on an album.

(25)	33.	Iko Iko-Dr. John-Atco
(39)	34.	How Can I Be Sure-David Cassidy-Bell
(38)	35.	(Last Night) I Didn't Get To Sleep At All-The 5th Dimension-Bell
(40)	36.	Diary-Bread-Elektra
(HB)	37.	Amazing Grace-Royal Scots Dragoon Guards-RCA
(HB)	38.	Living In A House Divided-Cher-Kapp
(HB)	39.	I Can Feel You-Addrisi Brothers-Columbia
(HB)	40.	Take It Easy-Eagles-Asylum

*Denotes Big 10 Climbers

WC
FL

ALL HIT MUSIC

IN THE MIDWEST

BIG 10 ALBUMS

1. First Take-Roberta Flack-Atlantic
2. Thick As A Brick-Jethro Tull-Reprise
3. America-America-WB
4. Harvest-Neil Young-Reprise
5. Nilsson Schmilsson-Nilsson-RCA
6. Fragile-Yes-Atlantic
7. Smokin'-Humble Pie-A & M
8. Eat A Peach-Allman Brothers Band-Capricorn
9. Graham Nash-David Crosby-Atlantic
10. Baby I'm-A Want You-Bread-Elektra

MEL TORME –Part Two
Sunday-9:00 PM
DICK BIONDI and FRIEND

BIG 10 GRADUATION DAY GOLD

*1. Graduation Day-The Beach Boys-(1967)
*2. School Is Out-Cat Stevens-(1966)
3. Cap & Gown-Marty Robbins-(1959)
4. Graduation Girl-The Letterman-(1964)
*5. I'll See You In September-The Happenings-(1966)
6. Graduation's Here-The Fleetwoods-(1962)
7. A White Sport Coat And A Pink Carnation-Marty Robbins-(1957)
8. Pomp & Circumstance-Adrian Kimberly-(1963)
*9. Be True To Your School-The Beach Boys-(1967)
10. Your Graduation Means Good-bye-The Cardigans-(1963)

*denotes currently available on a gold single or on an album.

WAS	IS	Week Ending May 25, 1972
(2)	1.	Back Off Boogaloo-Ringo Starr-Apple
(5)	2.	Morning Has Broken-Cat Stevens-A & M
(1)	3.	Suavecito-Malo-WB
(10)	*4.	Sylvia's Mother-Dr. Hook-Columbia
(6)	5.	In The Rain-Dramatics-Volt
(16)	*6.	Tumbling Dice-Rolling Stones-Rolling Stones
(4)	7.	Doctor My Eyes-Jackson Browne-Asylum
(18)	*8.	Oh Girl-Chi-Lites-Brunswick
(19)	*9.	Nice To Be With You-The Gallery-Sussex
(7)	10.	You Could Have Been A Lady-April Wine-Big Tree
(3)	11.	Vahevella-Ken Loggins/Jim Messina-Columbia
(8)	12.	Run Run Run-Jo Jo Gunne-Asylum
(17)	13.	Little Bitty Pretty One-The Jackson 5-Motown
(22)	*14.	Look What You Done For Me-Al Green-HI
(24)	*15.	Song Sung Blue-Neil Diamond-Uni
(21)	16.	Isn't Life Strange-The Moody Blues-Threshold
(25)	*17.	It's Going To Take Some Time-Carpenters-A & M
(15)	18.	Walk In The Night-Jr. Walker/All Stars-Soul
(9)	19.	Baby Blue-Badfinger-Apple
(11)	20.	Day Dreaming-Aretha Franklin-Atlantic
(12)	21.	Taurus-Dennis Coffey-Sussex
(26)	22.	I'll Take You There-The Staple Singers-Stax
(14)	23.	The Family Of Man-Three Dog Night-Dunhill
(29)	24.	Automatically Sunshine-The Supremes-Motown
(13)	25.	A Cowboy's Work Is Never Done-Sonny & Cher-Kapp
(32)	*26.	Walkin' In The Rain With The One I Love-Love Unlimited-Uni
(27)	27.	I Saw The Light-Todd Rundgren-Bearsville
(35)	*28.	(Last Night) I Didn't Get To Sleep At All-The 5th Dimension-Bell
(36)	*29.	Diary-Bread-Elektra
(34)	30.	How Can I Be Sure-David Cassidy-Bell
(20)	31.	Mister Can't You See-Buffy Sainte-Marie-Vanguard
(37)	32.	Amazing Grace-Royal Scots Dragoon Guards-RCA
(39)	33.	I Can Feel You-Addrisi Brothers-Columbia

(38)	34.	Living In A House Divided-Cher-Kapp
(40)	35.	Take It Easy-Eagles-Asylum
(HB)	36.	Outa-Space-Billy Preston-A & M
(HB)	37.	I Need You-America-WB
(HB)	38.	We're On Our Way-Chris Hodge-Apple
(HB)	39.	Ask Me What You Want-Millie Jackson-Spring
(HB)	40.	After Midnight-J.J. Cale-Shelter

*Denotes Big 10 Climbers

WC
FL

ALL HIT MUSIC

IN THE
MIDWEST

WAS IS Week Ending June 1, 1972

BIG 10 ALBUMS

1. Thick As A Brick-Jethro Tull-Reprise
2. First Take-Roberta Flack-Atlantic
3. Harvest-Neil Young-Reprise
4. America-America-WB
5. Graham Nash, David Crosby-Atlantic
6. Fragile-Yes-Atlantic
7. Smokin'-Humble Pie-A & M
8. Eat A Peach-Allman Brothers Band-Capricorn
9. Manassas-Stephen Stills-Atlantic
10. Nilsson Schmilsson-Nilsson-RCA

MEL TORME – Part Two
Sunday – 9:00 PM
DICK BIONDI and FRIEND

MIAMI'S BIG 10

1. In The Rain-Dramatics-Volt
2. I Gotcha-Joe Tex-Dial
3. First Time Ever I Saw-Roberta Flack-Atlantic
4. Roundabout-Yes-Atlantic
5. Family Of Man-Three Dog Night-Dunhill
6. A Cowboy's Work Is Never Done-Sonny & Cher-Kapp
7. Suavecito-Malo-WB
8. A Horse With No Name-America-WB
9. Rockin' Robin-Michael Jackson-Motown
10. Let's Stay Together-Al Green-HI

denotes:**(CAN'T READ THIS OR WHICH ONES TO MARK)**.

1. Morning Has Broken-Cat Stevens-A & M
2. Sylvia's Mother-Dr. Hook-Columbia
3. Tumbling Dice-Rolling Stones-Rolling Stones
4. Back Off Boogaloo-Ringo Starr-Apple
5. Nice To Be With You-The Gallery-Sussex
6. Song Sung Blue-Neil Diamond-Uni
7. Oh Girl-Chi-Lites-Brunswick
8. Isn't Life Strange-The Moody Blues-Threshold
9. Suavecito-Malo-WB
10. Little Bitty Pretty One-The Jackson 5-Motown
11. Look What You Done For Me-Al Green-HI
12. In The Rain-Dramatics-Volt
13. It's Going To Take Some Time-Carpenters-A & M
14. I'll Take You There-The Staple Singers-Stax
15. Automatically Sunshine-The Supremes-Motown
16. (Last Night) I Didn't Get To Sleep At All-The 5th Dimension-Bell
17. Doctor My Eyes-Jackson Browne-Asylum
18. You Could Have Been A Lady-April Wine-Big Tree
19. Amazing Grace-Royal Scots Dragoon Guards-RCA
20. Diary-Bread-Elektra
21. Run Run Run-Jo Jo Gunne-Asylum
22. How Can I Be Sure-David Cassidy-Bell
23. Walkin' In The Rain With The One I Love-Love Unlimited-Uni
24. Vahevella-Ken Loggins/Jim Messina-Columbia
25. I Saw The Light-Todd Rundgren-Bearsville
26. Walk In The Night-Jr. Walker/All Stars-Soul
27. Baby Blue-Badfinger-Apple
28. Living In A House Divided-Cher-Kapp
29. I Can Feel You-Addrisi Brothers-Columbia
30. Outa Space-Billy Preston-A & M
31. Ask Me What You Want-Millie Jackson-Spring
32. I Need You-America-WB
33. Take It Easy-Eagles-Asylum
34. After Midnight-J.J. Cale-Shelter
35. We're On Our Way-Chris Hodge-Apple
36. Daddy Don't You Walk So Fast-W. Newton-Chelsea

37. Troglodyte (Cave Man)-Jimmy Castor Bunch-RCA
38. In A Broken Dream-Python Lee Jackson-?? Crescendo
39. I Wanna Be Where You Are-Michael Jackson-Motown
40. Day By Day-Godspell-Bell

*Denotes Big 10 Climbers **(CAN'T READ WHICH ONES TO MARK**

WC
FL

ALL HIT MUSIC

IN THE
MIDWEST

WAS IS Week Ending June 8 1972

WAS	IS	
(2)	1.	Sylvia's Mother-Dr. Hook-Columbia
(3)	2.	Tumbling Dice-Rolling Stones-Rolling Stones
(5)	3.	Nice To Be With You-The Gallery-Sussex
(6)	4.	Song Sung Blue-Neil Diamond-Uni
(7)	5.	Oh Girl-Chi-Lites-Brunswick
(1)	6.	Morning Has Broken-Cat Stevens-A & M
(13)	*7.	It's Going To Take Some Time-Carpenters-A & M
(8)	8.	Isn't Life Strange-The Moody Bleus-Threshold
(11)	9.	Look What You Done For Me-Al Green-HI
(10)	10.	Little Bitty Pretty One-The Jackson 5-Motown
(16)	11.	(Last Night) I Didn't Get To Sleep At All-The 5th Dimension-Bell
(14)	12.	I'll Take you There-The Staple Singers-Stax
(4)	13.	Back Off Boogaloo-Ringo Starr-Stax
(15)	14.	Automatically Sunshine-The Supremes-Motown
(20)	15.	Diary-Bread-Elektra
(19)	16.	Amazing Grace-Royal Scots Dragoon Guards-RCA
(30)	*17.	Outa-Space-Billy Preston-A & M
(9)	18.	Suavecito-Malo-WB
(23)	19.	Walkin' In The Rain-Love Unlimited-Uni
(22)	20.	How Can I Be Sure-David Cassidy-Bell
(28)	*21.	Living In A House Divided-Cher Kapp
(12)	22.	In The Rain-Dramatics-Volt
(29)	*23.	I Can Feel You-Addrisi Brothers-Columbia
(32)	*24.	I Need You-America-WB
(31)	25.	Ask Me What You Want-Millie Jackson-Spring
(34)	*26.	After Midnight-J.J. Cale-Shelter
(18)	27.	You Could Have Been A Lady-April Wine-Big Tree
(25)	28.	I Saw The Light-Todd Rundgren-Bearsville
(33)	29.	Take It Easy-Eagles-Asylum
(37)	*30.	Troglodyte (Cave Man)-Jimmy Castor Bunch-RCA
(39)	*31.	I Wanna Be Where You Are-Michael Jackson-Motown
(35)	32.	Doctor My Eyes-Jackson Browne-Apple
(17)	33.	Doctor M Eyes-Jackson Browne-Asylum
(36)	34.	Daddy Don't You Walk So Fast-W. Newton-Chelsea
(40)	35.	Day By Day-Godspell-Bell

BIG 10 ALBUMS

1. Thick As A Brick-Jethro Tull-Reprise
2. Harvest-Neil Young-Reprise
3. First Take-Roberta Flack-Atlantic
4. Exile On Main St.-Rolling Stones-Rolling Stones
5. Graham Nash, David Crosby-Atlantic
6. America-America-WB
7. Fragile-Yes-Atlantic
8. Smokin'-Humble Pie-A & M
9. Eat A Peach-Allman Brothers Band-Capricorn
10. Manassas-Stephen Stills-Atlantic

THANK YOU CHICAGO
for your acceptance of BIG 10
as *the* ALL HIT MUSIC authority

BIG 10 SOUTHERN GOLD

*1. The Night They Drove Old Dixie Down-Joan Baez-(1971)
*2. Carolina On My Mind-James Taylor-(1971)
*3. Cottonfields-The Highwaymen-(1962)
*4. Rainy Night In Georgia-Brook Benton-(1970)
*5. Mississippi Queen-Mountain-(1970)
*6. Born On The Bayou-Credence Clearwater Revival-(1969)
*7. Shilo-Neil Diamond-(1970)
*8. Alabama-Neil Young-(1972)
*9. Louisiana Man-The Seekers-(1966)
10. Rebel Rouser-Duane Eddy-(1958)

*denotes currently available on a gold single or on an album.

(38)	36.	In A Broken Dream-Python Lee Jackson-Crescendo
(21)	37.	Run Run Run-Jo Jo Gunne-Asylum
(HB)	38.	Too Young-Donny Osmond-MGM
(HB)	39.	Little Bit O'Soul-Bullet-Big Tree
(HB)	40.	Too Late To Turn Back Now-Cornelius Brothers/Sister Rose-United Artists

*Denotes Big 10 Climbers

WC
FL

ALL HIT MUSIC

IN THE
MIDWEST

BIG 10 ALBUMS

1. Thick As A Brick-Jethro Tull-Reprise
2. Exile On Main St.-Rolling Stones-Rolling Stones
3. Harvest-Neil Young-Reprise
4. Graham Nash, David Crosby-Atlantic
5. First Take-Roberta Flack-Atlantic
6. America-America-WB
7. Manassas-Stephen Stills-Atlantic
8. Eat A Peach-Allman Brothers Band-Capricorn
9. Fragile-Yes-Atlantic
10. Smokin' Humble Pie-A & M

WCFL PRESENTS ELVIS!
Hundreds of Highest-Priced Elvis Tickets...
Hundreds of Elvis' Most Electrifying Hits...
ONLY on WCFL!

ELVIS PRESLEY GOLD

*1. In The Ghetto-(1969)
*2. Heartbreak Hotel-(1956)
*3. Can't Help Falling In Love With You-(1962)
*4. Teddy Bear-(1958)
*5. Don't Cry Daddy-(1970)
*6. All Shook Up-(1957)
*7. Are You Lonesome Tonight-(1961)
*8. Don't Be Cruel/Hound Dog-(1956)
*9. Suspicious Minds-(1969)
*10. Love Me Tender-(1956)

*Hear all of these and more during ELVIS PRESLEY WEEK on WCFL!

WAS	IS	Week Ending June 15, 1972
(3)	1.	Nice To Be With You-The Gallery-Sussex
(2)	2.	Tumbling Dice-Rolling Stones-Rolling Stones
(4)	3.	Song Sung Blue-Neil Diamond-Uni
(1)	4.	Sylvia's Mother-Dr. Hook-Columbia
(5)	5.	Oh Girl-Chi-Lites-Brunswick
(11)	6.	(Last Night) I Didn't Get To Sleep At All-The 5th Dimension-Bell
(14)	*7.	Automatically Sunshine-The Supremes-Motown
(15)	*8.	Diary-Bread-Elektra
(7)	9.	It's Going To Take Some Time-Carpenters-A & M
(16)	10.	Amazing Grace-Royal Scot Dragoon Guards-RCA
(17)	*11.	Outa-Space-Billy Preston-A & M
(12)	12.	I'll Take You There-The Staple Singers-Stax
(10)	13.	Little Bitty Pretty One-The Jackson 5-Motown
(30)	*14.	Troglodyte (Cave Man)-Jimmy Castor Bunch-RCA
(20)	15.	How Can I Be Sure-David Cassidy-Bell
(19)	16.	Walkin' In The Rain-Love Unlimited-Uni
(8)	17.	Isn't Life Strange-The Moody Blues-Threshold
(--)	18.	The Candy Man-Sammy Davis, Jr.-MGM
(21)	19.	Living In A House Divided-Cher-Kapp
(24)	20.	I Need You-America-WB
(23)	21.	I Can Feel You-Addrisi Brothers-Columbia
(9)	22.	Look What You Done For Me-Al Green-HI
(6)	23.	Morning Has Broken-Cat Stevens-A & M
(25)	24.	Ask Me What You Want-Millie Jackson-Spring
(31)	25.	I Wanna Be Where You Are-Michael Jackson-Motown
(26)	26.	After Midnight-J.J. Cale-Shelter
(29)	27.	Take It Easy-Eagles-Asylum
(13)	28.	Back Off Boogaloo-Ringo Starr-Apple
(32)	29.	We're On Our Way-Chris Hodge-Apple
(35)	30.	Day By Day-Godspell-Bell
(34)	31.	Daddy Don't You Walk So Fast-W. Newton-Chelsea
(36)	32.	In A Broken Dream-Python Lee Jackson-Crescendo
(18)	33.	Suavecito-Malo-WB
(39)	34.	Little Bit O'Soul-Bullet-Big Tree
(38)	35.	Too Young-Donny Osmond-MGM

(40)	36.	Too Late To Turn Back Now-Cornelius Brothers-Sister Rose-United Artists
(22)	37.	In The Rain-Dramatics-Volt
(--)	38.	Lean On Me-Bill Withers-Sussex
(--)	39.	I've Been Lonely For So Long-Fredrick Knight-Stax
(--)	40.	School's Out-Alice Cooper-WB

*Denotes Big 10 Climbers

WC FL — **ALL HIT MUSIC** — **IN THE MIDWEST**

WAS IS Week Ending June 22, 1972

WAS	IS	
(2)	1.	Tumbling Dice-Rolling Stones-Rolling Stones
(3)	2.	Song Sung Blue-Neil Diamond-Uni
(1)	3.	Nice To Be With You-The Gallery-Sussex
(6)	4.	I Didn't Get To Sleep At All-5th Dimension-Bell
(11)	5.	Outa-Space-Billy Preston-A & M
(7)	6.	Automatically Sunshine-The Supremes-Motown
(8)	7.	Diary-Bread-Elektra
(14)	*8.	Troglodyte (Cave Man)-Jimmy Castor Bunch-RCA
(18)	*9.	The Candy Man-Sammy Davis, Jr.-MGM
(10)	10.	Amazing Grace-Royal Scots Dragoon Guards-RCA
(5)	11.	Oh Girl-Chi-Lites-Brunswick
(9)	12.	It's Going To Take Some Time-Carpenters-A & M
(4)	13.	Sylvia's Mother-Dr. Hook-Columbia
(19)	14.	Living In A House Divided-Cher-Kapp
(20)	15.	I Need You-America-WB
(16)	16.	Walkin' In The Rain-Love Unlimited-Uni
(12)	17.	I'll Take You There-The Staple Singers-Stax
(13)	18.	Little Bitty Pretty One-The Jackson 5-Motown
(27)	*19.	Take It Easy-Eagles-Asylum
(21)	20.	I Can Feel You-Addrisi Brothers-Columbia
(35)	*21.	Too Young-Donny Osmond-MGM
(24)	22.	Ask Me What You Want-Millie Jackson-Spring
(25)	23.	I Wanna Be Where You Are-Michael Jackson-Motown
(29)	24.	We're On Our Way-Chris Hodge-Apple
(38)	*25.	Lean On Me-Bill Withers-Sussex
(30)	26.	Day By Day-Godspell-Bell
(31)	27.	Daddy Don't You Walk So Fast-W. Newton-Chelsea
(15)	28.	How Can I Be Sure-David Cassidy-Bell
(32)	29.	In A Broken Dream-Python Lee Jackson-Crescendo
(36)	30.	Too Late To Turn Back Now-Cornelius Brothers-UA
(22)	31.	Look What You Done For Me-Al Green-HI
(17)	32.	Isn't Life Strange-The Moody Blues-Threshold
(34)	33.	Little Bit O' Soul-Bullet-Big Tree
(40)	34.	School's Out-Alice Cooper-WB
(39)	35.	Been Lonely For So Long-

BIG 10 ALBUMS

1. Thick As A Brick-Jethro Tull-Reprise
2. Exile On Main St.-Rolling Stones-Rolling Stones
3. Harvest-Neil Young-Reprise
4. Graham Nash, David Crosby-Atlantic
5. First Take-Roberta Flack-Atlantic
6. Honky Chateau-Elton John-Uni
7. Manassas-Stephen Stills-Atlantic
8. America-America-WB
9. Eat A Peach-Allman Brothers Band-Capricorn
10. Joplin In Concert-Janis Joplin-Columbia

ART ROBERTS
in the morning!

CANADA'S BIG 10

1. Nice To Be With You-Gallery-Sussex
2. Sylvia's Mother-Dr. Hook-Columbia
3. Hot Rod Lincoln-Commander Cody-Paramount
4. Amazing Grace-Royal Scots Dragoons-RCA
5. Candy Man-Sammy Davis, Jr.-MGM
6. Song Sung Blue-Neil Diamond-Uni
7. Oh Girl-Chi-Lites-Brunswick

*8. Poor Little Fool-Frank Mills-Polydor
9. Rockin- Robin-Michael Jackson-Tamla Motown
*10. Wild Eyes-The Stampeders-MWC

*denotes Canadian talent
(courtesy: CHAM Radio-Hamilton)

		Fredrick Knight-Stax
(23)	36.	Morning Has Broken-Cat Stevens-A & M
(--)	37.	Happiest Girl In The Whole U.S.A.-Donna Fargo-Dot
(--)	38.	Where Is The Love-Flack/Hathaway-Atlantic
(--)	39.	Hold Her Tight-The Osmonds-MGM
(--)	40.	Alone Again (Naturally)-Gilbert O'Sullivan-MAM

*Denotes Big 10 Climbers

WC
FL

ALL HIT MUSIC IN THE MIDWEST

WAS	IS	Week Ending June 29, 1972
(5)	1.	Outa-Space-Billy Preston-A & M
(4)	2.	I Didn't Get To Sleep At All-5th Dimension-Bell
(9)	3.	The Candy Man-Sammy Davis, Jr.-MGM
(1)	4.	Tumbling Dice-Rolling Stones-Rolling Stones
(7)	5.	Diary-Bread-Elektra
(6)	6.	Automatically Sunshine-The Supremes-Motown
(8)	7.	Troglodyte (Cave Man0-Jimmy Castor Bunch-RCA
(2)	8.	Song Sung Blue-Neil Diamond-Uni
(14)	9.	Living In A House Divided-Cher-Kapp
(3)	10.	Nice To Be With You-The Gallery-Sussex
(15)	11.	I Need You-America-WB
(21)	*12.	Too Young-Donny Osmond-MGM
(19)	13.	Take It Easy-Eagles-Asylum
(25)	*14.	Lean On Me-Bill Withers-Sussex
(12)	15.	It's Going To Take Some Time-Carpenters-A & M
(23)	*16.	I Wanna Be Where You Are-Michael Jackson-Motown
(22)	17.	Ask Me What You Want-Millie Jackson-Spring
(10)	18.	Amazing Grace-Royal Scots Dragoon Guards-RCA
(27)	*19.	Daddy Don't You Walk So Fast-W. Newton-Chelsea
(30)	*20.	Too Late To Turn Back Now-Cornelius Brothers-UA
(11)	21.	Oh Girl-Chi-Lites-Brunswick
(26)	22.	Day By Day-Godspell-Bell
(24)	23.	We're On Our Way-Chris Hodge-Apple
(16)	24.	Walkin' In The Rain-Love Unlimited-Uni
(17)	25.	I'll Take You There-The Staple Singers-Stax
(34)	*26.	School's Out-Alice Cooper-WB
(33)	27.	Little Bit O' Soul-Bullet-Big Tree
(13)	28.	Sylvia's Mother-Dr. Hook-Columbia
(29)	29.	In A Broken Dream-Python Lee Jackson-Crescendo
(35)	30.	Been Lonely For So Long-Fredrick Knight-Stax
(18)	31.	Little Bitty Pretty One-The Jackson 5-Motown
(37)	32.	Happiest Girl In The Whole U.S.A.-Donna Fargo-Dot
(38)	33.	Where Is The Love-Flack/Hathaway-Atlantic
(40)	34.	Alone Again (Naturally)-Gilbert O'Sullivan-MAM
(39)	35.	Hold Her Tight-The Osmonds-MGM

BIG 10 ALBUMS

1. Thick As A Brick-Jethro Tull-Reprise
2. Exile On Main St.-Rolling Stones-Rolling Stones
3. Honky Chateau-Elton John-Uni
4. Harvest-Neil Young-Reprise
5. Manassas-Stephen Stills-Atlantic
6. Joplin In Concert-Janis Joplin-Columbia
7. America-America-WB
8. Graham Nash,David Crosby-Atlantic
9. First Take-Roberta Flack-Atlantic
10. Eat A Peach-Allman Brothers Band-Capricorn

ART ROBERTS
mornings – sunny side up

---

10 YEARS AGO TODAY

*1. The Stripper-David Rose-MGM
*2. I Can't Stop Loving You-Ray Charles-ABC
*3. The Man Who Shot Liberty Valance-Gene Pitney-Musicor
*4. Palisades Park-Freddy Cannon-Swan
*5. Roses Are Red-Bobby Vinton-Epic
6. Violetta-Ray Adams-Laurie
*7. Stranger On The Shore-Acker Bilk-Atco
8. Cindy's Birthday-Johnny Crawford- Del Fi
9. West Of The Wall-Toni Fisher-Big Top
10. It Keeps Right On A-Hurtin'-Johnny Tillotson-Cadence

*denotes currently available on a gold single or on an album.

(20)	36.	I Can Feel You-Addrisi Brothers-Columbia
(--)	37.	How Do You Do?-Mouth & MacNeal-Philips
(--)	38.	I Don't Want To Be Right-Luther Ingram-Koko
(--)	39.	Brandy-Looking Glass-Epic
(--)	40.	Jesahel-English Congregation-Signpost

*Denotes Big 10 Climbers

WC
FL

ALL HIT MUSIC IN THE MIDWEST

BIG 10 ALBUMS

1. Thick As A Brick-Jethro Tull-Reprise
2. Exile On Main Street-Rolling Stones-Rolling Stones
3. Honky Chateau-Elton John-Uni
4. Joplin In Concert-Janis Joplin-Columbia
5. America-America-WB
6. Harvest-Neil Young-Reprise
7. Manassas-Stephen Stills-Atlantic
8. Graham Nash,David Crosby-Atlantic
9. Roberta Flack & Donny Hathaway-Atlantic
10. First Take-Roberta Flack-Atlantic

SUPERSTAR
all day – every day
on WCFL!

DENVER'S BIG 10

*1. Nice To Be With You-The Gallery-Sussex
*2. Song Sung Blue-Neil Diamond-Uni
*3. The Candy Man-Sammy Davis, Jr.-MGM
*4. Diary-Bread-Elektra
*5. Sylvia's Mother-Dr. Hook-Columbia
*6. I Need You-America-WB
*7. Oh Girl-The Chi-Lites-Brunswick
*8. Outa-Space-Billy Preston-Asylum
*9. Last Night-5th Dimension-Bell
*10. Tumbling Dice-Rolling Stones-Rolling Stones

*denotes current or former WCFL hit in Chicago.
(courtesy: KTLK Radio-Denver

WAS	IS	Week Ending July 6, 1972
(7)	1.	Troglodyte (Cave Man)-Jimmy Castor Bunch-RCA
(2)	2.	I Didn't Get To Sleep At All-5th Dimension-Bell
(3)	3.	The Candy Man-Sammy Davis, Jr.-MGM
(9)	4.	Living In A House Divided-Cher-Kapp
(1)	5.	Outa-Space-Billy Preston-A & M
(13)	*6.	Take It Easy-Eagles-Asylum
(11)	7.	I Need You-America-WB
(14)	*8.	Lean On Me-Bill Withers-Sussex
(12)	9.	Too Young-Donny Osmond-MGM
(6)	10.	Automatically Sunshine-The Supremes-Motown
(16)	11.	I Wanna Be Where You Are-Michael Jackson-Motown
(4)	12.	Tumbling Dice-Rolling Stones-Rolling Stones
(19)	*13.	Daddy Don't You Walk So Fast-W. Newton-Chelsea
(5)	14.	Diary-Bread-Elektra
(26)	*15.	School's Out-Alice Cooper-WB
(20)	16.	Too Late To Turn Back Now-Cornelius Brothers-UA
(22)	17.	Dad By Day-Godspell-Bell
(8)	18.	Song Sung Blue-Neil Diamond-Uni
(17)	19.	Ask Me What You Want-Millie Jackson-Spring
(10)	20.	Nice To Be With You-The Gallery-Sussex
(23)	21.	We're On Our Way-Chris Hodge-Apple
(33)	*22.	Where Is The Love-Flack/Hathaway-Atlantic
(15)	23.	It's Going To Take Some Time-Carpenters-A & M
(34)	*24.	Alone Again (Naturally)-Gilbert O'Sullivan-MAM
(30)	25.	Been Lonely For So Long-Frederick Knight-Stax
(29)	26.	In A Broken Dream-Python Lee Jackson-Crescendo
(27)	27.	Little Bit O' Soul-Bullet-Big Tree
(32)	28.	Happiest Girl In The Whole U.S.A.-Donna Fargo-Dot
(35)	*29.	Hold Her Tight-The Osmonds-MGM
(18)	30.	Amazing Grace-Royal Scots Dragoon Guards-RCA
(37)	31.	How Do You Do?-Mouth & MacNeal-Philips
(21)	32.	Oh Girl-Chi-Lites-Brunswick
(38)	33.	I Don't Want To Be Right-Luther Ingram-Koko
(39)	34.	Brandy-Looking Glass-Epic
(40)	35.	Jesahel-English Congregation-Signpost
(25)	36.	I'll Take You There-The Staple Singers-Stax
(--)	37.	Conquistador-Procol Harum-A & M
(--)	38.	Long Cool Woman-The Hollies-Epic

(--)	39.	Small Beginnings-Flash-Capitol
(--)	40.	Rock And Roll Part 2-Gary Glitter-Bell

*Denotes Big 10 Climbers

WC FL — ALL HIT MUSIC — IN THE MIDWEST

WAS	IS	Week Ending July 15, 1972
(8)	1.	Lean On Me-Bill Withers-Sussex
(3)	2.	The Candy Man-Sammy Davis, Jr.-MGM
(9)	3.	Too Young-Donny Osmond-MGM
(6)	4.	Take It Easy-Eagles-Asylum
(7)	5.	I Need You-America-WB
(16)	*6.	Too Late To Turn Back Now-Cornelius Brothers-UA
(5)	7.	Outa-Space-Billy Preston-A & M
(13)	8.	Daddy Don't You Walk So Fast-W. Newton-Chelsea
(1)	9.	Troglodyte (Cave Man)-Jimmy Castor Bunch-RCA
(2)	10.	I Didn't Get To Sleep At All-5th Dimension-Bell
(15)	11.	School's Out-Alice Cooper-WB
(11)	12.	I Wanna Be Where You Are-Michael Jackson-Motown
(4)	13.	Living In A House Divided-Cher-Kapp
(24)	*14.	Alone Again (Naturally)-Gilbert O'Sullivan-MAM
(17)	15.	Day By Day-Godspell-Bell
(22)	16.	Where Is The Love-Flack/Hathaway-Atlantic
(18)	17.	Song Sung Blue-Neil Diamond-Uni
(31)	*18.	How Do You Do?-Mouth & MacNeal-Philips
(29)	*19.	Hold Her Tight-The Osmonds-MGM
(10)	20.	Automatically Sunshine-The Supremes-Motown
(20)	21.	Nice To Be With You-The Gallery Sussex
(12)	22.	Tumbling Dice-Rolling Stones-Rolling Stones
(34)	*23.	Brandy-Looking Glass-Epic
(21)	24.	We're On Our Way-Chris Hodge-Apple
(38)	*25.	Long Cool Woman-The Hollies-Epic
(28)	26.	Happiest Girl In The Whole U.S.A.-Donna Fargo-Dot
(33)	27.	I Don't Want To Be Right-Luther Ingram-Koko
(37)	28.	Conquistador-Procol Harum-A & M
(19)	29.	Ask Me What You Want-Millie Jackson-Spring
(--)	30.	Happy-Rolling Stones-Rolling Stones
(25)	31.	Been Lonely For So Long-Frederick Knight-Stax
(30)	32.	Amazing Grace-Royal Scots Dragoon Guards-RCA
(40)	*33.	Rock And Roll Part 2-Gary Glitter-Bell
(35)	34.	Jesahel-English Congregation-

BIG 10 ALBUMS

1. Thick As A Brick-Jethro Tull-Reprise
2. Honky Chateau-Elton John-Uni
3. Exile On Main Street-Rolling Stones-Rolling Stones
4. Elvis As Recorded At Madison Square Garden-Elvis Presley-RCA
5. School's Out-Alice Cooper-WB
6. Simon & Garfunkel's Greatest Hits-Columbia
7. Joplin In Concert-Janis Joplin-Columbia
8. A Song For You-Carpenters-A & M
9. Harvest-Neil Young-Reprise
10. Manassas-Stephen Stills-Atlantic

---

the greatest hits of the
SUPERSTARS!
Win them...from Super 'CFL!

SUPER SUMMER GOLD

*1. Summer In The City-Lovin' Spoonful-(1968)
*2. Summer Nights-Marianne Faithfull-(1966)
*3. Theme From A Summer Place-Percy Faith-(1959)
*4. Summer Rain-The Youngbloods-(1970)
*5. Summer Side Of Life-Gordon Lightfoot-(1970)
*6. A Summer Song-Chad & Jeremy-(1967)
*7. Summertime, Summertime-The Jamies-(1967)
8. Summer Means Fun-Jan & Dean-(1967)
*9. Summer Sand-Daawn-(1971)
10. Summertime-Billy Stewart-(1967)

*denotes currently available on a gold single or on an album.

		Signpost
(39)	35.	Small Beginnings-Flash-Capitol
(32)	36.	Oh Girl-Chi-Lites-Brunswick
(--)	37.	The Runway-The Grass Roots-Dunhill
(26)	38.	In A Broken Dream-Python Lee Jackson-Crescendo
(36)	39.	I'll Take You There-The Staple Singers-Stax
(--)	40.	Goodbye To Love-Carpenters-A & M

*Denotes Big 10 Climbers

WC
FL

ALL HIT MUSIC **IN THE MIDWEST**

WAS	IS	Week Ending July 22, 1972
(1)	1.	Lean On Me-Bill Withers-Sussex
(4)	2.	Take It Easy-Eagles-Asylum
(6)	3.	Too Late To Turn Back Now-Cornelius Brothers-UA
(8)	4.	Daddy Don't You Walk So Fast-W. Newton-Chelsea
(11)	*5.	School's Out-Alice Cooper-WB
(12)	6.	I Wanna Be Where You Are-Michael Jackson-Motown
(14)	*7.	Alone Again (Naturally)-Gilbert O'Sullivan-MAM
(2)	8.	The Candy Man-Sammy Davis, Jr.-MGM
(16)	*9.	Where Is The Love-Flack/Hathaway-Atlantic
(3)	10.	Too Young-Donny Osmond-MGM
(18)	*11.	How Do You Do?-Mouth & MacNeal-Philips
(15)	12.	Day By Day-Godspell-Bell
(23)	*13.	Brandy-Looking Glass-Epic
(19)	14.	Hold Her Tight-The Osmonds-MGM
(7)	15.	Outa-Space-Billy Prestone-A & M
(25)	*16.	Long Cool Woman-The Hollies-Epic
(5)	17.	I Need You-America-WB
(28)	*18.	Conquistador-Procol Harum-A & M
(9)	19.	Troglodyte (Cave Man)-Jimmy Castor Bunch-RCA
(10)	20.	I Didn't Get To Sleep At All-5th Dimension-Bell
(26)	21.	Happiest Girl In The Whole U.S.A.-Donna Fargo-Dot
(17)	22.	Song Sung Blue-Neil Diamond-Uni
(27)	23.	I Don't Want To Be Right-Luther Ingram-Koko
(30)	*24.	Happy-Rolling Stones-Rolling Stones
(13)	25.	Living In A House Divided-Cher-Kapp
(34)	*26.	Jesahel-English Congregation-Signpost
(33)	*27.	Rock And Roll Part 2-Gary Glitter-Bell
(21)	28.	Nice To Be With You-The Gallery-Sussex
(24)	29.	We're On Our Way-Chris Hodge-Apple
(35)	30.	Small Beginnings-Flash-Capitol
(37)	*31.	The Runway-The Grass Roots-Dunhill
(22)	32.	Tumbling Dice-Rolling Stones-Rolling Stones
(40)	*33.	Goodbye To Love-Carpenters-A & M
(--)	34.	I'm Coming Home-Stories-Kama Sutra
(20)	35.	Automatically Sunshine-The Supremes-Motown
(--)	36.	Back Stabbers-O'Jays-Philadelphia International

BIG 10 ALBUMS

1. Thick As A Brick-Jethro Tull-Reprise
2. Exile On Main Street-Rolling Stones-Rolling Stones
3. Honky Chateau-Elton John-Uni
4. Elvis As Recorded At Madison Square Garden-Elvis Presley-RCA
5. Simon & Garfunkel Greatest Hits-Columbia
6. School's Out-Alice Cooper-WB
7. Procol Harum Live-Procol Harum-A & M
8. A Song For You-Carpenters-A & M
9. Joplin In Concert-Janis Joplin-Columbia
10. Moods-Neil Diamond-Uni

the greatest hits of the
SUPERSTARS!
Win them...from Super 'CFL!

AUSTRALIA'S BIG 10

1. Nice To Be With You-The Gallery-Sussex
2. Baby Blue-Badfinger-Apple
3. Sylvia's Mother-Dr. Hook-Epic

*4. Marshall's Portable Music Machine-Robin Jolley-Fable

5. Everything I Own-Bread-Elektra
6. Beautiful Sunday-Daniel Boone-Mercury
7. Down By The Lazy River-The Osmonds-MGM
8. Hurting Each Other-The Carpenters-A & M
9. Open Up Your Heart-G. Wayne Thomas-NL
10. First Time Ever I Saw Your Face-Roberta Flack-Atlantic

*denotes Australian talent
(courtesy: 4BC Radio-Brisbane)

(29)	37.	Ask Me What You Want-Millie Jackson-Spring
(--)	38.	Summer Sun-Jamestown Massacre-WB
(--)	39.	Hold Your Head Up-Argent-Epic
(--)	40.	Motorcycle Mama-Sailcat-Elektra

*Denotes Big 10 Climbers

WC
FL

ALL HIT MUSIC IN THE MIDWEST

WAS	IS	Week Ending July 29, 1972
(1)	1.	Lean On Me-Bill Withers-Sussex
(3)	2.	Too Late To Turn Back Now-Cornelius Brothers-UA
(7)	3.	Alone Again (Naturally)-Gilbert O'Sullivan-MAM
(5)	4.	School's Out-Alice Cooper-WB
(4)	5.	Daddy Don't You Walk So Fast-W. Newton-Chelsea
(9)	6.	Where Is The Love-Flack/Hathaway-Atlantic
(11)	7.	How Do You Do?-Mouth & MacNeal-Philips
(13)	8.	Brandy-Looking Glass-Epic
(2)	9.	Take It Easy-Eagles-Asylum
(16)	*10.	Long Cool Woman-The Hollies-Epic
(6)	11.	I Wanna Be Where You Are-Michael Jackson-Motown
(14)	12.	Hold Her Tight-The Osmonds-MGM
(18)	*13.	Conquistador-Procol Harum-A & M
(8)	14.	The Candy Man-Sammy Davis, Jr.-MGM
(10)	15.	Too Young-Donny Osmond-MGM
(12)	16.	Day By Day-Godspell-Bell
(21)	17.	Happiest Girl In The Whole U.S.A.-Donna Fargo-Dot
(24)	*18.	Happy-Rolling Stones-Rolling Stones
(27)	*19.	Rock And Roll Part 2-Gary Glitter-Bell
(23)	20.	I Don't Want To Be Right-Luther Ingram-Koko
(15)	21.	Outa-Space-Billy Preston-A & M
(31)	*22.	The Runway-The Grass Roots-Dunhill
(26)	23.	Jesahel-English Congregation-Signpost
(17)	24.	I Need You-America-WB
(33)	*25.	Goodbye To Love-Carpenters-A & M
(19)	26.	Troglodyte (Cave Man)-Jimmy Castor Bunch-RCA
(20)	27.	I Didn't Get To Sleep At All-5th Dimension-Bell
(30)	28.	Small Beginnings-Flash-Capitol
(34)	*29.	I'm Coming Home-Stories-Kama Sutra
(22)	30.	Song Sung Blue-Neil Diamond-Uni
(39)	*31.	Hold Your Head Up-Argent-Epic
(36)	32.	Back Stabbers-O'Jays-Philadelphia International
(--)	33.	I'm Still In Love With You-Al Green-HI
(25)	34.	Living In A House Divided-Cher-Kapp

BIG 10 ALBUMS

1. Honky Chateau-Elton John-Uni
2. Thick As A Brick-Jethro Tull-Reprise
3. Elvis As Recorded At Madison Square Garden-Elvis Presley-RCA
4. Simon And Garfunkel's Greatest Hits-Columbia
5. Exile On Main St.-Rolling Stones-Rolling Stones
6. School's Out-Alice Cooper-WB
7. A Song For You-Carpenters-A & M
8. Procol Harum Live-Procol Harum-A & M
9. Joplin In Concert-Janis Joplin-Columbia
10. Moods-Neil Diamond-Uni

the greatest hits of the
SUPERSTARS!
Win them...from Super 'CFL!

MONTRÉAL'S BIG 10

*1. Wild Eyes-Stampeders-Quality
2. Jambalya-Nitty Gritty Dirt Band-United Artists
**3. Rain Showers-Pagliaro-Much
4. Sylvia's Mother-Dr. Hook & The Medicine Show-Col.
5. The Candy Man-Sammy Davis, Jr.-MGM
6. Daddy Don't You Walk So Fast-Wayne Newton-RCA
7. How Do You Do-Mouth & MacNeal-Philips
8. Troglodyte-Jimmy Castor Bunch-RCA
9. How Can I Be Sure-David Cassidy-Bell
10. Rocket Man-Elton John-Uni

*denotes Canadian talent-
**French Canadian
(courtesy: Les Palmares-CJMS-Montreal)

(38)	35.	Summer Sun-Jamestown Massacre-WB
(--)	36.	The Guitar Man-Bread-Elektra
(28)	37.	Nice To Be With You-The Gallery-Sussex
(--)	38.	You Don't Mess Around With Jim-Jim Croce-ABC
(40)	39.	Motorcycle Mama-Sailcat-Elektra
(29)	40.	We're On Our Way-Chris Hodge-Apple

*Denotes Big 10 Climbers

WC
FL

ALL HIT MUSIC **IN THE MIDWEST**

WAS	IS	Week Ending August 5, 1972
(3)	1.	Alone Again (Naturally)-Gilbert O'Sullivan-MAM
(4)	2.	School's Out-Alice Cooper-WB
(6)	3.	Where Is The Love-Flack/Hathaway-Atlantic
(7)	4.	How Do You Do?-Mouth & MacNeal-Philips
(8)	5.	Brandy-Looking Glass-Epic
(1)	6.	Lean On Me-Bill Withers-Sussex
(10)	7.	Long Cool Woman-The Hollies-Epic
(13)	*8.	Conquistador-Procol Harum-A & M
(2)	9.	Too Late To Turn Back Now-Cornelius Brothers-UA
(12)	10.	Hold Her Tight-The Osmonds-MGM
(5)	11.	Daddy Don't You Walk So Fast-W. Newton-Chelsea
(18)	*12.	Happy-Rolling Stones-Rolling Stones
(19)	*13.	Rock And Roll Part 2-Gary Glitter-Bell
(9)	14.	Take It Easy-Eagles-Asylum
(11)	15.	I Wanna Be Where You Are-Michael Jackson-Motown
(17)	16.	Happiest Girl In The Whole U.S.A.-Donna Fargo-Dot
(23)	*17.	Jesahel-English Congregation-Signpost
(14)	18.	The Candy Man-Sammy Davis, Jr.-MGM
(22)	19.	The Runway-The Grass Roots-Dunhill
(20)	20.	I Don't Want To Be Right-Luther Ingram-Koko
(25)	21.	Goodbye To Love-Carpenters-A & M
(31)	*22.	Hold Your Head Up-Argent-Epic
(15)	23.	Too Young-Donny Osmond-MGM
(28)	24.	Small Beginnings-Flash-Capitol
(16)	25.	Day By Day-Godspell-Bell
(32)	*26.	Back Stabbers-O'Jays-Philadelphia International
(29)	27.	I'm Coming Home-Stories-Kama Sutra
(35)	*28.	Summer Sun-Jamestown Massacre-WB
(21)	29.	Outa-Space-Billy Preston-A & M
(33)	30.	I'm Still In Love With You-Al Green-HI
(36)	*31.	The Guitar Man-Bread-Elektra
(39)	*32.	Motorcycle Mama-Sailcat-Elektra
(24)	33.	I Need You-America-WB
(38)	34.	You Don't Mess Around With Jim-Jim Croce-ABC
(--)	35.	Saturday In The Park-Chicago-Columbia
(26)	36.	Trogslodyte (Cave Man)-Jimmy Castor Bunch-RCA
(--)	37.	Power Of Love-Joe Simon-Spring

BIG 10 ALBUMS

1. Honky Chateau-Elton John-Uni
2. Simon & Garfunkel's Greatest Hits-Columbia
3. School's Out-Alice Cooper-WB
4. Thick As A Brick-Jethro Tull-Reprise
5. Elvis As Recorded At Madison Square Garden-Elvis Presley-RCA
6. Chicago V-Chicago-Columbia
7. Procol Harum Live-Procol Harum-A & M
8. Exile On Main St.-Rolling Stones-Rolling Stones
9. A Song For You-Carpenters-A & M
10. Moods-Neil Diamond-Uni

LARRY LUJACK
takes on all comers
Afternoons...on Super 'CFL!

BEATLES' BIG 10

1. Hey Jude-(1968)
2. I Want To Hold Your Hand-(1964)
3. She Loves You-(1964)
4. Let It Be-(1970)
5. A Hard Day's Night-(1964)
6. Can't Buy Me Love-(1964)
7. Help-(1965)
8. All You Need Is Love-(1967)
9. Come Together/Something-(1969)
10. Yesterday-(1965)

(--)	38.	Speak To The Sky-Rich Springfield-Capitol
(--)	39.	Run To Me-Bee Gees-Atco
(--)	40.	Join Together-The Who-Decca

*Denotes Big 10 Climbers

WC
FL

ALL HIT MUSIC

IN THE
MIDWEST

WAS	IS	Week Ending August 12, 1972
(1)	1.	Alone Again (Naturally)-Gilbert O'Sullivan-MAM
(3)	2.	Where Is The Love-Flack/Hathaway-Atlantic
(5)	3.	Brandy-Looking Glass-Epic
(7)	4.	Long Cool Woman-The Hollies-Epic
(8)	5.	Conquistador-Procol Harum-A & M
(13)	*6.	Rock And Roll Part 2-Gary Glitter-Bell
(2)	7.	School's Out-Alice Cooper-WB
(12)	8.	Happy-Rolling Stones-Rolling Stones
(10)	9.	Hold Her Tight-The Osmonds-MGM
(20)	*10.	I Don't Want To Be Right-Luther Ingram-Koko
(6)	11.	Lean On Me-Bill Withers-Sussex
(19)	*12.	The Runway-The Grass Roots-Dunhill
(4)	13.	How Do You Do?-Mouth & MacNeal-Philips
(22)	*14.	Hold Your Head Up-Argent-Epic
(21)	15.	Goodbye To Love-Carpenters-A & M
(16)	16.	Happiest Girl In The Whole U.S.A.-Donna Fargo-Dot
(17)	17.	Jesahel-English Congregation-Signpost
(26)	*18.	Back Stabbers-O'Jays-Philadelphia International
(30)	*19.	I'm Still In Love With You-Al Green-HI
(34)	*20.	You Don't Mess Around With Jim-Jim Croce-ABC
(9)	21.	Too Late To Turn Back Now-Cornelius Brothers-UA
(27)	22.	I'm Coming Home-Stories-Kama Sutra
(24)	23.	Small Beginnings-Flash-Capitol
(11)	24.	Daddy Don't You Walk So Fast-W. Newton-Chelsea
(28)	25.	Summer Sun-Jamestown Massacre-WB
(31)	*26.	The Guitar Man-Bread-Elektra
(14)	27.	Take It Easy-Eagles-Asylum
(32)	28.	Motorcycle Mama-Sailcat-Elektra
(35)	*29.	Saturday In The Park-Chicago-Columbia
(37)	*30.	Power Of Love-Joe Simon-Spring
(40)	*31.	Join Together-The Who-Decca
(--)	32.	Baby Let Me Take You-Detroit Emeralds-Westbound
(39)	33.	Run To Me-Bee Gees-Atco
(38)	34.	Speak To The Sky-Rick Springfield-Capitol

BIG 10 ALBUMS

1. Honky Chateau-Elton John-Uni
2. Simon & Garfunkel's Greatest Hits-Columbia
3. Chicago V-Chicago-Columbia
4. School's Out-Alice Cooper-WB
5. Elvis As Recorded At Madison Square Garden-Elvis Presley-RCA
6. A Song For You-Carpenters-A & M
7. Thick As A Brick-Jethro Tull-Reprise
8. Procol Harum Live-Procol Harum-A & M
9. Moods-Neil Diamond-Uni
10. Exile On Main St.-Rolling Stones-Rolling Stones

---

the greatest hits of the
SUPERSTARS
Win them...from Super 'CFL!

PITTSBURGH'S BIG 10

*1. Lean On Me-Bill Withers-Sussex
*2. School's Out-Alice Cooper-Warner Brothers
*3. Alone Again (Naturally)-Gilbert O'Sullivan-MAM
*4. Daddy, Don't You Walk So Fast-Wayne Newton-Chelsea
*5. Too Late To Turn Back Now-Cornelius Brothers-UA
*6. Brandy-Looking Glass-Epic
*7. Rocket Man-Elton John-Uni
*8. Where Is The Love?Flack/Hathaway-Atlantic
*9. How Do You Do-Mouth & MacNeal-Phillips
10. People Make The World Go 'Round-Stylistics-Avco

*denotes former or current WCFL hit in Chicago.
(courtesy: KQV Radio-Pittsburgh)

(15)	35.	I Wanna Be Where You Are-Michael Jackson-Motown
(--)	36.	Baby Don't Get Hooked On Me-Mac Davis-Columbia
(18)	37.	The Candy Man-Sammy Davis, Jr.-MGM
(23)	38.	Too Young-Donny Osmond-MGM
(--)	39.	Put It Where You Want It-The Crusaders-Blue Thumb
(--)	40.	Go All The Way-Raspberries-Capitol

*Denotes Big 10 Climbers

WCFL | ALL HIT MUSIC | IN THE MIDWEST

WAS IS Week Ending August 19, 1972

BIG 10 ALBUMS

1. Chicago V-Chicago-Columbia
2. Honky Chateau-Elton John-Uni
3. Simon & Garfunkel's Greatest Hits-Columbia
4. School's Out-Alice Cooper-WB
5. Thick As A Brick-Jethro Tull-Reprise
6. A Song For You-Carpenters-A & M
7. Elvis As Recorded At Madison Square Garden-Elvis Presley-RCA
8. Trilogy-Emerson, Lake & Palmer-Cotillion
9. Never A Dull Moment-Rod Stewart-Mercury
10. Exile On Main St.-Rolling Stones-Rolling Stones

LARRY LUJACK is a SUPERJOCK
He'll tell you so...
afternoons...on Super 'CFL!

EUROPE'S BIG 10

1. A Whiter Shade Of Pale-Procol Harum-Maxicube
2. Oh Babe What Would You Say-Hurricane Smith-Columbia
3. Hell-A-Mouth & McNeal-Philips
4. Un Canto A Galicia (Spanish) *A Song For Galicia*-Julio Iglesias-Decca
5. Meisjes Met Rode Haren (Dutch)-*Red-Haired Girls*-Arne Jansen-Imperial
6. Let's Dance-Cats-Imperial
7. Summertrain-Sandy Coast-Polydor
8. Matrimony-Gilbert O'Sullivan-MAM
9. Jumbo-Dizzy Man's Band-Imperial
10. Smoke Gets In Your Eyes-Blue Haze-A & M

WAS	IS	
(3)	1.	Brandy-Looking Glass-Epic
(4)	2.	Long Cool Woman-The Hollies-Epic
(5)	3.	Conquistador-Procol Harum-A & M
(1)	4.	Alone Again (Naturally)-Gilbert O'Sullivan-MAM
(6)	5.	Rock And Roll Part 2-Gary Glitter-Bell
(10)	6.	I Don't Want To Be Right-Luther Ingram-Koko
(8)	7.	Happy-Rolling Stones-Rolling Stones
(9)	8.	Hold Her Tight-The Osmonds-MBM
(2)	9.	Where Is The Love-Flack/Hathaway-Atlantic
(12)	10.	The Runway-The Grass Roots-Dunhill
(14)	11.	Hold Your Head Up-Argent-Epic
(15)	12.	Goodbye To Love-Carpenters-A & M
(20)	13.	You Don't Mess Around With Jim-Jim Croce-ABC
(18)	14.	Back Stabbers-O'Jays-Philadelphia International
(19)	15.	I'm Still In Love With You-Al Green-HI
(7)	16.	School's Out-Alice Cooper-WB
(13)	17.	How Do You Do?-Mouth & MacNeal-Philips
(29)	18.	Saturday In The Park-Chicago-Columbia
(11)	19.	Lean On Me-Bill Withers-Sussex
(17)	20.	Jesahel-English Congregation-Signpost
(26)	21.	The Guitar Man-Bread-Elektra
(25)	22.	Summer Sun-Jamestown Massacre-WB
(23)	23.	Small Beginnings-Flash-Capitol
(32)	24.	Baby Let Me Take You-Detroit Emeralds-Westbound
(28)	25.	Motorcycle Mama-Sailcat-Elektra
(30)	26.	Power Of Love-Joe Simon-Spring
(16)	27.	Happiest Girl In The Whole U.S.A.-Donna Fargo-Dot
(31)	28.	Join Together-The Who-Decca
(21)	29.	Too Late To Turn Back Now-Cornelius Brothers-UA
(33)	30.	Run To Me-Bee Gees-Atco
(34)	31.	Speak To The Sky-Rick Springfield-Capitol
(36)	32.	Baby Don't Get Hooked On Me-Mac Davis-Columbia

(Source: RADIO NOORDZEE, a radio ship broadcasting to Europe from the North Sea)

(22)	33.	I'm Coming Home-Stories-Kama Sutra
(39)	34.	Put It Where You Want It-Crusaders-Blue Thumb
(40)	35.	Go All The Way-Raspberries-Capitol
(--)	36.	America-Yes-Atlantic
(24)	37.	Daddy Don't You Walk So Fast-W. Newton-Chelsea
(--)	38.	Black & White-Three Dog Night-Dunhill
(--)	39.	Rock Me On The Water-Jackson Browne-Asylum
(--)	40.	You're Still A Young Man-Tower Of Power-WB

*Denotes Big 10 Climbers

WC FL — ALL HIT MUSIC — IN THE MIDWEST

Week Ending August 26, 1972

WAS	IS	
(1)	1.	Brandy-Looking Glass-Epic
(6)	2.	I Don't Want To Be Right-Luther Ingram-Koko
(2)	3.	Long Cool Woman-The Hollies-Epic
(5)	4.	Rock And Roll Part 2-Gary Glitter-Bell
(3)	5.	Conquistador-Procol Harum-A & M
(11)	6.	Hold Your Head Up-Argent-Epic
(13)	7.	You Don't Mess Around with Jim-Jim Croce-ABC
(12)	8.	Goodbye To Love-Carpenters-A & M
(14)	9.	Back Stabbers-O'Jays-Philadelphia International
(4)	10.	Alone Again (Naturally)-Gilbert O'Sullivan-MAM
(15)	11.	I'm Still In Love With You-Al Green-HI
(18)	*12.	Saturday In The Park-Chicago-Columbia
(7)	13.	Happy-Rolling Stones-Rolling Stones
(24)	*14.	Baby Let Me Take You-Detroit Emeralds-Westbound
(8)	15.	Hold Her Tight-The Osmonds-MGM
(21)	16.	The Guitar Man-Bread-Elektra
(9)	17.	Where Is The Love-Flack/Hathaway-Atlantic
(10)	18.	The Runway-The Grass Roots-Dunhill
(25)	19.	Motorcycle Mama-Sailcat-Elektra
(22)	20.	Summer Sun-Jamestown Massacre-WB
(26)	21.	Power Of Love-Joe Simon-Spring
(16)	22.	School's Out-Alice Cooper-WB
(32)	*23.	Baby Don't Get Hooked On Me-Mac Davis-Columbia
(28)	24.	Join Together-The Who-Decca
(30)	25.	Run To Me-Bee Gees-Atco
(17)	26.	How Do You Do?-Mouth & MacNeal-Columbia
(35)	*27.	Go All The Way-Raspberries-Capitol
(19)	28.	Lean On Me-Bill Withers-Sussex
(34)	29.	Put It Where You Want It-Crusaders-Blue Thumb
(31)	30.	Speak To The Sky-Rick Springfield-Capitol
(38)	*31.	Black & White-Three Dog Night-Dunhill
(36)	32.	America-Yes-Atlantic
(--)	33.	Play Me-Neil Diamond-Uni
(39)	34.	Rock Me On The Water-Jackson Browne-Asylum
(--)	35.	Burning Love-Elvis Presley-RCA

BIG 10 ALBUMS

1. Chicago V-Chicago-Columbia
2. Honky Chateau-Elton John-Uni
3. Simon & Garfunkel's Greatest Hits-Columbia
4. A Song For You-Carpenters-A & M
5. School's Out-Alice Cooper-WB
6. Never A Dull Moment-Rod Stewart-Mercury
7. Moods-Neil Diamond-Uni
8. Trilogy-Emerson, Lake And Palmer-Cotillion
9. Big Bambu-Cheech & Chong-Ode
10. Thick As A Brick-Jethro Tull-Reprise

Win SUPERSTAR hits
from the SUPERJOCKS
all day long...on SUPER 'CFL!

NEW ORLEANS' BIG 10

*1. Long Cool Woman-The Hollies-Epic
*2. Honky Cat-Elton John-Uni
*3. Saturday In The Park-Chicago-Columbia
*4. Layla-Derek & The Dominoes-Atco
5. Ben-Michael Jackson-Motown
6. In The Ghetto-Candi Staton-Fame
*7. Alone Again-Gilbert O'Sullivan-MAM
8. Looking Through A Window-Jackson 5-Motown
9. Stairway To Heaven-Led Zeppelin-Atlantic
*10. You Don't Mess Around With Jim-Jim Croce-ABC

*denotes former or current WCFL hit in Chicago
(courtesy: WTIX Radio-New Orleans)

(40)	36.	You're Still A Young Man-Tower Of Power-WB
(20)	37.	Jesahel-English Congregation-Signpost
(--)	38.	Pop That Thang-Isley Brothers-T-Neck
(--)	39.	Beautiful Sunday-Daniel Boone-Mercury
(--)	40.	Loving You Just Crossed My Mind-Sam Neely-Capitol

*Denotes Big 10 Climbers

FL WC ALL HIT MUSIC IN THE MIDWEST

BIG 10 ALBUMS

1. Chicago V-Chicago-Columbia
2. Simon & Garfunkel's Greatest Hits-Columbia
3. Honky Chateau-Elton John-Uni
4. Trilogy-Emerson, Lake And Palmer-Cotillion
5. School's Out-Alice Cooper-WB
6. Never A Dull Moment-Rod Stewart-Mercury
7. A Song For You-Carpenters-A&M
8. Big Bambu-Cheech & Chong-Ode
9. Thick As A Brick-Jethro Tull-Reprise
10. Elvis As Recorded At Madison Square Garden-Elvis Presley-RCA

HOLLAND'S BIG 10

1. Un Canto A Galicia-*A Song For Galicia*-Julio Iglesias-Decca
2. Popcorn-Various Versions-Basart
3. A Whiter Shade Of Pake-Procol Harum-Maxicube
4. Sing Along-Mac & Katie Kisoon-Young Blood
5. Sealed With A Kiss-Bobby Vinton-Altona
6. Oh Babe, What Would You Say-Hurricane Smith-Columbia
7. Rock & Roll Part I-Gary Glitter-Bell
8. Smoke Gets In Your Eyes-Blue Haze-A&M
9. Mario-Hanny En De Rekels-11 Provincien
10. Zomerzon-*Summer Sun*-DeMakkers-11 Provincien

Courtesy: Radio Veronica

WAS	IS	Week Ending September 2, 1972
(4)	1.	Rock And Roll Part 2-Gary Glitter-Bell
(7)	2.	You Don't Mess Around With Him-Jim Croce-ABC
(1)	3.	Brandy-Looking Glass-Epic
(8)	4.	Goodbye To Love-Carpenters-A&M
(6)	5.	Hold Your Head Up-Argent-Epic
(9)	6.	Back Stabbers-O'Jays-Philadelphia International
(12)	7.	Saturday In The Park-Chicago-Columbia
(14)	*8.	Baby Let Me Take You-Detroit Emeralds-Westbound
(2)	9.	I Don't Want To Be Right-Luther Ingram-Koke
(11)	10.	I'm Still In Love With You-Al Green-Hi
(3)	11.	Long Cool Woman-The Hollies-Epic
(5)	12.	Conquistador-Procol Harum-A&M
(16)	13.	The Guitar Man-Bread-Elektra
(10)	14.	Alone Again (Naturally)-Gilbert O'Sullivan-MAM
(19)	15.	Motorcycle Mama-Sailcat-Elektra
(23)	*16.	Baby Don't Get Hooked On Me-Mac Davis-Columbia
(21)	17.	Power Of Love-Joe Simon-Spring
(31)	*18.	Black & White-Three Dog Night-Dunhill
(27)	*19.	Go All The Way-Raspberries-Capitol
(25)	20.	Run To Me-Bee Gees-Atco
(13)	21.	Happy-Rolling Stones-Rolling Stones
(24)	22.	Join Together-The Who-Decca
(29)	*23.	Put It Where You Want It-Crusaders-Blue Thumb
(15)	24.	Hold Her Tight-The Osmonds-MGM
(32)	25.	America-Yes-Atlantic
(34)	*26.	Rock Me On The Water-Jackson Browne-Asylum
(17)	27.	Where Is The Love-Flack/Hathaway-Atlantic
(35)	*28.	Burning Love-Elvis Presley-RCA
(33)	29.	Play Me-Neil Diamond-Uni
(30)	30.	Speak To The Sky-Rick Springfield-Capitol
(36)	31.	You're Still A Young Man-Tower Of Power-WB
(18)	32.	The Runway-The Grass Roots-Dunhill
(39)	33.	Beautiful Sunday-Daniel Boone-Mercury
(--)	34.	Ben-Michael Jackson-Motown
(29)	35.	Summer Sun-Jamestown Massacre-WB
(38)	36.	Pop That Thang-Isley Brotehrs-T-Neck
(--)	37.	Honky Cat-Elton John-Utu
(40)	38.	Loving You Just Crossed My Mind-Sam Neely-Capitol
(--)	39.	Popcorn-Hot Butter-Musicor
(--)	40.	A Piece Of Paper-Gladstone-ABC

*Denotes Big 10 Climbers

WC
FL

ALL HIT MUSIC

IN THE
MIDWEST

WAS	IS	Week Ending September 9, 1972
(2)	1.	You Don't Mess Around With Jim-Jim Croce-ABC
(4)	2.	Goodbye To Love-Carpenters-A & M
(5)	3.	Hold Your Head Up-Argent-Epic
(6)	4.	Back Stabbers-O'Jays-Philadelphia International
(7)	5.	Saturday In The Park-Chicago-Columbia
(1)	6.	Rock And Roll Part 2-Gary Glitter-Bell
(8)	7.	Baby Let Me Take You-Detroit Emeralds-Westbound
(3)	8.	Brandy-Looking Glass-Epic
(10)	9.	I'm Still In Love With You-Al Green-HI
(18)	*10.	Black & White-Three Dog Night-Dunhill
(15)	11.	Motorcycle Mama-Sailcat-Elektra
(13)	12.	The Guitar Man-Bread-Elektra
(16)	13.	Baby Don't Get Hooked On Me-Mac Davis-Columbia
(19)	14.	Go All The Way-Raspberries-Columbia
(9)	15.	I Don't Want To Be Right-Luther Ingram-Koko
(28)	*16.	Burning Love-Elvis Presley-RCA
(17)	17.	Power Of Love-Joe Simon-Spring
(11)	18.	Long Cool Woman-The Hollies-Epic
(20)	19.	Run To Me-Bee Gees-Atco
(14)	20.	Alone Again (Naturally)-Gilbert O'Sullivan-MAM
(23)	21.	Put It Where You Want It-Crusaders-Blue Thumb
(22)	22.	Join Together-The Who-Decca
(29)	*23.	Play Me-Neil Diamond-Uni
(33)	*24.	Beautiful Sunday-Daniel Boone-Mercury
(12)	25.	Conquistador-Procol Harum-A & M
(21)	26.	Happy-Rolling Stones-Rolling Stones
(31)	27.	You're Still A Young Man-Tower Of Power-WB
(36)	*28.	Pop That Thang-Isley Brothers-T-Neck
(34)	29.	Ben-Michael Jackson-Motown
(25)	30.	America-Yes-Atlantic
(24)	31.	Hold Her Tight-The Osmonds-MGM
(38)	*32.	Loving You Just Crossed My Mind-Sam Neely-Capitol
(--)	33.	Tight Rope-Leon Russell-Shelter
(37)	34.	Honky Cat-Elton John-Uni
(39)	35.	Popcorn-Hot Butter-Musicor
(26)	36.	Rock Me On The Water-Jackson Browne-Asylum
(--)	37.	Everybody Plays The Fool-The Main Ingredient

BIG 10 ALBUMS

1. Chicago V-Chicago-Columbia
2. Trilogy-Emerson, Lake And Palmer-Cotillion
3. Big Bambu-Cheech & Chong-Ode
4. Honky Chateau-Elton John-Uni
5. Simon & Garfunkel's Greatest Hits-Columbia
6. School's Out-Alice Cooper-WB
7. A Song For You-Carpenters-A & M
8. Never A Dull Moment-Rod Stewart-Mercury
9. Thick As A Brick-Jethro Tull-Reprise
10. Moods-Neil Diamond-Uni

LARRY LUJACK says 'neat' things...
like "Oh Bleep" and "Cosmic"
Afternoons...on Super 'CFL!

BIG 10 LABOR DAY GOLD

*1. The Work Song-Herb Alpert-(1966)
*2. A Cowboy's Work Is Never Done-Sonny & Cher-(1972)
*3. Someday I'll Be A Farmer-Melanie-(1972)
*4. Working In A Coal Mine-Lee Dorsey-(1966)
*5. If I Were A Carpenter-4 Tops-(1968)
*6. Sky Pilot-Eric Burdon-(1969)
*7. Working On A Groovy Thing-5th Dimension-(1968)
8. Truck Drivin' Man-Grass Roots-(1969)
9. Butchers And Bakers-4 Jacks & A Jill-(1968)
10. Working Man Blues-Jerry Lee Lewis-(1970)

*denotes currently available on a gold single or on an album.

(40)	38.	A Piece Of Paper-Gladstone-ABC
(--)	39.	Good Time Charlie's Got The Blues-Danny O'Keefe-Signpost
(--)	40.	Nights In White Satin-The Moody Blues-Deram

*Denotes Big 10 Climbers

WC
FL

ALL HIT MUSIC

IN THE MIDWEST

WAS	IS	Week Ending September 16, 1972
(1)	1.	You Don't Mess Around With Jim-Jim Croce-ABC
(3)	2.	Hold Your Head Up-Argent-Epic
(4)	3.	Back Stabbers-O'Jays-Philadelphia International
(5)	4.	Saturday In The Park-Chicago-Columbia
(2)	5.	Goodbye To Love-Carpenters-A & M
(7)	6.	Baby Let Me Take You-Detroit Emeralds-Westbound
(10)	7.	Black & White-Three Dog Night-Dunhill
(11)	8.	Motorcycle Mama-Sailcat-Elektra
(13)	9.	Baby Don't Get Hooked On Me-Mac Davis-Columbia
(12)	10.	The Guitar Man-Bread-Elektra
(6)	11.	Rock And Roll Part 2-Gary Glitter-Bell
(14)	12.	Go All The Way-Raspberries-Capitol
(16)	13.	Burning Love-Elvis Presley-RCA
(8)	14.	Brandy-Looking Glass-Epic
(17)	15.	Power Of Love-Joe Simon-Spring
(19)	16.	Run To Me-Bee Gees-Atco
(24)	*17.	Beautiful Sunday-Daniel Boone-Mercury
(22)	18.	Join Together-The Who-Decca
(23)	19.	Play Me-Neil Diamond-Uni
(29)	*20.	Ben-Michael Jackson-Motown
(9)	21.	I'm Still In Love With You-Al Green-HI
(28)	*22.	Pop That Thang-Isley Brothers-T-Neck
(27)	23.	You're Still A Young Man-Tower Of Power-WB
(15)	24.	I Don't Want To Be Right-Luther Ingram-Koko
(20)	25.	Alone Again (Naturally)-Gilbert O'Sullivan-MAM
(35)	*26.	Popcorn-Hot Butter-Musicor
(18)	27.	Long Cool Woman-The Hollies-Epic
(32)	28.	Loving You Just Crossed My Mind-Sam Neely-Capitol
(33)	29.	Tight Rope-Leon Russell-Shelter
(34)	30.	Honky Cat-Elton John-Uni
(21)	31.	Pat It Where You Want It-Crusaders-Blue Thumb
(38)	*32.	A Piece Of Paper-Gladstone-ABC
(--)	33.	My Ding-A-Ling-Chuck Berry-Chess
(37)	34.	Everybody Plays The Fool-The Main Ingredient-RCA
(30)	35.	America-Yes-Atlantic
(39)	36.	Good Time Charlie's Got The Blues-Danny O'Keefe-Signpost

BIG 10 ALBUMS

1. Chicago V-Chicago-Columbia
2. Trilogy-Emerson, Lake And Palmer-Cotillion
3. Big Bambu-Cheech & Chong-Ode
4. Honky Chateau-Elton John-Uni
5. Simon & Garfunkel's Greatest Hits-Columbia
6. Never A Dull Moment-Rod Stewart-Mercury
7. School's Out-Alice Cooper-WB
8. A Song For You-Carpenters-A & M
9. Moods-Neil Diamond-Uni
10. Carney-Leon Russell-Shelter

Win SUPERSTAR hits
from the SUPERJOCKS
all day long...on SUPER 'CFL!

CHATTANOOGA'S BIG 10

*1. My Ding-A-Ling-Chuck Berry-Chess
*2. Alone Again (Naturally)-Gilbert O'Sullivan-MAM
*3. Baby Don't Get Hooked On Me-Mac Davis-Columbia
*4. Brandy-Looking Glass-Epic
*5. If Loving You Is Wrong-Luther Ingram-Koko
*6. Black And White-3 Dog Night-Dunhill
*7. I'm Still In Love With You-Al Green-HI
8. That's The Way God Planned It-Billy Preston-Apple
*9. Pop That Thang-Isley Brothers-T-Neck
*10. Long Cool Woman-The Hollies-Epic--

*denotes former of current WCFL hit in Chicago.
(courtesy: WFLI Radio-Chattanooga)

(40)	37.	Nights In White Satin-The Moody Blues-Deram
(--)	38.	I'll Be Around-The Spinners-Atlantic
(--)	39.	Listen To The Music-The Doobie Brothers-WB
(25)	40.	Conquistador-Procol Harum-A & M

*Denotes Big 10 Climbers

WC FL **ALL HIT MUSIC** **IN THE MIDWEST**

Week Ending September 23, 1972

WAS	IS	
(3)	1.	Back Stabbers-O'Jays-Philadelphia International
(7)	2.	Black & White-Three Dog Night-Dunhill
(4)	3.	Saturday In The Park-Chicago-Columbia
(9)	4.	Baby Don't Get Hooked On Me-Mac Davis-Columbia
(1)	5.	You Don't Mess Around With Jim-Jim Croce-ABC
(13)	6.	Burning Love-Elvis Presley-RCA
(12)	7.	Go All The Way-Raspberries-Capitol
(2)	8.	Hold Your Head Up-Argent-Epic
(17)	*9.	Beautiful Sunday-Daniel Boone-Mercury
(5)	10.	Goodbye To Love-Carpenters-A & M
(19)	*11.	Play Me-Neil Diamond-Uni
(20)	*12.	Ben-Michael Jackson-Motown
(33)	*13.	My Ding-A-Ling-Chuck Berry-Chess
(6)	14.	Baby Let Me Take You-Detroit Emeralds-Westbound
(22)	*15.	Pop That Thang-Isley Brothers-T-Neck
(23)	16.	You're Still A Young Man-Tower Of Power-WB
(8)	17.	Motorcycle Mama-Sailcat-Elektra
(26)	*18.	Popcorn-Hot Butter-Musicor
(10)	19.	The Guitar Man-Bread-Elektra
(18)	20.	Join Together-The Who-Decca
(29)	*21.	Tight Rope-Leon Russell-Shelter
(34)	*22.	Everybody Plays The Fool-The Main Ingredient-RCA
(28)	23.	Loving You Just Crossed My Mind-Sam Neely-Capitol
(30)	24.	Honky Cat-Elton John-Uni
(11)	25.	Rock And Roll Part 2-Gary Glitter-Bell
(37)	*26.	Nights In White Satin-The Moody Blues-Deram
(14)	27.	Brandy-Looking Glass-Epic
(15)	28.	Power Of Love-Joe Simon-Spring
(32)	29.	A Piece Of Paper-Gladstone-ABC
(25)	30.	Alone Again (Naturally)-Gilbert O'Sullivan-MAM
(36)	31.	Good Time Charlie's-Danny O'Keefe-Signpost
(16)	32.	Run To Me-Bee Gees-Atco
(38)	33.	I'll Be Around-The Spinners-Atlantic
(21)	34.	I'm Still In Love With You-Al Green-HI
(--)	35.	Starting All Over Again-Mel and Tim-Stax

BIG 10 ALBUMS

1. Chicago V-Chicao-Columbia
2. Big Bambu-Cheech And Chong-Ode
3. Trilogy-Emerson, Lake And Palmer-Cotillion
4. Never A Dull Moment-Rod Stewart-Mercury
5. Honky Chateau-Elton John-Uni
6. Moods-Neil Diamond-Uni
7. Simon & Garfunkel's Greatest Hits-Columbia
8. School's Out-Alice Cooper-WB
9. Carney-Leon Russell-Shelter
10. A Song For You-Carpenters-A & M

In the entire universe there is only ONE SUPERJOCK!!!
LARRY LUJACK...on Super 'CFL!

SWITZERLAND'S BIG 10

*1. Pop Corn-Hot Butter-Barclay
*2. Song Sung Blue-Neil Diamond-Uni
3. En Frappant Dans Nos Mains-Mireille Mathieu-Barclay
4. I've Found My Freedom-Mac & Katie Kissoon-Youngblood
5. Little Willy-The Sweet-RCA
6. Bonsoir Clara-Michel Sardou-Phonogram
7. Hello-A-Mouth & MacNeal-Philips
8. Ich Hab' Die Liebe Gesch'n-Vicky Leandros-Philips
*9. Join Together-The Who-Track
10. Let's Dance-The Cats-Columbia

*denotes current or former WCFL hit in Chicago
(courtesy: Radio Suisse Romande & German Swiss Broadcasting Service)

(39)	36.	Listen To The Music-The Doobie Brothers-WB
(--)	37.	Use Me-Bill Withers-Sussex
(24)	38.	I Don't Want To Be Right-Luther Ingram-Koko
(--)	39.	Garden Party-Rick Nelson-Decca
(--)	40.	I'd Love You To Want Me-Lobo-Big Tree

*Denotes Big 10 Climbers

WC
FL

ALL HIT MUSIC IN THE MIDWEST

WAS	IS	Week Ending September 30, 1972
(2)	1.	Black & White-Three Dog Night-Dunhill
(13)	2.	My Ding-A-Ling-Chuck Berry-Chess
(7)	3.	Go All The Way-Raspberries-Capitol
(4)	4.	Baby Don't Get Hooked On Me-Mac Davis-Columbia
(6)	5.	Burning Love-Elvis Presley-RCA
(9)	6.	Beautiful Sunday-Daniel Boone-Mercury
(3)	7.	Saturday In The Park-Chicago-Columbia
(12)	8.	Ben-Michael Jackson-Motown
(11)	9.	Play Me-Neil Diamond-Uni
(18)	*10.	Popcorn-Hot Butter-Musicor
(1)	11.	Back Stabbers-O'Jays-Philadelphia International
(22)	*12.	Everybody Plays The Fool-The Main Ingredient-RCA
(5)	13.	You Don't Mess Around With Jim-Jim Croce-ABC
(16)	14.	You're Still A Young Man-Tower Of Power-WB
(15)	15.	Pop That Thang-Isley Brothers-T-Neck
(21)	16.	Tight Rope-Leon Russell-Shelter
(8)	17.	Hold Your Head Up-Argent-Epic
(14)	18.	Baby Let Me Take You-Detroit Emeralds-Westbound
(26)	*19.	Nights In White Satin-The Moody Blues-Deram
(10)	20.	Goodbye To Love-Carpenters-A & M
(23)	21.	Loving You Just Crossed My Mind-Sam Neely-Capitol
(24)	22.	Honky Cat-Elton John-Uni
(33)	*23.	I'll Be Around-The Spinners-Atlantic
(36)	*24.	Listen To The Music-The Doobie Brothers-WB
(29)	25.	A Piece Of Paper-Gladstone-ABC
(17)	26.	Motorcycle Mama-Sailcat-Elektra
(31)	27.	Good Time Charlie's –Danny O'Keefe-Signpost
(19)	28.	The Guitar Man-Bread-Elektra
(20)	29.	Join Together-The Who-Decca
(37)	*30.	Use Me-Bill Withers-Sussex
(25)	31.	Rock And Roll Part 2-Gary Glitter-Bell
(39)	*32.	Garden Party-Rick Nelson-Decca
(35)	33.	Starting All Over Again-Mel And Tim-Stax
(40)	34.	I'd Love You To Want Me-Lobo-Big Tree
(--)	35.	Why-Donny Osmond-MGM
(--)	36.	Thunder And Lightning-Chi Coltrane-Columbia

BIG 10 ALBUMS

1. Chicago V-Chicago-Columbia
2. Trilogy-Emerson, Lake And Palmer-Cotillion
3. Honky Chateau-Elton John-Uni
4. Big Bambu-Cheech And Chong-Ode
5. Moods-Neil Diamond-Uni
6. Never A Dull Moment-Rod Stewart-Mercury
7. Simon & Garfunkel's Greatest Hits-Columbia
8. Carney-Leon Russell-Shelter
9. Days Of Future Passed-The Moody Blues-Deram
10. A Song For You-Carpenters-A & M

DIANE MATSON of Chicago...
Super 'CFL's first SUPERCAR winner!
Now it's YOUR turn!

BOISE'S BIG 10

*1. Baby Don't Get Hooked On Me-Mac Davis-Columbia
*2. Saturday In The Park-Chicago-Columbia
*3. Go All The Way-Raspberries-Capitol
*4. Back Stabbers-O'Jays-Philadelphia International
*5. Black And White-Three Dog Night-Dunhill
*6. Popcorn-Hot Butter-Musicor
7. Spaceman-Nilsson-RCA
8. To Do With Love-The Hollies-Epic
*9. Burning Love-Elvis Presley-RCA
*10. Play Me-Neil Diamond-Uni

*denotes former or current WCFL hit in Chicago
(courtesy: KFXD Radio-Boise, Idaho)

(27)	37.	Brandy-Looking Glass-Epic
(--)	38.	If I Could Reach You-The 5th Dimension-Bell
(--)	39.	You Wear It Well-Rod Stewart-Mercury
(--)	40.	Don't Ever Be Lonely-Cornelius Brothers-UA

*Denotes Big 10 Climbers

WC
FL

ALL HIT MUSIC

IN THE
MIDWEST

WAS IS Week Ending October 7, 1972

WAS	IS	
(2)	1.	My Ding-A-Ling-Chuck Berry-Chess
(3)	2.	Go All The Way-Raspberries-Capitol
(5)	3.	Burning Love-Elvis Presley-RCA
(9)	4.	Play Me-Neil Diamond-Uni
(6)	5.	Beautiful Sunday-Daniel Boone-Mercury
(8)	6.	Ben-Michael Jackson-Motown
(1)	7.	Black & White-Three Dog Night-Dunhill
(4)	8.	Baby Don't Get Hooked On Me-Mac Davis-Columbia
(12)	9.	Everybody Plays The Fool-The Main Ingredient-RCA
(10)	10.	Popcorn-Hot Butter-Musicor
(14)	11.	You're Still A Young Man-Tower Of Power-WB
(7)	12.	Saturday In The Park-Chicago-Columbia
(19)	*13.	Nights In White Satin-The Moody Blues-Deram
(16)	14.	Tight Rope-Leon Russell-Shelter
(23)	*15.	I'll Be Around-The Spinners-Atlantic
(13)	16.	You Don't Mess Around With Jim-Jim Croce-ABC
(11)	17.	Back Stabbers-O'Jays-Philadelphia International
(22)	18.	Honky Cat-Elton John-Uni
(21)	19.	Loving You Just Crossed My Mind-Sam Neely-Capitol
(24)	20.	Listen To The Music-The Doobie Brothers-WB
(32)	*21.	Garden Party-Rick Nelson-Decca
(27)	22.	Good Time Charlie's-Danny O'Keefe-Signpost
(30)	*23.	Use Me-Bill Withers-Sussex
(15)	24.	Pop That Thang-Isley Brothers-T-Neck
(25)	25.	A Piece Of Paper-Gladstone-ABC
(18)	26.	Baby Let Me Take You-Detroit Emeralds-Westbound
(33)	*27.	Starting All Over Again-Mel and Tim-Stax
(17)	28.	Hold Your Head Up-Argent-Epic
(34)	29.	I'd Love You To Want Me-Lobo-Big Tree
(35)	30.	Why-Donny Osmond-MGM
(36)	31.	Thunder And Lightning-Chi Coltrane-Columbia
(20)	32.	Goodbye To Love-Carpenters-A & M
(39)	33.	You Wear It Well-Rod Stewart-Mercury
(38)	34.	If I Could Reach You-The 5th Dimension-Bell
(26)	35.	Motorcycle Mama-Sailcat-Elektra

BIG 10 ALBUMS

1. Big Bambu-Cheech And Chong-Ode
2. Trilogy-Emerson, Lake And Palmer-Cotillion
3. Chicago V-Chicago-Columbia
4. Never A Dull Moment-Rod Stewart-Mercury
5. Honky Chateau-Elton John-Uni
6. Carney-Leon Russell-Shelter
7. Days Of Future Passed-Moody Blues-Deram
8. Moods-Neil Diamond-Uni
9. Simon & Garfunkel's Greatest Hits-Columbia
10. Superfly-Curtis Mayfield-Curtom

SUPERCARS...SUPERSTARS
You're Chance to Win
every day...on Super 'CFL!

CANADA'S BIG 10

1. My Ding-A-Ling-Chuck Berry-Chess
2. Black And White-Three Dog Night-Dunhill
3. Go All The Way-Raspberries-Capitol
4. Saturday In The Park-Chicago-Columbia
5. Honky Cat-Elton John-Uni
*6. Concrete Sea-Terry Jacks-London
7. Run To Me-Bee Gees-Atco
8. Play Me-Neil Diamond-Uni
9. Beautiful Sunday-Daniel Boone-Mercury
10. Baby Don't Get Hooked On Me-Mac Davis-Columbia

*denotes Canadian talent
(courtesy: CKRC Radio-Winnipeg)

(--)	36.	Freddie's Dead-Curtis Mayfield-Curtom
(28)	37.	The Guitar Man-Bread-Elektra
(40)	38.	Don't Ever Be Lonely-Cornelius Brothers-UA
(--)	39.	Witchy Woman-Eagles-Asylum
(--)	40.	I Can See Clearly Now-Johnny Nash-Epic

*Denotes Big 10 Climbers

WC
FL | ALL HIT MUSIC | IN THE MIDWEST

WAS IS Week Ending October 14, 1972

WAS	IS	
(1)	1.	My Ding-A-Ling-Chuck Berry-Chess
(3)	2.	Burning Love-Elvis Presley-RCA
(5)	3.	Beautiful Sunday-Daniel Boone-Mercury
(4)	4.	Play Me-Neil Diamond-Uni
(6)	5.	Ben-Michael Jackson-Motown
(2)	6.	Go All The Way-Raspberries-Capitol
(9)	7.	Everybody Plays The Fool-The Main Ingredient-RCA
(10)	8.	Popcorn-Hot Butter-Musicor
(13)	9.	Nights In White Satin-The Moody Blues-Deram
(15)	10.	I'll Be Around-The Spinners-Atlantic
(14)	11.	Tight Rope-Leon Russell-Shelter
(7)	12.	Black & White-Three Dog Night-Dunhill
(21)	*13.	Garden Party-Rick Nelson-Decca
(19)	14.	Loving You Just Crossed My Mind-Sam Neely-Capitol
(20)	15.	Listen To The Music-The Doobie Brothers-WB
(23)	*16.	Use Me-Bill Withers-Sussex
(8)	17.	Baby Don't Get Hooked On Me-Mac Davis-Columbia
(11)	18.	You're Still A Young Man-Tower Of Power-WB
(22)	19.	Good Time Charlie's-Danny O'Keefe-Signpost
(18)	20.	Honky Cat-Elton John-Uni
(12)	21.	Saturday In The Park-Chicago-Columbia
(25)	22.	A Piece Of Paper-Gladstone-ABC
(30)	*23.	Why-Donny Osmond-MGM
(29)	24.	I'd Love You To Want Me-Lobo-Big Tree
(27)	25.	Starting All Over Again-Mel And Tim-Stax
(31)	26.	Thunder And Lightning-Chi Coltrane-Columbia
(16)	27.	You Don't Mess Around With Jim-Jim Croce-ABC
(33)	28.	You Wear It Well-Rod Stewart-Mercury
(34)	29.	If I Could Reach You-The 5th Dimension-Bell
(36)	30.	Freddie's Dead-Curtis Mayfield-Curtom
(17)	31.	Back Stabbers-O'Jays-Philadelphia International
(38)	32.	Don't Ever Be Lonely-Cornelius Brothers-UA
(39)	33.	Witchy Woman-Eagles-Asylum
(24)	34.	Pop That Thang-Isley Brothers-T-Neck
(40)	35.	I Can See Clearly Now-Johnny Nash-Epic

BIG 10 ALBUMS

1. Trilogy-Emerson, Lake And Palmer-Cotillion
2. Chicago V-Chicago-Columbia
3. Days Of Future Past-Moody Blues-Deram
4. Big Bambu-Cheech And Chong-Ode
5. Never A Dull Moment-Rod Stewart-Mercury
6. Carney-Leon Russell-Shelter
7. Close To The Edge-Yes-Atlantic
8. Honky Chateau-Elton John-Uni
9. Moods-Neil Diamond-Uni
10. Simon & Garfunkel's Greatest Hits-Columbia

SUPERSCORE
more SUPERfun
from SUPER 'CFL!

SEATTLE'S BIG 10

*1. Ben-Michael Jackson-Motown
*2. I Can See Clearly Now-Johnny Nash-Epic
*3. Burning Love-Elvis Presley-RCA
4. Dancing In The Moonlight-King Harvest-Four
*5. Black And White-3 Dog Night-Dunhill
*6. I Believe In Music-Gallery-Sussex
*7. Saturday In The Park-Chicago-Columbia
*8. Listen To The Music-Doobie Brothers-WB
*9. Everybody Plays The Fool-Main Ingredient-RCA
*10. Loving You Just Crossed My Mind-Sam Neely-Capitol

*denotes current or former WCFL hit in Chicago
(courtesy: KJR Radio-Seattle)

(26)	36.	Baby Let Me Take You-Detroit Emeralds-Westbound
(28)	37.	Hold Your Head Up-Argent-Epic
(--)	38.	I Believe In Music-Gallery-Sussex
(--)	39.	Elected-Alice Cooper-WB
(--)	40.	I Am Woman-Helen Reddy-Capitol

*Denotes Big 10 Climbers

WC
FL

ALL HIT MUSIC

IN THE MIDWEST

Week Ending October 21, 1972

WAS	IS	
(3)	1.	Beautiful Sunday-Daniel Boone-Mercury
(5)	2.	Ben-Michael Jackson-Motown
(7)	3.	Everybody Plays The Fool-The Main Ingredient-RCA
(9)	4.	Nights In White Satin-The Moody Blues-Deram
(1)	5.	My Ding-A-Ling-Chuck Berry-Chess
(10)	6.	I'll Be Around-The Spinners-Atlantic
(2)	7.	Burning Love-Elvis Presley-RCA
(13)	8.	Garden Party-Rick Nelson-Decca
(4)	9.	Play Me-Neil Diamond-Uni
(15)	10.	Listen To The Music-The Doobie Brothers-WB
(16)	11.	Use Me-Bill Withers-Sussex
(8)	12.	Popcorn-Hot Butter-Musicor
(26)	*13.	Thunder And Lightning-Chi Coltrane-Columbia
(11)	14.	Tight Rope-Leon Russell-Shelter
(24)	*15.	I'd Love You To Want Me-Lobo-Big Tree
(6)	16.	Go All The Way-Raspberries-Capital
(19)	17.	Good Time Charlie's-Danny O'Keefe-Signpost
(30)	*18.	Freddie's Dead-Curtis Mayfield-Curtom
(25)	19.	Starting All Over Again-Mel And Tim-Stax
(28)	*20.	You Wear It Well-Rod Stewart-Mercury
(23)	21.	Why-Donny Osmond-MGM
(14)	22.	Loving You Just Crossed My Mind-Sam Neely-Capitol
(29)	*23.	If I Could Reach You-The 5th Dimension-Bell
(12)	24.	Black & White-Three Dog Night-Dunhill
(33)	*25.	Witchy Woman-Eagles-Asylum
(17)	26.	Baby Don't Get Hooked On Me-Mac Davis-Columbia
(35)	*27.	I Can See Clearly Now-Johnny Nash-Epic
(18)	28.	You're Still A Young Man-Tower Of Power-WB
(32)	29.	Don't Ever Be Lonely-Cornelius Brothers-UA
(20)	30.	Honky Cat-Elton John-Uni
(38)	*31.	I Believe In Music-Gallery-Sussex
(21)	32.	Saturday In The Park-Chicago-Columbia
(40)	*33.	I Am Woman-Helen Reddy-Capitol
(22)	34.	A Piece Of Paper-Gladstone-ABC
(39)	35.	Elected-Alice Cooper-WB
(--)	36.	Roberta-Bones-Signpost
(--)	37.	From The Beginning-Emerson, Lake & Palmer-Cotillion
(27)	38.	You Don't Mess Around With Jim-Jim Croce-ABC

BIG 10 ALBUMS

1. Days Of Future Past-Moody Blues-Deram
2. Trilogy-Emerson, Lake And Palmer-Cotillion
3. Chicago V-Chicago-Columbia
4. Big Bambu-Cheech And Chong-Ode
5. Never A Dull Moment-Rod Stewart-Mercury
6. Superfly-Curtis Mayfield-Curtom
7. Honky Chateau-Elton John-Uni
8. Close To The Edge-Yes-Atlantic
9. Moods-Neil Diamond-Uni
10. Carney-Leon Russell-Shelter

LARRY LUJACK
gets Chicago ecstatic
Afternoons on Super 'CFL!

TORONTO'S BIG 10

1. My Ding-A-Ling-Chuck Barry-Chess
2. Ben-Michael Jackson-Motown
3. Popcorn-Hot Butter-Musicor
4. Black And White-Three Dog Night-Dunhill
5. Garden Party-Rick Nelson-Decca
6. Saturday In The Park-Chicago-Columbia
7. I'd Love You To Want Me-Lobo-Big Tree
8. Burning Love-Elvis Presley-RCA
9. I Am Woman-Helen Reddy-Capitol
10. Nights In White Satin-Moody Blues-Deram

(Courtesy: CHUM Radio-Toronto)

(--)	39.	Rockin' Pneumonia-Johnny Rivers-UA
(--)	40.	Ventura Highway-America-WB

*Denotes Big 10 Climbers

WC FL — **ALL HIT MUSIC** — **IN THE MIDWEST**

WAS	IS	Week Ending October 28, 1972
(3)	1.	Everybody Plans The Fool-The Main Ingredient-RCA
(4)	2.	Nights In White Satin-The Moody Blues-Deram
(6)	3.	I'll Be Around-The Spinners-Altantic
(2)	4.	Ben-Michael Jackson-Motown
(8)	5.	Garden Party-Rick Nelson-Decca
(11)	6.	Use Me-Bill Withers-Sussex
(1)	7.	Beautiful Sunday-Daniel Boone-Mercury
(10)	8.	Listen To The Music-The Doobie Brothers-WB
(7)	9.	Burning Love-Elvis Presley-RCA
(13)	10.	Thunder And Lightning-Chi Coltrane-Columbia
(15)	11.	I'd Lose You To Want Me-Lobo-Big Tree
(17)	12.	Good Time Charlie's-Danny O'Keefe-Signpost
(21)	*13.	Why-Donny Osmond-MGM
(18)	14.	Freddie's Dead-Curtis Mayfield-Curtom
(5)	15.	My Ding-A-Ling-Chuck Berry-Chess
(12)	16.	Popcorn-Hot Butter-Musicor
(14)	17.	Tight Rope-Leon Russell-Shelter
(20)	18.	You Wear It Well-Rod Stewart-Mercury
(19)	19.	Starting All Over Again-Mel and Tim-Stax
(23)	20.	If I Could Reach You-The 5th Dimension-Bell
(27)	*21.	I Can See Clearly Now-Johnny Nash-Epic
(31)	22.	I Believe In Music-Gallery-Sussex
(25)	23.	Witchy Woman-Eagles-Asylum
(9)	24.	Play Me-Neil Diamond-Uni
(16)	25.	Go All The Way-Raspberries-Capitol
(29)	26.	Don't Ever Be Lonely-Cornelius Brothers-UA
(22)	27.	Loving You Just Crossed My Mind-Sam Neely-Capitol
(33)	28.	I Am Woman-Helen Reddy-Capitol
(35)	*29.	Elected-Alice Cooper-WB
(24)	30.	Black & White-Three Dog Night-Dunhill
(26)	31.	Baby Don't Get Hooked On Me-Mac Davis-Columbia
(37)	32.	From The Beginning-Emerson, Lake & Palmer-Cotillion
(--)	33.	Convention '72-The Delegates-Mainstream
(36)	34.	Roberta-Bones-Signpost
(28)	35.	You're Still A Young Man-Tower Of Power-WB
(40)	36.	Ventura Highway-America-WB

BIG 10 ALBUMS

1. Days Of Future Past-Moody Blues-Deram
2. Trilogy-Emerson, Lake And Palmer-Cotillion
3. Chicago V-Chicago-Columbia
4. Catch Bull At Four-Cat Stevens-A & M
5. Never A Dull Moment-Rod Stewart-Mercury
6. Big Bambu-Cheech And Chong-Ode
7. Close To The Edge-Yes-Atlantic
8. Black Sabbath, Vol. 4-Warner Brothers
9. Superfly-Curtis Mayfield-Curtom
10. Moods-Neil Diamond-Uni

SUPER 'CFL's exclusive
SUPERSCORE
YOUR chance to win-every hour!

---

10 YEARS AGO TODAY

*1. Monster Mash-Bobby Pickett-Garpax
*2. Only Love Can Break A Heart-Gene Pitney-Musicor
*3. He's A Rebel-The Crystals-Philles
4. Patches-Dickey Lee-Smash
*5. Gina-Johnny Mathis-Columbia
*6. Big Girls Don't Cry-Four Seasons-Vee-Jay
*7. Surfin' Safari/409-Beach Boys-Capitol
*8. Let's Dance-Chris Montez-Monogram
9. Close To Cathy-Mike Clifford-UA
*10. Next Door To An Angel-Neil Sedaka-RCA

36

*denotes: if you remember these hits you're getting up there!

(39)	37.	Rockin' Pneumonia-Johnny Rivers-UA
(--)	38.	Papa Was A Rollin' Stone-The Temptations-Gordy
(--)	39.	Something's Wrong With Me-Austin Roberts-Chelsea
(--)	40.	I Didn't Know I Loved You-Gary Glitter-Bell

*Denotes Big 10 Climbers

WC FL **ALL HIT MUSIC** **IN THE MIDWEST**

WAS	IS	Week Ending November 4, 1972
(2)	1.	Nights In White Satin-The Moody Blues-Deram
(3)	2.	I'll Be Around-The Spinners-Atlantic
(8)	3.	Listen To The Music-The Doobie Brothers-WB
(5)	4.	Garden Party-Rick Nelson-Decca
(1)	5.	Everybody Plays The Fool-The Main Ingredient-RCA
(11)	6.	I'd Love You To Want Me-Lobo-Big Tree
(14)	*7.	Freddie's Dead-Curtis Mayfield-Curtom
(12)	8.	Good Time Charlie's-Danny O'Keefe-Signpost
(10)	9.	Thunder And Lightning-Chi Coltrane-Columbia
(4)	10.	Ben-Michael Jackson-Motown
(13)	11.	Why-Donny Osmond-MGM
(21)	*12.	I Can See Clearly Now-Johnny Nash-Epic
(22)	*13.	I Believe In Music-Gallery-Sussex
(6)	14.	Use Me-Bill Withers-Sussex
(33)	*15.	Convention '72-The Delegates-Mainstream
(18)	16.	You Wear It Well-Rod Stewart-Mercury
(20)	17.	If I Could Reach You-The 5th Dimension-Bell
(23)	18.	Witchy Women-Eagles-Asylum
(7)	19.	Beautiful Sunday-Daniel Boone-Mercury
(32)	*20.	From The Beginning-Emerson, Lake & Palmer-Cotillion
(28)	21.	I Am Woman-Helen Reddy-Capitol
(9)	22.	Burning Love-Elvis Presley-RCA
(37)	*23.	Rockin' Pneumonia-Johnny Rivers-UA
(38)	*24.	Papa Was A Rollin' Stone-The Temptations-Gordy
(26)	25.	Don't Ever Be Lonely-Cornelius Brothers-UA
(29)	26.	Elected-Alice Cooper-WB
(36)	*27.	Ventura Highway-America-WB
(34)	28.	Roberta-Bones-Signpost
(19)	29.	Starting All Over Again-Mel And Tim-Stax
(17)	30.	Tight Rope-Leon Russell-Shelter
(16)	31.	Popcorn-Hot Butter-Musicor
(39)	32.	Something's Wrong With Me-Austin Roberts-Chelsea
(15)	33.	My Ding-A-Ling-Chuck Berry-Chess

BIG 10 ALBUMS

1. Days Of Future Past-Moody Blules-Deram
2. Trilogy-Emerson, Lake And Palmer-Cotillion
3. Chicago V-Chicago-Columbia
4. Superfly-Curtis Mayfield-Curtom
5. Catch Bull At Four-Cat Stevens-A & M
6. Close To The Edge-Yes-Atlantic
7. Big Bambu-Cheech And Chong-Ode
8. Never A Dull Moment-Rod Stewart-Mercury
9. Black Sabbath, Vol 4.-Warner Brothers
10. Moods-Neil Diamond-Uni

LARRY LUJACK...carves pumpkins and "klunk" letters... HALLOWEEN and every day on Super 'CFL!

BIG 10 HALLOWEEN HITS

1. Join Together-Dr. Y. Frankenstein-Mucelage
2. You Don't Mess Around With Drac-The Victims-Type A
3. A Hearse With No Name-Transylvania-ONA
4. Hold Your Head Up-Ichabod Crane-Decapitate
5. One Bad Apple-Snow White-Respite
6. The Mephisto Waltz-Liberace-Dischord
7. Smiling Faces Sometimes-Jack O'Lantern-Carve
8. Bier Barrel Polka-Chuck Bury-Chest
9. The First Time Ever I Saw Your Fangs-Mama Caskett-RIP
10. After Midnight-The Wolfman-Metamorphosis
11. Look What You Done For Me-Vincent Price-EA Poe
12. Living In A House Divided-Dr. Jekyl/Mr. Hyde-1/2
13. Drac Stabbers-The Witch Hunters-Stake

(--)	34.	Dialogue (Part I & II)-Chicago-Columbia
(40)	35.	I Didn't Know I Loved You-Gary Glitter-Bell
(24)	36.	Play Me-Neil Diamond-Uni
(--)	37.	Summer Breeze-Seals & Crofts-WB
(--)	38.	Operator-Jim Croce-ABC
(27)	39.	Loving You Just Crossed My Mind-Sam Neely-Capitol
(25)	40.	Go All The Way-Raspberries-Capitol

*Denotes Big 10 Climbers

HAPPY HALLOWEEN FROM THE GUYS AND GHOULS AT SUPER 'CFL!

WC
FL

ALL HIT MUSIC

IN THE
MIDWEST

WAS IS Week Ending November 11, 1972

WAS	IS	
(2)	1.	I'll Be Around-The Spinners-Atlantic
(4)	2.	Garden Party-Rick Nelson-Decca
(7)	3.	Freddie's Dead-Curtis Mayfield-Curtom
(6)	4.	I'd Love You To Want Me-Lobo-Big Tree
(15)	*5.	Convention '72-The Delegates-Mainstream
(8)	6.	Good Time Charlie's Danny O'Keefe-Signpost
(9)	7.	Thunder And Lightning-Chi Coltrane-Columbia
(11)	8.	Why-Donny Osmond-MGM
(12)	9.	I Can See Clearly Now-Johnny Nash-Epic
(1)	10.	Nights In White Satin-The Moody Blues-Deram
(5)	11.	Everybody Plays The Fool-The Main Ingredient-RCA
(13)	12.	I Believe In Music-Gallery-Sussex
(3)	13.	Listen To The Music-The Doobie Brothers-WB
(17)	14.	If I Could Reach You-The 5th Dimension-Bell
(21)	*15.	I Am Woman-Helen Reddy-Capitol
(16)	16.	You Wear It Well-Rod Stewart-Mercury
(18)	17.	Witchy Woman-Eagles-Asylum
(24)	*18.	Papa Was A Rollin' Stone-The Temptations-Gordy
(20)	19.	From The Beginning-Emerson-Lake & Palmer-Cotillion
(10)	20.	Ben-Michael Jackson-Motown
(23)	21.	Rockin' Pneumonia-Johnny Rivers-UA
(26)	22.	Elected-Alice Cooper-WB
(14)	23.	Use Me-Bill Withers-Sussex
(27)	24.	Ventura Highway-America-WB
(25)	25.	Don't Ever Be Lonely-Cornelius Brothers-UA
(28)	26.	Roberta-Bones-Signpost
(35)	*27.	I Didn't Know I Loved You-Gary Glitter-Bell
(19)	28.	Beautiful Sunday-Daniel Boone-Mercury
(32)	29.	Something's Wrong With Me-Austin Roberts-Chelsea
(29)	30.	Starting All Over Again-Mel and Tim-Stax
(22)	31.	Burning Love-Elvis Presley-RCA
(34)	32.	Dialogue (Part I & II)-Chicago-Columbia
(38)	33.	Operator-Jim Croce-ABC
(37)	34.	Summer Breeze-Seals & Crofts-WB
(--)	35.	Corner Of The Sky-The Jackson 5-Motown
(30)	36.	Tight Rope-Leon Russell-Shelter

BIG 10 ALBUMS

1. Days Of Future Past-Moody Blues-Deram
2. Catch Bull At Four-Cat Stevens-A & M
3. Trilogy-Emerson, Lake And Palmer-Cotillion
4. Superfly-Curtis Mayfield-Curtom
5. Close To The Edge-Yes-Atlantic
6. Black Sabbath, Vol. 4-Warner Brothers
7. Chicago V-Chicago-Columbia
8. Big Bambu-Cheech And Chong-Ode
9. Never A Dull Moment-Rod Stewart-Mercury
10. Caravanserai-Santana-Columbia

SUPER "CFL exclusive
SUPERSCORE
YOUR chance to win-every hour!

ENGLAND'S TOP 10

*1. Mouldy Old Dough-Lieutenant Pigeon-Decca
*2. Donna-10cc-UK-St. Annes
*3. You're A Lady-Peter Skellern-Decca
4. How Can I Be Sure-David Bell-Sparta Florida
5. I Didn't Know I Loved You-Gary Glitter-Bell
6. In A Broken Dream-Python Lee Jackson-Youngblood
7. Burning Love-Elvis Presley-RCA
*8. Wig-Wam Bam-Sweet-RCA
9. Elected-Alice Cooper-Warner Brothers
*10. Children Of The Revolution-Emi-Wizard

*denotes British talent
(courtesy: Music Week)

(--)	37.	You Ought To Be With Me-Al Green-HI
(--)	38.	Rock 'N Roll Soul-Grand Funk Railroad-Capitol
(--)	39.	Clair-Gilbert O'Sullivan-MAM
(31)	40.	Popcorn-Hot Butter-Musicor

*Denotes Big 10 Climbers

WC
FL

ALL HIT MUSIC

IN THE
MIDWEST

WAS	IS	Week Ending November 18, 1972
(1)	1.	I'll Be Around-The Spinners-Atlantic
(4)	2.	I'd Love You To Want Me-Lobo-Big Tree
(3)	3.	Freddie's Dead-Curtis Mayfield-Curtom
(7)	4.	Thunder And Lightning-Chi Coltrane-Columbia
(9)	5.	I Can See Clearly Now-Johnny Nash-Epic
(2)	6.	Garden Party-Rick Nelson-Decca
(8)	7.	Why-Donny Osmond-MGM
(12)	8.	I Believe In Music-Gallery-Sussex
(18)	*9.	Papa Was A Rollin' Stone-The Temptations-Gordy
(14)	10.	If I Could Reach You-The 5th Dimension-Bell
(15)	11.	I Am Woman-Helen Reddy-Capitol
(5)	12.	Convention '72-The Delegates-Mainstream
(17)	13.	Witchy Woman-Eagles-Asylum
(21)	*14.	Rockin' Pneumonia-Johnny Rivers-UA
(19)	15.	From The Beginning-Emerson, Lake & Palmer-Cotillion
(6)	16.	Good Time Charlie's-Danny O'Keefe-Signpost
(10)	17.	Nights In White Satin-The Moody Blues-Deram
(34)	*18.	Summer Breeze-Seals & Crofts-WB
(24)	19.	Ventura Highway-America-WB
(26)	20.	Roberta-Bones-Signpost
(32)	*21.	Dialogue (Part I & II)-Chicago-Columbia
(22)	22.	Elected-Alice Cooper-WB
(11)	23.	Everybody Plays The Fool-The Main Ingredient-RCA
(27)	24.	I Didn't Know I Loved You-Gary Glitter-Bell
(29)	25.	Something's Wrong With Me-Austin Roberts-Chelsea
(33)	*26.	Operator-Jim Croce-ABC
(13)	27.	Listen To The Music-The Doobie Brothers-WB
(16)	28.	You Wear It Well-Rod Stewart-Mercury
(35)	29.	Corner Of The Sky-The Jackson 5-Motown
(20)	30.	Ben-Michael Jackson-Motown
(37)	31.	You Ought To Be With Me-Al Green-HI
(23)	32.	Use Me-Bill Withers-Sussex
(38)	33.	Rock 'N Roll Soul-Grand Funk Railroad-Capitol
(39)	34.	Clair-Gilbert O'Sullivan-MAM
(25)	35.	Don't Ever Be Lonely-Cornelius

BIG 10 ALBUMS

1. Days Of Future Past-Moody Blues-Deram
2. Catch Bull At Four-Cat Stevens-A & M
3. Trilogy-Emerson, Lake And Palmer-Cotillion
4. Black Sabbath, Vol. 4-Warner Brothers
5. Seventh Sojourn-Moody Blues-Threshold
6. Superfly-Curtis Mayfield-Curtom
7. Never A Dull Moment-Rod Stewart-Mercury
8. Chicago V-Chicago-Columbia
9. Rhymes & Reasons-Carole King-Ode
10. Caravanserai-Santana-Columbia

LARRY LUJACK
wears it well
Afternoons...on Super 'CFL!

NEW MEXICO'S BIG 10

*1. You Wear It Well-Rod Stewart-Mercury
*2. From The Beginning-Emerson, Lake & Palmer-Cotillion
*3. Don't Ever Be Lonely-Cornelius Brothers-UA
*4. Use Me-Bill Withers-Sussex
*5. Good Time Charlie's-Danny O'Keefe-Signpost
6. Let It Rain-Eric Clapton-Polydor
*7. Ben-Michael Jackson-Motown
*8. Operator-Jim Croce-ABC
9. Happier Than The Morning Sun-B.J. Thomas-Scepter
*10. If I Could Reach You-5th Dimension-Bell

*denotes current or former WCFL hit in Chicago.
(courtesy: KRWG Radio-Las Cruces)

		Brothers-UA
(28)	36.	Beautiful Sunday-Daniel Boone-Mercury
(--)	37.	It Never Rains In So. Calif.-Albert Hammond-Mums
(--)	38.	Keeper Of The Castle-The Four Tops-Dunhill
(--)	39.	I'm Stone In Love With You-The Stylistics-Avco
(--)	40.	I Wanna Be With You-Raspberries-Capitol

*Denotes Big 10 Climbers

WC FL | ALL HIT MUSIC | IN THE MIDWEST

WAS IS Week Ending November 25, 1972

WAS	IS	Title
(2)	1.	I'd Love You To Want Me-Lobo-Big Tree
(5)	2.	I Can See Clearly Now-Johnny Nash-Epic
(11)	*3.	I Am Woman-Helen Reddy-Capitol
(4)	4.	Thunder And Lightning-Chi Coltrane-Columbia
(9)	5.	Papa Was A Rollin' Stone-The Temptations-Gordy
(10)	6.	If I Could Reach You-The 5th Dimension-Bell
(1)	7.	I'll Be Around-The Spinners-Atlantic
(8)	8.	I Believe In Music-Gallery-Sussex
(13)	9.	Witchy Woman-Eagles-Asylum
(3)	10.	Freddie's Dead-Curtis Mayfield-Curtom
(18)	*11.	Summer Breeze-Seals & Crofts-WB
(14)	12.	Rockin' Pneumonia-Johnny Rivers-UA
(6)	13.	Garden Party-Rick Nelson-Decca
(15)	14.	From The Beginning-Emerson, Lake & Palmer-Cotillion
(19)	15.	Ventura Highway-America-WB
(7)	16.	Why-Donny Osmond-MGM
(12)	17.	Convention '72-The Delegates-Mainstream
(21)	18.	Dialogue (Part I & II)-Chicago-Columbia
(20)	19.	Roberta-Bones-Signpost
(26)	*20.	Operator-Jim Croce-ABC
(24)	21.	I Didn't Know I Loved You-Gary Glitter-Bell
(25)	22.	Something's Wrong With Me-Austin Roberts-Chelsea
(29)	*23.	Corner Of The Sky-The Jackson 5-Motown
(31)	*24.	You Ought To Be With Me-Al Green-HI
(16)	25.	Good Time Charlie's-Danny \ O'Keefe-Signpost
(33)	*26.	Rock 'N Roll Soul-Grand Funk Railroad-Capitol
(34)	*27.	Clair-Gilbert O'Sullivan-MAM
(17)	28.	Nights In White Satin-The Moody Blues-Deram
(37)	*29.	It Never Rains In So. Calif.-Albert Hammond-Mums
(22)	30.	Elected-Alice Cooper-WB
(40)	*31.	I Wanna Be With You-Raspberries-Capitol
(23)	32.	Everybody Plays The Fool-The Main Ingredient-RCA
(--)	33.	If You Don't Know Me By Now-Harold Melvin-Philadelphia International
(38)	34.	Keeper Of The Castle-The Four

BIG 10 ALBUMS

1. Catch Bull At Four-Cat Stevens-A & M
2. Seventh Sojourn-Moody Blues-Threshold
3. Superfly-Curtis Mayfield-Curtom
4. Days Of Future Past-Moody Blues-Deram
5. Trilogy-Emerson, Lake And Palmer-Cotillion
6. Black Sabbath, Vol. 4-Warner Brothers
7. Living In The Past-Jethro Tull-Chrysalis
8. Rhymes & Reasons-Carole King-Ode
9. Caravanserai-Santana-Columbia
10. Close To The Edge-Yes-Atlantic

LARRY LUJACK
has a little something for your ears only
Afternoons on Super 'CFL

DAYTON'S BIG 10

*1. I Am Woman-Helen Reddy-Capitol
*2. Nights In White Satin-Moody Blues-Deram
*3. Witchy Woman-Eagles-Asylum
*4. Good Time Charlie's-Danny O'Keefe-Signpost
*5. If I Could Reach You-5th Dimension-Bell
*6. I Can See Clearly Now-Johnny Nash-Epic
*7. Listen To The Music-Doobie Brothers-WB
*8. Don't Ever Be Lonely-Cornelius Brothers-UA
*9. Everybody Plays The Fool-Main Ingredient-RCA
*10. I'd Love You To Want Me-Lobo-Big Tree

*denotes current or former WCFL hit in Chicago.
(courtesy: WING-Dayton)

		Tops-Dunhill
(39)	35.	I'm Stone In Love With You-The Stylistics-Avco
(--)	36.	Crazy Horses-The Osmonds-MGM
--	37.	Sweet Surrender-Bread-Elektra
--	38.	Sitting-Cat Stevens-A & M
28	39.	You Wear It Well-Rod Stewart-Mercury
27	40.	Listen To The Music-The Doobie Brothers-WB

*Denotes Big 10 Climbers

WC
FL

ALL HIT MUSIC

IN THE
MIDWEST

WAS IS Week Ending December 2, 1972

WAS	IS	
(3)	1.	I Am Woman-Helen Reddy-Capitol
(2)	2.	I Can See Clearly Now-Johnny Nash-Epic
(5)	3.	Papa Was A Rollin' Stone-The Temptations-Gordy
(1)	4.	I'd Love You To Want Me-Lobo-Big Tree
(6)	5.	If I Could Reach You-The 5th Dimension-Bell
(9)	6.	Witchy Woman-Eagles-Asylum
(8)	7.	I Believe In Music-Gallery-Sussex
(11)	8.	Summer Breeze-Seals & Crofts-WB
(12)	9.	Rockin' Pneumonia-Johnny Rivers-UA
(4)	10.	Thunder And Lightning-Chi Coltrane-Columbia
(15)	11.	Ventura Highway-America-WB
(7)	12.	I'll Be Around-The Spinners-Atlantic
(18)	13.	Dialogue (Part I & II)-Chicago-Columbia
(20)	14.	Operator-Jim Croce-ABC
(23)	*15.	Corner Of The Sky-Jackson 5-Motown
(27)	*16.	Clair-Gilbert O'Sullivan-MAM
(29)	*17.	It Never Rains In So. Calif.-Albert Hammond-Mums
(10)	18.	Freddie's Dead-Curtis Mayfield-Curtom
(22)	19.	Something's Wrong With Me-Austin Roberts-Chelsea
(21)	20.	I Didn't Know I Loved You-Gary Glitter-Bell
(24)	21.	You Ought To Be With Me-Al Green-HI
(31)	*22.	I Wanna Be With You-Raspberries-Capitol
(13)	23.	Garden Party-Rick Nelson-Decca
(26)	24.	Rock 'N Roll Soul-Grand Funk Railroad-Capitol
(14)	25.	From The Beginning-Emerson, Lake & Palmer-Cotillion
(16)	26.	Why-Donny Osmond-MGM
(35)	27.	I'm Stone In Love With You-The Stylistics-Avco
(34)	28.	Keeper Of The Castle-The Four Tops-Dunhill
(33)	29.	If You Don't Know ME By Now-Harold Melvin-Philadelphia International
(19)	30.	Roberta-Bones-Signpost
(38)	*31.	Sitting-Cat Stevens-A & M
(36)	32.	Crazy Horses-The Osmonds-MGM
(17)	33.	Convention '72-The Delegates-Mainstream
(37)	34.	Sweet Surrender-Bread-Elektra
(--)	35.	Me And Mrs. Jones-Billy Paul-

BIG 10 ALBUMS

1. Seventh Sojourn-Moody Blues-Threshold
2. Catch Bull At Four-Cat Stevens-A & M
3. Superfly-Curtis Mayfield-Curtom
4. Trilogy-Emerson, Lake And Palmer-Cotillion
5. Days Of Future Past-Moody Blues-Deram
6. Living In The Past-Jethro Tull-Chrysalis
7. Caravanserai-Santana-Columbia
8. Rhymes & Reasons-Carole King-Ode
9. Black Sabbath, Vol. 4-Warner Brothers
10. Never A Dull Moment-Rod Stewart-Mercury

LARRY LUJACK
has a little something for your ears only
Afternoons on Super 'CFL

STONE'S BIG 10

1. Satisfaction-(1965)
2. Jumping Jack Flash-(1968)
3. Honky Tonk Woman-(1969)
4. Brown Sugar-(1971)
5. Sympathy For The Devil-(1969)
6. You Can't Always Get What You Want-(1969)
7. Time Is On My Side-(1965)
8. Wild Horses-(1971)
9. Ruby Tuesday-(1967)
10. Get Off My Cloud-(1965)

		Philadelphia International
(25)	36.	Good Time Charlie's-Danny O'Keefe-Signpost
(28)	37.	Nights In White Satin-The Moody Blues-Deram
(--)	38.	No-Bulldog-Decca
(30)	39.	Elected-Alice Cooper-WB
(--)	40.	Funny Face-Donna Fargo-Dot

*Denotes Big 10 Climbers

WC
FL

ALL HIT MUSIC

IN THE
MIDWEST

WAS IS Week Ending December 9, 1972

WAS	IS	
(3)	1.	Papa Was A Rollin' Stone-The Temptations-Gordy
(1)	2.	I Am Woman-Helen Reddy-Capitol
(9)	3.	Rockin' Pneumonia-Johnny Rivers-UA
(8)	4.	Summer Breeze-Seals & Crofts-WB
(2)	5.	I Can See Clearly Now-Johnny Nash-Epic
(16)	*6.	Clair-Gilbert O'Sullivan-MAM
(11)	7.	Ventura Highway-America-WB
(14)	*8.	Operator-Jim Croce-ABC
(17)	*9.	It Never Rains In So. Calif.-Albert Hammond-Mums
(15)	10.	Corner Of The Sky-Jackson 5-Motown
(13)	11.	Dialogue (Part I & II)-Chicago-Columbia
(4)	12.	I'd Love You To Want Me-Lobo-Big Tree
(7)	13.	I Believe In Music-Gallery-Sussex
(19)	14.	Something's Wrong With Me-Austin Roberts-Chelsea
(22)	*15.	I Wanna Be With You-Raspberries-Capitol
(5)	16.	If I Could Reach You-The 5th Dimension-Bell
(21)	17.	You Ought To Be With Me-Al Green-HI
(28)	*18.	Keeper Of The Castle-The Four Tops-Dunhill
(20)	19.	I Didn't Know I Loved You-Gary Glitter-Bell
(24)	20.	Rock 'N Roll Soul-Grand Funk Railroad-Capitol
(6)	21.	Witchy Woman-Eagles-Asylum
(27)	*22.	I'm Stone In Love With You-The Stylistics-Avco
(29)	*23.	If You Don't Know Me By Now-Harold Melvin-Philadelphia International
(32)	*24.	Crazy Horses-The Osmonds-MGM
(10)	25.	Thunder And Lightning-Chi Coltrane-Columbia
(31)	*26.	Sitting-Cat Stevens-A & M
(35)	*27.	Me And Mrs. Jones-Billy Paul-Philadelphia International
(12)	28.	I'll Be Around-The Spinners-Atlantic
(34)	*29.	Sweet Surrender-Bread-Elektra
(18)	30.	Freddie's Dead-Curtis Mayfield-Curtom
(--)	31.	Why Can't We Live Together-Timmy Thomas-Glades
(23)	32.	Garden Party-Rick Nelson-Decca
(40)	33.	Funny Face-Donna Fargo-Dot

BIG 10 ALBUMS

1. Seventh Sojourn-Moody Blues-Threshold
2. Catch Bull At Four-Cat Stevens-A & M
3. Living In The Past-Jethro Tull-Chrysalis
4. Superfly-Curtis Mayfield-Curtom
5. Trilogy-Emerson, Lake And Palmer-Cotillion
6. Rhymes & Reasons-Carole King-Ode
7. Black Sabbath, Vol. 4-Warner Brothers
8. Homecoming-America-Warner Brothers
9. Caravanserai-Santana-Columbia
10. Days Of Future Past-Moody Blues-Deram

LARRY LUJACK
has a little something for your ears only
Afternoons on Super 'CFL

WYOMING'S BIG 10

*1. Use Me-Bill Withers-Sussex
*2. Tight Rope-Leon Russell-Shelter
*3. Summer Breeze-Seals & Crofts-WB
*4. Listen To The Music-Doobie Brothers-WB
*5. Garden Party-Rick Nelson-Decca
*6. I Am Woman-Helen Reddy-Capitol
7. Stringman-Kenny Rankin-Little David
8. The Girl You Think You See-Carly Simon-Elektra
*9. Freddie's Dead-Curtis Mayfield-Buddah
*10. Witchy Woman-Eagles-Asylum

*denotes current or former WCFL hit in Chicago
(courtesy: KUWR Radio, Laramie)

(38)	34.	No-Bulldog-Decca
(26)	35.	Why-Donny Osmond-MGM
(--)	36.	Living In The Past-Jethro Tull-Chrysalis
(--)	37.	Pieces Of April-Three Dog Night-Dunhill
(25)	38.	From The Beginning-Emerson, Lake & Palmer-Cotillion
(33)	39.	Convention '72-The Delegates-Mainstream
(--)	40.	Don't Let Me Be Lonely Tonight-James Taylor-WB

*Denotes Big 10 Climbers

WC FL **ALL HIT MUSIC** **IN THE MIDWEST**

WAS IS Week Ending December 16, 1972

WAS	IS	
(1)	1.	Papa Was A Rollin' Stone-The Temptations-Gordy
(4)	2.	Summer Breeze-Seals & Crofts-WB
(3)	3.	Rockin' Pneumonia-Johnny Rivers-UA
(6)	4.	Clair-Gilbert O'Sullivan-MAM
(7)	5.	Ventura Highway-America-WB
(9)	6.	It Never Rains In So. Calif.-Albert Hammond-Mums
(8)	7.	Operator-Jim Croce-ABC
(10)	8.	Corner Of The Sky-Jackson 5-Motown
(11)	9.	Dialogue (Part I & II)-Chicago-Columbia
(2)	10.	I Am Woman-Helen Reddy-Capitol
(15)	11.	I Wanna Be With You-Raspberries-Capitol
(14)	12.	Something's Wrong With Me-Austin Roberts-Chelsea
(5)	13.	I Can See Clearly Now-Johnny Nash-Epic
(17)	14.	You Ought To Be With Me-Al Green-HI
(18)	15.	Keeper Of The Castle-The Four Tops-Dunhill
(27)	*16.	Me And Mrs. Jones-Billy Paul-Philadelphia International
(23)	17.	If You Don't Know Me By Now-Harold Melvin-Philadelphia International
(20)	18.	Rock 'N Roll Soul-Grand Funk Railroad-Capitol
(22)	19.	I'm Stone In Love With You-The Stylistics-Avco
(24)	20.	Crazy Horses-The Osmonds-MGM
(13)	21.	I Believe In Music-Gallery-Sussex
(26)	22.	Sitting-Cat Stevens-A & M
(29)	23.	Sweet Surrender-Bread-Elektra
(12)	24.	I'd Love You To Want Me-Lobo-Big Tree
(31)	*25.	Why Can't We Live Together-Timmy Thomas-Glades
(33)	*26.	Funny Face-Donna Fargo-Dot
(16)	27.	If I Could Reach You-The 5th Dimension-Bell
(19)	28.	I Didn't Know I Loved You-Gary Glitter-Bell
(21)	29.	Witchy Woman-Eagles-Asylum
(34)	30.	No-Bulldog-Decca
(36)	31.	Living In The Past-Jethro Tull-Chrysalis
(--)	32.	Superstition-Stevie Wonder-Tamla
(37)	33.	Pieces Of April-Three Dog Night-Dunhill
(--)	34.	You're So Vain-Carly Simon-Elektra

BIG 10 ALBUMS

1. Seventh Sojourn-Moody Blues-Threshold
2. Catch Bull At Four-Cat Stevens-A & M
3. Living In The Past-Jethro Tull-Chrysalis
4. Homecoming-America-Warner Brothers
5. Rhymes & Reasons-Carole King-Ode
6. Superfly-Curtis Mayfield-Curtom
7. Summer Breeze-Seals & Crofts-Warner Brothers
8. Caravanserai-Santana-Columbia
9. Trilogy-Emerson, Lake And Palmer-Cotillion
10. Black Sabbath, Vol. 4-Warner Brothers

ART ROBERTS
makes mid-days meaningful
on SUPER CFL!

SPAIN'S BIG 10

1. Love Theme From "The Godfather"-Andy Williams-CBS
2. Alone Again (Naturally)-Gilbert O'Sullivan-CBS
3. Rocket Man-Elton John-EMI
4. When I'm A Kid-Demis Rousos-Phillips-F

*5. Algo De Mi-Camilo Sesto-Ariola

6. Song Sung Blue-Neil Diamond-Phillips-F
7. Popcorn-The Popcorn Makers-Movieplay
8. Beautiful Sunday-Daniel Boone-Belter
9. Rock & Roll-Gary Glitter-Polydor
10. Sealed With A Kiss-Bobby Vinton-CBS

*denotes Spanish talent
(courtesy: "El Musical")

(40)	35.	Don't Let Me Be Lonely Tonight-James Taylor-WB
(25)	36.	Thunder And Lightning-Chi Coltrane-Columbia
(--)	37.	Jambalaya-The Blue Ridge Rangers-Fantasy
(32)	38.	Garden Party-Rick Nelson-Decca
(--)	39.	Your Mama Don't Dance-Loggins/Messina-Columbia
(--)	40.	Oh Babe, What Would You Say-Hurricane Smith-Capitol

*Denotes Big 10 Climbers

WC
FL

ALL HIT MUSIC

IN THE
MIDWEST

WAS IS Week Ending December 23, 1972

WAS	IS	
(4)	1.	Clair-Gilbert O'Sullivan-MAM
(3)	2.	Rockin' Pneumonia-Johnny Rivers-UA
(2)	3.	Summer Breeze-Seals & Crofts-WB
(6)	4.	It Never Rains In So. Calif.-Albert Hammond-Mums
(1)	5.	Papa Was A Rollin' Stone-The Temptations-Gordy
(16)	*6.	Me And Mrs. Jones-Billy Paul-Philadelphia International
(5)	7.	Ventura Highway-America-WB
(9)	8.	Dialogue (Part I & II)-Chicago-Columbia
(15)	9.	Keeper Of The Castle-The Four Tops-Dunhill
(11)	10.	I Wanna Be With You-Raspberries-Capitol
(7)	11.	Operator-Jim Croce-ABC
(17)	12.	If You Don't Know Me By Now-Harold Melvin-Philadelphia International
(12)	13.	Something's Wrong With Me-Austin Roberts-Chelsea
(14)	14.	You Ought To Be With Me-Al Green-HI
(25)	*15.	Why Can't We Live Together-Timmy Thomas-Glades
(19)	16.	I'm Stone In Love With You-The Stylistics-Avco
(22)	17.	Sitting-Cat Stevens-A & M
(8)	18.	Corner Of The Sky-Jackson 5-Motown
(20)	19.	Crazy Horses-The Osmonds-MGM
(26)	20.	Funny Face-Donna Fargo-Dot
(23)	21.	Sweet Surrender-Bread-Elektra
(10)	22.	I Am Woman-Helen Reddy-Capitol
(31)	*23.	Living In The Past-Jethro Tull-Chrysalis
(13)	24.	I Can See Clearly Now-Johnny Nash-Epic
(33)	*25.	Pieces Of April-Three Dog Night-Dunhill
(18)	26.	Rock 'N Roll Soul-Grand Funk Railroad-Capitol
(34)	*27.	You're So Vain-Carly Simon-Elektra
(30)	28.	No-Bulldog-Decca
(35)	29.	Don't Let Me BE Lonely Tonight-James Taylor-WB
(32)	30.	Superstition-Stevie Wonder-Tamla
(21)	31.	I Believe In Music-Gallery-Sussex
(37)	32.	Jambalaya-The Blue Ridge Rangers-Fantasy

BIG 10 ALBUMS

1. Seventh Sojourn-Moody Blues-Threshold
2. Catch Bull At Four-Cat Stevens-A & M
3. Living In The Past-Jethro Tull-Chrysalis
4. Rhymes & Reasons-Carole King-Ode
5. Homecoming-America-Warner Brothers
6. No Secrets-Carly Simon-Elektra
7. Superfly-Curtis Mayfield-Curtom
8. Tommy-Various Artists-Ode
9. Caravanserai-Santana-Columbia
10. Summer Breeze-Seals & Crofts-Warner Brothers

BIG 10 CHRISTMAS '72

1. Can You Fix The Way I Talk For Christmas-Vincent & Pesci-Mainstream
2. The Little Drummer Boy-Royal Scots Dragoon Guards-RCA
3. Christmas Song-Shawn Phillips-A & M
4. Home For Christmas-Heaven & Earth-Ovation
5. Christmas in Chicago-Leon Russell-Shelter
6. Green Chritma-Stan Freberg-Capitol
7. Hotel Christmas-David Woeller-Playboy
8. It's Christmas Time-Louis Paul-Enterprise
9. How I Love Those Christmas Songs-The Country Squirrels-Metromedia Country
10. The Gift Of Giving-Bill Withers-Sussex

(24)	33.	I'd Love You To Want Me-Lobo-Big Tree
(28)	34.	I Didn't Know I Loved You-Gary Glitter-Bell
(--)	35.	Crocodile Rock-Elton John-MCA
(39)	36.	Your Mama Don't Dance-Loggins/Messina-Columbia
(40)	37.	Oh Babe-Hurricane Smith-Capitol
(27)	38.	If I Could Reach You-The 5th Dimension-Bell
(--)	39.	Do It Again-Steely Dan-ABC
(--)	40.	You Turn Me On, I'm A Radio-Joni Mitchell-Asylum

*Denotes Big 10 Climbers

WC FL ALL HIT MUSIC IN THE MIDWEST

WAS	IS	Week Ending December 30, 1972
(1)	1.	Clair-Gilbert O'Sullivan-MAM
(6)	2.	Me And Mrs. Jones-Billy Paul-Philadelphia International
(4)	3.	It Never Rains In So. Calif.-Albert Hammond-Mums
(9)	4.	Keeper Of The Castle-The Four Tops-Dunhill
(2)	5.	Rockin' Pneumonia-Johnny Rivers-UA
(10)	6.	I Wanna Be With You-Raspberries-Capitol
(15)	7.	Why Can't We Live Together-Timmy Thomas-Glades
(3)	8.	Summer Breeze-Seals & Crofts-WB
(27)	*9.	You're So Vain-Carly Simon-Elektra
(12)	10.	If You Don't Know Me By Now-Harold Melvin-Philadelphia International
(16)	11.	I'm Stone In Love With You-The Stylistics-Avco
(5)	12.	Papa Was A Rollin' Stone-The Temptations-Gordy
(23)	*13.	Living In The Past-Jethro Tull-Chrysalis
(19)	14.	Crazy Horses-The Osmonds-MGM
(17)	15.	Sitting-Cat Stevens-A & M
(21)	16.	Sweet Surrender-Bread-Elektra
(30)	*17.	Superstition-Stevie Wonder-Tamla
(7)	18.	Ventura Highway-America-WB
(25)	19.	Pieces Of April-Three Dog Night-Dunhill
(20)	20.	Funny Face-Donna Fargo-Dot
(28)	*21.	No-Bulldog-Decca
(8)	22.	Dialogue (Part I & II)-Chicago-Columbia
(32)	*23.	Jambalaya-The Blue Ridge Rangers-Fantasy
(29)	24.	Don't Let Me Be Lonely Tonight-James Taylor-WB
(11)	25.	Operator-Jim Croce-ABC
(38)	26.	Your Mama Don't Dance-Loggins/Messina-Columbia
(14)	27.	You Ought To Be With Me-Al Green-HI
(13)	28.	Something's Wrong With Me-Austin Roberts-Chelsea
(18)	29.	Corner Of The Sky-Jackson 5-Motown
(22)	30.	I Am Woman-Helen Reddy-Capitol
(39)	31.	Do It Again-Steely Dan-ABC
(35)	32.	Crocodile Rock-Elton John-MCA
(37)	33.	Oh Babe-Hurricane Smith-Capitol
(24)	34.	I Can See Clearly Now-Johnny Nash-Epic
(40)	35.	You Turn Me On, I'm A Radio-Joni Mitchell-Asylum

BIG 10 ALBUMS

1. Seventh Sojourn-Moody Blues-Threshold
2. Catch Bull At Four-Cat Stevens-A & M
3. Living In The Past-Jethro Tull-Chrysalis
4. No Secrets-Carly Simon-Elektra
5. Summer Breeze-Seals & Crofts-Warner Brothers
6. Tommy-Various Artists-Ode
7. Rhymes & Reasons-Carole King-Ode
8. Caravanseral-Santana-Columbia
9. Homecoming-America-Warner Brothers
10. Superfly-Curtis Mayfield-Curtom

BIG 10 CHRISTMAS '72

1. Christmas Song-Jethro Tull-Chrysalis
2. Green Chritma-Stan Freberg-Capitol
3. A Silent Night-Adam Perle & Wesley Crow-Atlantic
4. The Gift Of Giving-Bill Withers-Sussex
5. Home For Christmas-Heaven & Earth-Ovation
6. Christmas In Chicago-Leon Russell-Shelter
7. Little Drummer Boy-Royal Scots Dragoon Guards-RCA
8. 12 Days Of Christmas-Monday Morning Quarterback-WB
9. Christmas Song-Shawn Phillips-A & M
10. Can You Fix The Way I Talk For Christmas-Vincent & Pesci-Mainstream

(26)	36.	Rock 'N Roll Soul-Grand Funk Railroad-Capitol
(31)	37.	I Believe In Music-Gallery-Sussex
(--)	38.	Superfly-Curtis Mayfield-Curtom
(33)	39.	I'd Love You To Want Me-Lobo-Big Tree
(34)	40.	I Didn't Know I Loved You-Gary Glitter-Bell

*Denotes Big 10 Climbers

1. American Pie-Don McLean
2. A Cowboy's Work Is Never Done-Sonny & Cher
3. Down by the Lazy River-The Osmonds
4. Alone Again (Naturally)-Gilbert Sullivan
5. Bandy-Looking Glass
6. The First Ever I saw Your Face-Roberta Flack
7. You Don't Mess Around With Jim-Jim Croce
8. Lean On Me-Bill Withers
9. Stay With Me-Faces
10. Heart Of Gold-Neil Young
11. Doctor My Eyes-Jackson Browne-Asylum
12. Tumbling Dice-Rolling Stones-Rolling Stones
13. Nice To BE With You-Gallery-Sussex
14. I Can See Clearly Now-Johnny Nash-Epic
15. I'd Love You To Want Me-Lobo-Big Tree
16. Joy-Apollo 100-Mega
17. Run Run Run-Jo Jo Gunne-Asylum
18. Back Stabbers-O'Jays-Philadelphia International
19. Back Off Boogaloo-Ringo Starr-Apple
20. Song Sung Blue-Neil Diamond-Uni
21. Burning Love-Elvis Presley-RCA
22. A Horse With No Name-America-WB
23. Ben-Michael Jackson-Motown
24. Rockin' Robin-Michael Jackson-Motown
25. Goodbye To Love-Carpenters-A & M
26. Hold Your Head Up-Argent-Epic
27. I Am Woman-Helen Reddy-Capitol
28. Nights In White Satin-The Moody Blues-Deram
29. Outa-Space-Billy Preston-A & M
30. Without You-Nilsson-RCA
31. Black And White-3 Dog Night-Dunhill
32. Papa Was A Rollin' Stone-The Temptations-Gordy
33. I'll Be Around-The Spinners-Atlantic
34. Mother And Child Reunion-Paul Simon-Columbia
35. Suavecito-Malo-WB
36. The Candy Man-Sammy Davis, Jr.-MGM
37. Morning Has Broken-Cat Stevens-A & M
38. Black Dog-Led Zeppelin-Atlantic
39. Ring The Living Bell-Melanie-Neighborhood
40 Sunshine-Jonathan Edwards-Capricorn

BIG 10 ALBUMS '72

1. American Pie-Don McLean-United Artists
2. Thick As A Brick-Jethro Tull-Reprise
3. Honky Chateau-Elton John-Uni
4. Harvest-Neil Young-Reprise
5. America-America-Warner Brothers
6. Fragile-Yes-Atlantic
7. Trilogy-Emerson, Lake & Palmer-Cotillion
8. Chicago V-Chicago-Columbia
9. First Take-Roberta Flack-Atlanta
10. Chicago At Carnegie Hall-Chicago-Columbia

1973

SUPER CFL SURVEY JAN 13

wcfl wcfl wcfl wcfl wcfl wcfl wcfl wcfl wcfl wcfl wcfl wcfl wcfl wcfl

WAS	IS	
(2)	1.	Me And Mrs. Jones-Billy Paul-Philadelphia International
(4)	2.	Keeper Of The Castle-The Four Tops-Dunhill
(7)	3.	Why Can't We Live Together-Timmy Thomas-Glades
(9)	4.	You're So Vain-Carly Simon-Elektra
(3)	5.	It Never Rains In So. Calif.-Albert Hammond-Mums
(1)	6.	Clair-Gilbert O'Sullivan-MAM
(10)	7.	If You Don't Know Me By Now-Harold Melvin-Philadelphia International
(5)	8.	Rockin' Pneumonia-Johnny Rivers-UA
(13)	9.	Living In The Past-Jethro Tull-Chrysalis
(11)	10.	I'm Stone In Love With You-The Stylistics-Avco
(19)	11.	Pieces Of April-Three Dog Night-Dunhill
(17)	12.	Superstition-Stevie Wonder-Tamla
(15)	13.	Sitting-Cat Stevens-A & M
(16)	14.	Sweet Surrender-Bread-Elektra
(6)	15.	I Wanna Be With You-Raspberries-Capitol
(8)	16.	Summer Breeze-Seals & Crofts-WB
(21)	17.	No-Bulldog-Decca
(23)	18.	Jambalaya-The Blue Ridge Rangers-Fantasy
(20)	19.	Funny Face-Donna Fargo-Dot
(24)	20.	Don't Let Me Be Lonely Tonight-James Taylor-WB
(32)	*21.	Crocodile Rock-Elton John-MCA
(26)	22.	Your Mama Don't Dance-Loggins/Messina-Columbia
(31)	*23.	Do It Again-Steely Dan-ABC
(35)	*24.	You Turn Me On, I'm A Radio-Joni Mitchell-Asylum
(12)	25.	Papa Was A Rollin' Stone-The Temptations-Gordy
(14)	26.	Crazy Horses-The Osmonds-MGM
(18)	27.	Venture Highway-America-WB
(33)	28.	Oh Babe-Hurricane Smith-Capitol
(22)	29.	Dialogue (Part I & II)-Chicago-Columbia
(38)	30.	Superfly-Curtis Mayfield-Curtom
(25)	31.	Operator-Jim Croce-ABC
(--)	32.	Dancing In The Moonlight-King Harvest-Perception
(27)	33.	You Ought To Be With Me-Al Green-HI
(--)	34.	The Cover Of "Rolling Stone"-Dr. Hook-Columbia
(--)	35.	Don't Expect Me To Be Your Friend-Lobo-Big Tree

SUPER 10 ALBUMS

1. No Secrets-Carly Simon-Elektra
2. Living In The Past-Jethro Tull-Chrysalis
3. Seventh Sojourn-Moody Blues-Threshold
4. Catch Bull At Four-Cat Stevens-A & M
5. Tommy-Various Artists-Ode
6. Summer Breeze-Seals & Crofts-Warner Brothers
7. Rhymes & Reasons-Carole King-Ode
8. Homecoming-America-Warner Brothers
9. Superfly-Curtis Mayfield-Curtom
10. All Directions-Temptations-Gordy

'73 NEW YEAR'S RESOLUTION
Right on Super CFL
all year long!

ENGLAND'S SUPER HITS

1. My Ding-A-Ling-Chuck Berry-Chess
2. Gudbye T' Jane-Slade-Polydor
3. Crazy Horse-The Osmonds-MGM
*4. Solid Gold Easy Action-T. Rex-EMI
5. Long Haired Lover From Liverpool-Little Jimmy Osmond-MGM
6. Why-Donny Osmond-MGM
*7. Crocodile Rock-Elton John-DJM
8. Ben-Michael Jackson-Tamla/Motown
*9. Angel/What Made Milwaukee Famous-Rod Stewart-Mercury
*10. Shotgun Wedding-Roy C.-UK

*denotes British talent

(--)	36.	Last Song-Edward Bear-Capitol
(--)	37.	Smoke Gets In Your Eyes-Blue Haze-A & M
(--)	38.	Trouble Man-Marvin Gaye-Tamla
(30)	39.	I Am Woman-Helen Reddy-Capitol
(28)	40.	Something's Wrong With Me-Austin Roberts-Chelsea

*CFL Super Hits

collector's issue number ninety-seven

SUPER CFL SURVEY JAN 20

wcfl wcfl wcfl wcfl wcfl wcfl wcfl wcfl wcfl wcfl wcfl wcfl wcfl wcfl

SUPER 10 ALBUMS

1. No Secrets-Carly Simon-Elektra
2. Living In The Past-Jethro Tull-Chrysalis
3. Tommy-Various Artists-Ode
4. Seventh Sojourn-Moody Blues-Threshold
5. Catch Bull At Four-Cat Stevens-A & M
6. Chicago V-Chiago-Columbia
7. Summer Breeze-Seals & Crofts-Warner Brothers
8. Rhymes & Reasons-Carole King-Ode
9. Homecoming-America-Warner Brothers
10. The World Is A Ghetto-War-United Artists

LARRY LUJACK
has a little something for your ears only
Afternoons on Super CFL

LOUISVILLE'S SUPER HITS

1. Slow Motion-Johnny Williams-PI
2. Rockin' Pneumonia-Johnny Rivers-UA
3. I'm Stone In Love With You-The Stylistics-Avco
4. You're So Vain-Carly Simon-Elektra
5. Something's Wrong With Me-Austin Roberts-Chelsea
6. Me & Mrs. Jones-Billy Paul-PI
7. Papa Was A Rolling Stone-Temptations-Gordy
8. Rocky Mountain High-John Denver-RCA
9. Sweet Surrender-Bread-Elektra
10. Little Willy-The Sweet-Bell

WAS	IS	
(4)	1.	You're So Vain-Carly Simon-Elektra
(3)	2.	Why Can't We Live Together-Timmy Thomas-Glades
(1)	3.	Me And Mrs. Jones-Billy Paul-Philadelphia International
(9)	4.	Living In The Past-Jethro Tull-Chrysalis
(11)	5.	Pieces Of April-Three Dog Night-Dunhill
(12)	6.	Superstition-Stevie Wonder-Tamla
(2)	7.	Keeper Of The Castle-The Four Tops-Dunhill
(13)	8.	Sitting-Cat Stevens-A & M
(10)	9.	I'm Stone In Love With You-The Stylistics-Avco
(14)	10.	Sweet Surrender-Bread-Elektra
(5)	11.	It Never Rains In So. Calif.-Albert Hammond-Mums
(18)	12.	Jambalaya-The Blue Ridge Rangers-Fantasy
(21)	*13.	Crocodile Rock-Elton John-MCA
(6)	14.	Clair-Gilbert O'Sullivan-MAM
(30)	*15.	Superfly-Curtis Mayfield-Curtom
(20)	16.	Don't Let Me Be Lonely Tonight-James Taylor-WB
(19)	17.	Funny Face-Donna Fargo-Dot
(22)	18.	Your Mama Don't Dance-Loggins/ Messina-Columbia
(23)	19.	Do It Again-Steely Dan-ABC
(7)	20.	If You Don't Know Me By Now-Harold Melvin-Philadelphia International
(24)	21.	You Turn Me On, I'm A Radio-Joni Mitchell-Asylum
(17)	22.	No-Bulldog-Decca
(15)	23.	I Wanna Be With You-Raspberries-Capitol
(28)	24.	Oh Babe-Hurricane Smith-Capitol
(32)	25.	Dancing In The Moonlight-King Harvest-Perception
(35)	*26.	Don't Expect Me To Be Your Friend-Lobo-Big Tree
(8)	27.	Rockin' Pneumonia-Johnny Rivers-UA
(36)	*28.	Last Song-Edward Bear-Capitol
(34)	29.	The Cover Of "Rolling Stone"-Dr. Hook-Columbia
(25)	30.	Papa Was A Rolling' Stone-The Temptations-Gordy
(37)	31.	Smoke Gets In Your Eyes-Blue Haze-A & M
(--)	32.	Could It Be I'm Falling In Love-The Spinners-Atlantic
(16)	33.	Summer Breeze-Seals & Crofts-WB
(38)	34.	Trouble Man-Marvin Gaye-Tamla
(--)	35.	Love Jones-Brighter Side Of

		Darkness-20th Century
(26)	36.	Crazy Horses-The Osmonds-MGM
(--)	37.	Living Together, Growing Together-5th Dimension-Bell
(--)	38.	Peaceful Easy Feeling-Eagles-Asylum
(--)	39.	Rocky Mountain High-John Denver-RCA
(--)	40.	Hi, Hi, Hi-Wings-Apple

*CFL Super Hits

collector's issue number ninety-eight

SUPER CFL SURVEY JAN 27

wcfl wcfl wcfl wcfl wcfl wcfl wcfl wcfl wcfl wcfl wcfl wcfl wcfl wcfl

WAS	IS	
1	1.	You're So Vain-Carly Simon-Elektra
4	2.	Living In The Past-Jethro Tull-Chrysalis
5	3.	Pieces Of April-Three Dog Night-Dunhill
6	4.	Superstition-Stevie Wonder-Tamla
2	5.	Why Can't We Live Together-Timmy Thomas-Glades
8	6.	Sitting-Cat Stevens-A & M
13	7.	Crocodile Rock-Elton John-MCA
3	8.	Me And Mrs. Jones-Billy Paul-Philadelphia International
12	9.	Jambalaya-The Blue Ridge Rangers-Fantasy
7	10.	Keeper Of The Castle-The Four Tops-Dunhill
16	11.	Don't Let Me Be Lonely Tonight-James Taylor-WB
15	12.	Superfly-Curtis Mayfield-Curtom
18	13.	Your Mama Don't Dance-Loggins-Messina-Columbia
17	14.	Funny Face-Donna Fargo-Dot
21	15.	You Turn Me On, I'm A Radio-Joni Mitchell-Asylum
19	16.	Do It Again-Steely Dan-ABC
25	*17.	Dancing In The Moonlight-King Harvest-Perception
26	*18.	Don't Expect Me To Be Your Friend-Lobo-Big Tree
10	19.	Sweet Surrender-Bread-Elektra
28	*20.	Last Song-Edward Bear-Capitol
24	21.	Oh Babe-Hurricane Smith-Capitol
29	*22.	The Cover Of "Rolling Stone"-Dr. Hook-Columbia
9	23.	I'm Stone In Love With You-The Stylistics-Avco
34	*24.	Trouble Man-Marvin Gaye-Tamla
11	25.	It Never Rains In So. Calif.-Albert Hammond-Mums
14	26.	Clair-Gilbert O'Sullivan-MAM
31	27.	Smoke Gets In Your Eyes-Blue Haze-A & M
20	28.	If You Don't Know Me By Now-Harold Melvin-Philadelphia International
32	29.	Could It Be I'm Falling In Love-The Spinners-Atlantic
23	30.	I Wanna Be With You-Raspberries-Capitol
37	31.	Living Together, Growing Together-5th Dimension-Bell
38	32.	Peaceful Easy Feeling-Eagles-Asylum
40	33.	Hi, Hi, Hi-Wings-Apple
35	34.	Love Jones-Brighter Side Of Darkness-20th Century

SUPER 10 ALBUMS

1. No Secrets-Carly Simon-Elektra
2. Living In The Past-Jethro Tull-Chrysalis
3. Seventh Sojourn-Moody Blues-Threshold
4. Tommy-Various Artists-Ode
5. Catch Bull At Four-Cat Stevens-A & M
6. Rhymes & Reasons-Carole King-Ode
7. Homecoming-America-Warner Brothers
8. Summer Breeze-Seals & Crofts-Warner Brothers
9. One Man Dog-James Taylor-Warner Brothers
10. Superfly-Curtis Mayfield-Curtom

SUPER CFL
Where you win thousands and thousands and thousands of dollars in prizes in
THE LAST CONTEST!

CANADA'S SUPER HITS

1. Rockin' Pneumonia-Johnny Rivers-UA

*2. The Last Song-Edward Bear-Capitol

3. Me And Mrs. Jones-Billy Paul-Philadelphia International
4. I Wanna Be With You-Raspberries-Capitol
5. Long Dark Road-The Hollies-Epic
6. Funny Face-Donna Fargo-Dot
7. Your Mama Don't Dance-Loggins/Messina-Columbia
8. Sunny Days-Lighthouse-Evolution
9. You're So Vain-Carly Simon-Elektra
10. Sitting-Cat Stevens-A & M

*denotes Canadian talent
(courtesy: CKRC-Winnipeg)

22	35. No-Bulldog-Decca
--	36. Daddy's Home-Jermaine Jackson-Motown
39	37. Rocky Mountain High-John Denver-RCA
27	38. Rockin' Pneumonia-Johnny Rivers-UA
--	39. Don't Cross The River-America-WB
30	40. Papa Was A Rollin' Stone-The Temptations-Gordy

*CFL Super Hits

wcfl wcfl wcfl wcfl wcfl wcfl wcfl wcfl wcfl wcfl wcfl wcfl

collector's issue number ninety-nine

SUPER CFL SURVEY FEB 3

wcfl wcfl wcfl wcfl wcfl wcfl wcfl wcfl wcfl wcfl wcfl wcfl wcfl wcfl

WAS	IS	
1	1.	You're So Vain-Carly Simon-Elektra
3	2.	Pieces Of April-Three Dog Night-Dunhill
4	3.	Superstition-Stevie Wonder-Tamla
2	4.	Living In The Past-Jethro Tull-Chrysalis
7	5.	Crocodile Rock-Elton John-MCA
18	*6.	Don't Expect Me To Be Your Friend-Lobo-Big Tree
11	7.	Don't Let Me Be Lonely Tonight-James Taylor-WB
9	8.	Jambalaya-The Blue Ridge Rangers-Fantasy
17	*9.	Dancing In The Moonlight-King Harvest-Perception
6	10.	Sitting-Cat Stevens-A & M
12	11.	Superfly-Curtis Mayfield-Curtom
13	12.	Your Mama Don't Dance-Loggins/Messina-Columbia
16	13.	Do It Again-Steely Dan-ABC
20	*14.	Last Song-Edward Bear-Capitol
15	15.	You Turn Me On, I'm A Radio-Joni Mitchell-Asylum
22	*16.	The Cover Of "Rolling Stone"-Dr. Hook-Columbia
5	17.	Why Can't We Live Together-Timmy Thomas-Glades
21	18.	Oh Babe-Hurricane Smith-Capitol
8	19.	Me And Mrs. Jones-Billy Paul-Philadelphia International
29	*20.	Could It Be I'm Falling In Love-The Spinners-Atlantic
27	21.	Smoke Gets In Your Eyes-Blue Haze-A & M
10	22.	Keeper Of The Castle-The Four Tops-Dunhill
24	23.	Trouble Man-Marvin Gaye-Tamla
14	24.	Funny Face-Donna Fargo-Dot
31	25.	Living Together, Growing Together-5th Dimension-Bell
34	26.	Love Jones-Brighter Side Of Darkness-20th Century
37	*27.	Rocky Mountain High-John Denver-RCA
32	28.	Peaceful Easy Feeling-Eagles-Asylum
19	29.	Sweet Surrender-Bread-Elektra
33	30.	Hi, Hi, Hi-Wings-Apple
--	31.	Dueling Banjos-Deliverance-WB
23	32.	I'm Stone In Love With You-The Stylistics-Avco
39	33.	Don't Cross The River-America-WB
25	34.	It Never Rains In So. Calif.-Albert Hammond-Mums
36	35.	Daddy's Home-Jermaine Jackson-Motown
--	36.	The World Is A Ghetto-War-UA
26	37.	Clair-Gilbert O'Sullivan-MAM

SUPER 10 ALBUMS

1. No Secrets-Carly Simon-Elektra
2. Living In The Past-Jethro Tull-Chrysalis
3. Seventh Sojourn-Moody Blues-Threshold
4. Catch Bull At Four-Cat Stevens-A & M
5. Tommy-Various Artists-Ode
6. Rhymes & Reasons-Carole King-Ode
7. Homecoming-America-Warner Brothers
8. The World Is A Ghetto-War-United Artists
9. Who Do We Think We Are-Deep Purple-Warner Brothers
10. Chicago V-Chicago-Columbia

Over $5,000 in prizes to
COOKIE BERNACKE
first winner of "The Last Contest"
on SUPER CFL!

EL PASO'S SUPER HITS

1. Why Can't We Live Together-Timmy Thomas-Glades
2. Superstition-Stevie Wonder-Tamla
3. Do It Again-Steely Dan-ABC
4. Crocodile Rock-Elton John-MCA
5. Don't Expect Me To Be Your Friend-Lobo-Big Tree
6. You're So Vain-Carly Simon-Elektra
7. Your Mama Don't Dance-Loggins/Messina-Columbia
8. Oh Babe-Hurricane Smith-Capitol
9. Hi, Hi, Hi-Wings-Apple
10. Peaceful Easy Feeling-Eagles-Asylum

(Courtesy: KINT Radio)

28	38. If You Don't Know Me By Now-Harold Melvin-Philadelphia International
--	39. Killing Me Softly-Roberta Flack-Atlantic
--	40. I'm Just A Singer-The Moody Blues-Threshold

*CFL Super Hits

collector's issue number one hundred

SUPER CFL SURVEY FEB 10

wcfl wcfl wcfl wcfl wcfl wcfl wcfl wcfl wcfl wcfl wcfl wcfl wcfl wcfl

WAS	IS	
5	1.	Crocodile Rock-Elton John-MCA
1	2.	You're So Vain-Carly Simon-Elektra
9	*3.	Dancing In The Moonlight-King Harvest-Perception
6	4.	Don't Expect Me To Be Your Friend-Lobo-Big Tree
8	5.	Jambalaya-The Blue Ridge Rangers-Fantasy
7	6.	Don't Let ME Be Lonely Tonight-James Taylor-WB
2	7.	Pieces Of April-Three Dog Night-Dunhill
12	8.	Your Mama Don't Dance-Loggins/Messina-Columbia
14	9.	Last Song-Edward Bear-Capitol
16	*10.	The Cover Of "Rolling Stone"-Dr. Hook-Columbia
11	11.	Superfly-Curtis Mayfield-Curtom
13	12.	Do It Again-Steely Dan-ABC
15	13.	You Turn Me On, I'm A Radio-Joni Mitchell-Asylum
3	14.	Superstition-Stevie Wonder-Tamla
20	15.	Could It Be I'm Falling In Love-The Spinners-Atlantic
4	16.	Living In The Past-Jethro Tull-Chrysalis
18	17.	Oh Babe-Hurricane Smith-Capitol
23	18.	Trouble Man-Marvin Gaye-Tamla
10	19.	Sitting-Cat Stevens-A & M
21	20.	Smoke Gets In Your Eyes-Blue Haze-A & M
27	21.	Rocky Mountain High-John Denver-RCA
25	22.	Living Together, Growing Together-5th Dimension-Bell
31	*23.	Dueling Banjos-Deliverance-WB
28	24.	Peaceful Easy Feeling-Eagles-Asylum
26	25.	Love Jones-Brighter Side Of Darkness-20th Century
17	26.	Why Can't We Live Together-Timmy Thomas-Glades
33	27.	Don't Cross The River-America-WB
22	28.	Keeper Of The Castle-The Four Tops-Dunhill
35	29.	Daddy's Home-Jermaine Jackson-Motown
30	30.	Hi, Hi, Hi-Wings-Apple
24	31.	Funny Face-Donna Fargo-Dot
39	*32.	Killing Me Softly-Roberta Flack-Atlantic
36	33.	The World Is A Ghetto-War-UA
19	34.	Me And Mrs. Jones-Billy Paul-Philadelphia International
40	35.	I'm Just A Singer-The Moody Blues-Threshold

SUPER 10 ALBUMS

1. No Secrets-Carly Simon-Elektra
2. Living In The Past-Jethro Tull-Chrysalis
3. Seventh Sojourn-Moody Blues-Threshold
4. Catch Bull At Four-Cat Stevens-A & M
5. Rhymes & Reasons-Carole King-Ode
6. Tommy-Various Artists-Ode
7. One Man Dog-James Taylor-Warner Brothers
8. Homecoming-America-Warner Brothers
9. Shoot Out At The Fantasy Factory-Traffic-Island
10. Rocky Mountain High-John Denver-RCA

AUSTRALIA'S SUPER HITS

1. Ben-Michael Jackson-Tamla/Motown
2. You're A Lady-Peter Skellern-Decca
3. Popcorn-Hot Butter-Interfusion
4. I Can See Clearly Now-Johnny Nash-CBS
5. Baby Don't Get Hooked On Me-Mac Davis-CBS
6. Happiest Girl In The USA-Donna Fargo-Dot
7. Go All The Way-Raspberries-Capitol
8. Burning Love-Elvis Presley-RCA

*9. Wings Of An Eagle-Russell Morris-HMV

*10. Sugar Me-Lynsey De Paul-MAM

*denotes Australian talent
(courtesy: 2SN-Sydney)

29	36.	Sweet Surrender-Bread-Elektra
32	37.	I'm Stone In Love With You-The Stylistics-Avco
--	38.	Little Willy-The Sweet-Bell
--	39.	Love Train-O'Jays-Philadelphia International
34	40.	It Never Rains in So. Calif.-Albert Hammond-Mums

*CFL Super Hits

wcfl wcfl wcfl wcfl wcfl wcfl wcfl wcfl wcfl wcfl wcfl wcfl

collector's issue number one hundred-one

SUPER CFL SURVEY FEB 17

wcfl wcfl wcfl wcfl wcfl wcfl wcfl wcfl wcfl wcfl wcfl wcfl wcfl wcfl

WAS	IS	
1	1.	Crocodile Rock-Elton John-MCA
4	2.	Don't Expect Me To Be Your Friend-Lobo-Big Tree
3	3.	Dancing In The Moonlight-King Harvest-Perception
9	4.	Last Song-Edward Bear-Capitol
6	5.	Don't Let Me Be Lonely Tonight-James Taylor-WB
2	6.	You're So Vain-Carly Simon-Elektra
10	7.	The Cover Of "Rolling Stone"-Dr. Hook-Columbia
12	8.	Do It Again-Steely Dan-ABC
15	9.	Could It Be I'm Falling In Love-The Spinners-Atlantic
8	10.	Your Mama Don't Dance-Loggins/ Messina-Columbia
23	*11.	Dueling Banjos-Deliverance-WB
5	12.	Jambalaya-The Blue Ridge Rangers-Fantasy
21	*13.	Rocky Mountain High-John Denver-RCA
20	14.	Smoke Gets In Your Eyes-Blue Haze-A & M
11	15.	Superfly-Curtis Mayfield-Curtom
13	16.	You Turn Me On, I'm A Radio-Joni Mitchell-Asylum
32	*17.	Killing Me Softly-Roberta Flack-Atlantic
18	18.	Trouble Man-Marvin Gaye-Tamla
33	*19.	The World Is A Ghetto-War-UA
25	20.	Love Jones-Brighter Side of Darkness-20th Century
22	21.	Living Together, Growing Together-5th Dimension-Bell
24	22.	Peaceful Easy Feeling-Eagles-Asylum
14	23.	Superstition-Stevie Wonder-Tamla
29	24.	Daddy's Home-Jermaine Jackson-Motown
7	25.	Pieces Of April-Three Dog Night-Dunhill
27	26.	Don't Cross The River-America-WB
35	*27.	I'm Just A Singer-The Moody Blues-Threshold
16	28.	Living In The Past-Jethro Tull-Chrysalis
17	29.	Oh Babe-Hurricane Smith-Capitol
19	30.	Sitting-Cat Stevens-A & M
39	31.	Love Train-O'Jays-Philadelphia International
26	32.	Why Can't We Live Together-Timmy Thomas-Glades
38	33.	Little Willy-The Sweet-Bell

SUPER 10 ALBUMS

1. No Secrets-Carly Simon-Elektra
2. Seventh Sojourn-Moody Blues-Threshold
3. Don't Shoot Me I'm Only The Piano Player-Elton John-MCA
4. Catch Bull At Four-Cat Stevens-A & M
5. Tommy-Various Artists-Ode
6. Shoot Out At The Fantasy Factory-Traffic-Island
7. Living In The Past-Jethro Tull-Chrysalis
8. The World Is A Ghetto-War-United Artists
9. Homecoming-America-Warner Brothers
10. Rocky Mountain High-John Denver-RCA

Coming Soon:
SUPERJOCK BUMPER STICKERS
from Super CFL!

"LOVE IS..." SUPER CFL

1. Love Is All Around-The Troggs-(1968)
2. Love Is Blue-Paul Mauriat-(1968)
3. Love Is Here-The Supremes-(1967)
4. Love Is Good-The Mindbenders-(1965)
5. Love Is The Only Thing-Petula Clark-(1971)
6. Love Is Strange-Mickey & Sylvia-(1957)
7. Love Is Surrender-The Carpenters-(1971)
8. Love Is The Sweetest Thing-Mary Hopkin-(1968)
9. Love Is Plentiful-Staple Singers-(1971)
10. Love Is Like An Itchin' In My Heart-Supremes-(1966)

28	34.	Keeper Of The Castle-The Four Tops-Dunhill
31	35.	Funny Face-Donna Fargo-Dot
--	36.	Big City Miss Ruth Ann-Gallery-Sussex
34	37.	Me And Mrs. Jones-Billy Paul-Philadelphia International
--	38.	Kissing My Love-Bill Withers-Sussex
--	39.	One Man Band-Ronnie Dyson-Columbia
30	40.	Hi, Hi, Hi-Wings-Apple

*CFL Super Hits

collector's issue number one hundred and two

SUPER CFL SURVEY FEB 24

wcfl wcfl wcfl wcfl wcfl wcfl wcfl wcfl wcfl wcfl wcfl wcfl wcfl wcfl

SUPER 10 ALBUMS

1. No Secrets-Carly Simon-Elektra
2. Rocky Mountain High-John Denver-RCA
3. Don't Shoot Me I'm Only The Piano Player-Elton John-MCA
4. Seventh Sojourn-Moody Blues-Threshold
5. Shoot Out At The Fantasy Factory-Traffic-Island
6. Deliverance Sound Track-Warner Brothers
7. Catch Bull At Four-Cat Stevens-A & M
8. Tommy-Various Artists-Ode
9. Living In The Past-Jethro Tull-Chrysalis
10. Homecoming-America-Warner Brothers

LARRY LUJACK
has a little something for your ears only
Afternoons on Super CFL

ATLANTA'S SUPER HITS

1. Killing Me Softly-Roberta Flack-Atlantic
2. Dueling Banjos-Deliverance-WB
3. Could It Be I'm Falling In Love-The Spinners-Atlantic
4. Do It Again-Steely Dan-ABC
5. Your Mama Don't Dance-Loggins/Messina-Columbia
6. Do You Want To Dance-Bette Midler-Atlantic
7. Love Train-O'Jays-Philadelphia International
8. Crocodile Rock-Elton John-MCA
9. Why Can't We Live Together-Timmy Thomas-Glades
10. The World Is A Ghetto-War-UA

(courtesy: WQXI-Atlanta

WAS	IS	
3	1.	Dancing In The Moonlight-King Harvest-Perception
4	2.	Last Song-Edward Bear-Capitol
2	3.	Don't Expect Me To Be Your Friend-Lobo-Big Tree
9	4.	Could It Be I'm Falling In Love-The Spinners-Atlantic
1	5.	Crocodile Rock-Elton John-MCA
7	6.	The Cover Of "Rolling Stone"-Dr. Hook-Columbia
11	7.	Dueling Banjos-Deliverance-WB
17	*8.	Killing Me Softly-Roberta Flack-Atlantic
13	9.	Rocky Mountain High-John Denver-RCA
6	10.	You're So Vain-Carly Simon-Elektra
14	11.	Smoke Gets In Your Eyes-Blue Haze-A & M
8	12.	Do It Again-Steely Dan-ABC
5	13.	Don't Let Me Be Lonely Tonight-James Taylor-WB
10	14.	Your Mama Don't Dance-Loggins/Messina-Columbia
19	15.	The World Is A Ghetto-War-UA
12	16.	Jambalaya-The Blue Ridge Rangers-Fantasy
27	*17.	I'm Just A Singer-The Moody Blues-Threshold
20	18.	Love Jones-Brighter Side of Darkness-20th Century
21	19.	Living Together, Growing Together-5th Dimension-Bell
31	*20.	Love Train-O'Jays-Philadelphia International
16	21.	You Turn Me On, I'm A Radio-Joni Mitchell-Asylum
24	22.	Daddy's Home-Jermaine Jackson-Motown
15	23.	Superfly-Curtis Mayfield-Curtom
18	24.	Trouble Man-Marvin Gaye-Tamla
33	*25.	Little Willy-The Sweet-Bell
26	26.	Don't Cross The River-America-WB
22	27.	Peaceful Easy Feeling-Eagles-Asylum
36	*28.	Big City Miss Ruth Ann-Gallery-Sussex
38	*29.	Kissing My Love-Bill Withers-Sussex
23	30.	Superstition-Stevie Wonder-Tamla
25	31.	Pieces Of April-Three Dog Night-Dunhill
39	32.	One Man Band-Ronnie Dyson-Columbia
--	33.	Also Sprach Zarathustra (2001)-Deodato-CTI
--	34.	Ain't No Woman (Like The One I've

		Got)-4 Tops-Dunhill
29	35.	Oh Babe-Hurricane Smith-Capitol
28	36.	Living In The Past-Jethro Tull-Chrysalis
30	37.	Sitting-Cat Stevens-A & M
32	38.	Why Can't We Live Together-Timmy Thomas-Glades
--	39.	Danny's Song-Anne Murray-Capitol
--	40.	River Road-Uncle Dog-MCA

*CFL Super Hits

wcfl wcfl wcfl wcfl wcfl wcfl wcfl wcfl wcfl wcfl wcfl wcfl

collector's issue number one hundred and three

SUPER CFL SURVEY MAR 3

wcfl wcfl wcfl wcfl wcfl wcfl wcfl wcfl wcfl wcfl wcfl wcfl wcfl wcfl

WAS	IS	
8	*1.	Killing Me Softly-Roberta Flack-Atlantic
2	2.	Last Song-Edward Bear-Capitol
7	3.	Dueling Banjos-Deliverance-WB
4	4.	Could It Be I'm Falling In Love-The Spinners-Atlantic
6	5.	The Cover Of "Rolling Stone"-Dr. Hook-Columbia
1	6.	Dancing In The Moonlight-King Harvest-Perception
9	7.	Rocky Mountain High-John Denver-RCA
3	8.	Don't Expect Me To Be Your Friend-Lobo-Big Tree
15	*9.	The World Is A Ghetto-War-UA
5	10.	Crocodile Rock-Elton John-MCA
17	*11.	I'm Just A Singer-The Moody Blues-Threshold
10	12.	You're So Vain-Carly Simon-Elektra
20	*13.	Love Train-O'Jays-Philadelphia International
11	14.	Smoke Gets In Your Eyes-Blue Haze-A & M
18	15.	Love Jones-Brighter Side Of Darkness-20th Century
12	16.	Do It Again-Steely Dan-ABC
22	17.	Daddy's Home-Jermaine Jackson-Motown
25	*18.	Little Willy-The Sweet-Bell
19	19.	Living Together, Growing Together-5th Dimension-Bell
13	20.	Don't Let Me Be Lonely Tonight-James Taylor-WB
28	*21.	Big City Miss Ruth Ann-Gallery-Sussex
14	22.	Your Mama Don't Dance-Loggins/Messina-Columbia
29	*23.	Kissing My Love-Bill Withers-Sussex
16	24.	Jambalaya-The Blue Ridge Rangers-Fantasy
32	*25.	One Man Band-Ronnie Dyson-Columbia
27	26.	Peaceful Easy Feeling-Eagles-Asylum
33	*27.	Also Sprach Zarathustra (2001)-Deodato-CTI
26	28.	Don't Cross The River-America-WB
21	29.	You Turn Me On, I'm A Radio-Joni Mitchell-Asylum
34	30.	Ain't No Woman (Like The One I've Got)-4 Tops-Dunhill
24	31.	Trouble Man-Marvin Gaye-Tamla
39	32.	Danny's Song-Anne Murray-Capitol
23	33.	Superfly-Curtis Mayfield-Curtom
--	34.	Hummingbird-Seals & Crofts-WB
30	35.	Superstition-Stevie Wonder-Tamla
--	36.	Peaceful-Helen Reddy-Capitol
40	37.	River Road-Uncle Dog-MCA
--	38.	Break Up To Make Up-The Stylistics-Avco

SUPER 10 ALBUMS

1. No Secrets-Carly Simon-Elektra
2. Don't Shoot Me I'm Only The Piano Player-Elton John-MCA
3. Rocky Mountain High-John Denver-RCA
4. Shoot Out At The Fantasy Factory-Traffic-Island
5. The World Is A Ghetto-War-United Artists
6. Deliverance Sound Track-Warner Brothers
7. Seventh Sojourn-Moody Blues-Threshold
8. Catch Bull At Four-Cat Stevens-A & M
9. Aloha From Hawaii-Elvis Presley-RCA
10. Prelude-Deodato-CTI

MILWAUKEE'S SUPER HITS

1. Dueling Banjos-Deliverance-WB
2. Killing Me Softly-Roberta Flack-Atlantic
3. You're So Vain-Carly Simon-Elektra
4. Last Song-Edward Bear-Capitol
5. Crocodile Rock-Elton John-MCA
6. Rocky Mountain High-John Denver-RCA
7. Don't Expect Me To Be Your Friend-Lobo-Big Tree
8. The Cover Of "Rolling Stone"-Dr. Hook-Columbia
9. Superstition-Stevie Wonder-Tamla
10. Could It Be I'm Falling In Love-The Spinners-Atlantic

(courtesy: WOKY-Milwaukee

--	39.	Dead Skunk-Loudon Wainwright III-Columbia
--	40.	Names, Tags, Numbers & Labels-Association-Mums

*CFL Super Hits

collector's issue number one hundred and four

SUPER CFL SURVEY MAR 10

wcfl wcfl wcfl wcfl wcfl wcfl wcfl wcfl wcfl wcfl wcfl wcfl wcfl wcfl

WAS	IS	
1	1.	Killing Me Softly-Robert Flack-Atlantic
3	2.	Dueling Banjos-Deliverance-WB
4	3.	Could It Be I'm Falling In Love-The Spinners-Atlantic
7	4.	Rocky Mountain High-John Denver-RCA
9	5.	The World Is A Ghetto-War-UA
11	6.	I'm Just A Singer-The Moody Blues-Threshold
13	7.	Love Train-O'Jays-Philadelphia International
2	8.	Last Song-Edward Bear-Capitol
5	9.	The Cover Of "Rolling Stone"-Dr. Hook-Columbia
15	10.	Love Jones-Brighter Side Of Darkness-20th Century
18	*11.	Little Willy-The Sweet-Bell
6	12.	Dancing In The Moonlight-King Harvest-Perception
21	*13.	Big City Miss Ruth Ann-Gallery-Sussex
17	14.	Daddy's Home-Jermaine Jackson-Motown
27	*15.	Also Sprach Zarathustra (2001)-Deodato-CTI
8	16.	Don't Expect Me To Be Your Friend-Lobo-Big Tree
23	*17.	Kissing My Love-Bill Withers-Sussex
25	18.	One Man Band-Ronnie Dyson-Columbia
10	19.	Crocodile Rock-Elton John-MCA
12	20.	You're So Vain-Carly Simon-Elektra
16	21.	Do It Again-Steely Dan-ABC
14	22.	Smoke Gets In Your Eyes-Blue Haze-A & M
30	*23.	Ain't No Woman (Like The One I've Got)-4 Tops-Dunhill
19	24.	Living Together, Growing Together-5th Dimension-Bell
32	*25.	Danny's Song-Anne Murray-Capitol
22	26.	Your Mama Don't Dance-Loggins/Messina-Columbia
24	27.	Jambalaya-The Blue Ridge Rangers-Fantasy
34	28.	Hummingbird-Seals & Crofts-WB
36	29.	Peaceful-Helen Reddy-Capitol
20	30.	Don't Let Me Be Lonely Tonight-James Taylor-WB
37	31.	River Road-Uncle Dog-MCA
38	32.	Break Up To Make Up-The Stylistics-Avco

SUPER 10 ALBUMS

1. No Secrets-Carly Simon-Elektra
2. Deliverance Sound Track-Warner Brothers
3. Rocky Mountain High-John Denver-RCA
4. Don't Shoot Me I'm Only The Piano Player-Elton John-MCA
5. Shoot Out At The Fantasy Factory-Traffic-Island
6. The World Is A Ghetto-War-United Artists
7. Seventh Sojourn-Moody Blues-Threshold
8. Aloha From Hawaii-Elvis Presley-RCA
9. Homecoming-America-Warner Brothers
10. They Only Come Out At Night-Edgar Winter Group-Epic

LARRY LUJACK
has a little something for your ears only
Afternoons on Super 'CFL

SASKATOON'S SUPER HITS

1. Don't Expect Me To Be Your Friend-Lobo-Big Tree
2. Superstition-Stevie Wonder-Tamla
3. Crocodile Rock-Elton John-MCA
4. Dancing In The Moonlight-King Harvest-Perception
5. Your Mama Don't Dance-Loggins/Messina-Columbia
6. Stoned In Love With You-The Stylistics-Avco

*7. You Girl-Lighthouse-Evolution

8. Oh Babe, What Would You Say-Hurricane Smith-Capitol

*9. Control Of Me-Les Emerson-Canada

*10. Gypsy-Abraham's Children-Canada

*denotes Canadian talent (courtesy: CKOM Radio-Saskatoon)

--	33.	Neither One Of Us-Gladys Knight/ Pips-Soul
39	34.	Dead Skunk-Loudon Wainwright III-Columbia
40	35.	Names, Tags, Numbers Labels-Association-Mums
--	36.	Stir It Up-Johnny Nash-Epic
--	37.	Wildflower-Skylark-Capitol
--	38.	Tie A Yellow Ribbon-Dawn-Bell
--	39.	Stuck In The Middle With You-Stealers Wheel-A & M
--	40.	Call Me (Come Back Home)-Al Green-HI

*CFL Super Hits

wcfl wcfl wcfl wcfl wcfl wcfl wcfl wcfl wcfl wcfl wcfl wcfl

collector's issue number one hundred and five

SUPER CFL SURVEY MAR 17

wcfl wcfl wcfl wcfl wcfl wcfl wcfl wcfl wcfl wcfl wcfl wcfl wcfl wcfl

WAS	IS	
1	1.	Killing Me Softly-Roberta Flack-Atlantic
2	2.	Dueling Banjos-Deliverance-WB
6	3.	I'm Just A Singer-The Moody Blues-Threshold
7	4.	Love Train-O'Jays-Philadelphia International
15	*5.	Also Sprach Zarathustra (2001)-Deodato-CTI
3	6.	Could It Be I'm Falling In Love-The Spinners-Atlantic
4	7.	Rocky Mountain High-John Denver-RCA
10	8.	Love Jones-Brighter Side Of Darkness-20th Century
11	9.	Little Willy-The Sweet-Bell
13	10.	Big City Miss Ruth Ann-Gallery-Sussex
5	11.	The World Is A Ghetto-War-UA
8	12.	Last Song-Edward Bear-Capitol
23	*13.	Ain't No Woman (Like The One I've Got)-4 Tops-Dunhill
14	14.	Daddy's Home-Jermaine Jackson-Motown
17	15.	Kissing My Love-Bill Withers-Sussex
18	16.	One Man Band-Ronnie Dyson-Columbia
9	17.	The Cover Of "Rolling Stone"-Dr. Hook-Columbia
12	18.	Dancing In The Moonlight-King Harvest-Perception
25	*19.	Danny's Song-Anne Murray-Capitol
28	*20.	Hummingbird-Seals & Crofts-WB
16	21.	Don't Expect Me To Be Your Friend-Lobo-Big Tree
19	22.	Crocodile Rock-Elton John-MCA
20	23.	You're So Vain-Carly Simon-Elektra
29	24.	Peaceful-Helen Reddy-Capitol
34	*25.	Dead Skunk-Loudon Wainwright III-Columbia
32	26.	Break Up To Make Up-The Stylistics-Avco
21	27.	Do It Again-Steely Dan-ABC
33	*28.	Neither One Of Us-Gladys Knight/Pips-Soul
35	29.	Names, Tags, Numbers & Labels-Association-Mums
31	30.	River Road-Uncle Dog-MCA
22	31.	Smoke Gets In Your Eyes-Blue Haze-A & M
40	32.	Call Me (Come Back Home)-Al Green-HI

SUPER 10 ALBUMS

1. Deliverance Sound Track-Warner Brothers
2. No Secrets-Carly Simon-Elektra
3. Don't Shoot Me I'm Only The Piano Player-Elton John-MCA
4. Rocky Mountain High-John Denver-RCA
5. Shoot Out At The Fantasy Factory-Traffic-Island
6. Aloha From Hawaii-Elvis Presley-RCA
7. The World Is A Ghetto-War-United Artists
8. Prelude-Deodato-CTI
9. They Only Come Out At Night-Edgar Winter Group-Epic
10. Seventh Sojourn-Moody Blues-Threshold

LARRY LUJACK
has a little something for your ears only
Afternoons on Super 'CFL

QUÉBEC'S SUPER HITS

1. You're So Vain-Carly Simon-Elektra
2. Superstition-Stevie Wonder-Motown
3. Why Can't We Live Together-Timmy Thomas-Glades
*4. Can't Let You Walk/Everyday-Reno-Parrot
**5. Revolution-Pagliaro-Much
6. Do It Again-Steely Dan-ABC
7. Dancing In The Moonlight-King Harvest-Perception
8. Crocodile Rock-Elton John-MCA
*9. You're Still The One-Copper Penny-Sweet Plum
10. Do You Want To Dance-Bette Midler-Atlantic

*denotes Canadian talent
**French Canadien
(courtesy: CJRP-Quebec

26	33.	Your Mama Don't Dance-Loggins/ Messina-Columbia
39	34.	Stuck In The Middle With You- Stealers Wheel-A & M
36	35.	Stir It Up-Johnny Nash-Epic
37	36.	Wildflower-Skylark-Capitol
38	37.	Tie A Yellow Ribbon-Dawn-Bell
24	38.	Living Together, Growing Together- 5th Dimension-Bell
--	39.	Sing-Carpenters-A & M
--	40.	Masterpiece-The Temptations-Gordy

*CFL Super Hits

collector's issue number one hundred and six

SUPER CFL SURVEY MAR 24

wcfl wcfl wcfl wcfl wcfl wcfl wcfl wcfl wcfl wcfl wcfl wcfl wcfl wcfl

SUPER 10 ALBUMS

1. Deliverance Sound Track-Warner Brothers
2. Don't Shoot Me I'm Only The Piano Player-Elton John-MCA
3. No Secrets-Carly Simon-Elektra
4. Rocky Mountain High-John Denver-RCA
5. They Only Come Out At Night-Edgar Winter Group-Epic
6. The World Is A Ghetto-War-United Artists
7. Billion Dollar Babies-Alice Cooper-Warner Brothers
8. Shoot Out At The Fantasy Factory-Traffic-Island
9. Aloha From Hawaii-Elvis Presley-RCA
10. The Dark Side Of The Moon-Pink Floyd-Harvest

THE WORLD FAMOUS
TOM MURPHY SHOW...all yours
mornings...on Super CFL!

SAN DIEGO'S SUPER HITS

1. Killing Me Softly-Roberta Flack-Atlantic
2. You're So Vain-Carly Simon-Elektra
3. Fortune Teller-Rolling Stones-London
4. Crocodile Rock-Elton John-MCA
5. Dueling Banjos-Deliverance-WB
6. Cover Of "Rolling Stone"-Dr. Hook-Columbia
7. Also Sprach Zarathustra (2001)-Deodato-CTI
8. Oh Babe, What Would You Say-Hurricane Smith-Capitol
9. Reelin' And Rockin'-Chuck Berry-Chess
10. Hummingbird-Seals And Crofts-WB

WAS	IS	
5	1.	Also Sprach Zarathustra (2001)-Deodato-CTI
1	2.	Killing Me Softly-Roberta Flack-Atlantic
3	3.	I'm Just A Singer-The Moody Blues-Threshold
4	4.	Love Train-O'Jays-Philadelphia International
9	5.	Little Willy-The Sweet-Bell
2	6.	Dueling Banjos-Deliverance-WB
15	*7.	Kissing My Love-Bill Withers-Sussez
10	8.	Big City Miss Ruth Ann-Gallery-Sussex
13	9.	Ain't No Woman (Like The One I've Got)-4Tops-Dunhill
6	10.	Could It Be I'm Falling In Love-The Spinners-Atlantic
20	*11.	Hummingbird-Seals & Crofts-WB
19	*12.	Danny's Song-Anne Murray-Capitol
7	13.	Rocky Mountain High-John Denver-RCA
16	14.	One Man Band-Ronnie Dyson-Columbia
8	15.	Love Jones-Brighter Side Of Darkness-20th Century
24	*16.	Peaceful-Helen Reddy-Capitol
11	17.	The World Is A Ghetto-War-UA
25	*18.	Dead Skunk-Loudon Wainwright III-Columbia
12	19.	Last Song-Edward Bear-Capitol
14	20.	Daddy's Home-Jermaine Jackson-Motown
26	21.	Break Up To Make Up-The Stylistics-Avco
17	22.	The Cover Of "Rolling Stone"-Dr. Hook-Columbia
28	23.	Neither One Of Us-Gladys Knight/Pips-Soul
29	24.	Names, Tags, Numbers & Labels-Association-Mums
18	25.	Dancing In The Moonlight-King Harvest-Perception
23	26.	You're So Vain-Carly Simon-Elektra
32	27.	Call Me (Come Back Home)-Al Green-HI
34	28.	Stuck In The Middle With You-Stealers Wheel-A & M
30	29.	River Road-Uncle Dog-MCA
35	30.	Stir It Up-Johnny Nash-Epic
21	31.	Don't Expect Me To Be Your Friend-Lobo-Big Tree
37	32.	Tie A Yellow Ribbon-Dawn-Bell
36	33.	Wildflower-Skylark-Capitol
22	34.	Crocodile Rock-Elton John-MCA
39	35.	Sing-Carpenters-A & M

40	36.	Masterpiece-The Temptations-Gordy
27	37.	Do It Again-Steely Dan-ABC
--	38.	The Night The Lights Went Out In Georgia-Vicki Lawrence-Bell
--	39.	The Cisco Kid-War-UA
--	40.	Drift Away-Dobie Gray-Decca

*CFL Super Hits

wcfl wcfl wcfl wcfl wcfl wcfl wcfl wcfl wcfl wcfl wcfl wcfl

collector's issue number one hundred and seven

SUPER CFL SURVEY MAR 31

wcfl wcfl wcfl wcfl wcfl wcfl wcfl wcfl wcfl wcfl wcfl wcfl wcfl wcfl

WAS	IS	
1	1.	Also Sprach Zarathustra (2001)-Deodato-CTI
4	2.	Love Train-O'Jays-Philadelphia International
7	3.	Kissing My Love-Bill Withers-Sussex
5	4.	Little Willy-The Sweet-Bell
9	5.	Ain't No Woman (Like The One I've Got)-4 Tops-Dunhill
3	6.	I'm Just A Singer-The Moody Blues-Threshold
11	7.	Hummingbird-Seals & Crofts-WB
8	8.	Big City Miss Ruth Ann-Gallery-Sussex
2	9.	Killing Me Softly-Roberta Flack-Atlantic
12	10.	Danny's Song-Anne Murray-Capitol
16	*11.	Peaceful-Helen Reddy-Capitol
6	12.	Dueling Banjos-Deliverance-WB
14	13.	One Man Band-Ronnie Dyson-Columbia
24	*14.	Names, Tags, Numbers & Labels-Association-Mums
18	15.	Dead Skunk-Loudon Wainwright III-Columbia
10	16.	Could It Be I'm Falling In Love-The Spinners-Atlantic
23	17.	Neither One Of Us-Gladys Knight/Pips-Soul
13	18.	Rocky Mountain High-John Denver-RCA
21	19.	Break Up To Make Up-The Stylistics-Avco
15	20.	Love Jones-Brighter Side Of Darkness-20th Century
27	*21.	Call Me (Come Back Home)-Al Green-HI
30	*22.	Stir It Up-Johnny Nash-Epic
19	23.	Last Song-Edward Bear-Capitol
28	24.	Stuck In The Middle With You-Stealers Wheel-A & M
20	25.	Daddy's Home-Jermaine Jackson-Motown
35	*26.	Sing-Carpenters-A & M
22	27.	The Cover Of "Rolling Stone"-Dr. Hook-Columbia
32	28.	Tie A Yellow Ribbon-Dawn-Bell
26	29.	You're So Vain-Carly Simon-Elektra
36	30.	Masterpiece-The Temptations-Gordy
33	31.	Wildflower-Skylark-Capitol
29	32.	River Road-Uncle Dog-MCA
38	33.	The Night The Lights Went Out In Georgia-Vicki Lawrence-Bell
25	34.	Dancing In The Moonlight-King

SUPER 10 ALBUMS

1. Deliverance Sound Track-Warner Brothers
2. Billion Dollar Babies-Alice Cooper-Warner Brothers
3. Rocky Mountain High-John Denver-RCA
4. Don't Shoot Me I'm Only The Piano Player-Elton John-MCA
5. No Secrets-Carly Simon-Elektra
6. Prelude-Deodato-CTI
7. They Only Come Out At Night-Edgar Winter Group-Epic
8. Shoot Out At The Fantasy Factory-Traffic-Island
9. The Dark Side Of The Moon-Pink Floyd-Harvest
10. Aloha From Hawaii-Elvis Presley-RCA

LARRY LUJACK
has a little something for your ears only
Afternoons on Super 'CFL

AUSTRALIA'S SUPER HITS

1. You're So Vain-Carly Simon-Elektra
2. I'd Love You To Want Me-Lobo-Phillips
3. Crocodile Rock-Elton John-DJM
*4. I Am Woman-Helen Reddy-Capitol
5. Nights In White Satin-Moody Blues-Dream
*6. Dreams Are Ten A Penny-Kincade-Penny Farthing
7. I've Got To Have You-Carly Simon-Elektra
*8. Your Mama Don't Dance-Bootleg Family-Bootleg
9. Mouldy Old Dough-Lieutenant Pigeon-Decca
10. Ben-Michael Jackson-Tamla/Motown

*denotes Australian talent (courtesy: 3XY-Melbourne)

		Harvest-Perception
39	35.	The Cisco Kid-War-UA
--	36.	Cherry Cherry-Neil Diamond-MCA
17	37.	The World Is A Ghetto-War-UA
40	38.	Drift Away-Dobie Gray-Decca
--	39.	The Twelfth Of Never-Donny Osmond-MGM
--	40.	Frankenstein-The Edgar Winter Group-Epic

*CFL Super Hits

collector's issue number one hundred and eight

SUPER CFL SURVEY APR 7

WCFL WCFL WCFL WCFL WCFL WCFL WCFL

SUPER 10 ALBUMS

1. They Only Come Out At Night-Edgar Winter Group-Epic
2. Deliverance Sound Track-Warner Brothers
3. Billion Dollar Babies-Alice Cooper-Warner Brothers
4. The Dark Side Of The Moon-Pink Floyd-Harvest
5. Don't Shoot Me I'm Only The Piano Player-Elton John-MCA
6. Rocky Mountain High-John Denver-RCA
7. No Secrets-Carly Simon-Elektra
8. The World Is A Ghetto-War-United Artists
9. Aloha From Hawaii-Elvis Presley-RCA
10. Shoot Out At The Fantasy Factory-Traffic-Island

LARRY LUJACK
has a little something for your ears only
Afternoons on Super 'CFL

73's SUPER HITS

1. You're So Vain-Carly Simon-(January)
2. Killing Me Softly-Roberta Flack-(March)

WAS	IS	
3	1.	Kissing My Love-Bill Withers-Sussex
4	2.	Little Willy-The Sweet-Bell
1	3.	Also Sprach Zarathustra (2001)-Deodato-CTI
5	4.	Ain't No Woman (Like The One I've Got)-4 Tops-Dunhill
10	5.	Danny's Song-Anne Murray-Capitol
7	6.	Hummingbird-Seals & Crofts-WB
8	7.	Big City Miss Ruth Ann-Gallery-Sussex
11	8.	Peaceful-Helen Reddy-Capitol
2	9.	Love Train-O'Jays-Philadelphia International
15	10.	Dead Skunk-London Wainwright III-Columbia
13	11.	One Man Band-Ronnie Dyson-Columbia
14	12.	Names, Tags, Numbers & Labels-Association-Mums
6	13.	I'm Just A Singer-The Moody Blues-Threshold
22	*14.	Stir It Up-Johnny Nash-Epic
33	*15.	The Night The Lights Went Out to Georgia Vicki Lawrence-Bell
17	16.	Neither One Of Us-Gladys Knight/Pips-Soul
24	*17.	Stuck in The Middle With You-Stealers Wheel-A & M
19	18.	Break Up To Make Up-The Stylistics-Avco
21	19.	Call Me (Come Back Home)-

3. Rockin' Pneumonia-Johnny Rivers-(January)
4. Crocodile Rock-Elton John-(February)
5. Dueling Banjos-Deliverance-(March)
6. Also Sprach Zarathustra (2001)-Deodato-(March)
7. Last Song-Edward Bear-(February)
8. Dancing In The Moonlight-King Harvest-(February)
9. Keeper Of The Castle-Four Tops-(January)
10. Don't Expect Me To Be Your Friend-Lobo-(February)

		Al Green-Hi
26	20.	Sing-Carpenters-A & M
28	*21.	Tie A Yellow Ribbo-Dawn-Bell
12	22.	Dueling Banjos-Deliverance-WB
9	23.	Killing Me Softly-Roberta Flack-Atlantic
35	*24.	The Cisco Kid-War-UA
16	25.	Could It Be I'm Falling In Love-The Spinners-Atlantic
18	26.	Rocky Mountain High-John Denver-RCA
30	27.	Masterpiece-The Temptations-Gordy
23	28.	Last Song-Edward Bear-Capitol
31	29.	Wildflower-Skylark-Capitol
20	30.	Love Jones-Brighter Side Of Darkness-20th Century
38	31.	Drift Away-Dobie Gray-Decca
39	32.	The Twelfth Of Never-Donny Osmond-MGM
27	33.	The Cover Of "Rolling Stone"-Dr. Hook-Columbia
36	34.	Cherry Cherry-Neil Diamond-MCA
29	35.	You're So Vain-Carly Simon-Elektra
40	36.	Frankenstein-The Edgar Winter Group-Epic
--	37.	Hocus Pocus-Focus-Sire
25	38.	Daddy's Home-Jermaine Jackson-Motown
--	39.	You Are The Sunshine Of My Life-Stevie Wonder-Tamla
--	40.	Reeling In The Years-Steely Dan-ABC

* CFL Super Hits

WCFL WCFL WCFL WCFL WCFL WCFL

SUPER CFL SURVEY APR 14

wcfl wcfl wcfl wcfl wcfl wcfl wcfl wcfl wcfl wcfl wcfl wcfl wcfl wcfl

WAS	IS	
2	1.	Little Willy-The Sweet-Bell
8	2.	Peaceful-Helen Reddy-Capitol
15	*3.	The Night The Lights Went Out In Georgia-Vicki Lawrence-Bell
6	4.	Hummingbird-Seals & Crofts-WB
5	5.	Danny's Song-Anne Murray-Capitol
4	6.	Ain't No Woman (Like The One I've Got)-4 Tops-Dunhill
1	7.	Kissing My Love-Bill Withers-Sussex
10	8.	Dead Skunk-Loudon Wainwright III-Columbia
3	9.	Also Sprach Zarathustra (2001)-Deodato-CTI
14	10.	Stir It Up-Johnny Nash-Epic
7	11.	Big City Miss Ruth Ann-Gallery-Sussex
17	12.	Stuck In The Middle With You-Stealers Wheel-A & M
16	13.	Neither One Of Us-Gladys Knight/Pips-Soul
9	14.	Love Train-O'Jays-Philadelphia International
11	15.	One Man Band-Ronnie Dyson-Columbia
21	16.	Tie A Yellow Ribbon-Dawn-Bell
19	17.	Call Me (Come Back Home)-Al Green-HI
20	18.	Sing-Carpenters-A & M
12	19.	Names, Tags, Numbers & Labels-Association-Mums
24	20.	The Cisco Kid-War-UA
13	21.	I'm Just A Singer-The Moody Blues-Threshold
36	*22.	Frankenstein-The Edgar Winter Group-Epic
18	23.	Break Up To Make Up-The Stylistics-Avco
27	24.	Masterpiece-The Temptations-Gordy
29	25.	Wildflower-Skylark-Capitol
32	26.	The Twelfth Of Never-Donny Osmond-MGM
22	27.	Dueling Banjos-Deliverance-WB
34	28.	Cherry Cherry-Neil Diamond-MCA
37	29.	Hocus Pocus-Focus-Sire
31	30.	Drift Away-Dobie Gray-Decca
23	31.	Killing Me Softly-Roberta Flack-Atlantic
25	32.	Could It Be I'm Falling In Love-The Spinners-Atlantic
26	33.	Rocky Mountain High-John Denver-RCA
40	34.	Reeling In The Years-Steely Dan-ABC
39	35.	You Are The Sunshine Of My Life-Stevie Wonder-Tamla

SUPER 10 ALBUMS

1. The Dark Side Of The Moon-Pink Floyd-Harvest
2. Billion Dollar Babies-Alice Cooper-Warner Brothers
3. They Only Come Out At Night-Edgar Winter Group-Epic
4. Deliverance Sound Track-Warner Brothers
5. Eat It-Humble Pie-A & M
6. Don't Shoot Me I'm Only The Piano Player-Elton John-MCA
7. Prelude-Deodato-CTI
8. Rocky Mountain High-John Denver-RCA
9. Aloha From Hawaii-Elvis Presley-RCA
10. Shoot Out At The Fantasy Factory-Traffic-Island

LARRY LUJACK
has a little something for your ears only
Afternoons on Super 'CFL

NASHVILLE'S SUPER HITS

1. Tie A Yellow Ribbon-Dawn-Bell
2. Drift Away-Dobie Gray-Decca
3. The Twelfth Of Never-Donny Osmond-MGM
4. Sing-The Carpenters-A & M
5. Neither One Of Us-Gladys Knight/Pips-Soul
6. Danny's Song-Anne Murray-Capitol
7. Stir It Up-Johnny Nash-Epic
8. Playground Of My Mind-Clint Holmes-Epic
9. Love Train-O'Jays-Philadelphia International
10. Break Up To Make Up-The Stylistics-Avco

(Courtesy: WMAK-Nashville

28	36.	Last Song-Edward Bear-Capitol
--	37.	Pillow Talk-Sylvia-Vibration
--	38.	Give It To Me-The J. Geils Band-Atlantic
--	39.	It Sure Took A Long- Long Time-Lobo-Big Tree
--	40.	Drinking Wine Sop-Dee O'Dee-Jerry Lee Lewis-Mercury

*CFL Super Hits

collector's issue number one hundred and ten

SUPER CFL SURVEY APR 21

wcfl wcfl wcfl wcfl wcfl wcfl wcfl wcfl wcfl wcfl wcfl wcfl wcfl wcfl

SUPER 10 ALBUMS

1. The Dark Side Of The Moon-Pink Floyd-Harvest
2. They Only Come Out At Night-Edgar Winter Group-Epic
3. Billion Dollar Babies-Alice Cooper-Warner Brothers
4. Houses Of The Holy-Led Zeppelin-Atlantic
5. Aloha From Hawaii-Elvis Presley-RCA
6. Deliverance Sound Track-Warner Brothers
7. The Best of Bread-Bread-Elektra
8. Don't Shoot Me I'm Only The Piano Player-Elton John-MCA
9. Moving Waves-Focus-Sire
10. Shoot Out At The Fantasy Factory-Traffic-Island

CHUCK KNAPP entertains
while you sweat out the midnight deadline on Super CFL!

SUPER CFL "I.R.S." GOLD

1. Taxman-The Beatles-Apple
2. Nothing To Hide-Tommy James-Roulette
3. Stick-up-Honey Cone-Hot Wax
4. Liar-Three Dog Night-Dunhill
5. Whatcha See Is Whatcha Get-Dramatics-Volt
6. Hang Onto Your Life-Guess Who-RCA
7. If You Could Read My Mind-Gordon Lightfoot-Reprise
8. We Can Work It Out-Stevie Wonder-Tamla

*9. Free-Chicago-Columbia

10. Cherish What Is Dear To You-Freda Payne-Invictus

*denotes nondeductible

WAS	IS	
3	1.	The Night The Lights Went Out In Georgia-Vicki Lawrence-Bell
5	2.	Danny's Song-Anne Murray-Capitol
2	3.	Peaceful-Helen Reddy-Capitol
1	4.	Little Willy-The Sweet-Bell
10	5.	Stir It Up-Johnny Nash-Epic
12	6.	Stuck In The Middle With You-Stealers Wheel-A & M
16	*7.	Tie A Yellow Ribbon-Dawn-Bell
8	8.	Dead Skunk-Loudon Wainwright III-Columbia
4	9.	Hummingbird-Seals & Crofts-WB
18	*10.	Sing-Carpenters-A & M
20	*11.	The Cisco Kid-War-UA
6	12.	Ain't No Woman (Like The One I've Got)-4 Tops-Dunhill
13	13.	Neither One Of Us-Gladys Knight/Pips-Soul
22	*14.	Frankenstein-The Edgar Winter Group-Epic
11	15.	Big City Miss Ruth Ann-Gallery-Sussex
17	16.	Call Me (Come Back Home)-Al Green-HI
26	*17.	The Twelfth Of Never-Donny Osmond-MGM
29	*18.	Hocus Pocus-Focus-Sire
7	19.	Kissing My Love-Bill Withers-Sussex
25	20.	Wildflower-Skylark-Capitol
9	21.	Also Sprach Zarathustra (2001)-Deodato-CTI
28	22.	Cherry Cherry-Neil Diamond-MCA
24	23.	Masterpiece-The Temptations-Gordy
14	24.	Love Train-O'Jays-Philadelphia International
19	25.	Names, Tags, Numbers & Labels-Association-Mums
34	*26.	Reeling In The Years-Steely Dan-ABC
35	27.	You Are The Sunshine Of My Life-Stevie Wonder-Tamla
30	28.	Drift Away-Dobie Gray-Decca
15	29.	One Man Band-Ronnie Dyson-Columbia
37	*30.	Pillow Talk-Sylvia-Vibration
21	31.	I'm Just A Singer-The Moody Blues-Threshold
39	32.	It Sure Took A Long, Long Time-Lobo-Big Tree
--	33.	Walk On The Wild Side-Lou Reed-RCA
--	34.	The Right Thing To Do-Carly Simon-Elektra
38	35.	Give It To Me-The J. Geils Band-

		Atlantic
23	36.	Break Up To Make Up-The Stylistics-Avco
40	37.	Drinking Wine Spo-Dee O'Dee-Jerry Lee Lewis-Mercury
--	38.	Funky Worm-Ohio Players-Westbound
27	39.	Dueling Banjos-Deliverance-WB
--	40.	You Can't Always Get What You Want-The Rolling Stones-London

*CFL Super Hits

collector's issue number one hundred and eleven

SUPER CFL SURVEY APR 28

wcfl wcfl wcfl wcfl wcfl wcfl wcfl wcfl wcfl wcfl wcfl wcfl wcfl wcfl

SUPER 10 ALBUMS

1. The Dark Side Of The Moon-Pink Floyd-Harvest
2. They Only Come Out At Night-Edgar Winter Group-Epic
3. Houses Of The Holy-Led Zeppeliln-Atlantic
4. Billion Dollar Babies-Alice Cooper-Warner Brothers
5. Aloha From Hawaii-Elvis Presley-RCA
6. The Best Of Bread-Bread-Elektra
7. Moving Waves-Focus-Sire
8. The Beatles/1967-1970-Apple
9. The Beatles/1962-1966-Apple
10. Don't Shoot Me I'm Only The Piano Player-Elton John-MCA

LARRY LUJACK
has a little something for your ears only
Afternoons on Super 'CFL

MINNEAPOLIS' SUPER HITS

1. Night The Lights Went Out-Vicki Lawrence-Bell
2. Little Willy-The Sweet-Bell
3. Tie A Yellow Ribbon-Dawn-Bell
4. The Twelfth Of Never-Donny Osmond-MGM
5. Drift Away-Dobie Gray-Decca
6. The Cisco Kid-War-UA
7. Wildflower-Skylark-Capitol
8. Stuck In The Middle With You-Stealers Wheel-A & M
9. Sing-Carpenters-A & M
10. Daisy A Day-Jud Strunk-MGM

(courtesy: WDGY-Minneapolis/St. Paul)

WAS	IS	
1	1.	The Night The Lights Went Out In Georgia-Vicki Lawrence-Bell
6	2.	Stuck In The Middle With You-Stealers Wheel-A & M
10	*3.	Sing-Carpenters-A & M
3	4.	Peaceful-Helen Reddy-Capitol
7	5.	Tie A Yellow Ribbon-Dawn-Bell
5	6.	Stir It Up-Johnny Nash-Epic
2	7.	Danny's Song-Anne Murray-Capitol
11	8.	The Cisco Kid-War-UA
14	9.	Frankenstein-The Edgar Winter Group-Epic
8	10.	Dead Skunk-Loudon Wainwright III-Columbia
18	*11.	Hocus Pocus-Focus-Sire
17	12.	The Twelfth Of Never-Donny Osmond-MGM
20	13.	Wildflower-Skylark-Capitol
22	14.	Cherry Cherry-Neil Diamond-MCA
4	15.	Little Willy-The Sweet-Bell
27	*16.	You Are The Sunshine Of My Life-Stevie Wonder-Tamla
26	*17.	Reeling In The Years-Steely Dan-ABC
16	18.	Call Me (Come Back Home)-Al Green-HI
30	*19.	Pillow Talk-Sylvia-Vibration
23	20.	Masterpiece-The Temptations-Gordy
32	*21.	It Sure Took A Long, Long Time-Lobo-Big Tree
9	22.	Hummingbird-Seals & Crofts-WB
28	23.	Drift Away-Dobie Gray-Decca
12	24.	Ain't No Woman (Like The One I've Got)-4 Tops-Dunhill
35	*25.	Give It To Me-J. Geils Band-Atlantic
33	*26.	Walk On The Wild Side-Lou Reed-RCA
34	27.	The Right Thing To do-Carly Simon-Elektra
37	28.	Drinking Wine Spo-Dee O'Dee-Jerry Lee Lewis-Mercury
13	29.	Neither One Of Us-Gladys Knight/Pips-Soul
38	30.	Funky Worm-Ohio Players-Westbound
15	31.	Big City Miss Ruth Ann-Gallery-Sussex
40	32.	You Can't Always Get What You Want-The Rolling Stones-London
--	33.	Daniel-Elton John-MCA
--	34.	Let's Pretend-Raspberries-Capitol
19	35.	Kissing My Love-Bill Withers-Sussex
--	36.	Teddy Bear Song-Barbara Fairchild-Columbia
--	37.	No More Mr. Nice Guy-Alice Cooper-WB

--	38.	I'm Gonna Love You-Barry White-20th Century
--	39.	Back When My Hair Was Short-Gunhill Road-Kama Sutra
--	40.	Wild About My Lovin'-Adrian Smith-MCA

*CFL Super Hits

wcfl wcfl wcfl wcfl wcfl wcfl wcfl wcfl wcfl wcfl wcfl wcfl

collector's issue number one hundred and twelve

SUPER CFL SURVEY MAY 5

wcfl wcfl wcfl wcfl wcfl wcfl wcfl wcfl wcfl wcfl wcfl wcfl wcfl wcfl

SUPER 10 ALBUMS

1. The Dark Side Of The Moon-Pink Floyd-Harvest
2. They Only Come Out At Night-Edgar Winter Group-Epic
3. Billion Dollar Babies-Alice Cooper-Warner Brothers
4. Houses Of The Holy-Led Zeppelin-Atlantic
5. Moving Waves-Focus-Sire
6. The Best Of Bread-Bread-Elektra
7. Aloha From Hawaii-Elvis Presley-RCA
8. The Beatles/1967-1970-Apple
9. The Beatles/1962-1966-Apple
10. Don't Shoot Me I'm Only The Piano Player-Elton John-MCA

LARRY LUJACK
has a little something for your ears only
Afternoons on Super 'CFL

MONTANA'S SUPER HITS

1. Tie A Yellow Ribbon-Dawn-Bell
2. Playground In My Mind-Clint Holmes-Epic
3. The Lights Went Out-Vicki Lawrence-Bell
4. Let's Pretend-Raspberries-Capitol
5. Drift Away-Dobie Gray-Decca
6. Wildflower-Skylark-Capitol
7. Cherry Cherry-Neil Diamond-MCA
8. Reeling In The Years-Steely Dan-ABC
9. Stuck In The Middle-Stealers Wheel-A & M
10. Daisy A Day-Jud Strunk-MGM

(courtesy: KEIN Radio-Great Falls)

WAS	IS	
1	1.	The Night The Lights Went Out In Georgia-Vicke Lawrence-Bell
3	2.	Sing-Carpenters-A & M
5	3.	Tie A Yellow Ribbon-Dawn-Bell
9	*4.	Frankenstein-The Edgar Winter Group-Epic
8	5.	The Cisco Kid-War-UA
2	6.	Stuck In The Middle With You-Stealers Wheel-A & M
11	*7.	Hocus Pocus-Focus-Sire
12	8.	The Twelfth Of Never-Donny Osmond-MGM
7	9.	Danny's Song-Anne Murray-Capitol
6	10.	Stir It Up-Johnny Nash-Epic
14	11.	Cherry Cherry-Neil Diamond-MCA
13	12.	Wildflower-Skylark-Capitol
16	13.	You Are The Sunshine Of My Life-Stevie Wonder-Tamla
17	14.	Reeling In The Years-Steely Dan-ABC
19	15.	Pillow Talk-Sylvia-Vibration
26	*16.	Walk On The Wild Side-Lou Reed-RCA
4	17.	Peaceful-Helen Reddy-Capitol
21	18.	It Sure Took A Long, Long Time-Lobo-Big Tree
10	19.	Dead Skunk-Loudon Wainwright III-Columbia
23	20.	Drift Away-Dobie Gray-Decca
27	*21.	The Right Thing To Do-Carly Simon-Elektra
25	22.	Give It To Me-J. Geils Band-Atlantic
18	23.	Call Me (Come Back Home)-Al Green-HI
30	*24.	Funky Worm-Ohio Players-Westbound
15	25.	Little Willy-The Sweet-Bell
32	*26.	You Can't Always Get What You Want-The Rolling Stones-London
20	27.	Masterpiece-The Temptations-Gordy
28	28.	Drinking Wine Spo-Dee O'Dee-Jerry Lee Lewis-Mercury
33	29.	Daniel-Elton John-MCA
34	30.	Let's Pretend-Raspberries-Capitol
22	31.	Hummingbird-Seals & Crofts-WB
37	32.	No More Mr. Nice Guy-Alice Cooper-WB
39	33.	Back When My Hair Was Short-Gunhill Road-Kama Sutra
36	34.	Teddy Bear Song-Barbara Fairchild-Columbia
38	35.	I'm Gonna Love You-Barry White-20th Century
24	36.	Ain't No Woman (Like The One I've Got)-4 Tops-Dunhill

40	37.	Wild About My Lovin'-Adrian Smith-MCA
--	38.	My Love-Paul McCartney-Apple
--	39.	One Of A Kind (Love Affair)-Spinners-Atlantic
29	40.	Neither One Of Us-Gladys Knight/Pips-Soul

*CFL Super Hits

collector's issue number one hundred and thirteen

SUPER CFL SURVEY MAY 12

wcfl wcfl wcfl wcfl wcfl wcfl wcfl wcfl wcfl wcfl wcfl wcfl wcfl wcfl

SUPER 10 ALBUMS

1. The Dark Side Of The Moon-Pink Floyd-Harvest
2. Houses Of The Holy-Led Zeppelin-Atlantic
3. They Only Come Out At Night-Edgar Winter Group-Epic
4. Moving Waves-Focus-Sire
5. Billion Dollar Babies-Alice Cooper-Warner Brothers
6. The Best Of Bread-Bread-Elektra
7. The Beatles/1967-1970-Apple
8. The Beatles/1962-1966-Apple
9. Aloha From Hawaii-Elvis Presley-RCA
10. Diamond Girl-Seals & Crofts-Warner Brothers

LARRY LUJACK
Afternoons on Super 'CFL

ENGLAND'S SUPER HITS

1. Tie A Yellow Ribbon-Dawn-Bell
2. Hello Hello I'm Back Again-Gary Glitter-Bell
3. Get Down-Gilbert O'Sullivan-MAM
4. Tweedlee Dee-Jimmy Osmond-MGM
5. I'm A Clown/Some Kind Of Summer-David Cassidy-Bell
6. 12th Of Never-Donny Osmond-MGM
7. Power To All Our Friends-Cliff Richard-EMI
8. Drive-In Saturday-David Bowie-RCA
9. Never Never Never-Shirley Bassey-UA
10. Pyjamarama-Roxy Music-Island

WAS	IS	
2	1.	Sing-Carpenteers-A & M
4	2.	Frankenstein-The Edgar Winter Group-Epic
7	3.	Hocus Pocus-Focus-Sire
1	4.	Night The Lights Went Out-Vicki Lawrence-Bell
12	*5.	Wildflower-Skylark-Capitol
8	6.	The Twelfth Of Never-Donny Osmond-MGM
14	*7.	Reeling In The Years-Steely Dan-ABC
3	8.	Tie A Yellow Ribbon-Dawn-Bell
11	9.	Cherry Cherry-Neil Diamond-MCA
13	10.	You Are The Sunshine Of My Life-Stevie Wonder-Tamla
15	11.	Pillow Talk-Sylvia-Vibration
6	12.	Stuck In The Middle With You-Stealers Wheel-A & M
16	13.	Walk On The Wild Side-Lou Reed-RCA
5	14.	The Cisco Kid-War-UA
20	15.	Drift Away-Dobie Gray-Decca
18	16.	It Sure Took A Long, Long Time-Lobo-Big Tree
22	17.	Give It To Me-J. Geils Band-Atlantic
9	18.	Danny's Song-Anne Murray-Capitol
24	19.	Funky Worm-Ohio Players-Westbound
21	20.	The Right Thing To Do-Carly Simon-Elektra
26	21.	You Can't Always Get What You Want-The Rolling Stones-London
10	22.	Stir It Up-Johnny Nash-Epic
29	23.	Daniel-Elton John-MCA
30	*24.	Let's Pretend-Raspberries-Capitol
17	25.	Peaceful-Helen Reddy-Capitol
32	26.	No More Mr. Nice Guy-Alice Cooper-WB
33	27.	Back When My Hair Was Short-Gunhill Road-Kama Sutra
19	28.	Dead Skunk-Loudon Wainwright III-Columbia
34	29.	Teddy Bear Song-Barbara Fairchild-Columbia
35	30.	I'm Gonna Love You-Barry White-20th Century
38	31.	My Love-Paul McCartney-Apple
37	32.	Wild About My Lovin'-Adrian Smith-MCA
23	33.	Call Me (Come Back Home)-Al Green-HI
25	34.	Little Willy-The Sweet-Bell
39	35.	One Of A Kind (Love Affair)-Spinners-Atlantic

28	36.	Drinking Wine Spo-Dee O'Dee-Jerry Lee Lewis-Mercury
27	37.	Masterpiece-The Temptations-Gordy
N	38.	Bad, Bad Leroy Brown-Jim Croce-ABC
N	39.	Right Place Wrong Time-Dr. John-Atco
N	40.	Thinking Of You-Loggins & Messina-Columbia

*CFL Super Hits

wcfl wcfl wcfl wcfl wcfl wcfl wcfl wcfl wcfl wcfl wcfl wcfl

collector's issue number one hundred and fourteen

SUPER CFL SURVEY MAY 19

wcfl wcfl wcfl wcfl wcfl wcfl wcfl wcfl wcfl wcfl wcfl wcfl wcfl wcfl

SUPER 10 ALBUMS

1. The Dark Side Of The Moon-Pink Floyd-Harvest
2. Billion Dollar Babies-Alice Cooper-Warner Brothers
3. They Only Come Out At Night-Edgar Winter Group-Epic
4. Houses Of The Holy-Led Zeppelin-Atlantic
5. The Beatles/1967-1970-Apple
6. The Best Of Bread-Bread-Elektra
7. Moving Waves-Focus-Sire
8. The Beatles/1962-1966-Apple
9. Aloha From Hawaii-Elvis Presley-RCA
10. Ooh-La-La-Faces-Warner Brothers

TOM MURPHY
mornings on Super CFL!

MONTRÉAL'S SUPER HITS

(English)

1. Also Sprach Zarathustra-Deodato-CTI
2. Love Train-O'Jays-Columbia
3. Daddy's Home-Jermaine Jackson-Motown
*4. Oh My Lady-The Stampeders-MWC
*5. First Cut Is The Deepest-Keith Hampshire-A & M
6. Sing-Carpenters-A & M
7. The Night The Lights Went Out-Vicki Lawrence-Bell
8. Twelfth Of Never-Donny Osmond-MGM
9. Break Up To Make Up-The Stylistics-Avco
10. I'm Just A Singer-Moody Blues-London

*Canadian talent (courtesy: CJMS-Montreal)

WAS	IS	
2	1.	Frankenstein-The Edgar Winter Group-Epic
3	2.	Hocus Pocus-Focus-Sire
6	3.	The Twelfth Of Never-Donny Osmond-MGM
5	4.	Wildflower-Skylark-Capitol
7	5.	Reeling In The Years-Steely Dan-ABC
1	6.	Sing-Carpenters-A & M
13	*7.	Walk On The Wild Side-Lou Reed-RCA
9	8.	Cherry Cherry-Neil Diamond-MCA
11	9.	Pillow Talk-Sylvia-Vibration
10	10.	You Are The Sunshine Of My Life-Stevie Wonder-Tamla
15	11.	Drift Away-Dobie Gray-Decca
4	12.	Night The Lights Went Out-Vicki Lawrence-Bell
20	*13.	The Right Thing To Do-Carly Simon-Elektra
16	14.	It Sure Took A Long, Long Time-Lobo-Big Tree
17	15.	Give It To Me-J. Geils Band-Atlantic
19	16.	Funky Worm-Ohio Players-Westbound
8	17.	Tie A Yellow Ribbon-Dawn-Bell
12	18.	Stuck In The Middle With You-Stealers Wheel-A & M
21	19.	You Can't Always Get What You Want-The Rolling Stones-London
23	20.	Daniel-Elton John-MCA
24	21.	Let's Pretend-Raspberries-Capitol
26	22.	No More Mr. Nice Guy-Alice Cooper-WB
31	*23.	My Love-Paul McCartney-Apple
27	24.	Back When My Hair Was Short-Gunhill Road-Kama Sutra
14	25.	The Cisco Kid-War-UA
30	26.	I'm Gonna Love You-Barry White-20th Century
18	27.	Danny's Song-Anne Murray-Capitol
29	28.	Teddy Bear Song-Barbara Fairchild-Columbmia
32	29.	Wild About My Lovin'-Adrian Smith-MCA
22	30.	Stir It Up-Johnny Nash-Epic
35	31.	One Of A Kind (Love Affair) Spinners-Atlantic
25	32.	Peaceful-Helen Reddy-Capitol
40	33.	Thinking Of You-Loggins & Messina-Columbia
28	34.	Dead Skunk-Loudon Wainwright III-Columbia
39	35.	Right Place Wrong Time-Dr. John-Atco

38	36.	Bad, Bad Leroy Brown-Jim Croce-ABC
--	37.	Give Me Love-George Harrison-Apple
34	38.	Little Willy-The Sweet-Bell
--	39.	Kodachrome-Paul Simon-Columbia
--	40.	Long Train Runnin'-The Doobie Brothers-WB

*CFL Super Hits

collector's issue number one hundred and fifteen

SUPER CFL SURVEY MAY 26

wcfl wcfl wcfl wcfl wcfl wcfl wcfl wcfl wcfl wcfl wcfl wcfl wcfl wcfl

WAS	IS	
2	1.	Hocus Pocus-Focus-Sire
1	2.	Frankenstein-The Edgar Winter Group-Epic
5	3.	Reeling In The Years-Steely Dan-ABC
7	4.	Walk On The Wild Side-Lou Reed-RCA
9	5.	Pillow Talk-Sylvia-Vibration
4	6.	Wildflower-Skylark-Capitol
11	7.	Drift Away-Dobie Gray-Decca
8	8.	Cherry Cherry-Neil Diamond-MCA
3	9.	The Twelfth Of Never-Donny Osmond-MGM
13	10.	The Right Thing To Do-Carly Simon-Elektra
6	11.	Sing-Carpenters-A & M
20	*12.	Daniel-Elton John-MCA
15	13.	Give It To Me-J. Geils Band-Atlantic
16	14.	Funky Worm-Ohio Players-Westbound
10	15.	You Are The Sunshine Of My Life-Stevie Wonder-Tamla
23	*16.	My Love-Paul McCartney-Apple
22	17.	No More Mr. Nice Guy-Alice Cooper-WB
21	18.	Let's Pretend-Raspberries-Capitol
14	19.	It Sure Took A Long, Long Time-Lobo-Big Tree
26	20.	I'm Gonna Love You-Barry White-20th Century
24	21.	Back When My Hair Was Short-Gunhill Road-Kama Sutra
19	22.	You Can't Always Get What You Want-The Rolling Stones-London
12	23.	Night The Lights Went Out-Vicki Lawrence-Bell
17	24.	Tie A Yellow Ribbon-Dawn-Bell
29	25.	Wild About My Lovin'-Adrian Smith-MCA
31	26.	One Of A Kind (Love Affair)-Spinners-Atlantic
33	27.	Thinking Of You-Loggins & Messina-Columbia
18	28.	Stuck In The Middle With You-Stealers Wheel-A & M
36	29.	Bad, Bad Leroy Brown-Jim Croce-ABC
35	30.	Right Place Wrong Time-Dr. John-Atco
37	31.	Give Me Love-George Harrison-Apple
25	32.	The Cisco Kid-War-UA
--	33.	Will It Go Round In Circles-Billy Preston-A & M
40	34.	Long Train Runnin'-The Doobie Brothers-WB
28	35.	Teddy Bear Song-Barbara Fairchild-Columbia

SUPER 10 ALBUMS

1. The Dark Side Of The Moon-Pink Floyd-Harvest
2. Houses Of The Holy-Led Zeppelin-Atlantic
3. They Only Come Out At Night-Edgar Winter Group-Epic
4. The Beatles/1967-1970-Apple
5. Billion Dollar Babies-Alice Cooper-Warner Brothers
6. The Beatles/1962-1966-Apple
7. The Best Of Bread-Bread-Elektra
8. Red Rose Speedway-Paul McCartney-Apple
9. Moving Waves-Focus-Sire
10. Diamond Girl-Seals & Crofts-Warner Brothers

LARRY LUJACK
Afternoons on Super 'CFL

ARIZONA'S SUPER HITS

1. My Love-Paul McCartney-Apple
2. Reeling In The Years-Steely Dan-ABC
3. Frankenstein-Edgar Winter Group-Epic
4. Daniel-Elton John-MCA
5. The Right Thing To Do-Carly Simon-Elektra
6. Stuck In The Middle-Stealers Wheel-A & M
7. Cisco Kid-War-UA
8. Right Place Wrong Time-Dr. John-Atco
9. You Are The Sunshine Of My Life-Stevie Wonder-Tamla
10. Peaceful-Helen Reddy-Capitol

(courtesy: KIKX-Tucson)

--	36.	I'm Doin' Fine Now-New York City-Chelsea
39	37.	Kodachrome-Paul Simon-Columbia
--	38.	Fool-Elvis Presley-RCA
--	39.	Money-Pink Floyd-Harvest
--	40.	Shambala-Three Dog Night-Dunhill

*CFL Super Hits

wcfl wcfl wcfl wcfl wcfl wcfl wcfl wcfl wcfl wcfl wcfl wcfl

collector's issue number one hundred and sixteen

SUPER CFL SURVEY JUNE 2

wcfl wcfl wcfl wcfl wcfl wcfl wcfl wcfl wcfl wcfl wcfl wcfl wcfl wcfl

SUPER 10 ALBUMS

1. The Dark Side Of The Moon-Pink Floyd-Harvest
2. Houses Of The Holy-Led Zeppelin-Atlantic
3. The Beatles/1967-1970-Apple
4. Billion Dollar Babies-Alice Cooper-Warner Brothers
5. They Only Come Out At Night-Edgar Winter Group-Epic
6. Red Rose Speedway-Paul McCartney-Aple
7. The Best Of Bread-Bread-Elektra
8. Moving Waves-Focus-Sire
9. The Beatles/1962-1966-Apple
10. These Goges Rhymin' Simon-Paul Simon-Columbia

SUPER CFL salutes
PEPSI-COLA's 75th anniversary
in THE LAST CONTEXT

NEW BRUNSWICK'S SUPER HITS

1. Lilly Willy-The Sweet-Bell
2. Tie A Yellow Ribbon-Dawn-Bell
3. Stuck In The Middle-Stealers Wheel-A & M
4. Twelfth Of Never-Donny Osmond-MGM

*5. First Cut Is The Deepest-Keith Hampshire-A & M

6. Ain't No Woman-Four Tops-Dunhill
7. Out Of The Question-Gilbert O'Sullivan-MAM
8. Walk On The Wild Side-Lou Reed-RCA
9. Cisco Kid-War-UA
10. Masterpiece-The Temptations-Gordy

*denotes Canadian talent (courtesy: CFBC-St. John

WAS	IS	
5	1.	Pillow Talk-Sylvia-Vibration
1	2.	Hocus Pocus-Focus-Sire
10	*3.	The Right Thing To Do-Carly Simon-Elektra
4	4.	Walk On The Wild Side-Lou Reed-RCA
2	5.	Frankenstein-The Edgar Winter Group-Epic
16	*6.	My Love-Paul McCartney-Apple
7	7.	Drift Away-Dobie Gray-Decca
3	8.	Reeling In The Years-Steely Dan-ABC
6	9.	Wildflower-Skylark-Capitol
12	10.	Daniel-Elton John-MCA
13	11.	Give It To Me-J. Gelis Band-Atlantic
8	12.	Cherry Cherry-Neil Diamond-MCA
18	13.	Let's Pretend-Raspberries-Capitol
20	14.	I'm Gonna Love You-Barry White-20th Century
17	15.	No More Mr. Nice Guy-Alice Cooper-WB
21	16.	Back When My Hair Was Short-Gunhill Road-Kama Sutra
9	17.	The Twelfth Of Never-Donny Osmond-MGM
14	18.	Funky Worm-Ohio Players-Westbound
11	19.	Sing-Carpenters-A & M
25	20.	Wild About My Lovin'n-Adrian Smith-MCA
26	21.	One Of A Kind (Love Affair)-Spinners-Atlantic
31	*22.	Give Me Love-George Harrison-Apple
27	23.	Thinking Of You-Loggins & Messina-Columbia
15	24.	You Are The Sunshine Of My Life-Stevie Wonder-Tamla
30	25.	Right Place Wrong Time-Dr. John-Atco
37	*26.	Kodachrome-Paul Simon-Columbia
29	27.	Bad, Bad Leroy Brown-Jim Croce-ABC
34	28.	Long Train Runnin'-The Doobie Brothers-WB
36	29.	I'm Doin' Fine Now-New York City-Chelsea
19	30.	It Sure Took A Long, Long Time-Lobo-Big Tree
33	31.	Will It Go Round In Circles-Billy Preston-A & M
22	32.	You Can't Always Get-The Rolling Stones-London
40	33.	Shambala-Three Dog Night-Dunhill
38	34.	Fool-Elvis Presley-RCA
23	35.	Night The Lights Went Out-Vicki Lawrence-Bell

39	36.	Money-Pink Floyd-Harvest
24	37.	Tie A Yellow Ribbon-Dawn-Bell
--	38.	Monster Mash-Bobby (Boris) Pickett-Parrot
--	39.	So Very Hard To Go-Tower Of Power-WB
--	40.	Smoke On The Water-Deep Purple-WB

*CFL Super Hits

wcfl wcfl wcfl wcfl wcfl wcfl wcfl wcfl wcfl wcfl wcfl wcfl

collector's issue number one hundred and seventeen

SUPER CFL SURVEY JUNE 9

wcfl wcfl wcfl wcfl wcfl wcfl wcfl wcfl wcfl wcfl wcfl wcfl wcfl wcfl

SUPER 10 ALBUMS

1. The Dark Side Of The Moon-Pink Floyd-Harvest
2. Billion Dollar Babies-Alice Cooper-Warner Brothers
3. The Beatles/1967-1970-Apple
4. Houses Of The Holy-Led Zeppelin-Atlantic
5. Red Horse Speedway-Paul McCartney-Apple
6. They Only Come Out At Night-Edgar Winter Group-Epic
7. Moving Waves-Focus-Sire
8. The Best Of Bread-Bread-Elektra
9. The Beatles/1962-1966-Apple
10. There Goes Rhymin' Simon-Paul Simon-Columbia

LARRY LUJACK
Afternoons on Super 'CFL

SYRACUSE'S SUPER HITS

1. Frankenstein-Edgar Winter Group-Epic
2. You Are The Sunshine Of My Life-Stevie Wonder-Tamla
3. My Love-Paul McCartney-Apple
4. Reeling In The Years-Steely Dan-ABC
5. Little Willy-The Sweet-Bell
6. Cisco Kid-War-UA
7. Drift Away-Dobie Gray-Decca
8. Steamroller Blues-Elvis Presley-RCA
9. Daniel-Elton John-MCA
10. The Night The Lights Went Out In Georgia-Vickie Lawrence-Bell

(courtesy: WNDR-Syracuse

WAS	IS	
1	1.	Pillow Talk-Sylvia-Vibration
3	2.	The Right Thing To Do-Carly Simon-Elektra
4	3.	Walk On The Wild Side-Lou Reed-RCA
6	4.	My Love-Paul McCartney-Apple
10	5.	Daniel-Elton John-MCA
7	6.	Drift Away-Dobie Gray-Decca
2	7.	Hocus Pocus-Focus-Sire
11	8.	Give It To Me-J. Geils Band-Atlantic
14	9.	I'm Gonna Love You-Barry White-20th Century
13	10.	Let's Pretend-Raspberries-Capitol
5	11.	Frankenstein-The Edgar Winter Group-Epic
15	12.	No More Mr. Nice Guy-Alice Cooper-WB
16	13.	Back When My Hair Was Short-Gunhill Road-Kama Sutra
8	14.	Reeling In The Years-Steely Dan-ABC
22	15.	Give Me Love-George Harrison-Apple
9	16.	Wildflower-Skylark-Capitol
23	17.	Thinking Of You-Loggins & Messina-Columbia
26	*18.	Kodachrome-Paul Simon-Columbia
20	19.	Wild About My Lovin'-Adrian Smith-MCA
21	20.	One Of A Kind (Love Affair)-Spinners-Atlantic
25	21.	Right Place Wrong Time-Dr. John-Atco
12	22.	Cherry Cherry-Neil Diamond-MCA
27	23.	Bad, Bad Leroy Brown-Jim Croce-ABC
31	*24.	Will It Go Round In Circles-Billy Preston-A & M
28	25.	Long Train Runnin'-The Doobie Brothers-WB
29	26.	I'm Doin' Fine Now-New York City-Chelsea
17	27.	The Twelfth Of Never-Donny Osmond-MGM
34	28.	Fool-Elvis Presley-RCA
33	29.	Shambala-Three Dog Night-Dunhill
18	30.	Funky Worm-Ohio Players-Westbound
36	31.	Money-Pink Floyd-Harvest
19	32.	Sing-Carpenters-A & W
38	33.	Monster Mash-Bobby (Boris) Pickett-Parrot
24	34.	You Are The Sunshine Of My Life-Stevie Wonder-Tamla
--	35.	Natural High-Bloodstone-London

40	36.	Smoke On The Water-Deep Purple-WB
39	37.	So Very Hard To Go-Tower Of Power-WB
--	38.	There's No Me Without You-Manhattans-Columbia
30	39.	It Sure Took A Long- Long Time-Lobo-Big Tree
--	40.	Over The Hills And Far Away-Led Zeppelin-Atlantic

*CFL Super Hits

collector's issue number one hundred and eighteen

SUPER CFL SURVEY JUNE 16

wcfl wcfl wcfl wcfl wcfl wcfl wcfl wcfl wcfl wcfl wcfl wcfl wcfl wcfl

SUPER 10 ALBUMS

1. The Dark Side Of The Moon-Pink Floyd-Harvest
2. Billion Dollar Babies-Alice Cooper-Warner Brothers
3. Red Rose Speedway-Paul McCartney-Apple
4. Houses Of The Holy-Led Zeppellin-Atlantic
5. The Beatles/1967-1970-Apple
6. They Only Come Out At Night-Edgar Winter Group-Epic
7. There Goes Rhymin' Simon-Paul Simon-Columbia
8. The Best Of Bread-Bread-Elektra
9. The Beatles/1962-1966-Apple
10. Moving Waves-Focus-Sire

World Famous
TOM MURPHY
mornings...on SUPER CFL!

SASKATCHEWAN'S SUPER HITS

1. Frankenstein-Edgar Winter Group-Epic
2. Drift Away-Dobie Gray-Decca
3. My Love-Paul McCartney-Apple
4. Daniel-Elton John-MCA
5. Reelin' In The Years-Seely Dan-ABC
6. Stuck In The Middle With You-Stealers Wheel-A & M
7. Hocus Pocus-Focus-Sire
8. Playground In My Mind-Clint Holmes-Epic
9. Daisy A Day-Jud Strunk-MGM

*10. First Cut Is The Deepest-Keith Hampshire-A & M

*Canadian talent (courtesy: CJME-Regina)

WAS	IS	
2	1.	The Right Thing To Do-Carly Simon-Elektra
4	2.	My Love-Paul McCartney-Apple
1	3.	Pillow Talk-Sylvia-Vibration
5	4.	Daniel-Elton John-MCA
9	5.	I'm Gonna Love You-Barry White-20th Century
10	6.	Let's Pretend-Raspberries-Capitol
8	7.	Give It To Me-J. Geils Band-Atlantic
3	8.	Walk On The Wild Side-Lou Reed-RCA
15	*9.	Give Me Love-George Harrison-Apple
12	10.	No More Mr. Nice Guy-Alice Cooper-WB
18	*11.	Kodachrome-Paul Simon-Columbia
7	12.	Hocus Pocus-Focus-Sire
6	13.	Drift Away-Dobie Gray-Decca
20	14.	One Of A Kind (Love Affair)-Spinners-Atlantic
17	15.	Thinking Of You-Loggins & Messina-Columbia
19	16.	Wild About My Lovin'-Adrian Smith-MCA
13	17.	Back When My Hair Was Short-Gunhill Road-Kama Sustra
21	18.	Right Place Wrong Time-Dr. John-Atco
25	*19.	Long Train Runnin'-The Doobie Brothers-WB
24	20.	Will It Go Round In Circles-Billy Preston-A & M
11	21.	Frankenstein-The Edgar Winter Group-Epic
26	22.	I'm Doin' Fine New-New York City-Chelsea
23	23.	Bad, Bad Leroy Brown-Jim Croce-ABC
14	24.	Reeling In The Years-Steely Dan-ABC
16	25.	Wildflower-Skylark-Capitol
29	26.	Shambala-Three Dog Night-Dunhill
28	27.	Fool-Elvis Presley-RCA
33	*28.	Monster Mash-Bobby (Boris) Pickett-Parrot
31	29.	Money-Pink Floyd-Harvest
22	30.	Cherry Cherry-Neil Diamond-MCA
36	31.	Smoke On The Water-Deep Purple-WB
35	32.	Natural High-Bloodstone-London
37	33.	So Very Hard To Go-Tower Of Power-WB
--	34.	Standing Ovation-The Dells-Cadet
38	35.	There's No Me Without You-Manhattans-Columbia
40	36.	Over The Hills And Far Away-Led Zeppellin-Atlantic

--	37.	Daddy Could Swear-Gladys Knight & The Pips-Soul
--	38.	Get Down-Gilbert O'Sullivan-MAM
--	39.	Uneasy Rider-Charlie Daniels-Kama Sutra
--	40.	Boogie Woogie Bugle Boy-Bette Midler-Atlantic

*CFL Super Hits

wcfl wcfl wcfl wcfl wcfl wcfl wcfl wcfl wcfl wcfl wcfl wcfl

collector's issue number one hundred and nineteen

SUPER CFL SURVEY JUNE 23

wcfl wcfl wcfl wcfl wcfl wcfl wcfl wcfl wcfl wcfl wcfl wcfl wcfl wcfl

SUPER 10 ALBUMS

1. The Dark Side Of The Moon-Pink Floyd-Harvest
2. Living In The Material World-George Harrison-Apple
3. Red Rose Speedway-Paul McCartney-Apple
4. Billion Dollar Babies-Alice Cooper-Warner Brothers
5. Houses Of The Holy-Led Zeppelin-Atlantic
6. There Goes Rhymin' Simon-Paul Simon-Columbia
7. The Beatles/1967-1970-Apple
8. The Best Of Bread-Bread-Elektra
9. The Only Come Out At Night-Edgar Winter Group-Epic
10. Diamond Girl-Seals And Crofts-Warner Brothers

WORLD FAMOUS
TOM MURPHY
Mornings...on Super CFL!

SIMON & GARFUNKEL'S SUPER HITS

1. Bridge Over Troubled Water-(1970)
2. Mrs. Robinson-(1968)
3. Sounds Of Silence-(1965)
4. I Am A Rock-(1966)
5. Mother And Child Reunion-(1972)
6. Cecelia-(1970)
7. Homeward-Bound-(1966)
8. The Boxer-(1967)
9. At The Zoo-(1967)
10. Scarborough Fair-(1966)

WAS	IS	
2	1.	My Love-Paul McCartney-Apple
1	2.	The Right Thing To Do-Carly Simon-Elektra
9	3.	Give Me Love-George Harrison-Apple
6	4.	Let's Pretend-Raspberries-Capitol
5	5.	I'm Gonna Love You-Barry White-20th Century
11	*6.	Kodachrome-Paul Simon-Columbia
10	7.	No More Mr. Nice Guy-Alice Cooper-WB
4	8.	Daniel-Elton John-MCA
7	9.	Give It To Me-J. Geils Band-Atlantic
15	10.	Thinking Of You-Loggins & Messina-Columbia
3	11.	Pillow Talk-Sylvia-Vibration
14	12.	One Of A Kind (Love Affair)-Spinners-Atlantic
16	13.	Wild About My Lovin'-Adrian Smith-MCA
18	14.	Right Place Wrong Time-Dr. John-Atco
20	15.	Will It Go Round In Circles-Billy Preston-A & M
23	16.	Bad, Bad Leroy Brown-Jim Croce-ABC
19	17.	Long Train Runnin'-The Doobie Brothers-WB
28	*18.	Monster Mash-Bobby (Boris) Pickett-Parrot
22	19.	I'm Doin' Fine Now-New York City-Chelsea
29	*20.	Money-Pink Floyd-Harvest
8	21.	Walk On The Wild Side-Lou Reed-RCA
31	*22.	Smoke On The Water-Deep Purple-WB
26	23.	Shambala-Three Dog Night-Dunhill
27	24.	Fool-Elvis Presley-RCA
12	25.	Hocus Pocus-Focus-Sire
32	26.	Natural High-Bloodstone-London
36	27.	Over The Hills And Far Away-Led Zeppelin-Atlantic
17	28.	Back When My Hair Was Short-Gunhill Road-Kama Sutra
33	29.	So Very Hard to Go-Tower Of Power-WB
13	30.	Drift Away-Dobie Gray-Decca
34	31.	Standing Ovation-The Dells-Cadet
40	*32.	Boggie Woogie Bugle Boy-Bette Midler-Atlantic
35	33.	There's No Me Without You-Manhattans-Columbia
21	34.	Frankenstein-The Edgar Winter Group-Epic
--	35.	Yesterday Once More-Carpenters-A & M

37	36.	Daddy Could Swear-Gladys Knight & The Pips-Soul
38	37.	Get Down-Gilbert O'Sullivan-MAM
39	38.	Uneasy Rider-Charlie Daniels-Kama Sutra
24	39.	Reeling In The Years-Steely Dan-ABC
--	40.	Brother Louie-Stories-Kama Sutra

*CFL Super Hits

collector's issue number one hundred and twenty

SUPER CFL SURVEY JUN 30

wcfl wcfl wcfl wcfl wcfl wcfl wcfl wcfl wcfl wcfl wcfl wcfl wcfl wcfl

SUPER 10 ALBUMS

1. The Dark Side Of The Moon-Pink Floyd-Harvest
2. Living In The Material World-George Harrison-Apple
3. Red Rose Speedway-Paul McCartney-Apple
4. Houses Of The Holy-Led Zeppelin-Atlantic
5. There Goes Rhymin' Simon-Paul Simon-Columbia
6. Billion Dollar Babies-Alice Cooper-Warner Brothers
7. Made In Japan-Deep Purple-Warner Brothers
8. The Beatles/1967-1970-Apple
9. They Only Come Out At Night-Edgar Winter Group-Epic
10. The Best Of Bread-Bread-Elektra

LARRY LUJACK
Afternoons on Super 'CFL

OMAHA'S SUPER HITS

1. Kodachrome-Paul Simon-Columbia
2. Back When My Hair Was Short-Gunhill Road-Kama Sutra
3. Pillow Talk-Sylvia-Vibration
4. Playground In My Mind-Clint Holmes-Epic
5. Will It Go Round In Circles-Billy Preston-A & M
6. Give Me Love-George Harrison-Apple
7. My Love-Paul McCartney-Apple
8. Diamond Girl-Seals & Crofts-WB
9. You Are The Sunshine Of My Life-Stevie Wonder-Tamla
10. Give It To Me-J. Geils Band-Atlantic

courtesy: KOIL Omaha

WAS	IS	
3	1.	Give Me Love-George Harrison-Apple
6	*2.	Kodachrome-Paul Simon-Columbia
1	3.	My Love-Paul McCartney-Apple
5	4.	I'm Gonna Love You-Barry White-20th Century
2	5.	The Right Thing To Do-Carly Simon-Elektra
7	6.	No More Mr. Nice Guy-Alice Cooper-WB
17	*7.	Long Train Runnin'-The Doobie Brothers-WB
4	8.	Let's Pretend-Raspberries-Capitol
14	9.	Right Place Wrong Time-Dr. John-Atco
15	10.	Will It Go Round In Circles-Billy Preston-A & M
12	11.	One Of A Kind (Love Affair)-Spinners-Atlantic
18	*12.	Monster Mash-Bobby (Boris) Pickett-Parrot
13	13.	Wild About My Lovin'-Adrian Smith-MCA
16	14.	Bad, Bad Leroy Brown-Jim Croce-ABC
10	15.	Thinking Of You-Loggins & Messina-Columbia
19	16.	I'm Doin' Fine Now-New York City-Chelsea
8	17.	Daniel-Elton John-MCA
20	18.	Money-Pink Floyd-Harvest
23	19.	Shambala-Three Dog Night-Dunhill
22	20.	Smoke On The Water-Deep Purple-WB
26	*21.	Natural High-Bloodstone-London
27	*22.	Over The Hills And Far Away-Led Zeppelin-Atlantic
24	23.	Fool-Elvis Presley-RCA
9	24.	Give It To Me-J. Geils Band-Atlantic
32	*25.	Boogie Woogie Bugle Boy-Bette Midler-Atlantic
11	26.	Pillow Talk-Sylvia-Vibration
31	27.	Standing Ovation-The Dells-Cadet
29	28.	So Very Hard To Go-Tower Of Power-WB
33	29.	There's No Me Without You-Manhattans-Columbia
35	30.	Yesterday Once More-Carpenters-A & M
37	31.	Get Down-Gilbert O'Sullivan-MAM
21	32.	Walk On The Wild Side-Lou Reed-RCA
36	33.	Daddy Could Swear-Gladys Knight & The Pips-Soul
25	34.	Hocus Pocus-Focus-Sire
38	35.	Uneasy Rider-Charlie Daniels-Kama Sutra

28	36.	Back When My Hair Was Short-Gunhill Road-Kama Sutra
40	37.	Brother Louie-Stories-Kama Sutra
30	38.	Drift Away-Dobie Gray-Decca
--	39.	The Morning After-Maureen McGovern-20th Century
--	40.	Blockbuster-The Sweet-Bell

*CFL Super Hits

wcfl wcfl wcfl wcfl wcfl wcfl wcfl wcfl wcfl wcfl wcfl wcfl

collector's issue number one hundred and twenty-one

SUPER CFL SURVEY JUL 7

wcfl wcfl wcfl wcfl wcfl wcfl wcfl wcfl wcfl wcfl wcfl wcfl wcfl wcfl

WAS	IS	
1	1.	Give Me Love-George Harrison-Apple
2	2.	Kodachrome-Paul Simon-Columbia
9	3.	Right Place Wrong Time-Dr. John-Atco
12	*4.	Monster Mash-Bobby (Boris) Pickett-Parrot
7	5.	Long Train Runnin'-The Doobie Brothers-WB
10	6.	Will It Go Round In Circles-Billy Preston-A & M
3	7.	My Love-Paul McCartney-Apple
14	8.	Bad, Bad Leroy Brown-Jim Croce-ABC
11	9.	One Of A Kind (Love Affair)-Spinners-Atlantic
25	*10.	Boogie Woogie Bugle Boy-Bette Midler-Atlantic
4	11.	I'm Gonna Love You-Barry White-20th Century
18	*12.	Money-Pink Floyd-Harvest
16	13.	I'm Doin' Fine Now-New York City-Chelsea
20	14.	Smoke On The Water-Deep Purple-WB
19	15.	Shambala-Three Dog Night-Dunhill
30	*16.	Yesterday Once More-Carpenters-A & M
5	17.	The Right Thing To Do-Carly Simon-Elektra
22	18.	Over The Hills And Far Away-Led Zeppelin-Atlantic
21	19.	Natural High-Bloodstone-London
6	20.	No More Mr. Nice Guy-Alice Cooper-WB
--	*21.	Playground In My Mind-Clint Holmes-Epic
8	22.	Let's Pretend-Raspberries-Capitol
27	23.	Standing Ovation-The Dells-Cadet
31	24.	Get Down-Gilbert O'Sullivan-MAM
13	25.	Wild About My Lovin'-Adrian Smith-MCA
28	26.	So Very Hard To Go-Tower Of Power-WB
29	27.	There's No Me Without You-Manhattans-Columbia
35	28.	Uneasy Rider-Charlie Daniels-Kama Sutra
17	29.	Daniel-Elton John-MCA
33	30.	Daddy Could Swear-Gladys Knight & The Pips-Soul
37	31.	Brother Louie-Stories-Kama Sutra
39	32.	The Morning After-Maureen McGovern-20th Century
--	33.	Touch Me In The Morning-Diana Ross-Motown
15	34.	Thinking Of You-Loggins & Messina-Columbia
40	35.	Blockbuster-The Sweet-Bell

SUPER 10 ALBUMS

1. The Dark Side Of The Moon-Pink Floyd-Harvest
2. Living In The Material World-George Harrison-Apple
3. Houses Of The Holy-Led Zeppelin-Atlantic
4. Billion Dollar babies-Alice Cooper-Warner Brothers
5. There Goes Rhymin' Simon-Paul Simon-Columbia
6. Red Rose Speedway-Paul McCartney-Apple
7. Diamond Girl-Seals And Crofts-Warner Brothers
8. Made In Japan-Deep Purple-Warner Brothers
9. The Captain And Me-Doobie Brothers-Warner Brothers
10. Yessongs-Yes-Atlantic

WORLD FAMOUS
TOM MURPHY
mornings...on Super CFL!

73's SUPER HITS

1. You're So Vain-Carly Simon-(January)
2. Little Willy-The Sweeet-(April)
3. Hocus Pocus-Focus-(May)
4. Frankenstein-Edgar Winter Group-(May)
5. Rockin' Pneumonia-Johnny Rivers-(January)
6. Last Song-Edward Bear-(February)
7. Crocodile Rock-Elton John-(February)
8. The Right Thing To Do-Carly Simon-(June)
9. Danny's Song-Anne Murray-(April)
10. Sing-Carpenters-(May)

--	36. Feelin' Stronger Every Day-Chicago-Columbia
23	37. Fool-Elvis Presley-RCA
--	38. Roll Over Beethoven-Electric Light Orch.-UA
24	39. Give It To Me-J. Geils Band-Atlantic
--	40. Behind Closed Doors-Charlie Rich-Epic

*CFL Super Hits

collector's issue number one hundred and twenty-two

SUPER CFL SURVEY JUL 14

wcfl wcfl wcfl wcfl wcfl wcfl wcfl wcfl wcfl wcfl wcfl wcfl wcfl wcfl

SUPER 10 ALBUMS

1. The Dark Side Of The Moon-Pink Floyd-Harvest
2. Living In The Material World-George Harrison-Apple
3. Made In Japan-Deep Purple-Warner Brothers
4. There Goes Rhymin' Simon-Paul Simon-Columbia
5. Houses Of The Holy-Led Zeppelin-Atlantic
6. Red Rose Speedway-Paul McCartney-Apple
7. Billion Dollar Babies-Alice Cooper-Warner Brothers
8. The Captain And Me-Doobie Brothers-Warner Brothers
9. The Beatles/1967-1970-Apple
10. Diamond Girl-Seals And Crofts-Warner Brothers

World Famous
TOM MURPHY
mornings...on SUPER CFL!

SOUTH FLORIDA'S SUPER HITS

1. Yesterday Once More-Carpenters-A & M
2. Right Place, Wrong Time-Dr. John-Atco
3. I'm Gonna Love You-Barry White-20th Century
4. Playground In My Mind-Clint Holmes-Epic
5. Will It Go Round In Circles-Billy Preston-A & M
6. My Love-Paul McCartney-Apple
7. Give Me Love-George Harrison-Apple
8. Pillowtalk-Sylvia-Vibration
9. One Of A Kind-Spinners-Atlantic
10. Long Train Runnin'-The Double Brothers-WB

courtesy: -Miami

WAS	IS	
2	1.	Kodachrome-Paul Simon-Columbia
4	2.	Monster Mash-Bobby (Boris) Pickett-Parrot
5	3.	Long Train Runnin'-The Doobie Brothers-WB
6	4.	Will It Go Round In Circles-Billy Preston-A & M
8	5.	Bad, Bad Leroy Brown-Jim Croce-ABC
1	6.	Give Me Love-George Harrison-Apple
10	7.	Boogie Woogie Bugle Boy-Bette Midler-Atlantic
3	8.	Right Place Wrong Time-Dr. John-Atco
15	*9.	Shambala-Three Dog Night-Dunhill
16	*10.	Yesterday Once More-Carpenters-A & M
14	11.	Smoke On The Water-Deep Purple-WB
12	12.	Money-Pink Floyd-Harvest
13	13.	I'm Doin' Fine Now-New York City-Chelsea
9	14.	One Of A Kind (Love Affair)-Spinners-Atlantic
7	15.	My Love-Paul McCartney-Apple
21	16.	Playground In My Mind-Clint Holmes-Epic
18	17.	Over The Hills And Far Away-Led Zeppelin-Atlantic
19	18.	Natural High-Bloodstone-London
24	19.	Get Down-Gilbert O'Sullivan-MAM
11	20.	I'm Gonna Love You-Barry White-20th Century
23	21.	Standing Ovation-The Dells-Cadet
32	*22.	The Morning After-Maureen McGovern-20th Century
28	23.	Uneasy Rider-Charlie Daniels-Kama Sutra
26	24.	So Very Hard To Go-Tower Of Power-WB
31	25.	Brother Louie-Stories-Kama Sutra
27	26.	There's No Me Without You-Manhattans-Columbia
17	27.	The Right Thing To Do-Carly Simon-Elektra
30	28.	Daddy Could Swear-Gladys Knight & The Pips-Soul
33	29.	Touch Me In The Morning-Diana Ross-Motown
20	30.	No More Mr. Nice Guy-Alice Cooper-WB
35	31.	Blockbuster-The Sweet-Bell
22	32.	Let's Pretend-The Raspberries-Capitol
36	33.	Feelin' Stronger Every Day-Chicago-Columbia
40	34.	Behind Closed Doors-Charlie Rich-Epic

38	35.	Roll Over Beethoven-Electric Light Orch.-UA
25	36.	Wild About My Lovin'-Adrian Smith-MCA
--	37.	I Believe In You-Johnnie Taylor-Stax
--	38.	How Can I Tell Her-Lobo-Big Tree
--	39.	Delta Dawn-Helen Reddy-Capitol
--	40.	Say, Has Anybody Seen My Sweet Gypsy Rose-Dawn-Bell

*CFL Super Hits

wcfl wcfl wcfl wcfl wcfl wcfl wcfl wcfl wcfl wcfl wcfl wcfl

collector's issue number one hundred and twenty-three

SUPER CFL SURVEY JUL 21

wcfl wcfl wcfl wcfl wcfl wcfl wcfl wcfl wcfl wcfl wcfl wcfl wcfl wcfl

SUPER 10 ALBUMS

1. The Dark Side Of The Moon-Pink Floyd-Harvest
2. Chicago VI-Chicago-Columbia
3. Living In The Material World-George Harrison-Apple
4. Houses Of The Holy-Led Zeppelin-Atlantic
5. There Goes Rhymin' Simon-Paul Simon-Columbia
6. Made In Japan-Deep Purple-Warner Brothers
7. Red Rose Speedway-Paul McCartney-Apple
8. The Captain And Me-Doobie Brothers-Warner Brothers
9. Diamond Girl-Seals And Crofts-Warner Brothers
10. Love Devotion Surrender-Santana & McLaughlin-Columbia

BIG RON O'BRIEN
evenings...on SUPER CFL!

NORTH CAROLINA'S SUPER HITS

1. I'm Gonna Love You-Barry White-20th Century
2. So Very Hard To Go-Tower Of Power-WB
3. Long Train Runnin'-Doobie Brothers-WB
4. Give Me Love-George Harrison-Apple
5. Bad, Bad Leroy Brown-Jim Croce-ABC
6. Will It Go Round In Circles-Billy Preston-A & M
7. One Of A Kind-Spinners-Atlantic
8. Shambala-3 Dog Night-Dunhill
9. Let's Pretend-Raspberries-Capitol
10. Boogie Woogie Bugle Boy-Bette Midler-Atlantic

courtesy: WSTP - Salisbury

WAS	IS	
7	*1.	Boogie Woogie Bugle Boy-Bette Midler-Atlantic
2	2.	Monster Mash-Bobby (Boris) Pickett-Parrot
10	*3.	Yesterday Once More-Carpenters-A & M
1	4.	Kodachrome-Paul Simon-Columbia
16	*5.	Playground In My Mind-Clint Holmes-Epic
11	6.	Smoke On The Water-Deep Purple-WB
9	7.	Shambala-Three Dog Night-Dunhill
12	8.	Money-Pink Floyd-Harvest
3	9.	Long Train Runnin'-The Doobie Brothers-WB
8	10.	Right Place Wrong Time-Dr. John-Atco
4	11.	Will It Go Round In Circles-Billy Preston-A & M
19	12.	Get Down-Gilbert O'Sullivan-MAM
5	13.	Bad, Bad Leroy Brown-Jim Croce-ABC
6	14.	Give Me Love-George Harrison-Apple
21	15.	Standing Ovation-The Dells-Cadet
22	16.	The Morning After-Maureen McGovern-20th Century
25	17.	Brother Louie-Stories-Kama Sutra
18	18.	Natural High-Bloodstone-London
29	*19.	Touch Me In The Morning-Diana Ross-Motown
13	20.	I'm Doin' Fine Now-New York City-Chelsea
23	21.	Uneasy Rider-Charlie Daniels-Kama Sutra
15	22.	My Love-Paul McCartney-Apple
33	*23.	Feelin' Stronger Every Day-Chicago-Columbia
24	24.	So Very Hard To Go-Tower Of Power-WB
14	25.	One Of A Kind (Love Affair)-Spinners-Atlantic
17	26.	Over The Hills And Far Away-Led Zeppelin-Atlantic
28	27.	Daddy Could Swear-Gladys Knight & The Pips-Soul
31	28.	Blockbuster-The Sweet-Bell
35	29.	Roll Over Beethoven-Electric Light Orch.-UA
20	30.	I'm Gonna Love You-Barry White-20th Century
34	31.	Behind Closed Doors-Charlie Rich-Epic
37	32.	I Believe In You-Johnnie Taylor-Stax
--	33.	Diamond Girl-Seals & Crofts-WB
39	34.	Delta Dawn-Helen Reddy-Capitol
38	35.	How Can I Tell Her-Lobo-Big Tree

26	36.	There's No Me Without You-Manhattans-Columbia
--	37.	Let's Get It ON-Marvin Gaye-Tamla
40	38.	Say, Has Anybody Seen My Sweet Gypsy Rose-Dawn-Bell
--	39.	Live And Let Die-Wings-Apple
--	40.	Gypsy Man-War-UA

*CFL Super Hits

wcfl wcfl wcfl wcfl wcfl wcfl wcfl wcfl wcfl wcfl wcfl wcfl

collector's issue number one hundred and twenty-four

SUPER CFL SURVEY JUL 28

wcfl wcfl wcfl wcfl wcfl wcfl wcfl wcfl wcfl wcfl wcfl wcfl wcfl wcfl

SUPER 10 ALBUMS

1. Chicago VI-Chicago-Columbia
2. The Dark Side Of The Moon-Pink Floyd-Harvest
3. Houses Of The Holy-Led Zeppellin-Atlantic
4. Made In Japan-Deep Purple-Warner Brothers
5. Living In The Material World-George Harrison-Apple
6. There Goes Rhymin' Simon-Paul Simon-Columbia
7. Diamond Girl-Seals And Crofts-Warner Brothers
8. Love Devotion Surrender-Santana & McLaughlin-Columbia
9. The Captain And Me-Dobbie Brothers-Warner Brothers
10. Red Rose Speedway-Paul McCartney-Apple

World Famous
TOM MURPHY
mornings...on SUPER CFL!

IOWA'S SUPER HITS

1. Brother Louie-Stories-Kama Sutra
2. Smoke On The Water-Deep Purple-WB
3. Bad Bad Leroy Brown-Jim Croce-ABC
4. Back When My Hair Was Short-Gunhill Road-Kama Sutra
5. Uneasy Rider-Charlie Daniels-Kama Sutra
6. Boogie Woogie Bugle Boy-Bette Midler-Atlantic
7. Yesterday Once More-Carpenters-A & M
8. Swamp Witch-Jim Stafford-MGM
9. Kodachrome-Paul Simon-Columbia
10. The Morning After-Maureen McGovern-20th Century

courtesy: HSTT - Davenport

WAS	IS	
3	*1.	Yesterday Once More-Carpenters-A & M
5	2.	Playground In My Mind-Clint Holmes-Epic
2	3.	Monster Mash-Bobby (Boris) Pickett-Parrot
6	4.	Smoke On The Water-Deep Purple-WB
12	*5.	Get Down-Gilbert O'Sullivan-MAM
1	6.	Boogie Woogie Bugle Boy-Bette Midler-Atlantic
16	*7.	The Morning After-Maureen McGovern-20th Century
7	8.	Shambala-Three Dog Night-Dunhill
4	9.	Kodachrome-Paul Simon-Columbia
17	*10.	Brother Louie-Stories-Kama Sutra
8	11.	Money-Pink Floyd-Harvest
19	*12.	Touch Me In The Morning-Diana Ross-Motown
9	13.	Long Train Runnin'-The Doobie Brothers-WB
15	14.	Standing Ovation-The Dells-Cadet
11	15.	Will It Go Round In Circles-Billy Preston-A & M
21	16.	Uneasy Rider-Charlie Daniels-Kama Sutra
23	17.	Feelin' Stronger Every Day-Chicago-Columbia
13	18.	Bad, Bad Leroy Brown-Jim Croce-ABC
10	19.	Right Place Wrong Time-Dr. John-Atco
14	20.	Give Me Love-George Harrison-Apple
29	21.	Roll Over Beethoven-Electric Light Orch.-UA
28	22.	Blockbuster-The Sweet-Bell
33	23.	Diamond Girl-Seals & Crofts-WB
18	24.	Natural High-Bloodstone-London
31	25.	Behind Closed Doors-Charlie Rich-Epic
32	26.	I Believe In You-Johnnie Taylor-Stax
34	27.	Delta Dawn-Helen Reddy-Capitol
20	28.	I'm Doin' Fine Now-New York City-Chelsea
37	29.	Let's Get It On-Marvin Gaye-Tamla
39	30.	Live And Let Die-Wings-Apple
35	31.	How Can I Tell Her-Lobo-Big Tree
22	32.	My Love-Paul McCartney-Apple
38	33.	Say, Has Anybody Seen My Sweet Gypsy Rose-Dawn-Bell

24	34.	So Very Hard To Go-Tower Of Power-WB
40	35.	Gypsy Man-War-UA
--	36.	Are You Man Enough-Four Tops-Dunhill
27	37.	Daddy Could Swear-Gladys Knight & The Pips-Soul
--	38.	We're An American Band-Grand Funk-Capitol
--	39.	Saturday Night's Alright For Fighting-Elton John-MCA
--	40.	Sweet Charlie Babe-Jackie Moore-Atlantic

*CFL Super Hits

collector's issue number one hundred and twenty-five

SUPER CFL SURVEY AUG 4

wcfl wcfl wcfl wcfl wcfl wcfl wcfl wcfl wcfl wcfl wcfl wcfl wcfl wcfl

SUPER 10 ALBUMS

1. Chicago VI-Chicago-Columbia
2. The Dark Side Of The Moon-Pink Floyd-Harvest
3. Houses Of The Holy-Led Zeppelin-Atlantic
4. Made In Japan-Deep Purple-Warner Brothers
5. A Passion Play-Jethro Tull-Chrysalis
6. Diamond Girl-Seals And Crofts-Warner Brothers
7. The Captain And Me-Doobie Brothers-Warner Brothers
8. Living In The Material World-George Harrison-Apple
9. Love Devotion Surrender-Santana & McLaughlin-Columbia
10. There Goes Rhymin' Simon-Paul Simon-Columbia

LARRY LUJACK
Afternoons on Super 'CFL

ENGLAND'S SUPER HITS

*1. Skweeze Me Pleeze Me-Slade-Polydor
*2. Welcome Home-Peters & Lee-Phillips
*3. Rubber Bullets-10cc-UK
*4. Snoopy vs. The Red Baron-Hot Shots-Mooncrest
*5. Albatross-Fleetwood Mac-CBS
*6. Born To Be With You-Dave Edmunds-Rockfield
*7. Groover-T. Tex-EMI
*8. Life On Mars-David Bowie-RCA
*9. Give Me Love-George Harrison-Apple
*10. Live And Let Die-Wings-Apple

*denote British talent

WAS	IS	
1	1.	Yesterday Once More-Carpenters-A & M
2	2.	Playground In My Mind-Clint Holmes-Epic
4	3.	Smoke On The Water-Deep Purple-WB
5	4.	Get Down-Gilbert O'Sullivan-MAM
7	5.	The Morning After-Maureen McGovern-20th Century
10	6.	Brother Louie-Stories-Kama Sutra
12	7.	Touch Me In The Morning-Diana Ross-Motown
3	8.	Monster Mash-Bobby (Boris) Pickett-Parrot
6	9.	Boogie Woogie Bugle Boy-Bette Midler-Atlantic
17	10.	Feelin' Stronger Every Day-Chicago-Columbia
8	11.	Shambala-Three Dog Night-Dunhill
23	*12.	Diamond Girl-Seals & Crofts-WB
16	13.	Uneasy Rider-Charlie Daniels-Kama Sutra
9	14.	Kodachrome-Paul Simon-Columbia
21	15.	Roll Over Beethoven-Electric Light Orch.-UA
11	16.	Money-Pink Floyd-Harvest
14	17.	Standing Ovation-The Dells-Cadet
30	*18.	Live And Let Die-Wings-Apple
13	19.	Long Train Runnin'-The Doobie Brothers-WB
26	20.	I Believe In You-Johnnie Taylor-Stax
27	21.	Delta Dawn-Helen Reddy-Capitol
25	22.	Behind Closed Doors-Charlie Rich-Epic
15	23.	Will It Go Round In Circles-Billy Preston-A & M
29	24.	Let's Get It On-Marvin Gaye-Tamla
18	25.	Bad, Bad Leroy Brown-Jim Croce-ABC
31	26.	How Can I Tell Her-Lobo-Big Tree
22	27.	Blockbuster-The Sweet-Bell
19	28.	Right Place Wrong Time-Dr. John-Atco
35	29.	Gypsy Man-War-UA
33	30.	Say, Has Anybody Seen My Sweet Gypsy Rose-Dawn-Bell
20	31.	Give Me Love-George Harrison-Apple
36	32.	Are You Man Enough-Four Tops-Dunhill
24	33.	Natural High-Bloodsotne-London
38	34.	We're An American Band-Grand Funk-Capitol
39	35.	Saturday Night's Alright For

		Fighting-Elton John-MCA
28	36.	I'm Doin' Fine Now-New York City-Chelsea
40	37.	Sweet Charlie Babe-Jackie Moore-Atlantic
--	38.	Here I am-Al Green-HI
32	39.	My Love-Paul McCartney-Apple
--	40.	Angel-Aretha Franklin-Atlantic

*CFL Super Hits

wcfl wcfl wcfl wcfl wcfl wcfl wcfl wcfl wcfl wcfl wcfl wcfl

collector's issue number one hundred and twenty-six

SUPER CFL SURVEY AUG 11

wcfl wcfl wcfl wcfl wcfl wcfl wcfl wcfl wcfl wcfl wcfl wcfl wcfl wcfl

SUPER 10 ALBUMS

1. Chicago VI-Chicago-Columbia
2. A Passion Play-Jethro Tull-Chrysalis
3. Made In Japan-Deep Purple-Warner Brothers
4. Diamond Girl-Seals And Crofts-Warner Brothers
5. The Dark Side Of The Moon-Pink Floyd-Harvest
6. Foreigner-Cat Stevens-A & M
7. Houses Of The Holy-Led Zeppelin-Atlantic
8. Love Devotion Surrender-Santana & McLaughlin-Columbia
9. Living In The Material World-George Harrison-Apple
10. The Captain And Me-Doobie Brothers-Warner Brothers

Sailboats, Tennis Racquets, summer shirts, beach towels, concert tickets, bicycles plus many other prizes yet to come in THE LAST CONTEST from SUPER CFL!

OKLAHOMA'S SUPER HITS

1. Bad, Bad Leroy Brown-Jim Croce-ABC
2. Smoke On The Water-Deep Purple-WB
3. Live And Let Die-Wings-Apple
4. Monster Mash-Bobby (Boris) Pickett-Parrot
5. The Morning After-Maureen McGovern-20th Century
6. Brother Louie-Stories-Kama Sutra
7. Uneasy Rider-Charlie Daniels-Kama Sutra
8. Yesterday Once More-Carpenters-A & M
9. Back When My Hair Was Short-Gunhill Road-Kama Sutra
10. Behind Closed Doors-Charlie Rich-Epic

courtesy: Oklahoma City

WAS	IS	
4	1.	Get Down-Gilbert O'Sullivan-MAM
1	2.	Yesterday Once More-Carpenters-A & M
5	3.	The Morning After-Maureen McGovern-20th Century
6	4.	Brother Louie-Stories-Kama Sutra
12	*5.	Diamond Girl-Seals & Crofts-WB
7	6.	Touch Me In The Morning-Diana Ross-Motown
18	*7.	Live And Let Die-Wings-Apple
10	8.	Feelin' Stronger Every Day-Chicago-Columbia
2	9.	Playground In My Mind-Clint Holmes-Epic
13	10.	Uneasy Rider-Charlie Daniels-Kama Sutra
3	11.	Smoke On The Water-Deep Purple-WB
15	12.	Roll Over Beethoven-Electric Light Orch.-UA
8	13.	Monster Mash-Bobby (Boris) Pickett-Parrot
9	14.	Boogie Woogie Bugle Boy-Bette Midler-Atlantic
21	15.	Delta Dawn-Helen Reddy-Capitol
11	16.	Shambala-Three Dog Night-Dunhill
22	17.	Behind Closed Doors-Charlie Rich-Epic
20	18.	I Believe In You-Johnnie Taylor-Stax
24	19.	Let's Get It On-Marvin Gaye-Tamla
14	20.	Kodachrome-Paul Simon-Columbia
26	21.	How Can I Tell Her-Lobo-Big Tree
16	22.	Money-Pink Floyd-Harvest
19	23.	Long Train Runnin'-The Doobie Brothers-WB
32	24.	Are You Man Enough-Four Tops-Dunhill
30	25.	Say, Has Anybody Seen My Sweet Gypsy Rose-Dawn-Bell
34	26.	We're An American Band-Grand Funk-Capitol
29	27.	Gypsy Man-War-UA
23	28.	Will It Go Round In Circles-Billy Preston-A & M
17	29.	Standing Ovation-The Dells-Cadet
35	30.	Saturday Night's Alright For Fighting-Elton John-MCA
25	31.	Bad, Bad Leroy Brown-Jim Croce-ABC
37	32.	Sweet Charlie Babe-Jackie Moore-Atlantic
27	33.	Blockbuster-The Sweet-Bell
38	34.	Here I Am-Al Green-HI
28	35.	Right Place Wrong Time-Dr. John-

		Atco
40	36.	Angel-Aretha Franklin-Atlantic
31	37.	Give Me Love-George Harrison-Apple
--	38.	My Maria-B.W. Stevenson-RCA
--	39.	Loves Me Like A Rock-Paul Simon-Columbia
33	40.	Natural High-Bloodstone-London

*CFL Super Hits

collector's issue number one hundred and twenty-seven

SUPER CFL SURVEY AUG 18

wcfl wcfl wcfl wcfl wcfl wcfl wcfl wcfl wcfl wcfl wcfl wcfl wcfl wcfl

WAS	IS	
4	1.	Brother Louie-Stories-Kama Sutra
3	2.	The Morning After-Maureen McGovern-20th Century
5	3.	Diamond Girl-Seals & Crofts-WB
8	4.	Feelin' Stronger Every Day-Chicago-Columbia
6	5.	Touch Me In The Morning-Diana Ross-Motown
7	6.	Live And Let Die-Wings-Apple
1	7.	Get Down-Gilbert O'Sullivan-MAM
2	8.	Yesterday Once More-Carpenters-A & M
12	9.	Roll Over Beethoven-Electric Light Orch.-UA
10	10.	Uneasy Rider-Charlie Daniels-Kama Sutra
15	11.	Delta Dawn-Helen Reddy-Capitol
9	12.	Playground In My Mind-Clint Holmes-Epic
18	13.	I Believe In You-Johnnie Taylor-Stax
17	14.	Behind Closed Doors-Charlie Rich-Epic
19	15.	Let's Get It On-Marvin Gaye-Tamla
11	16.	Smoke On The Water-Deep Purple-WB
26	*17.	We're An American Band-Grand Funk-Capitol
13	18.	Monster Mash-Bobby (Boris) Pickett-Parrot
21	19.	How Can I Tell Her-Lobo-Big Tree
24	20.	Are You Man Enough-Four Tops-Dunhill
25	21.	Say, Has Anybody Seen My Sweet Gypsy Rose-Dawn-Bell
30	*22.	Saturday Night's Alright for Fighting-Elton John-MCA
14	23.	Boogie Woogie Bugle Boy-Bette Midler-Atlantic
27	24.	Gypsy Man-War-UA
16	25.	Shambala-Three Dog Night-Dunhill
23	26.	Long Train Runnin'-The Doobie Brother-WB
20	27.	Kodachrome-Paul Simon-Columbia
32	28.	Sweet Charlie Babe-Jackie Moore-Atlantic
34	29.	Here I am-Al Green-HI
22	30.	Money-Pink Floyd-Harvest
36	31.	Angel-Aretha Franklin-Atlantic
28	32.	Will It Go Round In Circles-Billy Preston-A & M
39	33.	Loves Me Like A Rock-Paul Simon-Columbia
38	34.	My Maria-B.W. Stevenson-RCA
31	35.	Bad, Bad Leroy Brown-Jim Croce-ABC
29	36.	Standing Ovation-The Dells-Cadet
--	37.	The Hurt-Cat Stevens-A & M
--	38.	Half-Breed-Cher-MCA

SUPER 10 ALBUMS

1. Chicago VI-Chicago-Columbia
2. A Passion Play-Jethro Tull-Chrysalis
3. Diamond Girl-Seals And Crofts-Warner Brothers
4. Foreigner-Cat Stevens-A & M
5. The Dark Side Of The Moon-Pink Floyd-Harvest
6. Houses Of The Holy-Led Zeppelin-Atlantic
7. Made In Japan-Deep Purple-Warner Brothers
8. Love Devotion Surrender-Santana & McLaughlin-Columbia
9. Living In The Material World-George Harrison-Apple
10. The Captain And Me-Doobie Brothers-Warner Brothers

World Famous
TOM MURPHY
mornings...on SUPER CFL!

LOS ANGELES' SUPER HITS

1. Brother Louie-Stories-Kama Sutra
2. Shambala-Three Dog Night-Dunhill
3. Live And Let Die-Wings-Apple
4. One Tin Soldier-Coven-MBM
5. Smoke On The Water-Deep Purple-WB
6. So Very Hard To Go-Tower Of Power-WB
7. Natural High-Bloodstone-London
8. Bad, Bad Leroy Brown-Jim Croce-ABC
9. Monster Mash-Bobby (Boris) Pickett-Parrot
10. Diamond Girl-Seals & Crofts-WB

courtesy: 93 KHJ – Los Angeles

--	39.	Higher Ground-Stevie Wonder-Tamla	
--	40.	Ramblin Man-The Allman Brothers Band-Capricorn	

*CFL Super Hits

wcfl wcfl wcfl wcfl wcfl wcfl wcfl wcfl wcfl wcfl wcfl wcfl

collector's issue number one hundred and twenty-eight

SUPER CFL SURVEY AUG 25

wcfl wcfl wcfl wcfl wcfl wcfl wcfl wcfl wcfl wcfl wcfl wcfl wcfl wcfl

SUPER 10 ALBUMS

1. Chicago VI-Chicago-Columbia
2. A Passion Play-Jethro Tull-Chrysalis
3. Diamond Girl-Seals And Crofts-Warner Brothers
4. Foreigner-Cat Stevens-A & M
5. We're An American Band-Grand Funk-Capitol
6. Made In Japan-Deep Purple-Warner Brothers
7. Houses Of The Holy-Led Zeppelin-Atlantic
8. The Dark Side Of The Moon-Pink Floyd-Harvest
9. The Captain And Me-Doobie Brothers-Warner Brothers
10. Love Devotion Surrender-Santana & McLaughlin-Columbia

LARRY LUJACK
Afternoons on Super 'CFL

CHICAGO'S SUPER HITS

1. 25 Or 6 to 4-(1970)
2. Does Anybody Really Know What Time It Is?-(1970)
3. Make Me Smile-(1970)
4. Beginnings-(1971)
5. Saturday In The Park-(1972)
6. Free-(1971)
7. Colour My World-(1971)
8. Dialogue (Part I & 2)-(1973)
9. Feelin' Stronger Every Day-(1973)

WAS	IS	
2	1.	The Morning After-Maureen McGovern-20th Century
3	2.	Diamond Girl-Seals & Crofts-WB
4	3.	Feelin' Stronger Every Day-Chicago-Columbia
6	4.	Live And Let Die-Wings-Apple
11	*5.	Delta Dawn-Helen Reddy-Capitol
1	6.	Brother Louie-Stories-Kama Sutra
9	7.	Roll Over Beethoven-Electric Light Orch.-UA
5	8.	Touch Me In The Morning-Diana Ross-Motown
7	9.	Get Down-Gilbert O'Sullivan-MAM
15	10.	Let's Get It On-Marvin Gaye-Tamla
17	*11.	We're An American Band-Grand Funk-Capitol
13	12.	I Believe In You-Johnnie Taylor-Stax
14	13.	Behind Closed Doors-Charlie Rich-Epic
8	14.	Yesterday Once More-Carpenters-A & M
19	15.	How Can I Tell Her-Lobo-Big Tree
10	16.	Uneasy Rider-Charlie Daniels-Kama Sutra
21	17.	Say, Has Anybody Seen My Sweet Gypsy Rose-Dawn-Bell
12	18.	Playground In My Mind-Clint Holmes-Epic
20	19.	Are You Man Enough-Four Tops-Dunhill
22	20.	Saturday Night's Alright For Fighting-Elton John-MCA
16	21.	Smoke On The Water-Deep Purple-WB
24	22.	Gypsy Man-War-UA
33	*23.	Loves Me Like A Rock-Paul Simon-Columbia
28	24.	Sweet Charlie Babe-Jackie Moore-Atlantic
29	25.	Here I Am-Al Green-HI
34	*26.	My Maria-B.W. Stevenson-RCA
18	27.	Monster Mash-Bobby (Boris) Pickett-Parrot
23	28.	Boogie Woogie Bugle Boy-Bette Midler-Atlantic
31	29.	Angel-Aretha Franklin-Atlantic
37	30.	The Hurt-Cat Stevens-A & M
25	31.	Shambala-Three Dog Night-Dunhill
38	32.	Half-Breed-Cher-MCA
26	33.	Long Train Runnin'-The Doobie Brothers-WB
39	34.	Higher Ground-Stevie Wonder-Tamla
--	35.	A Million To One-Donny Osmond-MGM

40	36.	Ramblin' Man-The Allman Brothers Band-Capricorn
27	37.	Kodachrome-Paul Simon-Columbia
--	38.	China Grove-The Doobie Brothers-WB
30	39.	Money-Pink Floyd-Harvest
--	40.	Shady Lady-Shepstone & Dibbens-Buddah

*CFL Super Hits

wcfl wcfl wcfl wcfl wcfl wcfl wcfl wcfl wcfl wcfl wcfl wcfl

collector's issue number one hundred and twenty-nine

SUPER CFL SURVEY SEP 1

wcfl wcfl wcfl wcfl wcfl wcfl wcfl wcfl wcfl wcfl wcfl wcfl wcfl wcfl

WAS	IS	
1	1.	The Morning After-Maureen McGovern-20th Century
3	2.	Feelin' Stronger Every Day-Chicago-Columbia
5	3.	Delta Dawn-Helen Reddy-Capitol
4	4.	Live And Let Die-Wings-Apple
2	5.	Diamond Girl-Seals & Crofts-WB
7	6.	Roll Over Beethoven-Electric Light Orch.-UA
10	7.	Let's Get It On-Marvin Gaye-Tamla
11	8.	We're An American Band-Grand Funk-Capitol
6	9.	Brother Louie-Stories-Kama Sutra
12	10.	I Believe In You-Johnnie Taylor-Stax
8	11.	Touch Me In The Morning-Diana Ross-Motown
13	12.	Behind Closed Doors-Charlie Rich-Epic
15	13.	How Can I Tell Her-Lobo-Big Tree
20	*14.	Saturday Night's Alright For Fighting-Elton John-MCA
17	15.	Say, Has Anybody Seen My Sweet Gypsy Rose-Dawn-Bell
19	16.	Are You Man Enough-Four Tops-Dunhill
23	*17.	Loves Me Like A Rock-Paul Simon-Columbia
22	18.	Gypsy Man-War-UA
9	19.	Get Down-Gilbert O'Sullivan-MAM
14	20.	Yesterday Once More-Carpenters-A & M
24	21.	Sweet Charlie Babe-Jackie Moore-Atlantic
26	22.	My Maria-B.W. Stevenson-RCA
30	*23.	The Hurt-Cat Stevens-A & M
16	24.	Uneasy Rider-Charlie Daniels-Kama Sutra
29	25.	Angel-Aretha Franklin-Atlantic
32	26.	Half-Breed-Cher-MCA
18	27.	Playground In My Mind-Clint Holmes-Epic
36	*28.	Ramblin' Man-The Allman Brothers Band-Capricorn
25	29.	Here I Am-Al Green-HI
34	30.	Higher Ground-Stevie Wonder-Tamla
35	31.	A Million To One-Donny Osmond-MGM
21	32.	Smoke On The Water-Deep Purple-WB
27	33.	Monster Mash-Bobby (Boris) Pickett-Parrot
38	34.	China Grove-The Doobie Brothers-WB

SUPER 10 ALBUMS

1. Chicago VI-Chicago-Columbia
2. A Passion Play-Jethro Tull-Chrysalis
3. Diamond Girl-Seals And Crofts-Warner Brothers
4. Foreigner-Cat Stevens-A & M
5. We're An American Band-Grand Funk-Capitol
6. Brothers And Sisters-Allman Brothers Band-Capricorn
7. The Dark Side Of The Moon-Pink Floyd-Harvest
8. Made In Japan-Deep Purple-Warner Brothers
9. Houses Of The Holy-Led Zeppelin-Atlantic
10. The Captain And Me-Doobie Brothers-Warner Brothers

World Famous
TOM MURPHY
mornings...on SUPER CFL!

10 YEARS AGO TODAY

1. My Boyfriend's Back-The Angels-Smash
2. Hello Muddah Hello Faddah-Alan Sherman-WB
3. Martian Hop-The Randalls-Chairman
4. Blowin' In The Wind-Peter, Paul & Mary-WB
5. Then He Kissed Me-Crystals-Philles
6. Abilene-George Hamilton IV-RCA
7. Denise-Randy & The Rainbows-Rust
8. Detroit City-Bobby Bare-RCA
9. Wipe Out-Surfaris-Dot
10. Blue Velvet-Bobby Vinton-Epic

28	35.	Boogie Woogie Bugle Boy-Bette Midler-Atlantic
31	36.	Shambala-Three Dog Night-Dunhill
--	37.	Jimmy Loves Mary-Anne-Looking Glass-Epic
40	38.	Shady Lady-Shepstone & Dibbens-Buddah
--	39.	Free Ride-The Edgar Winter Group-Epic
--	40.	

*CFL Super Hits

collector's issue number one hundred and thirty

SUPER CFL SURVEY SEP 8

wcfl wcfl wcfl wcfl wcfl wcfl wcfl wcfl wcfl wcfl wcfl wcfl wcfl wcfl

SUPER 10 ALBUMS

1. Chicago VI-Chicago-Columbia
2. We're An American Band-Grand Funk-Capitol
3. Brothers And Sisters-Allman Brothers Band-Capricorn
4. Diamond Girl-Seals And Crofts-Warner Brothers
5. Foreigner-Cat Stevens-A & M
6. A Passion Play-Jethro Tull-Chrysalis
7. The Captain And Me-Doobie Brothers-Warner Brothers
8. Houses Of The Holy-Led Zeppelin-Atlantic
9. The Dark Side Of The Moon-Pink Floyd-Harvest
10. Made In Japan-Deep Purple-Warner Brothers

LARRY LUJACK works out...
"Labor Day" and every afternoon
on SUPER CFL!

LABOR DAY
WOMEN'S LIB SUPERSONGS

1. I Am Woman-Helen Reedy-Capitol
2. Working Together-Ike & Tina Turner-Liberty
3. Work Me, Lord-Janis Joplin-Columbia
4. I've Waited So Long-Carly Simon-Elektra
5. It's About Time-Linda Rondstadt-Capitol
6. United We Stand-Anne Murray-Capitol
7. Don't Mess With A Woman-Helen Reddy-Capitol
8. Working Girl-Merilee Rush-Bell
9. Roller Derby Queen-Leon Russell-Shelter
10. Someday I'll Be A Farmer-Melanie-Neighborhood

WAS	IS	
3	1.	Delta Dawn-Helen Reddy-Capitol
1	2.	The Morning After-Maureen McGovern-20th Century
7	3.	Let's Get It On-Marvin Gaye-Tamla
2	4.	Feelin' Stronger Every Day-Chicago-Columbia
4	5.	Live And Let Die-Wings-Apple
8	6.	We're An American Band-Grand Funk-Capitol
6	7.	Roll Over Beethoven-Electric Light Orch.-UA
17	*8.	Loves Me Like A Rock-Paul Simon-Columbia
14	9.	Saturday Night's Alright For Fighting-Elton John-MCA
10	10.	I Believe In You-Johnnie Taylor-Stax
5	11.	Diamond Girl-Seals & Crofts-WB
23	*12.	The Hurt-Cat Stevens-A & M
15	13.	Say, Has Anybody Seen My Sweet Gypsy Rose-Dawn-Bell
22	14.	My Maria-B.W. Stevenson-RCA
16	15.	Are You Man Enough-Four Tops-Dunhill
18	16.	Gypsy Man-War-UA
13	17.	How Can I Tell Her-Lobo-Big Tree
9	18.	Brother Louie-Stories-Kama Sutra
21	19.	Sweet Charlie Babe-Jackie Moore-Atlantic
26	20.	Half-Breed-Cher-MCA
28	*21.	Ramblin' Man-The Allman Brothers Band-Capricorn
11	22.	Touch Me In The Morning-Diana Ross-Motown
34	*23.	China Grove-The Doobie Brothers-WB
25	24.	Angel-Aretha Franklin-Atlantic
12	25.	Behind Closed Doors-Charlie Rich-Epic
31	26.	A Million To One-Donny Osmond-MGM
30	27.	Higher Ground-Stevie Wonder-Tamla
19	28.	Get Down-Gilbert O'Sullivan-MGM
20	29.	Yesterday Once More-Carpenters-A & M
37	30.	Jimmy Loves Mary-Anne-Looking Glass-Epic
24	31.	Uneasy Rider-Charlie Daniels-Kama Sutra
38	32.	Shady Lady-Shepstone & Dibbens-Buddah
27	33.	Playground In My Mind-Clint Holmes-Epic
--	34.	That Lady-Isley Brothers-T-Neck
39	35.	Free Ride-The Edgar Winter Group-Epic
40	36.	In The Midnight Hour-Cross Country-Atco

29 37. Here I Am-Al Green-HI
32 38. Smoke On The Water-Deep Purple-WB
-- 39. Angie-Rolling Stones-Rolling Stones
-- 40. Heartbeat, It's A Lovebeat-The DeFranco Family-20th Century

* CFL Super Hits

wcfl wcfl wcfl wcfl wcfl wcfl wcfl wcfl wcfl wcfl wcfl wcfl

collector's issue number one hundred and thirty-one

SUPER CFL SURVEY SEP 15

wcfl wcfl wcfl wcfl wcfl wcfl wcfl wcfl wcfl wcfl wcfl wcfl wcfl wcfl

SUPER 10 ALBUMS

1. Chicago VI-Chicago-Columbia
2. Brothers And Sisters-Allman Brothers Band-Capricorn
3. We're An American Band-Grand Funk-Capitol
4. Diamond Girl-Seals And Crofts-Warner Brothers
5. Houses Of The Holy-Led Zeppelin-Atlantic
6. Foreigner-Cat Stevens-A & M
7. A Passion Play-Jethro Tull-Chrysalis
8. The Dark Side Of The Moon-Pink Floyd-Harvest
9. The Captain And Me-Doobie Brothers-Warner Brothers
10. Los Cochinos @ #!!*-Cheech Y Chong-Ode

World Famous TOM MURPHY
Morning's Much Better
on Super CFL!

KNOM

ALASKA'S SUPER HITS

1. I Believe In Humanity-Carole King-Ode
2. Maybe Baby-Gallery-Sussex
3. Say, Has Anybody Seen My Sweet Gypsy Rose-Dawn-Bell
4. Live And Let Die-Wings-Apple
5. Uneasy Rider-Charlie Daniels-Kama Sutra
6. Swamp Witch-Jim Stafford-MGM
7. Don't Want To Be Lonely-Ronnie Dyson-Columbia
8. Show Biz-Kids-Steely Dan-ABC
9. One Tin Soldier-Coven-MGM
10. Baby, Missing You-The Independents-Wand

Courtesy: KNOM-Nome

WAS	IS	
1	1.	Delta Dawn-Helen Reddy-Capitol
3	2.	Let's Get It On-Marvin Gaye-Tamla
6	3.	We're An American Band-Grand Funk-Capitol
8	*4.	Loves Me Like A Rock-Paul Simon-Columbia
9	5.	Saturday Night's Alright For Fighting-Elton John-MCA
2	6.	The Morning After-Maureen McGovern-20th Century
7	7.	Roll Over Beethoven-Electric Light Orch.-UA
12	8.	The Hurt-Cat Stevens-A & M
15	9.	Are You Man Enough-Four Tops-Dunhill
14	10.	My Maria-B.W. Stevenson-RCA
5	11.	Live And Let Die-Wings-Apple
13	12.	Say, Has Anybody Seen My Sweet Gypsy Rose-Dawn-Bell
16	13.	Gypsy Man-War-UA
20	*14.	Half-Breed-Cher-MCA
4	15.	Feelin' Stronger Every Day-Chicago-Columbia
19	16.	Sweet Charlie Babe-Jackie Moore-Atlantic
21	17.	Ramblin' Man-The Allman Brothers Band-Capricorn
10	18.	I Believe In You-Johnnie Taylor-Stax
17	19.	How Can I Tell Her-Lobo-Big Tree
23	20.	China Grove-The Doobie Brothers-WB
26	21.	A Million To One-Donny Osmond-MGM
11	22.	Diamond Girl-Seals- & Crofts-WB
27	23.	Higher Ground-Stevie Wonder-Tamla
24	24.	Angel-Aretha Franklin-Atlantic
30	25.	Jimmy Loves Mary-Anne-Looking Glass-Epic
18	26.	Brother Louie-Stories-Kama Sutra
35	27.	Free Ride-The Edgar Winter Group-Epic
22	28.	Touch Me In The Morning-Diana Ross-Motown
32	29.	Shady Lady-Shepstone & Dibbens-Buddah
34	30.	That Lady-Isley Brothers-T-Neck
36	31.	In The Midnight Hour-Cross Country-Atco
--	32.	Keep On Truckin'-Eddie Kendricks-Tamla
25	33.	Behind Closed Doors-Charlie Rich-Epic
39	34.	Angie-Rolling Stones-Rolling Stones
--	35.	Rubber Bullets-10 C.C.-UK
29	36.	Yesterday Once More-Carpenters-A & M

40	37.	Heartbeat, It's A Lovebeat-DeFranco Family-20th Century
--	38.	Theme From Cleopatra Jones-Joe Simon-Spring
--	39.	Let Me In-The Osmonds-MGM
--	40.	You Got Me Anyway-Sutherland Brothers-Island

* CFL Super Hits

wcfl wcfl wcfl wcfl wcfl wcfl wcfl wcfl wcfl wcfl wcfl wcfl

collector's issue number one hundred and thirty-two

SUPER CFL SURVEY SEP 22

wcfl wcfl wcfl wcfl wcfl wcfl wcfl wcfl wcfl wcfl wcfl wcfl wcfl wcfl

SUPER ALBUMS

1. Brothers And Sisters-Allman Brothers Band-Capricorn
2. We're An American Band-Grand Funk-Capitol
3. Chicago VI-Chicago-Columbia
4. Foreigner-Cat Stevens-A & M
5. Los Cochinos @ #!!*-Cheech Y Chong-Ode
6. Diamond Girl-Seals And Crofts-Warner Brothers
7. The Captain And Me-Doobie Brothers-Warner Brothers
8. A Passion Play-Jethro Tull-Chrysalis
9. Killing Me Softly-Roberta Flack-Atlantic
10. Houses Of The Holy-Led Zeppelin-Atlantic
11. Let's Get It On-Marvin Gaye-Tamla
12. Deliver The Word-War-United Artists
13. The Dark Side Of The Moon-Pink Floyd-Harvest
14. There Goes Rhymin' Simon-Paul Simon-Columbia
15. Made In Japan-Deep Purple-Warner Brothers
16. American Graffiti Sound Track-MCA
17. Innervisions-Stevie Wonder-Tamla
18. 3 + 3-The Isley Brothers-T-Neck
19. Fresh-Sly And The Family Stone-Epic
20. Long Hard Climb-Helen Reddy-Capitol

wcfl wcfl wcfl wcfl wcfl

LARRY LUJACK

WAS	IS	
4	1.	Loves Me Like A Rock-Paul Simon-Columbia
2	2.	Let's Get It On-Marvin Gaye-Tamla
3	3.	We're An American Band-Grand Funk-Capitol
5	4.	Saturday Night's Alright For Fighting-Elton John-MCA
8	5.	The Hurt-Cat Stevens-A & M
1	6.	Delta Dawn-Helen Reddy-Capitol
10	7.	My Maria-B.W. Stevenson-RCA
12	8.	Say, Has Anybody Seen My Sweet Gypsy Rose-Dawn-Bell
14	9.	Half-Breed-Cher-MCA
17	10.	Ramblin' Man-The Allman Brothers Band-Capricorn
6	11.	The Morning After-Maureen McGovern-20th Century
9	12.	Are You Man Enough-Four Tops-Dunhill
20	13.	China Grove-The Doobie Brothers-WB
16	14.	Sweet Charlie Babe-Jackie Moore-Atlantic
7	15.	Roll Over Beethoven-Electric Light Orch.-UA
21	16.	A Million T One-Donny Osmond-MGM
11	17.	Live And Let Die-Wings-Apple
13	18.	Gypsy Man-War-UA
25	19.	Jimmy Loves Mary-Anne-Looking Glass-Epic
15	20.	Feelin' Stronger Every Day-Chicago-Columbia
23	21.	Higher Ground-Stevie Wonder-Tamla
27	22.	Free Ride-The Edgar Winter Group-Epic
30	23.	That Lady-Isley Brothers-T-Neck
34	24.	Angie-Rolling Stones-Rolling Stones
18	25.	I Believe In You-Johnnie Taylor-Stax
29	26.	Shady Lady-Shepstone & Dibbens-Buddah
19	27.	How Can I Tell Her-Lobo-Big Tree
31	28.	In The Midnight Hour-Cross Country-Atco
32	29.	Keep On Truckin'-Eddie Kendricks-Tamla
37	30.	Heartbeat, It's A Lovebeat-DeFranco Family-20th Century
35	31.	Rubber Bullets-10 C.C.-UK
24	32.	Angel-Aretha Franklin-Atlantic
38	33.	Theme From Cleopatra Jones-Joe Simon-Spring
40	34.	You Got Me Anyway-Sutherland Brothers-Island
39	35.	Let Me In-The Osmonds-MGM
--	36.	Yes We Can Can-The Pointer Sisters-Blue Thumb

Afternoons on SUPER 'CFL	22	37.	Diamond Girl-Seals & Crofts-WB
	--	38.	All I Know-Art Garfunkel-Columbia
	--	39.	Midnight Train To Georgia-Gladys Knight And The Pips-Buddah
	--	40.	Rocky Mountain Way-Joe Walsh-Dunhill

collector's issue number one hundred and thirty-three

SUPER CFL SURVEY SEP 29

wcfl wcfl wcfl wcfl wcfl wcfl wcfl wcfl wcfl wcfl wcfl wcfl wcfl wcfl

SUPER ALBUMS

1. Brothers And Sisters-Allman Brothers Band-Capricorn
2. We're An American Band-Grand Funk-Capitol
3. Los Cochinos @ #!!*-Cheech Y Chong-Ode
4. Chicago VI-Chicago-Columbia
5. The Captain And Me-Doobie Brothers-Warner Brothers
6. Diamond Girl-Seals And Crofts-Warner Brothers
7. A Passion Play-Jethro Tull-Chrysalis
8. Foreigner-Cat Stevens-A & M
9. Houses Of The Holy-Led Zeppelin-Atlantic
10. Deliver The Word-War-United Artists
11. The Dark Side Of The Moon-Pink Floyd-Harvest
12. Killing Me Softly-Roberta Flack-Atlantic
13. Innervisions-Stevie Wonder-Tamla
14. Let's Get It On-Marvin Gaye-Tamla
15. Long Hard Climb-Helen Reddy-Capitol
16. American Graffiti Sound Track-MCA
17. There Goes Rhymin' Simon-Paul Simon-Columbia
18. Made In Japan-Deep Purple-Warner Brothers
19. Love Devotion Surrender-Santana/McLaughlin-Columbia
20. Sufficiently Breathless-Captain Beyond-Capricorn

wcfl wcfl wcfl wcfl wcfl

WAS	IS	
2	1.	Let's Get It On-Marvin Gaye-Tamla
1	2.	Loves Me Like A Rock-Paul Simon-Columbia
5	3.	The Hurt-Cat Stevens-A & M
9	4.	Half-Breed-Cher-MCA
7	5.	My Maria-B.W. Stevenson-RCA
3	6.	We're An American Band-Grand Funk-Capitol
13	7.	China Grove-The Doobie Brothers-WB
10	8.	Ramblin' Man-The Allman Brothers Band-Capricorn
4	9.	Saturday Night's Alright For Fighting-Elton John-MCA
24	10.	Angie-Rolling Stones-Rolling Stones
21	11.	Higher Ground-Stevie Wonder-Tamla
19	12.	Jimmy Loves Mary-Anne-Looking Glass-Epic
14	13.	Sweet Charlie Babe-Jackie Moore-Atlantic
6	14.	Delta Dawn-Helen Reddy-Capitol
22	15.	Free Ride-The Edgar Winter Group-Epic
8	16.	Say, Has Anybody Seen My Sweet Gypsy Rose-Dawn-Bell
16	17.	A Million To One-Donny Osmond-MGM
23	18.	That Lady-The Isley Brothers-T-Neck
29	19.	Keep On Truckin'-Eddie Kendricks-Tamla
12	20.	Are You Man Enough-Four Tops-Dunhill
26	21.	Shady Lady-Shepstone & Dibbens-Buddah
31	22.	Rubber Bullets-10 C.C.-UK
30	23.	Heartbeat, It's A Lovebeat-DeFranco Family-20th Century
34	24.	You Got Me Anyway-Sutherland Brothers-Island
28	25.	In The Midnight Hour-Cross Country-Atco
11	26.	The Morning After-Maureen McGovern-20th Century
35	27.	Let Me In-The Osmonds-MGM
33	28.	Theme From Cleopatra Jones-Joe Simon-Spring
15	29.	Roll Over Beethoven-Electric Light Orch.-UA
38	30.	All I Know-Art Garfunkel-Columbia
36	31.	Yes We Can Can-The Pointer Sisters-Blue Thumb
17	32.	Live And Let Die-Wings-Apple
40	33.	Rocky Mountain Way-Joe Walsh-Dunhill
--	34.	Basketball Jones-Cheech Y Chong-

		Ode
39	35.	Midnight Train To Georgia-Gladys Knight/Pips-Buddah
--	36.	Paper Roses-Marie Osmond-MGM
18	37.	Gypsy Man-War-UA
--	38.	We May Never Pass This Way (Again)-Seals & Crofts-WB
--	39.	Jesse-Roberta Flack-Atlantic
--	40.	Just You 'N' Me-Chicago-Columbia

BIG RON O'BRIEN
evenings...on SUPER CFL!

wcfl wcfl wcfl wcfl wcfl wcfl wcfl wcfl wcfl wcfl wcfl wcfl

collector's issue number one hundred and thirty-four

SUPER CFL SURVEY OCT 6

wcfl wcfl wcfl wcfl wcfl wcfl wcfl wcfl wcfl wcfl wcfl wcfl wcfl wcfl

SUPER ALBUMS

1. Brothers And Sisters-Allman Brothers Band-Capricorn
2. Los Cochinos @ #!!*-Cheech Y Chong-Ode
3. Goats Head Soup-The Rolling Stones-Rolling Stones
4. We're An American Band-Grand Funk-Capitol
5. Foreigner-Cat Stevens-A & M
6. Chicago VI-Chicago-Columbia
7. The Captain And Me-Doobie Brothers-Warner Brothers
8. Killing Me Softly-Roberta Flack-Atlantic
9. Let's Get It On-Marvin Gaye-Tamla
10. Angel Clare-Garfunkel-Columbia
11. Rainbow Concert-Eric Clapton-RSO
12. A Passion Play-Jethro Tull-Chrysalis
13. Deliver The Word-War-United Artists
14. Houses Of The Holy-Led Zeppelin-Atlantic
15. Diamond Girl-Seals And Crofts-Warner Brothers
16. Innervisions-Stevie Wonder-Tamla
17. The Dark Side Of The Moon-Pink Floyd-Harvest
18. There Goes Rhymin' Simon-Paul Simon-Columbia
19. Long Hard Climb-Helen Reddy-Capitol
20. 3 + 3-The Isley Brothers-T-Neck

wcfl wcfl wcfl wcfl wcfl

WAS	IS	
3	1.	The Hurt-Cat Stevens-A & M
4	2.	Half-Breed-Cher-MCA
5	3.	My Maria-B.W. Stevenson-RCA
2	4.	Loves Me Like A Rock-Paul Simon-Columbia
7	5.	China Grove-The Doobie Brothers-WB
8	6.	Ramblin' Man-The Allman Brothers Band-Capricorn
1	7.	Let's Get It On-Marvin Gaye-Tamla
10	8.	Angie-Rolling Stones-Rolling Stones
12	9.	Jimmy Loves Mary-Anne-Looking Glass-Epic
15	10.	Free Ride-The Edgar Winter Group-Epic
11	11.	Higher Ground-Stevie Wonder-Tamla
9	12.	Saturday Night's Alright For Fighting-Elton John-MCA
13	13.	Sweet Charlie Babe-Jackie Moore-Atlantic
6	14.	We're An American Band-Grand Funk-Capitol
18	15.	That Lady-The Isley Brothers-T-Neck
23	16.	Heartbeat, It's A Lovebeat-DeFranco Family-20th Century
19	17.	Keep On Truckin'-Eddie Kendricks-Tamla
14	18.	Delta Dawn-Helen Reddy-Capitol
21	19.	Shady Lady-Shepstone & Dibbens-Buddah
22	20.	Rubber Bullets-10 C.C.-UK
17	21.	A Million To One-Donny Osmond-MGM
30	22.	All I Know-Art Garfunkel-Columbia
24	23.	You Got Me Anyway-Sutherland Brothers-Island
25	24.	In The Midnight Hour-Cross Country-Atco
27	25.	Let Me IN-The Osmonds-MGM
28	26.	Theme From Cleopatra Jones-Joe Simon-Spring
31	27.	Yes We Can Can-The Pointer Sisters-Blue Thumb
34	28.	Basketball Jones-Cheech Y Chong-Ode
16	29.	Say, Has Anybody Seen My Sweet Gypsy Rose-Dawn-Bell
35	30.	Midnight Train To Georgia-Gladys Knight/Pips-Buddah
33	31.	Rocky Mountain Way-Joe Walsh-Dunhill
36	32.	Paper Roses-Marie Osmond-MGM
20	33.	Are You Man Enough-Four Tops-Dunhill
39	34.	Jesse-Roberta Flack-Atlantic
38	35.	We May Never Pass This Way (Again)-Seals & Crofts-WB
40	36.	Just You 'N' Me-Chicago-Columbia

World Famous TOM MURPHY
Morning's Much Better
on Super CFL!

26	37.	The Morning After-Maureen McGovern-20th Century
--	38.	Knockin' On Heaven's Door-Bob Dylan-Columbia
--	39.	You're A Special Part Of Me-Diana Ross/Marvin Gaye-Motown
--	40.	Ooh Baby-Gilbert O'Sullivan-MAM

collector's issue number one hundred and thirty-five

SUPER CFL SURVEY OCT 13

wcfl wcfl wcfl wcfl wcfl wcfl wcfl wcfl wcfl wcfl wcfl wcfl wcfl wcfl

SUPER ALBUMS

1. Brothers And Sisters-Allman Brothers Band-Capricorn
2. Los Cochinos @ #!!*-Cheech Y Chong-Ode
3. Goats Head Soup-The Rolling Stones-Rolling Stones
4. We're An American Band-Grand Funk-Capitol
5. The Captain And Me-Doobie Brothers-Warner Brothers
6. Chicago VI-Chicago-Columbia
7. Let's Get It On-Marvin Gaye-Tamla
8. Foreigner-Cat Stevens-A & M
9. Killing Me Softly-Roberta Flack-Atlantic
10. Angel Clare-Garfunkel-Columbia
11. Innervisions-Stevie Wonder-Tamla
12. Sweet Freedom-Uriah Heep-Warner Brothers
13. A Passion Play-Jethro Tull-Chrysalis
14. Houses Of The Holy-Led Zeppelin-Atlantic
15. Rainbow Concert-Eric Clapton-RSO
16. Diamond Girl-Seals And Crofts-Warner Brothers
17. The Dark Side Of The Moon-Pink Floyd-Harvest
18. The Smoker You Drink, The Player You Get-Joe Walsh-Dunhill
19. Deliver The Word-War-United Artists
20. Sufficiently Breathless-Captain Beyond-Capricorn

wcfl wcfl wcfl wcfl wcfl

WAS	IS	
2	1.	Half-Breed-Cher-MCA
5	2.	China Grove-The Doobie Brothers-WB
8	3.	Angie-Rolling Stones-Rolling Stones
6	4.	Ramblin' Man-The Allman Brothers Band-Capricorn
9	5.	Jimmy Loves Mary-Anne-Looking Glass-Epic
10	6.	Free Ride-The Edgar Winter Group-Epic
3	7.	My Maria-B.W. Stevenson-RCA
11	8.	Higher Ground-Stevie Wonder-Tamla
16	9.	Heartbeat, It's A Lovebeat-DeFranco Family-20th Century
4	10.	Loves Me Like A Rock-Paul Simon-Columbia
1	11.	The Hurt-Cat Stevens-A & M
15	12.	That Lady-The Isley Brothers-T-Neck
17	13.	Keep On Truckin'-Eddie Kendricks-Tamla
7	14.	Let's Get It On-Marvin Gaye-Tamla
23	15.	You Got Me Anyway-Sutherland Brothers-Island
22	16.	All I Know-Art Garfunkel-Columbia
20	17.	Rubber Bullets-10 C.C.-UK
12	18.	Saturday Night's Alright For Fighting-Elton John-MCA
28	19.	Basketball Jones-Cheech Y Chong-Ode
24	20.	In The Midnight Hour-Cross Country-Atco
27	21.	Yes We Can Can-The Pointer Sisters-Blue Thumb
25	22.	Let Me In-The Osmonds-MGM
13	23.	Sweet Charlie Babe-Jackie Moore-Atlantic
14	24.	We're An American Band-Grand Funk-Capitol
30	25.	Midnight Train To Georgia-Gladys Knight/Pips-Buddah
32	26.	Paper Roses-Marie Osmond-MGM
18	27.	Delta Dawn-Helen Reddy-Capitol
31	28.	Rocky Mountain Way-Joe Walsh-Dunhill
35	29.	We May Never Pass This Way (Again)-Seals & Crofts-WB
19	30.	Shady Lady-Shepstone & Dibbens-Buddah
34	31.	Jesse-Roberta Flack-Atlantic
36	32.	Just You 'N' Me-Chicago-Columbia
--	33.	Photograph-Ringo Starr-Apple
38	34.	Knockin' On Heaven's Door-Bob Dylan-Columbia

21	35.	A Million To One-Donny Osmond-MGM
39	36.	You're A Special Part Of Me-Diana Ross/Marvin Gaye-Motown
40	37.	Ooh Baby-Gilbert O'Sullivan-MAM
--	38.	I Got A Name-Jim Croce-ABC
--	39.	Top Of The World-Carpenters-A & M
--	40.	D'yer Mak'er-Led Zeppelin-Atlantic

LARRY LUJACK
Afternoons on SUPER 'CFL

wcfl wcfl wcfl wcfl wcfl wcfl wcfl wcfl wcfl wcfl wcfl wcfl

collector's issue number one hundred and thirty-six

SUPER CFL SURVEY OCT 20

wcfl wcfl wcfl wcfl wcfl wcfl wcfl wcfl wcfl wcfl wcfl wcfl wcfl wcfl

SUPER ALBUMS

1. Goats Head Soup-The Rolling Stones-Rolling Stones
2. Los Cochinos @ #!!*-Cheech Y Chong-Ode
3. Brothers And Sisters-Allman Brothers Band-Capricorn
4. We're An American Band-Grand Funk-Capitol
5. Let's Get It On-Marvin Gaye-Tamla
6. The Captain And Me-Doobie Brothers-Warner Brothers
7. Angel Clare-Garfunkel-Columbia
8. Innervisions-Stevie Wonder-Tamla
9. Sweet Freedom-Uriah Heep-Warner Brothers
10. Chicago VI-Chicago-Columbia
11. Rainbow Concert-Eric Clapton-RSO
12. Foreigner-Cat Stevens-A & M
13. Deliver The Word-War-United Artists
14. Killing Me Softly-Roberta Flack-Atlantic
15. Goodbye Yellow Brick Road-Elton John-MCA
16. Houses Of The Holy-Led Zeppelin-Atlantic
17. 3 + 3-The Isley Brothers-T-Neck
18. Over-Nite Sensation-Mothers-Discreet/WB
19. A Passion Play-Jethro Tull-Chrysalis
20. Life And Times-Jim Croce-ABC

wcfl wcfl wcfl wcfl wcfl

World Famous

WAS	IS	
1	1.	Half-Breed-Cher-MCA
3	2.	Angie-Rolling Stones-Rolling Stones
4	3.	Ramblin' Man-The Allman Brothers Band-Capricorn
9	4.	Heartbeat, It's A Lovebeat-DeFranco Family-20th Century
6	5.	Free Ride-The Edgar Winter Group-Epic
5	6.	Jimmy Loves Mary-Anne-Looking Glass-Epic
2	7.	China Grove-The Doobie Brothers-WB
8	8.	Higher Ground-Stevie Wonder-Tamla
19	9.	Basketball Jones-Cheech Y Chong-Ode
12	10.	That Lady-The Isley Brothers-T-Neck
13	11.	Keep On Truckin'-Eddie Kendricks-Tamla
16	12.	All I Know-Art Garfunkel-Columbia
7	13.	My Maria-B.W. Stevenson-RCA
15	14.	You Got Me Anyway-Sutherland Brothers-Island
17	15.	Rubber Bullets-10 C.C.-UK
21	16.	Yes We Can Can-The Pointer Sisters-Blue Thumb
10	17.	Loves Me Like A rock-Paul Simon-Columbia
22	18.	Let Me In-The Osmonds-MGM
26	19.	Paper Roses-Marie Osmond-MGM
20	20.	In The Midnight Hour-Cross Country-Atco
11	21.	The Hurt-Cat Stevens-A & M
14	22.	Let's Get It On-Marvin Gaye-Tamla
25	23.	Midnight Train To Georgia-Gladys Knight/Pips-Buddah
18	24.	Saturday Night's Alright For Fighting-Elton John-MCA
29	25.	We May Never Pass This Way (Again)-Seals & Crofts-WB
28	26.	Rocky Mountain Way-Joe Walsh-Dunhill
33	27.	Photograph-Ringo Starr-Apple
32	28.	Just You 'N' Me-Chicago-Columbia
31	29.	Jesse-Roberta Flack-Atlantic
34	30.	Knockin' On Heaven's Door-Bob Dylan-Columbia
23	31.	Sweet Charlie Babe-Jackie Moore-Atlantic
37	32.	Ooh Baby-Gilbert O'Sullivan-MAM
36	33.	You're A Special Part Of Me-Diana Ross/Marvin Gays-Motown
24	34.	We're An American Band-Grand Funk-Capitol
39	35.	Top Of The World-Carpenters-A & M
38	36.	I Got A Name-Jim Croce-ABC
40	37.	D'yer Mak'er-Led Zeppelin-Atlantic

TOM MURPHY mornings...on SUPER CFL!	--	38.	Be-Neil Diamond-Columbia
	--	39.	Hello It's Me-Todd Rundgren-Bearsville
	--	40.	Rock 'N Roll-Kevin Johnson-Mainstream

collector's issue number one hundred and thirty-seven

SUPER CFL SURVEY OCT 27

wcfl wcfl wcfl wcfl wcfl wcfl wcfl wcfl wcfl wcfl wcfl wcfl wcfl wcfl

SUPER ALBUMS

1. Goats Head Soup-The Rolling Stones-Rolling Stones
2. Los Cochinos @ #!!*-Cheech Y Chong-Ode
3. Brothers And Sisters-Allman Brothers Band-Capricorn
4. Let's Get It On-Marvin Gaye-Tamla
5. Chicago VI-Chicago-Columbia
6. We're An American Band-Grand Funk-Capitol
7. Goodbye Yellow Brick Road-Elton John-MCA
8. Sweet Freedom-Uriah Heep-Warner Brothers
9. Angel Clare-Garfunkel-Columbia
10. The Captain And Me-Doobie Brothers-Warner Brothers
11. Innervisions-Stevie Wonder-Tamla
12. Killing Me Softly-Roberta Flack-Atlantic
13. The Smoker You Drink, The Player You Get-Joe Walsh-Dunhill
14. Deliver The Word-War-United Artists
15. Foreigner-Cat Stevens-A & M
16. Rainbow Concert-Eric Clapton-RSO
17. Houses Of The Holy-Led Zeppelin-Atlantic
18. Life And Times-Jim Croce-ABC
19. American Graffiti Sound Track-MCA
20. A Passion Play-Jethro Tull-Chrysalis

wcfl wcfl wcfl wcfl wcfl

World Famous

WAS	IS	
2	1.	Angie-Rolling Stones-Rolling Stones
4	2.	Heartbeat, It's a Lovebeat-DeFranco Family-20th Century
5	3.	Free Ride-The Edgar Winter Group-Epic
1	4.	Half-Breed-Cher-MCA
12	5.	All I Know-Art Garfunkel-Columbia
9	6.	Basketball Jones-Cheech Y Chong-Ode
3	7.	Ramblin' Man-The Allman Brothers Band-Capricorn
11	8.	Keep On Truckin'-Eddie Kendricks-Tamla
6	9.	Jimmy Loves Mary-Anne-Looking Glass-Epic
15	10.	Rubber Bullets-10 C.C.-UK
23	11.	Midnight Train To Georgia-Gladys Knight/Pips-Buddah
7	12.	China Grove-The Doobie Brothers-WB
16	13.	Yes We Can Can-The Pointer Sisters-Blue Thumb
19	14.	Paper Roses-Marie Osmond-MGM
10	15.	That Lady-The Isley Brothers-T-Neck
18	16.	Let Me In-The Osmonds-MGM
27	17.	Photograph-Ringo Starr-Apple
8	18.	Higher Ground-Stevie Wonder-Tamla
14	19.	You Got Me Anyway-Sutherland Brothers-Island
25	20.	We May Never Pass This Way (Again)-Seals & Crofts-WB
13	21.	My Maria-B.W. Stevenson-RCA
26	22.	Rocky Mountain Way-Joe Walsh-Dunhill
28	23.	Just You 'N' Me-Chicago-Columbia
17	24.	Loves Me Like A rock-Paul Simon-Columbia
32	25.	Ooh Baby-Gilbert O'Sullivan-MAM
36	26.	I Got A Name-Jim Croce-ABC
30	27.	Knockin' On Heaven's Door-Bob Dylan-Columbia
29	28.	Jesse-Roberta Flack-Atlantic
22	29.	Let's Get It On-Marvin Gaye-Tamla
35	30.	Top Of The World-Carpenters-A & M
33	31.	You're A Special Part Of Me-Diana Ross/Marvin Gaye-Motown
37	32.	D'yer Mak'er-Led Zeppelin-Atlantic
21	33.	The Hurt-Cat Stevens-A & M
38	34.	Be-Neil Diamond-Columbia
39	35.	Hello It's Me-Todd Rundgren-Bearsville
20	36.	In The Midnight Hour-Cross Country-Atco

TOM MURPHY
mornings...on SUPER CFL!

40	37.	Rock 'N Roll-Kevin Johnson-Mainstream
--	38.	Space Race-Billy Preston-A & M
--	39.	Mammy Blue-Stories-Kama Sutra
--	40.	Goodbye Yellow Brick Road-Elton John-MCA

wcfl wcfl wcfl wcfl wcfl wcfl wcfl wcfl wcfl wcfl wcfl wcfl

collector's issue number one hundred and thirty-eight

SUPER CFL SURVEY NOV 3

wcfl wcfl wcfl wcfl wcfl wcfl wcfl wcfl wcfl wcfl wcfl wcfl wcfl wcfl

SUPER ALBUMS

1. Goats Head Soup-The Rolling Stones-Rolling Stones
2. Los Cochinos @ #!!*-Cheech Y Chong-Ode
3. Brothers And Sisters-Allman Brothers Band-Capricorn
4. Sweet Freedom-Uriah Heep-Warner Brothers
5. Goodbye Yellow Brick Road-Elton John-MCA
6. Let's Get It On-Marvin Gaye-Tamla
7. Angel Clare-Garfunkel-Columbia
8. We're An American Band-Grand Funk-Capitol
9. The Captain And Me-Doobie Brothers-Warner Brothers
10. Life And Times-Jim Croce-ABC
11. 3 + 3-The Isley Brothers-T-Neck
12. The Smoker You Drink, The Player You Get-Joe Walsh-Dunhill
13. Time Fades Away-Neil Young-Reprise
14. Innervisions-Stevie Wonder-Tamla
15. Rainbow Concert-Eric Clapton-RSO
16. Jonathan Livingston Seagull-Neil Diamond-Columbia
17. Chicago VI-Chicago-Columbia
18. Killing Me Softly-Roberta Flack-Atlantic
19. Ol' Blue Eyes Is Back-Frank Sinatra-Reprise
20. Wake Of The Flood-Grateful Dead-Grateful Dead

wcfl wcfl wcfl wcfl wcfl

WAS	IS	
1	1.	Angie-Rolling Stones-Rolling Stones
2	2.	Heartbeat, It's A Lovebeat-DeFranco Family-20th Century
6	3.	Basketball Jones-Cheech Y Chong-Ode
5	4.	All I Know-Art Garfunkel-Columbia
3	5.	Free Ride-The Edgar Winter Group-Epic
8	6.	Keep On Truckin'-Eddie Kendricks-Tamla
10	7.	Rubber Bullets-10 C.C.-UK
4	8.	Half-Breed-Cher-MCA
11	9.	Midnight Train To Georgia-Gladys Knight/Pips-Buddah
14	10.	Paper Roses-Marie Osmond-MGM
13	11.	Yes We Can Can-The Pointer Sisters-Blue Thumb
17	12.	Photograph-Ringo Starr-Apple
7	13.	Ramblin' Man-The Allman Brothers Band-Capricorn
16	14.	Let Me In-The Osmonds-MGM
20	15.	We May Never Pass This Way (Again)-Seals & Crofts-WB
9	16.	Jimmy Loves Mary-Anne-Looking Glass-Epic
12	17.	China Grove-The Doobie Brothers-WB
23	18.	Just You 'N' Me-Chicago-Columbia
25	19.	Ooh Baby-Gilbert O'Sullivan-MAM
15	20.	That Lady-The Isley Brothers-T-Neck
26	21.	I Got A Name-Jim Croce-ABC
27	22.	Knockin' On Heaven's Door-Bob Dylan-Columbia
35	23.	Hello It's Me-Todd Rundgren-Bearsville
18	24.	Higher Ground-Stevie Wonder-Tamla
19	25.	You Got Me Anyway-Sutherland Brothers-Island
32	26.	D'yer Mak'er-Led Zeppelin-Atlantic
30	27.	Top Of The World-Carpenters-A & M
31	28.	You're A Special Part Of Me-Diana Ross/Marvin Gaye-Motown
34	29.	Be-Neil Diamond-Columbia
22	30.	Rocky Mountain Way-Joe Walsh-Dunhill
21	31.	My Maria-B.W. Stevenson-RCA
37	32.	Rock 'N Roll-Kevin Johnson-Mainstream
28	33.	Jesse-Roberta Flack-Atlantic
38	34.	Space Race-Billy Preston-A & M
40	35.	Goodbye Yellow Brick Road-Elton John-MCA

LARRY LUJACK Afternoons on SUPER 'CFL	39	36. Mammy Blue-Stories-Kama Sutra
	24	37. Loves Me Like A rock-Paul Simon-Columbia
	--	38. Let Me Serenade You-Three Dog Night-Dunhill
	--	39. Painted Ladies-Ian Thomas-Janus
	--	40. Leave Me Alone-Helen Reddy-Capitol

collector's issue number one hundred and thirty-nine

SUPER CFL SURVEY NOV 10

wcfl wcfl wcfl wcfl wcfl wcfl wcfl wcfl wcfl wcfl wcfl wcfl wcfl wcfl

SUPER ALBUMS

1. Goats Head Soup-The Rolling Stones-Rolling Stones
2. Goodbye Yellow Brick Road-Elton John-MCA
3. Los Cochinos @ #!!*-Cheech Y Chong-Ode
4. Brothers And Sisters-Allman Brothers Band-Capricorn
5. Jonathan Livingston Seagull-Neil Diamond-Columbia
6. Sweet Freedom-Uriah Heep-Warner Brothers
7. Angel Clare-Garfunkel-Columbia
8. Let's Get It On-Marvin Gaye-Tamla
9. Life And Times-Jim Croce-ABC
10. Wake Of The Flood-Grateful Dead-Grateful Dead
11. Chicago VI-Chicago-Columbia
12. The Smoker You Drink-The Player You Get-Joe Walsh-Dunhill
13. Time Fades Away-Neil Young-Reprise
14. Ol' Blue Eyes Is Back-Frank Sinatra-Reprise
15. Quadrophenia-The Who-MCA
16. We're An American Band-Grand Funk-Capitol
17. The Captain And Me-Doobie Brothers-Warner Brothers
18. Innervisions-Stevie Wonder-Tamla
19. Killing Me Softly-Roberta Flack-Atlantic
20. Houses Of The Holy-Led Zeppelin-Atlantic

wcfl wcfl wcfl wcfl wcfl

WAS	IS	
2	1.	Heartbeat, It's A Lovebeat-DeFranco Family-20th Century
9	2.	Midnight Train to Georgia-Gladys Knight/Pips-Buddah
6	3.	Keep On Truckin'-Eddie Kendricks-Tamla
4	4.	All I Know-Art Garfunkel-Columbia
1	5.	Angie-Rolling Stones-Rolling Stones
12	6.	Photograph-Ringo Starr-Apple
3	7.	Basketball Jones-Cheech Y Chong-Ode
10	8.	Paper Roses-Marie Osmond-MGM
11	9.	Yes We Can Can-The Pointer Sisters-Blue Thumb
5	10.	Free Ride-The Edgar Winter Group-Epic
15	11.	We May Never Pass This Way (Again)-Seals & Crofts-WB
7	12.	Rubber Bullets-10 C.C.-UK
19	13.	Ooh Baby-Gilbert O'Sullivan-MAM
21	14.	I Got A Name-Jim Croce-ABC
8	15.	Half Breed-Cher-MCA
18	16.	Just You 'N' Me-Chicago-Columbia
27	17.	Top Of The World-Carpenters-A & M
14	18.	Let Me In-The Osmonds-MGM
22	19.	Knockin' On Heaven's Door-Bob Dylan-Columbia
23	20.	Hello It's Me-Todd Rundgren-Bearsville
29	21.	Be-Neil Diamond-Columbia
13	22.	Ramblin' Man-The Allman Brothers Band-Capricorn
26	23.	D'yer Mak'er-Led Zeppelin-Atlantic
16	24.	Jimmy Loves Mary-Anne-Looking Glass-Epic
17	25.	China Grove-The Doobie Brothers-WB
35	26.	Goodbye Yellow Brick Road-Elton John-MCA
32	27.	Rock 'N Roll-Kevin Johnson-Mainstream
34	28.	Space Race-Billy Preston-A & M
20	29.	That Lady-The Isley Brothers-T-Neck
36	30.	Mammy Blue-Stories-Kama Sutra
28	31.	You're A Special Part Of Me-Diana Ross/Marvin Gaye-Motown
38	32.	Let Me Serenade You-Three Dog Night-Dunhill
24	33.	Higher Ground-Stevie Wonder-Tamla

	--	34. Mind Games-John Lennon-Apple
	39	35. Painted Ladies-Ian Thomas-Janus
	40	36. Leave Me Alone-Helen Reddy-Capitol
LARRY LUJACK Afternoons on SUPER 'CFL	25	37. You Got Me Anyway-Sutherland Brothers-Island
	--	38. Smokin' In The Boy's Room-Brownsville Station-Big Tree
	--	39. Show And Tell-Al Wilson-Rocky Road
	30	40. Rocky Mountain Way-Joe Walsh-Dunhill

thank you for collecting the Super CFL survey!

SUPER CFL SURVEY NOV 17

wcfl wcfl wcfl wcfl wcfl wcfl wcfl wcfl wcfl wcfl wcfl wcfl wcfl wcfl

SUPER ALBUMS

1. Goats Head Soup-The Rolling Stones-Rolling Stones
2. Goodbye Yellow Brick Road-Elton John-MCA
3. Los Cochinos @ #!!*-Cheech Y Chong-Ode
4. Jonathan Livingston Seagull-Neil Diamond-Columbia
5. Quadrophenia-The Who-MCA
6. Sweet Freedom-Uriah Heep-Warner Brothers
7. Brothers And Sisters-Allman Brothers Band-Capricorn
8. Life And Time-Jim Croce-ABC
9. Angel Clare-Garfunkel-Columbia
10. The Smoker You Drink, The Player You Get-Joe Walsh-Dunhill
11. Full Sail-Loggins And Messina-Columbia
12. Time Fades Away-Neil Young-Reprise
13. Wake Of The Flood-Grateful Dead-Grateful Dead
14. Hat Trick-America-Warner Brothers
15. We're An American Band-Grand Funk-Capitol
16. Let's Get It On-Marvin Gaye-Tamla
17. Ol' Blue Eyes Is Back-Frank Sinatra-Reprise
18. The Captain And Me-Doobie Brothers-Warner Brothers
19. Innervisions-Stevie Wonder-Tamla
20. You Don't Mess Around With Jim-Jim Croce-ABC

wcfl wcfl wcfl wcfl wcfl

WAS	IS	
2	1.	Midnight Train To Georgia-Gladys Knight/Pips-Buddah
1	2.	Heartbeat, It's A Lovebeat-DeFranco Family-20th Century
6	3.	Photograph-Ringo Starr-Apple
3	4.	Keep On Truckin'-Eddie Kendricks-Tamla
14	5.	I Got A Name-Jim Croce-ABC
4	6.	All I Know-Art Garfunkel-Columbia
16	7.	Just You 'N' Me-Chicago-Columbia
8	8.	Paper Roses-Marie Osmond-MGM
11	9.	We May Never Pass This Way (Again)-Seals & Crofts-WB
13	10.	Ooh Baby-Gilbert O'Sullivan-MAM
5	11.	Angie-Rolling Stones-Rolling Stones
17	12.	Top Of The World-Carpenters-A & M
9	13.	Yes We Can Can-The Pointer Sisters-Blue Thumb
20	14.	Hello It's Me-Todd Rundgren-Bearsville
7	15.	Basketball Jones-Cheech Y Chong-Ode
19	16.	Knockin' On Heaven's Door-Bob Dylan-Columbia
21	17.	Be-Neil Diamond-Columbia
28	18.	Space Race-Billy Preston-A & M
23	19.	D'yer Mak'er-Led Zeppelin-Atlantic
10	20.	Free Ride-The Edgar Winter Group-Epic
26	21.	Goodbye Yellow Brick Road-Elton John-MCA
12	22.	Rubber Bullets-10 C.C.-UK
15	23.	Half-Breed-Cher-MCA
27	24.	Rock 'N Roll-Kevin Johnson-Mainstream
18	25.	Let Me In-The Osmonds-MGM
30	26.	Mammy Blue-Stories-Kama Sutra
32	27.	Let Me Serenade You-Three Dog Night-Dunhill
36	28.	Leave Me Alone-Helen Reddy-Capitol
34	29.	Mind Games-John Lennon-Apple
35	30.	Painted Ladies-Ian Thomas-Janus
22	31.	Ramblin' Man-The Allman Brothers Band-Capricorn
39	32.	Show And Tell-Al Wilson-Rocky Road
25	33.	China Grove-The Doobie Brothers-WB
38	34.	Smokin' In The Boy's Room-Brownsville Station-Big Tree
24	35.	Jimmy Loves Mary-Anne-Looking Glass-Epic

LARRY LUJACK
Afternoons on SUPER 'CFL

--	36.	Helen Wheels-Paul McCartney-Apple
--	37.	If You're Ready (Come Go With Me)-The Staple Singers-Stax
--	38.	Who's In The Strawberry Patch With Sally-Dawn-Bell
--	39.	The Joker-Steve Miller Band-Capitol
--	40.	Tell Her She's Lovely-El Chicano-MCA

wcfl wcfl wcfl wcfl wcfl wcfl wcfl wcfl wcfl wcfl wcfl wcfl

collector's issue number one hundred and forty-one

SUPER CFL SURVEY NOV 24

wcfl wcfl wcfl wcfl wcfl wcfl wcfl wcfl wcfl wcfl wcfl wcfl wcfl wcfl

SUPER ALBUMS

1. Goodbye Yellow Brick Road-Elton John-MCA
2. Goats Head Soup-The Rolling Stones-Rolling Stones
3. Jonathan Livingston Seagull-Neil Diamond-Columbia
4. Los Cochinos @ #!!*-Cheech Y Chong-Ode
5. Ringo-Ringo Starr-Apple
6. Quadrophenia-The Who-MCA
7. Sweet Freedom-Uriah Heep-Warner Brothers
8. Life And Times-Jim Croce-ABC
9. Brothers And Sisters-Allman Brothers Band-Capricorn
10. Let's Get It On-Marvin Gaye-Tamla
11. Mind Games-John Lennon-Apple
12. Angel Clare-Garfunkel-Columbia
13. Hat Trick-America-Warner Brothers
14. Wake Of The Flood-Grateful Dead-Grateful Dead
15. The Smoker You Drink, The Player You Get-Joe Walsh-Dunhill
16. Time Fades Away-Neil Young-Reprise
17. Ol' Blue Eyes Is Back-Frank Sinatra-Reprise
18. The Captain And Me-Doobie Brothers-Warner Brothers
19. You Don't Mess Around With Jim-Jim Croce-ABC
20. Chicago VI-Chicago-Columbia

wcfl wcfl wcfl wcfl wcfl

WAS	IS	
1	1.	Midnight Train to Georgia-Gladys Knight/Pips-Buddah
3	2.	Photograph-Ringo Starr-Apple
5	3.	I Got A Name-Jim Croce-ABC
7	4.	Just You 'N Me-Chicago-Columbia
2	5.	Heartbeat, It's A Lovebeat-DeFranco Family-20th Century
4	6.	Keep On Truckin'-Eddie Kendricks-Tamla
14	7.	Hello It's Me-Todd Rundgren-Bearsville
10	8.	Ooh Baby-Gilbert O'Sullivan-MAM
12	9.	Top Of The World-Carpenters-A & M
6	10.	All I Know-Art Garfunkel-Columbia
8	11.	Paper Roses-Marie Osmond-MGM
21	12.	Goodbye Yellow Brick Road-Elton John-MCA
9	13.	We May Never Pass This Way (Again)-Seals & Crofts-WB
18	14.	Space Race-Billy Preston-A & M
13	15.	Yes We Can Can-The Pointer Sisters-Blue Thumb
17	16.	Be-Neil Diamond-Columbia
19	17.	D'yer Mak'er-Led Zeppelin-Atlantic
26	18.	Mammy Blue-Stories-Kama Sutra
11	19.	Angie-Rolling Stones-Rolling Stones
16	20.	Knockin' On Heaven's Door-Bob Dylan-Columbia
29	21.	Mind Games-John Lennon-Apple
30	22.	Painted Ladies-Ian Thomas-Janus
27	23.	Let Me Serenade You-Three Dog Night-Dunhill
28	24.	Leave Me Alone-Helen Reddy-Capitol
32	25.	Show And Tell-Al Wilson-Rocky Road
15	26.	Basketball Jones-Cheech Y Chongl-Ode
20	27.	Free Ride-The Edgar Winter Group-Epic
24	28.	Rock 'N Roll-Kevin Johnson-Mainstream
34	29.	Smokin' In The Boy's Room-Brownsville Station-Big Tree
22	30.	Rubber Bullets- 10 C.C.-UK
36	31.	Helen Wheels-Paul McCartney-Apple
39	32.	The Joker-Steve Miller Band-Capitol
--	33.	The Most Beautiful Girl-Charlie Rich-Epic
37	34.	If You're Ready (Come Go With Me)-The Staple Singers-Stax
23	35.	Half-Breed-Cher-MCA

BIG RON O'BRIEN
evenings...on SUPER CFL!

38	36.	Who's In The Strawberry Patch With Sally-Dawn-Bell
40	37.	Tell Her She's Lovely-El Chicano-MCA
25	38.	Let Me In-The Osmonds-MGM
--	39.	Living For The City-Stevie Wonder-Tamla
--	40.	Blue Collar-Bachman-Turner Overdrive-Mercury

wcfl wcfl wcfl wcfl wcfl wcfl wcfl wcfl wcfl wcfl wcfl wcfl

collector's issue number one hundred and forty-two

SUPER CFL SURVEY DEC 1

wcfl wcfl wcfl wcfl wcfl wcfl wcfl wcfl wcfl wcfl wcfl wcfl wcfl wcfl

SUPER ALBUMS

1. Goodbye Yellow Brick Road-Elton John-MCA
2. Jonathan Livingston Seagull-Neil Diamond-Columbia
3. Quadrophenia-The Who-MCA
4. Goats Head Soup-The Rolling Stones-Rolling Stones
5. Ringo-Ringo Starr-Apple
6. Los Cochinos @ #!!*-Cheech Y Chong-Ode
7. Brothers And Sisters-Allman Brothers Band-Capricorn
8. Life And Times-Jim Croce-ABC
9. Sweet Freedom-Uriah Heep-Warner Brothers
10. Mind Games-John Lennon-Apple
11. You Don't Mess Around With Jim-Jim Croce-ABC
12. Wake Of The Flood-Grateful Dead-Grateful Dead
13. Angel Clare-Garfunkel-Columbia
14. Full Sail-Loggins And Messina-Columbia
15. Ol' Blue Eyes Is Back-Frank Sinatra-Reprise
16. The Smoker You Drink, The Player You Get-Joe Walsh-Dunhill
17. Hat Trick-America-Warner Brothers
18. Innervisions-Stevie Wonder-Tamla
19. Laid Back-Gregg Allman-Capricorn
20. The Joker-Steve Miller Band-Capitol

wcfl wcfl wcfl wcfl wcfl

WAS	IS	
2	1.	Photograph-Ringo Starr-Apple
3	2.	I Got A Name-Jim Croce-ABC
7	3.	Hello It's Me-Todd Rundgren-Bearsville
9	4.	Top Of The World-Carpenters-A & M
4	5.	Just You 'N Me-Chicago-Columbia
8	6.	Ooh Baby-Gilbert O' Sullivan-MAM
12	7.	Goodbye Yellow Brick Road-Elton John-MCA
1	8.	Midnight Train To Georgia-Gladys Knight/Pips-Buddah
14	9.	Space Race-Billy Preston-A & M
5	10.	Heartbeat, It's A Lovebeat-DeFranco Family-20th Century
6	11.	Keep On Truckin'-Eddie Kendricks-Tamla
17	12.	D'yer Mak'er-Led Zeppelin-Atlantic
21	13.	Mind Games-John Lennon-Apple
18	14.	Mammy Blue-Stories-Kama Sutra
11	15.	Paper Roses-Marie Osmond-MGM
23	16.	Let Me Serenade You-Three Dog Night-Dunhill
13	17.	We May Never Pass This Way (Again)-Seals & Crofts-WB
16	18.	Be-Neil Diamond-Columbia
22	19.	Painted Ladies-Ian Thomas-Janus
24	20.	Leave Me Alone-Helen Reddy-Capitol
25	21.	Show And Tell-Al Wilson-Rocky Road
10	22.	All I Know-Art Garfunkel-Columbia
29	23.	Smokin' In The Boy's Room-Brownsville Station-Big Tree
15	24.	Yes We Can Can-The Pointer Sisters-Blue Thumb
31	25.	Helen Wheels-Paul McCartney-Apple
19	26.	Angie-Rolling Stones-Rolling Stones
32	27.	The Joker-Steve Miller Band-Capitol
34	28.	If You're Ready-The Staple Singers-Stax
33	29.	The Most Beautiful Girl-Charlie Rich-Epic
20	30.	Knockin' On Heaven's Door-Bob Dylan-Columbia
36	31.	Who's In The Strawberry Patch With Sally-Dawn-Bell
26	32.	Basketball Jones-Cheech Y Chong-Ode
37	33.	Tell Her She's Lovely-El Chicano-MCA
39	34.	Living For The City-Stevie Wonder-Tamla
27	35.	Free Ride-The Edgar Winter Group-

LARRY LUJACK
afternoons...on SUPER CFL!

		Epic
40	36.	Blue Collar-Bachman-Turner Overdrive-Mercury
--	37.	Rockin' Roll Baby-Stylistics-Avco
--	38.	Time In A Bottle-Jim Croce-ABC
28	39.	Rock 'N Roll-Kevin Johnson-Mainstream
30	40.	Rubber Bullets-10 C.C.-UK

wcfl wcfl wcfl wcfl wcfl wcfl wcfl wcfl wcfl wcfl wcfl wcfl

collector's issue number one hundred and forty-three

SUPER CFL SURVEY DEC 8

wcfl wcfl wcfl wcfl wcfl wcfl wcfl wcfl wcfl wcfl wcfl wcfl wcfl wcfl

SUPER ALBUMS

1. Goodbye Yellow Brick Road-Elton John-MCA
2. Jonathan Livingston Seagull-Neil Diamond-Columbia
3. Ringo-Ringo Starr-Apple
4. Quadrophenia-The Who-MCA
5. Goats Head Soup-The Rolling Stones-Rolling Stones
6. Life And Times-Jim Croce-ABC
7. Mind Games-John Lennon-Apple
8. The Joker-Steve Miller Band-Capitol
9. Los Cochinos @ #!!*-Cheech Y Chong-Ode
10. You Don't Mess Around With Jim-Jim Croce-ABC
11. Ol' Blue Eyes Is Back-Frank Sinatra-Reprise
12. The Singles-Carpenters-A & M
13. Brothers And Sisters-Allman Brothers Band-Capricorn
14. Sweet Freedom-Uriah Heep-Warner Brothers
15. Full Sail-Loggins And Messina-Columbia
16. Laid Back-Gregg Allman-Capricorn
17. Welcome-Santana-Columbia
18. Hat Trick-America-Warner Brothers
19. Live Dates-WishboneAsh-MCA
20. Wake Of The Flood-Grateful Dead-Grateful Dead

wcfl wcfl wcfl wcfl wcfl

World Famous
TOM MURPHY

WAS	IS	
3	1.	Hello It's Me-Todd Rundgren-Bearsville
4	2.	Top Of The World-Carpenters-A & M
2	3.	I Got A Name-Jim Croce-ABC
7	4.	Goodbye Yellow Brick Road-Elton John-MCA
9	5.	Space Race-Billy Preston-A & M
1	6.	Photograph-Ringo Starr-Apple
13	7.	Mind Games-John Lennon-Apple
12	8.	D'yer Mak'er-Zeppelin-Atlantic
14	9.	Mammy Blue-Stories-Kama Sutra
6	10.	Ooh Baby-Gilbert O'Sullivan-MAM
16	11.	Let Me Serenade You-Three Dog Night-Dunhill
5	12.	Just You 'N Me-Chicago-Columbia
19	13.	Painted Ladies-Ian Thomas-Janus
20	14.	Leave Me Alone-Helen Reddy-Capitol
10	15.	Heartbeat, It's A Lovebeat-DeFranco Family-20th Century
21	16.	Show And Tell-Al Wilson-Rocky Road
8	17.	Midnight Train To Georgia-Gladys Knight/Pips-Buddah
23	18.	Smokin' In The Boy's Room-Brownsville Station-Big Tree
27	19.	The Joker-Steve Miller Band-Capitol
25	20.	Helen Wheels-Paul McCartney-Apple
11	21.	Keep On Truckin'-Eddie Kendricks-Tamla
29	22.	The Most Beautiful Girl-Charlie Rich-Epic
15	23.	Paper Roses-Marie Osmond-MGM
28	24.	If You're Ready-The Staple Singers-Stax
18	25.	Be-Neil Diamond-Columbia
17	26.	We May Never Pass This Way (Again)-Seals & Crofts-WB
31	27.	Who's In The Strawberry Patch With Sally-Dawn-Bell
34	28.	Living For The City-Stevie Wonder-Tamla
38	29.	Time In A Bottle-Jim Croce-ABC
33	30.	Tell Her She's Lovely-El Chicano-MCA
22	31.	All I Know-Art Garfunkel-Columbia
36	32.	Blue Collor-Bachman-Turner Overdrive-Mercury
24	33.	Yes We Can Can-The Pointer Sisters-Blue Thumb
37	34.	Rockin' Roll Baby-Stylistics-Avco
30	35.	Knockin' On Heaven's Door-Bob Dylan-Columbia
26	36.	Angie-Rolling Stones-Rolling Stones

mornings...on SUPER CFL!

--	37.	Let Me Be There-Olivia Newton-John-MCA
32	38.	Basketball Jones-Cheech & Chong-Ode
35	39.	Free Ride-The Edgar Winter Group-Epic
40	40.	Rubber Bullets-10 C.C.-UK

collector's issue number one hundred and forty-four

SUPER CFL SURVEY DEC 15

wcfl wcfl wcfl wcfl wcfl wcfl wcfl wcfl wcfl wcfl wcfl wcfl wcfl wcfl

WAS	IS	
2	1.	Top Of The World-Carpenters-A & M
1	2.	Hello It's Me-Todd Rundgren-Bearsville
4	3.	Goodbye Yellow Brick Road-Elton John-MCA
8	4.	D'yer Mak'er-Led Zeppelin-Atlantic
5	5.	Space Race-Billy Preston-A & M
7	6.	Mind Games-John Lennon-Apple
9	7.	Mammy Blue-Stories-Kama Sutra
11	8.	Let Me Serenade You-Three Dog Night-Dunhill
3	9.	I Got A Name-Jim Croce-ABC
19	10.	The Joker-Steve Miller Band-Capitol
13	11.	Painted Ladies-Ian Thomas-Janus
14	12.	Leave Me Alone-Helen Reddy-Capitol
18	13.	Smokin' In The Boy's Room-Brownsville Station-Big Tree
16	14.	Show And Tell-Al Wilson-Rocky Road
6	15.	Photograph-Ringo Starr-Apple
22	16.	The Most Beautiful Girl-Charlie Rich-Epic
20	17.	Helen Wheels-Paul McCartney-Apple
10	18.	Ooh Baby-Gilbert O'Sullivan- MAM
12	19.	Just You 'N Me-Chicago-Columbia
29	20.	Time In A Bottle-Jim Croce-ABC
24	21.	If You're Ready-The Staple Singers-Stax
27	22.	Who's In The Strawberry Patch With Sally-Dawn-Bell
15	23.	Heartbeat, It's A Lovebeat-DeFranco Family-20th Century
28	24.	Living For The City-Stevie Wonder-Tamla
30	25.	Tell Her She's Lovely-El Chicano-MCA
--	26.	One Tin Soldier-Coven-MGM/WB
17	27.	Midnight Train To Georgia-Gladys Knight/Pips-Buddah
34	28.	Rockin' Roll Baby-Stylistics-Avco
32	29.	Blue Collar-Bachman-Turner Overdrive-Mercury
--	30.	Sister Mary Elephant-Cheech & Chong-Ode
37	31.	Let Me Be There-Olivia Newton-John-MCA
23	32.	Paper Roses-Marie Osmond-MGM
--	33.	Never, Never Gonna Give Ya Up-Barry White-20th Century
25	34.	Be-Neil Diamond-Columbia
--	35.	You're Sixteen-Ringo Starr-Apple

SUPER ALBUMS

1. Goodbye Yellow Brick Road-Elton John-MCA
2. Quadrophenia-The Who-MCA
3. Jonathan Livingston Seagull-Neil Diamond-Columbia
4. Ringo-Ringo Starr-Apple
5. I Got A Name-Jim Croce-ABC
6. The Joker-Steve Miller Band-Capitol
7. The Singles-Carpenters-A & M
8. Brain Salad Surgery-Emerson, Lake & Palmer-Manticore
9. You Don't Mess Around With Jim-Jim Croce-ABC
10. Muscle Of Love-Alice Cooper-Warner Brothers
11. Mind Games-John Lennon-Apple
12. Life And Times-Jim Croce-ABC
13. Bette Midler-Better Midler-Atlantic
14. Los Cochinos @ #!!*-Cheech Y Chong-Ode
15. Goats Head Soup-The Rolling Stones-Rolling Stones
16. Sweet Freedom-Uriah Heep-Warner Brothers
17. The Serpent Is Rising-Styx-Wooden Nickel
18. Brothers And Sisters-Allman Brothers Band-Capricorn
19. Ol' Blue Eyes Is Back-Frank Sinatra-Reprise
20. The Smoker You Drink, The Player You Get-Joe Walsh-Dunhill

wcfl wcfl wcfl wcfl wcfl

World Famous TOM MURPHY mornings...on SUPER CFL!	21	36.	Keep On Truckin'-Eddie Kendricks-Tamla
	26	37.	We May Never Pass This Way (Again)-Seals & Crofts-WB
	--	38.	Love's Theme-Love Unlimited Orch.-20th Century
	--	39.	Walk Like A Man-Grand Funk-Grand Funk
	--	40.	Midnight Rider-Gregg Allman-Capricorn

wcfl wcfl wcfl wcfl wcfl wcfl wcfl wcfl wcfl wcfl wcfl wcfl

collector's issue number one hundred and forty-five

SUPER CFL SURVEY DEC 22

wcfl wcfl wcfl wcfl wcfl wcfl wcfl wcfl wcfl wcfl wcfl wcfl wcfl wcfl

WAS	IS	
1	1.	Top Of The World-Carpenters-A & M
6	2.	Mind Games-John Lennon-Apple
5	3.	Space Race-Billy Preston-A & M
10	4.	The Joker-Steve Miller Band-Capitol
11	5.	Painted Ladies-Ian Thomas-Janus
3	6.	Goodbye Yellow Brick Road-Elton John-MCA
14	7.	Show And Tell-Al Wilson-Rocky Road
8	8.	Let ME Serenade You-Three Dog Night-Dunhill
2	9.	Hello It's Me-Todd Rundgren-Bearsville
13	10.	Smokin' In The Boy's Room-Brownsville Station-Big Tree
12	11.	Leave Me Alone-Helen Reddy-Capitol
16	12.	The Most Beautiful Girl-Charlie Rich-Epic
20	13.	Time In A Bottle-Jim Croce-ABC
4	14.	D'yer Mak'er-Led Zeppelin-Atlantic
17	15.	Helen Wheels-Paul McCartney-Apple
7	16.	Mammy Blue-Stories-Kama Sutra
9	17.	I Got A Name-Jim Croce-ABC
21	18.	If You're Ready-The Staple Singers-Stax
24	19.	Living For The City-Stevie Wonder-Tamla
22	20.	Who's In The Strawberry Patch With Sally-Dawn-Bell
26	21.	One Tin Soldier-Coven MGM/WB
25	22.	Tell Her She's Lovely-El Chicano-MCA
30	23.	Sister Mary Elephant-Cheech & Chong-Ode
15	24.	Photograph-Ringo Starr-Apple
31	25.	Let Me Be There-Olivia Newton-John-MCA
28	26.	Rockin' Roll Baby-Stylistics-Avco
29	27.	Blue Collar-Bachman-Turner Overdrive-Mercury
18	28.	Ooh Baby-Gilbert O'Sullivan-MAM
33	29.	Never, Never Gonna Give Ya Up-Barry White-20th Century
35	30.	You're Sixteen-Ringo Starr-Apple
19	31.	Just You 'N Me-Chicago-Columbia
23	32.	Heartbeat, It's A Lovebeat-DeFranco Family-20th Century
--	33.	The Way We Were-Barbra Streisand-Columbia
40	34.	Midnight Rider-Gregg Allman-Capricorn
39	35.	Walk Like A Man-Grand Funk-Grand Funk

SUPER ALBUMS

1. Goodbye Yellow Brick Road-Elton John-MCA
2. I Got A Name-Jim Croce-ABC
3. Jonathan Livingston Seagull-Neil Diamond-Columbia
4. Quadrophenia-The Who-MCA
5. The Singles-Carpenters-A & M
6. The Joker-Steve Miller Band-Capitol
7. Ringo-Ringo Starr-Apple
8. Brian Salad Surgery-Emerson, Lake & Palmer-Manticore
9. You Don't Mess Around With Jim-Jim Croce-ABC
10. Life And Times-Jim Croce-ABC
11. Los Cochinos @ #!!*-Cheech Y Chong-Ode
12. Bette Midler-Bette Midler-Atlantic
13. Muscle Of Love-Alice Cooper-Warner Brothers
14. Mind Games-John Lennon-Apple
15. Sweet Freedom-Uriah Heep-Warner Brothers
16. Bright White-Shawn Phillips-A & M
17. Brothers And Sisters-Allman Brothers Band-Capricorn
18. Live Dates-Wishbone Ash-MCA
19. Goats Head Soup-The Rolling Stones-Rolling Stones
20. Welcome-Santana-Columbia

wcfl wcfl wcfl wcfl wcfl

World Famous

TOM MURPHY

mornings...on SUPER CFL!

38	36.	Love's Theme-Love Unlimited Orch.-20th Century
27	37.	Midnight Train To Georgia-Gladys Knight/Pips-Buddah
32	38.	Paper Roses-Marie Osmond-MGM
--	39.	Teenage Lament '74-Alice Cooper-WB
34	40.	Be-Neil Diamond-Columbia

collector's issue number one hundred and forty-six

SUPER CFL SURVEY DEC 29

wcfl wcfl wcfl wcfl wcfl wcfl wcfl wcfl wcfl wcfl wcfl wcfl wcfl wcfl

SUPER ALBUMS

1. I Got A Name-Jim Croce-ABC
2. The Singles-Carpenters-A & M
3. Goodbye Yellow Brick Road-Elton John-MCA
4. Jonathan Livingston Seagull-Neil Diamond-Columbia
5. Quadrophenia-The Who-MCA
6. The Joker-Steve Miller Band-Capitol
7. Brain Salad Surgery-Emerson, Lake & Palmer-Manticore
8. You Don't Mess Around With Jim-Jim Croce-ABC
9. Muscle Of Love-Alice Cooper-Warner Brothers
10. Ringo-Ringo Starr-Apple
11. Bette Midler-Bette Midler-Atlantic
12. "Billy Jack"-Sound Track-Warner Brothers
13. Los Cochinos @ #!!*-Cheech Y Chong-Ode
14. Life And Times-Jim Croce-ABC
15. The Serpent Is Rising-Styx-Wooden Nickel
16. Mind Games-John Lennon-Apple
17. Band On The Run-Paul McCartney & Wings-Apple
18. Bright White-Shawn Phillips-A & M
19. "Sunshine"-Sound Track-MCA
20. Dylan-Bob Dylan-Columbia

wcfl wcfl wcfl wcfl wcfl

BOB DEARBORN

WAS	IS	
2	1.	Mind Games-John Lennon-Apple
4	2.	The Joker-Steve Miller Band-Capitol
1	3.	Top Of The World-Carpenters-A & M
5	4.	Painted Ladies-Ian Thomas-Janus
10	5.	Smokin' In The Boy's Room-Brownsville Station-Big Tree
7	6.	Show and Tell-Al Wilson-Rocky Road
13	7.	Time In A Bottle-Jim Croce-ABC
3	8.	Space Race-Billy Preston-A & M
21	9.	One Tin Soldier-Coven-MGM-WB
11	10.	Leave Me Alone-Helen Reddy-Capitol
6	11.	Goodbye Yellow Brick Road-Elton John-MCA
12	12.	The Most Beautiful Girl-Charlie Rich-Epic
23	13.	Sister Mary Elephant-Cheech & Chong-Ode
15	14.	Helen Wheels-Paul McCartney-Apple
9	15.	Hello It's Me-Todd Rundgren-Bearsville
18	16.	If You're Ready-The Staple Singers-Stax
19	17.	Living For The City-Stevie Wonder-Tamla
25	18.	Let Me Be There-Olivia Newton-John-MCA
8	19.	Let Me Serenade You-Three Dog Night-Dunhill
22	20.	Tell Her She's Lovelly-El Chicano-MCA
14	21.	D'yer Mak'er-Led Zeppelin-Atlantic
26	22.	Rockin' Roll Baby-Stylistics-Avco
30	23.	You're Sixteen-Ringo Starr-Apple
20	24.	Who's In The Strawberry Patch With Sally-Dawn-Bell
29	25.	Never, Never Gonna Give Ya Up-Barry White-20th Century
17	26.	I Got A Name-Jim Croce-ABC
36	27.	Love's Theme-Love Unlimited Orch.-20th Century
16	28.	Mammy Blue-Stories-Kama Sutra
34	29.	Midnight Rider-Gregg Allman-Capricorn
33	30.	The Way We Were-Barbra Stsreisand-Columbia
35	31.	Walk Line A Man-Grand Funk-Grand Funk
27	32.	Blue Collar-Bachman-Turner Overdrive-Mercury
24	33.	Photograph-Ringo Starr-Apple
39	34.	Teenage Lament'74-Alice Cooper-WB
--	35.	Abra-Ca-Dabra-DeFranko Family-20th Century
--	36.	I've Got To Use My Imagination-Gladys Knight/Pips-Buddah
28	37.	Ooh Baby-Gilbert O'Sullivan-MAM

...mid-day on SUPER CFL

--	38.	Jim Dandy-Black Oak Arkansas-Atco
31	39.	Just You 'N Me-Chicago-Columbia
--	40.	I Love-Tom T. Hall-Mercury

collector's issue number one hundred and forty-seven

SUPER HITS OF '73

wcfl wcfl wcfl wcfl wcfl wcfl wcfl wcfl wcfl wcfl wcfl wcfl wcfl wcfl

SUPER LP's OF '73

1. The Dark Side Of The Moon-Pink Floyd-Harvest
2. No Secrets-Carly Simon-Elektra
3. Catch Bull At Four-Cat Stevens-A & M
4. Houses Of The Holy-Led Zeppelin-Atlantic
5. Seventh Sojourn-Moody Blues-Deram
6. Chicago VI-Chicago-Columbia
7. Billion Dollar Babies-Alice Cooper-WB
8. They Only Come Out At Night-Edgar Winter Group-Epic
9. Brothers And Sisters-Allman Brothers Band-Capricorn
10. Los Cohinos @ #!!*J-Cheech Y Chong-Ode
11. Goats Head Soup-Rolling Stones-Rolling Stones
12. Living In The Past-Jethro Tull-Chrysalis
13. Deliverance Sound Track-Warner Brothers
14. We're An American Band-Grand Funk-Capitol
15. Don't Shoot Me, I'm Only The Piano Player-Elton John-MCA
16. A Passion Play-Jethro Tull-Chrysalis
17. Living In The Material World-George Harrison-Apple
18. Rocky Mountain High-John Denver-RCA
19. Foreigner-Cat Stevens-A & M
20. Tommy-Various Artists-Ode

1. You're So Vain-Carly Simon-Elektra
2. Heartbeat, It's A Lovebeat-DeFranco Family-20th Century
3. The Morning After-Maureen McGovern-20th Century
4. The Right Thing To Do-Carly Simon-Elektra
5. Little Willy-The Sweet-Bell
6. Half Breed-Cher-MCA
7. Monster Mash-Bobby (Boris) Pickett-Parrot
8. Crocodile Rock-Elton John-MCA
9. Frankenstein-The Edgar Winter Group-Epic
10. My Love-Paul McCartney-Apple
11. We're An American Band-Grand Funk-Capitol
12. Delta Dawn-Helen Reddy-Capitol
13. Lets Get It On-Marvin Gaye-Tamla
14. Tie A Yellow Ribbon-Dawn-Bell
15. Pillow Talk-Sylvia-Vibration
16. Angie-Rolling Stones-Rolling Stones
17. Top Of The World-Carpenters-A & M
18. Hocus Pocus-Focus-Sire
19. Killing Me Softly With His Song-Roberta Flack-Atlantic
20. The Night The Lights Went Out In Georgia-Vicki Lawrence-Bell
21. Last Song-Edward Bear-Capitol
22. Smoke On The Water-Deep Purple-WB
23. Sing-Carpenters-A & M
24. Mind Games-John Lennon-Apple
25. Yesterday Once More-Carpenters-A & M
26. Loves Me Like A Rock-Paul Simon-Columbia
27. Playground In My Mind-Clint Holmes-Epic
28. Kodachrome-Paul Simon-Columbia
29. Feeling Stronger Every Day-Chicago-Columbia
30. Danny's Song-Anne Murray-Capitol
31. Free Ride-The Edgar Winter Group-Epic
32. Give Me Love-George Harrison-Apple
33. Summer Breeze-Seals & Crofts-WB
34. Long Train Runnin'-Doobie Brothers-WB
35. Dancing In The Moonlight-King Harvest-Perception
36. Clair-Gilbert O'Sullivan-MAM

37. My Maria-B.W. Stevenson-RCA
38. Goodbye Yellow Brick Road-
    Elton John-MCA
39. Photograph-Ringo Star-Apple
40. Get Down-Gilbert O'Sullivan-MAM

1974

SUPER CFL SURVEY JAN 12

wcfl wcfl wcfl wcfl wcfl wcfl wcfl wcfl wcfl wcfl wcfl wcfl wcfl wcfl

WAS	IS	
2	1.	The Joker-Steve Miller Band-Capitol
5	2.	Smokin' In The Boy's Room-Brownsville Station-Big Tree
9	3.	One Tin Soldier-Coven-MGM-WB
13	4.	Sister Mary Elephant-Cheech & Chong-Ode
6	5.	Show And Tell-Al Wilson-Rocky Road
7	6.	Time In A Bottle-Jim Croce-ABC
1	7.	Mind Games-John Lennon-Apple
10	8.	Leave Me Alone-Helen Reddy-Capitol
12	9.	The Most Beautiful Girl-Charlie Rich-Epic
3	10.	Top Of The World-Carpenters-A & M
16	11.	If You're Ready-The Staple Singers-Stax
18	12.	Let Me Be There-Olivia Newton-John-MCA
23	13.	You're Sixteen-Ringo Starr-Apple
17	14.	Living For The City-Stevie Wonder-Tamla
14	15.	Helen Wheels-Paul McCartney-Apple
4	16.	Painted Ladies-Ian Thomas-Janus
22	17.	Rockin' Roll Baby-Stylistics-Avco
30	18.	The Way We Were-Barbra Streisand-Columbia
25	19.	Never, Never Gonna Give Ya Up-Barry White-20th Century
11	20.	Goodbye Yellow Brick Road-Elton John-MCA
27	21.	Love's Theme-Love Unlimited Orch.-20th Century
8	22.	Space Race-Billy Preston-A & M
29	23.	Midnight Rider-Gregg Allman-Capricorn
19	24.	Let Me Serenade You-Three Dog Night-Dunhill
31	25.	Walk Like A Man-Grand Funk-Grand Funk
15	26.	Hello It's Me-Todd Rundgren-Bearsville
34	27.	Teenage Lament '74-Alice Cooper-WB
20	28.	Tell Her She's Lovely-El Chicano-MCA
36	29.	I've Got To Use My Imagination-Gladys Knight/Pips-Buddah
--	30.	Americans-Byron MacGregor-Westbound
35	31.	Abra-Ca-Dabra-DeFranco Family-20th Century
21	32.	D'yer Mak'er-Led Zeppelin-Atlantic
38	33.	Jim Dandy-Black Oak Arkansas-Atco
26	34.	I Got A Name-Jim Croce-ABC
28	35.	Mammy Blue-Stories-Kama Sutra
40	36.	I Love-Tom T. Hall-Mercury

SUPER ALBUMS

1. I Got A Name-Jim Croce-ABC
2. You Don't Mess Around With Jim-Jim Croce-ABC
3. Goodbye Yellow Brick Road-Elton John-MCA
4. Jonathan Livingston Seagull-Neil Diamond-Columbia
5. The Singles-Carpenters-A & M
6. Brain Salad Surgery-Emerson, Lake & Palmer-Manticore
7. Quadrophenia-The Who-MCA
8. The Joker-Steve Miller Band-Capitol
9. Los Cochinos @ #!!*-Cheech Y Chong-Ode
10. Ringo-Ringo Starr-Apple
11. "Billy Jack"-Sound Track-Warner Brothers
12. Life And Times-Jim Croce-ABC
13. Muscle Of Love-Alive Cooper-Warner Brothers
14. Big Bambu-Cheech & Chong-Ode
15. Bright White-Shawn Phillips-A & M
16. Bette Midler-Bette Midler-Atlantic
17. Laid Back-Gregg Allman-Capricon
18. Band On The Run-Paul McCartney & Wings-Apple
19. Dylan-Bob Dylan-Columbia
20. Welcome-Santana-Columbia

wcfl wcfl wcfl wcfl wcfl

BOB DEARBORN	--	37.	Until You Come Back To Me-Aretha Franklin-Atlantic
...mid-day on SUPER CFL	24	38.	Who's In The Strawberry Patch With Sally-Dawn-Bell
	--	39.	Seasons In The Sun-Terry Jacks-Bell
	--	40.	Doo Doo Doo Doo Doo (Heartbreaker)-Rolling Stones-Rolling Stones

collector's issue number one hundred and forty-nine

SUPER CFL SURVEY JAN 19

wcfl wcfl wcfl wcfl wcfl wcfl wcfl wcfl wcfl wcfl wcfl wcfl wcfl wcfl

WAS	IS	
1	1.	The Joker-Steve Miller Band-Capitol
4	2.	Sister Mary Elephant-Cheech & Chong-Ode
3	3.	One Tin Soldier-Coven-MGM/WB
2	4.	Smokin' In The Boy's Room-Brownsville Station-Big Tree
12	5.	Let Me Be There-Olivia Newton-John-MCA
6	6.	Time In A Bottle-Jim Croce-ABC
13	7.	You're Sixteen-Ringo Starr-Apple
9	8.	The Most Beautiful Girl-Charlie Rich-Epic
5	9.	Show And Tell-Al Wilson-Rocky Road
14	10.	Living For The City-Stevie Wonder-Tamla
18	11.	The Way We Were-Barbra Streisand-Columbia
7	12.	Mind Games-John Lennon-Apple
21	13.	Love's Theme-Love Unlimited Orch.-20th Century
17	14.	Rockin' Roll Baby-Stylistics-Avco
19	15.	Never, Never Gonna Give Ya Up-Barry White-20th Century
10	16.	Top Of The World-Carpenters-A & M
30	17.	Americans-Byron MacGregor-Westbound
23	18.	Midnight Rider-Gregg Allman-Capricorn
8	19.	Leave Me Alone-Helen Reddy-Capitol
27	20.	Teenage Lament '74-Alice Cooper-WB
15	21.	Helen Wheels-Paul McCartney-Apple
25	22.	Walk Like A Man-Grand Funk-Grand Funk
11	23.	If You're Ready-The Staple Singers-Stax
31	24.	Abra-Ca-Dabra-DeFranco Family-20th Century
29	25.	I've Got To Use My Imagination-Gladys Knight/Pips-Buddah
20	26.	Goodbye Yellow Brick Road-Elton John-MCA
16	27.	Painted Ladies-Ian Thomas-Janus
33	28.	Jim Dandy-Black Oak Arkansas-Atco
22	29.	Space Race-Billy Preston-A & M
37	30.	Until You Come Back To Me-Aretha Franklin-Atlantic
39	31.	Seasons In The Sun-Terry Jacks-Bell
40	32.	Doo Doo Doo Doo Doo (Heartbreaker)-Rolling Stones-Rolling Stones
--	33.	Boogie Down-Eddie Kendricks-Tamla

SUPER ALBUMS

1. You Don't Mess Around With Jim-Jim Croce-ABC
2. The Singles-Carpenters-A & M
3. Goodbye Yellow Brick Road-Elton John-MCA
4. I Got A Name-Jim Croce-ABC
5. Jonathan Livingston Seagull-Neil Diamond-Columbia
6. Quadrophenia-The Who-MCA
7. The Joker-Steve Miller Band-Capitol
8. Brain Salad Surgery-Emerson, Lake & Palmer-Manticore
9. Muscle Of Love-Alice Cooper-Warner Brothers
10. Los Cochinos @ #!!*-Cheech Y Chong-Ode
11. Ringo-Ringo Starr-Apple
12. "Billy Jack"-Sound Track-Warner Brothers
13. Life And Times-Jim Croce-ABC
14. Bette Midler-Bette Midler-Atlantic
15. Big Bambu-Cheech & Chong-Ode
16. John Denver's Greatest Hits-RCA
17. Laid Back-Gregg Allman-Capricorn
18. Under The Influence Of Love Unlimited-Love Unlimited-20th Century
19. Stone Gon'-Barry White-20th Century
20. "American Graffiti"-Sound Track-MCA

36	34.	I Love-Tom T. Hall-Mercury
--	35.	Dark Lady-Cher-MCA
24	36.	Let Me Serenade You-Three Dog Night-Dunhill
--	37.	I Miss You-The Dells-Cadet
--	38.	Rock On-David Essex-Columbia
26	39.	Hello It's Me-Todd Rundgren-Bearsville
--	40.	In The Mood-Bette Midler-Atlantic

wcfl wcfl wcfl wcfl wcfl wcfl wcfl wcfl wcfl wcfl wcfl wcfl

collector's issue number one hundred and fifty

SUPER CFL SURVEY JAN 26

wcfl wcfl wcfl wcfl wcfl wcfl wcfl wcfl wcfl wcfl wcfl wcfl wcfl wcfl

SUPER ALBUMS

1. You Don't Mess Around With Jim-Jim Croce-ABC
2. Goodbye Yellow Brick Road-Elton John-MCA
3. I Got A Name-Jim Croce-ABC
4. Brain Salad Surgery-Emerson, Lake & Palmer-Manticore
5. The Singles-Carpenters-A & M
6. Quadrophenia-The Who-MCA
7. Jonathan Livingston Seagull-Neil Diamond-Columbia
8. John Denver's Greatest Hits-RCA
9. Ringo-Ringo Starr-Apple
10. The Joker-Steve Miller Band-Capitol
11. Los Cochinos @ #!!*-Cheech Y Chong-Ode
12. Life And Times-Jim Croce-ABC
13. Muscle Of Love-Alice Cooper-Warner Brothers
14. Laid Back-Gregg Allman-Capricorn
15. Band On The Run-Paul McCartney & Wings-Apple
16. Dylan-Bob Dylan-Columbia
17. "Billy Jack"-Sound Track-Warner Brothers
18. Bette Midler-Bette Midler-Atlantic
19. Under The Influence Of Love Unlimited-Love Unlimited-20th Century
20. "American Graffiti"-Sound Track-MCA

WAS	IS	
3	1.	One Tin Soldier-Coven-MGM/WB
2	2.	Sister Mary Elephant-Cheech & Chong-Ode
7	3.	You're Sixteen-Ringo Starr-Apple
5	4.	Let Me Be There-Olivia Newton-John-MCA
8	5.	The Most Beautiful Girl-Charlie Rich-Epic
1	6.	The Joker-Steve Miller Band-Capitol
6	7.	Time In A Bottle-Jim Croce-ABC
13	8.	Love's Theme-Love Unlimited Orch.-20th Century
11	9.	The Way We Were-Barbra Streisand-Columbia
17	10.	Americans-Byron MacGregor-Westbound
15	11.	Never, Never Gonna Give Ya Up-Barry White-20th Century
14	12.	Rockin' Roll Baby-Stylistics-Avco
4	13.	Smokin' In The Boy's Room-Brownsville Station-Big Tree
10	14.	Living For The City-Stevie Wonder-Tamla
18	15.	Midnight Rider-Gregg Allman-Capricorn
9	16.	Show And Tell-Al Wilson-Rocky Road
20	17.	Teenage Lament '74-Alice Cooper-WB
12	18.	Mind Games-John Lennon-Apple
22	19.	Walk Like A Man-Grand Funk-Grand Funk
24	20.	Abra-Ca-Dabra-DeFranco Family-20th Century
16	21.	Top Of The World-Carpenters-A & M
25	22.	I've Got To Use My Imagination-Gladys Knight/Pips-Buddah
19	23.	Leave Me Alone-Helen Reddy-Capitol
28	24.	Jim Dandy-Black Oak Arkansas-Atco
31	25.	Seasons In The Sun-Terry Jacks-Bell
30	26.	Until You Come Back To Me-Aretha Franklin-Atlantic
23	27.	If You're Ready-The Staple Singers-Stax
32	28.	Doo Doo Doo Doo Doo (Heartbreaker)-Rolling Stones-Rolling Stones
34	29.	I Love-Tom T. Hall-Mercury
33	30.	Boogie Down-Eddie Kendricks-Tamla
21	31.	Helen Wheels-Paul McCartney-Apple
35	32.	Dark Lady-Cher-MCA
38	33.	Rock On-David Essex-Columbia
26	34.	Goodbye Yellow Brick Road-Elton John-MCA

37	35.	I Miss You-The Dells-Cadet
--	36.	Last Kiss-Wednesday-Sussex
40	37.	In The Mood-Bette Midler-Atlantic
--	38.	Sunshine On My Shoulders-John Denver-RCA
27	39.	Painted Ladies-Ian Thomas-Janus
29	40.	Space Race-Billy Preston-A & M

collector's issue number one hundred and fifty-one

SUPER CFL SURVEY FEB 2

wcfl wcfl wcfl wcfl wcfl wcfl wcfl wcfl wcfl wcfl wcfl wcfl wcfl wcfl

WAS	IS	
3	1.	You're Sixteen-Ringo Starr-Apple
4	2.	Let Me Be There-Olivia Newton-John-MCA
1	3.	One Tin Soldier-Coven-MGM/WB
9	4.	The Way We Were-Barbra Streisand-Columbia
8	5.	Love's Theme-Love Unlimited Orch.-20th Century
2	6.	Sister Mary Elephant-Cheech & Chong-Ode
11	7.	Never, Never Gonna Give Ya Up-Barry White-20th Century
10	8.	Americans-Byron MacGregor-Westbound
25	9.	Seasons In The Sun-Terry Jacks-Bell
6	10.	The Joker-Steve Miller Band-Capitol
5	11.	The Most Beautiful Girl-Charlie Rich-Epic
20	12.	Abra-Ca-Dabra-DeFranco Family-20th Century
17	13.	Teenage Lament '74-Alice Cooper-WB
12	14.	Rockin' Roll Baby-Stylistics-Avco
15	15.	Midnight Rider-Gregg Allman-Capricorn
19	16.	Walk Like A Man-Grand Funk-Grand Funk
24	17.	Jim Dandy-Black Oak Arkansas-Atco
7	18.	Time In A Bottle-Jim Croce-ABC
22	19.	I've Got To Use My Imagination-Gladys Knight/Pips-Buddah
13	20.	Smokin' In The Boy's Room-Brownsville Station-Big Tree
30	21.	Boogie Down-Eddie Kendricks-Tamla
14	22.	Living For The City-Stevie Wonder-Tamla
26	23.	Until You Come Back To Me-Aretha Franklin-Atlantic
28	24.	Doo Doo Doo Doo Doo (Heartbreaker)-Rolling Stones-Rolling Stones
32	25.	Dark Lady-Cher-MCA
16	26.	Show And Tell-Al Wilson-Rocky Road
29	27.	I Love-Tom T. Hall-mercury
33	28.	Rock ON-David Essex-Columbia
18	29.	Mind Games-John Lennon-Apple
35	30.	I Miss You-The Dells-Cadet
37	31.	In The Mood-Bette Midler-Atlantic
36	32.	Last Kiss-Wednesday-Sussex
21	33.	Top Of The World-Carpenters-A & M
38	34.	Sunshine On My Shoulders-John Denver-RCA
23	35.	Leave Me Alone-Helen Reedy-Capitol

SUPER ALBUMS

1. You Don't Mess Around With Jim-Jim Croce-ABC
2. I Got A Name-Jim Croce-ABC
3. Brain Salad Surgery-Emerson, Lake & Palmer-Manticore
4. Goodbye Yellow Brick Road-Elton John-MCA
5. The Singles-Carpenters-A & M
6. Quadrophenia-The Who-MCA
7. Sabbath, Bloody Sabbath-Black Sabbath-Warner Brothers
8. John Denver's Greatest Hits-RCA
9. The Joker-Steve Miller Band-Capitol
10. Muscle Of Love-Alice Cooper-Warner Brothers
11. Under The Influence Of Love Unlimited-Love Unlimited-20th Century
12. Los Cochinos @ #!!*-Cheech Y Chong-Ode
13. Laid Back-Gregg Allman-Capricorn
14. Ringo-Ringo Starr-Apple
15. "American Graffiti"-Sound Track-MCA
16. "Billy Jack"-Sound Track-Warner Brothers
17. Jonathan Livingston Seagull-Neil Diamond-Columbia
18. Band On The Run-Paul McCartney & Wings-Apple
19. Big Bambu-Cheech & Chong-Ode
20. Bette Midler-Bette Midler-Atlantic

--	36. Jungle Boogie-Kool/Gang-De-Lite
--	37. Love Song-Anne Murray-Capitol
--	38. Can This Be Real-Natural Four-Curtom
27	39. If You're Ready-The Staple Singers-Stax
31	40. Helen Wheels-Paul McCartney-Apple

wcfl wcfl wcfl wcfl wcfl wcfl wcfl wcfl wcfl wcfl wcfl wcfl

collector's issue number one hundred and fifty-three

SUPER CFL SURVEY FEB 9

wcfl wcfl wcfl wcfl wcfl wcfl wcfl wcfl wcfl wcfl wcfl wcfl wcfl wcfl

SUPER ALBUMS

1. You Don't Mess Around With Jim-Jim Croce-ABC
2. Brain Salad Surgery-Emerson, Lake & Palmer-Manticore
3. John Denver's Greatest Hits-RCA
4. Goodbye Yellow Brick Road-Elton John-MCA
5. The Singles-Carpenters-A & M
6. I Got A Name-Jim Croce-ABC
7. Quadrophenia-The Who-MCA
8. The Joker-Steve Miller Band-Capitol
9. Muscle Of Love-Alice Cooper-Warner Brothers
10. Band On The Run-Paul McCartney & Wings-Apple
11. Sabbath, Bloody Sabbath-Black Sabbath-Warner Brothers
12. Under The Influence Of Love Unlimited-Love Unlimited-20th Century
13. Laid Back-Gregg Allman-Capricorn
14. Hotcakes-Carly Simon-Elektra
15. "American Graffiti"-Sound Track-MCA
16. Los Cochinos @ #!!*-Cheech Y Chong-Ode
17. Jonathan Livingston Seagull-Neil Diamond-Columbia
18. Tales From Topographic Oceans-Yes-Atlantic
19. Ringo-Ringo Starr-Apple
20. Planet Waves-Bob Dylan-Asylum

WAS	IS	
4	1.	The Way We Were-Barbra Streisand-Columbia
5	2.	Love's Theme-Love Unlimited Orch.-20th Century
1	3.	You're Sixteen-Ringo Starr-Apple
9	4.	Seasons In The Sun-Terry Jacks-Bell
7	5.	Never, Never Gonna Give Ya Up-Barry White-20th Century
2	6.	Let Me Be There-Olivia Newton-John-MCA
3	7.	One Tin Soldier-Coven-MGM-WB
12	8.	Abra-Ca-Dabra-DeFranco Family-20th Century
13	9.	Teenage Lament '74-Alice Cooper-WB
19	10.	I've Got To Use My Imagination-Gladys Knight/Pips-Buddah
6	11.	Sister Mary Elephant-Cheech & Chong-Ode
17	12.	Jim Dandy-Black Oak Arkansas-Atco
25	13.	Dark Lady-Cher-MCA
10	14.	The Joker-Steve Miller Band-Capitol
21	15.	Boogie Down-Eddie Kendricks-Tamla
11	16.	The Most Beautiful Girl-Charlie Rich-Epic
28	17.	Rock On-David Essex-Columbia
8	18.	Americans-Byron MacGregor-Westbound
23	19.	Until You Come Back To Me-Aretha Franklin-Atlantic
14	20.	Rockin' Roll Baby-Stylistics-Avco
18	21.	Time In A Bottle-Jim Croce-ABC
24	22.	Doo Doo Doo Doo Doo (Heartbreaker)-Rolling Stones-Rolling Stones
15	23.	Midnight Rider-Gregg Allman-Capricorn
34	24.	Sunshine On My Shoulders-John Denver-RCA
32	25.	Last Kiss-Wednesday-Sussex
27	26.	I Love-Tom T. Hall-Mercury
30	27.	I Miss You-The Dells-Cadet
16	28.	Walk Like A Man-Grand Funk-Grand Funk
31	29.	In The Mood-Bette Midler-Atlantic
36	30.	Jungle Boogie-Kool/Gang-De-Lite
20	31.	Smokin' In The Boy's Room-Brownsville Station-Big Tree
37	32.	Love Song-Anne Murray-Capitol
22	33.	Living For The City-Stevie Wonder-Tamla
38	34.	Can This Be Real-Natural Four-Curtom
--	35.	Sexy Mama-Moments-Stang
26	36.	Show And Tell-Al Wilson-Rocky

		Road
--	37.	Last Time I Saw Him-Diana Ross-Motown
33	38.	Top Of The World-Carpenters-A & M
--	39.	Mockingbird-Carly Simon & James Taylor-Elektra
29	40.	Mind Games-John Lennon-Apple

collector's issue number one hundred and fifty-four

SUPER CFL SURVEY FEB 16

wcfl wcfl wcfl wcfl wcfl wcfl wcfl wcfl wcfl wcfl wcfl wcfl wcfl wcfl

SUPER ALBUMS

1. You Don't Mess Around With Jim-Jim Croce-ABC
2. Brain Salad Surgery-Emerson-Lake & Palmer-Manticore
3. John Denver's Greatest Hits-RCA
4. Sabbath, Bloody Sabbath-Black Sabbath-Warner Brothers
5. I Got A Name-Jim Croce-ABC
6. Court And Spark-Joni Mitchell-Asylum
7. Under The Influence Of Love Unlimited-Love Unlimited-20th Century
8. The Singles-Carpenters-A & M
9. Hotcakes-Carly Simon-Elektra
10. Planet Waves-Bob Dylan-Asylum
11. Goodbye Yellow Brick Road-Elton John-MCA
12. Muscle Of Love-Alice Cooper-Warner Brothers
13. "The Way We Were"-Sound Track-Columbia
14. Tales From Topographic Oceans-Yes-Atlantic
15. "American Graffiti"-Sound Track-MCA
16. Laid Back-Gregg Allman-Capricorn
17. Band On The Run-Paul McCartney & Wings-Apple
18. The Joker-Steve Miller Band-Capitol
19. Quadrophenia-The Who-MCA
20. Ringo-Ringo Starr-Apple

WAS	IS	
1	1.	The Way We Were-Barbra Streisand-Columbia
4	2.	Seasons In The Sun-Terry Jacks-Bell
2	3.	Love's Theme-Love Unlimited Orch.-20th Century
8	4.	Abra-Ca-Dabra-DeFranco Family-20th Century
3	5.	You're Sixteen-Ringo Starr-Apple
10	6.	I've Got To Use My Imagination-Gladys Knight/Pips-Buddah
12	7.	Jim Dandy-Black Oak Arkansas-Atco
5	8.	Never, Never Gonna Give Ya Up-Barry White-20th Century
15	9.	Boogie Down-Eddie Kendricks-Tamla
13	10.	Dark Lady-Cher-MCA
6	11.	Let Me Be There-Olivia Newton-John-MCA
17	12.	Rock On-David Essex-Columbia
9	13.	Teenage Lament '74-Alice Cooper-WB
24	14.	Sunshine On My Shoulders-John Denver-RCA
7	15.	One Tin Soldier-Coven-MGM/WB
25	16.	Last Kiss-Wednesday-Sussex
19	17.	Until You Come Back To Me-Aretha Franklin-Atlantic
22	18.	Doo Doo Doo Doo Doo (Heartbreaker)-Rolling Stones-Rolling Stones
14	19.	The Joker-Steve Miller Band-Capitol
30	20.	Jungle Boogie-Kool/Gang-De-Lite
16	21.	The Most Beautiful Girl-Charlie Rich-Epic
11	22.	Sister Mary Elephant-Cheech & Chong-Ode
32	23.	Love Song-Anne Murray-Capitol
29	24.	In The Mood-Bette Midler-Atlantic
27	25.	I Miss You-The Dells-Cadet
26	26.	I Love-Tom T. Hall-Mercury
21	27.	Time In A Bottle-Jim Croce-ABC
34	28.	Can This Be Real-Natural Four-Curtom
20	29.	Rockin' Roll Baby-Stylistics-Avco
23	30.	Midnight Rider-Gregg Allman-Capricorn
35	31.	Sexy Mama-Moments-Stang
37	32.	Last Time I Saw Him-Diana Ross-Motown
18	33.	Americans-Byron MacGregor-Westbound
39	34.	Mockingbird-Carly Simon & James Taylor-Elektra
--	35.	Energy Crisis '74-Dickie Goodman-Rainy Wednesday
--	36.	Jet-Paul McCartney-Apple
--	37.	Bennie And The Jets-Elton John-MCA

28	38.	Walk Like A Man-Grand Funk-Grand Funk
--	39.	I Like To Live The Love-B.B. King-ABC
--	40.	Hooked On A Feeling-Blue Swede-EMI

wcfl wcfl wcfl wcfl wcfl wcfl wcfl wcfl wcfl wcfl wcfl wcfl

collector's issue number one hundred and fifty-five

SUPER CFL SURVEY FEB 23

wcfl wcfl wcfl wcfl wcfl wcfl wcfl wcfl wcfl wcfl wcfl wcfl wcfl wcfl

SUPER ALBUMS

1. Court And Spark-Joni Mitchell-Asylum
2. Sabbath, Bloody Sabbath-Black Sabbath-Warner Brothers
3. You Don't Mess Around With Jim-Jim Croce-ABC
4. Brain Salad Surgery-Emerson, Lake & Palmer-Manticore
5. John Denver's Greatest Hits-RCA
6. Hotcakes-Carly Simon-Elektra
7. Planet Waves-Bob Dylan-Asylum
8. I Got A Name-Jim Croce-ABC
9. Tales From Topographic Oceans-Yes-Atlantic
10. The Singles-Carpenters-A & M
11. Under The Influence Of Love Unlimited-Love Unlimited-20th Century
12. The Way We Were-Barbra Streisand-Columbia
13. Laid Back-Gregg Allman-Capricorn
14. "The Way We Were"-Sound Track-Columbia
15. The Joker-Steve Miller Band-Capitol
16. Quadrophenia-The Who-MCA
17. Muscle of Love-Alice Cooper-Warner Brothers
18. Goodbye Yellow Brick Road-Elton John-MCA
19. Band On The Run-Paul McCartney & Wings-Apple
20. "American Graffiti"-Sound Track-MCA

WAS	IS	
1	1.	The Way We Were-Barbra Streisand-Columbia
2	2.	Seasons In The Sun-Terry Jacks-Bell
4	3.	Abra-Ca-Dabra-DeFranco Family-20th Century
10	4.	Dark Lady-Cher-MCA
3	5.	Love's Theme-Love Unlimited Orch.-20th Century
7	6.	Jim Dandy-Black Oak Arkansas-Atco
12	7.	Rock On-David Essex-Columbia
9	8.	Boogie Down-Eddie Kendricks-Tamla
14	9.	Sunshine On My Shoulders-John Denver-RCA
6	10.	I've Got To Use My Imagination-Gladys Knight/Pips-Buddah
5	11.	You're Sixteen-Ringo Starr-Apple
16	12.	Last Kiss-Wednesday-Sussex
8	13.	Never, Never Gonna Give Ya Up-Barry White-20th Century
17	14.	Until You Come Back To Me-Aretha Franklin-Atlantic
23	15.	Love Song-Anne Murray-Capitol
18	16.	Doo Doo Doo Doo Doo (Heartbreaker)-Rolling Stones-Rolling Stones
13	17.	Teenage Lament '74-Alice Cooper-WB
20	18.	Jungle Boogie-Kool/Gang-De-Lite
11	19.	Let Me Be There-Olivia Newton-John-MCA
25	20.	I Miss You-The Dells-Cadet
15	21.	One Tin Soldier-Coven-MGM/WB
31	22.	Sexy Mama-Moments-Stang
32	23.	Last Time I Saw Him-Diana Ross-Motown
28	24.	Can This Be Real-Natural Four-Curtom
34	25.	Mockingbird-Carly Simon & James Taylor-Elektra
21	26.	The Most Beautiful Girl-Charlie Rich-Epic
24	27.	In The Mood-Bette Midler-Atlantic
26	28.	I Love-Tom T. Hall-Mercury
39	29.	Jet-Paul McCartney-Apple
35	30.	Energy Crisis '74-Dickie Goodman-Rainy Wednesday
19	31.	The Joker-Steve Miller Band-Capitol
22	32.	Sister Mary Elephant-Cheech & Chong-Ode
39	33.	I Like To Live The Love-B.B. King-ABC
37	34.	Bennie And The Jets-Elton John-MCA
27	35.	Time In A Bottle-Jim Croce-ABC
40	36.	Hooked On A Feeling-Blue Swede-

		EMI
--	37.	Eres Tu (Touch The Wind)- Mocedades-Tara
29	38.	Rockin' Roll Baby-Stylistics-Avco
--	39.	Rock And Roll, Hoochie Koo-Rick Derringer-Blue Sky
--	40.	Trying To Hold On To My Woman- Lamont Dozier-ABC

collector's issue number one hundred and fifty-six

SUPER CFL SURVEY MAR 2

wcfl wcfl wcfl wcfl wcfl wcfl wcfl wcfl wcfl wcfl wcfl wcfl wcfl wcfl

WAS	IS	
2	1.	Seasons In The Sun-Terry Jacks-Bell
4	2.	Dark Lady-Cher-MCA
9	3.	Sunshine On My Shoulders-John Denver-RCA
1	4.	The Way We Were-Barbra Streisand-Columbia
6	5.	Jim Dandy-Black Oak Arkansas-Atco
7	6.	Rock On-David Essex-Columbia
3	7.	Abra-Ca-Dabra-DeFranco Family-20th Century
8	8.	Boogie Down-Eddie Kendricks-Tamla
12	9.	Last Kiss-Wednesday-Sussex
15	10.	Love Song-Anne Murray-Capitol
5	11.	Love's Theme-Love Unlimited Orch.-20th Century
18	12.	Jungle Boogie-Kool/Gang-De-Lite
11	13.	You're Sixteen-Ringo Starr-Apple
30	14.	Energy Crisis '74-Dickie Goodman-Rainy Wednesday
10	15.	I've Got To Use My Imagination-Gladys Knight/Pips-Buddah
16	16.	Doo Doo Doo Doo Doo (Heartbreaker)-Rolling Stones-Rolling Stones
22	17.	Sexy Mama-Moments-Stang
24	18.	Can This Be Real-Natural Four-Curtom
25	19.	Mockingbird-Carly Simon & James Taylor-Elektra
29	20.	Jet-Paul McCartney-Apple
23	21.	Last Time I Saw Him-Diana Ross-Motown
14	22.	Until You Come Back To Me-Aretha Franklin-Atlantic
13	23.	Never, Never Gonna Give Ya Up-Barry White-20th Century
19	24.	Let Me Be There-Olivia Newton-John-MCA
36	25.	Hooked On A Feeling-Blue Swede-EMI
20	26.	I Miss You-The Dells-Cadet
34	27.	Bennie And The Jets-Elton John-MCA
17	28.	Teenage Lament '74-Alice Cooper-WB
--	29.	Spiders & Snakes-Jim Stafford-MGM
33	30.	I Like To Live The Love-B.B. King-ABC
21	31.	One Tin Soldier-Coven-MGM/WB
37	32.	Eres Tu (Touch The Wind)-Mocedades-Tara
26	33.	The Most Beautiful Girl-Charlie Rich-Epic

SUPER ALBUMS

1. John Denver's Greatest Hits-RCA
2. The Way We Were-Barbra Streisand-Columbia
3. Hotcakes-Carly Simon-Elektera
4. Court And Spark-Joni Mitchell-Asylum
5. Sabbath, Bloody Sabbath-Black Sabbath-Warner Brothers
6. Planet Waves-Bob Dylan-Asylum
7. You Don't Mess Around With Jim-Jim Croce-ABC
8. Brain Salad Surgery-Emerson, Lake & Palmer-Manticore
9. Under The Influence Of Love Unlimited-Love Unlimited-20th Century
10. Tales From Topographic Oceans-Yes-Atlantic
11. I Got A Name-Jim Croce-ABC
12. Laid Back-Gregg Allman-Capricorn
13. "The Way We Were"-Sound Track-Columbia
14. Band On The Run-Paul McCartney & Wings-Apple
15. Goodbye Yellow Brick Road-Elton John-MCA
16. Tubular Bells-Mike Oldfield-Virgin
17. The Singles-Carpenters-A & M
18. "American Graffiti"-Sound Track-MCA
19. Quadrophenia-The Who-MCA
20. Love Is The Message-MFSB-Philadelphia International

40	34.	Trying To Hold On To My Woman-Lamont Dozier-ABC
--	35.	The Lord's Prayer-Sister Janet Mead-A & M
39	36.	Rock And Roll, Hoochie Koo-Rick Derringer-Blue Sky
27	37.	In The Mood-Bette Midler-Atlantic
--	38.	Just Don't Want To Be Lonely-The Main Ingredient-RCA
--	39.	Virginia-Bill Amesbury-Casablanca
28	40.	I Love-Tom T. Hall-Mercury

wcfl wcfl wcfl wcfl wcfl wcfl wcfl wcfl wcfl wcfl wcfl wcfl

collector's issue number one hundred and fifty-seven

SUPER CFL SURVEY MAR 9

wcfl wcfl wcfl wcfl wcfl wcfl wcfl wcfl wcfl wcfl wcfl wcfl wcfl wcfl

WAS	IS	
1	1.	Seasons In The Sun-Terry Jacks-Bell
3	2.	Sunshine On My Shoulders-John Denver-RCA
6	3.	Rock On-David Essex-Columbia
2	4.	Dark Lady-Cher-MCA
8	5.	Boogie Down-Eddie Kendricks-Tamla
10	6.	Love Song-Anne Murray-Capitol
9	7.	Last Kiss-Wednesday-Sussex
4	8.	The Way We Were-Barbra Streisand-Columbia
19	9.	Mockingbird-Carly Simon & James Taylor-Elektra
12	10.	Jungle Boogie-Kool/Gang-De-Lite
5	11.	Jim Dandy-Black Oak Arkansas-Atco
14	12.	Energy Crisis '74-Dickie Goodman-Rainy Wednesday
17	13.	Sexy Mama-Moments-Stang
7	14.	Abra-Ca-Dabra-DeFranco Family-20th Century
11	15.	Love's Theme-Love Unlimited Orch.-20th Century
21	16.	Last Time I Saw Him-Diana Ross-Motown
16	17.	Doo Doo Doo Doo Doo (Heartbreaker)-Rolling Stones-Rolling Stones
18	18.	Can This Be Real-Natural Four-Curtom
20	19.	Jet-Paul McCartney-Apple
27	20.	Bennie And The Jets-Elton John-MCA
25	21.	Hooked On A Feeling-Blue Swede-EMI
15	22.	I've Got To Use My Imagination-Gladys Knight/Pips-Buddah
13	23.	You're Sixteen-Ringo Starr-Apple
32	24.	Eres Tu (Touch The Wind)-Mocedades-Tara
29	25.	Spiders & Snakes-Jim Stafford-MGM
30	26.	I Like To Live The Love-B.B. King-ABC
22	27.	Until You Come Back To Me-Aretha Franklin-Atlantic
36	28.	Rock And Roll, Hoochie Koo-Rick Derringer-Blue Sky
23	29.	Never, Never Gonna Give Ya Up-Barry White-20th Century
35	30.	The Lord's Prayer-Sister Janet Mead-A & M
34	31.	Trying To Hold On To My Woman-Lamont Dozier-ABC
24	32.	Let Me Be There-Olivia Newton-John-MCA
38	33.	Just Don't Want To Be Lonely-The

SUPER ALBUMS

1. The Way We Were-Barbra Streisand-Columbia
2. Court And Spark-Joni Mitchell-Asylum
3. John Denver's Greatest Hits-RCA
4. Hotcakes-Carly Simon-Elektra
5. Sabbath, Bloody Sabbath-Black Sabbath-Warner Brothers
6. Planet Waves-Bob Dylan-Asylum
7. Brain Salad Surgery-Emerson, Lake & Palmer-Manticore
8. You Don't Mess Around With Jim-Jim Croce-ABC
9. Goodbye Yellow Brick Road-Elton John-MCA
10. "American Graffiti"-Sound Track-MCA
11. I Got A Name-Jim Croce-ABC
12. Tales From Topographic Oceans-Yes-Atlantic
13. Band On The Run-Paul McCartney & Wings-Apple
14. Under The Influence Of Love Unlimited-Love Unlimited-20th Century
15. Tubular Bells-Mike Oldfield-Virgin
16. Burn-Deep Purple-Warner Brothers
17. "The Way We Were"-Sound Track-Columbia
18. Laid Back-Gregg Allman-Capricorn
19. Unborn Child-Seals & Crofts-Warner Brothers
20. All American Boy-Rick Derringer-Blue Sky

		Main Ingredient-RCA
26	34.	I Miss You-The Dells-Cadet
39	35.	Virginia-Bill Amesbury-Casablanca
28	36.	Teenage Lament '74-Alice Cooper-WB
--	37.	I'll Have To Say I Love You In A Song-Jim Croce-ABC
--	38.	Oh My My-Ringo Starr-Apple
--	39.	The Loco-motion-Grand Funk-Grand Funk
--	40.	Let It Ride-Bachman-Turner Overdrive-Mercury

collector's issue number one hundred and fifty-eight

SUPER CFL SURVEY MAR 16

wcfl wcfl wcfl wcfl wcfl wcfl wcfl wcfl wcfl wcfl wcfl wcfl wcfl wcfl

WAS	IS	
1	1.	Seasons In The Sun-Terry Jacks-Bell
2	2.	Sunshine On My Shoulders-John Denver-RCA
3	3.	Rock On-David Essex-Columbia
9	4.	Mockingbird-Carly Simon & James Taylor-Elektra
7	5.	Last Kiss-Wednesday-Sussex
6	6.	Love Song-Anne Murray-Capitol
4	7.	Dark Lady-Cher-MCA
10	8.	Jungle Boogie-Kool/Gang-De-Lite
5	9.	Boogie Down-Eddie Kendricks-Tamla
21	10.	Hooked On A Feeling-Blue Swede-EMI
13	11.	Sexy Mama-Moments-Stang
19	12.	Jet-Paul McCartney-Apple
8	13.	The Way We Were-Barbra Streisand-Columbia
20	14.	Bennie And The Jets-Elton John-MCA
16	15.	Last Time I Saw Him-Diana Ross-Motown
25	16.	Spiders & Snakes-Jim Stafford-MGM
12	17.	Energy Crisis '74-Dickie Goodman-Rainy Wednesday
24	18.	Eres Tu (Touch The Wind)-Mocedades-Tara
11	19.	Jim Dandy-Black Oak Arkansas-Atco
30	20.	The Lord's Prayer-Sister Janet Mead-A & M
17	21.	Doo Doo Doo Doo Doo (Heartbreaker)-Rolling Stones-Rolling Stones
15	22.	Love's Theme-Love Unlimited Orch.-20th Century
26	23.	I Like To Live The Love-B.B. King-ABC
28	24.	Rock And Roll, Hoochie Koo-Rick Derringer-Blue Sky
14	25.	Abra-Ca-Dabra-DeFranco Family-20th Century
18	26.	Can This Be Real-Natural Four-Curtom
31	27.	Trying To Hold On To My Woman-Lamont Dozier-ABC
22	28.	I've Got To Use My Imagination-Gladys Knight/Pips-Buddah
33	29.	Just Don't Want To Be Lonely-The Main Ingredient-RCA
35	30.	Virginia-Bill Amesbury-Casablanca
23	31.	You're Sixteen-Ringo Starr-Apple
37	32.	I'll Have To Say I Love You In A Song-Jim Croce-ABC
--	33.	Best Thing That Ever Happened To Me-Gladys Knight/Pips-Buddah

SUPER ALBUMS

1. The Way We Were-Barbra Streisand-Columbia
2. John Denver's Greatest Hits-RCA
3. Court And Spark-Joni Mitchell-Asylum
4. Hotcakes-Carly Simon-Elektra
5. Tubular Bells-Mike Oldfield-Virgin
6. Planet Waves-Bob Dylan-Asylum
7. Sabbath, Bloody Sabbath-Black Sabbath-Warner Brothers
8. Brain Salad Surgery-Emerson, Lake & Palmer-Manticore
9. Burn-Deep Purple-Warner Brothers
10. Unborn Child-Seals & Crofts-Warner Brothers
11. Tales From Topographic Oceans-Yes-Atlantic
12. Goodbye Yellow Brick Road-Elton John-MCA
13. Saints & Sinners-Johnny Winter-Columbia
14. Band On The Run-Paul McCartney & Wings-Apple
15. You Don't Mess Around With Jim-Jim Croce-ABC
16. Under The Influence Of Love Unlimited-Love Unlimited-20th Century
17. Love Is The Message-MFSB-Philadelphia International
18. "American Graffiti"-Sound Track-MCA
19. Rock N Roll Animal-Lou Reed-RCA
20. Ship Ahoy-The O'Jays-Philadelphia International

27	34.	Until You Came Back To Me-Aretha Franklin-Atlantic
--	35.	Tubular Bells-Mike Oldfield-Virgin
38	36.	Oh My My-Ringo Starr-Apple
40	37.	Let It Ride-Bachman-Turner Overdrive-Mercury
39	38.	The Loco-motion-Grand Funk-Grand Funk
--	39.	Simone-Henry Gross-A & M
--	40.	The Show Must Go On-Three Dog Night-Dunhill

wcfl wcfl wcfl wcfl wcfl wcfl wcfl wcfl wcfl wcfl wcfl wcfl

collector's issue number one hundred and fifty-nine

SUPER CFL SURVEY MAR 23

wcfl wcfl wcfl wcfl wcfl wcfl wcfl wcfl wcfl wcfl wcfl wcfl wcfl wcfl

SUPER ALBUMS

1. John Denver's Greatest Hits-RCA
2. Court And Spark-Joni Mitchell-Asylum
3. Hotcakes-Carly Simon-Elektra
4. Tubular Bells-Mike Oldfield-Virgin
5. The Way We Were-Barbra Streisand-Columbia
6. Tales From Topographic Oceans-Yes-Atlantic
7. Burn-Deep Purple-Warner Brothers
8. Unborn Child-Seals & Crofts-Warner Brothers
9. What Were Once Vices Are Now Habits-Doobie Brothers-Warner Brothers
10. You Don't Mess Around With Jim-Jim Croce-ABC
11. Sabbath, Bloody Sabbath-Black Sabbath-Warner Brothers
12. Goodbye Yellow Brick Road-Elton John-MCA
13. "The Way We Were"-Sound Track-Columbia
14. Brain Salad Surgery-Emerson, Lake & Palmer-Manticore
15. Band On The Run-Paul McCartney & Wings-Apple
16. Planet Waves-Bob Dylan-Asylum
17. "American Graffiti"-Sound Track-MCA
18. All American Boy-Rick Derringer-Blue Sky
19. Saints & Sinners-Johnny Winter-Columbia
20. I Got A Name-Jim Croce-ABC

WAS	IS	
4	1.	Mockingbird-Carly Simon & James Taylor-Elektra
2	2.	Sunshine On My Shoulders-John Denver-RCA
1	3.	Seasons In The Sun-Terry Jacks-Bell
3	4.	Rock On-David Essex-Columbia
10	5.	Hooked On A Feeling-Blue Swede-EMI
8	6.	Jungle Boogie-Kool/Gang-De-Lite
6	7.	Love Song-Anne Murray-Capitol
14	8.	Bennie And The Jets-Elton John-MCA
16	9.	Spiders & Snakes-Jim Stafford-MGM
12	10.	Jet-Paul McCartney-Apple
11	11.	Sexy Mama-Moments-Stang
5	12.	Last Kiss-Wednesday-Sussex
15	13.	Last Time I Saw Him-Diana Ross-Motown
18	14.	Eres Tu (Touch The Wind)-Mocedades-Tara
7	15.	Dark Lady-Cher-MCA
20	16.	The Lord's Prayer-Sister Janet Mead-A & M
9	17.	Boogie Down-Eddie Kendricks-Tamla
29	18.	Just Don't Want To Be Lonely-The Main Ingredient-RCA
13	19.	The Way We Were-Barbra Streisand-Columbia
23	20.	I Like To Live The Love-B.B. King-ABC
32	21.	I'll Have To Say I Love You In A Song-Jim Croce-ABC
30	22.	Virginia-Bill Amesbury-Casablanca
17	23.	Energy Crisis '74-Dickie Goodman-Rainy Wednesday
27	24.	Trying To Hold On To My Woman-Lamont Dozier-ABC
33	25.	Best Thing That Ever Happened To Me-Gladys Knight/Pips-Buddah
19	26.	Jim Dandy-Black Oak Arkansas-Atco
36	27.	Oh My My-Ringo Starr-Apple
21	28.	Doo Doo Doo Doo Doo (Heartbreaker)-Rolling Stones-Rolling Stones
37	29.	Let It Ride-Bachman-Turner Overdrive-Mercury
35	30.	Tubular Bells-Mike Oldfield-Virgin
22	31.	Love's Theme-Love Unlimited Orch.-20th Century
24	32.	Rock And Roll, Hoochie Koo-Rich Derringer-Blue Sky
38	33.	The Loco-motion-Grand Funk-Grand Funk

--	34.	TSOP-MFSB-Philadelphia International
39	35.	Simone-Henry Gross-A & M
40	36.	The Show Must Go On-Three Dog Night-Dunhill
25	37.	Abra-Ca-Dabra-DeFranco Family-20th Century
--	38.	I Am What I Am-Lois Fletcher-Playboy
--	39.	Searchin' So Long-Chicago-Columbia
26	40.	Can This Be Real-Natural Four-Curtom

collector's issue number one hundred and sixty

SUPER CFL SURVEY MAR 30

wcfl wcfl wcfl wcfl wcfl wcfl wcfl wcfl wcfl wcfl wcfl wcfl wcfl wcfl

SUPER ALBUMS

1. John Denver's Greatest Hits-RCA
2. Tubular Bells-Mike Oldfield-Virgin
3. Hotcakes-Carly Simon-Elektra
4. The Way We Were-Barbra Streisand-Columbia
5. Court And Spark-Joni Mitchell-Asylum
6. Burn-Deep Purple-Warner Brothers
7. Unborn Child-Seals And Crofts-Warner Brothers
8. Goodbye Yellow Brick Road-Elton John-MCA
9. Sabbath, Bloody Sabbath-Black Sabbath-Warner Brothers
10. What Were Once Vices Are Now Habits-Doobie Brothers-Warner Brothers
11. Tales From Topographic Oceans-Yes-Atlantic
12. You Don't Mess Around With Jim-Jim Croce-ABC
13. Band On The Run-Paul McCartney-Apple
14. Planet Waves-Bob Dylan-Asylum
15. Ship Ahoy-O'Jays-Philadelphia International
16. Brain Salad Surgery-Emerson, Lake & Palmer-Manticore
17. "American Graffiti"-Sound Track-MCA
18. "Sting"-Sound Track-MCA
19. Live Rhymin'-Paul Simon-Columbia
20. Saints & Sinners-Johnny Winter-Columbia

WAS	IS	
1	1.	Mockingbird, Carly Simon & James Taylor-Elektra
5	2.	Hooked On A Feeling-Blue Swede-EMI
8	3.	Bennie And The Jets-Elton John-MCA
3	4.	Seasons In The Sun-Terry Jacks-Bell
9	5.	Spiders & Snakes-Jim Stafford-MGM
2	6.	Sunshine On My Shoulders-John Denver-RCA
10	7.	Jet-Paul McCartney-Apple
4	8.	Rock On-David Essex-Columbia
16	9.	The Lord's Prayer-Sister Janet Mead-A & M
14	10.	Eres Tu (Touch The Wind)-Mocedadaes-Tara
6	11.	Jungle Boogie-Kool/Gang-De-Lite
7	12.	Love Song-Anne Murray-Capitol
18	13.	Just Don't Want To Be Lonely-The Main Ingredient-RCA
21	14.	I'll Have To Say I Love You In A Song-Jim Croce-ABC
11	15.	Sexy Mama-Moments-Stang
12	16.	Last Kiss-Wednesday-Sussex
25	17.	Best Thing That Ever Happened To Me-Gladys Knight/Pips-Buddah
13	18.	Last Time I Saw Him-Diana Ross-Motown
30	19.	Tubular Bells-Mike Oldfield-Virgin
22	20.	Virginia-Bill Amesbury-Casablanca
27	21.	Oh My My-Ringo Starr-Apple
17	22.	Boogie Down-Eddie Kendricks-Tamla
15	23.	Dark Lady-Cher-MCA
20	24.	I Like To Live The Love-B.B. King-ABC
33	25.	The Loco-motion-Grand Funk-Grand Funk
29	26.	Let It Ride-Bachman-Turner Overdrive-Mercury
19	27.	The Way We Were-Barbra Streisand-Columbia
36	28.	The Show Must Go On-Three Dog Night-Dunhill
34	29.	TSOP-MFSB-Philadelphia International
--	30.	Come And Get Your Love-Redbone-Epic
24	31.	Trying To Hold On To My Woman-Lamont Dozier-ABC
35	32.	Simone-Henry Gross-A & M
38	33.	I Am What I Am-Lois Fletcher-Playboy
--	34.	Keep On Singing-Helen Reddy-

	Capitol
39	35. Searchin' So Long-Chicago-Columbia
26	36. Jim Dandy-Black Oak Arkansas-Atco
--	37. The Entertainer-Marvin Hamlisch-MCA
23	38. Energy Crisis '74-Dickie Goodman-Rainy Wednesday
--	39. Standing At The End Of The Line-Lobo-Big Tree
--	40. Lookin' For A Love-Bobby Womack-UA

wcfl wcfl wcfl wcfl wcfl wcfl wcfl wcfl wcfl wcfl wcfl wcfl

collector's issue number one hundred and sixty-one

SUPER CFL SURVEY APR 6

wcfl wcfl wcfl wcfl wcfl wcfl wcfl wcfl wcfl wcfl wcfl wcfl wcfl wcfl

SUPER ALBUMS

1. John Denver's Greatest Hits-RCA
2. Tubular Bells-Mike Oldfield-Virgin
3. Chicago VII-Columbia
4. Hotcakes-Carly Simon-Elektra
5. Court And Spark-Joni Mitchell-Asylum
6. Goodbye Yellow Brick Road-Elton John-MCA
7. What Were Once Vices Are Now Habits-Doobie Brothers-Warner Brothers
8. Burn-Deep Purple-Warner Brothers
9. Band On The Run-Paul McCartney-Apple
10. Unborn Child-Seals And Crofts-Warner Brothers
11. "The Sting"-Sound Track-MCA
12. The Way We Were-Barbra Streisand-Columbia
13. Brain Salad Surgery-Emerson, Lake & Palmer-Manticore
14. Shinin' On-Grand Funk-Grand Funk
15. You Don't Mess Around With Jim-Jim Croce-ABC
16. Tales From Topographic Oceans-Yes-Atlantic
17. Sabbath, Bloody Sabbath-Black Sabbath-Warner Brothers
18. "American Graffiti"-Sound Track-MCA
19. Live Rhymin'-Paul Simon-Columbia
20. Love Is The Message-MFSB-Philadelphia International

WAS	IS	
1	1.	Mockingbird-Carly Simon & James Taylor-Elektra
2	2.	Hooked On A Feeling-Blue Swede-EMI
3	3.	Bennie And The Jets-Elton John-MCA
5	4.	Spiders & Snakes-Jim Stafford-MGM
7	5.	Jet-Paul McCartney-Apple
10	6.	Eres Tu (Tough The Wind)-Mocedades-Tara
6	7.	Sunshine On My Shoulders-John Denver-RCA
9	8.	The Lord's Prayer-Sister Janet Mead-A & M
13	9.	Just Don't Want To Be Lonely-The Main Ingredient-RCA
14	10.	I'll Have To Say I Love You In A Song-Jim Croce-ABC
4	11.	Seasons In The Sun-Terry Jacks-Bell
17	12.	Best Thing That Ever Happened To Me-Gladys Knight/Pips-Buddah
19	13.	Tubular Bells-Mike Oldfield-Virgin
21	14.	Oh My My-Ringo Starr-Apple
20	15.	Virginia-Bill Amesbury-Casablanca
12	16.	Love Song-Anne Murray-Capitol
29	17.	TSOP-MFSB-Philadelphia International
8	18.	Rock On-David Essex-Columbia
28	19.	The Show Must Go On-Three Dog Night-Dunhill
25	20.	The Loco-motion-Grand Funk-Grand Funk
11	21.	Jungle Boogie-Kool/Gang/De-Lite
26	22.	Let It Ride-Bachman-Turner Overdrive-Mercury
15	23.	Sexy Mama-Moments-Stang
30	24.	Come And Get Your Love-Redbone-Epic
16	25.	Last Kiss-Wednesday-Sussex
35	26.	Searchin' So Long-Chicago-Columbia
32	27.	Simone-Henry Gross-A & M
33	28.	I Am What I Am-Lois Fletcher-Playboy
18	29.	Last Time I Saw Him-Diana Ross-Motown
34	30.	Keep On Singing-Helen Reddy-Capitol
22	31.	Boogie Down-Eddie Kendricks-Tamla
37	32.	The Entertainer-Marvin Hamlisch-MCA
23	33.	Dark Lady-Cher-MCA
39	34.	Standing At The End Of The Line-Lobo-Big Tree

27	35.	The Way We Were-Barbra Streisand-Columbia
40	36.	Lookin' For A Love-Bobby Womack-UA
24	37.	I Like To Live The Love-B.B. King-ABC
--	38.	Dancing Machine-The Jackson 5-Motown
31	39.	Trying To Hold On To My Woman-Lamont Dozier-ABC
--	40.	Star Baby-The Guess Who-RCA

collector's issue number one hundred and sixty-two

SUPER CFL SURVEY APR 13

wcfl wcfl wcfl wcfl wcfl wcfl wcfl wcfl wcfl wcfl wcfl wcfl wcfl wcfl

SUPER ALBUMS

1. Tubular Bells-Mike Oldfield-Virgin
2. John Denver's Greatest Hits-RCA
3. Chicago VII-Columbia
4. "The Sting"-Sound Track-MCA
5. Court And Spark-Joni Mitchell-Asylum
6. Goodbye Yellow Brick Road-Elton John-MCA
7. Burn-Deep Purple-Warner Brothers
8. Unborn Child-Seats And Crofts-Warner Brothers
9. Hotcakes-Carly Simon-Elektra
10. Sabbath, Bloody Sabbath-Black Sabbath-Warner Brothers
11. The Way We Were-Barbra Streisand-Columbia
12. What Were Once Vices Are Now Habits-Doobie Brothers-Warner Brothers
13. Band On The Run-Paul McCartney-Apple
14. Shinin' On-Grand Funk-Grand Funk
15. Love Is The Message-MFSB-Philadelphia International
16. Buddha And The Chocolate Box-Cat Stevens-A & M
17. Saints & Sinners-Johnny Winter-Columbia
18. Live Rhymin'-Paul Simon-Columbia
19. You Don't Mess Around With Jim-Jim Croce-ABC
20. "American Graffiti"-Sound Track-MCA

WAS	IS	
2	1.	Hooked On A Feeling-Blue Swede-EMI
3	2.	Bennie And The Jets-Elton John-MCA
8	3.	The Lord's Prayer-Sister Janet Mead-A & M
4	4.	Spiders & Snakes-Jim Stafford-MGM
5	5.	Jet-Paul McCartney-Apple
6	6.	Eres Tu (Touch The Wind)-Mocedades-Tara
10	7.	I'll Have To Say I Love You In A Song-Jim Croce-ABC
9	8.	Just Don't Want To Be Lonely-The Main Ingredient-RCA
12	9.	Best Thing That Ever Happened To Me-Gladys Knight/Pips-Buddah
1	10.	Mockingbird-Carly Simon & James Taylor-Elektra
13	11.	Tubular Bells-Mike Oldfield-Virgin
15	12.	Virginia-Bill Amesbury-Casablanca
14	13.	Oh My My-Ringo Starr-Apple
24	14.	Come And Get Your Love-Redbone-Epic
17	15.	TSOP-MFSB-Philadelphia International
19	16.	The Show Must Go On-Three Dog Night-Dunhill
7	17.	Sunshine On My Shoulders-John Denver-RCA
20	18.	The Loco-motion-Grand Funk-Grand Funk
26	19.	Searchin' So Long-Chicago-Columbia
11	20.	Seasons In The Sun-Terry Jacks-Bell
22	21.	Let It Ride-Bachman-Turner Overdrive-Mercury
16	22.	Love Song-Anne Murray-Capitol
30	23.	Keep On Singing-Helen Reddy-Capitol
28	24.	I Am What I Am-Lois Fletcher-Playboy
27	25.	Simone-Henry Gross-A & M
32	26.	The Entertainer-Marvin Hamlisch-MCA
18	27.	Rock On-David Essex-Columbia
34	28.	Standing At The End Of The Line-Lobo-Big Tree
21	29.	Jungle Boogie-Kool/Gang-De-Lite
36	30.	Lookin' For A Love-Bobby Womack-UA
38	31.	Dancing Machine-The Jackson 5-Motown
23	32.	Sexy Mama-Moments-Stang
40	33.	Star Baby-The Guess Who-RCA
25	34.	Last Kiss-Wednesday-Sussex
--	35.	Midnight At The Oasis-Maria Muldaur-Reprise

SMALL PRINT DEPARTMENT: The Super CFL survey represents the relative popularity of records based on industry tabulations and record sales. So there. WCFL, the

Voice of Labor, is o	29	36. Last Time I Saw Him-Diana Ross-Motown
	--	37. Touch A Hand, Make A Friend-The Staple Singers-Stax
	--	38. I'm A Train-Albert Hammond-Mums
	31	39. Boogie Down-Eddie Kendricks-Tamla
	33	40. Dark Lady-Cher-MCA

the long awaited collector's issue number one hundred and fifty-two

SUPER CFL SURVEY APR 20

wcfl wcfl wcfl wcfl wcfl wcfl wcfl wcfl wcfl wcfl wcfl wcfl wcfl wcfl

SUPER ALBUMS

1. Tubular Bells-Mike Oldfield-Virgin
2. John Denver's Greatest Hits-RCA
3. "The Sting"-Sound Track-MCA
4. Chicago VII-Columbia
5. Court And Spark-Joni Mitchell-Asylum
6. Goodbye Yellow Brick Road-Elton John-MCA
7. Shinin' On-Grand Funk-Grand Funk
8. Band On The Run-Paul McCartney-Apple
9. Buddha And The Chocolate Box-Cat Stevens-A & M
10. Burn-Deep Purple-Warner Brothers
11. Hotcakes-Carly Simon-Elektra
12. Love Is The Message-MFSB-Philadelphia International
13. What Were Once Vices Are Now Habits-Doobie Brothers-Warner Brothers
14. Unborn Child-Seals And Crofts-Warner Brothers
15. "American Graffiti"-Sound Track-MCA
16. Brain Salad Surgery-Emerson, Lake & Palmer-Manticore
17. The Way We Were-Barbra Streisand-Columbia
18. Tales From Topographic Oceans-Yes-Atlantic
19. Sabbath, Bloody Sabbath-Black Sabbath-Warner Brothers
20. Rock n Roll Animal-Lou Reed-RCA

WAS	IS	
1	1.	Hooked On A Feeling-Blue Swede-EMI
2	2.	Bennie And The Jets-Elton John-MCA
4	3.	Spiders & Snakes-Jim Stafford-MGM
6	4.	Eres Tu (Touch The Wind)-Mocedades-Tara
14	5.	Come And Get Your Love-Redbone-Epic
8	6.	Just Don't Want To Be Lonely-The Main Ingredient-RCA
7	7.	I'll Have To Say I Love You In A Song-Jim Croce-ABC
9	8.	Best Thing That Ever Happened To Me-Gladys Knight/Pips-Buddah
11	9.	Tubular Bells-Mike Oldfield-Virgin
15	10.	TSOP-MFSB-Philadelphia International
12	11.	Virginia-Bill Amesbury-Casablanca
16	12.	The Show Must Go On-Three Dog Night-Dunhill
13	13.	Oh My My-Ringo Starr-Apple
18	14.	The Loco-motion-Grand Funk-Grand Funk
3	15.	The Lord's Prayer-Sister Janet Mead-A & M
19	16.	Searchin' So Long-Chicago-Columbia
5	17.	Jet-Paul McCartney-Apple
10	18.	Mockingbird-Carly Simon & James Taylor-Elektra
26	19.	The Entertainer-Marvin Hamlisch-MCA
24	20.	I Am What I Am-Lois Fletcher-Playboy
23	21.	Keep On Singing-Helen Reddy-Capitol
17	22.	Sunshine On My Shoulders-John Denver-RCA
33	23.	Star Baby-The Guess Who-RCA
28	24.	Standing At The End Of The Line-Lobo-Big Tree
21	25.	Let It Ride-Bachman-Turner Overdrive-Mercury
31	26.	Dancing Machine-The Jackson 5-Motown
30	27.	Lookin' For A Love-Bobby Womack-UA
22	28.	Love Song-Anne Murray-Capitol
35	29.	Midnight At The Oasis-Maria Muldaur-Reprise
--	30.	The Streak-Ray Stevens-Barnaby
25	31.	Simone-Henry Gross-A & M
37	32.	Touch A Hand, Make A Friend-The Staple Singers-Stax
20	33.	Seasons In The Sun-Terry Jacks-Bell
38	34.	I'm A Train-Albert Hammond-Mums
--	35.	I Won't Last A Day Without You-

		Carpenters-A & M
27	36.	Rock On-David Essex-Columbia
29	37.	Jungle Boogie-Kool/Gang-De-Lite
--	38.	You Make Me Feel Brand New- The Stylistics-Avco
34	39.	Last Kiss-Wednesday-Susex
--	40.	Billy, Don't Be A Hero-Bo Donaldson And The Heywoods- ABC

wcfl wcfl wcfl wcfl wcfl wcfl wcfl wcfl wcfl wcfl wcfl wcfl

collector's issue number one hundred and sixty-three

SUPER CFL SURVEY APR 27

wcfl wcfl wcfl wcfl wcfl wcfl wcfl wcfl wcfl wcfl wcfl wcfl wcfl wcfl

SUPER ALBUMS

1. "The Sting"-Sound Track-MCA
2. Tubular Bells-Mike Oldfield-Virgin
3. Chicago VII-Columbia
4. John Denver's Greatest Hits-RCA
5. Buddha And The Chocolate Box-Cat Stevens-A & M
6. Goodbye Yellow Brick Road-Elton John-MCA
7. Court And Spark-Joni Mitchell-Asylum
8. Band On The Run-Paul McCartney-Apple
9. Unborn Child-Seals And Crofts-Warner Brothers
10. Shinin' On-Grand Funk-Grand Funk
11. Burn-Deep Purple-Warner Brothers
12. The Way We Were-Barbra Streisand-Columbia
13. Bachman-Turner Overdrive II-Mercury
14. Behind Closed Doors-Charlie Rich-Epic
15. What Were Once Vices Are Now Habits-Doobie Brothers-Warner Brothers
16. Apostrophe'-Frank Zappa-Discreet
17. Hotcakes-Carly Simon-Elektra
18. "Maria Muldaur"-Maria Muldaur-Reprise
19. Sabbath, Bloody Sabbath-Black Sabbath-Warner Brothers
20. Love Is The Message-MFSB-Philadelphia International

WAS	IS	
5	1.	Come And Get Your Love-Redbone-Epic
1	2.	Hooked On A Feeling-Blue Swede-EMI
9	3.	Tubular Bells-Mike Oldfield-Virgin
7	4.	I'll Have To Say I Love You In A Song-Jim Croce-ABC
2	5.	Bennie And The Jets-Elton John-MCA
6	6.	Just Don't Want To Be Lonely-The Main Ingredient-RCA
8	7.	Best Thing That Ever Happened To Me-Gladys Knight/Pips-Buddah
14	8.	The Loco-motion-Grand Funk-Grand Funk
10	9.	TSOP-MFSB-Philadelphia International
12	10.	The Show Must Go On—Three Dog Night-Dunhill
11	11.	Virginia-Bill Amesbury-Casablanca
16	12.	Searchin' So Long-Chicago-Columbia
19	13.	The Entertainer-Marvin Hamlisch-MCA
4	14.	Eres Tu (Touch The Wind)-Mocedadesl-Tara
23	15.	Star Baby-The Guess Who-RCA
21	16.	Keep On Singing-Helen Reddy-Capitol
13	17.	On My My-Ringo Starr-Apple
20	18.	I Am What I Am-Lois Fletcher-Playboy
3	19.	Spiders & Snakes-Jim Stafford-MGM
30	20.	The Streak-Ray Stevens-Barnaby
24	21.	Standing At The End Of The Line-Lobo-Big Tree
29	22.	Midnight At The Oasis-Maria Muldaur-Reprise
15	23.	The Lord's Prayer-Sister Janet Mead-A & M
26	24.	Dancing Machine-The Jackson 5-Motown
27	25.	Lookin' For A Love-Bobby Womack-UA
17	26.	Jet-Paul McCartney-Apple
32	27.	Touch A Hand, Make A friend-The Staple Singers-Stax
18	28.	Mockingbird-Carly Simon & James Taylor-Elektra
34	29.	I'm A Train-Albert Hammond-Mums
35	30.	I Won't Last A Day Without You-Carpenters-A & M
38	31.	You Make Me Feel Brand New-The Stylistics-Avco
22	32.	Sunshine On My Shoulders-John Denver-RCA
25	33.	Let It Ride-Bachman-Turner Overdrive-Mercury

--	34.	Band On The Run-Paul McCartney-Apple
40	35.	Billy, Don't Be A Hero-Bo Donaldson/Heywoods-ABC
28	36.	Love Song-Anne Murray-Capitol
--	37.	Don't You Worry 'Bout A Thing-Stevie Wonder-Tamla
31	38.	Simone-Henry Gross-A & M
--	39.	Save The Last Dance For Me-The DeFranco Family-20th Century
33	40.	Seasons In The Sun-Terry Jacks-Bell
--	99	The Ballad Of The Mad Streaker-Larry Lujack "Superjock" Curtom

collector's issue number one hundred and sixty-four

SUPER CFL SURVEY MAY 4

wcfl wcfl wcfl wcfl wcfl wcfl wcfl wcfl wcfl wcfl wcfl wcfl wcfl wcfl

WAS	IS	
1	1.	Come And Get Your Love-Redbone-Epic
3	2.	Tubular Bells-Mike Oldfield-Virgin
6	3.	Just Don't Want To Be Lonely-The Main Ingredient-RCA
8	4.	The Loco-motion-Grand Funk-Grand Funk
20	5.	The Streak-Ray Stevens-Barnaby
7	6.	Best Thing That Ever Happened To Me-Gladys Knight/Pips-Buddah
9	7.	TSOP-MFSB-Philadelphia International
10	8.	The Show Must Go On-Three Dog Night-Dunhill
13	9.	The Entertainer-Marvin Hamlisch-MCA
12	10.	Searchin' So Long-Chicago-Columbia
2	11.	Hooked On A Feeling-Blue Swede-EMI
4	12.	I'll Have To Say I Love You In A Song-Jim Croce-ABC
15	13.	Star Baby-The Guess Who-RCA
18	14.	I Am What I Am-Lois Fletcher-Playboy
16	15.	Keep On Singing-Helen Reddy-Capitol
5	16.	Bennie And The Jets-Elton John-MCA
22	17.	Midnight At The Oasis-Maria Muldaur-Reprise
21	18.	Standing At The End Of The Line-Lobo-Big Tree
24	19.	Dancing Machine-The Jackson 5-Motown
11	20.	Virginia-Bill Amesbury-Casablanca
14	21.	Eres Tu (Touch The Wind)-Mocedades-Tara
25	22.	Lookin' For A Love-Bobby Womack-UA
17	23.	Oh My My-Ringo Starr-Apple
30	24.	I Won't Last A Day Without You-Carpenters-A & M
27	25.	Touch A Hand, Make A Friend-The Staple Singers-Stax
29	26.	I'm A Train-Albert Hammond-Mums
19	27.	Spiders & Snakes-Jim Stafford-MGM
31	28.	You Make Me Feel Brand New-The Stylistics-Avco
34	29.	Band On The Run-Paul McCartney-Apple
35	30.	Billy, Don't Be A Here-Bo Donaldson/Heywoods-ABC
23	31.	The Lord's Prayer-Sister Janet Mead-A & M
37	32.	Don't You Worry 'Bout A Thing-Stevie Wonder-Tamla
26	33.	Jet-Paul McCartney-Apple
39	34.	Save The Last Dance For Me-The DeFranco Family-20th Century
--	35.	Oh Very Young-Cat Stevens-A & M

SUPER ALBUMS

1. "The Sting"-Sound Track-MCA
2. Tubular Bells-Mike Oldfield-Virgin
3. Chicago VII-Columbia
4. John Denver's Greatest Hits-RCA
5. Goodbye Yellow Brick Road-Elton John-MCA
6. Buddha And The Chocolate Box-Cat Stevens-A & M
7. Shinin' On-Grand Funk-Grand Funk
8. Court And Spark-Joni Mitchell-Asylum
9. Band On The Run-Paul McCartney-Apple
10. Burn-Deep Purple-Warner Brothers
11. "Maria Muldaur"-Maria Muldaur-Reprise
12. Behind Closed Doors-Charlie Rich-Epic
13. Love Is The Message-MFSB-Philadelphia International
14. Unborn Child-Seals And Crofts-Warner Brothers
15. Hotcakes-Carly Simon-Elektra
16. Starless And Bible Black-King Crimson-Atlantic
17. Sabbath, Bloody Sabbath-Black Sabbath-Warner Brothers
18. Apostrophe'-Frank Zappa-Discreet
19. Bridge Of Sighs-Robin Trower-Chrysalis
20. Bachman-Turner Overdrive II-Mercury

--	36.	Sundown-Gordon Lightfoot-Reprise
28	37.	Mockingbird-Carly Simon & James Taylor-Elektra
--	38.	My Girl Bill-Jim Stafford-MGM
32	39.	Sunshine On My Shoulders-John Denver-RCA
36	40.	Love Song-Anne Murray-Capitol
99	98	The Ballad Of The Mad Streaker-Larry Lujack "Superjock"-Curtom

wcfl wcfl wcfl wcfl wcfl wcfl wcfl wcfl wcfl wcfl wcfl wcfl

collector's issue number one hundred and sixty-five

SUPER CFL SURVEY MAY 11

wcfl wcfl wcfl wcfl wcfl wcfl wcfl wcfl wcfl wcfl wcfl wcfl wcfl wcfl

SUPER ALBUMS

1. "The Sting"-Sound Track-MCA
2. Tubular Bells-Mike Oldfield-Virgin
3. John Denver's Greatest Hits-RCA
4. Chicago VII-Columbia
5. Band On The Run-Paul McCartney-Apple
6. Buddha And The Chocolate Box-Cat Stevens-A & M
7. Goodbye Yellow Brick Road-Elton John-MCA
8. Shinin' On-Grand Funk-Grand Funk
9. Court And Spark-Joni Mitchell-Asylum
10. Love Is The Message-MFSB-Philadelphia International
11. Bridge Of Sighs-Robin Trower-Chrysalis
12. Bachman-Turner Overdrive II-Mercury
13. "Maria Muldaur"-Maria Muldaur-Reprise
14. What Were Once Vices Are Now Habits-Doobie Brothers-Warner Brothers
15. Apostrophe'-Frank Zappa-Discreet
16. Road Food-Guess Who-RCA
17. Unborn Child-Seals And Crofts-Warner Brothers
18. Secret Treaties-Blue Oyster Cult-Columbia
19. Starless And Bible Black-King Crimson-Atlantic
20. Hotcakes-Carly Simon-Elektra

WAS	IS	
1	1.	Come And Get Your Love-Redbone-Epic
4	2.	The Loco-motion-Grand Funk-Grand Funk
5	3.	The Streak-Ray Stevens-Barnaby
2	4.	Tubular Bells-Mike Oldfield-Virgin
7	5.	TSOP-MSFB-Philadelphia International
6	6.	Best Thing That Ever Happened To Me-Gladys Knight/Pips-Buddah
9	7.	The Entertainer-Marvin Hamlisch-MCA
8	8.	The Show Must Go On-Three Dog Night-Dunhill
10	9.	Searchin' So Long-Chicago-Columbia
13	10.	Star Baby-The Guess Who-RCA
3	11.	Just Don't Want To Be Lonely-The Main Ingredient-RCA
15	12.	Keep On Singing-Helen Reddy-Capitol
17	13.	Midnight At The Oasis-Maria Muldaur-Reprise
14	14.	I Am What I Am-Lois Fletcher-Playboy
18	15.	Standing At The End Of The Line-Lobo-Big Tree
12	16.	I'll Have To Say I Love You In A Song-Jim Croce-ABC
19	17.	Dancing Machine-The Jackson 5-Motown
24	18.	I Won't Last A Day Without You-Carpenters-A & M
29	19.	Band On The Run-Paul McCartney-Apple
11	20.	Hooked On A Feeling-Blue Swede-EMI
27	21.	You Make Me Feel Brand New-The Stylistics-Avco
16	22.	Bennie And The Jets-Elton John-MCA
30	23.	Billy, Don't Be A Hero-Bo Donaldson/Heywoods-ABC
26	24.	I'm A Train-Albert Hammond-Mums
20	25.	Virginia-Bill Amesbury-Casablanca
32	26.	Don't You Worry 'Bout A Thing-Stevie Wonder-Tamla
21	27.	Eres Tu (Touch The Wind)-Mocedades-Tara
34	28.	Save The Last Dance-DeFranco Family-20th Century
22	29.	Lookin' For A Love-Bobby Womack-UA
36	30.	Sundown-Gordon Lightfoot-Reprise
35	31.	Oh Very Young-Cat Stevens-A & M
23	32.	Oh My My-Ringo Starr-Apple
25	33.	Touch A Hand, Make A Friend-The Staple Singers-Stax
28	34.	My Girl Bill-Jim Stafford-MGM
--	35.	Help Me-Joni Mitchell-Asylum
27	36.	Spiders & Snakes-Jim Stafford-

	MGM
31	37. The Lord's Prayer-Sister Janet Mead-A & M
--	38. Another Park, Another Sunday-The Doobie Brothers-WB
33	39. Jet-Paul McCartney-Apple
--	40. Son Of Sagittarius-Eddie Kendricks-Tamla
98	97. The Ballad Of The Mad Streaker-Larry Lujack "Superjock"-Curtom

collector's issue number one hundred and sixty-six

SUPER CFL SURVEY MAY 18

wcfl wcfl wcfl wcfl wcfl wcfl wcfl wcfl wcfl wcfl wcfl wcfl wcfl wcfl

SUPER ALBUMS

1. "The Sting"-Sound Track-MCA
2. Band On The Run-Paul McCartney-Apple
3. Tubular Bells-Mike Oldfield-Virgin
4. Chicago VII-Columbia
5. John Denver's Greatest Hits-RCA
6. Goodbye Yellow Brick Road-Elton John-MCA
7. Shinin' On-Grand Funk-Grand Funk
8. Buddha And The Chocolate Box-Cat Stevens-A & M
9. Hotcakes-Carly Simon-Elektra
10. Court And Spark-Joni Mitchell-Asylum
11. "Maria Muldaur"-Maria Muldaur-Reprise
12. Bachman-Turner Overdrive II-Mercury
13. Unborn Child-Seals And Crofts-Warner Brothers
14. Love Is The Message-MFSB-Philadelphia International
15. Apostrophe'-Frank Zappa-Discreet
16. Behind Closed Doors-Charlie Rich-Epic
17. Road Food-Guess Who-RCA
18. Starless And Bible Black-King Crimson-Atlantic
19. Bridge Of Sighs-Robin Trower-Chrysalis
20. Burn-Deep Purple-Warner Brothers

WAS	IS	
3	1.	The Streak-Ray Stevens-Barnaby
2	2.	The Loco-motion-Grand Funk-Grand Funk
8	3.	The Show Must Go ON-Three Dog Night-Dunhill
1	4.	Come And Get Your Love-Redbone-Epic
6	5.	Best Thing That Ever Happened To Me-Gladys Knight/Pips-Buddah
7	6.	The Entertainer-Marvin Hamlisch-MCA
9	7.	Searchin' So Long-Chicago-Columbia
10	8.	Star Baby-The Guess Who-RCA
5	9.	TSOP-MFSB-Philadelphia International
13	10.	Midnight At The Oasis-Maria Muldaur-Reprise
4	11.	Tubular Bells-Mike Oldfield-Virgin
17	12.	Dancing Machine-The Jackson 5-Motown
19	13.	Band On The Run-Paul McCartney-Apple
15	14.	Standing At The End Of The Line-Lobo-Big Tree
18	15.	I Won't Last A Day Without You-Carpenters-A & M
12	16.	Keep On Singing-Helen Reddy-Capitol
14	17.	I Am What I Am-Lois Fletcher-Playboy
23	18.	Billy, Don't Be A Hero-Bo Donaldson/Heywoods-ABC
21	19.	You Make Me Feel Brand New-The Stylistics-Avco
11	20.	Just Don't Want To Be Lonely-The Main Ingredient-RCA
30	21.	Sundown-Gordon Lightfoot-Reprise
28	22.	Save The Last Dance-DeFranco Family-20th Century
16	23.	I'll Have To Say I Love You In A Song-Jim Croce-ABC
26	24.	Don't You Worry 'Bout A Thing-Stevie Wonder-Tamla
31	25.	Oh Very Young-Cat Stevens-A & M
34	26.	My Girl Bill-Jim Stafford-MGM
24	27.	I'm A Train-Albert Hammond-Mums
20	28.	Hooked On A Feeling-Blue Swede-EMI
22	29.	Bennie And The Jets-Elton John-MCA
35	30.	Help Me-Joni Mitchell-Asylum
27	31.	Eres Tu (Touch The Wind)-Mocedades-Tara
29	32.	Lookin' For A Love-Bobby Womack-UA
38	33.	Another Park, Another Sunday-The Doobie Brothers-WB
--	34.	For The Love Of Money-O'Jays-Philadelphia International
40	35.	Son Of Sagittarius-Eddie Kendricks-

		Tamla
32	36.	Oh My My-Ringo Starr-Apple
33	37.	Touch A Hand, Make A Friend- The Staple Singers-Stax
--	38.	La Grange-ZZ Top-London
25	39.	Virginia-Bill Amesbury-Casablanca
--	40.	I'm The Leader Of The Gang- Brownsville Station-Big Tree
97	96	The Ballad Of The Mad Streaker- Larry Lujack "Superjock"-Curtom

wcfl wcfl wcfl wcfl wcfl wcfl wcfl wcfl wcfl wcfl wcfl wcfl

collector's issue number one hundred and sixty-seven

SUPER CFL SURVEY MAY 25

wcfl wcfl wcfl wcfl wcfl wcfl wcfl wcfl wcfl wcfl wcfl wcfl wcfl wcfl

SUPER ALBUMS

1. "The Sting"-Sound Track-MCA
2. Band On The Run-Paul McCartney-Apple
3. John Denver's Greatest Hits-RCA
4. Buddha And The Chocolate Box-Cat Stevens-A & M
5. Chicago VII-Columbia
6. Tubular Bells-Mike Oldfield-Virgin
7. Shinin' On-Grand Funk-Grand Funk
8. Bachman-Turner Overdrive II-Mercury
9. Road Food-Guess Who-RCA
10. Goodbye Yellow Brick Road-Elton John-MCA
11. "Maria Muldaur"-Maria Muldaur-Reprise
12. Court And Spark-Joni Mitchell-Asylum
13. Love Is The Message-MFSB-Philadelphia International
14. Bridge Of Sighs-Robin Trower-Chrysalis
15. Apostrophe'-Frank Zappa-Discreet
16. Unborn Child-Seals And Crofts-Warner Brothers
17. Secret Treaties-Blue Oyster Cult-Columbia
18. Hotcakes-Carly Simon-Elektra
19. On Stage-Loggins And Messina-Columbia
20. Starless And Bible Black-King Crimson-Atlantic

WAS	IS	
1	1.	The Streak-Ray Stevens-Barnaby
2	2.	The Loco-motion-Grand Funk-Grand Funk
6	3.	The Entertainer-Marvin Hamlisch-MCA
5	4.	Best Thing That Ever Happened To Me-Gladys Knight/Pips-Buddah
10	5.	Midnight At The Oasis-Maria Muldaur-Reprise
8	6.	Star Baby-The Guess Who-RCA
7	7.	Searchin' So Long-Chicago-Columbia
3	8.	The Show Must Go On-Three Dog Night-Dunhill
13	9.	Band On The Run-Paul McCartney-Apple
12	10.	Dancing Machine-The Jackson 5-Motown
15	11.	I Won't Last A Day Without You-Carpenters-A & M
14	12.	Standing At The End Of The Line-Lobo-Big Tree
21	13.	Sundown-Gordon Lightfoot-Reprise
4	14.	Come And Get Your Love-Redbone-Epic
18	15.	Billy, Don't Be A Hero-Bo Donaldson/Heywoods-ABC
19	16.	You Make Me Feel Brand New-The Stylistics-Avco
9	17.	TSOP-MFSB-Philadelphia International
26	18.	My Girl Bill-Jim Stafford-MGM
22	19.	Save The Last Dance-DeFranco Family-20th Century
24	20.	Don't You Worry 'Bout A Thing-Stevie Wonder-Tamla
11	21.	Tubular Bells-Mike Oldfield-Virgin
30	22.	Help Me-Joni Mitchell-Asylum
25	23.	Oh Very Young-Cat Stevens-A & M
16	24.	Keep On Singing-Helen Reddy-Capitol
33	25.	Another Park,, Another Sunday-The Doobie Brothers-WB
17	26.	I Am What I Am-Lois Fletcher-Playboy
20	27.	Just Don't Want To be Lonely-The Main Ingredient-RCA
34	28.	For The Love Of Money-O'Jays-Philadelphia International
23	29.	I'll Have To Say I Love You In A Song-Jim Croce-ABC
35	30.	Son Of Sagittarius-Eddie Kendricks-Tamla
27	31.	I'm A Train-Albert Hammond-Mums
--	32.	You Won't See Me-Anne Murray-Capitol
38	33.	La Grange-ZZ Top-London
28	34.	Hooked On A Feeling-Blue Swede-

		EMI
40	35.	I'm The Leader Of The Gang-Brownsville Station-Big Tree
29	36.	Bennie And The Jets-Elton John-MCA
31	37.	Eres Tu (Touch The Wind)-Mocedades-Tara
32	38.	Lookin' For A Love-Bobby Womack-UA
--	39.	Waterloo-Abba-Atlantic
36	40.	Oh My My-Ringo Starr-Apple

collector's issue number one hundred and sixty-eight

SUPER CFL SURVEY JUNE 1

wcfl wcfl wcfl wcfl wcfl wcfl wcfl wcfl wcfl wcfl wcfl wcfl wcfl wcfl

SUPER ALBUMS

1. "The Sting"-Sound Track-MCA
2. Band On The Run-Paul McCartney-Apple
3. Buddha And The Chocolate Box-Cat Stevens-A & M
4. John Denver's Greatest Hits-RCA
5. Chicago VII-Columbia
6. Bachman-Turner Overdrive II-Mercury
7. Road Food-Guess Who-RCA
8. Goodbye Yellow Brick Road-Elton John-MCA
9. Shinin' On-Grand Funk-Grand Funk
10. "Maria Muldaur"-Maria Muldaur-Reprise
11. Tubular Bells-Mike Oldfield-Virgin
12. Court And Spark-Joni Mitchell-Asylum
13. Bridge Of Sighs-Robin Trower-Chrysalis
14. Apostrophe'-Frank Zappa-Discreete
15. Sundown-Gordon Lightfoot-Reprise
16. Unborn Child-Seals And Crofts-Warner Brothers
17. Hotcakes-Carly Simon-Elektra
18. Best Of Bread-Volume II-Elektra
19. On Stage-Loggins And Messina-Columbia
20. Love Is The Message-MFSB-Philadelphia International

WAS	IS	
1	1.	The Streak-Ray Stevens-Barnaby
3	2.	The Entertainer-Marvin Hamlisch-MCA
6	3.	Star Baby-The Guess Who-RCA
5	4.	Midnight At The Oasis-Maria Mauldaur-Reprise
9	5.	Band On The Run-Paul McCartney-Apple
13	6.	Sundown-Gordon Lightfoot-Reprise
10	7.	Dancing Machine-The Jackson 5-Motown
11	8.	I Won't Last A Day Without You-Carpenters-A & M
15	9.	Billy, Don't Be A Hero-Bo Donaldson/Heywoods-ABC
4	10.	Best Thing That Ever Happened To Me-Gladys Knight/Pips-Buddah
7	11.	Searchin' So Long-Chicago-Columbia
22	12.	Help Me-Joni Mitchell-Asylum
18	13.	My Girl Bill-Jim Stafford-MGM
2	14.	The Loco-motion-Grand Funk-Grand Funk
8	15.	The Show Must Go On-Three Dog Night-Dunhill
20	16.	Don't You Worry 'Bout A Thing-Stevie Wonder-Tamla
23	17.	Oh Very Young-Cat Stevens-A & M
28	18.	For The Love Of Money-O'Jays-Philadelphia International
25	19.	Another Park, Another Sunday-The Doobie Brothers-WB
12	20.	Standing At The End Of The Line-Lobo-Big Tree
16	21.	You Make Me Feel Brand New-The Stylistics-Avco
14	22.	Come And Get Your Love-Redbone-Epic
33	23.	La Grange-ZZ Top-London
17	24.	TSOP-MFSB-Philadelphia International
30	25.	Son Of Sagittarius-Eddie Kendricks-Tamla
19	26.	Save The Last Dance-DeFranco Family-20th Century
32	27.	You Won't See Me-Anne Murray-Capitol
21	28.	Tubular Bells-Mike Oldfield-Virgin
24	29.	Keep On Singing-Helen Reddy-Capitol
35	30.	I'm The Leader Of The Gang-Brownsville Station-Big Tree
27	31.	Just Don't Want To Be Lonely-The Main Ingredient-RCA
29	32.	I'll Have To Say I Love You In A Song-Jim Croce-ABC
26	33.	I Am What I Am-Lois Fletcher-Playboy
39	34.	Waterloo-Abba-Atlantic
31	35.	I'm A Train-Albert Hammond-Mums

--	36.	Train Of Thought-Cher-MCA
--	37.	Rock The Boat-The Hues Corporation-RCA
34	38.	Hooked On A Feeling-Blue Swede-EMI
--	39.	Rock And Roll Heaven-The Righteous Brothers-Capitol
36	40.	Bennie And The Jets-Elton John-MCA

wcfl wcfl wcfl wcfl wcfl wcfl wcfl wcfl wcfl wcfl wcfl wcfl

collector's issue number one hundred and sixty-nine

SUPER CFL SURVEY JUNE 8

wcfl wcfl wcfl wcfl wcfl wcfl wcfl wcfl wcfl wcfl wcfl wcfl wcfl wcfl

WAS	IS	
4	1.	Midnight At The Oasis-Maria Muldaur-Reprise
5	2.	Band On The Run-Paul McCartney-Apple
3	3.	Star Baby-The Guess Who-RCA
6	4.	Sundown-Gordon Lightfoot-Reprise
9	5.	Billy, Don't Be A Hero-Bo Donaldson/Heywoods-ABC
7	6.	Dancing Machine-The Jackson 5-Motown
1	7.	The Streak-Ray Stevens-Barnaby
8	8.	I Won't Last A Day Without You-Carpenters-A & M
12	9.	Help Me-Joni Mitchell-Asylum
13	10.	My Girl Bill-Jim Stafford-MGM
2	11.	The Entertainer-Marvin Hamlisch-MCA
18	12.	For The Love Of Money-O'Jays-Philadelphia International
10	13.	Best Thing That Ever Happened To Me-Gladys Knight/Pips-Buddah
23	14.	La Grange-ZZ Top-London
16	15.	Don't You Worry 'Bout A Thing-Stevie Wonder-Tamla
19	16.	Another Park, Another Sunday-The Doobie Brothers-WB
11	17.	Searchin' So Long-Chicago-Columbia
17	18.	Oh Very Young-Cat Stevens-A & M
14	19.	The Loco-motion-Grand Funk-Grand Funk
27	20.	You Won't See Me-Anne Murray-Capitol
15	21.	The Show Must Go On-Three Dog Night-Dunhill
25	22.	Son Of Sagittarius-Eddie Kendricks-Tamla
21	23.	You Make Me Feel Brand New-The Stylistics-Avco
30	24.	I'm The Leader Of The Gang-Brownsville Station-Big Tree
20	25.	Standing At The End Of The Line-Lobo-Big Tree
22	26.	Come And Get Your Love-Redbone-Epic
34	27.	Waterloo-Abba-Atlantic
26	28.	Save The Last Dance-DeFranco Family-20th Century
--	29.	If You Love Me (Let Me Know)-Olivia Newton-John-MCA
36	30.	Train Of Thought-Cher-MCA
24	31.	TSOP-MFSB-Philadelphia International
--	32.	Rikki Don't Lose That Number-Steely Dan-ABC
37	33.	Rock The Boat-The Hues Corporation-RCA
28	34.	Tubular Bells-Mike Oldfield-Virgin
39	35.	Rock And Roll Heaven-The Righteous Brothers-Capitol
--	36.	Hollywood Swinging-Kool & The

SUPER ALBUMS

1. Band On The Run-Paul McCartney-Apple
2. "The Sting"-Sound Track-MCA
3. Buddha And The Chocolate Box-Cat Stevens-A & M
4. Bachman-Turner Overdrive II-Mercury
5. Shinin' On-Grand Funk-Grand Funk
6. Sundown-Gordon Lightfoot-Reprise
7. Chicago VII-Columbia
8. John Denver's Greatest Hits-RCA
9. Road Food-Guess Who-RCA
10. Goodbye Yellow Brick Road-Elton John-MCA
11. Bridge Of Sighs-Robin Trower-Chrysalis
12. Tres Hombres-ZZ Top-London
13. "Maria Muldaur"-Maria Muldaur-Reprise
14. Apostrophe'-Frank Zappa-Discreet
15. Court And Spark-Joni Mitchell-Asylum
16. Tubular Bells-Mike Oldfield-Virgin
17. On Stage-Loggins And Messina-Columbia
18. Queen II-Elektra
19. Unborn Child-Seals And Crofts-Warner Brothers
20. Hotcakes-Carly Simon-Elektra

	Gang-De-Lite
29	37. Keep On Singing-Helen Reddy-Capitol
32	38. I'll Have To Say I Love You In A Song-Jim Croce-ABC
31	39. Just Don't Want To Be Lonely-The Main Ingredient-RCA
33	40. I Am What I Am-Lois Fletcher-Playboy

collector's issue number one hundred and seventy

SUPER CFL SURVEY JUNE 15

wcfl wcfl wcfl wcfl wcfl wcfl wcfl wcfl wcfl wcfl wcfl wcfl wcfl wcfl

SUPER ALBUMS

1. Band On The Run-Paul McCartney-Apple
2. "The Sting"-Sound Track-MCA
3. Sundown-Gordon Lightfoot-Reprise
4. Buddha And The Chocolate Box-Cat Stevens-A & M
5. Tres Hombres-ZZ Top-London
6. Bachman-Turner Overdrive II-Mercury
7. Road Food-Guess Who-RCA
8. Chicago VII-Columbia
9. Bridge Of Sighs-Robin Trower-Chrysalis
10. John Denver's Greatest Hits-RCA
11. "Claudine"-Sound Track-Gladys Knight & The Pips-Buddah
12. Shinin' On-Grand Funk-Grand Funk
13. Goodbye Yellow Brick Road-Elton John-MCA
14. "Maria Muldaur"-Maria Muldaur-Reprise
15. Court And Spark-Joni Mitchell-Asylum
16. Apostrophe'-Frank Zappa-Discreet
17. Open Our Eyes-Earth, Wind & Fire-Columbia
18. +'Justments-Bill Withers-Sussex
19. On Stage-Loggins And Messina-Columbia
20. Best Of Bread Volume II-Bread-Elektra
21. Mighty Love-The Spinners-Atlantic
22. Hotcakes-Carly Simon-Elektra
23. Street Lady-Donald Byrd-UA
24. Head Hunters-Herbie Hancock-Columbia
25. Queen II-Elektra

WAS	IS	
2	1.	Band On The Run-Paul McCartney-Apple
5	2.	Billy, Don't Be A Hero-Bo Donaldson/Heywoods-ABC
4	3.	Sundown-Gordon Lightfoot-Reprise
9	4.	Help Me-Joni Mitchell-Asylum
1	5.	Midnight At The Oasis-Maria Muldaur-Reprise
3	6.	Star Baby-The Guess Who-RCA
14	7.	La Grange-ZZ Top-London
6	8.	Dancing Machine-The Jackson 5-Motown
12	9.	For The Love Of Money-O'Jays-Philadelphia International
10	10.	My Girl Bill-Jim Stafford-MGM
15	11.	Don't You Worry 'Bout A Thing-Stevie Wonder-Tamla
16	12.	Another Park, Another Sunday-The Doobie Brothers-WB
8	13.	I Won't Last A Day Without You-Carpenters-A & M
20	14.	You Won't See Me-Anne Murray-Capitol
7	15.	The Streak-Ray Stevens-Barnaby
11	16.	The Entertainer-Marvin Hamlisch-MCA
24	17.	I'm The Leader Of The Gang-Brownsville Station-Big Tree
27	18.	Waterloo-Abba-Atlantic
29	19.	If You Love Me (Let Me Know)-Olivia Newton-John-MCA
13	20.	Best Thing That Ever Happened To Me-Gladys Knight/Pips-Buddah
22	21.	Son Of Sagittarius-Eddie Kendricks-Tamla
18	22.	Oh Very Young-Cat Stevens-A & M
30	23.	Train Of Thought-Cher-MCA
33	24.	Rock The Boat-The Hues Corporation-RCA
17	25.	Searchin' So Long-Chicago-Columbia
35	26.	Rock And Roll Heaven-The Righteous Brothers-Capitol
32	27.	Rikki Don't Lose That Number-Steely Dan-ABC
19	28.	The Loco-motion-Grand Funk-Grand Funk
23	29.	You Make Me Feel Brand New-The Stylistics-Avco
36	30.	Hollywood Swinging-Kool & The Gang-De-Lite
21	31.	The Show Must Go On-Three Dog Night-De-Lite
--	32.	Annie's Song-John Denver-RCA
25	33.	Standing At The End Of The Line-Lobo-Big Tree
28	34.	Save The Last Dance-DeFranco Family-20th Century

--	35.	Takin' Care Of Business-Bachman-Turner Overdrive-Mercury
26	36.	Come And Get Your Love-Redbone-Epic
--	37.	This Heart-Gene Redding-Haven
--	38.	Finally Got Myself Together-The Impressions-Curtom
31	39.	TSOP-MFSB-Philadelphia International
34	40.	Tubular Bells-Mike Oldfield-Virgin

collector's issue number one hundred and seventy-one

SUPER CFL SURVEY JUNE 22

wcfl wcfl wcfl wcfl wcfl wcfl wcfl wcfl wcfl wcfl wcfl wcfl wcfl wcfl

SUPER ALBUMS

1. Band On The Run-Paul McCartney-Apple
2. Sundown-Gordon Lightfoot-Reprise
3. "The Sting"-Sound Track-MCA
4. Tres Hombres-ZZ Top-London
5. Bachman-Turner Overdrive II-Mercury
6. Buddha And The Chocolate Box-Cat Stevens-A & M
7. Shinin' On-Grand Funk-Grand Funk
8. Road Food-Guess Who-RCA
9. "Claudine"-Sound Track-Gladys Knight & The Pips-Buddah
10. John Denver's Greatest Hits-RCA
11. "Maria Muldaur"-Maria Muldaur-Reprise
12. Bridge Of Sighs-Robin Trower-Chrysalis
13. Chicago VII-Columbia
14. Goodbye Yellow Brick Road-Elton John-MCA
15. +'Justments-Bill Withers-Sussex
16. Diamond Dogs-David Bowie-RCA
17. Journey To The Centre Of The Earth-Rick Wakeman-A & M
18. Court And Spark-Joni Mitchell-Asylum
19. Apostrophe'-Frank Zappa-Discreet
20. Sweet Exorcist-Curtis Mayfield-Curtom
21. Street Lady-Donald Byrd-UA
22. Head Hunters-Herbie Hancock-Columbia
23. Open Our Eyes-Earth, Wind & Fire-Columbia
24. Queen II-Elektra
25. Body Heat-Quincy Jones-A & M

WAS	IS	
1	1.	Band On The Run-Paul McCartney-Apple
2	2.	Billy, Don't Be A Hero-Bo Donaldson/Heywoods-ABC
3	3.	Sundown-Gordon Lightfoot-Reprise
7	4.	La Grange-ZZ Top-London
4	5.	Help Me-Joni Mitchell-Asylum
9	6.	For The Love Of Money-O'Jays-Philadelphia International
19	7.	If You Love Me (Let Me Know)-Olivia Newton-John-MCA
10	8.	My Girl Bill-Jim Stafford-MGM
14	9.	You Won't See Me-Anne Murray-Capitol
12	10.	Another Park, Another Sunday-The Doobie Brothers-WB
5	11.	Midnight At The Oasis-Maria Muldaur-Reprise
8	12.	Dancing Machine-The Jackson 5-Motown
18	13.	Waterloo-Abba-Atlantic
17	14.	I'm The Leader Of The Gang-Brownsville Station-Big Tree
6	15.	Star Baby-The Guess Who-RCA
11	16.	Don't You Worry 'Bout A Thing-Stevie Wonder-Tamla
24	17.	Rock The Boat-The Hues Corporation-RCA
13	18.	I Won't Last A Day Without You-Carpenters-A & M
23	19.	Train Of Thought-Cher-MCA
27	20.	Rikki Don't Lose That Number-Steely Dan-ABC
15	21.	The Streak-Ray Stevens-Barnaby
26	22.	Rock And Roll Heaven-The Righteous Brothers-Capitol
21	23.	Son Of Sagittarius-Eddie Kendricks-Tamla
30	24.	Hollywood Swinging-Kool & The Gang-De-Lite
22	25.	Oh Very Young-Cat Stevens-A & M
16	26.	The Entertainer-Marvin Hamlisch-MCA
20	27.	Best Thing That Ever Happened To Me-Gladys Knight/Pips-Buddah
32	28.	Annie's Song-John Denver-RCA
35	29.	Takin' Care Of Business-Bachman-Turner Overdrive-Mercury
25	30.	Searchin' So Long-Chicago-Columbia
29	31.	You Make Me Feel Brand New-The Stylistics-Avco
38	32.	Finally Got Myself Together-The Impressions-Curtom
--	33.	Rock Your Baby-George McCrae-T.K.
28	34.	The Loco-motion-Grand Funk-Grand

	Funk
37	35. This Heart-Gene Redding-Haven
--	36. Already Gone-Eagles-Asylum
31	37. The Show Must Go On-Three Dog Night-Dunhill
--	38. Don't Let The Sun Go Down On Me-Elton John-MCA
34	39. Save The Last Dance-DeFranco Family-20th Century
33	40. Standing At The End Of The Line-Lobo-Big Tree

collector's issue number one hundred and seventy-two

SUPER CFL SURVEY JUN 29

wcfl wcfl wcfl wcfl wcfl wcfl wcfl wcfl wcfl wcfl wcfl wcfl wcfl wcfl

SUPER ALBUMS

1. Band On The Run-Paul McCartney-Apple
2. Sundown-Gordon Lightfoot-Reprise
3. Tres Hombres-ZZ Top-London
4. "The Sting"-Sound Track-MCA
5. Bachman-Turner Overdrive II-Mercury
6. Buddha And The Chocolate Box-Cat Stevens-A & M
7. Bridge Of Sighs-Robin Trower-Chrysalis
8. Road Food-Guess Who-RCA
9. Journey To The Centre Of The Earth-Rick Wakeman-A & M
10. John Denver's Greatest Hits-RCA
11. Court And Spark-Joni Mitchell-Asylum
12. "Claudine"-Sound Track-Gladys Knight & The Pips-Buddah
13. Chicago VII-Columbia
14. Shinin' On-Grand Funk-Grand Funk
15. Goodbye Yellow Brick Road-Elton John-MCA
16. "Maria Muldaur"-Maria Muldaur-Reprise
17. Open Our Eyes-Earth, Wind & Fire-Columbia
18. +'Justments-Bill Withers-Sussex
19. Queen II-Elektra
20. Street Lady-Donald Byrd-UA
21. Head Hunters-Herbie Hancock-Columbia
22. On Stage-Loggins And Messina-Columbia
23. Sweet Exorcist-Curtis Mayfield-Curtom
24. Apostrophe'-Frank Zappa-Discreet
25. Diamond Dogs-David Bowie-RCA

WAS	IS	
3	1.	Sundown-Gordon Lightfoot-Reprise
4	2.	La Grange-ZZ Top-London
1	3.	Band On The Run-Paul McCartney-Apple
7	4.	If You Love Me (Let Me Know)-Olivia Newton-John-MCA
2	5.	Billy, Don't Be A Hero-Bo Donaldson/Heywoods-ABC
9	6.	You Won't See Me-Anne Murray-Capitol
5	7.	Help Me-Joni Mitchell-Asylum
13	8.	Waterloo-Abba-Atlantic
6	9.	For The Love Of Money-O'Jays-Philadelphia International
14	10.	I'm The Leader Of The Gang-Brownsville Station-Big Tree
17	11.	Rock The Boat-The Hues Corporation-RCA
8	12.	My Girl Bill-Jim Stafford-MGM
19	13.	Train Of Thought-Cher-MCA
20	14.	Rikki Don't Lose That Number-Steely Dan-ABC
11	15.	Midnight At The Oasis-Maria Muldaur-Reprise
22	16.	Rock And Roll Heaven-The Righteous Brothers-Capitol
28	17.	Annie's Song-John Denver-RCA
10	18.	Another Park, Another Sunday-The Doobie Brothers-WB
24	19.	Hollywood Swinging-Kool & The Gang-De-Lite
15	20.	Star Baby-The Guess Who-RCA
29	21.	Takin' Care Of Business-Bachman-Turner Overdrive-Mercury
12	22.	Dancing Machine-The Jackson 5-Motown
16	23.	Don't You Worry 'Bout A Thing-Stevie Wonder-Tamla
33	24.	Rock Your Baby-George McCrae-T.K.
18	25.	I Won't Last A Day Without You-Carpenters-A & M
35	26.	This Heart-Gene Redding-Haven
21	27.	The Streak-Ray Stevens-Barnaby
36	28.	Already Gone-Eagles-Asylum
32	29.	Finally Got Myself Together-The Impressions-Curtom
--	30.	Radar Love-Golden Earring-MCA
38	31.	Don't Let The Sun Go down On Me-Elton John-MCA
25	32.	Oh Very Young-Cat Stevens-A & M
--	33.	Workin' At The Car Wash Blues-Jim Croce-ABC
26	34.	The Entertainer-Marvin Hamlisch-MCA
23	35.	Son Of Sagittarius-Eddie Kendricks-Tamla

--	36.	Wild Thing-Fancy-Big Tree
--	37.	Hang On In There Baby-Johnny Bristol-MGM
31	38.	You Make Me Feel Brand New-The Stylistics-Avco
30	39.	Searchin' So Long-Chicago-Columbia
27	40.	Best Thing That Ever Happened To Me-Gladys Knight/Pips-Buddah

collector's issue number one hundred and seventy-three

SUPER CFL SURVEY JUL 6

wcfl wcfl wcfl wcfl wcfl wcfl wcfl wcfl wcfl wcfl wcfl wcfl wcfl wcfl

SUPER ALBUMS

1. Sundown-Gordon Lightfoot-Reprise
2. Band On The Run-Paul McCartney-Apple
3. Tres Hombres-ZZ Top-London
4. Bachman-Turner Overdrive II-Mercury
5. "The Sting"-Sound Track-MCA
6. "Claudine"-Sound Track-Gladys Knight & The Pips-Buddah
7. Buddha And The Chocolate Box-Cat Stevens-A & M
8. Bridge Of Sighs-Robin Trower-Chrysalis
9. John Denver's Greatest Hits-RCA
10. Road Food-Guess Who-RCA
11. Caribou-Elton John-MCA
12. Chicago VII-Columbia
13. Goodbye Yellow Brick Road-Elton John-MCA
14. Journey To The Centre Of The Earth-Rick Wakeman-A & M
15. Shinin' On-Grand Funk-Grand Funk
16. +'Justments-Bill Withers-Sussex
17. Court And Spark-Joni Mitchell-Asylum
18. Head Hunters-Herbie Hancock-Columbia
19. Back Home Again-John Denver-RCA
20. Diamond Dogs-David Bowie-RCA
21. "Maria Muldaur"-Maria Muldaur-Reprise
22. Apostophe'-Frank Zappa-Discreet
23. Sweet Exorcist-Curtis Mayfield-Curtom
24. Street Lady-Donald Byrd-UA
25. Body Heat-Quincy Jones-A & M

4	1.	If You Love Me-Olivia Newton-John-MCA
2	2.	La Grange-ZZ Top-London
1	3.	Sundown-Gordon Lightfoot-Reprise
6	4.	You Won't See Me-Anne Murray-Capitol
8	5.	Waterloo-Abba-Atlantic
3	6.	Band On The Run-Paul McCartney-Apple
11	7.	Rock The Boat-The Hues Corporation-RCA
14	8.	Rikki Don't Lose That Number-Steely Dan-ABC
17	9.	Annie's Song-John Denver-RCA
10	10.	I'm The Leader Of The Gang-Brownsville Station-Big Tree
5	11.	Billy, Don't Be A Hero-Bo Donaldson/Heywoods-ABC
24	12.	Rock Your Baby-George McCrae-T.K.
7	13.	Help Me-Joni Mitchell-Asylum
13	14.	Train Of Thought-Cher-MCA
19	15.	Holly Swinging-Kool & The Gang-De-Lite
16	16.	Rock And Roll Heaven-The Righteous Brothers-Haven
21	17.	Takin' Care Of Business-Bachman-Turner Overdrive-Mercury
12	18.	My Girl Bill-Jim Stafford-MGM
9	19.	For The Love Of Money-O'Jays-Phil. Intl.
28	20.	Already Gone-Eagles-Asylum
31	21.	Don't Let The Sun Go Down On Me-Elton John-MCA
15	22.	Midnight At The Oasis-Maria Muldaur-Reprise
18	23.	Another Park, Another Sunday-The Doobie Brothers-WB
26	24.	This Heart-Gene Redding-Haven
33	25.	Workin' At The Car Wash Blues-Jim Croce-ABC
30	26.	Radar Love-Golden Earring-MCA
20	27.	Star Baby-The Guess Who-RCA
29	28.	Finally Got Myself Together-The Impressions-Curtom
22	29.	Dancing Machine-The Jackson 5-Motown
36	30.	Wild Thing-Fancy-Big Tree
23	31.	Don't You Worry 'Bout A Thing-Stevie Wonder-Tamla
37	32.	Hang On In There Baby-Johnny Bristol-MGM
--	33.	Sideshow-Blue Magic-Atco
--	34.	Wildwood Weed-Jim Stafford-MGM
25	35.	I Won't Last A Day Without You-Carpenters-A & M
--	36.	The Night Chicago Died-Paper Lace-Mercury
32	37.	Oh Very Young-Cat Stevens-A & M
27	38.	The Streak-Ray Stevens-Barnaby
34	39.	The Entertainer-Marvin Hamlisch-

		MCA
35	40.	Son Of Sagittarius-Eddie Kendricks-Tamla

collector's issue number one hundred and seventy-four

SUPER CFL SURVEY JUL 13

wcfl wcfl wcfl wcfl wcfl wcfl wcfl wcfl wcfl wcfl wcfl wcfl wcfl wcfl

SUPER ALBUMS

1. Band On The Run-Paul McCartney-Apple
2. Sundown-Gordon Lightfoot-Reprise
3. Tres Hombres-ZZ Top-London
4. Back Home Again-John Denver-RCA
5. Caribou-Elton John-MCA
6. Bachman-Turner Overdrive II-Mercury
7. "Claudine"-Sound Track-Gladys Knight & The Pips-Buddah
8. Bridge Of Sighs-Robin Trower-Chrysalis
9. "The Sting"-Sound Track-MCA
10. Road Food-The Guess Who-RCA
11. Journey To The Centre Of The Earth-Rick Wakeman-A & M
12. Buddha And The Chocolate Box-Cat Stevens-A & M
13. Court And Spark-Joni Mitchell-Asylum
14. +'Justments-Bill Withers-Sussex
15. Shiniin' On-Grand Funk-Grand Funk
16. Chicago VII-Columbia
17. Open Our Eyes-Earth, Wind & Fire-Columbia
18. John Denver's Greatest Hits-RCA
19. Body Heat-Quincy Jones-A & M
20. Goodbye Yellow Brick Road-Elton John-MCA
21. Head Hunters-Herbie Hancock-Columbia
22. Diamond Dogs-David Bowie-RCA
23. On Stage-Loggins And Messina-Columbia
24. "Maria Muldaur"-Maria Muldaur-Reprise
25. Apostrophe'-Frank Zappa-Discreet

1	1.	If You Love Me-Olivia Newton-John-MCA
4	2.	You Won't See Me-Anne Murray-Capitol
5	3.	Waterloo-Abba-Atlantic
7	4.	Rock The Boat-The Hues Corporation-RCA
3	5.	Sundown-Gordon Lightfoot-Reprise
8	6.	Rikki Don't Lose That Number-Steely Dan-ABC
9	7.	Annie's Song-John Denver-RCA
2	8.	La Grange—ZZ Top-London
12	9.	Rock Your Baby-George McCrae-T.K.
6	10.	Band On The Run-Paul McCartney-Apple
10	11.	I'm The Leader Of The Gang-Brownsville Station-Big Tree
16	12.	Rock And Roll Heaven-The Righteous Brothers-Haven
17	13.	Takin' Care Of Business-Bachman-Turner Overdrive-Mercury
20	14.	Already Gone-Eagles-Asylum
15	15.	Hollywood Swinging-Kool & The Gang-De-Lite
21	16.	Don't Let The Sun Go Down On Me-Elton John-MCA
11	17.	Billy, Don't Be A Hero-Bo Donaldson/Heywoods-ABC
26	18.	Radar Love-Golden Earring-MCA
24	19.	This Heart-Gene Redding-Haven
13	20.	Help Me-Joni Mitchell-Asylum
25	21.	Workin' At The Car Wash Blues-Jim Croce-ABC
14	22.	Train Of Thought-Cher-MCA
19	23.	For The Love Of Money-O'Jays-Phil. Intl.
30	24.	Wild Thing-Fancy-Big Tree
18	25.	My Girl Bill-Jim Stafford-MGM
28	26.	Finally Got Myself Together-The Impressions-Curtom
32	27.	Hang On In There Baby-Johnny Bristol-MGM
22	28.	Midnight At The Oasis-Maria Muldaur-Reprise
33	29.	Sideshow-Blue Magic-Atco
--	30.	On And On-Gladys Knight And The Pips-Buddah
23	31.	Another Park Another Sunday-The Doobie Brothers-WB
34	32.	Wildwood Weed-Jim Stafford-MGM
36	33.	The Night Chicago Died-Paper Lace-Mercury
27	34.	Star Baby-The Guess Who-RCA
--	35.	You And Me Against The World-Helen Reddy-Capitol

29	36.	Dancing Machine-The Jackson 5-Motown
31	37.	Don't You Worry 'Bout A Thing-Stevie Wonder-Tamla
35	38.	I Won't Last A Day Without You-Carpenters-A & M
39	39.	The Entertainer-Marvin Hamlisch-RCA
37	40.	Oh Very Young-Cat Stevens-A & M

collector's issue number one hundred and seventy-five

SUPER CFL SURVEY JUL 20

wcfl wcfl wcfl wcfl wcfl wcfl wcfl wcfl wcfl wcfl wcfl wcfl wcfl wcfl

SUPER ALBUMS

1. Caribou-Elton John-MCA
2. Sundown-Gordon Lightfoot-Reprise
3. Band On The Run-Paul McCartney-Apple
4. Tres Hombres-ZZ Top-London
5. Back Home Again-John Denver-RCA
6. Bachman-Turner Overdrive II-Mercury
7. "Claudine"-Sound Track-Gladys Knight & The Pips-Buddah
8. Bridge Of Sighs-Robin Trower-Chrysalis
9. Before The Flood-Bob Dylan/Band-Asylum
10. Mars Hotel-Grateful Dead-Grateful Dead
11. Buddha And The Chocolate Box-Cat Stevens-A & M
12. John Denver's Greatest Hits-RCA
13. Journey To The Centre Of The Earth-Rick Wakeman-A & M
14. +'Justments-Bill Withers-Sussex
15. "The Sting"-Sound Track-MCA
16. Goodbye Yellow Brick Road-Elton John-MCA
17. Head Hunters-Herbie Hancock-Columbia
18. Shinin' On-Grand Funk-Grand Funk
19. Body Heat-Quincy Jones-A & M
20. Diamond Dogs-David Bowie-RCA
21. Chicago VII-Columbia
22. Open Our Eyes-Earth, Wind & Fire-Columbia
23. Court And Spark-Joni Mitchell-Asylum
24. Sweet Exorcist-Curtis Mayfield-Curtom
25. On The Border-Eagles-Asylum

4	1.	Rock The Boat-The Hues Corporation-RCA
3	2.	Waterloo-Abba-Atlantic
1	3.	If You Love Me-Olivia Newton-John-MCA
7	4.	Annie's Song-John Denver-RCA
6	5.	Rikki Don't Lose That Number-Steely Dan-ABC
9	6.	Rock Your Baby-George McCrae-T.K.
2	7.	You Won't See Me-Anne Murray-Capitol
5	8.	Sundown-Gordon Lightfoot-Reprise
16	9.	Don't Let The Sun Go Down On Me-Elton John-MCA
8	10.	La Grange ZZ Top-London
13	11.	Takin' Care Of Business-Bachman-Turner Overdrive-Mercury
18	12.	Radar Love-Golden Earring-MCA
12	13.	Rock And Roll Heaven-The Righteous Brothers-Haven
14	14.	Already Gone-Eagles-Asylum
11	15.	I'm The Leader Of The Gang-Brownsville Station-Big Tree
21	16.	Workin' At The Car Wash Blues-Jim Croce-ABC
19	17.	This Heart-Gene Redding-Haven
10	18.	Band On The Run-Paul McCartney-Apple
24	19.	Wild Thing-Fancy-Big Tree
14	20.	Hollywood Swinging-Kool & The Gang-De-Lite
--	21.	The Air That I Breathe-The Hollies-Epic
17	22.	Billy, Don't Be A Hero-Bo Donaldson/Heywoods-ABC
27	23.	Hang On In There Baby-Johnny Bristol-MGM
29	24.	Sideshow-Blue Magic-Atco
26	25.	Finally Got Myself Together-The Impressions—Curtom
--	26.	Machine Gun-The Commodores-Motown
30	27.	On And On-Gladys Knight & The Pips-Buddah
32	28.	Wildwood Weed-Jim Stafford-MGM
20	29.	Help Me-Joni Mitchell-Asylum
33	30.	The Night Chicago Died-Paper Lace-Mercury
22	31.	Train Of Thought-Cher-MCA
35	32.	You And Me Against The World-Helen Reddy-Capitol
--	33.	Keep On Smilin'-Wet Willie-Capricorn
23	34.	For The Love Of Money-O'Jays-Phil. Intl.
25	35.	My Girl Bill-Jim Stafford-MGM
--	36.	I'm Leaving It Up To You-Donny &

	Marie Osmond-MGM
28	37. Midnight At The Oasis-Maria Muldaur-Reprise
--	38. Tell Me Something Good-Rufus-ABC
31	39. Another Park, Another Sunday-The Doobie Brothers-WB
34	40. Star Baby-The Guess Who-RCA

SUPER CFL SURVEY JUL 27

wcfl wcfl wcfl wcfl wcfl wcfl wcfl wcfl wcfl wcfl wcfl wcfl wcfl wcfl

SUPER ALBUMS

1. Caribou-Elton John-MCA
2. Back Home Again-John Denver-RCA
3. Band On The Run-Paul McCartney-Apple
4. Tres Hombres-ZZ Top-London
5. Sundown-Gordon Lightfoot-Reprise
6. Bachman-Turner Overdrive II-Mercury
7. Before The Flood-Bob Dylan/Band-Asylum
8. "Claudine"-Sound Track-Gladys Knight & The Pips-Buddah
9. 461 Ocean Boulevard-Eric Clapton-RSO
10. "The Sting"-Sound Track-MCA
11. Journey To The Centre Of The Earth-Rick Wakeman-A & M
12. Moontan-Golden Earring-MCA
13. Chicago VII-Columbia
14. Bridge Of Sighs-Robin Trower-Chrysalis
15. Body Heat-Quincy Jones-A & M
16. Buddha And The Chocolate Box-Cat Stevens-A & M
17. Mars Hotel-Grateful Dead-Grateful Dead
18. Open Our Eyes-Earth, Wind & Fire-Columbia
19. Diamond Dogs-David Bowie-RCA
20. Shinin' On-Grand Funk-Grand Funk
21. Court And Spark-Joni Mitchell-Asylum
22. Head Hunters-Herbie Hancock-Columbia
23. Sweeet Exorcist-Curtis Mayfield-Curtom
24. +'Justments-Bill Withers-Sussex
25. Goodbye Yellow Brick Road-Elton John-MCA

6	1.	Rock Your Baby-George McCrae-T.K.
1	2.	Rock The Boat-The Hues Corporation-RCA
4	3.	Annie's Song-John Denver-RCA
5	4.	Rikki Don't Lose That Number-Steely Dan-ABC
2	5.	Waterloo-Abba-Atlantic
9	6.	Don't Let The Sun Go Down On Me-Elton John-MCA
7	7.	You Won't See Me-Anne Murray-Capitol
12	8.	Radar Love-Golden Earring-MCA
11	9.	Takin' Care Of Business-Bachman-Turner Overdrive-Mercury
3	10.	If You Love Me-Olivia Newton-John-MCA
8	11.	Sundown-Gordon Lightfoot-Reprise
16	12.	Workin' At The Car Wash Blues-Jim Croce-ABC
19	13.	Wild Thing-Fancy-Big Tree
13	14.	Rock And Roll Heaven-The Righteous Brothers-Haven
10	15.	La Grange-ZZ Top-London
21	16.	The Air That I Breathe-The Hollies-Epic
17	17.	This Heart-Gene Redding-Haven
24	18.	Sideshow-Blue Magic-Atco
23	19.	Hang On In There Baby-Johnny Bristol-MGM
14	20.	Already Gone-Eagles-Asylum
28	21.	Wildwood Weed-Jim Stafford-MGM
30	22.	The Night Chicago Died-Paper Lace-Mercury
27	23.	On And ON-Gladys Knight & The Pips-Buddah
26	24.	Machine Gun-The commodores-Motown
15	25.	I'm The Leader Of The Gang-Brownsville Station-Big Tree
32	26.	You And Me Against The World-Helen Reddy-Capitol
25	27.	Finally Got Myself Together-The Impressions-Curtom
20	28.	Hollywood Swinging-Kool & The Gang-De-Lite
18	29.	Band On The Run-Paul McCartney-Apple
33	30.	Keep On Smilin'-Wet Willie-Capricorn
22	31.	Billy, Don't Be A Hero-Bo Donaldson/Heywoods-ABC
38	32.	Tell Me Something Good-Rufus-ABC
36	33.	I'm Leaving It Up To You-Donny & Marie Osmond-MGM
--	34.	Feel Like Makin' Love-Roberta

		Flack-Atlantic
--	35.	Sure As I'm Sittin' Here-Three Dog Night-Dunhill
29	36.	Help Me-Joni Mitchell-Asylum
--	37.	Captain Howdy-Simon Stokes-Casablanca
31	38.	Train Of Thought-Cher-MCA
35	39.	My Girl Bill-Jim Stafford-MGM
34	40.	For The Love Of Money-O'Jays-Phil. Intl.

SUPER CFL SURVEY AUG 3

wcfl wcfl wcfl wcfl wcfl wcfl wcfl wcfl wcfl wcfl wcfl wcfl wcfl wcfl

SUPER ALBUMS

1. Caribou-Elton John-MCA
2. Back Home Again-John Denver-RCA
3. Bachman-Turner Overdrive II-Mercury
4. Sundown-Gordon Lightfoot-Reprise
5. Band On The Run-Paul McCartney-Apple
6. "Claudine"-Sound Track-Gladys Knight & The Pips-Buddah
7. Tres Hombres-ZZ Top-London
8. 461 Ocean Boulevard-Eric Clapton-RSO
9. Before The Flood-Bob Dylan/ Band-Asylum
10. Moontan-Golden Earring-MCA
11. Bridge Of Sighs-Robin Trower-Chrysalis
12. Journey To The Centre Of The Earth-Rick Wakeman-A & M
13. Mars Hotel-Grateful Dead-Grateful Dead
14. "The Sting"-Sound Track-MCA
15. John Denver's Greatest Hits-RCA
16. +'Justments-Bill Withers-Sussex
17. Body Heat-Quincy Jones-A & M
18. Perfect Angel-Minnie Riperton-Epic
19. Open Our Eyes-Earth, Wind & Fire-Columbia
20. The Souther, Hillman, Furay Band-Asylum
21. Goodbye Yellow Brick Road-Elton John-MCA
22. Head Hunters-Herbie Hancock-Columbia
23. On The Beach-Neil Young-Reprise
24. Blue Magic-Atco
25. Diamond Dogs-David Bowie-RCA

3	1.	Annie's Song-John Denver-RCA
6	2.	Don't Let The Sun Go Down On Me-Elton John-MCA
1	3.	Rock Your Baby-George McCrae-T.K.
4	4.	Rikki Don't Lose That Number-Steely Dan-ABC
16	5.	The Air That I Breathe-The Hollies-Epic
8	6.	Radar Love-Golden Earring-MCA
2	7.	Rock The Boat-The Hues Corporation-RCA
9	8.	Takin' Care Of Business-Bachman-Turner Overdrive-Mercury
5	9.	Waterloo-Abba-Atlantic
13	10.	Wild Thing-Fancy-Big Tree
12	11.	Workin' At The Car Wash Blues-Jim Croce-ABC
10	12.	If You Love Me-Olivia Newton-John-MCA
7	13.	You Won't See Me-Anne Murray-Capitol
24	14.	Machine Gun-The Commodores-Motown
18	15.	Sideshow-Blue Magic-Atco
23	16.	On And On-Gladys Knight & The Pips-Buddah
19	17.	Hang On In There Baby-Johnny Bristol-MGM
21	18.	Wildwood Weed-Jim Stafford-MGM
11	19.	Sundown-Gordon Lightfoot-Reprise
22	20.	The Night Chicago Died-Paper Lace-Mercury
17	21.	This Heart-Gene Redding-Haven
26	22.	You And Me Against The World-Helen Reddy-Capitol
14	23.	Rock And Roll Heaven-The Righteous Brothers-Haven
30	24.	Keep On Smilin'-Wet Willie-Capricorn
15	25.	La Grange-ZZ Top-London
32	26.	Tell Me Something Good-Rufus-ABC
34	27.	Feel Like Makin' Love-Roberta Flack-Atlantic
20	28.	Already Gone-Eagles-Asylum
35	29.	Sure As I'm Sittin' Here-Three Dog Night-Dunhill
33	30.	I'm Leaving It Up To You-Donny & Marie Osmond-MGM
25	31.	I'm The Leader Of The Gang-Brownsville Station-Big Tree
--	32.	(You're) Having My Baby-Paul Anka-UA
27	33.	Finally Got Myself Together-The Impressions-Cartom
37	34.	Captain Howdy-Simon Stokes-Casablanca

--	35.	I Shot The Sheriff-Eric Clapton-RSO
29	36.	Band On The Run-Paul McCartney-Apple
--	37.	Can't Get Enough Of Your Love, Babe-Barry White-20th Century
28	38.	Hollywood Swinging-Kool & The Gang-De-Lite
31	39.	Billy, Don't Be A Hero-Bo Donaldson/Heywoods-ABC
36	40.	Help Me-Joni Mitchell-Asylum

SUPER CFL SURVEY AUG 10

wcfl wcfl wcfl wcfl wcfl wcfl wcfl wcfl wcfl wcfl wcfl wcfl wcfl wcfl

SUPER ALBUMS

1. Caribou-Elton John-MCA
2. Back Home Again-John Denver-RCA
3. Bachman-Turner Overdrive II-Mercury
4. 461 Ocean Boulevard-Eric Clapton-RSO
5. Moontan-Golden Earring-MCA
6. Before The Flood-Bob Dylan/Band-Asylum
7. Band On The Run-Paul McCartney-Apple
8. Tres Hombres-ZZ Top-London
9. "Claudine"-Sound Track-Gladys Knight & The Pips-Buddah
10. Sundown-Gordon Lightfoot-Reprise
11. Mars Hotel-Grateful Dead-Grateful Dead
12. Journey To The Centre Of The Earth-Rick Wakeman-A & M
13. Bridge Of Sighs-Robin Trower-Chrysalis
14. The Souther, Hillman, Furay Band-Asylum
15. Open Our Eyes-Earth, Wind & Fire-Columbia
16. On The Beach-Neil Young-Reprise
17. Head Hunters-Herbie Hancock-Columbia
18. +'Justments-Bill Withers-Sussex
19. Diamond Dogs-David Bowie-RCA
20. Buddha And The Chocolate Box-Cat Stevens-A & M
21. Body Heat-Quincy Jones-A & M
22. Perfect Angel-Minnie Riperton-Epic
23. You're Gonna Love Yourself In The Morning-Bonnie Koloc-Ovation
24. On Stage-Loggins And Messina-Columbia
25. Walking Man-James Taylor-Warner Brothers

1	1.	Annie's Song-John Denver-RCA
2	2.	Don't Let The Sun Go Down On Me-Elton John-MCA
5	3.	The Air That I Breathe-The Hollies-Epic
6	4.	Radar Love-Golden Earring-MCA
8	5.	Takin' Care Of Business-Bachman-Turner Overdrive-Mercury
10	6.	Wild Thing-Fancy-Big Tree
3	7.	Rock Your Baby-George McCrae-T.K.
20	8.	The Night Chicago Died-Paper Lace-Mercury
4	9.	Rikki Don't Lose That Number-Steely Dan-ABC
7	10.	Rock The Beat-The Hues Corporation-RCA
18	11.	Wildwood Weed-Jim Stafford-MGM
14	12.	Machine Gun-The Commodores-Motown
17	13.	Hang On In There Baby-Johnny Bristol-MGM
22	14.	You And Me Against The World-Helen Reddy-Capitol
9	15.	Waterloo-Abba-Atlantic
16	16.	On And On-Gladys Knight & The Pips-Buddah
11	17.	Workin' At The Car Wash Blues-Jim Croce-ABC
15	18.	Sideshow-Blue Magic-Atco
27	19.	Feel Like Makin' Love-Roberta Flack-Atlantic
13	20.	You Won't See Me-Anne Murray-Capitol
24	21.	Keep On Smilin'-Wet Willie-Capricorn
26	22.	Tell Me Something Good-Rufus-ABC
29	23.	Sure As I'm Sittin' Here-Three Dog Night-Dunhill
12	24.	If You Love Me-Olivia Newton-John-MCA
30	25.	I'm Leaving It Up To You-Donny & Marie Osmond-MGM
19	26.	Sundown-Gordon Lightfoot-Reprise
32	27.	(You're) Having My Baby-Paul Anka-UA
34	28.	Captain Howdy-Simon Stokes-Casablanca
21	29.	This Heart-Gene Redding-Haven
35	30.	I Shot The Sheriff-Eric Clapton-RSO
--	31.	Rock Me Gently-Andy Kim-Capitol
37	32.	Can't Get Enough Of Your Love-Babe-Barry White-20th Century
23	33.	Rock And Roll Heaven-The Righteous Brothers-Haven

1--	34.	It's Only Rock 'N Roll-Rolling Stones-Rolling Stones
25	35.	La Grange-ZZ Top-London
28	36.	Already Gone-Eagles-Asylum
31	37.	I'm The Leader Of The Gang-Brownsville Station-Big Tree
33	38.	Finally Got Myself Together-The Impressions-Curtom
36	39.	Band On The Run-Paul McCartney-Apple
38	40.	Hollywood Swinging-Kool & The Gang-De-Lite

SUPER CFL SURVEY AUG 17

wcfl wcfl wcfl wcfl wcfl wcfl wcfl wcfl wcfl wcfl wcfl wcfl wcfl wcfl

SUPER ALBUMS

1. Caribou-Elton John-MCA
2. Back Home Again-John Denver-RCA
3. 461 Ocean Boulevard-Eric Clapton-RSO
4. Bachman-Turner Overdrive II-Mercury
5. Moontan-Golden Earring-MCA
6. Before The Flood-Bob Dylan/Band-Asylum
7. "Claudine"-South Track-Gladys Knight & The Pips-Buddah
8. Band On The Run-Paul McCartney-Apple
9. On The Beach-Neil Young-Reprise
10. Journey To The Centre Of The Earth-Rick Wakeman-A & M
11. Tres Hombres-ZZ Top-London
12. Sundown-Gordon Lightfoot-Reprise
13. Perfect Angel-Minnie Riperton-Epic
14. Body Heat-Quincy Jones-A & M
15. Mars Hotel-Grateful Dead-Grateful Dead
16. Bridge Of Sighs-Robin Trower-Chrysalis
17. +'Justments-Bill Withers-Sussex
18. Fulfillingness' First Finale-Stevie Wonder-Tamla
19. Open Our Eyes-Earth, Wind & Fire-Columbia
20. The Souther, Hillman, Furay Band-Asylum
21. Diamond Dogs-David Bowie-RCA
22. You're Gonna Love Yourself In The Morning-Bonnie Koloc-Ovation
23. On Stage-Loggins And Messina-Columbia
24. Head Hunters-Herbie Hancock-Columbia
25. Buddha And The Chocolate Box-Cat Stevens-A & M

2	1.	Don't Let The Sun Go Down On Me-Elton John-MCA
8	2.	The Night Chicago Died-Paper Lace-Mercury
3	3.	The Air That I Breathe-The Hollies-Epic
1	4.	Annie's Song-John Denver-RCA
6	5.	Wild Thing-Fancy-Big Tree
4	6.	Radar Love-Golden Earring-MCA
5	7.	Takin' Care Of Business-Bachman-Turner Overdrive-Mercury
16	8.	On And On-Gladys Knight & The Pips-Buddah
11	9.	Wildwood Weed-Jim Stafford-MGM
19	10.	Feel Like Makin' Love-Roberta Flack-Atlantic
7	11.	Rock Your Baby-George McCrae-T.K.
14	12.	You And Me Against The World-Helen Reddy-Capitol
10	13.	Rock The Boat-The Hues Corporation-RCA
12	14.	Machine Gun-The Commodores-Motown
22	15.	Tell Me Something Good-Rufus-ABC
9	16.	Rikki Don't Lose That Number Steely Dan-ABC
27	17.	(You're) Having My Baby-Paul Anka-UA
21	18.	Keep On Smilin'-Wet Willie-Capricorn
23	19.	Sure As I'm Sittin' Here-Three Dog Night-Dunhill
30	20.	I Shot The Sheriff-Eric Clapton-RSO
13	21.	Hang On In There Baby-Johnny Bristol-MGM
25	22.	I'm Leaving It Up To You-Donny & Marie Osmond-MGM
28	23.	Captain Howdy-Simon Stokes-Casablanca
15	24.	Waterloo-Abba-Atlantic
32	25.	Can't Get Enough Of Your Love, Babe-Barry White-20th Century
17	26.	Workin' At The Car Wash Blues-Jim Croce-ABC
31	27.	Rock Me Gently-Andy Kim-Capitol
18	28.	Sideshow-Blue Magic-Atco
34	29.	It's Only Rock 'N Roll-Rolling Stones-Rolling Stones
--	30.	Then Came You-Dionne Warwicke/Spinners-Atlantic
--	31.	Nothing From Nothing-Billy Preston-A & M
24	32.	If You Love Me-Olivia Newton-John-MCA

--	33.	Who Do You Think You Are-Bo Donaldson/Heywoods-ABC
20	34.	You Won't See Me-Anne Murray-Capitol
--	35.	Beach Baby-First Class-UK
26	36.	Sundown-Gordon Lightfoot-Reprise
35	37.	La Grange-ZZ Top-London
36	38.	Already Gone-Eagles-Asylum
33	39.	Rock And Roll Heaven-The Righteous Brothers-Haven
29	40.	This Heart-Gene Redding-Haven

SUPER CFL SURVEY AUG 24

wcfl wcfl wcfl wcfl wcfl wcfl wcfl wcfl wcfl wcfl wcfl wcfl wcfl wcfl

SUPER ALBUMS

1. Caribou-Elton John-MCA
2. Back Home Again-John Denver-RCA
3. 461 Ocean Boulevard-Eric Clapton-RSO
4. Bachman-Turner Overdrive II-Mercury
5. Moontan-Golden Earring-MCA
6. On The Beach-Neil Young-Reprise
7. "Claudine" Sound Track-Gladys Knight & The Pips-Buddah
8. Bridge Of Sighs-Robin Trower-Chrysalis
9. Before The Flood-Bob Dylan/ Band-Asylum
10. Band On The Run-Paul McCartney-Apple
11. Perfect Angel-Minnie Riperton-Epic
12. Rags To Rufus-Rufus-ABC
13. Fulfillingness' First Finale-Stevie Wonder-Tamla
14. Sundown-Gordon Lightfoot-Reprise
15. Mars Hotel-Grateful Dead-Grateful Dead
16. Journey To The Centre Of The Earth-Rich Wakeman-A & M
17. Body Heat-Quincy Jones-A & M
18. Open Our Eyes-Earth, Wind & Fire-Columbia
19. "Marvin Gaye Live"-Tamla
20. Tres Hombres-ZZ Top-London
21. The Souther, Hillman, Furay Band-Asylum
22. You're Gonna Love Yourself In The Morning-Bonnie Koloc-Ovation
23. "Santana's Greatest Hits"-Columbia
24. Walking Man-James Taylor-Warner Brothers
25. Holiday-America-WB

2	1.	The Night Chicago Died-Paper Lace-Mercury
3	2.	The Air That I Breathe-The Hollies-Epic
10	3.	Feel Like Makin' Love-Roberta Flack-Atlantic
1	4.	Don't Let The Sun Go Down On Me-Elton John-MCA
9	5.	Wildwood Weed-Jim Stafford-MGM
5	6.	Wild Thing-Fancy-Big Tree
8	7.	On And On-Gladys Knight & The Pips-Buddah
12	8.	You And Me Against The World-Helen Reddy-Capitol
15	9.	Tell Me Something Good-Rufus-ABC
17	10.	(You're) Having My Baby-Paul Anka-UA
7	11.	Takin' Care Of Business-B.T.O.-Mercury
4	12.	Annie's Song-John Denver-RCA
20	13.	I Shot The Sheriff-Eric Clapton-RSO
18	14.	Keep On Smilin'-Wet Willie-Capricorn
19	15.	Sure As I'm Sittin' Here-Three Dog Night-Dunhill
27	16.	Rock Me Gently-Andy Kim-Capitol
22	17.	I'm Leaving It Up To You-Donny & Marie Osmond-MGM
6	18.	Radar Love-Golden Earring-MCA
25	19.	Can't Get Enough Of Your Love-Barry White-20th Century
23	20.	Captain Howdy-Simon Stokes-Casablanca
13	21.	Rock The Boat-The Hues Corporation-RCA
29	22.	It's Only "Rock 'N Roll-Rolling Stones-Rolling Stones
11	23.	Rock Your ?Baby-George McCrae-T.K.
21	24.	Hang On In There Baby-Johnny Bristol-MGM
30	25.	Then Came You-Dionne Warwicke/ Spinners-Atlantic
14	26.	Machine Gun-Commodores-Motown
31	27.	Nothing From Nothing-Billy Preston-A & M
16	28.	Rikki Don't Lose That Number-Steely Dan-ABC
--	29.	You Haven't Done Nothin'-Stevie Wonder-Tamla
33	30.	Who Do You Think You Are-Bo Donaldson/Heywoods-ABC
24	31.	Waterloo-Abba-Atlantic
--	32.	I Honestly Love You-Olivia Newton-John-MCA

35	33.	Beach Baby-First Class-UK
28	34.	Sideshow-Blue Magic-Atco
26	35.	Workin' At The Car Wash Blues-Jim Croce-ABC
--	36.	The Need To be-Jim Weatherly-Buddah
34	37.	You Won't See Me-Anne Murray-Capitol
36	38.	Sundown-Gordon Lightfoot-Reprise
32	39.	If You Love Me-Olivia Newton-John-MCA
37	40.	La Grange-ZZ Top-London

SUPER CFL SURVEY AUG 31

wcfl wcfl wcfl wcfl wcfl wcfl wcfl wcfl wcfl wcfl wcfl wcfl wcfl wcfl

SUPER ALBUMS

1. 461 Ocean Boulevard-Eric Clapton-RSO
2. Caribou-Elton John-MCA
3. Bachman-Turner Overdrive II-Mercury
4. Back Home Again-John Denver-RCA
5. Moontan-Golden Earring-MCA
6. Fulfillingness' First Finale-Stevie Wonder-Tamla
7. On The Beach-Neil Young-Reprise
8. Before The Flood-Bob Dylan/Band-Asylum
9. "Claudine" Sound Track-Gladys Knight & The Pips-Buddah
10. Band On The Run-Paul McCartney-Apple
11. Journey To The Centre Of The Earth-Rich Wakeman-A & M
12. "Marvin Gaye Live"-Tamla
13. Bridge Of Sighs-Robin Trower-Chrysalis
14. Rags To Rufus-Rufus-ABC
15. Perfect Angel-Minnie Riperton-Epic
16. The Souther, Hillman, Furay Band-Asylum
17. Open Our Eyes-Earth, Wind & Fire-Columbia
18. You're Gonna Love Yourself In The Morning-Bonnie Koloc-Ovation
19. "Bad Company"-Bad Company-Swan Song
20. Sundown-Gordon Lightfoot-Reprise
21. Diamond Dogs-David Bowie-RCA
22. Body Heat-Quincy Jones-A & M
23. "Santana's Greatest Hits"-Columbia
24. Tres Hombres-ZZ Top-London
25. Mars Hotel-Grateful Dead-Grateful Dead

3	1.	Feel Like Makin' Love-Roberta Flack-Atlantic
10	2.	(You're) Having My Baby-Paul Anka-UA
1	3.	The Night Chicago Died-Paper Lace-Mercury
4	4.	Don't Let The Sun Go Down On Me-Elton John-MCA
9	5.	Tell Me Something Good-Rufus-ABC
2	6.	The Air That I Breathe-The Hollies-Epic
5	7.	Wildwood Weed-Jim Stafford-MGM
13	8.	I Shot The Sheriff-Eric Clapton-RSO
8	9.	You And Me Against The World-Helen Reddy-Capitol
16	10.	Rock Me Gently-Andy Kim-Capitol
15	11.	Sure As I'm Sittin' Here-Three Dog Night-Dunhill
7	12.	On And On-Gladys Knight & The Pips-Buddah
19	13.	Can't Get Enough Of Your Love-Barry White-20th Century
6	14.	Wild Thing-Fancy-Big Tree
17	15.	I'm Leaving It Up To You-Donny & Marie Osmond-MGM
12	16.	Annie's Song-John Denver-RCA
14	17.	Keep On Smilin'-Wet Willie-Capricorn
22	18.	It's Only Rock 'N Roll-Rolling Stones-Rolling Stones
11	19.	Takin' Care Of Business-B.T.O.-Mercury
20	20.	Captain Howdy-Simon Stokes-Casablanca
25	21.	Then Came You-Dionne Warwicke/Spinners-Atlantic
27	22.	Nothing From Nothing-Billy Preston-A & M
33	23.	Beach Baby-First Class-UK
18	24.	Radar Love-Golden Earring-MCA
30	25.	Who Do You Think You Are-Bo Donaldson/Heywoods-ABC
29	26.	You Haven't Done Nothin'-Stevie Wonder-Tamla
23	27.	Rock Your Baby-George McCrae-T.K.
32	28.	I Honestly Love You-Olivia Newton-John-MCA
--	29.	Sweet Home Alabama-Lynyrd Skynyrd-MCA
21	30.	Rock The Boat-The Hues Corporation-RCA
36	31.	The Need To Be-Jim Weatherly-Buddah
--	32.	Earache My Eye-Cheech & Chong-Ode
--	33.	It Could Have Been Me-Sami Jo-MGM
26	34.	Machine Gun-Commodores-Motown

28	35.	Rikki Don't Lose That Number-Steely Dan-ABC
24	36.	Hang On In There Baby-Johnny Bristol-MGM
31	37.	Waterloo-Abba-Atlantic
34	38.	Sideshow-Blue Magic-Atco
37	39.	You Won't See Me-Anne Murray-Capitol
35	40.	Workin' At The Car Wash Blues-Jim Croce-ABC

SUPER CFL SURVEY SEP 7

wcfl wcfl wcfl wcfl wcfl wcfl wcfl wcfl wcfl wcfl wcfl wcfl wcfl wcfl

SUPER ALBUMS

1. 461 Ocean Boulevard-Eric Clapton-RSO
2. Caribou-Elton John-MCA
3. Bachman-Turner Overdrive II-Mercury
4. Back Home Again-John Denver-RCA
5. Rags To Rufus-Rufus-ABC
6. Fulfillingness' First Finale-Stevie Wonder-Tamla
7. Moontan-Golden Earring-MCA
8. Not Fragile-Bachman-Turner Overdrive-Mercury
9. On The Beach-Neil Young-Reprise
10. Bridge Of Sighs-Robin Trower-Chrysalis
11. More, More, More-Latimore-Glades
12. "Bad Company"-Bad Company-Swan Song
13. Highway Call-Richard Betts-Capricorn
14. "Marvin Gaye Live"-Tamla
15. "Claudine" Sound Track-Gladys Knight & The Pips-Buddah
16. Before The Flood-Bob Dylan/Band-Asylum
17. Perfect Angel-Minnie Riperton-Epic
18. Band On The Run-Paul McCartney-Apple
19. The Souther, Hillman, Furay Band-Asylum
20. Sundown-Gordon Lightfoot-Reprise
21. "Santana's Greatest Hits"-Columbia
22. Open Our Eyes-Earth, Wind & Fire-Columbia
23. Mars Hotel-Grateful Dead-Grateful Dead
24. Anka-Paul Anka-United Artists
25. Body Heat-Quincy Jones-A & M

2	1.	(You're) Having My Baby-Paul Anka-UA
1	2.	Feel Like Makin' Love-Roberta Flack-Atlantic
5	3.	Tell Me Something Good-Rufus-ABC
8	4.	I Shot The Sheriff-Eric Clapton-RSO
10	5.	Rock Me Gently-Andy Kim-Capitol
3	6.	The Night Chicago Died-Paper Lace-Mercury
13	7.	Can't Get Enough Of Your Love-Barry White-20th Century
28	8.	I Honestly Love You-Olivia Newton-John-MCA
15	9.	I'm Leaving It Up To You-Donny & Marie Osmond-MGM
4	10.	Don't Let The Sun Go Down On Me-Elton John-MCA
18	11.	It's Only Rock 'N Roll-Rolling Stones-Rolling Stones
6	12.	The Air That I Breathe-The Hollies-Epic
9	13.	You And Me Against The World-Helen Reddy-Capitol
23	14.	Beach Baby-First Class-UK
21	15.	Then Came You-Dionne Warwicke/Spinners-Atlantic
7	16.	Wildwood Weed-Jim Stafford-MGM
22	17.	Nothing From Noting-Billy Preston-A & M
26	18.	You Haven't Done Nothin'-Stevie Wonder-Tamla
11	19.	Sure As I'm Sittin' Here-Three Dog Night-Dunhill
25	20.	Who Do You Think You Are-Bo Donaldson/Heywoods-ABC
12	21.	On And On-Gladys Knight & The Pips-Buddah
20	22.	Captain Howdy-Simon Stokes-Casablanca
16	23.	Annie's Song-John Denver-RCA
32	24.	Earache My Eye-Cheech & Chong-Ode
29	25.	Sweet Home Alabama-Lynyrd Skynyrd-MCA
14	26.	Wild Thing-Fancy-Big Tree
31	27.	The Need To be-Jim Weatherly-Buddah
17	28.	Keep On Smilin'-Wet Willie-Capricorn
33	29.	It Could Have Been Me-Sami Jo-MGM
--	30.	Let's Put It All Together-The Stylistics-Avco
19	31.	Takin' Care Of Business-B.T.O.-Mercury
--	32.	Stop And Smell The Roses-Mac Davis-Columbia
24	33.	Radar Love-Golden Earring-MCA
--	34.	Never My Love-Blue Swede-EMI

--	35.	Kings Of The Party-Brownsville Station-Big Tree
30	36.	Rock The Boat-The Hues Corporation-RCA
27	37.	Rock Your Baby-George McCrae-T.K.
36	38.	Hang On In There Baby-Johnny Bristol-MGM
35	39.	Rikki Don't Lose That Number-Steely Dan-ABC
34	40.	Machine Gun-Commodores-Motown

SUPER CFL SURVEY SEP 14

wcfl wcfl wcfl wcfl wcfl wcfl wcfl wcfl wcfl wcfl wcfl wcfl wcfl wcfl

SUPER ALBUMS

1. 461 Ocean Boulevard-Eric Clapton-RSO
2. Caribou-Elton John-MCA
3. Fulfillingness' First Finale-Stevie Wonder-Tamla
4. Bachman-Turner Overdrive II-Mercury
5. Back Home Again-John Denver-RCA
6. Not Fragile-Bachman-Turner Overdrive-Mercury
7. Rags To Rufus-Rufus-ABC
8. On The Beach-Neil Young-Reprise
9. Skin Tight-Ohio Players-Mercury
10. "Bad Company"-Bad Company-Swan Song
11. More, More, More-Latimore-Glades
12. Moontan-Golden Earring-MCA
13. "Marvin Gaye Live"-Tamla
14. Perfect Angel-Minnie Riperton-Epic
15. Bridge Of Sighs-Robin Trower-Chrysalis
16. Body Heat-Quincy Jones-A & M
17. Band On The Run-Paul McCartney-Apple
18. Highway Call-Richard Betts-Capricorn
19. One-Bob James-CTI
20. Journey To The Centre Of The Earth-Rich Wakeman-A & M
21. You're Gonna Love Yourself In The Morning-Bonnie Koloc-Ovation
22. Mars Hotel-Grateful Dead-Grateful Dead
23. Before The Flood-Bob Dylan/Band-Asylum
24. The Souther, Hillman Furay Band-Asylum
25. Can't Get Enough-Barry White-20th Century

1	1.	(You're) Having My Baby-Paul Anka-UA
5	2.	Rock Me Gently-Andy Kim-Capitol
4	3.	I Shot The Sheriff-Eric Clapton-RSO
3	4.	Tell Me Something Good-Rufus-ABC
7	5.	Can't Get Enough Of Your Love-Barry White-20th Century
8	6.	I Honestly Love You-Olivia Newton-John-MCA
14	7.	Beach Baby-First Class-UK
2	8.	Feel Like Makin' Love-Roberta Flack-Atlantic
9	9.	I'm Leaving It Up To You-Donny & Marie Osmond-MGM
11	10.	It's Only Rock 'N Roll-Rolling Stones-Rolling Stones
15	11.	Then Came You-Dionne Warwicke/Spinners-Atlantic
17	12.	Nothing From Nothing-Billy Preston-A & M
6	13.	The Night Chicago Died-Paper Lace-Mercury
24	14.	Earache My Eye-Cheech & Chong-Ode
10	15.	Don't Let The Sun Go Down On Me-Elton John-MCA
18	16.	You Haven't Done Nothin'-Stevie Wonder-Tamla
20	17.	Who Do You Think You Are-Bo Donaldson/Heywoods-ABC
25	18.	Sweet Home Alabama-Lynyrd Skynyrd-MCA
12	19.	The Air That I Breathe-The Hollies-Epic
13	20.	You And Me Against The World-Helen Reddy-Capitol
27	21.	The Need To Be-Jim Weatherly-Buddah
19	22.	Sure As I'm Sittin' Here-Three Dog Night-Dunhill
29	23.	It Could Have Been Me-Sami Jo-MGM
30	24.	Let's Put It All Together-The Stylistics-Avco
34	25.	Never My Love-Blue Swede-EMI
21	26.	On And On-Gladys Knight & The Pips-Buddah
32	27.	Stop And Smell The Roses-Mac Davis-Columbia
--	28.	Can't Get Enough-Bad Company-Swan Song
35	29.	Kings Of The Party-Brownsville Station-Big Tree
--	30.	You Little Trustmaker-The Tymes-RCA
22	31.	Captain Howdy-Simon Stokes-Casablanca
--	32.	Love Me For A Reason-The Osmonds-MGM

23	33.	Annie's Song-John Denver-RCA
26	34.	Wild Thing-Fancy-Big Tree
--	35.	After The Goldrush-Prelude-Island
28	36.	Keep On Smilin'-Wet Willie-Capricorn
31	37.	Takin' Care Of Business-B.T.O.-Mercury
16	38.	Wildwood Weed-Jim Stafford-MGM
33	39.	Radar Love-Golden Earring-MCA
37	40.	Rock Your Baby-George McCrae-T.K.

SUPER CFL SURVEY SEP 21

wcfl wcfl wcfl wcfl wcfl wcfl wcfl wcfl wcfl wcfl wcfl wcfl wcfl wcfl

SUPER ALBUMS

1. 461 Ocean Boulevard-Eric Clapton-RSO
2. Caribou-Elton John-MCA
3. Fulfillingness' First Finale-Stevie Wonder-Tamla
4. Not Fragile-Bachman-Turner Overdrive-Mercury
5. Bachman-Turner Overdrive II-Mercury
6. Rags To Rufus-Rufus-ABC
7. "Bad Company"-Bad Company-Swan Song
8. Back Home Again-John Denver-RCA
9. Welcome Back-Emerson, Lake & Palmer-Manticore
10. On The Beach-Neil Young-Reprise
11. Bridge Of Sighs-Robin Trower-Chrysalis
12. Can't Get Enough-Barry White-20th Century
13. So Far-Crosby, Stills, Nash & Young-Atlantic
14. More, More, More-Latimore-Glades
15. Skin Tight-Ohio Players-Mercury
16. Anka-Paul Anka-United Artists
17. Moontan-Golden Earring-MCA
18. Perfect Angel-Minnie Riperton-Epic
19. Endless Summer-The Beach Boys-Capitol
20. "Marvin Gaye Live"-Tamla
21. Before The Flood-Bob Dylan/Band-Asylum
22. Highway Call-Richard Betts-Capricorn
23. "Santana's Greatest Hits"-Columbia
24. The Souther, Hillman, Furay Band-Asylum
25. Body Heat-Quincy Jones-A & M

3	1.	I Shot The Sheriff-Eric Clapton-RSO
1	2.	(You're) Having My Baby-Paul Anka-UA
6	3.	I Honestly Love You-Olivia Newton-John-MCA
7	4.	Beach Baby-First Class-UK
2	5.	Rock Me Gently-Andy Kim-Capitol
12	6.	Nothing From Nothing-Billy Preston-A & M
5	7.	Can't Get Enough Of Your Love-Barry White-20th Century
11	8.	Then Came You-Dionne Warwicke/Spinners-Atlantic
14	9.	Earache My Eye-Cheech & Chong-Ode
4	10.	Tell Me Something Good-Rufus-ABC
9	11.	I'm Leaving It Up To You-Donny & Marie Osmond-MGM
17	12.	Who Do You Think You Are-Bo Donaldson/Heywoods-ABC
10	13.	It's Only Rock 'N Roll-Rolling Stones-Rolling Stones
16	14.	You Haven't Done Nothin'-Stevie Wonder-Tamla
18	15.	Sweet Home Alabama-Lynyrd Skynyrd-MCA
8	16.	Feel Like Makin' Love-Roberta Flack-Atlantic
21	17.	The Need To Be-Jim Weatherly-Buddah
24	18.	Let's Put It All Together-The Sty;listics-Avco
13	19.	The Night Chicago Died-Paper Lace-Mercury
23	20.	It Could Have Been Me-Sami Jo-MGM
25	21.	Never My Love-Blue Swede-EMI
27	22.	Stop And Smell The Roses-Mac Davis-Columbia
19	23.	The Air That I Breathe-The Hollies-Epic
28	24.	Can't Get Enough-Bad Company-Swan Song
15	25.	Don't Let The Sun Go Down On Me-Elton John-MCA
29	26.	Kings Of The Party-Brownsville Station-Big Tree
20	27.	You And Me Against The World-Helen Reddy-Capitol
30	28.	You Little Trustmaker-The Tymes-RCA
--	29.	Another Saturday Night-Cat Stevens-A & M
32	30.	Love Me For A Reason-The Osmonds-MGM
--	31.	Do It Baby-The Miracles-Tamla
22	32.	Sure As I'm Sittin' Here-Three Dog Night-Dunhill
35	33.	After The Goldrush-Prelude-Island
--	34.	Clap For The Wolfman-The Guess

		Who-RCA
26	35.	On And On-Gladys Knight & The Pips-Buddah
--	36.	Skin Tight-Ohio Players-Mercury
34	37.	Wild Thing-Fancy-Big Tree
36	38.	Keep On Smilin'-Wet Willie-Capricorn
33	39.	Annie's Song-John Denver-RCA
31	40.	Captain Howdy-Simon Stokes-Casablanca

SUPER CFL SURVEY SEP 28

wcfl wcfl wcfl wcfl wcfl wcfl wcfl wcfl wcfl wcfl wcfl wcfl wcfl wcfl

SUPER ALBUMS

1. 461 Ocean Boulevard-Eric Clapton-RSO
2. Not Fragile-Bachman-Turner Overdrive-Mercury
3. Caribou-Elton John-MCA
4. Fulfillingness' First Finale-Stevie Wonder-Tamla
5. "Bad Company"-Bad Company-Swan Song
6. Bachman-Turner Overdrive II-Mercury
7. Rags To Rufus-Rufus-ABC
8. Skin Tight-Ohio Players-Mercury
9. Welcome Back-Emerson, Lake & Palmer-Manticore
10. Perfect Angel-Minnie Riperton-Epic
11. Can't Get Enough-Barry White-20th Century
12. Back Home Again-John Denver-RCA
13. So Far-Crosby-Stills, Nash & Young-Atlantic
14. On The Beach-Neil Young-Reprise
15. Endless Summer-Beach Boys-Capitol
16. Bridge Of sighs-Robin Trower-Chrysalis
17. Anka-Paul Anka-United Artists
18. One-Bob James-CTI
19. More, More, More-Latimore-Glades
20. "Marvin Gaye Live"-Tamla
21. Highway Call-Richard Betts-Capricorn
22. Before The Flood-Bob Dylan/Band-Asylum
23. Mars Hotel-Grateful Dead-Grateful Dead
24. Moontan-Golden Earring-MCA
25. If You Love Me, Let Me Know-Olivia Newton-John-MCA

3	1.	I Honestly Love You-Olivia Newton-John-MCA
4	2.	Beach Baby-First Class-UK
1	3.	I Shot The Sheriff-Eric Clapton-RSO
6	4.	Nothing From Nothing-Billy Preston-A & M
2	5.	(You're) Having My Baby-Paul Anka-UA
9	6.	Earache My Eye-Cheech & Chong-Ode
8	7.	Then Came You-Dionne Warwicke/Spinners-Atlantic
5	8.	Rock Me Gently-Andy Kim-Capitol
15	9.	Sweet Home Alabama-Lynyrd Skynyrd-MCA
12	10.	Who Do You Think You Are-Bo Donaldson/Heywoods-ABC
7	11.	Can't Get Enough Of Your Love=Barry White-20th Century
14	12.	You Haven't Done Nothin'-Stevie Wonder-Tamla
18	13.	Let's Put It All Together-The Stylistics-Avco
17	14.	The Need To be-Jim Weatherly-Buddah
11	15.	I'm Leaving It Up To You-Donny & Marie Osmond-MGM
21	16.	Never My Love-Blue Swede-EMI
22	17.	Stop And Smell The Roses-Mac Davis-Columbia
13	18.	It's Only Rock 'N Roll-Rolling Stones-Rolling Stones
24	19.	Can't Get Enough-Bad Company-Swan Song
20	20.	It Could Have Been Me-Sami Jo-MGM
10	21.	Tell Me Something Good-Rufus-ABC
26	22.	Kings Of The Party-Brownsville Station-Big Tree
16	23.	Feel Like Makin' Love-Roberta Flack-Atlantic
28	24.	You Little Trustmaker-The Tymes-RCA
30	25.	Love Me For A Reason-The Osmonds-MGM
29	26.	Another Saturday Night-Cat Stevens-A & M
33	27.	After The Goldrush-Prelude-Island
31	28.	Do It Baby-The Miracles-Tamla
19	29.	The Night Chicago Died-Paper Lace-Mercury
34	30.	Clap For The Wolfman-The Guess Who-RCA
23	31.	The Air That I Breathe-The Hollies-Epic
36	32.	Skin Tight-Ohio Players-Mercury
--	33.	The Bitch Is Back-Elton John-MCA
27	34.	You And Me Against The World-Helen Reddy-Capitol

25	35.	Don't Let The Sun Go Down On Me-Elton John-MCA
32	36.	Sure As I'm Sittin' Here-Three Dog Night-Dunhill
35	37.	On And On-Gladys Knight & The Pips-Buddah
37	38.	Wild Thing-Fancy-Big Tree
39	39.	Annie's Song-John Denver-RCA
38	40.	Keep On Smilin'-Wet Willie-Capricorn

SUPER CFL SURVEY OCT 5

wcfl wcfl wcfl wcfl wcfl wcfl wcfl wcfl wcfl wcfl wcfl wcfl wcfl wcfl

SUPER ALBUMS

1. Not Fragile-Bachman-Turner Overdrive-Mercury
2. 461 Ocean Boulevard-Eric Clapton-RSO
3. Caribou-Elton John-MCA
4. Fulfillingness' First Finale-Stevie Wonder-Tamla
5. "Bad Company"-Bad Company-Swan Song
6. Bachman-Turner Overdrive II-Mercury
7. Can't Get Enough-Barry White-20th Century
8. Rags To Rufus-Rufus-ABC
9. Welcome Back-Emerson, Lake & Palmer-Manticore
10. Back Home Again-John Denver-RCA
11. So Far-Crosby-Stills Nash & Young-Atlantic
12. Anka-Paul Anka-United Artists
13. Endless Summer-Beach Boys-Capitol
14. Skin Tight-Ohio Players-Mercury
15. When The Eagle Flies-Traffic-Asylum
16. If You Love Me. Let Me Know-Olivia Newton-John-MCA
17. Live It Up-Isley Brothers-T-Neck
18. On The Beach-Neil Young-Reprise
19. Perfect Angel-Minnie Riperton-Epic
20. Bridge Of Sighs-Robin Trower-Chrysalis
21. More, More, More-Latimore-Glades
22. Second Helping-Lynyrd Skynyrd-MCA
23. "Marvin Gaye Live"-Tamla
24. Wrap Around Joy-Carole King-Ode
25. Thrust-Herbie Hancock-Columbia

1	1.	I Honestly Love You-Olivia Newton-John-MCA
4	2.	Nothing From Nothing-Billy Preston-A & M
2	3.	Beach Baby-First Class-UK
3	4.	I Shot The Sheriff-Eric Clapton-RSO
7	5.	Then Came You-Dionne Warwicke/ Spinners-Atlantic
6	6.	Earache My Eye-Cheech & Chong-Ode
9	7.	Sweet Home Alabama-Lynyrd Skynyrd-MCA
5	8.	(You're) Having My Baby-Paul Anka-UA
19	9.	Can't Get Enough-Bad Company-Swan Song
13	10.	Let's Put It All Together-The Stylistics-Avco
10	11.	Who Do You Think You Are-Bo Donaldson/Heywoods-ABC
14	12.	The Need To Be-Jim Weatherly-Buddah
8	13.	Rock Me Gently-Andy Kim-Capitol
17	14.	Stop And Smell The Roses-Mac Davis-Columbia
16	15.	Never My Love-Blue Swede-EMI
26	16.	Another Saturday Night-Cat Stevens-A & M
12	17.	You Haven't Done Nothin'-Stevie Wonder-Tamla
22	18.	Kings Of The Party-Brownsville Station—Big Tree
11	19.	Can't Get Enough Of Your Love-Barry White-20th Century
24	20.	You Little Trustmaker-The Tymes-RCA
25	21.	Love Me For A Reason-The Osmonds-MGM
15	22.	I'm Leaving It Up To You-Donny & Marie Osmond-MGM
28	23.	Do It Baby-The Miracles-Tamla
27	24.	After The Goldrush-Prelude-Island
30	25.	Clap For The Wolfman-The Guess Who-RCA
18	26.	It's Only Rock 'N Roll-Rolling Stones-Rolling Stones
32	27.	Skin Tight-Ohio Players-Mercury
20	28.	It Could Have Been Me-Sami Jo-MGM
23	29.	Feel Like Makin' Love-Roberta Flack-Atlantic
33	30.	The Bitch Is Back-Elton John-MCA
--	31.	Steppin' Out (Gonna Boogie Tonight)-Dawn-Bell
21	32.	Tell Me Something Good-Rufus-ABC
--	33.	Give It To The People-The Righteous Brothers-Haven

--	34.	Straight Shootin' Woman-Steppenwolf-Mums
29	35.	The Night Chicago Died-Paper Lace-Mercury
31	36.	The Air That I Breathe-The Hollies-Epic
--	37.	Everlasting Love-Carl Carlton-Back Beat
34	38.	You And Me Against The World-Helen Reddy-Capitol
35	39.	Don't Let The Sun Go Down On Me-Elton John-MCA
36	40.	Sure As I'm Sittin' Here-Three Dog Night-Dunhill

SUPER CFL SURVEY OCT 12

wcfl wcfl wcfl wcfl wcfl wcfl wcfl wcfl wcfl wcfl wcfl wcfl wcfl wcfl

SUPER ALBUMS

1. Not Fragile-Bachman-Turner Overdrive-Mercury
2. "Bad Company"-Bad Company-Swan Song
3. 461 Ocean Boulevard-Eric Clapton-RSO
4. Caribou-Elton John-MCA
5. Can't Get Enough-Barry White-20th Century
6. Fulfillingness' First Finale-Stevie Wonder-Tamla
7. Welcome Back-Emerson, Lake & Palmer-Manticore
8. Skin Tight-Ohio Players-Mercury
9. Bachman-Turner Overdrive II-Mercury
10. So Far-Crosby, Stills, Nash & Young-Atlantic
11. Endless Summer-Beach Boys-Capitol
12. Anka-Paul Anka-United Artists
13. If You Love Me, Let Me Know, Olivia Newton-John-MCA
14. Back Home Again-John Denver-RCA
15. Rags To Rufus-Rufus-ABC
16. Perfect Angel-Minnie Riperton-Epic
17. Photographs And Memories-Jim Croce-ABC
18. When The Eagle Flies-Traffic-Asylum
19. On The Beach-Neil Young-Reprise
20. Wrap Around Joy-Carole King-Ode1
21. Second Helping-Lynyrd Skynyrd-MCA
22. Live It Up-Isley Brothers-T-Neck
23. Body Heat-Quincy Jones-A & M
24. One-Bob James-CTI
25. All In Love Is Fair-Nancy Wilson-Capitol

1	1.	I Honestly Love You-Olivia Newton-John-MCA
2	2.	Nothing From Nothing-Billy Preston-A & M
5	3.	Then Came You-Dionne Warwicke/ Spinners-Atlantic
7	4.	Sweet Home Alabama-Lynyrd Skynyrd-MCA
3	5.	Beach Baby-First Class-UK
9	6.	Can't Get Enough-Bad Company-Swan Song
6	7.	Earache My Eye-Cheech & Chong-Ode
4	8.	I Shot The Sheriff-Eric Clapton-RSO
14	9.	Stop And Smell The Roses-Mac Davis-Columbia
10	10.	Let's Put It All Together-The Stylistics-Avco
15	11.	Never My Love-Blue Swede-EMI
12	12.	The Need To be-Jim Weatherly-Buddah
16	13.	Another Saturday Night-Cat Stevens-A & M
8	14.	(You're) Having My Baby-Paul Anka-UA
11	15.	Who Do You Think You Are-Bo Donaldson/Heywoods-ABC
21	16.	Love Me For A Reason-The Osmonds-MGM
20	17.	You Little Trustmaker-The Tymes-RCA
18	18.	Kings Of The Party-Brownsville Station-Big Tree
24	19.	After The Goldrush-Prelude-Island
30	20.	The Bitch Is Back-Elton John-MCA
23	21.	Do It Baby-The Miracles-Tamla
13	22.	Rock Me Gently-Andy Kim-Capitol
25	23.	Clap For The Wolfman-The Guess Who-RCA
17	24.	You Haven't Done Nothin'-Stevie Wonder-Tamla
27	25.	Skin Tight-Ohio Players-Mercury
19	26.	Can't Get Enough Of Your Love-Barry White-20th Century
31	27.	Steppin' Out (Gonna Boogie Tonight)-Dawn-Bell
33	28.	Give It To The People-The Righteous Brothers-Haven
--	29.	You Ain't Seen Nothing Yet-Bachman-Turner Overdrive-Mercury
22	30.	I'm Leaving It Up To You-Donny & Marie Osmond-MGM
34	31.	Straight Shootin' Woman-Steppenwolf-Mums
--	32.	Back Home Again-John Denver-RCA

37	33.	Everlasting Love-Carl Carlton-Back Beat
--	34.	Life Is A Rock-Reunion-RCA
26	35.	It's Only Rock "n Roll-Rolling Stones-Rolling Stones
--	36.	Longfellow Serenade-Neil Diamond-Columbia
32	37.	Tell Me Something Good-Rufus-ABC
29	38.	Feel Like Makin' Love-Roberta Flack-Atlantic
35	39.	The Night Chicago Died-Paper Lace-Mercury
28	40.	It Could Have Been Me-Sami Jo-MGM

SUPER CFL SURVEY OCT 19

wcfl wcfl wcfl wcfl wcfl wcfl wcfl wcfl wcfl wcfl wcfl wcfl wcfl wcfl

SUPER ALBUMS

1. Not Fragile-Bachman-Turner Overdrive-Mercury
2. "Bad Company"-Bad Company-Swan Song
3. Caribou-Elton John-MCA
4. 461 Ocean Boulevard-Eric Clapton-RSO
5. Endless Summer-Beach Boys-Capitol
6. Welcome Back-Emerson, Lake & Palmer-Manticore
7. Fulfillingness' First Finale-Stevie Wonder-Tamla
8. So Far-Crosby, Stills, Nash & Young-Atlantic
9. If You Love Me, Let Me Know-Olivia Newton-John-MCA
10. Can't Get Enough-Barry White-20th Century
11. Bachman-Turner Overdrive II-Mercury
12. Back Home Again-John Denver-RCA
13. Skin Tight-Ohio Players-Mercury
14. Rags To Rufus-Rufus-ABC
15. Wrap Around Joy-Carole King-Ode
16. Perfect Angel-Minnie Riperton-Epic
17. Photographs And Memories-Jim Croce-ABC
18. When The Eagle Flies-Traffic-Asylum
19. Anka-Paul Anka-United Artists
20. Second Helping-Lynyrd Skynyrd-MCA
21. Roxy & Elsewhere-Zappa/Mothers-Discreet
22. Walls And Bridges-John Lennon-Apple
23. One-Bob James-CTI
24. Body Heat-Quincy Jones-A & M
25. Cheech & Chong's Wedding Album-Cheech & Chong-Ode

4	1.	Sweet Home Alabama-Lynyrd Skynyrd-MCA
6	2.	Can't Get Enough-Bad Company-Swan Song
1	3.	I Honestly Love You-Olivia Newton-John-MCA
2	4.	Nothing From Nothing-Billy Preston-A & M
9	5.	Stop And Smell The Roses-Mac Davis-Columbia
13	6.	Another Saturday Night-Cat Stevens-A & M
3	7.	Then Came You-Dionne Warwicke/Spinners-Atlantic
5	8.	Beach Baby-First Class-UK
17	9.	You Little Trustmaker-The Tymes-RCA
7	10.	Earache My Eye-Cheech & Chong-Ode
11	11.	Never My Love-Blue Swede-EMI
16	12.	Love Me For A Reason-The Osmonds-MGM
10	13.	Let's Put It All Together-The Stylistics-Avco
8	14.	I Shot The Sheriff-Eric Clapton-RSO
20	15.	The Bitch Is Back-Elton John-MCA
23	16.	Clap For The Wolfman-The Guess Who-RCA
19	17.	After The Goldrush-Prelude-Island
25	18.	Skin Tight-Ohio Players-Mercury
14	19.	(You're) Having My Baby-Paul Anka-UA
27	20.	Steppin' Out (Gonna Boogie Tonight)-Dawn-Bell
12	21.	The Need To be-Jim Weatherly-Buddah
29	22.	You Ain't Seen Nothing Yet-B-T Overdrive-Mercury
15	23.	Who Do You ?Think You Are-Bo Donaldson/Heywoods-ABC
31	24.	Straight Shootin' Woman-Steppenwolf-Mums
28	25.	Give It To The People-The Righteous Brothers-Haven
21	26.	Do It Baby-The Miracles-Tamla
18	27.	Kings Of The Party-Brownsville Station-Big Tree
32	28.	Back Home Again-John Denver-RCA
--	29.	Jazzman-Carole King-Ode
33	30.	Everlasting Love-Carl Carlton-Back Beat
--	31.	My Melody Of Love-Bobby Vinton-ABC
34	32.	Life Is A Rock-Reunion-RCA
22	33.	Rock Me Gently-Andy Kim-Capitol
36	34.	Longfellow Serenade-Neil Diamond-Columbia
24	35.	You Haven't Done Nothin'-Stevie

		Wonder-Tamla
--	36.	Kung Fu Fighting-Carl Douglas-20th Century
26	37.	Can't Get Enough Of Your Love-Barry Whtie-20th Century
--	38.	I've Got The Music In Me-The Kiki Dee Band-MCA
30	39.	I'm Leaving It Up To You-Donny & Marie Osmond-MGM
35	40.	It's Only Rock 'N Roll-Rolling Stones-Rolling Stones

SUPER CFL SURVEY OCT 26

wcfl wcfl wcfl wcfl wcfl wcfl wcfl wcfl wcfl wcfl wcfl wcfl wcfl wcfl

SUPER ALBUMS

1. Not Fragile-Bachman-Turner Overdrive-Mercury
2. "Bad Company"-Bad Company-Swan Song
3. Caribou-Elton John-MCA
4. If You Love Me, Let Me Know-Olivia Newton-John-MCA
5. Can't Get Enough-Barry White-20th Century
6. So Far-Crosby, Stills, Nash & Young-Atlantic
7. Welcome Back-Emerson, Lake & Palmer-Manticore
8. Photographs And Memories-Jim Croce-ABC
9. Skin Tight-Ohio Players-Mercury
10. Endless Summer-Beach Boys-Capitol
11. When The Eagle Flies-Traffic-Asylum
12. Wrap Around Joy-Carole King-Ode
13. Fulfillingness' First Finale-Stevie Wonder-Tamla
14. Bachman-Turner Overdrive II-Mercury
15. Perfect Angel-Minnie Riperton-Epic
16. 461 Ocean Boulevard-Eric Clapton-RSO
17. Second Helping-Lynyrd Skynyrd-MCA
18. Back Home Again-John Denver-RCA
19. Cheech & Chong's Wedding Album-Cheech & Chong-Ode
20. One-Bob James-CTI
21. Anka-Paul Anka-United Artists
22. Late For The Sky-Jackson Browne-Asylum
23. Live It Up-Isley Brothers-T-Neck
24. Body Heat-Quincy Jones-A & M
25. "Live"-Marvin Gaye-Tamla

1	1.	Sweet Home Alabama-Lynyrd Skynyrd-MCA
2	2.	Can't Get Enough-Bad Company-Swan Song
5	3.	Stop And Smell The Roses-Mac Davis-Columbia
3	4.	I Honestly Love You-Olivia Newton-John-MCA
6	5.	Another Saturday Night-Cat Stevens-A & M
4	6.	Nothing From Nothing-Billy Preston-A & M
9	7.	You Little Trustmaker-The Tymes-RCA
15	8.	The Bitch Is Back-Elton John-MCA
12	9.	Love Me For A Reason-The Osmonds-MGM
7	10.	Then Came You-Dionne Warwicke/Spinners-Atlantic
22	11.	You Ain't Seen Nothing Yet-B-T Overdrive-Mercury
17	12.	After The Goldrush-Prelude-Island
8	13.	Beach Baby-First Class-UK
16	14.	Clap For The Wolfman-The Guess Who-RCA
18	15.	Skin Tight-Ohio Players-Mercury
10	16.	Earache My Eye-Cheech & Chong-Ode
20	17.	Steppin' Out (Gonna Boogie Tonight)-Dawn-Bell
28	18.	Back Home Again-John Denver-RCA
11	19.	Never My Love-Blue Swede-EMI
13	20.	Let's Put It All Together-The Stylistics-Avco
24	21.	Straight Shootin' Women-Steppenwolf-Mums
14	22.	I Shot The Sheriff-Eric Clapton-RSO
25	23.	Give It To The People-The Righteous Brothers-Haven
31	24.	My Melody Of Love-Bobby Vinton-ABC
29	25.	Jazzman-Carole King-Ode
32	26.	Life Is A Rock-Reunion-RCA
34	27.	Longfellow Serenade-Neil Diamond-Columbia
30	28.	Everlasting Love-Carl Carlton-Back Beat
19	29.	(You're) Having My Baby-Paul Anka-UA
21	30.	The Need To Be-Jim Weatherly-Buddah
--	31.	Whatever Gets You Thru The Night-John Lennon-Apple
26	32.	Do It Baby-The Miracles-Tamla
36	33.	Kung Fu Fighting-Carl Douglas-20th Century
--	34.	Angie Baby-Helen Reddy-Capitol
38	35.	I've Got The Music In Me-The Kiki

	Dee Band-MCA
23	36. Who Do You Think You Are-Bo Donaldson/Heywoods-ABC
--	37. Wishing You Were Here-Chicago-Columbia
33	38. Rock Me Gently-Andy Kim-Capitol
27	39. Kings Of The Party-Brownsville Station-Big Tree
35	40. You Haven't Done Nothin'-Stevie Wonder-Tamla

SUPER CFL SURVEY NOV 2

wcfl wcfl wcfl wcfl wcfl wcfl wcfl wcfl wcfl wcfl wcfl wcfl wcfl wcfl

SUPER ALBUMS

1. Not Fragile-Bachman-Turner Overdrive-Mercury
2. "Bad Company"-Bad Company-Swan Song
3. If You Love Me, Let Me Know-Olivia Newton-John-MCA
4. Caribou-Elton John-MCA
5. Welcome Back-Emerson, Lake & Palmer-Manticore
6. Wrap Around Joy-Carole King-Ode
7. Photographs And Memories-Jim Croce-ABC
8. Cheech & Chong's Wedding Album-Cheech & Chong-Ode
9. War Child-Jethro Tull-Chrysalis
10. Endless Summer-Beach Boys-Capitol
11. So Far-Crosby, Stills, Nash & Young-Atlantic
12. Can't Get Enough-Barry White-20th Century
13. Fulfillingness' First Finale-Stevie Wonder-Tamla
14. When The Eagle Flies-Traffic-Asylum
15. Walls And Bridges-John Lennon-Apple
16. Back Home Again-John Denver-RCA
17. Second Helping-Lynyrd Skynyrd-MCA
18. Skin Tight-Ohio Players-Mercury
19. 461 Ocean Boulevard-Eric Clapton-RSO
20. Anka-Paul Anka-United Artists
21. Bachman-Turner Overdrive II-Mercury
22. Smiler-Rod Stewart-Mercury
23. Odds & Sods-The Who-MCA
24. Live It Up-Isley Brothers-T-Neck
25. One-Bob James-CTI

2	1.	Can't Get Enough-Bad Company-Swan Song
5	2.	Another Saturday Night-Cat Stevens-A & M
3	3.	Stop And Smell The Roses-Mac Davis-Columbia
7	4.	You Little Trustmaker-The Tymes-RCA
1	5.	Sweet Home Alabama-Lynyrd Skynyrd-MCA
8	6.	The Bitch Is Back-Elton John-MCA
11	7.	You Ain't Seen Nothing Yet-B-T Overdrive-Mercury
4	8.	I Honestly Love You-Olivia Newton-John-MCA
9	9.	Love Me For A Reason-The Osmonds-MGM
12	10.	After The Goldrush-Prelude-Island
14	11.	Clap For The Wolfman-The Guess Who-RCA
6	12.	Nothing From Nothing-Billy Preston-A & M
25	13.	Jazzman-Carole King-Ode
17	14.	Steppin' Out (Gonna Boogie Tonight)-Dawn-Bell
24	15.	My Melody Of Love-Bobby Vinton-ABC
18	16.	Back Home Again-John Denver-RCA
10	17.	Then Came You-Dionne Warwicke/ Spinners-Atlantic
26	18.	Life Is A Rock-Reunion-RCA
15	19.	Skin Tight-Ohio Players-Mercury
21	20.	Straight Shootin' Woman-Steppenwolf-Mums
13	21.	Beach Baby-First Class-UK
28	22.	Everlasting Love-Carl Carlton-Back Beat
23	23.	Give It To The People-The Righteous Brothers-Haven
27	24.	Longfellow Serenade-Neil Diamond-Columbia
16	25.	Earache My Eye-Cheech & Chong-Ode
--	26.	Tin Man-America-WB
31	27.	Whatever Gets You Thru The Night-John Lennon-Apple
19	28.	Never My Love-Blue Swede-EMI
33	29.	Kung Fu Fighting-Carl Douglas-20th Century
22	30.	I Shot The Sheriff-Eric Clapton-RSO
35	31.	I've Got The Music In Me-The Kiki Dee Band-MCA
34	32.	Angie Baby-Helen Reddy-Capitol
--	33.	Cat's In The Cradle-Harry Chapin-Elektra
20	34.	Let's Put It All Together-The Stylistics-Avco
37	35.	Wishing You Were Here-Chicago-

		Columbia
--	36.	Fire, Baby I'm On Fire-Andy Kim-Capitol
29	37.	(You're) Having My Baby-Paul Anka-UA
30	38.	The Need To be-Jim Weatherly-Buddah
36	39.	Who Do You Think You Are-Bo Donaldson/Heywoods-ABC
--	40.	Must Of Got Lost-J Geils Band-Atlantic

SUPER CFL SURVEY NOV 9

wcfl wcfl wcfl wcfl wcfl wcfl wcfl wcfl wcfl wcfl wcfl wcfl wcfl wcfl

SUPER ALBUMS

1. Not Fragile-Bachman-Turner Overdrive-Mercury
2. "Bad Company"-Bad Company-Swan Song
3. Photographs And Memories-Jim Croce-ABC
4. War Child-Jethro Tull-Chrysalis
5. Caribou-Elton John-MCA
6. It's Only Rock 'N Roll-The Rolling Stones-Rolling Stones
7. If You Love Me, Let Me Know-Olivia Newton-John-MCA
8. Welcome Back-Emerson, Lake & Palmer-Manticore
9. Wrap Around Joy-Carole King-Ode
10. So Far-Crosby, Stills, Nash & Young-Atlantic
11. Can't Get Enough-Barry White-20th Century
12. Cheech & Chong's Wedding Album-Cheech & Chong-Ode
13. Endless Summer-Beach Boys-Capitol
14. Skin Tight-Ohio Players-Mercury
15. Second Helping-Lynyrd Skynyrd-MCA
16. Perfect Angel-Minnie Riperton-Epic
17. Bachman-Turner Overdrive II-Mercury
18. When The Eagle Flies-Traffic-Asylum
19. Odds & Sods-The Who-MCA
20. 461 Ocean Boulevard-Eric Clapton-RSO
21. Back Home Again-John Denver-RCA
22. Walls And Bridges-John Lennon-Apple
23. Live It Up-Isley Brothers-T-Neck
24. One-Bob James-CTI
25. Serenade-Neil Diamond-Columbia

6	1.	The Bitch Is Back-Elton John-MCA
7	2.	You Ain't Seen Nothing Yet-B-T Overdrive-Mercury
1	3.	Can't Get Enough-Bad Company-Swan Song
4	4.	You Little Trustmaker-The Tymes-RCA
2	5.	Another Saturday Night-Cat Stevens-A & M
13	6.	Jazzman-Carole King-Ode
15	7.	My Melody Of Love-Bobby Vinton-ABC
5	8.	Sweet Home Alabama-Lynyrd Skynyrd-MCA
10	9.	After The Goldrush-Prelude-Island
16	10.	Back Home Again-John Denver-RCA
8	11.	I Honestly Love You-Olivia Newton-John-MCA
18	12.	Life Is A Rock-Reunion-RCA
3	13.	Stop And Smell The Roses-Mac Davis-Columbia
14	14.	Steppin' Out (Gonna Boogie Tonight)-Dawn-Bell
24	15.	Longfellow Serenade-Neil Diamond-Columbia
11	16.	Clap For The Wolfman-The Guess Who-RCA
9	17.	Love Me For A Reason-The Osmonds-MGM
22	18.	Everlasting Love-Carl Carlton-Back Beat
12	19.	Nothing From Nothing-Billy Preston-A & M
27	20.	Whatever Gets You Thru The Night-John Lennon-Apple
19	21.	Skin Tight-Ohio Players-Mercury
26	22.	Tin Man-America-WB
29	23.	Kung Fu Fighting-Carl Douglas-20th Century
31	24.	I've Got The Music In Me-The Kiki Dee Band-MCA
17	25.	Then Came You-Dionne Warwicke/Spinners-Atlantic
32	26.	Angie Baby-Helen Reddy-Capitol
20	27.	Straight Shootin' Woman-Steppenwolf-Mums
33	28.	Cat's In The Cradle-Harry Chapin-Elektra
35	29.	Wishing You Were Here-Chicago-Columbia
21	30.	Beach Baby-First Class-UK
--	31.	So You Are A Star-Hudson Brothers-Casablanca
23	32.	Give It To The People-The Righteous Brothers-Haven
36	33.	Fire, Baby I'm On Fire-Andy Kim-Capitol
25	34.	Earache My Eye-Cheech & Chong-Ode

40	35.	Must Of Got Lost-J. Geils Band-Atlantic
--	36.	Junior's Farm-Paul McCartney-Apple
--	37.	One Man Woman-Paul Anka-UA
28	38.	Never My Love-Blue Swede-EMI
30	39.	I Shot The Sheriff-Eric Clapton-RSO
34	40.	Let's Put It All Together-The Stylistics-Avco

SUPER CFL SURVEY NOV 16

wcfl wcfl wcfl wcfl wcfl wcfl wcfl wcfl wcfl wcfl wcfl wcfl wcfl wcfl

SUPER ALBUMS

1. Not Fragile-Bachman-Turner Overdrive-Mercury
2. Wrap Around Joy-Carole King-Ode
3. "Bad Company"-Bad Company-Swan Song
4. War Child-Jethro Tull-Chrysalis
5. Caribou-Elton John-MCA
6. It's Only Rock 'N Roll-The Rolling Stones-Rolling Stones
7. Photographs And Memories-Jim Croce-ABC
8. If You Love Me, Let Me Know-Olivia Newton-John-MCA
9. Serenade-Neil Diamond-Columbia
10. Walls And Bridges-John Lennon-Apple
11. Cheech & Chong's Wedding Album-Cheech & Chong-Ode
12. So Far-Crosby, Stills, Nash & Young-Atlantic
13. Back Home Again-John Denver-RCA
14. Can't Get Enough-Barry White-20th Century
15. Welcome Back-Emerson, Lake & Palmer-Manticore
16. "Live"-Marvin Gaye-Tamla
17. When The Eagle Flies-Traffic-Asylum
18. Skin Tight-Ohio Players-Mercury
19. Second Helping-Lynyrd Skynyrd-MCA
20. Perfect Angel-Minnie Riperton-Epic
21. Holiday-America-Warner Brothers
22. David Live-David Bowie-RCA
23. Endless Summer-Beach Boys-Capitol
24. Smiler-Rod Stewart-Mercury
25. Fulfillingness' First Finale-Stevie Wonder-Tamla

2	1.	You Ain't Seen Nothing Yet-B-T Overdrive-Mercury
1	2.	The Bitch Is Back-Elton John-MCA
7	3.	My Melody Of Love-Bobby Vinton-ABC
6	4.	Jazzman-Carole King-Ode
3	5.	Can't Get Enough-Bad Company-Swan Song
12	6.	Life Is A Rock-Reunion-RCA
9	7.	After The Goldrush-Prelude-Island
5	8.	Another Saturday Night-Cat Stevens-A & M
10	9.	Back Home Again-John Denver-RCA
22	10.	Tin Man-America-WB
15	11.	Longfellow Serenade-Neil Diamond-Columbia
4	12.	You Little Trustmaker-The Tymes-RCA
11	13.	I Honestly Love you-Olivia Newton-John-MCA
20	14.	Whatever Gets You Thru The Night-John Lennon-Apple
18	15.	Everlasting Love-Carl Carlton-Back Beat
8	16.	Sweet Home Alabama-Lynyrd Skynyrd-MCA
28	17.	Cat's In The Cradle-Harry Chapin-Elektra
23	18.	Kung Fu Fighting-Carl Douglas-20th Century
13	19.	Stop And Smell The Roses-Mac Davis-Columbia
24	20.	I've Got The Music In Me-The Kiki Dee Band-MCA
26	21.	Angie Baby-Helen Reddy-Capitol
17	22.	Love Me For A Reason-The Osmonds-MGM
29	23.	Wishing You Were Here-Chicago-Columbia
16	24.	Clap For The Wolfman-The Guess Who-RCA
33	25.	Fire, Baby I'm On Fire-Andy Kim-Capitol
14	26.	Steppin' Out (Gonna Boogie Tonight)-Dawn-Bell
31	27.	So You Are A Star-Hudson Brothers-Casablanca
19	28.	Nothing From Nothing-Billy Preston-A & M
35	29.	Must Of Got Lost-J Geils Band-Atlantic
--	30.	Do It ('Til You're Satisfied)-B.T. Express-Scepter
36	31.	Junior's Farm-Paul McCartney-Apple
21	32.	Skin Tight-Ohio Players-Mercury
37	33.	One Man Woman-Paul Anka With Odia Coates-UA
--	34.	Sha-La-La-Al Green-HI
30	35.	Beach Baby-First Class-UK

25	36.	Then Came You-Dionne Warwicke/ Spinners-Atlantic
--	37.	Fairytale-The Pointer Sisters-Blue Thumb
27	38.	Straight Shootin' Woman-Steppenwolf-Mums
34	39.	Earache My Eye-Cheech & Chong-Ode
32	40.	Give It To The People-The Righteous Brothers-Haven

SUPER CFL SURVEY NOV 23

wcfl wcfl wcfl wcfl wcfl wcfl wcfl wcfl wcfl wcfl wcfl wcfl wcfl wcfl

1	1.	You Ain't Seen Nothing Yet-B-T Overdrive-Mercury
3	2.	My Melody Of Love-Bobby Vinton-ABC
4	3.	Jazzman-Carole King-Ode
6	4.	Life Is A Rock-Reunion-RCA
2	5.	The Bitch Is Back-Elton John-MCA
10	6.	Tin Man-America-WB
7	7.	After The Goldrush-Prelude-Island
11	8.	Longfellow Serenade-Neil Diamond-Columbia
14	9.	Whatever Gets You Thru The Night-John Lennon-Apple
5	10.	Can't Get Enough-Bad Company-Swan Song
15	11.	Everlasting Love-Carl Carlton-Back Beat
9	12.	Back Home Again-John Denver-RCA
18	13.	Kung Fu Fighting-Carl Douglas-20th Century
17	14.	Cat's In The Cradle-Harry Chapin-Elektra
8	15.	Another Saturday Night-Cat Stevens-A & M
20	16.	I've Got The Music In Me-The Kiki Dee Band-MCA
12	17.	You Little Trustmaker-The Tymes-RCA
21	18.	Angie Baby-Helen Reddy-Capitol
23	19.	Wishing You Were Here-Chicago-Columbia
13	20.	I Honestly Love You-Olivia Newton-John-MCA
25	21.	Fire, Baby I'm On Fire-Andy Kim-Capitol
27	22.	So You Are A Star-Hudson Brothers-Casablanca
16	23.	Sweet Home Alabama-Lynyrd Skynyrd-MCA
31	24.	Junior's Farm-Paul McCartney-Apple
39	25.	Must Of Got Lost-J. Geils Band-Atlantic
30	26.	Do It ('Til You're Satisfied)-B.T. Express-Scepter
--	27.	I Can Help-Billy Swan-Monument
19	28.	Stop And Smell The Roses-Mac Davis-Columbia
33	29.	One Man Woman-Paul Anka With Odia Coates-UA
34	30.	Sha-La-La-Al Green-HI
22	31.	Love Me For A Reason-The Osmonds-MGM
--	32.	Play Something Sweet-Three Dog Night-Dunhill
37	33.	Fairytale-The Pointer Sisters-Blue Thumb
--	34.	I Feel A Song (In My Heart)-Gladys

SUPER ALBUMS

1. Not Fragile-Bachman-Turner Overdrive-Mercury
2. War Child-Jethro Tull-Chrysalis
3. "Bad Company"-Bad Company-Swan Song
4. It's Only Rock 'N Roll-The Rolling Stones-Rolling Stones
5. Wrap Around Joy-Carole King-Ode
6. Caribou-Elton John-MCA
7. Photographs And Memories-Jim Croce-ABC
8. Serenade-Neil Diamond-Columbia
9. Cheech & Chong's Wedding Album-Cheech & Chong-Ode
10. Can't Get Enough-Barry White-20th Century
11. If You Love Me, Let Me Know-Olivia Newton-John-MCA
12. When The Eagle Flies-Traffic-Asylum
13. Skin Tight-Ohio Players-Mercury
14. Walls And Bridges-John Lennon-Apple
15. So Far-Crosby-Stills, Nash & Young-Atlantic
16. Back Home Again-John Denver-RCA
17. "Lilve"-Marvin Gaye-Tamla
18. Mother Lode-Loggins And Messina-Columbia
19. Perfect Angel-Minnie Riperton-Epic
20. Second Helping-Lynyrd Skynyrd-MCA
21. Live It Up-Isley Brothers-T-Neck
22. Borboletta-Santana-Columbia
23. Fulfillingness' First Finale-Stevie Wonder-Tamla
24. Greatest Hits-Elton John-MCA
25. Odds & Sods-The Who-MCA

		Knight And The Pips-Buddah
--	35.	Ain't Too Proud To Beg-The Rolling Stones-Rolling Stones
26	36.	Steppin' Out (Gonna Boggie Tonight)-Dawn-Bell
24	37.	Clap For The Wolfman-The Guess Who-RCA
32	38.	Skin Tight-Ohio Players-Mercury
28	39.	Nothing From Nothing-Billy Preston-A & M
35	40.	Beach Baby-First Class-UK

SUPER CFL SURVEY NOV 30

wcfl wcfl wcfl wcfl wcfl wcfl wcfl wcfl wcfl wcfl wcfl wcfl wcfl wcfl

SUPER ALBUMS

1. Greatest Hits-Elton John-MCA
2. Not Fragile-Bachman-Turner Overdrive-Mercury
3. Serenade-Neil Diamond-Columbia
4. Caribou-Elton John-MCA
5. War Child-Jethro Tull-Chrysalis
6. "Bad Company"-Bad Company-Swan Song
7. It's Only Rock 'N Roll-The Rolling Stones-Rolling Stones
8. Wrap Around Joy-Carole King-Ode
9. Photographs And Memories-Jim Croce-ABC
10. This Is The Moody Blues-Threshold
11. Walls And Bridges-John Lennon-Apple
12. Cheech & Chong's Wedding Album-Cheech & Chong-Ode
13. Skin Tight-Ohio Players-Mercury
14. Mother Lode-Loggins And Messina-Columbia
15. When The Eagle Flies-Traffic-Asylum
16. Holiday-America-Warner Brothers
17. If You Love Me, Let Me Know-Olivia Newton-John-MCA
18. Back Home Again-John Denver-RCA
19. Heavy-Stylistics-Avco
20. Odds & Sods-The Who-MCA
21. David Live-David Bowie-RCA
22. Welcome Back-Emerson, Lake & Palmer-Manticore
23. "Live"-Marvin Gaye-Tamla
24. Borboletta-Santana-Columbia
25. Butterfly-Barbra Streisand-Columbia

2	1.	My Melody Of Love-Bobby Vinton-ABC
6	2.	Tin Man-America-WB
3	3.	Jazzman-Carole King-Ode
4	4.	Life Is A Rock-Reunion-RCA
1	5.	You Ain't Seen Nothing Yet-B-T Overdrive-Mercury
8	6.	Longfellow Serenade-Neil Diamond-Columbia
9	7.	Whatever Gets You Thru The Night-John Lennon-Apple
14	8.	Cat's In The Cradle-Harry Chapin-Elektra
13	9.	Kung Fu Fighting-Carl Douglas-20th Century
11	10.	Everlasting Love-Carl Carlton-Back Beat
7	11.	After The Goldrush-Prelude-Island
5	12.	The Bitch Is Back-Elton John-MCA
16	13.	I've Got The Music In Me-The Kiki Dee Band-MCA
24	14.	Junior's Farm-Paul McCartney-Apple
18	15.	Angie Baby-Helen Reddy-Capitol
19	16.	Wishing You Were Here-Chicago-Columbia
27	17.	I Can Help-Billy Swan-Monument
21	18.	Fire, Baby I'm On Fire-Andy Kim-Capitol
22	19.	So You Are A Star-Hudson Brothers-Casablanca
26	20.	Do It ('Til You're Satisfied)-B.T. Express-Scepter
30	21.	Sha-La-La-Al Green-HI
25	22.	Must Of Got Lost-J. Geils Band-Atlantic
12	23.	Back Home Again-John Denver-RCA
15	24.	Another Saturday Night-Cat Stevens-A & M
29	25.	One Man Woman-Paul Anka With Odia Coates-UA
--	26.	When Will I See You Again-The Three Degrees-PI
10	27.	Can't Get Enough-Bad Company-Swan Song
33	28.	Fairytale-The Pointer Sisters-Blue Thumb
32	29.	Play Something Sweet-Three Dog Night-Dunhill
--	30.	You're The First-The Last, My Everything-Barry White-20th Century
34	31.	I Feel A Song (In My Heart)-Gladys King And The Pips-Buddah
17	32.	You Little Trustmaker-The Tymes-RCA
35	33.	Ain't Too Proud To Beg-The Rolling Stones-Rolling Stones
--	34.	Lucy In The Sky With Diamonds-

		Elton John-MCA
20	35.	I Honestly Love You-Olivia Newton-John-MCA
23	36.	Sweet Home Alabama-Lynyrd Skynyrd-MCA
28	37.	Stop And Smell The Roses-Mac Davis-Columbia
31	38.	Love Me For A Reason-The Osmonds-MGM
37	39.	Clap For The Wolfman-The Guess Who-RCA
39	40.	Nothing From Nothing-Billy Preston-A & M

SUPER CFL SURVEY DEC 7

wcfl wcfl wcfl wcfl wcfl wcfl wcfl wcfl wcfl wcfl wcfl wcfl wcfl wcfl

SUPER ALBUMS

1. Greatest Hits-Elton John-MCA
2. Not Fragile-Bachman-Turner Overdrive-Mercury
3. Serenade-Neil Diamond-Columbia
4. War Child-Jethro Tull-Chrysalis
5. It's Only Rock 'N Roll-The Rolling Stones-Rolling Stones
6. "Bad Company"-Bad Company-Swan Song
7. Wrap Around Joy-Carole King-Ode
8. Photographs and Memories-Jim Croce-ABC
9. This Is The Moody Blues-Threshold
10. Caribou-Elton John-MCA
11. Walls and Bridges-John Lennon-Apple
12. Cheech & Chong's Wedding Album-Cheech & Chong-Ode
13. Can't Get Enough-Barry White-20th Century
14. Mother Lode-Loggins And Messina-Columbia
15. Skin Tight-Ohio Players-Mercury
16. When The Eagle Flies-Traffic-Asylum
17. Al Green Explores Your Mind-HI
18. Back Home Again-John Denver-RCA
19. Fire-Ohio Players-Mercury
20. Butterfly-Barbra Streisand-Columbia
21. If You Love Me, Let Me Know-Olivia Newton-John-MCA
22. Fulfillingness' First Finale-Stevie Wonder-Tamla
23. Perfect Angel-Minnie Riperton-Epic
24. So Far-Crosby-Stills, Nash & Young-Atlantic
25. "Live"-Marvin Gaye-Tamla

1	1.	My Melody Of Love-Bobby Vinton-ABC
2	2.	Tin Man-America-WB
8	3.	Cat's In The Cradle-Happy Chapin-Elektra
9	4.	Kung Fu Fighting-Carl Douglas-20th Century
6	5.	Longfellow Serenade-Neil Diamond-Columbia
7	6.	Whatever Gets You Thru The Night-John Lennon-Apple
3	7.	Jazzman-Carole King-Ode
4	8.	Life Is A Rock-Reunion-RCA
14	9.	Junior's Farm-Paul McCartney-Apple
17	10.	I Can Help-Billy Swan-Monument
5	11.	You Ain't Seen Nothing Yet-B-T Overdrive-Mercury
15	12.	Angie Baby-Helen Reddy-Capitol
10	13.	Everlasting Love-Carl Carlton-Back Beat
20	14.	Do It ('Til You're Satisfied)-B.T. Express-Scepter
16	15.	Wishing You Were Here-Chicago-Columbia
19	16.	So You Are A Star-Hudson Brothers-Casablanca
18	17.	Fire, Baby I'm On Fire-Andy Kim-Capitol
13	18.	I've Got The Music In me-The Kiki Dee Band-MCA
21	19.	Sha-La-La-Al Green-HI
22	20.	Must Of Got Lost-J. Geils Band-Atlantic
25	21.	One Man Woman-Paul Anka With Odia Coates-UA
26	22.	When Will I See You Again-The Three Degrees-PI
11	23.	After The Goldrush-Prelude-Island
29	24.	Play Something Sweet-Three Dog Night-Dunhill
12	25.	The Bitch Is Back-Elton John-RCA
28	26.	Fairytale-The Pointer Sisters-Blue Thumb
31	27.	I Feel A Song (In My Heart)-Gladys King And The Pips-Buddah
30	28.	You're The First, The Last, My Everything-Barry White-20th Century
33	29.	Ain't Too Proud To Beg-The Rolling Stones-Rolling Stones
23	30.	Back Home Again-John Denver-RCA
34	31.	Lucy In The Sky With Diamonds-Elton John-MCA
24	32.	Another Saturday Night-Cat Stevens-A & M
--	33.	Laughter In The Rain-Neil Sedaka-MCA

27	34.	Can't Get Enough-Bad Company-Swan Song
35	35.	I Honestly Love You-Olivia Newton-John-MCA
32	36.	You Little Trustmaker-The Tymes-RCA
36	37.	Sweet Home Alabama-Lynyrd Skynyrd-MCA
38	38.	Love Me For A Reason-The Osmonds-MGM
37	39.	Stop And Smell The Roses-Mac Davis-Columbia
39	40.	Clap For The Wolfman-The Guess Who-RCA

SUPER CFL SURVEY DEC 14

wcfl wcfl wcfl wcfl wcfl wcfl wcfl wcfl wcfl wcfl wcfl wcfl wcfl wcfl

SUPER ALBUMS

1. Greatest Hits-Elton John-MCA
2. War Child-Jethro Tull-Chrysalis
3. Not Fragile-Bachman-Turner Overdrive-Mercury
4. This Is The Moody Blues-Threshold
5. Serenade-Neil Diamond-Columbia
6. "Bad Company"-Bad Company-Swan Song
7. It's Only Rock 'N Roll-The Rolling Stones-Rolling Stones
8. Wrap Around Joy-Carole King-Ode
9. Caribou-Elton John-MCA
10. Verities & Balderdash-Harry Chapin-Elektra
11. Walls And Bridges-John Lennon-Apple
12. Photographs And Memories-Jim Croce-ABC
13. Holiday-America-Warner Brothers
14. Fire-Ohio Players-Mercury
15. Butterfly-Barbra Streisand-Columbia
16. Al Green Explores Your Mind-HI
17. Cheech & Chong's Wedding Album-Cheech & Chong-Ode
18. When The Eagle Flies-Traffic-Asylum
19. Fulfullingness' First Finale-Stevie Wonder-Tamla
20. Mother Lode-Loggins And Messina-Columbia
21. Together For The First Time-B.B. King & Bobby Bland-ABC Dunhill
22. David Live-David Bowie-RCA
23. Caught Up-Millie Jackson-Spring
24. Goodnight Vienna-Ringo Starr-Apple
25. Borboletta-Santana-Columbia

3	1.	Cat's In The Cradle-Harry Chapin-Elektra
4	2.	Kung Fu Fighting-Carl Douglas-20th Century
1	3.	My Melody Of Love-Bobby Vinton-ABC
9	4.	Junior's Farm-Paul McCartney-Apple
10	5.	I Can Help-Billy Swan-Monument
2	6.	Tin Man-America-WB
5	7.	Longfellow Serenade-Neil Diamond-Columbia
12	8.	Angie Baby-Helen Reddy-Capitol
14	9.	Do It ('Til You're Satisfied)-B.T. Express-Scepter
6	10.	Whatever Gets You Thru The Night-John Lennon-Apple
22	11.	When Will I See You Again-The Three Degrees-PI
31	12.	Lucy In The Sky With Diamonds-Elton John-MCA
7	13.	Jazzman-Carole King-Ode
19	14.	Sha-La-La-Al Green-HI
28	15.	You're The First, The Last, My Everything-Barry White-20th Century
21	16.	One Man Woman-Paul Anka With Odia Coates-UA
20	17.	Must Of Got Lost-J. Geils Band-Atlantic
8	18.	Life Is A Rock-Reunion-RCA
11	19.	You Ain't Seen Nothing Yet-B-T Overdrive-Mercury
16	20.	So You Are A Star-Hudson Brothers-Casablanca
26	21.	Fairytale-The Pointer Sisters-Blue Thumb
24	22.	Play Something Sweet-Three Dog Night-Dunhill
27	23.	I Feel A Song (In My Heart)-Gladys King And The Pips-Buddah
13	24.	Everlasting Love-Carl Carlton-Back Beat
15	25.	Wishing You Were Here-Chicago-Columbia
29	26.	Ain't Too Proud To Beg-The Rolling Stones-Rolling Stones
17	27.	Fire, Baby I'm On Fire-Andy Kim-Capitol
--	28.	You Got The Love-Rufus-ABC
33	29.	Laughter In The Rain-Neil Sedaka-MCA
18	30.	I've Got The Music In Me-The Kiki Dee Band-MCA
--	31.	Only You-Ringo Starr-Apple
23	32.	After The Goldrush-Prelude-Island
--	33.	Best Of My Love-The Eagles-Asylum

25	34.	The Bitch Is Back-Elton John-MCA
30	35.	Back Home Again-John Denver-RCA
35	36.	I Honestly Love You-Olivia Newton-John-MCA
32	37.	Another Saturday Night-Cat Stevens-A & M
--	38.	When A Child Is Born-Michael Holm-Mercury
36	39.	You Little Trustmaker-The Tymes-RCA
34	40.	Can't Get Enough-Bad Company-Swan Song

SUPER CFL SURVEY DEC 21

wcfl wcfl wcfl wcfl wcfl wcfl wcfl wcfl wcfl wcfl wcfl wcfl wcfl wcfl

SUPER ALBUMS

1. Greatest Hits-Elton John-MCA
2. This Is The Moody Blues-Threshold
3. Not Fragile-Bachman-Turner Overdrive-Mercury
4. Serenade-Neil Diamond-Columbia
5. War Child-Jethro Tull-Chrysalis
6. "Bad Company"-Bad Company-Swan Song
7. It's Only Rock 'N Roll-The Rolling Stones-Rolling Stones
8. Wrap Around Joy-Carole King-Ode
9. Butterfly-Barbra Streisand-Columbia
10. Fire-Ohio Players-Mercury
11. Miles Of Aisles-Joni Mitchell-Asylum
12. Back Home Again-John Denver-RCA
13. Photographs And Memories-Jim Croce-ABC
14. Verities & Balderdash-Harry Chapin-Elektra
15. Mother Lode-Loggins And Messina-Columbia
16. Caribou-Elton John-MCA
17. Al Green Explores Your Mind-HI
18. Walls And Bridges-John Lennon-Apple
19. Can't Get Enough-Barry White-20th Century
20. Stormbringer-Deep Purple-Warner Brothers
21. Bobby Vinton With Love-Epic
22. Fulfillingness' First Finale-Stevie Wonder-Tamla
23. Goodnight Vienna-Ringo Starr-Apple
24. Caught Up-Millie Jackson-Spring
25. Cheech & Chong's Wedding Album-Cheech & Chong-Ode

2	1.	Kung Fu Fighting-Carl Douglas-20th Century
1	2.	Cat's In The Cradle-Harry Chapin-Elektra
4	3.	Junior's Farm-Paul McCartney-Apple
5	4.	I Can Help-Billy Swan-Monument
3	5.	My Melody Of Love-Bobby Vinton-ABC
9	6.	Do It ('Til You're Satisfied)-B.T. Express-Scepter
8	7.	Angie Baby-Helen Reddy-Capitol
11	8.	When Will I See You Again-The Three Degrees-PI
12	9.	Lucy In The Sky With Diamonds-Elton John-MCA
6	10.	Tin Man-America-WB
15	11.	You're The First, The Last, My Everything-Barry White-20th Century
14	12.	Sha-La-La-Al Green-HI
16	13.	One Man Woman-Paul Anka With Odia Coates-UA
7	14.	Longfellow Serenade-Neil Diamond-Columbia
10	15.	Whatever Gets You Thru The Night-John Lennon-Apple
17	16.	Must Of Got Lost-J. Geils Band-Atlantic
21	17.	Fairytale-The Pointer Sisters-Blue Thumb
13	18.	Jazzman-Carole King-Ode
23	19.	I Feel A Song (In My Heart)-Gladys Knight And The Pips-Buddah
22	20.	Play Something Sweet-Three Dog Night-Dunhill
29	21.	Laughter In The Rain-Neil Sedaka-MCA
26	22.	Ain't Too Proud To Beg-The Rolling Stones-Rolling Stones
18	23.	Life Is A Rock-Reunion-RCA
28	24.	You Got The Love-Rufus-ABC
19	25.	You Ain't Seen Nothing Yet-B-T Overdrive-Mercury
31	26.	Only You-Ringo Starr-Apple
--	27.	Boogie On Reggae Woman-Stevie Wonder-Tamla
20	28.	So You Are A Star-Hudson Brothers-Casablanca
25	29.	Wishing You Were Here-Chicago-Columbia
--	30.	Please Mr. Postman-Carpenters-A & M
33	31.	Best Of My Love-The Eagles-Asylum
24	32.	Everlasting Love-Carl Carlton-Back Beat
--	33.	Get Dancin'-Disco Tex & The Sex-O-Lettes-Chelsea
38	34.	When A Child Is Born-Michael Holm-Mercury

27	35.	Fire, Baby I'm On Fire-Andy Kim-Capitol
30	36.	I've Got The Music In Me-The Kiki Dee Band-MCA
--	37.	Some Kind Of Wonderful-Grand Funk-Grand Funk
32	38.	After The Goldrush-Prelude-Island
34	39.	The Bitch Is Back-Elton John-MCA
36	40.	I Honestly Love You-Olivia Newton-John-MCAt

SUPER CFL SURVEY DEC 28

wcfl wcfl wcfl wcfl wcfl wcfl wcfl wcfl wcfl wcfl wcfl wcfl wcfl wcfl

SUPER ALBUMS

1. Greatest Hits-Elton John-MCA
2. This Is The Moody Blues-Threshold
3. Not Fragile-Bachman-Turner Overdrive-Mercury
4. War Child-Jethro Tull-Chrysalis
5. Verities & Balderdash-Harry Chapin-Elektra
6. Serenade-Neil Diamond-Columbia
7. It's Only Rock 'N Roll-The Rolling Stones-Rolling Stones
8. Fire-Ohio Players-Mercury
9. "Bad Company"-Bad Company-Swan Song
10. Butterfly-Barbra Streisand-Columbia
11. Photographs And Memories-Jim Croce-ABC
12. Holiday-America-Warner Brothers
13. Back Home Again-John Denver-RCA
14. Bobby Vinton With Love-Epic
15. Together For The First Time-B.B. King & Bobby Bland-ABC Dunhill
16. Wrap Around Joy-Carole King-Ode
17. Al Green Explores Your Mind-HI
18. Miles Of Aisles-Joni Mitchell-Asylum
19. Caribou-Elton John-MCA
20. Walls And Bridges-John Lennon-Apple
21. Mother Lode-Loggins And Messina-Columbia
22. New And Improved-Spinners-Atlantic
23. Cheech & Chong's Wedding Album-Cheech & Chong-Ode
24. Caught Up-Millie Jackson-Spring
25. Dark Horse-George Harrison-Apple

4	1.	I Can Help-Billy Swan-Monument
1	2.	Kung Fu Fighting-Carl Douglas-20th Century
9	3.	Lucy In The Sky With Diamonds-Elton John-MCA
3	4.	Junior's Farm-Paul McCartney-Apple
6	5.	Do It ('Til You're Satisfied)-B.T. Express-Scepter
8	6.	When Will I See You Again-The Three Degrees-PI
11	7.	You're The First-The Last, My Everything-Barry White-20th Century
7	8.	Angie Baby-Helen Reddy-Capitol
2	9.	Cat's In The Cradle-Harry Chapin-Elektra
13	10.	One Man Woman-Paul Anka With Odia Coates-UA
5	11.	My Melody Of Love-Bobby Vinton-ABC
21	12.	Laughter In The Rain-Neil Sedaka-MCA
17	13.	Fairytale-The Pointer Sisters-Blue thumb
12	14.	Sha-La-La-Al Green-HI
10	15.	Tin Man-America-WB
19	16.	I Feel A Song (In My Heart)-Gladys Knight And The Pips-Buddah
24	17.	You Got The Love-Rufus-ABC
16	18.	Must Of Got Lost-J. Geils Band-Atlantic
14	19.	Longfellow Serenade-Neil Diamond-Columbia
26	20.	Only You-Ringo Starr-Apple
20	21.	Play Something Sweet-Three Dog Night-Dunhill
27	22.	Boogie On Reggae Woman-Stevie Wonder-Tamla
15	23.	Whatever Gets You Thru The Night-John Lennon-Apple
31	24.	Best Of My Love-The Eagles-Asylum
30	25.	Please Mr. Postman-Carpenters-A & M
18	26.	Jazzman-Carole King-Ode
22	27.	Ain't Too Proud To Beg-The Rolling Stones-Rolling Stones
--	28.	Promised Land-Elvis Presley-RCA
34	29.	When A Child Is Born-Michael Holm-Mercury
33	30.	Get Dancin'-Disco Tex & The Sex-O'Lettes-Chelsea
23	31.	Life Is A Rock-Reunion-RCA
--	32.	Mandy-Barry Manilow-Bell
37	33.	Some Kind Of Wonderful-Grand Funk-Grand Funk

28	34.	So You Are A Star-Hudson Brothers-Casablanca
--	35.	Morning Side Of The Mountain-Donny & Marie Osmond-MGM
25	36.	You Ain't Seen Nothing Yet-B.T. Overdrive-Mercury
--	37.	Doctor's Orders-Carol Douglas-Midland Int'l
29	38.	Wishing You Were Here-Chicago-Columbia
32	39.	Everlasting Love-Carl Carlton-Back Beat
35	40.	Fire, Baby I'm On Fire-Andy Kim-Capitol

1. I Honestly Love You-Olivia Newton John
2. Billy, Don't Be A Hero-Bo Donaldson & The Heywoods
3. Season's In The Sun-Terry Jacks
4. My Melody Of Love-Bobby Vinton
5. Sundown-Gordon Lightfoot
6. Cat's In The Cradle-Harry Chapin
7. Band On The Run-Paul McCartney & The Wings
8. You're Havinf My Baby-Paul Anka
9. The Way We Were-Barbara Streisand
10. Loco-Motion-Grand Funk
11. Rock The Boat-The Hues Corporation
12. Ennie & The Jets-Elton John
13. Hooked On A Feeling-Blue Swede-EMI
14. The Entertainer-Marvin Hamlisch-MCA
15. Star Baby-Guess Who-RCA
16. Annie's Song-John Denver-RCA
17. Beach Baby-First Class-UK
18. Sweet Home Alabama-Lynyrd Skynyrd-MCA
19. The Streak-Ray Stevens-Barnaby
20. I Can Help-Billy Swan-Monument
21. Mockingbird-Carly Simon & James Taylor-Elektra
22. Rock On-David Essex-Columbia
23. Sunshine On My Shoulders-John Denver-RCA
24. If You Love Me-Olivia Newton-John-MCA
25. Dancing Machine-Jackson 5-Motown
26. The Night Chicago Died-Paper Lace-Mercury
27. The Most Beautiful Girl-Charlie Rich-Epic
28. Nothing From Nothing-Billy Preston-A & M
29 Midnight At The Oasis-Maria Muldaur-Reprise
30. The Air That I Breathe-The Hollies-Epic
31. Takin' Care Of Business-Bachman-Turner-Overdrive-Mercury
32. Come And Get Your Love-Redbone-Epic
33. Rock Your Baby-George McCrae-TK
34. The Joker-Steve Miller Band-Capitol
35. Tin Man-America-WB
36. I Shot The Sheriff-Eric Clapton-RSO
37. Rikki, Don't Lose That Number-Steely Dan-ABC
38. You're 16-Ringo Starr-Apple
39. Love's Theme-Love Unlimited Orchestra-20th Century
40. Can't Get Enough-Bad Company-Swan Song

SMALL PRINT DEPARTMENT: This Super CFL survey repre-
sents the relative popularity of records during1974 based on sales reports and Super CFL's highly expert musical judgment. By the way, this is collector's issue number 200 for you clowns that are still counting! Thank you for reminding me, Mr. Lujack.

1975

SUPER CFL SURVEY JAN 11

wcfl wcfl wcfl wcfl wcfl wcfl wcfl wcfl wcfl wcfl wcfl wcfl wcfl wcfl

SUPER ALBUMS

1. Greatest Hits-Elton John-MCA
2. This Is The Moody Blues-Threshold
3. Not Fragile-Bachman-Turner Overdrive-Mercury
4. Verities And Balderdash-Harry Chapin-Elektera
5. Butterfly-Barbra Streisand-Columbia
6. War Child-Jethro Tull-Chrysalis
7. Serenade-Neil Diamond-Columbia
8. Fire-Ohio Players-Mercury
9. With Love-Bobby Vinton-Epic
10. Back Home Again-John Denver-RCA
11. Miles Of Aisles-Joni Mitchell-Asylum
12. Al Green Explores Your Mind-HI
13. Photographs And Memories-Jim Croce-ABC
14. Dark Horse-George Harrison-Apple
15. "Bad Company"-Bad Company-Swan Song
16. Relayer-Yes-Atlantic
17. Mother Lode-Loggins And Messina-Columbia
18. New And Improved-Spinners-Atlantic
19. Caught Up-Millie Jackson-Spring
20. Can't Get Enough-Barry White-20th Century
21. Fulfillingness' First Finale-Stevie Wonder-Tamla
22. It's Only Rock 'N Roll-The Rolling Stones-Rolling Stones
23. Dragon Fly-Jefferson Starship-Grunt
24. Holiday-America-Warner Brothers
25. Souvenirs-Dan Fogelberg-Epic

3	1.	Lucy In The Sky With Diamonds-Elton John-MCA
1	2.	I Can Help-Billy Swan-Monument
7	3.	You're The First-The Last-My Everything-Barry White-20th Century
6	4.	When Will I See You Again-The Three Degrees-PI
5	5.	Do It ('Til You're Satisfied)-B.T. Express-Scepter
10	6.	One Man Woman-Paul Anka With Odia Coates-UA
12	7.	Laughter In The Rain-Neil Sedaka-MCA
25	8.	Please Mr. Postman-Carpenters-A & M
2	9.	Kung Fu Fighting-Carl Douglas-20th Century
20	10.	Only You-Ringo Starr-Apple
17	11.	You Got The Love-Rufus-ABC
4	12.	Junior's Farm-Paul McCartney-Apple
22	13.	Boogie On Reggae Woman-Stevie Wonder-Tamla
9	14.	Cat's In The Cradle-Harry Chapin-Elektra
8	15.	Angie Baby-Helen Reddy-Capitol
24	16.	Best Of My Love-The Eagles-Asylum
11	17.	My Melody Of Love-Bobby Vinton-ABC
16	18.	I Feel A Song (In My Heart)-Gladys Knight And The Pips-Buddah
13	19.	Fairytale-The Pointer Sisters-Blue Thumb
32	20.	Mandy-Barry Manilow-Bell
14	21.	Sha-La-La-Al Green-HI
29	22.	When A Child Is Born-Michael Holm-Mercury
28	23.	Promised Land-Elvis Presley-RCA
30	24.	Get Dancin'-Disco Tex & Sex-O-Lettes-Chelsea
35	25.	Morning Side Of The Mountain-Donny & Marie Osmond-MGM
15	26.	Tin Man-America-WB
--	27.	Bungle In The Jungle-Jethro Tull-Chrysalis
33	28.	Some Kind Of Wonderful-Grand Funk-Grand Funk
19	29.	Longfellow Serenade-Neil Diamond-Columbia
--	30.	Never Can Say Goodbye-Gloria Gaynor-MGM
37	31.	Doctor's Orders-Carol Douglas-Midland Int'l
18	32.	Must Of Got Lost-J. Geils Band-Atlantic
--	33.	Pick Up The Pieces-AWB-Atlantic
--	34.	You're No Good-Linda Ronstadt-Capitol

21	35.	Play Something Sweet-Three Dog Night-Dunhill
--	36.	#9 Dream-John Lennon-Apple
--	37.	Your Bulldog Drinks Champagne-Jim Stafford-MGM
26	38.	Jazzman-Carole King-Ode
--	39.	Lady-Styx-Wooden Nickel
--	40.	It's All Right-Jim Capaldi-Island

SUPER CFL SURVEY JAN 18

WCFL WCFL WCFL WCFL WCFL WCFL WCFL WCFL WFCL

SUPER ALBUMS

1. Greatest Hits-Elton John-MCA
2. This Is The Moody Blues-Threshold
3. War Child-Jethro Tull-Chrysalis
4. Not Fragile-Bachman-Turner Overdirve-Mercury
5. Verities And Balderdash-Harry Chapin-Elektra
6. Butterfly-Barbra Streisand-Columbia
7. Melodies Of Love-Bobby Vinton-ABC
8. Fire-Ohio Players-Mercury
9. Serenade-Neil Diamond-Columbia
10. Dark Horse-George Harrison-Apple
11. Al Green Explores Your Mind-Hi
12. Miles Of Aisles-Joni Mitchell-Asylum
13. Back Home Again-John Denver-RCA
14. Photographs And Memories-Jim Croce-ABC
15. New And Improved-Spinners-Atlantic
16. Relayer-Yes-Atlantic
17. Styx II-Styx-Wooden Nickel
18. "Bad Company"-Bad Company-Swan Song
19. Mother Lode-Loggins And Messina-Columbia
20. Souvenirs-Dan Fogelberg-Epic
21. Stormbringer-Deep Purple-Warner Brothers
22. Heart Like A Wheel-Linda Ronstadt-Capitol
23. Average White Band-AWB-Atlantic
24. It's Only Rock 'N Roll-The Rolling Stones-Rolling Stones
25. Light Of Worlds-Kool & The Gang-De-Lite

WAS	IS	
1	1.	LUCY IN THE SKY WITH DIAMONDS-Elton John-MCA
7	2.	Laughter In The Rain-Neil Sedaka-MCA
8	3.	Please Mr. Postman-Carpenters-A&M
4	4.	When Will I See You Again-The Three Degrees-PI
6	5.	One Man Woman-Paul Anka With Odra Coates-UA
2	6.	I Can Help-Billy Swan-Monument
10	7.	Only You-Ringo Starr-Apple
3	8.	You're The First, The Last ,My Everything-Barry White-20th Century
13	9.	Boogie On Reggae Woman-Stevie Wonder-Tamla
11	10.	You Got The Love-Rufus-ABC
5	11.	Do It ('Till You're Satisfied)-B.T. Express-Scepter
16	12.	Best Of My Love-The Eagles-Asylum
9	13.	Kung Fu Fighting-Carl Douglas-20th Century
20	14.	Mandy-Barry Manilow-Bell
12	15.	Junior's Farm-Paul McCartney-Apple
25	16.	Morning Side Of The Mountain-Donny & Marie-MGM
23	17.	Promised Land-Elvis Presley-RCA
15	18.	Angie Baby-Helen Reddy-Capitol
22	19.	When A Child Is Born-Michael Helm-Mercury
27	20.	Bungle In The Jungle-Jethro Tail-Chrysalis
24	21.	Get Dancin'-Disco Tex & Sex 0-Lettes-Chelsea
39	22.	Lady-Styx-Wooden Nickel
14	23.	Cat's In The Cradle-Harry Chapin-Elektra
28	24.	Some Kind Of Wonderful-Grand Funk-Grand Funk
30	25.	Never Can Say Goodbye-Gloria Gaynor-MGM
17	26.	My Melody Of Love-Bobby Vinton-ABC
31	27.	Doctor's Orders-Carol Douglas-Midland Int'l
18	28.	I Feel A Song-Gladys Knight/Pips-Buddah
34	29.	You're No Good-Linda Ronstadt-Capitol
33	30.	Pick Up The Pieces-AWB-Atlantic
19	31.	Fairytale-The Pointer Sisters-Blue Thumb
36	32.	#9 Dream-John Lennon-Apple
26	33.	Tin Man-America-WH
37	34.	Your Bulldog Drinks Champagne-Jim Stafford-MGM
40	35.	It's All Right-Jim Capaldi-Island
21	36.	Sha-La-La-Al Green-HI
--	37.	Sweet Surrender-John Denver-RCA
--	38.	My Eyes Adored You-Frankie Valli-Private Stock
29	39.	Longfellow Serenade-Neil Diamond-Columbia
32	40.	Must Of Got Lost-J. Geils Band-Atlantic

SUPER CFL SURVEY JAN 25

wcfl wcfl wcfl wcfl wcfl wcfl wcfl wcfl wcfl wcfl wcfl wcfl wcfl wcfl

SUPER ALBUMS

1. Greatest Hits-Elton John-MCA
2. This Is The Moody Blues-Threshold
3. War Child-Jethro Tull-Chrysalis
4. Dark Horse-George Harrison-Apple
5. Not Fragile-Bachman-Turner Overdrive-Mercury
6. Fire-Ohio Players-Mercury
7. Verities And Balderdash-Harry Chapin-Elektra
8. Styx II-Styx-Wooden Nickel
9. Butterfly-Barbra Streisand-Columbia
10. Miles Of Aisles-Joni Mitchell-Asylum
11. Serenade-Neil Diamond-Columbia
12. Relayer-Yes-Atlantic
13. New And Improved-Spinners-Atlantic
14. Photographs And Memories-Jim Croce-ABC
15. Al Green Explores Your Mind-Hi
16. Melodies Of Love-Bobby Vinton-ABC
17. Joy To The World-Three Dog Night-Dunhill
18. Heart Like A Wheel-Linda Ronstadt-Capitol
19. Back Home Again-John Denver-RCA
20. So What-Joe Walsh-Dunhill
21. Mother Lode-Loggins And Messina-Columbia
22. Souvenirs-Dan Fogelberg-Epic
23. Average White Band-AWB-Atlantic
24. Caught Up-Millie Jackson-Spring
25. White Gold-The Love Unlimited Orchestra-20th Century

2	1.	Laughter In The Rain-Neil Sedaka-MCA
3	2.	Please Mr. Postman-Carpenters-A & M
1	3.	Lucy In The Sky With Diamonds-Elton John-MCA
14	4.	Mandy-Barry Manilow-Bell
4	5.	When Will I See You Again-The Three Degrees-PI
7	6.	Only You-Ringo Starr-Apple
9	7.	Boogie On Reggae Woman-Stevie Wonder-Tamla
12	8.	Best Of My Love-The Eagles-Asylum
5	9.	One Man Woman-Paul Anka With Odia Coates-UA
10	10.	You Got The Love-Rufus-ABC
8	11.	You're The First, The Last, My Everything-Barry White-20th Century
6	12.	I Can Help-Billy Swan-Monument
20	13.	Bungle In The Jungle-Jethro Tull-Chrysalis
16	14.	Morning Side Of The Mountain-Donny & Marie-MGM
22	15.	Lady-Styx-Wooden Nickel
17	16.	Promised Land-Elvis Presley-RCA
11	17.	Do It ('Til You're Satisfied)-B.T. Express-Scepter
25	18.	Get Dancin'-Disco Tex & Sex-O-Lettes-Chelsea
25	19.	Never Can Say Goodbye-Gloria Gaynor-MGM
19	20.	When A Child Is Born-Michael Holm-Mercury
24	21.	Some Kind Of Wonderful-Grand Funk-Grand Funk
13	22.	Kung Fu Fighting-Carl Douglas-20th Century
18	23.	Angie Baby-Helen Reddy-Capitol
27	24.	Doctor's Orders-Carol Douglas-Midland-Int'l
29	25.	You're No Good-Linda Ronstadt-Capitol
15	26.	Junior's Farm-Paul McCartney-Apple
--	27.	Fire-Ohio Players-Mercury
30	28.	Pick Up The Pieces-AWB-Atlantic
23	29.	Cat's In The Cradle-Harry Chaplin-Elektra
32	30.	#9 Dream-John Lennon-Apple
35	31.	It's All Right-Jim Capaldi-Island
34	32.	Your Bulldog Drinks Champagne-Jim Stafford-MGM
37	33.	Sweet Surrender-John Denver-RCA
26	34.	My Melody Of Love-Bobby Vinton-ABC
--	35.	Up In A Puff Of Smoke-Polly

		Brown-GTO
38	36.	My Eyes Adored You-Frankie Valli-Private Stock
28	37.	I Feel A Song-Gladys Knight/Pips-Buddah
--	38.	Part Of The Plan-Dan Fogelberg-Epic
31	39.	Fairytale-The Pointer Sisters-Blue Thumb
--	40.	Emma-Hot Chocolate-Big Tree

SUPER CFL SURVEY FEB 1

wcfl wcfl wcfl wcfl wcfl wcfl wcfl wcfl wcfl wcfl wcfl wcfl wcfl wcfl

SUPER ALBUMS

1. Greatest Hits-Elton John-MCA
2. War Child-Jethro Tull-Chrysalis
3. Styx II-Styx-Wooden Nickel
4. Fire-Ohio Players-Mercury
5. Not Fragile-Bachman-Turner Overdrive-Mercury
6. This Is The Moody Blues-Threshold
7. Heart Like A Wheel-Linda Ronstadt-Capitol
8. Miles Of Aisles-Joni Mitchell-Asylum
9. Melodies Of Love-Bobby Vinton-ABC
10. Butterfly-Barbra Streisand-Columbia
11. Relayer-Yes-Atlantic
12. Verities And Balderdash-Harry Chapin-Elektra
13. New And Improved-Spinners-Atlantic
14. Dark Horse-George Harrison-Apple
15. Al Green Explores Your Mind-Hi
16. Mother Lode-Loggins And Messina-Columbia
17. Souvenirs-Dan Fogelberg-Epic
18. Joy To The World-Three Dog Night-Dunhill
19. Back Home Again-John Denver-RCA
20. White Gold-The Love Unlimited Orchestra-20th Century
21. Photographs And Memories-Jim Croce-ABC
22. Serenade-Neil Diamond-Columbia
23. Barry Manilow II-Bell
24. Perfect Angel-Minnie Riperton-Epic
25. Average White Band-AWB-Atlantic

2	1.	Please Mr. Postman-Carpenters-A & M
1	2.	Laughter In The Rain-Neil Sedaka-MCA
4	3.	Mandy-Barry Manilow-Bell
8	4.	Best Of My Love-The Eagles-Asylum
3	5.	Lucy In The Sky With Diamonds-Elton John-MCA
7	6.	Boogie On Reggae Woman-Stevie Wonder-Tamla
14	7.	Morning Side Of The Mountain-Donny & Marie-MGM
6	8.	Only You-Ringo Starr-Apple
13	9.	Bungle In The Jungle-Jethro Tull-Chrysalis
15	10.	Lady-Styx-Wooden Nickel
5	11.	When Will I See You Again-The Three Degrees-PI
9	12.	One Man Woman-Paul Anka With Odia Coates-UA
19	13.	Never Can Say Goodbye-Gloria Gaynor-MGM
10	14.	You Got The Love-Rufus-ABC
21	15.	Some Kind Of Wonderful-Grand Funk-Grand Funk
24	16.	Doctor's Orders-Carol Douglas-Midland Int'l
18	17.	Get Dancin'-Disco Tex & Sex-O-Lettes-Chelsea
12	18.	I Can Help-Billy Swan-Monument
25	19.	You're No Good-Linda Ronstadt-Capitol
16	20.	Promised Land-Elvis Presley-RCA
11	21.	You're The First, The Last, My Everything-Barry White-20th Century
27	22.	Fire-Ohio Players-Mercury
28	23.	Pick Up The Pieces-AWB-Atlantic
33	24.	Sweet Surrender-John Denver-RCA
30	25.	#9 Dream-John Lennon-Apple
22	26.	Kung Fu Fighting-Carl Douglas-20th Century
31	27.	It's All Right-Jim Capaldi-Island
--	28.	Black Water-The Doobie Brothers-WB
32	29.	Your Bulldog Drinks Champagne-Jim Stafford-MGM
36	30.	My Eyes Adored You-Frankie Valli-Private Stock
--	31.	Roll On Down The Highway-Bachman-Turner Overdrive-Mercury
35	32.	Up In A Puff Of Smoke-Polly Brown-GTO
17	33.	Do It ('Til You're Satisfied)-B.T.

		Express-Scepter
38	34.	Part Of The Plan-Dan Fogelberg-Epic
--	35.	Have You Never Been Mellow-Olivia Newton-John-MCA
20	36.	When A Child Is Born-Michael Holm-Mercury
40	37.	Emma-Hot Chocolate-Big Tree
26	38.	Junior's Farm-Paul McCartney-Apple
23	39.	Angie Baby-Helen Reddy-Capitol
--	40.	I'll Still Love You-Jim Weatherly-Buddah

SUPER CFL SURVEY FEB 8

wcfl wcfl wcfl wcfl wcfl wcfl wcfl wcfl wcfl wcfl wcfl wcfl wcfl wcfl

SUPER ALBUMS

1. Greatest Hits-Elton John-MCA
2. Styx II-Styx-Wooden Nickel
3. War Child-Jethro Tull-Chrysalis
4. Fire-Ohio Players-Mercury
5. Heart Like A Wheel-Linda Ronstadt-Capitol
6. This Is The Moody Blues-Threshold
7. Not Fragile-Bachman-Turner Overdrive-Mercury
8. Relayer-Yes-Atlantic
9. Miles Of Aisles-Joni Mitchell-Asylum
10. Dark Horse-George Harrison-Apple
11. Melodies Of Love-Bobby Vinton-ABC
12. Average White Band-AWB-Atlantic
13. New And Improved-Spinners-Atlantic
14. Joy To The World-Three Dog Night-Dunhill
15. Verities And Balderdash-Harry Chapin-Elektra
16. Back Home Again-John Denver-RCA
17. So What-Joe Walsh-Dunhill
18. Souvenirs-Dan Fogelberg-Epic
19. Butterfly-Barbra Streisand-Columbia
20. Photographs And Memories-Jim Croce-ABC
21. Perfect Angel-Minnie Riperton-Epic
22. White Gold-The Love Unlimited Orchestra-20th Century
23. Serenade-Neil Diamond-Columbia
24. Barry Manilow II-Bell
25. Mother Lode-Loggins And Messina-Columbia

3	1.	Mandy-Barry Manilow-Bell
1	2.	Please Mr. Postman-Carpenters-A & M
4	3.	Best Of My Love-The Eagles-Asylum
6	4.	Boogie On Reggae Woman-Stevie Wonder-Tamla
2	5.	Laughter In The Rain-Neil Sedaka-MCA
10	6.	Lady-Styx-Wooden Nickel
9	7.	Bungle In The Jungle-Jethro Tull-Chrysalis
13	8.	Never Can Say Goodbye-Gloria Gaynor-MGM
7	9.	Morning Side Of The Mountain-Donny & Marie-MGM
5	10.	Lucy In The Sky With Diamonds-Elton John-MCA
15	11.	Some Kind Of Wonderful-Grand Funk-Grand Funk
19	12.	You're No Good-Linda Ronstadt-Capitol
16	13.	Doctor's Orders-Carol Douglas-Midland Int'l
22	14.	Fire-Ohio Players-Mercury
17	15.	Get Dancin'-Disco Tex & Sex-O-Lettes-Chelsea
11	16.	When Will I See You Again-The Three Degrees-PI
8	17.	Only You-Ringo Starr-Apple
23	18.	Pick Up The Pieces-AWB-Atlantic
12	19.	One Man Woman-Paul Anka With Odia Coates-UA
28	20.	Black Water-The Doobie Brothers-WB
25	21.	#9 Dream-John Lennon-Apple
18	22.	I Can Help-Billy Swan-Monument
24	23.	Sweet Surrender-John Denver-RCA
14	24.	You Got The Love-Rufus-ABC
27	25.	It's All Right-Jim Capaldi-Island
30	26.	My Eyes Adored You-Frankie Valli-Private Stock
21	27.	You're The First, The Last, My Everything-Barry White-20th Century
31	28.	Roll On Down The Highway-B-T Overddrive-Mercury
29	29.	Your Bulldog Drinks Champagne-Jim Stafford-MGM
32	30.	Up In A Puff Of Smoke-Polly Brown-GTO
34	31.	Part Of The Plan-Dan Fogelberg-Epic
35	32.	Have You Never Been Mellow-Olivia Newton-John-MCA
26	33.	Kung Fu Fighting-Carl Douglas-20th Century

20	34.	Promised Land-Elvis Presley-RCA
37	35.	Emma-Hot Chocolate-Big Tree
--	36.	Movin' On-Bad Company-Swan Song
--	37.	To The Door Of The Sun-Al Martino-Capitol
40	38.	I'll Still Love You-Jim Weatherly-Buddah
--	39.	I Am Love-The Jackson 5-Motown
--	40.	The South's Gonna Do It-The Charlie Daniels Band-Kama Sutra

SUPER CFL SURVEY FEB 15

wcfl wcfl wcfl wcfl wcfl wcfl wcfl wcfl wcfl wcfl wcfl wcfl wcfl wcfl

SUPER ALBUMS

1. Greatest Hits-Elton John-MCA
2. Fire-Ohio Players-Mercury
3. War Child-Jethro Tull-Chrysalis
4. Styx II-Styx-Wooden Nickel
5. Heart Like A Wheel-Linda Ronstadt-Capitol
6. Blood On The Tracks-Bob Dylan-Columbia
7. Relayer-Yes-Atlantic
8. Barry Manilow II-Bell
9. New And Improved-Spinners-Atlantic
10. Melodies Of Love-Bobby Vinton-ABC
11. Average White Band-AWB-Atlantic
12. Joy To The World-Three Dog Night-Dunhill
13. Souvenirs-Dan Fogelberg-Epic
14. Perfect Angel-Minnie Riperton-Epic
15. Dark Horse-George Harrison-Apple
16. Mother Lode-Loggins And Messina-Columbia
17. This Is The Moody Blues-Threshold
18. Not Fragile-Bachman-Turner Overdrive-Mercury
19. Miles of Aisles-Joni Mitchell-Asylum
20. So What-Joe Walsh-Dunhill
21. Butterfly-Barbra Streisand-Columbia
22. White Gold-The Love Unlimited Orchestra-20th Century
23. Never Can Say Goodbye-Gloria Gaynor-MGM
24. Photographs And Memories-Jim Croce-ABC
25. Al Green Explores Your Mind-Hi

3	1.	Best Of My Love-The Eagles-Asylum
1	2.	Mandy-Barry Manilow-Bell
6	3.	Lady-Styx-Wooden Nickel
8	4.	Never Can Say Goodbye-Gloria Gaynor-MGM
4	5.	Boogie On Reggae Woman-Stevie Wonder-Tamla
7	6.	Bungle In The Jungle-Jethro Tull-Chrysalis
14	7.	Fire-Ohio Players-Mercury
13	8.	Doctor's Orders-Carol Douglas-Midland Int'l
11	9.	Some Kind Of Wonderful-Grand Funk-Grand Funk
12	10.	You're No Good-Linda Ronstadt-Capitol
2	11.	Please Mr. Postman-Carpenters-A & M
20	12.	Black Water-The Doobie Brothers-WB
18	13.	Pick Up The Pieces-AWB-Atlantic
5	14.	Laughter In The Rain-Neil Sedaka-MCA
9	15.	Morning Side Of The Mountain-Donny & Marie-MGM
15	16.	Get Dancin'-Disco Tex & Sex-O-Lettes-Chelsea
10	17.	Lucy In The Sky With Diamonds-Elton John-MCA
21	18.	#9 Dream-John Lennon-Apple
26	19.	My Eyes Adored You-Frankie Valli-Private Stock
23	20.	Sweet Surrender-John Denver-RCA
28	21.	Roll On Down The Highway-B-T Overdrive-Mercury
25	22.	It's All Right-Jim Capaldi-Island
32	23.	Have You Never Been Mellow-Olivia Newton-John-MCA
16	24.	When Will I See You Again-The Three Degrees-PI
30	25.	Up In A Puff Of Smoke-Polly Brown-GTO
--	26.	Lonely People-America-WB
31	27.	Part Of The Plan-Dan Fogelberg-Epic
17	28.	Only You-Ringo Starr-Apple
--	29.	Can't Get It Out Of My Head-Electric Light Orchestra-UA
35	30.	Emma-Hot Chocolate-Big Tree
22	31.	I Can Help-Billy Swan-Monument
38	32.	I'll Still Love You-Jim Weatherly-Buddah
36	33.	Movin' On-Bad Company-Swan Song
--	34.	You Are So Beautiful-Joe Cocker-A & M
40	35.	South's Gonna Do It-Charlie Daniels Band-Kama Sutra

37	36.	To The Door Of The Sun-Al Martino-Capitol
39	37.	I Am Love-The Jackson 5-Motown
19	38.	One Man Woman-Paul Anka With Odia Coates-UA
27	39.	You're The First, The Last, My Everything-Barry White-20th Century
24	40.	You Got The Love-Rufus-ABC

SUPER CFL SURVEY FEB 22

wcfl wcfl wcfl wcfl wcfl wcfl wcfl wcfl wcfl wcfl wcfl wcfl wcfl wcfl

SUPER ALBUMS

1. Greatest Hits-Elton John-MCA
2. Styx II-Styx-Wooden Nickel
3. Fire-Ohio Players-Mercury
4. War Child-Jethro Tull-Chrysalis
5. Heart Line A Wheel-Linda Ronstadt-Capitol
6. Average White Band-AWB-Atlantic
7. Blood On The Tracks-Bob Dylan-Columbia
8. Barry Manilow II-Bell
9. New And Improved-Spinners-Atlantic
10. Perfect Angel-Minnie Riperton-Epic
11. So What-Joe Walsh-Dunhill
12. Souvenirs-Dan Fogelberg-Epic
13. Not Fragile-Bachman-Turner Overdrive-Mercury
14. Melodies Of Love-Bobby Vinton-ABC
15. Dark Horse-George Harrison-Apple
16. What Were Once Vices Are Now Habits-The Doobie Brothers-WB
17. Miles Of Aisles-Joni Mitchell-Asylum
18. This Is The Moody Blues-Threshold
19. Relayer-Yes-Atlantic
20. Joy To The World-Three Dog Night-Dunhill
21. Do It-B.T. Express-Scepter
22. All The Girls In The World Beware!!!-Grand Funk-Grand Funk
23. Back Home Again-John Denver-RCA
24. Mother Lode-Loggins And Messina-Columbia
25. White Gold-The Love Unlimited Orchestra-20th Century

1	1.	Best Of My Love-The Eagles-Asylum
4	2.	Never Can Say Goodbye-Gloria Gaynor-MGM
3	3.	Lady-Styx-Wooden Nickel
7	4.	Fire-Ohio Players-Mercury
9	5.	Some Kind of Wonderful-Grand Funk-Grand Funk
12	6.	Black Water-The Doobie Brothers-WB
10	7.	You're No Good-Linda Ronstadt-Capitol
2	8.	Mandy-Barry Manilow-Bell
13	9.	Pick Up The Pieces-AWB-Atlantic
8	10.	Doctor's Orders-Carol Douglas-Midland Int'l
8	11.	Bungle In The Jungle-Jethro Tull-Chrysalis
11	12.	Please Mr. Postman-Carpenters-A & M
21	13.	Roll On Down The Highway-B-T Overdrive-Mercury
18	14.	#9 Dream-John Lennon-Apple
5	15.	Boogie On Reggae Woman-Stevie Wonder-Tamla
19	16.	My Eyes Adored You-Frankie Valli-Private Stock
23	17.	Have You Never Been Mellow-Olivia Newton-John-MCA
14	18.	Laughter In The Rain-Neil Sedaka-MCA
25	19.	Up In A Puff Of Smoke-Polly Brown-GTO
27	20.	Part Of The Plan-Dan Fogelberg-Epic
26	21.	Lonely People-America-WB
15	22.	Morning Side Of The Mountain-Donny & Marie-MGM
33	23.	Movin' On-Bad Company-Swan Song
20	24.	Sweet Surrender-John Denver-RCA
29	25.	Can't Get It Out Of My Head-Electric Light Orchestra-UA
16	26.	Get Dancin'-Disco Tex & Sex-O-Lettes-Chelsea
30	27.	Emma-Hot Chocolate-Big Tree
22	28.	It's All Right-Jim Capaldi-Island
35	29.	South's Gonna Do It-Charlie Daniels Band-Kama Sutra
17	30.	Lucy In The Sky With Diamonds-Elton John-MCA
37	31.	I Am Love-The Jackson 5-Motown
--	32.	Don't Call Us, We'll Call You-Sugarloaf-Claridge
34	33.	You Are So Beautiful-Joe Cocker-A & M
36	34.	To The Door Of The Sun-Al Martino-Capitol

--	35.	Poetry Man-Phoebe Snow-Shelter
24	36.	When Will I See You Again-The Three Degrees-PI
--	37.	Lovin' You-Minnie Riperton-Epic
31	38.	I Can Help-Billy Swan-Monument
--	39.	No No Song-Ringo Starr-Apple
38	40.	One Man Woman-Paul Anka With Odia Coates-UA

SUPER CFL SURVEY MAR 1

wcfl wcfl wcfl wcfl wcfl wcfl wcfl wcfl wcfl wcfl wcfl wcfl wcfl wcfl

SUPER ALBUMS

1. Blood On The Tracks-Bob Dylan-Columbia
2. Heart Like A Wheel-Linda Ronstadt-Capitol
3. Greatest Hits-Elton John-MCA
4. Fire-Ohio Players-Mercury
5. Styx II-Styx-Wooden Nickel
6. Average White Band-AWB-Atlantic
7. War Child-Jethro Tull-Chrysalis
8. Barry Manilow II-Bell
9. Perfect Angel-Minnie Riperton-Epic
10. New And Improved-Spinners-Atlantic
11. So What-Joe Walsh-Dunhill
12. Souvenirs-Dan Fogelberg-Epic
13. Not Fragile-Bachman-Turner Overdrive-Mercury
14. Miles Of Aisles-Joni Mitchell-Asylum
15. Melodies Of Love-Bobby Vinton-ABC
16. What Were Once Vices Are Now Habits-The Doobie Brothers-WB
17. Relayer-Yes-Atlantic
18. Mother Lode-Loggins And Messina-Columbia
19. Have You Never Been Mellow-Olivia Newton-John-MCA
20. Do It-B.T. Express-Scepter
21. White Gold-The Love Unlimited Orchestra-20th Century
22. Dark Horse-George Harrison-Apple
23. Joy To The World-Three Dog Night-Dunhill
24. Never Can Say Goodbye-Gloria Gaynor-MGM
25. This Is The Moody Blues-Threshold

6	1.	Black Water-The Doobie Brothers-WB
2	2.	Never Can Say Goodbye-Gloria Gaynor-MGM
4	3.	Fire-Ohio Players-Mercury
7	4.	You're No Good-Linda Ronstadt-Capitol
1	5.	Best Of My Love-The Eagles-Asylum
3	6.	Lady-Styx-Wooden Nickel
9	7.	Pick Up The Pieces-AWB-Atlantic
5	8.	Some Kind Of Wonderful-Grand Funk-Grand Funk
13	9.	Roll On Down The Highway-B-T Overdrive-Mercury
16	10.	My Eyes Adored You-Frankie Valli-Private Stock
8	11.	Mandy-Barry Manilow-Bell
17	12.	Have You Never Been Mellow-Olivia Newton-John-MCA
14	13.	#9 Dream-John Lennon-Apple
21	14.	Lonely People-America-WB
10	15.	Doctor's Orders-Carol Douglas-Midland Int'l
20	16.	Part Of The Plan-Dan Fogelberg-Epic
19	17.	Up In A Puff Of Smoke-Polly Brown-GTO
11	18.	Bungle In The Jungle-Jethro Tull-Chrysalis
15	19.	Boogle On Reggae Woman-Stevie Wonder-Tamla
25	20.	Can't Get It Out Of My Head-Electric Light Orchestra-UA
23	21.	Movin' On-Bad Company-Swan Song
12	22.	Please Mr. Postman-Carpenters-A & M
27	23.	Emma-Hot Chocolate-Big Tree
24	24.	Sweet Surrender-John Denver-RCA
33	25.	You Are So Beautiful-Joe Cocker-A & M
29	26.	South's Gonna Do It-Charlie Daniels Band-Kama Sutra
18	27.	Laughter In The Rain-Neil Sedaka-MCA
31	28.	I Am Love-The Jackson 5-Motown
34	29.	To The Door Of The Sun-Al Martino-Capitol
32	30.	Don't Call Us, We'll Call You-Sugarloaf-Claridge
26	31.	Get Dancin'-Disco Tex & Sex-O-Lettes-Chelsea
37	32.	Lovin' You-Minnie Riperton-Epic
35	33.	Poetry Man-Phoebe Snow-Shelter
39	34.	No No Song-Ringo Starr-Apple
--	35.	Once You Get Started-Rufus-ABC
22	36.	Morning Side Of The Mountain-Donny & Marie-MGM

--	37.	Runaway-Charlie Kulis-Playboy
--	38.	Harry Truman-Chicago-Columbia
28	39.	It's All Right-Jim Capaldi-Island
30	40.	Lucy In The Sky With Diamonds- Elton John-MCA

SUPER CFL SURVEY MAR 8

wcfl wcfl wcfl wcfl wcfl wcfl wcfl wcfl wcfl wcfl wcfl wcfl wcfl wcfl

SUPER ALBUMS

1. Average White Band-AWB-Atlantic
2. Greatest Hits-Elton John-MCA
3. Blood On The Tracks-Bob Dylan-Columbia
4. Heart Like A Wheel-Linda Ronstadt-Capitol
5. Fire-Ohio Players-Mercury
6. Styx II-Styx-Wooden Nickel
7. War Child-Jethro Tull-Chrysalis
8. Barry Manilow II-Bell
9. What Were Once Vices Are Now Habits-The Doobie Brothers-WB
10. Perfect Angel-Minnie Riperton-Epic
11. For Earth Below-Robin Trower-Chrysalis
12. New And Improved-Spinners-Atlantic
13. Have You Never Been Mellow-Olivia Newton-John-MCA
14. So What-Joe Walsh-Dunhill
15. Cold On The Shoulder-Gordon Lightfoot-Reprise
16. Souvenirs-Dan Fogelberg-Epic
17. Relayer-Yes-Atlantic
18. Melodies Of Love-Bobby Vinton-ABC
19. Do It-B.T. Express-Scepter
20. Autobahn-Kraftwerk-Vertigo
21. White Gold-The Love Unlimited Orchestra-20th Century
22. Joy To The World-Three Dog Night-Dunhill
23. Not Fragile-Bachman-Turner Overdrive-Mercury
24. Miles Of Aisles-Joni Mitchell-Asylum
25. "Phoebe Snow"-Shelter

1	1.	Black Water-The Doobie Brothers-WB
3	2.	Fire-Ohio Players-Mercury
7	3.	Pick Up The Pieces-AWB-Atlantic
4	4.	You're No Good-Linda Ronstadt-Capitol
2	5.	Never Can Say Goodbye-Gloria Gaynor-MGM
10	6.	My Eyes Adored You-Frankie Valli-Private Stock
9	7.	Roll On Down The Highway-B-T Overdrive-Mercury
12	8.	Have You Never Been Mellow-Olivia Newton-John-MCA
14	9.	Lonely People-America-WB
5	10.	Best Of My Love-The Eagles-Asylum
8	11.	Some Kind Of Wonderful-Grand Funk-Grand Funk
17	12.	Up In A Puff Of Smoke-Polly Brown-GTO
16	13.	Part Of The Plan-Dan Fogelberg-Epic
21	14.	Movin' On-Bad Company-Swan Song
6	15.	Lady-Styx-Wooden Nickel
13	16.	#9 Dream-John Lennon-Apple
20	17.	Can't Get It Out Of My Head-Electric Light Orchestra-UA
11	18.	Mandy-Barry Manilow-Bell
25	19.	You Are So Beautiful-Joe Cocker-A & M
23	20.	Emma-Hot Chocolate-Big Tree
26	21.	South's Gonna Do It-Charlie Daniels Band-Kama Sutra
15	22.	Doctor's Orders-Carol Douglas-Midland Int'l
28	23.	I Am Love-The Jackson 5-Motown
32	24.	Lovin' You-Minnie Riperton-Epic
29	25.	To The Door Of The Sun-Al Martino-Capitol
33	26.	Poetry Man-Phoebe Snow-Shelter
30	27.	Don't Call Us, We'll Call You-Sugarloaf-Claridge
--	28.	Lady Marmalade-Labelle-Epic
22	29.	Please Mr. Postman-Carpenters-A & M
34	30.	No No Song-Ringo Starr-Apple
--	31.	Express-B.T. Express-Roadshow
18	32.	Bungle In The Jungle-Jethro Tull-Chrysalis
35	33.	Once You Get Started-Rufus-ABC
38	34.	Harry Truman-Chicago-Columbia
37	35.	Runaway-Charlie Kulis-Playboy
19	36.	Boogie On Reggae Woman-Stevie Wonder-Tamla
--	37.	Chevy Van-Sammy Johns-GRC

27	38.	Laughter In The Rain-Neil Sedaka-MCA
24	39.	Sweet Surrender-John Denver-RCA
31	40.	Get Dancin'-Disco Tex & Sex-O-Lettes-Chelsea

SUPER CFL SURVEY MAR 15

wcfl wcfl wcfl wcfl wcfl wcfl wcfl wcfl wcfl wcfl wcfl wcfl wcfl wcfl

2	1.	Fire-Ohio Players-Mercury
1	2.	Black Water-The Doobie Brothers-WB
6	3.	My Eyes Adored You-Frankie Valli-Private Stock
8	4.	Have You Never Been Mellow-Olivia Newton-John-MCA
7	5.	Roll On Down The Highway-B-T Overdrive-Mercury
4	6.	You're No Good-Linda Ronstadt-Capitol
9	7.	Lonely People-America-WB
3	8.	Pick Up The Pieces-AWB-Atlantic
5	9.	Never Can Say Goodbye-Gloria Gaynor-MGM
13	10.	Part Of The Plan-Dan Fogelberg-Epic
24	11.	Lovin' You-Minnie Riperton-Epic
12	12.	Up In A Puff Of Smoke-Polly Brown-GTO
14	13.	Movin' On-Bad Company-Swan Song
17	14.	Can't Get It Out Of My Head-Electric Light Orchestra-UA
20	15.	Emma-Hot Chocolate-Big Tree
19	16.	You Are So Beautiful-Joe Cocker-A & M
10	17.	Best Of My Love-The Eagles-Asylum
21	18.	South's Gonna Do It-Charlie Daniels Band-Kama Sutra
23	19.	I Am Love-The Jackson 5-Motown
27	20.	Don't Call Us, We'll Call You-Sugarloaf-Claridge
11	21.	Some Kind Of Wonderful-Grand Funk-Grand Funk
28	22.	Lady Marmalade-Labelle-Epic
26	23.	Poetry Man-Phoebe Snow-Shelter
15	24.	Lady-Styx-Wooden Nickel
25	25.	To The Door Of The Sun-Al Martino-Capitol
34	26.	Harry Truman-Chicago-Columbia
30	27.	No No Song-Ringo Starr-Apple
33	28.	Once You Get Started-Rufus-ABC
31	29.	Express-B.T. Express-Roadshow
18	30.	Mandy-Barry Manilow-Bell
--	31.	Philadelphia Freedom-The Elton John Band-MCA
35	32.	Runaway-Charlie Kulis-Playboy
16	33.	#9 Dream-John Lennon-Apple
--	34.	(Hey Won't You Play) Another Somebody Done Somebody Wrong Song-B.J. Thomas-ABC
37	35.	Chevy Van-Sammy Johns-GRC
22	36.	Doctor's Orders-Carol Douglas-Midland Int'l

SUPER ALBUMS

1. Blood On The Tracks-Bob Dylan-Columbia
2. Average White Band-AWB-Atlantic
3. Heart Like A Wheel-Linda Ronstadt-Capitol
4. Greatest Hits-Elton John-MCA
5. Fire-Ohio Players-Mercury
6. What Were Once Vices Are Now Habits-The Doobie Brothers-WB
7. Perfect Angel-Minnie Riperton-Epic
8. Have You Never Been Mellow-Olivia Newton-John-MCA
9. For Earth Below-Robin Trower-Chrysalis
10. Barry Manilow II-Bell
11. So What-Joe Walsh-Dunhill
12. Styx II-Styx-Wooden Nickel
13. War Child-Jethro Tull-Chrysalis
14. Physical Graffiti-Led Zeppelin-Swan Song
15. New And Improved-Spinners-Atlantic
16. Rufusized-Rufus-ABC
17. Souvenirs-Dan Fogelberg-Epic
18. Do It-B.T. Express-Scepter
19. Cold On The Shoulder-Gordon Lightfoot-Reprise
20. Autobahn-Kraftwerk-Vertigo
21. An Evening With John Denver-RCA
22. Melodies Of Love-Bobby Vinton-ABC
23. Not Fragile-Bachman-Turner Overdrive-Mercury
24. Never Can Say Goodbye-Gloria Gaynor-MGM
25. Mother Lode-Loggins And Messina-Columbia

--	37.	Bertha Butt Boogie-Jimmy Castor Bunch-Atlantic
29	38.	Please Mr. Postman-Carpenters-A & M
--	39.	Who's Sorry Now-Marie Osmond-MGM
--	40.	He Don't Love You (Like I Love You)-Tony Orlando & Dawn-Elektra

SUPER CFL SURVEY MAR 22

wcfl wcfl wcfl wcfl wcfl wcfl wcfl wcfl wcfl wcfl wcfl wcfl wcfl wcfl

SUPER ALBUMS

1. Physical Graffiti-Led Zeppelin-Swan Song
2. Blood On The Tracks-Bob Dylan-Columbia
3. Average White Band-AWB-Atlantic
4. What Were Once Vices Are Now Habits-The Doobie Brothers-WB
5. Greatest Hits-Elton John-MCA
6. Have You Never Been Mellow-Olivia Newton-John-MCA
7. Heart Like A Wheel-Linda Ronstadt-Capitol
8. Perfect Angel-Minnie Riperton-Epic
9. For Earth Below-Robin Trower-Chrysalis
10. Fire-Ohio Players-Mercury
11. War Child-Jethro Tull-Chrysalis
12. Barry Manilow II-Bell
13. New And Improved-Spinners-Atlantic
14. Styx II-Styx-Wooden Nickel
15. So What-Joe Walsh-Dunhill
16. Rufusized-Rufus-ABC
17. Souvenirs-Dan Fogelberg-Epic
18. Cold On The Shoulder-Gordon Lightfoot-Reprise
19. "Phoebe Snow"-Shelter
20. Do It-B.T. Express-Scepter
21. Rock 'N Roll-John Lennon-Apple
22. White Gold-The Love Unlimited Orchestra-20th Century
23. Autobahn-Kraftwerk-Vertigo
24. Nightbirds-Labelle-Epic
25. That's The Way Of The World-Earth, Wind & Fire-Columbia

3	1.	My Eyes Adored You-Frankie Valli-Private Stock
4	2.	Have You Never Been Mellow-Olivia Newton-John-MCA
1	3.	Fire-Ohio Players-Mercury
5	4.	Roll On Down The Highway-B-T Overdrive-Mercury
11	5.	Lovin' You-Minnie Riperton-Epic
7	6.	Lonely People-America-WB
2	7.	Black Water-The Doobie Brothers-WB
14	8.	Can't Get It Out Of My Head-Electric Light Orchestra-UA
6	9.	You're No Good-Linda Ronstadt-Capitol
10	10.	Part Of The Plan-Dan Fogelberg-Epic
16	11.	You Are So Beautiful-Joe Cocker-A & M
8	12.	Pick Up The Pieces-AWB-Atlantic
15	13.	Emma-Hot Chocolate-Big Tree
19	14.	I Am Love-The Jackson 5-Motown
22	15.	Lady Marmalade-Labelle-Epic
13	16.	Movin' On-Bad Company-Swan Song
20	17.	Don't Call Us, We'll Call You-Sugarloaf-Claridge
18	18.	South's Gonna Do It-Charlie Daniels Band-Kama Sutra
23	19.	Poetry Man-Phoebe Snow-Shelter
12	20.	Up In A Puff Of Smoke-Polly Brown-GTO
9	21.	Never Can Say Goodbye-Gloria Gaynor-MGM
29	22.	Express-B.T. Express-Roadshow
27	23.	No No Song-Ringo Star-Apple
31	24.	Philadelphia Freedom-The Elton John Band-MCA
26	25.	Harry Truman-Chicago-Columbia
28	26.	Once You Get Started-Rufus-ABC
25	27.	To The Door Of The sun-Al Martino-Capitol
32	28.	Runaway-Charlie Kulis-Playboy
17	29.	Best Of My Love-The Eagles-Asylum
35	30.	Chevy Van-Sammy Johns-GRC
34	31.	(Hey Won't You Play) Another Somebody Done Somebody Wrong Song-B.J. Thomas-ABC
37	32.	Bertha Butt Boogie-Jimmy Castor Bunch-Atlantic
21	33.	Some Kind Of Wonderful-Grand Funk-Grand Funk
39	34.	Who's Sorry Now-Marie Osmond-MGM
24	35.	Lady-Styx-Wooden Nickel
40	36.	He Don't Love You-Tony Orlando/Dawn-Elektra
--	37.	Supernatural Ting-Ben E. King-

		Atlantic
30	38.	Mandy-Barry Manilow-Bell
--	39.	Jackie Blue-Ozark Mountain Daredevils-A & M
--	40.	How Long-Ace-Anchor

SUPER CFL SURVEY MAR 29

wcfl wcfl wcfl wcfl wcfl wcfl wcfl wcfl wcfl wcfl wcfl wcfl wcfl wcfl

SUPER ALBUMS

1. Physical Graffiti-Led Zeppelin-Swan Song
2. Have You Never Been Mellow-Olivia Newton-John-MCA
3. Blood On The Tracks-Bob Dylan-Columbia
4. Average White Band-AWB-Atlantic
5. What Were Once Vices Are Now Habits-The Doobie Brothers-WB
6. Heart Like A Wheel-Linda Ronstadt-Capitol
7. For Earth Below-Robin Trower-Chrysalis
8. Greatest Hits-Elton John-MCA
9. Fire-Ohio Players-Mercury
10. Perfect Angel-Minnie Riperton-Epic
11. Rufusized-Rufus-ABC
12. New And Improved-Spinners-Atlantic
13. "An Evening With John Denver"-RCA
14. Crash Landing-Jimi Hendrix-Reprise
15. Barry Manilow II-Bell
16. So What-Joe Walsh-Dunhill
17. Autobahn-Kraftwerk-Vertigo
18. "Phoebe Snow"-Shelter
19. Styx II-Styx-Wooden Nickel
20. Sun Goddess-Ramsey Lewis And Earth, Wind & Fire-Columbia
21. Cold On The Shoulder-Gordon Lightfoot-Reprise
22. Do It-B.T. Express-Scepter
23. Welcome To My Nightmare-Alice Cooper-Atlantic
24. Yesterdays-Yes-Atlantic
25. War Child-Jethro Tull-Chrysalis

2	1.	Have You Never Been Mellow-Olivia Newton-John-MCA
5	2.	Lovin' You-Minnie Riperton-Epic
1	3.	My Eyes Adored You-Frankie Valli-Private Stock
6	4.	Lonely People-America-WB
4	5.	Roll On Down The Highway-B-T Overdrive-Mercury
15	6.	Lady Marmalade-Labelle-Epic
8	7.	Can't Get It Out Of My Head-Electric Light Orchestra-UA
11	8.	You Are So Beautiful-Joe Cocker-A & M
7	9.	Black Water-The Doobie Brothers-WB
3	10.	Fire-Ohio Players-Mercury
13	11.	Emma-Hot Chocolate-Big Tree
17	12.	Don't Call Us, We'll Call You-Sugarloaf-Claridge
14	13.	I Am Love-The Jackson 5-Motown
9	14.	You're No Good-Linda Ronstadt-Capitol
24	15.	Philadelphia Freedom-The Elton John Band-MCA
19	16.	Poetry Man-Phoebe Snow-Shelter
10	17.	Part Of The Plan-Dan Fogelberg-Epic
22	18.	Express-B.T. Express-Roadshow
23	19.	No No Song-Ringo Starr-Apple
25	20.	Harry Truman-Chicago-Columbia
18	21.	South's Gonna Do It-Charlie Daniels Band-Kama Sutra
26	22.	Once You Get Started-Rufus-ABC
12	23.	Pick Up The Pieces-AWB-Atlantic
28	24.	Runaway-Charlie Kulis-Playboy
31	25.	(Hey Won't You Play) Another Somebody Done Somebody Wrong Song-B.J. Thomas-ABC
30	26.	Chevy Van-Sammy Johns-GRC
--	27.	Shame, Shame, Shame-Shirley (And Company)-Vibration
32	28.	Bertha Butt Boogie-Jimmy Castor Bunch-Atlantic
16	29.	Movin' On-Bad Company-Swan Song
36	30.	He Don't Love you-Tony Orlando/Dawn-Elektra
34	31.	Who's Sorry Now-Marie Osmond-MGM
37	32.	Supernatural Ting-Ben E. King-Atlantic
20	33.	Up In A Puff Of Smoke-Polly Brown-GTO
--	34.	What Am I Gonna Do With You-Barry White-20th Century
39	35.	Jackie Blue-Ozark Mountain Daredevils-A & M
21	36.	Never Can Say Goodbye-Gloria

		Gaynor-MGM
40	37.	How Long-Ace-Anchor
--	38.	I Don't Like To Sleep Alone-Paul Anka-UA
27	39.	To The Door Of The Sun-Al Martino-Capitol
--	40.	Don't Tell Me Goodnight-Lobo-Big Tree

SUPER CFL SURVEY APR 5

wcfl wcfl wcfl wcfl wcfl wcfl wcfl wcfl wcfl wcfl wcfl wcfl wcfl wcfl

SUPER ALBUMS

1. Physical Graffiti-Led Zeppelin-Swan Song
2. Blood On The Tracks-Bob Dylan-Columbia
3. Have You Never Been Mellow-Olivia Newton-John-MCA
4. Average White Band-AWB-Atlantic
5. What Were Once Vices Are Now Habits-The Doobie Brothers-WB
6. For Earth Below-Robin Trower-Chrysalis
7. Heart Like A Wheel-Linda Ronstadt-Capitol
8. Greatest Hits-Elton John-MCA
9. Perfect Angel-Minnie Riperton-Epic
10. Rufusized-Rufus-ABC
11. Fire-Ohio Players-Mercury
12. New And Improved-Spinners-Atlantic
13. Crash Landing-Jimi Hendrix-Reprise
14. Young Americans-David Bowie-RCA
15. "Phoebe Snow"-Shelter
16. "An Evening With John Denver"-RCA
17. Autobahn-Kraftwerk-Vertigo
18. Nightbirds-Labelle-Epic
19. War Child-Jethro Tull-Chrysalis
20. Do It-B.T. Express-Scepter
21. Barry Manilow II-Bell
22. Styx II-Styx-Wooden Nickel
23. So What-Joe Walsh-Dunhill
24. That's The Way Of The World-Earth, Wind & Fire-Columbia
25. On Your Feet Or On Your Knees-Blue Oyster Cult-Columbia

2	1.	Lovin' You-Minnie Riperton-Epic
6	2.	Lady Marmalade-Labelle-Epic
1	3.	Have You Never Been Mellow-Olivia Newton-John-MCA
3	4.	My Eyes Adored You-Frankie Valli-Private Stock
4	5.	Lonely People-America-WB
8	6.	You Are So Beautiful-Joe Cocker-A & M
7	7.	Can't Get It Out Of My Head-Electric Light Orchestra-UA
13	8.	I Am Love-The Jackson 5-Motown
11	9.	Emma-Hot Chocolate-Big Tree
12	10.	Don't Call Us, We'll Call You-Sugarloaf-Claridge
15	11.	Philadelphia Freedom-The Elton John Band-MCA
5	12.	Roll On Down The Highway-B-T Overdrive-Mercury
18	13.	Express-B.T. Express-Roadshow
9	14.	Black Water-The Doobie Brothers-WB
19	15.	No No Song-Ringo Starr-Apple
16	16.	Poetery Man-Phoebe Snow-Shelter
10	17.	Fire-Ohio Players-Mercury
20	18.	Harry Truman-Chicago-Columbia
22	19.	Once You Get Started-Rufus-ABC
24	20.	Runaway-Charlie Kulis-Playboy
26	21.	Chevy Van-Sammy Johns-GRC
14	22.	You're No Good-Linda Ronstadt-Capitol
25	23.	(Hey Won't You Play) Another Somebody Done Somebody Wrong Song-B.J. Thomas-ABC
28	24.	Bertha Butt Boogie-Jimmy Castor Bunch-Atlantic
27	25.	Shame, Shame, Shame-Shirley (And Company)-Vibration
30	26.	He Don't Love You-Tony Orlando/Dawn-Elektra
17	27.	Part Of The Plan-Dan Fogelberg-Epic
31	28.	Who's Sorry Now-Marie Osmond-MGM
34	29.	What Am I Gonna Do With You-Barry White-20th Century
32	30.	Supernatural Thing-Ben E. King-Atlantic
21	31.	South's Gonna Do It-Charlie Daniels Band-Kama Sutra
37	32.	How Long-Ace-Anchor
35	33.	Jackie Blue-Ozark Mountain Daredevils-A & M
23	34.	Pick Up The Pieces-AWB-Atlantic
38	35.	I Don't Like To Sleep Alone-Paul Anka-UA

SMALL PRINT DEPARTMENT: The Super CFL survey represents the relative popularity of records based on industry tabulations and record sales. Hot diggity. WCFL, the Voice of Labor, is o

29	36.	Movin' On-Bad Company-Swan Song
--	37.	Stand By Me-John Lennon-Apple
40	38.	Don't Tell Me Goodnight-Lobo-Big Tree
--	39.	Amie-Pure Prairie League-RCA
--	40.	Autobahn-Kraftwerk-Vertigo

SUPER CFL SURVEY APR 12

wcfl wcfl wcfl wcfl wcfl wcfl wcfl wcfl wcfl wcfl wcfl wcfl wcfl wcfl

1	1.	Lovin' You-Minnie Riperton-Epic
2	2.	Lady Marmalade-Labelle-Epic
6	3.	You Are So Beautiful-Joe Cocker-A & M
3	4.	Have You Never Been Mellow-Olivia Newton-John-MCA
8	5.	I Am Love-The Jackson 5-Motown
11	6.	Philadelphia Freedom-The Elton John Band-MCA
10	7.	Don't Call Us, We'll Call you-Sugarloaf-Claridge
5	8.	Lonely People-America-WB
9	9.	Emma-Hot Chocolate-Big Tree
15	10.	No No Song-Ringo Starr-Apple
13	11.	Express-B.T. Express-Roadshow
7	12.	Can't Get It Out Of My Head-Electric Light Orchestra-UA
4	13.	My Eyes Adored You-Frankie Valli-Private Stock
18	14.	Harry Truman-Chicago-Columbia
20	15.	Runaway-Charlie Kulis-Playboy
25	16.	Shame, Shame, Shame-Shirley (And Company)-Vibration
19	17.	Once You Get Started-Rufus-ABC
21	18.	Chevy Van-Sammy Johns-GRC
23	19.	(Hey Won't You Play) Another Somebody Done Somebody Wrong Song-B.J. Thomas-ABC
16	20.	Poetry Man-Phoebe Snow-Shelter
14	21.	Black Water-The Doobie Brothers-WB
26	22.	He Don't Love You-Tony Orlando/ Dawn-Elektra
24	23.	Bertha Butt Boogie-Jimmy Castor Bunch-Atlantic
29	24.	What Am I Gonna Do With You-Barry White-20th Century
33	25.	Jackie Blue-Ozark Mountain Daredevils-A & M
28	26.	Who's Sorry Now-Marie Osmond-MGM
12	27.	Roll On Down The Highway-B-T Overdrive-Mercury
32	28.	How Long-Ace-Anchor
30	29.	Supernatural Thing-Ben E. King-Atlantic
35	30.	I Don't Like To Sleep Alone-Paul Anka-UA
17	31.	Fire-Ohio Players-Mercury
37	32.	Stand By Me-John Lennon-Apple
--	33.	Before The Next Teardrop Falls-Freddy Fender-ABC/Dot
--	34.	It's A Miracle-Barry Manilow-Arista
38	35.	Don't Tell Me Goodnight-Lobo-Big Tree
22	36.	You're No Good-Linda Ronstadt-Capitol
40	37.	Autobahn-Kraftwerk-Vertigo
39	38.	Amie-Pure Prairie League-RCA
--	39.	L-O-V-E (Love)-Al Green-HI
--	40.	Killer Queen-Queen-Elektra

SUPER ALBUMS

1. Physical Graffiti-Led Zeppelin-Swan Song
2. Have You Never Been Mellow-Olivia Newton-John-MCA
3. Blood On The Tracks-Bob Dylan-Columbia
4. Average White Band-AWB-Atlantic
5. Perfect Angel-Minnie Riperton-Epic
6. Crash Landing-Jimi Hendrix-Reprise
7. For Earth Below-Robin Trower-Chrysalis
8. What Were Once Vices Are Now Habits-The Doobie Brothers-WB
9. "An Evening With John Denver"-RCA
10. Greatest Hits-Elton John-MCA
11. Blue Jays-Justin Hayward and John Lodge-Threshold
12. Heart Like A Wheel-Linda Ronstadt-Capitol
13. Rufusized-Rufus-ABC
14. Fire-Ohio Players-Mercury
15. Autobahn-Kraftwerk-Vertigo
16. "Phoebe Snow"-Shelter
17. So What-Joe Walsh-Dunhill
18. That's The Way Of The World-Earth, Wind & Fire-Columbia
19. New And Improved-Spinners-Atlantic
20. Nightbirds-Labelle-Epic
21. Sun Goddess-Ramsey Lewis-Columbia
22. "Funny Lady'-Original Sound Track-Arista
23. Rock 'N Roll-John Lennon-Apple
24. Welcome To My Nightmare-Alice Cooper-Atlantic
25. Cold On The Shoulder-Gordon Lightfoot-Reprise

SUPER CFL SURVEY APR 19

wcfl wcfl wcfl wcfl wcfl wcfl wcfl wcfl wcfl wcfl wcfl wcfl wcfl wcfl

SUPER ALBUMS

1. Physical Graffiti-Led Zeppelin-Swan Song
2. Chicago VIII-Columbia
3. Have You Never Been Mellow-Olivia Newton-John-MCA
4. Blood On The Tracks-Bob Dylan-Columbia
5. Average White Band-AWB-Atlantic
6. For Earth Below-Robin Trower-Chrysalis
7. Crash Landing-Jimi Hendrix-Reprise
8. Perfect Angel-Minnie Riperton-Epic
9. What Were Once Vices Are Now Habits-The Doobie Brothers-WB
10. Greatest Hits-Elton John-MCA
11. That's The Way Of The World-Earth, Wind & Fire-Columbia
12. Rufusized-Rufus-ABC
13. "An Evening With John Denver"-RCA
14. There's One In Every Crowd-Eric Clapton-RSO
15. Blue Jays-Justin Hayward & John Lodge-Threshold
16. Autobahn-Kraftwerk-Vertigo
17. "Tommy"-Original Sound Track-Polydor
18. Blow By Blow-Jeff Beck-Epic
19. Nightbirds-Labelle-Epic
20. Nuthin' Fancy-Lynyrd Skynyrd-MCA
21. Heart Like A Wheel-Linda Ronstadt-Capitol
22. Katy Lied-Steely Dan-ABC
23. Do It-B.T. Express-Scepter
24. Young Americans-David Bowie-RCA
25. "Funny Lady"-Original Sound Track-Arista

2	1.	Lady Marmalade-Labelle-Epic
3	2.	You Are So Beautiful-Joe Cocker-A & M
6	3.	Philadelphia Freedom-The Elton John Band-MCA
7	4.	Don't Call Us, We'll Call You-Sugarloaf-Claridge
1	5.	Lovin' You-Minnie Riperton-Epic
10	6.	No No Song-Ringo Starr-Apple
5	7.	I Am Love-The Jackson 5-Motown
4	8.	Have You Never Been Mellow-Olivia Newton-John-MCA
11	9.	Express-B.T. Express-Roadshow
9	10.	Emma-Hot Chocolate-Big Tree
18	11.	Chevy Van-Sammy Johns-GRC
16	12.	Shame, Shame, Shame-Shirley (& Company)-Vibration
15	13.	Runaway-Charlie Kulis-Playboy
19	14.	(Hey Won't You Play) Another Somebody Done Somebody Wrong Song-B.J. Thomas-ABC
13	15.	My Eyes Adored You-Frankie Valli-Private Stock
8	16.	Lonely People-America-WB
22	17.	He Don't Love You-Tony Orlando/ Dawn-Elektra
12	18.	Can't Get It Out Of My Head-Electric Light Orchestra-UA
23	19.	Bertha Butt Boogie-Jimmy Castor Bunch-Atlantic
14	20.	Harry Truman-Chicago-Columbia
25	21.	Jackie Blue-Ozark Mountain Daredevils-A & M
24	22.	What Am I Gonna Do With You-Barry White-20th Century
28	23.	How Long-Ace-Anchor
17	24.	Once You ?Get Started-Rufus-ABC
30	25.	I Don't Like To Sleep Alone-Paul Anka-UA
20	26.	Poetry Man-Phoebe Snow-Shelter
32	27.	Stand By Me-John Lennon-Apple
34	28.	It's A Miracle-Barry Manilow-Arista
33	29.	Before The Next Teardrop Falls-Freddy Fender-ABC/Dot
26	30.	Who's Sorry Now-Marie Osmond-MGM
37	31.	Autobahn-Kraftwerk-Vertigo
21	32.	Black Water-The Doobie Brothers-WB
35	33.	Don't Tell Me Goodnight-Lobo-Big Tree
39	34.	L-O-V-E (Love)-Al Green-HI
38	35.	Amie-Pure Prairie League-RCA
40	36.	Killer Queen-Queen-Elektra
27	37.	Roll On Down The Highway-B-T Overdrive-Mercury
--	38.	Bad Time-Grand Funk-Capitol
29	39.	Supernatural Thing-Ben E. King-Atlantic
--	40.	Remember What I Told You To Forget-Tavares-Capitol

SUPER CFL SURVEY APR 26

wcfl wcfl wcfl wcfl wcfl wcfl wcfl wcfl wcfl wcfl wcfl wcfl wcfl wcfl

SUPER ALBUMS

1. Physical Graffiti-Led Zeppelin-Swan Song
2. Chicago VIII-Columbia
3. Have You Never Been Mellow-Olivia Newton-John-MCA
4. Perfect Angel-Minnie Riperton-Epic
5. Average White Band-AWB-Atlantic
6. Crash Landing-Jimi Hendrix-Reprise
7. Blood on the Tracks-Bob Dylan-Columbia
8. For Earth Below-Robin Trower-Chrysalis
9. That's The Way Of The World-Earth, Wind & Fire-Columbia
10. Blue Jays-Justin Hayward & John Lodge-Threshold
11. Autobahn-Kraftwerk-Vertigo
12. Greatest Hits-Elton John-MCA
13. "An Evening With John Denver"-RCA
14. What Were Once Vices Are Now Habits-The Doobie Brothers-WB
15. Blow By Blow-Jeff Beck-Epic
16. Rufusized-Rufus-ABC
17. Cold On The Shoulder-Gordon Lightfoot-Reprise
18. "Funny Lady"-Original Sound Track-Arista
19. Nightbirds-Labelle-Epic
20. There's One In Every Crowd-Eric Clapton-RSO
21. Straight Shooter-Bad Company-Swan Song
22. Fire-Ohio Players-Mercury
23. "Tommy"-Original Sound Track-Polydor
24. "Phoebe Snow"-Shelter
25. Hearts-America-WB

3	1.	Philadelphia Freedom-The Elton John Band-MCA
1	2.	Lady Marmalade-Labelle-Epic
6	3.	No No Song-Ringo Starr-Apple
2	4.	You Are So Beautiful-Joe Cocker-A & M
11	5.	Chevy Van-Sammy Johns-GRC
5	6.	Lovin' You-Minnie Riperton-Epic
12	7.	Shame, Shame, Shame-Shirley (& Company)-Vibration
13	8.	Runaway-Charlie Kulis-Playboy
4	9.	Don't Call Us, We'll Call You-Sugarloaf-Claridge
14	10.	(Hey Won't You Play) Another Somebody Done Somebody Wrong Song-B.J. Thomas-ABC
21	11.	Jackie Blue-Ozark Mountain Daredevils-A & M
9	12.	Express-B.T. Express-Roadshow
17	13.	He Don't Love You-Tony Orlando/Dawn-Elektra
7	14.	I Am Love-The Jackson 5-Motown
8	15.	Have You Never Been Mellow-Olivia Newton-John-MCA
22	16.	What Am I Gonna Do With You-Barry White-20th Century
10	17.	Emma-Hot Chocolate-Big Tree
23	18.	How Long-Ace-Anchor
19	19.	Bertha Butt Boogie-Jimmy Castor Bunch-Atlantic
28	20.	It's A Miracle-Barry Manilow-Arista
25	21.	I Don't Like To Sleep Alone-Paul Anka-UA
29	22.	Before The Next Teardrop Falls-Freddy Fender-ABC/Dot
15	23.	My Eyes Adored You-Frankie Valli-Private Stock
31	24.	Autobahn-Kraftwerk-Vertigo
27	25.	Stand By Me-John Lennon-Apple
20	26.	Harry Truman-Chicago-Columbia
16	27.	Lonely People-America-WB
33	28.	Don't Tell Me Goodnight-Lobo-Big Tree
--	29.	Long Gall Glasses-Leo Sayer-WB
35	30.	Amie-Pure Prairie League-RCA
34	31.	L-O-V-E (Love)-Al Green-HI
18	32.	Can't Get It Out Of My Head-Electric Light Orchestra-UA
36	33.	Killer Queen-Queen-Elektra
26	34.	Poetry Man-Phoebe Snow-Shelter
38	35.	Bad Time-Grand Funk-Capitol
40	36.	Remember What I Told You To Forget-Tavares-Capitol
24	37.	Once You Get Started-Rufus-ABC
--	38.	Hijack-Herbie Mann-Atlantic
--	39.	Only Women-Alice Cooper-Atlantic
30	40.	Who's Sorry Now-Marie Osmond-MGM

SUPER CFL SURVEY MAY 3

wcfl wcfl wcfl wcfl wcfl wcfl wcfl wcfl wcfl wcfl wcfl wcfl wcfl wcfl

SUPER ALBUMS

1. Physical Graffiti-Led Zeppelin-Swan Song
2. Chicago VIII-Columbia
3. Crash Landing-Jimi Hendrix-Reprise
4. Have You Never Been Mellow-Olivia Newton-John-MCA
5. That's The Way Of The World-Earth, Wind & Fire-Columbia
6. Average White Band-AWB-Atlantic
7. Blue Jays-Justin Hayward-John Lodge-Threshold
8. Autobahn-Kraftwerk-Vertigo
9. "Tommy"-Original Sound Track-Polydor
10. "An Evening With John Denver"-RCA
11. For Earth Below-Robin Trower-Chrysalis
12. Perfect Angel-Minnie Riperton-Epic
13. Straight Shooter-Bad Company-Swan Song
14. Greatest Hits-Elton John-MCA
15. Blood On The Tracks-Bob Dylan-Columbia
16. Blow By Blow-Jeff Beck-Epic
17. To Be True-Harold Melvin & The Blue Notes-PI
18. Hearts-America-WB
19. There's One In Every Crowd-Eric Clapton-RSO
20. Nightbirds-Labelle-Epic
21. Rufusized-Rufus-ABC
22. Just Another Way To Say I Love You-Barry White-20th Century
23. Nuthin' Fancy-Lynyrd Skynyrd-MCA
24. Katy Lied-Steely Dan-ABC
25. Judith-Judy Collins-Elektra

1	1.	Philadelphia Freedom-The Elton John Band-MCA
5	2.	Chevy Van-Sammy Johns-GRC
10	3.	(Hey Won't You Play) Another Somebody Done Somebody Wrong Song-B.J. Thomas-ABC
7	4.	Shame, Shame, Shame-Shirley (& Company)-Vibration
2	5.	Lady Marmalade-Labelle-Epic
11	6.	Jackie Blue-Ozark Mountain Daredevils-A & M
13	7.	He Don't Love you-Tony Orlando/ Dawn-Elektra
8	8.	Runaway-Charlie Kulis-Playboy
3	9.	No No 'Song-Ringo Starr-Apple
9	10.	Don't Call Us, We'll Call You-Sugarloaf-Claridge
4	11.	You Are So Beautiful-Joe Cocker-A & M
20	12.	It's A Miracle-Barry Manilow-Arista
18	13.	How Long-Ace-Anchor
16	14.	What Am I Gonna Do With You-Barry White-20th Century
6	15.	Lovin' You-Minnie Riperton-Epic
22	16.	Before The Next Teardrop Falls-Freddy Fender-ABC/Dot
12	17.	Express-B.T. Express-Roadshow
21	18.	I Don't Like To Sleep Alone-Paul Anka-UA
15	19.	Have You Never Been Mellow-Olivia Newton-John-MCA
24	20.	Autobahn-Kraftwerk-Vertigo
14	21.	I Am Love-The Jackson 5-Motown
29	22.	Long Tall Glasses-Leo Sayer-WB
30	23.	Amie-Pure Prairie League-RCA
25	24.	Stand By Me-John Lennon-Apple
28	25.	Don't Tell Me Goodnight-Lobo-Big Tree
33	26.	Killer Queen-Queen-Elektra
31	27.	L-O-V-E (Love)-Al Green-HI
19	28.	Bertha Butt Boogie-Jimmy Castor Bunch-Atlantic
23	29.	My Eyes Adored You-Frankie Valli-Private Stock
35	30.	Bad Time-Grand Funk-Capitol
--	31.	Only Yesterday-Carpenters-A & M
17	32.	Emma-Hot Chocolate-Big Tree
36	33.	Remember What I Told You To Forget-Tavaras-Capitol
--	34.	Thank God I'm A Country Boy-John Denver-RCA
39	35.	Only Women-Alice-Cooper-Atlantic
38	36.	Hijack-Herbie Mann-Atlantic
26	37.	Harry Truman-Chicago-Columbia
27	38.	Lonely People-America-WB
--	39.	I Wanna Dance Wit' Choo-Disco Tex & The Sex-O-Lettes-Chelsea
32	40.	Can't Get It Out Of My Head-Electric Light Orchestra-UA

SUPER CFL SURVEY MAY 10

wcfl wcfl wcfl wcfl wcfl wcfl wcfl wcfl wcfl wcfl wcfl wcfl wcfl wcfl

SUPER ALBUMS

1. Physical Graffiti-Led Zeppelin-Swan Song
2. Chicago VIII-Columbia
3. That's The Way Of The World-Earth, Wind & Fire-Columbia
4. Autobahn-Kraftwerk-Vertigo
5. Blue Jays-Justin Hayward & John Lodge-Threshold
6. Crash Landing-Jimi Hendrix-Reprise
7. Have You Never Been Mellow-Olivia Newton-John-MCA
8. Blow By Blow-Jeff Beck-Epic
9. "Tommy"-Original Sound Track-Polydor
10. Perfect Angel-Minine Riperton-Epic
11. For Earth Below-Robin Trower-Chrysalis
12. Average White Band-AWB-Atlantic
13. Greatest Hits-Elton John-MCA
14. Playing Possum-Carly Simon-Elektra
15. "An Evening With John Denver"-RCA
16. Just Another Way To Say I Love You-Barry White-20th Century
17. Straight Shooter-Bad Company-Swan Song
18. To BE True-Harold Melvin & The Blue Notes-PI
19. Katy Lied-Steely Dan-ABC
20. Hearts-America-WB
21. Blood On The Tracks-Bob Dylan-Columbia
22. Cold On The Shoulder-Gordon Lightfoot-Reprise
23. Rufusized-Rufus-ABC
24. Nuthin' Fancy-Lynyrd Skynyrd-MCA
25. There's One In Every Crowd-Eric Clapton-RSO

2	1.	Chevy Van-Sammy Johns-GRC
3	2.	(Hey Won't You Play) Another Somebody Done Somebody Wrong Song-B.J. Thomas-ABC
6	3.	Jackie Blue-Ozark Mountain Daredevils-A & M
7	4.	He Don't Love You-Tony Orlando/ Dawn-Elektra
1	5.	Philadelphia Freedom-The Elton John Band-MCA
12	6.	It's A Miracle-Barry Manilow-Arista
4	7.	Shame, Shame, Shame-Shirley (And Company)-Vibration
22	8.	Long Tall Glasses-Leo Sayer-WB
13	9.	How Long-Ace-Anchor
5	10.	Lady Marmalade-Labelle-Epic
9	11.	No No Song-Ringo Starr-Apple
18	12.	I Don't Like To Sleep Alone-Paul Anka-UA
16	13.	Before The Next Teardrop Falls-Freddy Fender-ABC/Dot
20	14.	Autobahn-Kraftwerk-Vertigo
8	15.	Runaway-Charlie Kulis-Playboy
11	16.	You Are So Beautiful-Joe Cocker-A & M
14	17.	What Am I Gonna Do With You-Barry White-20th Century
26	18.	Killer Queen-Queen-Elektra
10	19.	Don't Call Us, We'll Call You-Sugarloaf-Claridge
23	20.	Amie-Pure Prairie League-RCA
15	21.	Lovin' You-Minnie Riperton-Epic
19	22.	Have You Never Been Mellow-Olivia Newton-John-MCA
25	23.	Don't Tell Me Goodnight-Lobo-Big Tree
30	24.	Bad Time-Grand Funk-Capitol
17	25.	Express-B.T. Express-Roadshow
31	26.	Only Yesterday-Carpenters-A & M
27	27.	L-O-V-E (Love)-Al Green-HI
35	28.	Only Women-Alice Cooper-Atlantic
--	29.	Shining Star-Earth, Wind & Fire-Columbia
34	30.	Thank God I'm A Country Boy-John Denver-RCA
33	31.	Remember What I Told You To Forget-Tavares-Capitol
21	32.	I Am Love=The Jackson 5-Motown
--	33.	Sister Golden Hair-America-WB
36	34.	Hijack-Herbie Mann-Atlantic
24	35.	Stand By Me-John Lennon-Apple
--	36.	Shoeshine Boy-Eddie Kendricks-Tamla
39	37.	I Wanna Dance Wit' Choo-Disco Tex & The Sex-O-Lettes-Chelsea
--	38.	When Will I Be Loved-Linda Rondstadt-Capitol
29	39.	My Eyes Adored You-Frankie Valli-Private Stock
28	40.	Bertha Butt Boogie-Jimmy Castor Bunch-Atlantic

SUPER CFL SURVEY MAY 17

wcfl wcfl wcfl wcfl wcfl wcfl wcfl wcfl wcfl wcfl wcfl wcfl wcfl wcfl

SUPER ALBUMS

1. Chicago VIII-Columbia
2. Physical Graffiti-Led Zeppelin-Swan Song
3. That's The Way Of The World-Earth, Wind & Fire-Columbia
4. "Tommy"-Original Sound Track-Polydor
5. Autobahn-Kraftwerk-Vertigo
6. Crash Landing-Jimi Hendrix-Reprise
7. Blue Jays-Justin Hayward & John Lodge-Threshold
8. Have You Never Been Mellow-Olivia Newton-John-MCA
9. Blow By Blow-Jeff Beck-Epic
10. Straight Shooter-Bad Company-Swan Song
11. Average White Band-AWB-Atlantic
12. To Be True-Harold Melvin & The Blue Notes-PI
13. Greatest Hits-Elton John-MCA
14. For Earth Below-Robin Trower-Chrysalis
15. Playing Possum-Carly Simon-Elektra
16. Hearts-America-WB
17. Mister Magic-Grover Washington, Jr.-Motown
18. A Song For You-The Temptations-Gordy
19. Welcome To My Nightmare-Alice Cooper-Atlantic
20. "An Evening With John Denver"-RCA
21. Nightbirds-Labelle-Epic
22. Blood On The Tracks-Bob Dylan-Columbia
23. The Myths And Legends Of King Arthur And The Knights Of The Round Table-Rick Wakeman-A & M
24. Just Another Way To Say I Love You-Barry White-20th Century
25. Kath Lied-Steely Dan-ABC

2	1.	(Hey Won't You Play) Another Somebody Done Somebody Wrong Song-B.J. Thomas-ABC
4	2.	He Don't Love You-Tony Orlando/Dawn-Elektra
6	3.	It's A Miracle-Barry Manilow-Arista
3	4.	Jackie Blue-Ozark Mountain Daredevils-A & M
8	5.	Long Tall Glasses-Leo Sayer-WB
1	6.	Chevy Van-Sammy Johns-GRC
9	7.	How Long-Ace-Anchor
13	8.	Before The Next Teardrop Falls-Freddy Fender-ABC/Dot
12	9.	I Don't Like To Sleep Alone-Paul Anka-UA
18	10.	Killer Queen-Queen-Elektra
7	11.	Shame, Shame, Shame-Shirley (And Company)-Vibration
14	12.	Autobahn-Kraftwerk-Vertigo
5	13.	Philadelphia Freedom-The Elton John Band-MCA
10	14.	Lady Marmalade-Labelle-Epic
24	15.	Bad Time-Grand Funk-Capitol
11	16.	No No Song-Ringo Starr-Apple
26	17.	Only Yesterday-Carpenters-A & M
15	18.	Runaway-Charlie Kulis-Playboy
30	19.	Thank God I'm A Country Boy-John Denver-RCA
20	20.	Amie-Pure Prairie League-RCA
28	21.	Only Women-Alice Cooper-Atlantic
16	22.	You Are So Beautiful-Joe Cocker-A & M
23	23.	Don't Tell Me Goodnight-Lobo-Big Tree
29	24.	Shining Star-Earth-Wind & Fire-Columbia
19	25.	Don't Call Us, We'll Call You-\ Sugarloaf-Claridge
31	26.	Remember What I Told You To Forget-Tavares-Capitol
17	27.	What Am I Gonna Do With You-Barry White-20th Century
33	28.	Sister Golden Hair-America-WB
21	29.	Lovin' You-Minnie Riperton-Epic
34	30.	Hijack-Herbie Mann-Atlantic
37	31.	I Wanna Dance Wit' Choo-Disco Tex & The Sex-O-Lettes-Chelsea
27	32.	L-O-V-E (Love)-Al Green-HI
36	33.	Shoeshine Boy-Eddie Kendricks-Tamla
38	34.	When Will I Be loved-Linda Ronstadt-Capitol
22	35.	Have You Never Been Mellow-Olivia Newton-John-MCA
--	36.	Old Days-Chicago-Columbia
--	37.	Magic-Pilot-EMI
--	38.	Take Me In Your Arms-The Doobie Brothers-WB
--	39.	Hey You-Bachman-Turner Overdrive-Mercury
25	40.	Express-B.T. Express-Roadshow

SUPER CFL SURVEY MAY 24

wcfl wcfl wcfl wcfl wcfl wcfl wcfl wcfl wcfl wcfl wcfl wcfl wcfl wcfl

SUPER ALBUMS

1. Chicago VIII-Columbia
2. Physical Graffiti-Led Zeppelin-Swan Song
3. That's The Way Of The World-Earth, Wind & Fire-Columbia
4. "Tommy"-Original Sound Track-Polydor
5. Autobahn-Kraftwerk-Vertigo
6. Blue Jays-Justin Hayward & John Lodge-Threshold
7. Blow By Blow-Jeff Beck-Epic
8. Crash Landing-Jimi Hendrix-Reprise
9. Straight Shooter-Bad Company-Swan Song
10. Hearts-America-WB
11. To Be True-Harold Melvin & The Blue Notes-PI
12. Playing Possum-Carly Simon-Elektra
13. Mister Magic-Grover Washington, Jr.-Kudu
14. Stampede-The Doobie Brothers-WB
15. Crime Of The Century-Supertramp-A & M
16. Have You Never Been Mellow-Olivia Newton-John-MCA
17. A Song For You-The Temptations-Gordy
18. Katy Lied-Steely Dan-ABC
19. For Earth Below-Robin Trower-Chrysalis
20. Fandango-ZZ Top-London
21. Average White Band-AWB-Atlantic
22. Welcome To My Nightmare-Alice Cooper-Atlantic
23. Just Another Way To Say I Love You-Barry White-20th Century
24. "An Evening With John Denver"-RCA
25. Blood On The Tracks-Bob Dylan-Columbia

1	1.	(Hey Won't You Play) Another Somebody Done Somebody Wrong Song-B.J. Thomas-ABC
2	2.	He Don't Love You-Tony Orlando/Dawn-Elektra
3	3.	It's A Miracle-Barry Manilow-Arista
5	4.	Long Tall Glasses-Leo Sayer-WB
7	5.	How Long-Ace-Anchor
8	6.	Before The Next Teardrop Falls-Freddy Fender-ABC/Dot
10	7.	Killer Queen-Queen-Elektra
4	8.	Jackie Blue-Ozark Mountain Daredevils-A & M
15	9.	Bad Time-Grand Funk-Capitol
17	10.	Only Yesterday-Carpenters-A & M
9	11.	I Don't Like To Sleep Alone-Paul Anka-UA
6	12.	Chevy Van-Sammy Johns-GRC
12	13.	Autobahn-Kraftwerk-Vertigo
19	14.	Thank God I'm A Country Boy-John Denver-RCA
24	15.	Shining Star-Earth, Wind & Fire-Columbia
11	16.	Shame, Shame, Shame-Shirley (And Company)-Vibration
21	17.	Only Women-Alice Cooper-Atlantic
13	18.	Philadelphia Freedom-The Elton John Band-MCA
28	19.	Sister Golden Hair-America-WB
14	20.	Lady Marmalade-Labelle-Epic
26	21.	Remember What I Told You To Forget-Tavares-Capitol
20	22.	Amie-Pure Prairie League-RCA
16	23.	No No Song-Ringo Starr-Apple
34	24.	When Will I Be Loved-Linda Ronstadt-Capitol
23	25.	Don't Tell Me Goodnight-Lobo-Big Tree
31	26.	I Wanna Dance Wit' Choo-Disco Tex-Chelsea
30	27.	Hijack-Herbie Mann-Atlantic
22	28.	You Are So Beautiful-Joe Cocker-A & M
33	29.	Shoeshine Boy-Eddie Kendricks-Tamla
36	30.	Old Days-Chicago-Columbia
18	31.	Runaway-Charlie Kulis-Playboy
37	32.	Magic-Pilot-EMI
38	33.	Take Me In Your Arms-The Doobie Brothers-WB
--	34.	Love Won't Let Me Wait-Major Harris-Atlantic
39	35.	Hey You-Bachman-Turner Overdrive-Mercury
25	36.	Don't Call Us, We'll Call You-Sugarloaf-Claridge
--	37.	Wildfire-Michael Murphey-Epic

--	38.	The Way We Were/Try To Remember-Gladys Knight & The Pips-Buddah
27	39.	What Am I Gonna Do With You-Barry White-20th Century
--	40.	One Of These Nights-Eagles-Asylum

SUPER CFL SURVEY MAY 31

wcfl wcfl wcfl wcfl wcfl wcfl wcfl wcfl wcfl wcfl wcfl wcfl wcfl wcfl

SUPER ALBUMS

1. Chicago VIII-Columbia
2. That's The Way Of The World-Earth, Wind & Fire-Columbia
3. Physical Graffiti-Led Zeppelin-Swan Song
4. "Tommy"-Original Sound Track-Polydor
5. Hearts-America-WB
6. Straight Shooter-Bad Company-Swan Song
7. Autobahn-Kraftwerk-Vertigo
8. Playing Possum-Carly Simon-Elektra
9. Crash Landing-Jimi Hendrix-Reprise
10. Blow By Blow-Jeff Beck-Epic
11. Stampede-The Doobie Brothers-WB
12. To Be True-Harold Melvin & The Blue Notes-PI
13. Mister Magic-Grover Washington, Jr.-Kudu
14. Welcome To My Nightmare-Alice Cooper-Atlantic
15. "An Evening With John Denver"-RCA
16. Blue Jays-Justin Hayward & John Lodge-Threshold
17. Greatest Hits-Elton John-MCA
18. Have You Never Been Mellow-Olivia Newton-John-MCA
19. Crime Of The Century-Supertramp-A & M
20. Disco Baby-Van McCoy-Avco
21. Fandango-ZZ Top-London
22. Just A Boy-Leo Sayer-WB
23. "The Original Sound Track"-10cc-Mercury
24. Just Another Way To Say I Love You-Barry White-20th Century
25. Four Wheel Drive-Bachman-Turner Overdrive-Mercury

3	1.	It's A Miracle-Barry Manilow-Arista
7	2.	Killer Queen-Queen-Elektra
6	3.	Before The Next Teardrop Falls-Freddy Fender-ABC/Dot
4	4.	Long Tall Glasses-Leo Sayer-WB
1	5.	(Hey Won't You Play) Another Somebody Done Somebody Wrong Song-B.J. Thomas-ABC
10	6.	Only Yesterday-Carpenters-A & M
9	7.	Bad Time-Grand Funk-Capitol
14	8.	Thank God I'm A Country Boy-John Denver-RCA
2	9.	He Don't Love You-Tony Orlando/Dawn-Elektra
17	10.	Only Women-Alice Cooper-Atlantic
5	11.	How Long-Ace-Anchor
15	12.	Shining Star-Earth, Wind & Fire-Columbia
8	13.	Jackie Blue-Ozark Mountain Daredevils-A & M
19	14.	Sister Golden Hair-America-WB
11	15.	I Don't Like To Sleep Alone-Paul Anka-UA
13	16.	Autobahn-Kraftwerk-Vertigo
24	17.	When Will I Be Loved-Linda Ronstadt-Capitol
12	18.	Chevy Van-Sammy Johns-GRC
21	19.	Remember What I Told You To Forget-Tavares-Capitol
30	20.	Old Days-Chicago-Columbia
16	21.	Shame, Shame, Shame-Shirley (And Company)-Vibration
18	22.	Philadelphia Freedom-The Elton John Band-MCA
22	23.	Amie-Pure Prairie League-RCA
33	24.	Take Me In Your Arms-The Doobie Brothers-WB
20	25.	Lady Marmalade-Labelle-Epic
35	26.	Hey You-Bachman-Turner Overdrive-Mercury
27	27.	Hijack-Herbie Mann-Atlantic
32	28.	Magic-Pilot-EMI
29	29.	Shoeshine Boy-Eddie Kendricks-Tamla
34	30.	Love Won't Let Me Wait-Major Harris-Atlantic
37	31.	Wildfire-Michael Murphey-Epic
26	32.	I Wanna Dance Wit' Choo-Disco Tex-Chelsea
38	33.	The Way We Were/Try To Remember-Gladys Knight & The Pips-Buddah
--	34.	I'm Not Lisa-Jessi Colter-Capitol

--	35.	The Last Farewell-Roger Whittaker-RCA
23	36.	No No Song-Ringo Starr-Apple
40	37.	One Of These Nights-Eagles-Asylum
--	38.	The Hustle-Van McCoy-Avco
25	39.	Don't Tell Me Goodnight-Lobo-Big Tree
--	40.	It's All Up To You-Jim Capaldi-Island

SUPER CFL SURVEY JUN 7

wcfl wcfl wcfl wcfl wcfl wcfl wcfl wcfl wcfl wcfl wcfl wcfl wcfl wcfl

SUPER ALBUMS

1. Chicago VIII-Columbia
2. That's The Way Of The World-Earth-Wind & Fire-Columbia
3. Physical Graffiti-Led Zeppelin-Swan Song
4. "Tommy"-Original Sound Track-Polydor
5. Hearts-America-WB
6. Stampede-The Doobie Brothers-WB
7. Captain Fantastic-Elton John-MCA
8. Playing Possum-Carly Simon-Elektra
9. Straight Shooter-Bad Company-Swan Song
10. Crash Landing-Jimi Hendrix-Reprise
11. Blow By Blow-Jeff Beck-Epic
12. Blue Jays-Justin Hayward & John Lodge-Threshold
13. Mister Magic-Grover Washington, Jr.-Kudu
14. Four Wheel Drive-Bachman-Turner Overdrive-Mercury
15. Welcome To My Nightmare-Alice Cooper-Atlantic
16. Autobahn-Kraftwerk-Vertigo
17. Survival-O'Jays-PI
18. To Be True-Harold Melvin & The Blue Notes-PI
19. A Song For You-The Temptations-Gordy
20. Greatest Hits-Elton John-MCA
21. Have You Never Been Mellow-Olivia Newton-John-MCA
22. Just Another Way To Say I Love You-Barry White-20th Century
23. Disco Baby-Van McCoy-Avco
24. Fandango-ZZ Top-London
25. "An Evening With John Denver"-RCA

2	1.	Killer Queen-Queen-Elektra
4	2.	Long Tall Glasses-Leo Sayer-WB
6	3.	Only Yesterday-Carpenters-A & M
7	4.	Bad Time-Grand Funk-Capitol
1	5.	It's A Miracle-Barry Manilow-Arista
3	6.	Before The Next Teardrop Falls-Freddy Fender-ABC/Dot
8	7.	Thank God I'm A Country Boy-John Denver-RCA
10	8.	Only Women-Alice Cooper-Atlantic
12	9.	Shining Star-Earth, Wind & Fire-Columbia
14	10.	Sister Golden Hair-America-WB
5	11.	(Hey Won't You Play) Another Somebody Done Somebody Wrong Song-B.J. Thomas-ABC
17	12.	When Will I Be Loved-Linda Ronstadt-Capitol
9	13.	He Don't Love You-Tony Orlando/Dawn-Elektra
20	14.	Old Days-Chicago-Columbia
11	15.	How Long-Ace-Anchor
24	16.	Take Me In Your Arms-The Doobie Brothers-WB
13	17.	Jackie Blue-Ozark Mountain Daredevils-A & M
26	18.	Hey You-Bachman-Turner Overdrive-Mercury
28	19.	Magic-Pilot-EMI
19	20.	Remember What I Told You To Forget-Tavares-Capitol
15	21.	I Don't Like To Sleep Alone-Paul Anka-UA
18	22.	Chevy Van-Sammy Johns-GRC
30	23.	Love Won't Let Me Wait-Major Harris-Atlantic
16	24.	Autobahn-Kraftwerk-Vertigo
31	25.	Wildfire-Michael Murphey-Epic
33	26.	The Way We Were/Try To Remember-Gladys Knight & The Pips-Buddah
23	27.	Amie-Pure Prairie League-RCA
34	28.	I'm Not Lisa-Jessi Colter-Capitol
27	29.	Hijack-Herbie Mann-Atlantic
38	30.	The Hustle-Van McCoy-Avco
35	31.	The Last Farewell-Roger Whittaker-RCA
22	32.	Philadelphia Freedom-The Elton John Band-MCA
37	33.	One Of These Nights-Eagles-Asylum
25	34.	Lady Marmalade-Labelle-Epic
--	35.	I'm On Fire-Dwight Twilley Band-Shelter
40	36.	It's All Up To You-Jim Capaldi-Island

29	37.	Shoeshine Boy-Eddie Kendricks-Tamla
--	38.	Black Friday-Steely Dan-ABC
--	39.	Listen To What The Man Said-Paul McCartney-Capitol
21	40.	Shame, Shame, Shame-Shirley (And Company)-Vibration

SUPER CFL SURVEY JUN 14

wcfl wcfl wcfl wcfl wcfl wcfl wcfl wcfl wcfl wcfl wcfl wcfl wcfl wcfl

SUPER ALBUMS

1. Captain Fantastic-Elton John-MCA
2. That's The Way Of The World-Earth, Wind & Fire-Columbia
3. Chicago VIII-Columbia
4. Stampede-The Doobie Brothers-WB
5. "Tommy"-Original Sound Track-Polydor
6. Hearts-America-WB
7. Disco Baby-Van McCoy-Avco
8. Playing Possum-Carly Simon-Elektra
9. Four Wheel Drive-Bachman-Turner Overdrive-Mercury
10. Physical Graffiti-Led Zeppelin-Swan Song
11. Straight Shooter-Bad Company-Swan Song
12. Mister Magic-Grover Washington, Jr.-Kudu
13. Blow By Blow-Jeff Beck-Epic
14. Welcome To My Nightmare-Alice Cooper-Atlantic
15. Survival-O'Jays-PI
16. "The Original Sound Track"-10cc-Mercury
17. To BE True-Harold Melvin & The Blue Notes-PI
18. Lake Shore Drive-Aliotta, Haynes, Jeremiah-Big Foot
19. Fandango-ZZ Top-London
20. A Song For You-The Temptations-Gordy
21. Autobahn-Kraftwerk-Vertigo
22. Crash Landing-Jimi Hendrix-Reprise
23. "An Evening With John Denver"-RCA
24. Blue Jays-Justin Hayward & John Lodge-Threshold
25. A Quiet Storm-Smokey Robinson-Tamla

1	1.	Killer Queen-Queen-Elektra
3	2.	Only Yesterday-Carpenters-A & M
4	3.	Bad Time-Grand Funk-Capitol
7	4.	Thank God I'm A Country Boy-John Denver-RCA
10	5.	Sister Golden Hair-America-WB
8	6.	Only Women-Alice Cooper-Atlantic
12	7.	When Will I Be Loved-Linda Ronstadt-Capitol
9	8.	Shining Star-Earth, Wind & Fire-Columbia
14	9.	Old Days-Chicago-Columbia
2	10.	Long Tall Glasses-Leo Sayer-WB
16	11.	Take Me In Your Arms-The Doobie Brothers-WB
5	12.	It's A Miracle-Barry Manilow-Arista
18	13.	Hey You-Bachman-Turner Overdrive-Mercury
6	14.	Before The Next Teardrop Falls-Freddy Fender-ABC/Dot
19	15.	Magic-Pilot-EMI
25	16.	Wildfire-Michael Murphey-Epic
11	17.	(Hey Won't You Play) Another Somebody Done Somebody Wrong Song-B.J. Thomas-ABC
23	18.	Love Won't Let Me Wait-Major Harris-Atlantic
28	19.	I'm Not Lisa-Jessi Colter-Capitol
26	20.	The Way We Were/Try To Remember-Gladys Knight & The Pips-Buddah
15	21.	How Long-Ace-Anchor
30	22.	The Hustle-Van McCoy-Avco
13	23.	He Don't Love You-Tony Orlando/Dawn-Elektra
17	24.	Jackie Blue-Ozark Mountain Daredevils-A & M
31	25.	The Last Farewell-Roger Whittaker-RCA
21	26.	I Don't Like To Sleep Alone-Paul Anka-UA
--	27.	Love Will Keep Us Together-The Captain & Tennille-A & M
33	28.	One Of These Nights-Eagles-Asylum
22	29.	Chevy Van-Sammy Johns-GRC
35	30.	I'm On Fire-Dwight Twilley Band-Shelter
24	31.	Autobahn-Kraftwerk-Vertigo
39	32.	Listen To What The Man Said-Paul McCartney-Capitol
20	33.	Remember What I Told You To Forget-Tavares-Capitol
36	34.	It's All Up To You-Jim Capaldi-Island
38	35.	Black Friday-Steely Dan-ABC
--	36.	I'm Not In Love-10cc-Mercury

32	37.	Philadelphia Freedom-The Elton John Band-MCA
--	38.	Jive Talkin'-Bee Gees-RSO
29	39.	Hijack-Herbie Mann-Atlantic
27	40.	Amie – Pure Prairie League – RCA

SUPER CFL SURVEY JUN 21

wcfl wcfl wcfl wcfl wcfl wcfl wcfl wcfl wcfl wcfl wcfl wcfl wcfl wcfl

SUPER ALBUMS

1. Captain Fantastic-Elton John-MCA
2. That's The Way Of The World-Earth, Wind & Fire-Columbia
3. Chicago VIII-Columbia
4. Venus And Mars-Paul McCartney & Wings-Capitol
5. Stampede-The Doobie Brothers-WB
6. Hearts-America-WB
7. Playing Possum-Carly Simon-Elektra
8. Physical Graffiti-Led Zeppelin-Swan Song
9. Four Wheel Drive-Bachman-Turner Overdrive-Mercury
10. Disco Baby-Van McCoy-Avco
11. "Tommy"-Original Sound Track-Polydor
12. Mister Magic-Grover Washington, Jr.-Kudu
13. Blow By Blow-Jeff Beck-Epic
14. Straight Shooter-Bad Company-Swan Song
15. To Be True-Harold Melvin & The Blue Notes-PI
16. Survival-O'Jays-PI
17. Welcome To My Nightmare-Alice Cooper-Atlantic
18. Blue Jays-Justin Hayward & John Lodge-Threshold
19. Lake Shore Drive-Aliotta, Haynes, Jeremiah-Big Foot
20. Fandango-ZZ Top-London
21. Diamonds & Rust-Joan Baez-A & M
22. A Quiet Storm-Smokey Robinson-Tamla
23. Gorilla-James Taylor-WB
24. A Song For You-The Temptations-Gordy
25. Toys In The Attic-Aerosmith-Columbia

5	1.	Sister Golden Hair-America-WB
7	2.	When Will I Be Loved-Linda Ronstadt-Capitol
2	3.	Only Yesterday-Carpenters-A & M
4	4.	Thank God I'm A Country Boy-John Denver-RCA
1	5.	Killer Queen-Queen-Elektra
8	6.	Shining Star-Earth, Wind & Fire-Columbia
16	7.	Wildfire-Michael Murphey-Epic
11	8.	Take Me In Your Arms-The Doobie Brothers-WB
9	9.	Old Days-Chicago-Columbia
13	10.	Hey You-Bachman-Turner Overdrive-Mercury
3	11.	Bad Time-Grand Funk-Capitol
15	12.	Magic-Pilot-EMI
6	13.	Only Women-Alice Cooper-Atlantic
19	14.	I'm Not Lisa-Jessi Colter-Capitol
18	15.	Love Won't Let Me Wait-Major Harris-Atlantic
22	16.	The Hustle-Van McCoy-Avco
20	17.	The Way We Were/Try To Remember-Gladys Knight & The Pips-Buddah
10	18.	Long Tall Glasses-Leo Sayer-WB
25	19.	The Last Farewell-Roger Whittaker-RCA
27	20.	Love Will Keep Us Together-The Captain &y Tennille-A & M
12	21.	It's A Miracle-Barry Manilow-Arista
17	22.	(Hey Won't You Play) Another Somebody Done Somebody Wrong Song-B.J. Thomas-ABC
28	23.	One Of These Nights-Eagles-Asylum
32	24.	Listen To What The Man Said-Paul McCartney-Capitol
14	25.	Before The Next Teardrop Falls-Freddy Fender-ABC/Dot
30	26.	I'm On Fire-Dwight Twilley Band-Shelter
21	27.	How Long-Ace-Anchor
35	28.	Black Friday-Steely Dan-ABC
24	29.	Jackie Blue-Ozark Mountain Daredevils-A & M
36	30.	I'm Not In Love-10cc-Mercury
34	31.	It's All Up To You-Jim Capaldi-Island
--	32.	Why Can't We Be Friends?-War-UA
38	33.	Jive Talkin'-Bee Gees-RSO
26	34.	I Don't Like To Sleep Alone-Paul Anka-UA
--	35.	Rockin' Chair-Gwen McCrae-TK

29	36.	Chevy Van-Sammy Johns-GRC
--	37.	Please Mr. Please-Olivia Newton-John-MCA
37	38.	Philadelphia Freedom-The Elton John Band-MCA
31	39.	Autobahn-Kraftwerk-Vertigo
23	40.	He Don't Love You-Tony Orlando/ Dawn-Elektra

SUPER CFL SURVEY JUN 28

wcfl wcfl wcfl wcfl wcfl wcfl wcfl wcfl wcfl wcfl wcfl wcfl wcfl wcfl

1	1.	Sister Golden Hair-America-WB
2	2.	When Will I Be Loved-Linda Ronstadt-Capitol
7	3.	Wildfire-Michael Murphey-Epic
3	4.	Only Yesterday-Carpenters-A & M
20	5.	Love Will Keep Us Together-The Captain & Tennille-A & M
8	6.	Take Me In Your Arms-The Doobie Brothers-WB
10	7.	Hey You-Bachman-Turner Overdrive-Mercury
12	8.	Magic-Pilot-EMI
6	9.	Shining Star-Earth, Wind & Fire-Columbia
14	10.	I'm Not Lisa-Jessi Colter-Capitol
16	11.	The Hustle-Van McCoy-Avco
17	12.	Way We Were (Medley)-Gladys Knight/Pips-Buddah
15	13.	Love Won't Let Me Wait-Major Harris-Atlantic
4	14.	Thank God I'm A Country Boy-John Denver-RCA
19	15.	The Last Farewell-Roger Whittaker-RCA
9	16.	Old Days-Chicago-Columbia
5	17.	Killer Queen-Queen-Elektra
23	18.	One Of These Nights-Eagles-Asylum
11	19.	Bad Time-Grand Funk-Capitol
24	20.	Listen To What The Man Said-Paul McCartney-Capitol
13	21.	Only Women-Alice Cooper-Atlantic
30	22.	I'm Not In Love-10cc-Mercury
26	23.	I'm On Fire-Dwight Twilley Band-Shelter
18	24.	Long Tall Glasses-Leo Sayer-WB
33	25.	Jive Talkin'-Bee Gees-RSO
32	26.	Why Can't We Be Friends?-War-UA
21	27.	It's A Miracle-Barry Manilow-Arista
28	28.	Black Friday-Steely Dan-ABC
35	29.	Rockin' Chair-Gwen McCrae-TK
37	30.	Please Mr. Please-Olivia Newton-John-MCA
--	31.	Swearin' To God-Frankie Valli-Private Stock
22	32.	(Hey Won't You Play) Another Somebody Done Somebody Wrong Song-B.J. Thomas-ABC

SUPER ALBUMS

1. Captain Fantastic-Elton John-MCA
2. That's The Way Of The World-Earth, Wind & Fire-Columbia
3. Chicago VIII-Columbia
4. Venus And Mars-Paul McCartney & Wings-Capitol
5. Disco Baby-Van McCoy-Avco
6. Stampede-The Doobie Brothers-WB
7. Playing Possum-Carly Simon-Elektra
8. Hearts-America-WB
9. Four Wheel Drive-Bachman-Turner Overdrive-Mercury
10. "Tommy"-Original Sound Track-Polydor
11. Survival-O'Jays-PI
12. Mister Magic-Grover Washington, Jr.-Kudu
13. Physical Graffiti-Led Zeppelin-Swan Song
14. "The Original Sound Track"-10cc-Mercury
15. Welcome To My Nightmare-Alice Cooper-Atlantic
16. Straight Shooter-Bad Company-Swan Song
17. Metamorphosis-The Rolling Stones-ABKCO
18. Blow By Blow-Jeff Beck-Epic
19. Lake Shore Drive-Aliotta, Haynes, Jeremiah-Big Foot
20. Spirit Of America-The Beach Boys-Capitol
21. Fandango-ZZ Top-London
22. A Quiet Storm-Smokey Robinson-Tamla
23. To Be True-Harold Melvin & The Blue Notes-PI
24. The Myths And Legends Of King Arthur And The Knights Of The Round Table-Rick Wakeman-A & M
25. A Song For You-The Temptations-Gordy

31	33.	It's All Up To You-Jim Capaldi-Island
27	34.	How Long-Ace-Anchor
--	35.	Midnight Blue-Melissa Manchester-Arista
25	36.	Before The Next Teardrop Falls-Freddy Fender-ABC/Dot
29	37.	Jackie Blue-Ozark Mountain Daredevils-A & M
36	38.	Chevy Van-Sammy Johns-GRC
34	39.	I Don't Like To Sleep Alone-Paul Anka-UA
38	40.	Philadelphia Freedom-The Elton John Band-MCA

SUPER CFL SURVEY JUL 5

wcfl wcfl wcfl wcfl wcfl wcfl wcfl wcfl wcfl wcfl wcfl wcfl wcfl wcfl

SUPER ALBUMS

1. Captain Fantastic-Elton John-MCA
2. Venus And Mars-Paul McCartney & Wings-Capitol
3. That's The Way Of The World-Earth, Wind & Fire-Columbia
4. Chicago VIII-Columbia
5. One Of These Nights-Eagles-Asylum
6. Stampede-The Doobie Brothers-WB
7. Disco Baby-Van McCoy-Avco
8. Hearts-America-WB
9. Four Wheel Drive-Bachman-Turner Overdrive-Mercury
10. Physical Graffiti-Led Zeppelin-Swan Song
11. The Heat Is On-The Isley Brothers-T-Neck
12. Mister Magic-Grover Washington, Jr.-Kudu
13. Survival-O'Jays-PI
14. "Tommy"-Original Sound Track-Polydor
15. To Be True-Harold Melvin & The Blue Notes-PI
16. Playing Possum-Carly Simon-Elektra
17. Horizon-Carpenters-A & M
18. A Song For You-Temptations-Gordy
19. Welcome To My Nightmare-Alice Cooper-Atlantic
20. Love Will Keep Us Together-The Captain & Tennille-A & M
21. Blow By Blow-Jeff Beck-Epic
22. Gorilla-James Taylor-WB
23. Lake Shore Drive-Aliotta, Haynes, Jeremiah-Big Foot
24. A Quiet Storm-Smokey Robinson-Tamla
25. Blue Sky-Night Thunder-Michael Murphey-Epic

3	1.	Wildfire-Michael Murphey-Epic
5	2.	Love Will Keep Us Together-The Captain & Tennille-A & M
1	3.	Sister Golden Hair-America-WB
6	4.	Take Me In Your Arms-The Doobie Brothers-WB
8	5.	Magic-Pilot-EMI
11	6.	The Hustle-Van McCoy-Avco
10	7.	I'm Not Lisa-Jessi Colter-Capitol
2	8.	When Will I Be Loved-Linda Ronstadt-Capitol
4	9.	Only Yesterday-Carpenters-A & M
18	10.	One Of These Nights-Eagles-Asylum
7	11.	Hey You-Bachman-Turner Overdrive-Mercury
12	12.	Way We Were (Medley)-Gladys Knight/Pips-Buddah
20	13.	Listen To What The Man Said-Paul McCartney-Capitol
13	14.	Love Won't Let Me Wait-Major Harris-Atlantic
15	15.	The Last Farewell-Roger Whittaker-RCA
22	16.	I'm Not In Love-10cc-Mercury
9	17.	Shining Star-Earth, Wind & Fire-Columbia
30	18.	Please Mr. Please-Olivia Newton-John-MCA
25	19.	Jive Talkin'-Bee Gees-RSO
26	20.	Why Can't We Be Friends-War-UA
16	21.	Old Days-Chicago-Columbia
29	22.	Rockin' Chair-Gwen McCrae-TK
23	23.	I'm On Fire-Dwight Twilley Band-Shelter
17	24.	Killer Queen-Queen-Elektra
31	25.	Swearin' To God-Frankie Valli-Private Stock
21	26.	Only Women-Alice Cooper-Atlantic
14	27.	Thank God I'm A Country Boy-John Denver-RCA
35	28.	Midnight Blue-Melissa Manchester-Arista
19	29.	Bad Time-Grand Funk-Capitol
--	30.	Dynomite-Bazuka-A & M
24	31.	Long Tall Glasses-Leo Sayer-WB
28	32.	Black Friday-Steely Dan-ABC
--	33.	Disco Queen-Hot Chocolate-Big Tree
27	34.	It's A Miracle-Barry Manilow-Arista
--	35.	Mornin' Beautiful-Tony Orlando & Dawn-Elektra
32	36.	(Hey Won't You Play) Another Somebody Done Somebody Wrong Song-B.J. Thomas-ABC
--	37.	Someone Saved My Life Tonight-

		Elton John-MCA
--	38.	Rhinestone Cowboy-Glen Campbell-Capitol
34	39.	How Long-Ace-Anchor
40	40.	Philadelphia Freedom-The Elton John Band-MCA

SUPER CFL SURVEY JUL 12

wcfl wcfl wcfl wcfl wcfl wcfl wcfl wcfl wcfl wcfl wcfl wcfl wcfl wcfl

SUPER ALBUMS

1. Captain Fantastic-Elton John-MCA
2. Venus And Mars-Paul McCartney & Wings-Capitol
3. That's The Way Of The World-Earth, Wind & Fire-Columbia
4. Disco Baby-Van McCoy-Avco
5. Chicago VIII-Columbia
6. One Of These Nights-Eagles-Asylum
7. The Heat Is On-The Isley Brothers-T-Neck
8. Stampede-The Doobie Brothers-WB
9. Physical Graffiti-Led Zeppelin-Swan Song
10. Four Wheel Drive-Bachman-Turner Overdrive-Mercury
11. Mister Magic-Grover Washington, Jr.-Kudu
12. Hearts-America-WB
13. Survival-O'Jays-PI
14. Lake Shore Drive-Aliotta, Haynes, Jeremiah-Big Foot
15. Love Will Keep Us Together-The Captain & Tennille-A & M
16. Gorilla-James Taylor-WB
17. Fandango-ZZ Top-London
18. Spirit Of America-The Beach Boys-Capitol
19. Blow By Blow-Jeff Beck-Epic
20. Playing Possum-Carly Simon-Elektra
21. To BE True-Harold Melvin & The Blue Notes-PI
22. Horizon-Carpenters-A & M
23. Made In The Shade-Rolling Stones-Rolling Stones
24. Stills-Stephen Stills-Columbia
25. "The Original Sound Track"-10cc-Mercury

2	1.	Love Will Keep Us Together-The Captain & Tennille-A & M
1	2.	Wildfire-Michael Murphey-Epic
5	3.	Magic-Pilot-EMI
6	4.	The Hustle-Van McCoy-Avco
4	5.	Take Me In Your Arms-The Doobie Brothers-WB
7	6.	I'm Not Lisa-Jessi Colter-Capitol
10	7.	One Of These Nights-Eagles-Asylum
12	8.	Way We Were (Medley)-Gladys Knight/Pips-Buddah
13	9.	Listen To What The Man Said-Paul McCartney-Capitol
3	10.	Sister Golden Hair-America-WB
16	11.	I'm Not In Love-10cc-Mercury
19	12.	Jive Talkin'-Bee Gees-RSO
18	13.	Please Mr. Please-Olivia Newton-John-MCA
8	14.	When Will I Be Loved-Linda Ronstadt-Capitol
20	15.	Why Can't We Be Friends?-War-UA
14	16.	Love Won't Let Me Wait-Major Harris-Atlantic
11	17.	Hey You-Bachman-Turner Overdrive-Mercury
22	18.	Rockin' Chair-Gwen McCrae-TK
9	19.	Only Yesterday-Carpenters-A & M
25	20.	Swearin' To God-Frankie Valli-Private Stock
28	21.	Midnight Blue-Melissa Manchester-Arista
17	22.	Shining Star-Earth, Wind & Fire-Columbia
15	23.	The Last Farewell-Roger Whittaker-RCA
30	24.	Dynomite-Bazuka-A & M
21	25.	Old Days-Chicago-Columbia
23	26.	I'm On Fire-Dwight Twilley Band-Shelter
33	27.	Disco Queen-Hot Chocolate-Big Tree
35	28.	Mornin' Beautiful-Tony Orlando & Dawn-Elektra
24	29.	Killer Queen-Queen-Elektra
38	30.	Rhinestone Cowboy-Glen Campbell-Capitol
26	31.	Only Women-Alice Cooper-Atlantic
37	32.	Someone Saved My Life Tonight-Elton John-MCA
27	33.	Thank God I'm A Country Boy-John Denver-RCA
32	34.	Black Friday-Steely Dan-ABC
--	35.	Misty-Ray Stevens-Barnaby
29	36.	Bad Time-Grand Funk-Capitol
31	37.	Long Tall Glasses-Leo Sayer-WB
--	38.	'Til The World Ends-Three Dog Night-ABC
34	39.	It's A Miracle-Barry Manilow-Arista

39 | 40. How Long-Ace-Anchor

SUPER CFL SURVEY JULY 19

wcfl wcfl wcfl wcfl wcfl wcfl wcfl wcfl wcfl wcfl wcfl wcfl wcfl wcfl

SUPER ALBUMS

1. Captain Fantastic-Elton John-MCA
2. Venus And Mars-Paul McCartney & Wings-Capitol
3. One Of These Nights-Eagles-Asylum
4. Chicago VIII-Columbia
5. Love Will Keep Us Together-The Captain & Tennille-A & M
6. The Heat Is On-The Isley Brothers-T-Neck
7. That's The Way Of The World-Earth, Wind & Fire-Columbia
8. Four Wheel Drive-Bachman-Turner Overdrive-Mercury
9. Disco Baby-Van McCoy-Avco
10. Physical Graffiti-Led Zeppelin-Swan Song
11. Stills-Stephen Stills-Columbia
12. Mister Magic-Grover Washington, Jr.-Kudu
13. Gorilla-James Taylor-WB
14. Horizon-Carpenters-A & M
15. Stampede-The Doobie Brothers-WB
16. Survival-O'Jays PI
17. Hearts-America-WB
18. Playing Possum-Carly Simon-Elektra
19. Lake Shore Drive-Aliotta, Haynes, Jeremiah-Big Foot
20. Blow By Blow-Jeff Beck-Epic
21. Blue Sky-Night Thunder-Michael Murphey-Epic
22. "Tommy"-Original Sound Track-Polydor
23. Cut The Cake-AWB-Atlantic
24. Spirit Of America-The Beach Boys-Capitol
25. "The Original Sound Track"-10cc-Mercury

1	1.	Love Will Keep Us Together-The Captain & Tennille-A & M
3	2.	Magic-Pilot-EMI
4	3.	The Hustle-Van McCoy-Avco
7	4.	One Of These Nights-Eagles-Asylum
9	5.	Listen To What The Man Said-Paul McCartney-Capitol
6	6.	I'm Not Lisa-Jessi Colter-Capitol
13	7.	Please Mr. Please-Olivia Newton-John-MCA
11	8.	I'm Not In Love-10cc-Mercury
12	9.	Jive Talkin'-Bee Gees-RSO
2	10.	Wildfire-Michael Murphey-Epic
15	11.	Why Can't We Be Friends?-War-UA
5	12.	Take Me In Your Arms-The Doobie Brothers-WB
18	13.	Rockin' Chair-Gwen McCrae-TK
8	14.	Way We Were (Medley)-Gladys Knight/Pips-Buddah
10	15.	Sister Golden Hair-America-WB
20	16.	Swearin' To God-Frankie Valli-Private Stock
24	17.	Dynomite-Bazuka-A & M
14	18.	When Will I Be Loved-Linda Ronstadt-Capitol
21	19.	Midnight Blue-Melissa Manchester-Arista
27	20.	Disco Queen-Hot Chocolate-Big Tree
17	21.	Hey You-Bachman-Turner Overdrive-Mercury
28	22.	Mornin' Beautiful-Tony Orlando & Dawn-Elektra
30	23.	Rhinestone Cowboy-Glen Campbell-Capitol
19	24.	Only Yesterday-Carpenters-A & M
32	25.	Someone Saved My Life Tonight-Elton John-MCA
16	26.	Love Won't Let Me Wait-Major Harris-Atlantic
35	27.	Misty-Ray Stevens-Barnaby
22	28.	Shining Star-Earth, Wind & Fire-Columbia
--	29.	The Rockford Files-Mike Post-MGM
23	30.	The Last Farewell-Roger Whittaker-RCA
29	31.	Killer Queen-Queen-Elektra
38	32.	'Til The World Ends-Three Dog Night-ABC
--	33.	Fallin' In Love-Hamilton, Joe Frank & Reynolds-Playboy
25	34.	Old Days-Chicago-Columbia
31	35.	Only Women-Alice Cooper-Atlantic
26	36.	I'm On Fire-Dwight Twilley Band-Shelter
34	37.	Black Friday-Steely Dan-ABC

33	38.	Thank God I'm A Country Boy-John Denver-RCA
36	39.	Bad Time-Grand Funk-Capitol
37	40.	Long Tall Glasses-Leo Sayer-WB

SUPER CFL SURVEY JUL 26

wcfl wcfl wcfl wcfl wcfl wcfl wcfl wcfl wcfl wcfl wcfl wcfl wcfl wcfl

SUPER ALBUMS

1. Captain Fantastic-Elton John-MCA
2. Venus And Mars-Paul McCartney & Wings-Capitol
3. One Of These Nights-Eagles-Asylum
4. Love Will Keep Us Together-The Captain & Tennille-A & M
5. The Heat Is On-The Isley Brothers-T-Neck
6. That's The Way Of The World-Earth, Wind & Fire-Columbia
7. Disco Baby-Van McCoy-Avco
8. Stills-Stephen Stills-Columbia
9. Four Wheel Drive-Bachman-Turner Overdrive-Mercury
10. Chicago VIII-Columbia
11. Physical Graffiti-Led Zeppelin-Swan Song
12. Stampede-The Doobie Brothers-WB
13. Spirit Of America-The Beach Boys-Capitol
14. Survival-O'Jays-PI
15. Hearts-America-WB
16. Gorilla-James Taylor-WB
17. Mister Magic-Grover Washington, Jr.-Kudu
18. Tonight's The Night-Neil Young-Reprise
19. Greatest Hits-Cat Stevens-A & M
20. "The Original Sound Track"-10cc-Mercury
21. One Size Fits All-Frank Zappa-Discreet
22. Blow By Blow-Jeff Beck-Epic
23. Made In The Shade-Rolling Stones-Rolling Stones
24. Cut The Cake-AWB-Atlantic
25. Lake Shore Drive-Aliotta, Haynes-Jeremiah-Big Foot

1	1.	Love Will Keep Us Together-The Captain & Tennille-A & M
5	2.	Listen To What The Man Said-Paul McCartney-Capitol
4	3.	One Of These Nights-Eagles-Asylum
3	4.	The Hustle-Van McCoy-Avco
7	5.	Please Mr. Please-Olivia Newton-John-MCA
8	6.	I'm Not In Love-10cc-Mercury
9	7.	Jive Talkin'-Bee Gees-RSO
2	8.	Magic-Pilot-EMI
11	9.	Why Can't We Be Friends?-War-UA
13	10.	Rockin' Chair-Gwen McCrae-TK
6	11.	I'm Not Lisa-Jessi Colter-Capitol
16	12.	Swearin' To God-Frankie Valli-Private Stock
19	13.	Midnight Blue-Melissa Manchester-Arista
17	14.	Dynomite-Bazuka-A & M
10	15.	Wildfire-Michael Murphey-Epic
14	16.	Way We Were (Medley)-Gladys Knight/Pips-Buddah
25	17.	Someone Saved My Life Tonight-Elton John-MCA
20	18.	Disco Queen-Hot Chocolate-Big Tree
22	19.	Mornin' Beautiful-Tony Orlando & Dawn-Elektra
12	20.	Take Me In Your Arms-The Doobie Brothers-WB
23	21.	Rhinestone Cowboy-Glen Campbell-Capitol
15	22.	Sister Golden Hair-America-WB
27	23.	Misty-Ray Stevens-Barnaby
18	24.	When Will I Be Loved-Linda Ronstadt-Capitol
29	25.	The Rockford files-Mike Post-MGM
21	26.	Hey You-Bachman-Turner Overdrive-Mercury
32	27.	'Til The World Ends-Three Dog Night-ABC
26	28.	Love Won't Let Me Wait-Major Harris-Atlantic
33	29.	Fallin' In Love-Hamilton, Joe Frank & Reynolds-Playboy
24	30.	Only Yesterday-Carpenters-A & M
--	31.	Black Superman-Johnny Wakelin-Pye
28	32.	Shining Star-Earth, Wind & Fire-Columbia
--	33.	Tush-ZZ Top-London
30	34.	The Last Farewell-Roger Whittaker-RCA

31	35.	Killer Queen-Queen-Elektra
34	36.	Old Days-Chicago-Columbia
36	37.	I'm On Fire-Dwight Twilley Band-Shelter
35	38.	Only Women-Alice Cooper-Atlantic
37	39.	Black Friday-Steely Dan-ABC
38	40.	Thank God I'm A Country Boy-John Denver-RCA

SUPER CFL SURVEY AUG 2

wcfl wcfl wcfl wcfl wcfl wcfl wcfl wcfl wcfl wcfl wcfl wcfl wcfl wcfl

SUPER ALBUMS

1. Captain Fantastic-Elton John-MCA
2. Venus And Mars-Paul McCartney & Wings-Capitol
3. Love Will Keep Us Together-The Captain & Tennille-A & M
4. One Of These Nights-Eagles-Asylum
5. The Heat Is On-The Isley Brothers-T-Neck
6. Stills-Stephen Stills-Columbia
7. That's The Way Of The World-Earth, Wind & Fire-Columbia
8. Chicago VIII-Columbia
9. Four Wheel Drive-Bachman-Turner Overdrive-Mercury
10. Gorilla-James Taylor-WB
11. Spirit Of America-The Beach Boys-Capitol
12. Survival-O'Jays-PI
13. Disco Baby-Van McCoy-Avco
14. Red Octopus-Jefferson Starship-Grunt
15. Greatest Hits-Cat Stevens-A & M
16. Stampede-The Doobie Brothers-WB
17. "The Original Sound Track"-10cc-Mercury
18. Physical Graffiti-Led Zeppelin-Swan Song
19. One Size Fits All-Frank Zappa-Discreet
20. Tonight's The Night-Neil Young-Reprise
21. Latimore III-Glades
22. Blow By Blow-Jeff Beck-Epic
23. Cut The Cake-AWB-Atlantic
24. Horizon-Carpenters-A & M
25. Hearts-America-WB

2	1.	Listen To What The Man Said-Paul McCartney-Capitol
5	2.	Please Mr. Please-Olivia Newton-John-MCA
9	3.	Why Can't We Be Friends?-War-UA
6	4.	I'm Not In Love-10cc-Mercury
7	5.	Jive Talkin'-Bee Gees-RSO
1	6.	Love Will Keep Us Together-The Captain & Tennille-A & M
10	7.	Rockin' Chair-Gwen McCrae-TK
3	8.	One Of These Nights-Eagles-Asylum
4	9.	The Hustle-Van McCoy-Avco
13	10.	Midnight Blue-Melissa Manchester-Arista
12	11.	Swearin' To God-Frankie Valli-Private Stock
14	12.	Dynomite-Bazuka-A & M
17	13.	Someone Saved My Life Tonight-Elton John-MCA
8	14.	Magic-Pilot-EMI
19	15.	Mornin' Beautiful-Tony Orlando & Dawn-Elektra
18	16.	Disco Queen-Hot Chocolate-Big Tree
21	17.	Rhinestone Cowboy-Glen Campbell-Capitol
25	18.	The Rockford Files-Mike Post-MGM
23	19.	Misty-Ray Stevens-Barnaby
15	20.	Wildfire-Michael Murphey-Epic
11	21.	I'm Not Lisa-Jessi Colter-Capitol
27	22.	'Til The World Ends-Three Dog Night-ABC
16	23.	Way We Were (Medley)-Gladys Knight/Pips-Buddah
29	24.	Fallin' In Love-Hamilton, Joe Frank & Reynolds-Playboy
33	25.	Tush-ZZ Top-London
20	26.	Take Me In Your Arms-The Doobie Brothers-WB
22	27.	Sister Golden Hair-America-WB
31	28.	Black Superman-Johnny Wakelin-Pye
24	29.	When Will I Be Loved-Linda Ronstadt-Capitol
--	30.	At Seventeen-Janis Ian-Columbia
--	31.	Rendezvous-Hudson Brothers-Rocket
30	32.	Only Yesterday-Carpenters-A & M
26	33.	Hey You-Bachman-Turner Overdrive-Mercury
32	34.	Shining Star-Earth, Wind & Fire-Columbia
--	35.	Run Joey Run-David Geddes-Big Tree
35	36.	Killer Queen-Queen-Elektra

28	37.	Love Won't Let Me Wait-Major Harris-Atlantic
36	38.	Old Days-Chicago-Columbia
34	39.	The Last Farewell-Roger Whittaker-RCA
37	40.	I'm On Fire-Dwight Twilley Band-Shelter

SUPER CFL SURVEY AUG 9

wcfl wcfl wcfl wcfl wcfl wcfl wcfl wcfl wcfl wcfl wcfl wcfl wcfl wcfl

SUPER ALBUMS

1. Captain Fantastic-Elton John-MCA
2. Venus And Mars-Paul McCartney & Wings-Capitol
3. One Of These Nights-Eagles-Asylum
4. Love Will Keep Us Together-The Captain & Tennille-A & M
5. The Heat Is On-The Isley Brothers-T-Neck
6. Stills-Stephen Stills-Columbia
7. That's The Way Of The World-Earth, Wind & Fire-Columbia
8. Chicago VIII-Columbia
9. Four Wheel Drive-Bachman-Turner Overdrive-Mercury
10. Gorilla-James Taylor-WB
11. Disco Baby-Van McCoy-Avco
12. Survival-O'Jays-PI
13. Greatest hits-Cat Stevens-A & M
14. Physical Graffiti-Led Zeppelin-Swan Song
15. Made In The Shade-The Rolling Stones-Rolling Stones
16. Mister Magic-Grover Washington, Jr.-Kudu
17. Cut The Cake-AWB-Atlantic
18. Spirit Of America-The Beach Boys-Capitol
19. Red Octopus-Jefferson Starship-Grunt
20. "The Original Sound Track"-10cc-Mercury
21. The Basement Tapes-Bob Dylan & The Band-Columbia
22. Tonight's The Night-Neil Young-Reprise
23. Moving Violation-The Jackson 5-Motown
24. Blow By Blow-Jeff Beck-Epic
25. Fandango-ZZ Top-London

1	1.	Listen To What The Man Said-Paul McCartney-Capitol
3	2.	Why Can't We Be Friends?-War-UA
5	3.	Jive Talkin'-Bee Gees-RSO
2	4.	Please Mr. Please-Olivia Newton-John-MCA
7	5.	Rockin' Chair-Gwen McCrae-TK
10	6.	Midnight Blue-Melissa Manchester-Arista
6	7.	Love Will Keep Us Together-The Captain & Tennille-A & M
13	8.	Someone Saved My Life Tonight-Elton John-MCA
4	9.	I'm Not In Love-10cc-Mercury
12	10.	Dynomite-Bazuka-A & M
17	11.	Rhinestone Cowboy-Glen Campbell-Capitol
8	12.	One Of These Nights-Eagles-Asylum
11	13.	Swearin' To God-Frankie Valli-Private Stock
9	14.	The Hustle-Van McCoy-Avco
18	15.	The Rockford Files-Mike Post-MGM
16	16.	Disco Queen-Hot Chocolate-Big Tree
15	17.	Mornin' Beautiful-Tony Orlando & Dawn-Elektra
25	18.	Tush-ZZ Top-London
22	19.	'Til The World Ends-Three Dog Night-ABC
19	20.	Misty-Ray Stevens-Barnaby
24	21.	Fallin' In Love-Hamilton, Joe Frank & Reynolds-Playboy
14	22.	Magic-Pilot-EMI
28	23.	Black Superman-Johnny Wakelin-Pye
23	24.	Way We Were (Medley)-Gladys Knight/Pips-Buddah
30	25.	At Seventeen-Janis Ian-Columbia
20	26.	Wildfire-Michael Murphey-Epic
31	27.	Rendezvous-Hudson Brothers-Rocket
21	28.	I'm Not Lisa-Jessi Colter-Capitol
--	29.	Fell Like Makin' Love-Bad Company-Swan Song
27	30.	Sister Golden Hair-America-WB
--	31.	Third Rate Romance-The Amazing Rhythm Aces-ABC
35	32.	Run Joey Run-David Geddes-Big Tree
--	33.	(I Believe) There's Nothing Stronger Than Our Love-Paul Anka-UA
33	34.	Hey You-Bachman-Turner Overdrive-Mercury
26	35.	Take Me In Your Arms-The Doobie Brothers-WB
34	36.	Shining Star-Earth-Wind & Fire-Columbia

29	37.	When Will I Be Loved-Linda Ronstadt-Capitol
37	38.	Love Won't Let Me Wait-Major Harris-Atlantic
32	39.	Only Yesterday-Carpenters-A & M
36	40.	Killer Queen-Queen-Elektra

SUPER CFL SURVEY AUG 16

wcfl wcfl wcfl wcfl wcfl wcfl wcfl wcfl wcfl wcfl wcfl wcfl wcfl wcfl

2	1.	Why Can't We Be Friends?-War-UA
3	2.	Jive Talkin'-Bee Gees-RSO
5	3.	Rockin' Chair-Gwen McCrae-TK
8	4.	Someone Saved My Life Tonight-Elton John-MCA
1	5.	Listen To What The Man Said-Paul McCartney-Capitol
6	6.	Midnight Blue-Melissa Manchester-Arista
10	7.	Dynomite-Bazuka-A & M
4	8.	Please Mr. Please-Olivia Newton-John-MCA
9	9.	I'm Not In Love-10cc-Mercury
11	10.	Rhinestone Cowboy-Glen Campbell-Capitol
7	11.	Love Will Keep Us Together-The Captain & Tennille-A & M
18	12.	Tush-ZZ Top-London
15	13.	The Rockford Files-Mike Post-MGM
19	14.	'Til The World Ends-Three Dog Night-ABC
21	15.	Fallin' In Love-Hamilton, Joe Frank & Reynolds-Playboy
13	16.	Swearin' To God-Frankie Valli-Private Stock
25	17.	At Seventeen-Janis Ian-Columbia
23	18.	Black Superman-Johnny Wakelin-Pye
12	19.	One Of These Nights-Eagles-Asylum
14	20.	The Hustle-Van McCoy-Avco
17	21.	Mornin' Beautiful-Tony Orlando & Dawn-Elektra
27	22.	Rendezvous-Hudson Brothers-\ Rocket
16	23.	Disco Queen-Hot Chocolate-Big Tree
--	24.	Fight The Power-The Isley Brothers-T-Neck
32	25.	Run Joey Run-David Geddes-Big Tree
29	26.	Feel Like Makin' Love-Bad Company-Swan Song
20	27.	Misty-Ray Stevens-Barnaby
31	28.	Third Rate Romance-The Amazing Rhythm Aces-ABC
--	29.	Get Down Tonight-K.C. & The Sunshine Band-TK
33	30.	(I Believe) There's Nothing Stronger Than Our Love-Paul Anka-UA
22	31.	Magic-Pilot-EMI
--	32.	Ballroom Blitz-Sweet-Capitol
26	33.	Wildfire-Michael Murphey-Epic
30	34.	Sister Golden Hair-America-WB
24	35.	Way We Were (Medley)-Gladys Knight/Pips-Buddah
34	36.	Hey You-Bachman-Turner

SUPER ALBUMS

1. Captain Fantastic-Elton John-MCA
2. Venus And Mars-Paul McCartney & Wings-Capitol
3. The Heat Is On-The Isley Brothers-T-Neck
4. One Of These Nights-Eagles-Asylum
5. Love Will Keep Us Together-The Captain & Tennille-A & M
6. Gorilla-James Taylor-WB
7. That's The Way Of The World-Earth, Wind & Fire-Columbia
8. Stills-Stephen Stills-Columbia
9. Greatest Hits-Cat Stevens-A & M
10. Disco Baby-Van McCoy-Avco
11. Endless Summer-The Beach Boys-Capitol
12. Four Wheel Drive-Bachman-Turner Overdrive-Mercury
13. Survival-O'Jays-PI
14. Chicago VIII-Columbia
15. Spirit Of America-The Beach Boys-Capitol
16. Physical Graffiti-Led Zeppelin-Swan Song
17. Diamonds And Rust-Joan Baez-A & M
18. Why Can't We Be Friends?-War-UA
19. Fleetwood Mac-Fleetwood Mac-Reprise
20. Red Octopus-Jefferson Starship-Grunt
21. The Basement Tapes-Bob Dylan & The Band-Columbia
22. Cut The Cake-AWB-Atlantic
23. Made In The Shade-The Rolling Stones-Rolling Stones
24. Fandango-ZZ Top-London
25. Blow By Blow-Jeff Beck-Epic

		Overdrive-Mercury
28	37.	I'm Not Lisa-Jessi Colter-Capitol
35	38.	Take Me In Your Arms-The Doobie Brothers-WB
37	39.	When Will I Be Loved-Linda Ronstadt-Capitol
39	40.	Only Yesterday-Carpenters-A & M

SUPER CFL SURVEY AUG 23

wcfl wcfl wcfl wcfl wcfl wcfl wcfl wcfl wcfl wcfl wcfl wcfl wcfl wcfl

2	1.	Jive Talkin'-Bee Gees-RSO
4	2.	Someone Saved My Life Tonight-Elton John-MCA
1	3.	Why Can't We Be Friends?-War-UA
7	4.	Dynomite-Bazuka-A & M
10	5.	Rhinestone Cowboy-Glen Campbell-Capitol
12	6.	Tush-ZZ Top-London
3	7.	Rockin' Chair-Gwen McCrae-TK
15	8.	Fallin' In Love-Hamilton, Joe Frank & Reynolds-Playboy
6	9.	Midnight Blue-Melissa Manchester-Arista
13	10.	The Rockford Files-Mike Post-MGM
17	11.	At Seventeen-Janis Ian-Columbia
5	12.	Listen To What The Man Said-Paul McCartney-Capitol
11	13.	Love Will Keep Us Together-The Captain & Tennille-A & M
14	14.	'Til The World Ends-Three Dog Night-ABC
18	15.	Black Superman-Johnny Wakelin-Pye
9	16.	I'm Not In Love-10cc-Mercury
22	17.	Rendezvous-Hudson Brothers-Rocket
8	18.	Please Mr. Please-Olivia Newton-John-MCA
26	19.	Feel Like Makin' Love-Bad Company-Swan Song
25	20.	Run Joey Run-David Geddes-Big Tree
24	21.	Fight The Power-The Isley Brothers-T-Neck
16	22.	Swearin' To God-Frankie Valli-Private Stock
28	23.	Third Rate Romance-The Amazing Rhythm Aces-ABC
19	24.	One Of These Nights-Eagles-Asylum
30	25.	(I Believe) There's Nothing Stronger Than Our Love-Paul Anka-UA
29	26.	Get Down Tonight-K.C. & The Sunshine Band-TK
23	27.	Disco Queen-Hot Chocolate-Big Tree
32	28.	Ballroom Blitz-Sweet-Capitol
20	29.	The Hustle-Van McCoy-Avco
--	30.	Dance With Me-Orleans-Asylum
--	31.	Ain't No Way To Treat A Lady-Helen Reddy-Capitol
27	32.	Misty-Ray Stevens-Barnaby
21	33.	Mornin' Beautiful-Tony Orlando & Dawn-Elektra
31	34.	Magic-Pilot-EMI
33	35.	Wildfire-Michael Murphey-Epic
34	36.	Sister Golden Hair-America-WB

SUPER ALBUMS

1. Captain Fantastic-Elton John-MCA
2. The Heat Is On-The Isley Brothers-T-Neck
3. One Of These Nights-Eagles-Asylum
4. Venus And Mars-Paul McCartney & Wings-Capitol
5. Love Will Keep Us Together-The Captain & Tennille-A & M
6. Gorilla-James Taylor-WB
7. Stills-Stephen Stills-Columbia
8. That's The Way Of The World-Earth, Wind & Fire-Columbia
9. Greatest Hits-Cat Stevens-A & M
10. Red Octopus-Jefferson Starship-Grunt
11. Chicago VIII-Columbia
12. Endless Summer-The Beach Boys-Capitol
13. The Basement Tapes-Bob Dylan & The Band-Columbia
14. Toys In The Attic-Aerosmith-Columbia
15. Disco Baby-Van McCoy-Avco
16. Survival-O'Jays-PI
17. Four Wheel Drive-Bachman-Turner Overdrive-Mercury
18. Spirit Of America-The Beach Boys-Capitol
19. Fandango-ZZ Top-London
20. Physical Graffiti-Led Zeppelin-Swan Song
21. Steppin'-The Pointer Sisters-Blue Thumb
22. Fleetwood Mac-Fleetwood Mac-Reprise
23. Cut The Cake-AWB-Atlantic
24. Made In The Shade-The Rolling Stones-Rolling Stones
25. "The Original Sound Track"-10cc-Mercury

35	37.	Way We Were (Medley)-Gladys Knight/Pips-Buddah
37	38.	I'm Not Lisa-Jessi Colter-Capitol
38	39.	Take Me In Your Arms-The Doobie Brothers-WB
36	40.	Hey You-Bachman-Turner Overdrive-Mercury

SUPER CFL SURVEY AUG 30

wcfl wcfl wcfl wcfl wcfl wcfl wcfl wcfl wcfl wcfl wcfl wcfl wcfl wcfl

SUPER ALBUMS

1. Captain Fantastic-Elton John-MCA
2. One Of These Nights-Eagles-Asylum
3. The Heat Is On-The Isley Brothers-T-Neck
4. Venus And Mars-Paul McCartney & Wings-Capitol
5. Gorilla-James Taylor-WB
6. Love Will Keep Us Together-The Captain & Tennille-A & M
7. Stills-Stephen Stills-Columbia
8. That's The Way Of The World-Earth, Wind & Fire-Columbia
9. Endless Summer-The Beach Boys-Capitol
10. Greatest hits-Cat Stevens-A & M
11. Red Octopus-Jefferson Starship-Grunt
12. Why Can't We Be Friends?-War-UA
13. Fandango-ZZ Top-London
14. Chicago VIII-Columbia
15. Fleetwood Mac-Fleetwood Mac-Reprise
16. Steppin'-The Pointer Sisters-Blue Thumb
17. Pick Of The Litter-Spinners-Atlantic
18. Insides Out-Hamilton Bohannon-Dakar
19. Survival-O'Jays-PI
20. Cut The Cake-AWB-Atlantic
21. Spirit of America-The Beach Boys-Capitol
22. The Basement Tapes-Bob Dylan & The Band-Columbia
23. Diamonds & Rust-Joan Baez-A & M
24. Toys In The Attic-Aerosmith-Columbia
25. Disco Baby-Van McCoy-Avco

2	1.	Someone Saved My Life Tonight-Elton John-MCA
5	2.	Rhinestone Cowboy-Glen Campbell-Capitol
1	3.	Jive Talkin'-Bee Gees-RSO
6	4.	Tush-ZZ Top-London
8	5.	Fallin' In Love-Hamilton, Joe Frank & Reynolds-Playboy
3	6.	Why Can't We Be Friends?-War-UA
11	7.	At Seventeen-Janis Ian-Columbia
7	8.	Rockin' Chair-Gwen McCrae-TK
4	9.	Dynomite-Bazuka-A & M
10	10.	The Rockford Files-Mike Post-MGM
21	11.	Fight The Power-The Isley Brothers-T-Neck
15	12.	Black Superman-Johnny Wakelin-Pye
20	13.	Run Joey Run-David Geddes-Big Tree
26	14.	Get Down Tonight-K.C. & The Sunshine Band-TK
17	15.	Rendezvous-Hudson Brothers-Rocket
19	16.	Feel Like Makin' Love-Bad Company-Swan Song
9	17.	Midnight Blue-Melissa Manchester-Arista
16	18.	I'm Not In Love-10cc-Mercury
23	19.	Third Rate Romance-The Amazing Rhythm Aces-ABC
13	20.	Love Will Keep Us Together-The Captain & Tennille-A & M
18	21.	Please Mr. Please-Olivia Newton-John-MCA
25	22.	(I Believe) There's Nothing Stronger Than Our Love-Paul-Anka-UA
28	23.	Ballroom Blitz-Sweet-Capitol
12	24.	Listen To What The Man Said-Paul McCartney-Capitol
30	25.	Dance With Me-Orleans-Asylum
--	26.	How Long-The Pointer Sisters-Blue Thumb
14	27.	'Til The World Ends-Three Dog Night-ABC
31	28.	Ain't No Way To Treat A Lady-Helen Reddy-Capitol
--	29.	Games People Play-Spinners-Atlantic
--	30.	Wasted Days And Wasted Nights-Freddy Fender-ABC
24	31.	One Of These Nights-Eagles-Asylum
--	32.	Theme From "Jaws"-Original Sound Track—MCA
29	33.	The Hustle-Van McCoy-Avco
22	34.	Swearin' To God-Frankie Valli-Private Stock

27	35.	Disco Queen-Hot Chocolate-Big Tree
34	36.	Magic-Pilot-EMI
35	37.	Wildfire-Michael Murphey-Epic
33	38.	Mornin' Beautiful-Tony Orlando & Dawn-Elektra
32	39.	Misty-Ray Stevens-Barnaby
36	40.	Sister Golden Hair-America-WB

SUPER CFL SURVEY SEP 6

wcfl wcfl wcfl wcfl wcfl wcfl wcfl wcfl wcfl wcfl wcfl wcfl wcfl wcfl

SUPER ALBUMS

1. Captain Fantastic-Elton John-MCA
2. The Heat Is On-The Isley Brothers-T-Neck
3. One Of These Nights-Eagles-Asylum
4. Venus And Mars-Paul McCartney & Wings-Capitol
5. Gorilla-James Taylor-WB
6. Love Will Keep Us Together-The Captain & Tennille-A & M
7. Red Octopus-Jefferson Starship-Grunt
8. Greatest Hits-Cat Stevens-A & M
9. Stills-Stephen Stills-Columbia
10. Toys In The Attic-Aerosmith-Columbia
11. Endless Summer-The Beach Boys-Capitol
12. Fandango-ZZ Top-London
13. That's The Way Of The World-Earth, Wind & Fire-Columbia
14. Is It Something I Said?-Richard Pryor-Reprise
15. Steppin'-The Pointer Sisters-Blue Thumb
16. The Basement Tapes-Bob Dylan & The Band-Columbia
17. Fleetwood Mac-Fleetwood Mac-Reprise
18. Why Can't We Be Friends?-War-UA
19. Cut The Cake-AWB-Atlantic
20. Sabotage-Black Sabbath-WB
21. Insides Out-Hamilton Bohannon-Dakar
22. Honey-Ohio Players-Mercury
23. Made In The Shade-The Rolling Stones-Rolling Stones
24. Survival-O'Jays-PI
25. Pick Of The Litter-Spinners-Atlantic

2	1.	Rhinestone Cowboy-Glen Campbell-Capitol
5	2.	Fallin' In Love-Hamilton, Joe Frank & Reynolds-Playboy
4	3.	Tush—ZZ Top-London
1	4.	Someone Saved My Life Tonight-Elton John-MCA
7	5.	At Seventeen-Janis Ian-Columbia
3	6.	Jive Talkin'-Bee Gees-RSO
12	7.	Black Superman-Johnny Wakelin-Pye
11	8.	Fight The Power-The Isley Brothers-T-Neck
16	9.	Feel Like Makin' Love-Bad Company-Swan Song
13	10.	Run Joey Run-David Geddes-Big Tree
14	11.	Get Down Tonight-K.C. & The Sunshine Band-TK
6	12.	Why Can't We Be Friends?-War-UA
9	13.	Dynomite-Bazuka-A & M
15	14.	Rendezvous-Hudson Brothers-Rocket
19	15.	Third Rate Romance-The Amazing Rhythm Aces-ABC
8	16.	Rockin' Chair-Gwen McCrae-TK
23	17.	Ballroom Blitz-Sweet-Capitol
10	18.	The Rockford Files-Mike Post-MGM
28	19.	Ain't No Way To Treat A Lady-Helen Reddy-Capitol
25	20.	Dance With Me-Orleans-Asylum
18	21.	I'm Not In Love-10cc-Mercury
22	22.	(I Believe) There's Nothing Stronger Than Our Love-Paul Anka-UA
26	23.	How Long-The Pointer Sisters-Blue Thumb
29	24.	Games People Play-Spinners-Atlantic
--	25.	Fame-David Bowie-RCA
30	26.	Wasted Days And Wasted Nights-Freddy Fender-ABC
17	27.	Midnight Blue-Melissa Manchester-Arista
--	28.	That's The Way Of The World-Earth, Wind & Fire-Columbia
32	29.	Theme From "Jaws"-Original Sound Track-MCA
20	30.	Love Will Keep Us Together-The Captain & Tennille-A & M
--	31.	Mr. Jaws-Dickie Goodman-Cash
24	32.	Listen To What The Man Said-Paul McCartney-Capitol
27	33.	'Til The World Ends-Three Dog Night-ABC

21	34.	Please Mr. Please-Olivia Newton-John-MCA
31	35.	One Of These Nights-Eagles-Asylum
33	36.	The Hustle-Van McCoy-Avco
34	37.	Swearin' To God-Frankie Valli-Private Stock
36	38.	Magic-Pilot-EMI
37	39.	Wildfire-Michael Murphey-Epic
35	40.	Disco Queen-Hot Chocolate-Big Tree

SUPER CFL SURVEY SEP 13

wcfl wcfl wcfl wcfl wcfl wcfl wcfl wcfl wcfl wcfl wcfl wcfl wcfl wcfl

SUPER ALBUMS

1. Captain Fantastic-Elton John-MCA
2. One Of These Nights-Eagles-Asylum
3. The Heat Is On-Isley Brothers-T-Neck
4. Venus And Mars-Paul McCartney & Wings-Capitol
5. Gorilla-James Taylor-WB
6. Love Will Keep Us Together-The Captain & Tennille-A & M
7. Endless Summer-The Beach Boys-Capitol
8. Greatest Hits-Cat Stevens-A & M
9. Red Octopus-Jefferson Starship-Grunt
10. Fandango-ZZ Top-London
11. Is It Something I Said?-Richard Pryor-Reprise
12. Stills-Stephen Stills-Columbia
13. Honey-Ohio Players-Mercury
14. Pick Of The Litter-Spinners-Atlantic
15. That's The Way Of The World-Earth, Wind & Fire-Columbia
16. Blues For Allah-The Grateful Dead-UA
17. Fleetwood Mac-Fleetwood Mac-Reprise
18. Cut The Cake-AWB-Atlantic
19. Diamonds & Rust-Joan Baez-A & M
20. The Basement Tapes-Bob Dylan & The Band-Columbia
21. Toys In The Attic-Aerosmith-Columbia
22. Sabotage-Black Sabbath-WB
23. Between The Lines-Janis Ian-Columbia
24. Main Course-Bee Gees-RSO
25. Why Can't We Be Friends-War-UA

2	1.	Fallin' In Love-Hamilton, Joe Frank & Reynolds-Playboy
3	2.	Tush-ZZ Top-London
5	3.	At Seventeen-Janis Ian-Columbia
11	4.	Get Down Tonight-K.C. & The Sunshine Band-TK
8	5.	Fight The Power-The Isley Brothers-T-Neck
7	6.	Black Superman-Johnny Wakelin-Pye
10	7.	Run Joey Run-David Geddes-Big Tree
9	8.	Feel Like Makin' Love-Bad Company-Swan Song
1	9.	Rhinestone Cowboy-Glen Campbell-Capitol
4	10.	Someone Saved My Life Tonight-Elton John-MCA
17	11.	Ballroom Blitz-Sweet-Capitol
15	12.	Third Rate Romance-The Amazing Rhythm Aces-ABC
6	13.	Jive Talkin'-Bee Gees-RSO
20	14.	Dance With Me-Orleans-Asylum
19	15.	Ain't No Way To Treat A Lady-Helen Reddy-Capitol
12	16.	Why Can't We Be Friends?-War-UA
23	17.	How Long-The Pointer Sisters-Blue Thumb
14	18.	Rendezvous-Hudson Brothers-Rocket
24	19.	Games People Play-Spinners-Atlantic
26	20.	Wasted Days And Wasted Nights-Freddy Fender-ABC
29	21.	Theme From "Jaws'-Original Sound Track-MCA
25	22.	Fame-David Bowie-RCA
16	23.	Rockin' Chair-Gwen McCrae-TK
28	24.	That's The Way Of The World-Earth, Wind & Fire-Columbia
13	25.	Dynomite-Bazuka-A & M
31	26.	Mr. Jaws-Dickie Goodman-Cash
--	27.	It Only Takes A Minute-Tavares-Capitol
18	28.	The Rockford Files-Mike Post-MGM
22	29.	(I Believe) There's Nothing Stronger Than Our Love-Paul Anka-UA
--	30.	Eighteen With A Bullet-Pete Wingfield-Island
21	31.	I'm Not In Love-10cc-Mercury
--	32.	Lady Blue-Leon Russell-Shelter
--	33.	Brazil-The Richie Family-20th Century
30	34.	Love Will Keep Us Together-The Captain & Tennille-A & M
27	35.	Midnight Blue-Melissa Manchester-Arista

36	36.	The Hustle-Van McCoy-Avco
32	37.	Listen To What The Man Said-Paul McCartney-Capitol
34	38.	Please Mr. Please-Olivia Newton-John-MCA
35	39.	One Of These Nights-Eagles-Asylum
33	40.	'Til The World Ends-Three Dog Night-ABC

SUPER CFL SURVEY SEP 20

wcfl wcfl wcfl wcfl wcfl wcfl wcfl wcfl wcfl wcfl wcfl wcfl wcfl wcfl

SUPER ALBUMS

1. Captain Fantastic-Elton John-MCA
2. Red Octopus-Jefferson Starship-Grunt
3. One Of These Nights-Eagles-Asylum
4. The Heat Is On-Isley Brothers-T-Neck
5. Toys In The Attic-Aerosmith-Columbia
6. Venus And Mars-Paul McCartney & Wings-Capitol
7. Gorilla-James Taylor-WB
8. Is It Something I Said?-Richard Pryor-Reprise
9. Fandango-ZZ Top-London
10. Greatest Hits-Cat Stevens-A & M
11. Endless Summer-The Beach Boys-Capitol
12. Why Can't We Be Friends?-War-UA
13. Love Will Keep Us Together-The Captain & Tennille-A & M
14. Pick Of The Litter-Spinners-Atlantic
15. Honey-Ohio Players-Mercury
16. Sabotage-Black Sabbath-WB
17. Fleetwood Mac-Fleetwood Mac-Reprise
18. That's The Way Of the World-Earth, Wind & Fire-Columbia
19. Born To Run-Bruce Springsteen-Columbia
20. Steppin'-The Pointer Sisters-Blue Thumb
21. Between The Lines-Janis Ian-Columbia
22. Cut The Cake-AWB-Atlantic
23. Rhinestone Cowboy-Glen Campbell-Capitol
24. Stills-Stephen Stills-Columbia
25. Blues For Allah-The Grateful Dead-UA

4	1.	Get Down Tonight-K.C. & The Sunshine Band-TK
2	2.	Tush-ZZ Top-London
3	3.	At Seventeen-Janis Ian-Columbia
1	4.	Fallin' In Love-Hamilton, Joe Frank & Reynolds-Playboy
8	5.	Feel Like Makin' Love-Bad Company-Swan Song
5	6.	Fight The Power-The Isley Brothers-T-Neck
11	7.	Ballroom Blitz-Sweet-Capitol
6	8.	Black Superman-Johnny Wakelin-Pye
14	9.	Dance With Me-Orleans-Asylum
12	10.	Third Rate Romance-The Amazing Rhythm Aces-ABC
15	11.	Ain't No Way To Treat A Lady-Helen Reddy-Capitol
9	12.	Rhinestone Cowboy-Glen Campbell-Capitol
7	13.	Run Joey Run-David Geddes-Big Tree
19	14.	Games People Play-Spinners-Atlantic
17	15.	How Long-The Pointer Sisters-Blue Thumb
26	16.	Mr. Jaws-Dickie Goodman-Cash
22	17.	Fame-David Bowie-RCA
24	18.	That's The Way Of The World-Earth, Wind & Fire-Columbia
21	19.	Theme From "Jaws"-Original Sound Track-MCA
20	20.	Wasted Days And Wasted Nights-Freddy Fender-ABC
10	21.	Someone Saved My Life Tonight-Elton John-MCA
13	22.	Jive Talkin'-Bee Gees-RSO
27	23.	It Only Takes A Minute-Tavares-Capitol
16	24.	Why Can't We Be Friends?-War-UA
30	25.	Eighteen With A Bullet-Pete Wingfield-Island
--	26.	I'm Sorry-John Denver-RCA
32	27.	Lady Blue-Leon Russell-Shelter
18	28.	Rendezvous-Hudson Brothers-Rocket
--	29.	Bad Blood-Neil Sedaka-Rocket
33	30.	Brazil-The Ritchie Familiy-20th Century
25	31.	Dynomite-Bazuka-A & M
29	32.	(I Believe) There's Nothing Stronger Than Our Love-Paul Anka-UA
31	33.	I'm Not In Love-10cc-Mercury
28	34.	The Rockford Files-Mike Post-MGM
--	35.	I Got Stoned And I Missed It-Jim Stafford-MGM
23	36.	Rockin' Chair-Gwen McCrae-TK
36	37.	The Hustle-Van McCoy-Avco
34	38.	Love Will Keep Us Together-The

Captain & Tennille-A & M

37 | 39. Listen To What The Man Said-Paul McCartney-Capitol

35 | 40. Midnight Blue-Melissa Manchester-Arista

SUPER CFL SURVEY SEP 27

wcfl wcfl wcfl wcfl wcfl wcfl wcfl wcfl wcfl wcfl wcfl wcfl wcfl wcfl

SUPER ALBUMS

1. Captain Fantastic-Elton John-MCA
2. Red Octopus-Jefferson Starship-Grunt
3. The Heat Is On-Isley Brothers-T-Neck
4. One Of These Nights-Eagles-Asylum
5. Venus And Mars-Paul McCartney & Wings-Capitol
6. Is It Something I Said?-Richard Pryor-Reprise
7. Gorilla-James Taylor-WB
8. Toys In The Attic-Aerosmith-Columbia
9. Honey-Ohio Players-Mercury
10. Win, Lose Or Draw-The Allman Brothers Band-Capricorn
11. Pick Of The Litter-Spinners-Atlantic
12. Endless Summer-The Beach Boys-Capitol
13. Fandango-ZZ Top-London
14. Greatest Hits-Cat Stevens-A & M
15. Sabotage-Black Sabbath-WB
16. Love Will Keep Us Together-The Captain & Tennille-A & M
17. Atlantic Crossing-Rod Stewart-WB
18. Blues For Allah-The Grateful Dead-UA
19. Fleetwood Mac-Fleetwood Mac-Reprise
20. Why Can't We Be Friends?-War-UA
21. Steppin'-The Pointer Sisters-Blue Thumb
22. Between The Lines-Janis Ian-Columbia
23. Born To Run-Bruce Springsteen-Columbia
24. Rhinestone Cowboy-Glen Campbell-Capitol
25. E.C. Was Here-Eric Clapton-RSO

1	1.	Get Down Tonight-K.C. & The Sunshine Band-TK
3	2.	At Seventeen-Janis Ian-Columbia
7	3.	Ballroom Blitz-Sweet-Capitol
5	4.	Feel Like Makin' Love-Bad Company-Swan Song
9	5.	Dance With Me-Orleans-Asylum
4	6.	Fallin' In Love-Hamilton, Joe Frank & Reynolds-Playboy
11	7.	Ain't No Way To Treat A Lady-Helen Reddy-Capitol
16	8.	Mr. Jaws-Dickie Goodman-Cash
2	9.	Tush-ZZ Top-London
6	10.	Fight The Power-The Isley Brothers-T-Neck
10	11.	Third Rate Romance-The Amazing Rhythm Aces-ABC
15	12.	How Long-The Pointer Sisters-Blue Thumb
14	13.	Games People Play-Spinners-Atlantic
17	14.	Fame-David Bowie-RCA
8	15.	Black Superman-Johnny Wakelin-Pye
19	16.	Theme From "Jaws"-Original Sound Track-MCA
13	17.	Run Joey Run-David Geddes-Big Tree
18	18.	That's The Way Of The World-Earth, Wind & Fire-Columbia
23	19.	It Only Takes A Minute-Tavares-Capitol
12	20.	Rhinestone Cowboy-Glen Campbell-Capitol
25	21.	Eighteen With A Bullet-Pete Wingfield-Island
20	22.	Wasted Days And Wasted Nights-Freddy Fender-ABC
26	23.	I'm Sorry-John Denver-RCA
27	24.	Lady Blue-Leon Russell-Shelter
22	25.	Jive Talkin'-Bee Gees-RSO
30	26.	Brazil-The Ritchie Family-20th Century
29	27.	Bad Blood-Neil Sedaka-Rocket
--	28.	This Will Be-Natalie Cole-Capitol
--	29.	What A Difference A Day Makes-Esther Phillips-Kudu
21	30.	Someone Saved My Life Tonight-Elton John-MCA
--	31.	Lyin' Eyes-The Eagles-Asylum
35	32.	I Got Stoned And I Missed It-Jim Stafford-MGM
--	33.	SOS-Abba-Atlantic
24	34.	Why Can't We Be Friends?-War-UA
32	35.	(I Believe) There's Nothing Stronger Than Our Love-Paul Anka-UA
34	36.	The Rockford Files-Mike Post-MGM

36	37.	Rockin' Chair-Gwen McCrae-TK
28	38.	Rendezvous-Hudson Brothers-Rocket
33	39.	I'm Not In Love-10cc-Mercury
37	40.	The Hustle-Van McCoy-Avco

SUPER CFL SURVEY OCT 4

wcfl wcfl wcfl wcfl wcfl wcfl wcfl wcfl wcfl wcfl wcfl wcfl wcfl wcfl

SUPER ALBUMS

1. Captain Fantastic-Elton John-MCA
2. Red Octopus-Jefferson Starship-Grunt
3. Wish You Were Here-Pink Floyd-Columbia
4. The Heat Is On-Isley Brothers-T-Neck
5. Toys In The Attic-Aerosmith-Columbia
6. One Of These Nights-Eagles-Asylum
7. Is It Something I Said?-Richard Pryor-Reprise
8. Venus And Mars-Paul McCartney & Wings-Capitol
9. Gorilla-James Taylor-WB
10. Win, Lose Or Draw-The Allman Brothers Band-Capricorn
11. Minstrel In The Gallery-Jethro Tull-Chrysalis
12. Fandango-ZZ Top-London
13. Pick Of The Litter-Spinners-Atlantic
14. Honey-Ohio Players-Mercury
15. Greatest Hits-Cat Stevens-A & M
16. Why Can't We Be Friends?-War-UA
17. Born To Run-Bruce Springsteen-Columbia
18. Sabotage-Black Sabbath-WB
19. KC & The Sunshine Band-TK
20. Between The Lines-Janis Ian-Columbia
21. Blues For Allah-The Grateful Dead-UA
22. Fleetwood Mac-Reprise
23. Endless Summer-The Beach Boys-Capitol
24. Love Will Keep Us Together-The Captain & Tennille-A & M
25. Steppin'-The Pointer Sisters-Blue Thumb

3	1.	Ballroom Blitz-Sweet-Capitol
5	2.	Dance With Me-Orleans-Asylum
1	3.	Get Down Tonight-K.C. & The Sunshine Band-TK
4	4.	Feel Like Makin' Love-Bad Company-Swan Song
8	5.	Mr. Jaws-Dickie Goodman-Cash
7	6.	Ain't No Way To Treat A Lady-Helen Reddy-Capitol
14	7.	Fame-David Bowie-RCA
2	8.	At Seventeen-Janis Ian-Columbia
13	9.	Games People Play-Spinners-Atlantic
6	10.	Fallin' In Love-Hamilton, Joe Frank & Reynolds-Playboy
12	11.	How Long-The Pointer Sisters-Blue Thumb
23	12.	I'm Sorry-John Denver-RCA
9	13.	Tush-ZZ Top-London
11	14.	Third Rate Romance-The Amazing Rhythm Aces-ABC
19	15.	It Only Takes A Minute-Tavares-Capitol
21	16.	Eighteen With A Bullet-Pete Wingfield-Island
16	17.	Theme From "Jaws"-Original Sound Track-MCA
27	18.	Bad Blood-Neil Sedaka-Rocket
10	19.	Fight The Power-The Isley Brothers-T-Neck
24	20.	Lady Blue-Leon Russell-Shelter
18	21.	That's The Way Of The World-Earth, Wind & Fire-Columbia
22	22.	Wasted Days And Wasted Nights-Freddy Fender-ABC
26	23.	Brazil-The Ritchie Family-20th Century
20	24.	Rhinestone Cowboy-Glen Campbell-Capitol
28	25.	This Will Be-Natalie Cole-Capitol
31	26.	Lyin' Eyes-The Eagles-Asylum
29	27.	What A Difference A Day Makes-Esther Phillips-Kudu
32	28.	I Got Stoned And I Missed It-Jim Stafford-MGM
15	29.	Black Superman-Johnny Wakelin-Pye
--	30.	Heat Wave-Linda Ronstadt-Asylum
33	31.	SOS-Abba-Atlantic
17	32.	Run Joey Run-David Geddes-Big Tree
25	33.	Jive Talkin'-Bee Gees-RSO
--	34.	Born To Run-Bruce Springsteen-Columbia
--	35.	Nights On Broadway-Bee Gees-RSO

30	36.	Someone Saved My Life Tonight-Elton John-MCA
34	37.	Why Can't We Be Friends?-War-UA
36	38.	The Rockford Files-Mike Post-MGM
35	39.	(I Believe) There's Nothing Stronger Than Our Love-Paul Anka-UA
37	40.	Rockin' Chair-Gwen McCrae-TK

SUPER CFL SURVEY OCT 11

wcfl wcfl wcfl wcfl wcfl wcfl wcfl wcfl wcfl wcfl wcfl wcfl wcfl wcfl

SUPER ALBUMS

1. Wish You Were Here-Pink Floyd-Columbia
2. Red Octopus-Jefferson Starship-Grunt
3. Captain Fantastic-Elton John-MCA
4. Toys In The Attic-Aerosmith-Columbia
5. One Of These Nights-Eagles-Asylum
6. The Heat Is On-Isley Brothers-T-Neck
7. Win, Lose Or Draw-The Allman Brothers Band-Capricorn
8. Honey-Ohio Players-Mercury
9. Is It Something I Said?-Richard Pryor-Reprise
10. Venus And Mars-Paul McCartney & Wings-Capitol
11. Born To Run-Bruce Springsteen-Columbia
12. Pick Of The Litter-Spinners-Atlantic
13. Sabotage-Black Sabbath-WB
14. Gorilla-James Taylor-WB
15. Windsong-John Denver-RCA
16. Minstrel In The Gallery-Jethro Tull-Chrysalis
17. Fandango-ZZ Top-London
18. Fleetwood Mac-Reprise
19. Atlantic Crossing-Rod Stewart-WB
20. Prisoner In Disguise-Linda Ronstadt-Asylum
21. Save Me-Silver Convention-Midland Int'l
22. Endless Summer-The Beach Boys-Capitol
23. Greatest Hits-Cat Stevens-A & M
24. Mellow Madness-Quincy Jones-A & M
25. Between The Lines-Janis Ian-Columbia

1	1.	Ballroom Blitz-Sweet-Capitol
2	2.	Dance With Me-Orleans-Asylum
7	3.	Fame-David Bowie-RCA
6	4.	Ain't No Way To Treat A Lady-Helen Reddy-Capitol
5	5.	Mr. Jaws-Dickie Goodman-Cash
9	6.	Games People Play-Spinners-Atlantic
3	7.	Get Down Tonight-K.C. & The Sunshine Band-TK
12	8.	I'm Sorry-John Denver-RCA
4	9.	Feel Like Makin' Love-Bad Company-Swan Song
15	10.	It Only Takes A Minute-Tavares-Capitol
10	11.	Fallin' In Love-Hamilton- Joe Frank & Reynolds-Playboy
16	12.	Eighteen With A Bullet-Pete Wingfield-Island
18	13.	Bad Blood-Neil Sedaka-Rocket
8	14.	At Seventeen-Janis Ian-Columbia
20	15.	Lady Blue-Leon Russell-Shelter
11	16.	How Long-The Pointer Sisters-Blue Thumb
23	17.	Brazil-The Ritchie Family-20th Century
14	18.	Third Rate Romance-The Amazing Rhythm Aces-ABC
25	19.	This Will Be-Natalie Cole-Capitol
17	20.	Theme From "Jaws"-Original Sound Track-MCA
26	21.	Lyin' Eyes-The Eagles-Asylum
27	22.	What A Difference A Day Makes-Esther Phillips-Kudu
13	23.	Tush-ZZ Top-London
31	24.	SOS-Abba-Atlanta
--	25.	Miracles-Jefferson Starship-Grunt
30	26.	Heat Wave-Linda Ronstadt-Asylum
21	27.	That's The Way Of The World-Earth, Wind & Fire-Columbia
24	28.	Rhinestone Cowboy-Glen Campbell-Capitol
19	29.	Fight The Power-The Isley Brothers-T-Neck
--	30.	Island Girl-Elton John-MCA
34	31.	Born To Run-Bruce Springsteen-Columbia
22	32.	Wasted Days And Wasted Nights-Freddy Fender-ABC
35	33.	Nights On Broadway-Bee Gees-RSO
28	34.	I Got Stoned And I Missed It-Jim Stafford-MGM
32	35.	Run Joey Run-David Geddes-Big Tree

33	36.	Jive Talkin'-Bee Gees-RSO
36	37.	Someone Saved My Life Tonight-Elton John-MCA
38	38.	The Rockford Files-Mike Post-MGM
37	39.	Why Can't We Be Friends?-War-UA
29	40.	Black Superman-Johnny Wakelin-Pye

SUPER CFL SURVEY OCT 18

wcfl wcfl wcfl wcfl wcfl wcfl wcfl wcfl wcfl wcfl wcfl wcfl wcfl wcfl

SUPER ALBUMS

1. Wish You Were Here-Pink Floyd-Columbia
2. Red Octopus-Jefferson Starship-Grunt
3. Toys In The Attic-Aerosmith-Columbia
4. Captain Fantastic-Elton John-MCA
5. The Heat Is On-Isley Brothers-T-Neck
6. Minstrel In The Gallery-Jethro Tull-Chrysalis
7. One Of These Nights-Eagles-Asylum
8. Born To Run-Bruce Springsteen-Columbia
9. Win, Lose or Draw-The Allman Brothers Band-Capricorn
10. Extra Texture-George Harrison-Apple
11. Windsong-John Denver-RCA
12. Honey-Ohio Players-Mercury
13. Is It Something I Said?-Richard Pryor-Reprise
14. Fandango-ZZ Top-London
15. Sabotage-Black Sabbath-WB
16. Pick Of The Litter-Spinners-Atlantic
17. Fleetwood Mac-Reprise
18. Between The Lines-Janis Ian-Columbia
19. Save Me-Silver Convention-Midland Int'l
20. Greatest Hits-Cat Stevens-A & M
21. Gorilla-James Taylor-WB
22. K.C. & The Sunshine Band-TK
23. Venus And Mars-Paul McCartney & Wings-Capitol
24. Why Can't We Be Friends?-War-UA
25. Mellow Madness-Quincy Jones-A & M

3	1.	Fame-David Bowie-RCA
2	2.	Dance With Me-Orleans-Asylum
8	3.	I'm Sorry-John Denver-RCA
1	4.	Ballroom Blitz-Sweet-Capitol
6	5.	Games People Play-Spinners-Atlantic
13	6.	Bad Blood-Neil Sedaka-Rocket
4	7.	Ain't No Way To Treat A Lady-Helen Reddy-Capitol
12	8.	Eighteen With A Bullet-Pete Wingfield-Island
10	9.	It Only Takes A Minute-Tavares-Capitol
7	10.	Get Down Tonight-K.C. & The Sunshine Band-TK
9	11.	Feel Like Makin' Love-Bad Company-Swan Song
15	12.	Lady Blue-Leon Russell-Shelter
17	13.	Brazil-The Ritchie Family-20th Century
21	14.	Lyin' Eyes-The Eagles-Asylum
11	15.	Fallin' In Love-Hamilton, Joe Frank & Reynolds-Playboy
19	16.	This Will Be-Natalie Cole-Capitol
5	17.	Mr. Jaws-Dickie Goodman-Cash
24	18.	SOS-Abba-Atlantic
22	19.	What A Difference A Day Makes-Esther Phillips-Kudu
26	20.	Heat Wave-Linda Ronstadt-Asylum
25	21.	Miracles-Jefferson Starship-Grunt
16	22.	How Long-The Pointer Sisters-Blue Thumb
14	23.	At Seventeen-Janis Ian-Columbia
33	24.	Nights On Broadway-Bee Gees-RSO
18	25.	Third Rate Romance-The Amazing Rhythm Aces-ABC
30	26.	Island Girl-Elton John-MCA
--	27.	Who Loves You-Four Seasons-WB
31	28.	Born To Run-Bruce Springsteen-Columbia
20	29.	Theme From "Jaws"-Original Sound Track-MCA
--	30.	I Want'a Do Something Freaky To You-Leon Heywood-20th Century
27	31.	That's The Way Of The World-Earth, Wind & Fire-Columbia
--	32.	I'm On Fire-5000 Volts-Philips
28	33.	Rhinestone Cowboy-Glen Campbell-Capitol
23	34.	Tush-ZZ Top-London
29	35.	Fight The Power-The Isley Brothers-T-Neck
32	36.	Wasted Days And Wasted Nights-Freddy Fender-ABC
36	37.	Jive Talkin'-Bee Gees-RSO

34	38.	I Got Stoned And I Missed It-Jim Stafford-MGM
35	39.	Run Joey Run-David Geddes-Big Tree
37	40.	Someone Saved My Life Tonight-Elton John-MCA

SUPER CFL SURVEY OCT 25
75

wcfl wcfl wcfl wcfl wcfl wcfl wcfl wcfl wcfl wcfl wcfl wcfl wcfl wcfl

SUPER ALBUMS

1. Wish You Were Here-Pink Floyd-Columbia
2. Red Octopus-Jefferson Starship-Grunt
3. Toys In The Attic-Aerosmith-Columbia
4. Captain Fantastic-Elton John-MCA
5. Windsong-John Denver-RCA
6. Born To Run-Bruce Springsteen-Columbia
7. One Of These Nights-Eagles-Asylum
8. Honey-Ohio Players-Mercury
9. Win, Lose Or Draw-The Allman Brothers Band-Capricorn
10. Minstrel In The Gallery-Jethro Tull-Chrysalis
11. Prisoner In Disguise-Linda Ronstadt-Asylum
12. Fleetwood Mac-Reprise
13. Save Me-Silver Convention-Midland Int'l
14. Extra Texture-George Harrison-Apple
15. Venus And Mars-Paul McCartney & Wings-Capitol
16. The Heat Is On-Isley Brothers-T-Neck
17. Sabotage-Black Sabbath-WB
18. Mellow Madness-Quincy Jones-A & M
19. Pick Of The Litter-Spinners-Atlantic
20. Wind On The Water-David Crosby/Graham Nash-ABC
21. K.C. & The Sunshine Band-TK
22. Gorilla-James Taylor-WB
23. Fandango-ZZ Top-London
24. Boogie Down U.S.A.-People's Choice-TSOP
25. Is It Something I Said?-Richard Pryor-Reprise

1	1.	Fame-David Bowie-RCA
3	2.	I'm Sorry-John Denver-RCA
6	3.	Bad Blood-Neil Sedaka-Rocket
5	4.	Games People Play-Spinners-Atlantic
8	5.	Eighteen With A Bullet-Pete Wingfield-Island
2	6.	Dance With Me-Orleans-Asylum
9	7.	It Only Takes A Minute-Tavares-Capitol
4	8.	Ballroom Blitz-Sweet-Capitol
12	9.	Lady Blue-Leon Russell-Shelter
14	10.	Lyin' Eyes-The Eagles-Asylum
21	11.	Miracles-Jefferson Starship-Grunt
16	12.	This Will Be-Natalie Cole-Capitol
13	13.	Brazil-The Ritchie Family-20th Century
18	14.	SOS-Abba-Atlantic
7	15.	Ain't No Way To Treat A Lady-Helen Reddy-Capitol
20	16.	Heat Wave-Linda Ronstadt-Asylum
19	17.	What A Difference A Day Makes-Esther Phillips-Kudu
26	18.	Island Girl-Elton John-MCA
10	19.	Get Down Tonight-K.C. & The Sunshine Band-TK
15	20.	Fallin' In Love-Hamilton, Joe Frank & Reynolds-Playboy
24	21.	Nights On Broadway-Bee Gees-RSO
27	22.	Who Loves You-Four Seasons-WB
11	23.	Feel Like Makin' Love-Bad Company-Swan Song
--	24.	The Way I Want To Touch You-Captain & Tennille-A & M
17	25.	Mr. Jaws-Dickie Goodman-Cash
30	26.	I Want'a Do Something Freaky To You-Leon Haywood-20th Century
--	27.	Sky High-Jigsaw-Chelsea
23	28.	At Seventeen-Janis Ian-Columbia
32	29.	I'm On Fire-5000 Volts-Philips
--	30.	Welcome To My Nightmare-Alice Cooper-Atlantic
22	31.	How Long-The Pointer Sisters-Blue Thumb
28	32.	Born To Run-Bruce Springsteen-Columbia
25	33.	Third Rate Romance-The Amazing Rhythm Aces-ABC
33	34.	Rhinestone Cowboy-Glen Campbell-Capitol
29	35.	Theme From "Jaws"-Original Sound

		Track-MCA
31	36.	That's The Way Of The World-Earth, Wind & Fire-Columbia
34	37.	Tush-ZZ Top-London
37	38.	Jive Talkin'-Bee Gees-RSO
35	39.	Fight The Power-The Isley Brothers-T-Neck
36	40.	Wasted Days And Wasted Nights-Freddy Fender-ABC

SUPER CFL SURVEY NOV 1
75

wcfl wcfl wcfl wcfl wcfl wcfl wcfl wcfl wcfl wcfl wcfl wcfl wcfl wcfl

SUPER ALBUMS

1. Wish You Were Here-Pink Floyd-Columbia
2. Red Octopus-Jefferson Starship-Grunt
3. Windsong-John Denver-RCA
4. Toys In The Attic-Aerosmith-Columbia
5. Born To Run-Bruce Springsteen-Columbia
6. Minstrel In The Gallery-Jethro Tull-Chrysalis
7. Honey-Ohio Players-Mercury
8. Extra Texture-George Harrison-Apple
9. One Of These Nights-Eagles-Asylum
10. Save Me-Silver Convention-Midland Int'l
11. Captain Fantastic-Elton John-RCA
12. Win, Lose Or Draw-The Allman Brothers Band-Capricorn
13. The Heat Is On-Isley Brothers-T-Neck
14. The Who By Numbers-The Who-MCA
15. Pick Of The Litter-Spinners-Atlantic
16. Mellow Madness-Quincy Jones-A & M
17. Alive-Kiss-Casablanca
18. Still Crazy After All These Years-Paul Simon-Columbia
19. K.C. & The Sunshine Band-TK
20. Prisoner In Disguise-Linda Ronstadt-Asylum
21. Fleetwood Mac-Reprise
22. Sabotage-Black Sabbath-WB
23. Is It Something I Said?-Richard Pryor-Reprise
24. Young Americans-David Bowie-RCA
25. Between The Lines-Janis Ian-Columbia

2	1.	I'm Sorry-John Denver-RCA
3	2.	Bad Blood-Neil Sedaka-Rocket
4	3.	Games People Play-Spinners-Atlantic
1	4.	Fame-David Bowie-RCA
5	5.	Eighteen With A Bullet-Pete Wingfield-Island
9	6.	Lady Blue-Leon Russell-Shelter
10	7.	Lyin' Eyes-The Eagles-Asylum
7	8.	It Only Takes A Minute-Tavares-Capitol
11	9.	Miracles-Jefferson Starship-Grunt
18	10.	Island Girl-Elton John-MCA
12	11.	This Will BE-Natalie Cole-Capitol
14	12.	SOS-Abba-Atlantic
16	13.	Heat Wave-Linda Ronstadt-Asylum
22	14.	Who Loves You-Four Seasons-WB
6	15.	Dance With Me-Orleans-Asylum
21	16.	Nights On Broadway-Bee Gees-RSO
8	17.	Ballroom Blitz-Sweet-Capitol
15	18.	Ain't No Way To Treat A Lady-Helen Reddy-Capitol
24	19.	The Way I Want To Touch You-Captain & Tennille-A & M
13	20.	Brazil-The Ritchie Family-20th Century
19	21.	Get Down Tonight-K.C. & The Sunshine Band-TK
26	22.	I Want'a Do Something Freaky To You-Leon Haywood-20th Century
29	23.	I'm On Fire-5000 Volts-Philips
27	24.	Sky High-Jigsaw-Chelsea
17	25.	What A Difference A Day Makes-Esther Phillips-Kudu
30	26.	Welcome To My Nightmare-Alice Cooper-Atlantic
20	27.	Fallin' In Love-Hamilton, Joe Frank & Reynolds- Playboy
--	28.	Fly, Robin, Fly-Silver Convention-Midland Int'l
23	29.	Feel Like Makin' Love-Bad Company-Swan Song
--	30.	Operator-The Manhattan Transfer-Atlantic
--	31.	Winners And Losers-Hamilton, Joe Frank & Reynolds-Playboy
28	32.	At Seventeen-Janis Ian-Columbia
--	33.	You Sexy Thing-Hot Chocolate-Big Tree
31	34.	How Long-The Pointer Sisters-Blue Thumb

25	35.	Mr. Jaws-Dickie Goodman-Cash
34	36.	Rhinestone Cowboy-Glen Campbell-Capitol
33	37.	Third Rate Romance-The Amazing Rhythm Aces-ABC
32	38.	Born To Run-Bruce Springsteen-Columbia
35	39.	Theme From "Jaws"-Original Sound Track-MCA
36	40.	That's The Way Of The World-Earth, Wind & Fire-Columbia

SUPER CFL SURVEY NOV 8
75

wcfl wcfl wcfl wcfl wcfl wcfl wcfl wcfl wcfl wcfl wcfl wcfl wcfl wcfl

SUPER ALBUMS

1. Wish You Were Here-Pink Floyd-Columbia
2. Red Octopus-Jefferson Starship-Grunt
3. Windsong-John Denver-RCA
4. Toys In The Attic-Aerosmith-Columbia
5. Minstrel In The Gallery-Jethro Tull-Chrysalis
6. The Who By Numbers-The Who-MCA
7. One Of These Nights-Eagles-Asylum
8. Honey-Ohio Players-Mercury
9. Prisoner In Disguise-Linda Ronstadt-Asylum
10. Save Me-Silver Convention-Midland Int'l
11. Captain Fantastic-Elton John-MCA
12. Rock Of The Westies-Elton John-MCA
13. Fleetwood Mac-Reprise
14. Mellow Madness-Quincy Jones-A & M
15. Born To Run-Bruce Springsteen-Columbia
16. Still Crazy After All These Years-Paul Simon-Columbia
17. Extra Texture-George Harrison-Apple
18. Breakaway-Art Garfunkel-Columbia
19. Wind On The Water-David Crosby/Graham Nash-ABC
20. The Heat Is On-Isley Brothers-T-Neck
21. Win, Lose Or Draw-The Allman Brothers Band-Capricorn
22. K.C. & The Sunshine Band-TK
23. Boogie Down U.S.A.-People's Choice-TSOP
24. Pick Of The Litter-Spinners-Atlantic
25. Is It Something I Said?-Richard Pryor-Reprise

2.	1.	Bad Blood-Neil Sedaka-Rocket
1	2.	I'm Sorry-John Denver-RCA
5	3.	Eighteen With A Bullet-Pete Wingfield-Island
7	4.	Lyin' Eyes-The Eagles-Asylum
6	5.	Lady Blue-Leon Russell-Shelter
9	6.	Miracles-Jefferson Starship-Grunt
10	7.	Island Girl-Elton John-MCA
11	8.	This Will Be-Natalie Cole-Capitol
14	9.	Who Loves You-Four Seasons-WB
12	10.	SOS-Abba-Atlantic
13	11.	Heat Wave-Linda Ronstadt-Asylum
3	12.	Games People Play-Spinners-Atlantic
16	13.	Nights On Broadway-Bee Gees-RSO
19	14.	The Way I Want To Touch You-Captain & Tennille-A & M
4	15.	Fame-David Bowie-RCA
8	16.	It Only Takes A Minute-Tavares-Capitol
24	17.	Sky High-Jigsaw-Chelsea
22	18.	I Want'a Do Something Freaky To You-Leon Haywood-20th Century
17	19.	Ballroom Blitz-Sweet-Capitol
23	20.	I'm On Fire-5000 Volts-Philips
26	21.	Welcome To My Nightmare-Alice Cooper-Atlantic
15	22.	Dance With Me-Orleans-Asylum
28	23.	Fly, Robin, Fly-Silver Convention-Midland Int'l
18	24.	Ain't No Way To Treat A Lady-Helen Reddy-Capitol
30	25.	Operator-The Manhattan Transfer-Atlantic
20	26.	Brazil-The Ritchie Family-20th Century
31	27.	Winners And Losers-Hamilton, Joe Frank & Reynolds-Playboy
--	28.	That's The Way I Like It-K.C. & The Sunshine Band-TK
33	29.	You Sexy Thing-Hot Chocolate-Big Tree
--	30.	Venus And Mars Rock Show-Paul McCartney-Capitol
25	31.	What A Difference A Day Makes-Esther Phillips-Kudu
21	32.	Get Down Tonight-K.C. & The Sunshine Band-TK
27	33.	Fallin' In Love-Hamilton, Joe Frank & Reynolds-Playboy
29	34.	Feel Like Makin' Love-Bad

		Company-Swan Song
32	35.	At Seventeen-Janis Ian-Columbia
36	36.	Rhinestone Cowboy-Glen Campbell-Capitol
34	37.	How Long-The Pointer Sisters-Blue Thumb
35	38.	Mr. Jaws-Dickie Goodman-Cash
37	39.	Third Rate Romance-The Amazing Rhythm Aces-ABC
38	40.	Born To Run-Bruce Springsteen-Columbia

SUPER CFL SURVEY NOV 15

75

wcfl wcfl wcfl wcfl wcfl wcfl wcfl wcfl wcfl wcfl wcfl wcfl wcfl wcfl

SUPER ALBUMS

1. Red Octopus-Jefferson Starship-Grunt
2. Wish You Were Here-Pink Floyd-Columbia
3. Windsong-John Denver-RCA
4. Rock Of The Westies-Elton John-MCA
5. Toys In The Attic-Aerosmith-Columbia
6. Minstrel In The Gallery-Jethro Tull-Chrysalis
7. Still Crazy After All These Years-Paul Simon-Columbia
8. Save Me-Silver Convention-Midland Int'l
9. The Who By Numbers-The Who-MCA
10. Born To Run-Bruce Springsteen-Columbia
11. Breakaway-Art Garfunkel-Columbia
12. Honey-Ohio Players-Mercury
13. One Of These Nights-Eagles-Asylum
14. Tryin' To Get The Feeling-Barry Manilow-Arista
15. Alive-Kiss-Casablanca
16. Mellow Madness-Quincy Jones-A & M
17. K.C. & The Sunshine Band-TK
18. Captain Fantastic-Elton John-MCA
19. Prisoner In Disguise-Linda Ronstadt-Asylum
20. Fleetwood Mac-Reprise
21. Pick Of The Litter-Spinners-Atlantic
22. Extra Texture-George Harrison-Apple
23. The Heat Is On-Isley Brothers-T-Neck
24. Boogie Down U.S.A.-People's Choice-TSOP
25. Win, Lose Or Draw-The Allman Brothers Band-Capricorn

1	1.	Bad Blood-Neil Sedaka-Rocket
7	2.	Island Girl-Elton John-MCA
6	3.	Miracles-Jefferson Starship-Grunt
9	4.	Who Loves You-Four Seasons-WB
2	5.	I'm Sorry-John Denver-RCA
10	6.	SOS-Abba-Atlantic
3	7.	Eighteen With A Bullet-Pete Wingfield-Island
14	8.	The Way I Want To Touch You-Captain & Tennille-A & M
11	9.	Heat Wave-Linda Ronstadt-Asylum
13	10.	Nights On Broadway-Bee Gees-RSO
4	11.	Lyin' Eyes-The Eagles-Asylum
8	12.	This Will Be-Natalie Cole-Capitol
17	13.	Sky High-Jigsaw-Chelsea
5	14.	Lady Blue-Leon Russell-Shelter
23	15.	Fly, Robin, Fly-Silver Convention-Midland Int'l
18	16.	I Want'a Do Something Freaky To You-Leon Haywood-20th Century
20	17.	I'm On Fire-5000 Volts-Philips
12	18.	Games People Play-Spinners-Atlantic
21	19.	Welcome To My Nightmare-Alice Cooper-Atlantic
25	20.	Operator-The Manhattan Transfer-Atlantic
15	21.	Fame-David Bowie-RCA
27	22.	Winners And Losers-Hamilton, Joe Frank & Reynolds-Playboy
16	23.	It Only Takes A Minute-Tavares-Capitol
29	24.	You Sexy Thing-Hot Chocolate-Big Tree
28	25.	That's The Way I Like It-K.C. & The Sunshine Band-TK
30	26.	Venus And Mars Rock Show-Paul McCartney-Capitol
--	27.	Let's Do It Again-The Staple Singers-Curtom
22	28.	Dance With Me-Orleans-Asylum
--	29.	Fox On The Run-Sweet-Capitol
--	30.	I Write The Songs-Barry Manilow-Arista
--	31.	Blind Man in the Bleachers-David Geddes-Big Tree
19	32.	Ballroom Blitz-Sweet-Capitol
32	33.	Get Down Tonight-K.C. & The Sunshine Band-TK
24	34.	Ain't No Way To Treat A Lady-Helen Reddy-Capitol
26	35.	Brazil-The Ritchie Family-20th Century
33	36.	Fallin' In Love-Hamilton, Joe Frank & Reynolds-Playboy

31	37.	What A Difference A Day Makes-Esther Phillips-Kudu
34	38.	Feel Like Makin' Love-Bad Company-Swan Song
36	39.	Rhinestone Cowboy-Glen Campbell-Capitol
35	40.	At Seventeen-Janis Ian-Columbia

SUPER CFL SURVEY NOV 22
75

wcfl wcfl wcfl wcfl wcfl wcfl wcfl wcfl wcfl wcfl wcfl wcfl wcfl wcfl

SUPER ALBUMS

1. Red Octopus-Jefferson Starship-Grunt
2. Rock Of The Westies-Elton John-MCA
3. Wish You Were Here-Pink Floyd-Columbia
4. Windsong-John Denver-RCA
5. Still Crazy After All These Years-Paul Simon-Columbia
6. Toys In The Attic-Aerosmith-Columbia
7. The Who By Numbers-The Who-MCA
8. Minstrel In The Gallery-Jethro Tull-Chrysalis
9. Tryin' To Get The Feeling-Barry Manilow-Arista
10. Save Me-Silver Convention-Midland Int'l
11. Breakaway-Art Garfunkel-Columbia
12. Wind On The Water-David Crosby/Graham Nash-ABC
13. Prisoner In Disguise-Linda Ronstadt-Asylum
14. Fleetwood Mac-Reprise
15. K.C. & The Sunshine Band-TK
16. One Of These Nights-Eagles-Asylum
17. Mellow Madness-Quincy Jones-A & M
18. Honey-Ohio Players-Mercury
19. Born To Run-Bruce Springsteen-Columbia
20. Boogie Down U.S.A.-People's Choice-TSOP
21. Captain Fantastic-Elton John-MCA
22. Extra Texture-George Harrison-Apple
23. Let's Do It Again-The Staple Singers-Curtom
24. Al Green Is Love-HL
25. The Heat Is On-Isley Brothers-T-Neck

2	1.	Island Girl-Elton John-MCA
3	2.	Miracles-Jefferson Starship-Grunt
4	3.	Who Loves You-Four Seasons-WB
6	4.	SOS-Abba-Atlantic
1	5.	Bad Blood-Neil Sedaka-Rocket
8	6.	The Way I Want To Touch You-Captain & Tennille-A & M
10	7.	Nights On Broadway-Bee Gees-RSO
13	8.	Sky High-Jigsaw-Chelsea
9	9.	Heat Wave-Linda Ronstadt-Asylum
15	10.	Fly, Robin, Fly-Silver Convention-Midland Int'l
5	11.	I'm Sorry-John Denver-RCA
11	12.	Lyin' Eyes-The Eagles-Asylum
17	13.	I'm On Fire-5000 Volts-Phillips
7	14.	Eighteen With A Bullet-Peter Wingfield-Island
16	15.	I Want'a Do Something Freaky To You-Leon Haywood-20th Century
20	16.	Operator-The Manhattan Transfer-Atlantic
19	17.	Welcome To My Nightmare-Alice Cooper-Atlantic
25	18.	That's The Way I Like It-K.C. & The Sunshine Band-TK
22	19.	Winners And Losers-Hamilton, Joe Frank & Reynolds-Playboy
12	20.	This Will Be-Natalie Cole-Capitol
--	21.	Feelings-Morris Albert-RCA
24	22.	You Sexy Thing-Hot Chocolate-Big Tree
26	23.	Venus And Mars Rock Show-Paul McCartney-Capitol
14	24.	Lady Blue-Leon Russell-Shelter
27	25.	Let's Do It Again-The Staple Singers-Curtom
29	26.	Fox On The Run-Sweet-Capitol
--	27.	Saturday Night-Bay City Rollers-Arista
30	28.	I Write The Songs-Barry Manilow-Arista
31	29.	The Last Game Of The Season (A Blind Man In The Bleachers-David Geddes-Big Tree
--	30.	Secret Love-Freddy Fender-ABC
18	31.	Games People Play-Spinners-Atlantic
23	32.	It Only Takes A Minute-Tavares-Capitol
21	33.	Fame-David Bowie-RCA
32	34.	Ballroom Blitz-Sweet-Capitol

28	35.	Dance With Me-Orleans-Asylum
34	36.	Ain't No Way To Treat A Lady-Helen Reddy-Capitol
33	37.	Get Down Tonight-K.C. & The Sunshine Band-TK
38	38.	Feel Like Makin' Love-Bad Company-Swan Song
39	39.	Rhinestone Cowboy-Glen Campbell-Capitol
35	40.	Brazil-The Ritchie Family-20th Century

SUPER CFL SURVEY NOV 29
75

wcfl wcfl wcfl wcfl wcfl wcfl wcfl wcfl wcfl wcfl wcfl wcfl wcfl wcfl

SUPER ALBUMS

1. Rock Of The Westies-Elton John-MCA
2. Red Octopus-Jefferson Starship-Grunt
3. Wish You Were Here-Pink Floyd-Columbia
4. Windsong-John Denver-RCA
5. Still Crazy After All These Years-Paul Simon-Columbia
6. Toys In The Attic-Aerosmith-Columbia
7. The Who By Numbers-The Who-MCA
8. Minstrel In The Gallery-Jethro Tull-Chrysalis
9. Breakaway-Art Garfunkel-Columbia
10. Tryin' To Get The Feeling-Barry Manilow-Arista
11. Save Me-Silver Convention-Midland Int'l
12. Alive-Kiss-Casablanca
13. Honey-Ohio Players-Mercury
14. Prisoner In Disguise-Linda Ronstadt-Asylum
15. Let's Do It Again-The Staple Singers-Curtom
16. Fleetwood Mac-Reprise
17. K.C. & The Sunshine Band-TK
18. One Of These Nights-Eagles-Asylum
19. Boogie Down U.S.A.-People's Choice-TSOP
20. Love To Love You Baby-Donna Summer-Oasis
21. Wind On The Water-David Crosby/Graham Nash-ABC
22. Mellow Madness-Quincy Jones-A & M
23. Al Green Is Love-HL
24. Born To Run-Bruce Springsteen-Columbia
25. Pick Of The Litter-Spinners-Atlantic

3	1.	Who Loves You-Four Seasons-WB
6	2.	The Way I Want To Touch You-Captain & Tennille-A & M
1	3.	Island Girl-Elton John-MCA
7	4.	Nights On Broadway-Bee Gees-RSO
8	5.	Sky High-Jigsaw-Chelsea
4	6.	SOS-Abba-Atlantic
10	7.	Fly, Robin, Fly-Silver Convention-Midland Int'l
18	8.	That's The Way I Like It-K.C. & The Sunshine Band-TK
2	9.	Miracles-Jefferson Starship-Grunt
5	10.	Bad Blood-Neil Sedaka-Rocket
9	11.	Heat Wave-Linda Ronstadt-Asylum
13	12.	I'm On Fire-5000 Volts-Philips
21	13.	Feelings-Morris Albert-RCA
19	14.	Winners And Losers-Hamilton, Joe Frank & Reynolds-Playboy
25	15.	Let's Do It Again-The Staple Singers-Curtom
17	16.	Welcome To My Nightmare-Alice Cooper-Atlantic
12	17.	Lyin' Eyes-The Eagles-Asylum
23	18.	Venus And Mars Rock Show-Paul McCartney-Capitol
22	19.	You Sexy Thing-Hot Chocolate-Big Tree
26	20.	Fox On The Run-Sweet-Capitol
11	21.	I'm Sorry-John Denver-RCA
16	22.	Operator-The Manhattan Transfer-Atlantic
28	23.	I Write The Songs-Barry Manilow-Arista
27	24.	Saturday Night-Bay City Rollers-Arista
14	25.	Eighteen With A Bullet-Pete Wingfield-Island
29	26.	The Last Game Of T he Season (A Blind Man In The Bleachers)-David Geddes-Big Tree
--	27.	My Little Town-Simon And Garfunkel-Columbia
30	28.	Secret Love-Freddy Fender-ABC
--	29.	Love Rollercoaster-Ohio Players-Mercury
15	30.	I Want'a Do Something Freaky To You-Leon Haywood-20th Century
--	31.	Theme From Mahogany-Diana Ross-Motown
31	32.	Games People Play-Spinners-Atlantic

20	33.	This Will Be-Natalie Cole-Capitol
--	34.	Rock And Roll All Nite-Kiss-Casablanca
33	35.	Fame-David Bowie-RCA
24	36.	Lady Blue-Leon Russell-Shelter
32	37.	It Only Takes A Minute-Tavares-Capitol
37	38.	Get Down Tonight-K.C. & The Sunshine Band-TK
34	39.	Ballroom Blitz-Sweet-Capitol
35	40.	Dance With Me-Orleans-Asylum

SUPER CFL SURVEY DEC 6
75

wcfl wcfl wcfl wcfl wcfl wcfl wcfl wcfl wcfl wcfl wcfl wcfl wcfl wcfl

SUPER ALBUMS

1. Rock Of The Westies-Elton John-MCA
2. Red Octopus-Jefferson Starship-Grunt
3. Still Crazy After All These Years-Paul Simon-Columbia
4. Wish You Were Here-Pink Floyd-Columbia
5. Windsong-John Denver-RCA
6. The Who By Numbers-The Who-MCA
7. Save Me-Silver Convention-Midland Int'l
8. Toys In The Attic-Aerosmith-Columbia
9. Tryin' To Get The Feeling-Barry Manilow-Arista
10. Chicago's Greatest Hits-Columbia
11. Breakaway-Art Garfunkel-Columbia
12. Let's Do It Again-The Staple Singers-Curtom
13. Minstrel In The Gallery-Jethro Tull-Chrysalis
14. Fleetwood Mac-Reprise
15. Wind On The Water-David Crosby/Graham Nash-ABC
16. History-America's Greatest Hits-WB
17. Prisoner In Disguise-Linda Ronstadt-Asylum
18. K.C. & The Sunshine Band-TK
19. Honey-Ohio Players-Mercury
20. One Of These Nights-Eagles-Asylum
21. Love To Love You Baby-Donna Summer-Oasis
22. Boogie Down U.S.A.-People's Choice-TSOP
23. Seals & Crofts' Greatest Hits-WB
24. Zuma-Neil Young-Reprise
25. Feelings-Morris Albert-RCA

1	1.	Who Loves You-Four Seasons-WB
2	2.	The Way I Want To Touch You-Captain & Tennille-A & M
5	3.	Sky High-Jigsaw-Chelsea
7	4.	Fly, Robin, Fly-Silver Convention-Midland Int'l
8	5.	That's The Way I Like It-K.C. & The Sunshine Band-TK
13	6.	Feelings-Morris Albert-RCA
3	7.	Island Girl-Elton John-MCA
6	8.	SOS-Abba-Atlantic
14	9.	Winners And Losers-Hamilton, Joe Frank & Reynolds-Playboy
4	10.	Nights On Broadway-Bee Gees-RSO
15	11.	Let's Do It Again-The Staple Singers-Curtom
9	12.	Miracles-Jefferson Starship-Grunt
18	13.	Venus And Mars Rock Show-Paul McCartney-Capitol
10	14.	Bad Blood-Neil Sedaka-Rocket
24	15.	Saturday Night-Bay City Rollers-Arista
20	16.	Fox On The Run-Sweet-Capitol
19	17.	You Sexy Thing-Hot Chocolate-Big Tree
23	18.	I Write The Songs-Barry Manilow-Arista
11	19.	Heat Wave-Linda Ronstadt-Asylum
17	20.	Lyin' Eyes-The Eagles-Asylum
12	21.	I'm On Fire-5000 Volts-Philips
26	22.	The Last Game Of The Season (A Blind Man In The Bleachers)-David Geddes-Big Tree
16	23.	Welcome To My Nightmare-Alice Cooper-Atlantic
27	24.	My Little Town-Simon And Garfunkel-Columbia
22	25.	Operator-The Manhattan Transfer-Atlantic
28	26.	Secret Love-Freddy Fender-ABC
29	27.	Love Rollercoaster-Ohio Players-Mercury
--	28.	Convoy-C.W. McCall-MGM
31	29.	Theme From Mahogany-Diana Ross-Motown
21	30.	I'm Sorry-John Denver-RCA
34	31.	Rock And Roll All Nite-Kiss-Casablanca
--	32.	Walk Away From Love-David Ruffin-Motown
25	33.	Eighteen With A Bullet-Pete Wingfield-Island
--	34.	Country Boy-Glen Campbell-Capitol
30	35.	I Want'a Do Something Freaky To You-Leon Haywood-20th Century
32	36.	Games People Play-Spinners-

		Atlantic
36	37.	Lady Blue-Leon Russell-Shelter
33	38.	This Will Be-Natalie Cole-Capitol
35	39.	Fame-David Bowie-RCA
37	40.	It Only Takes A Minute-Tavares-Capitol

SUPER CFL SURVEY DEC 13
75

wcfl wcfl wcfl wcfl wcfl wcfl wcfl wcfl wcfl wcfl wcfl wcfl wcfl wcfl

SUPER ALBUMS

1. Rock Of The Westies-Elton John-MCA
2. Red Octopus-Jefferson Starship-Grunt
3. Still Crazy After All These Years-Paul Simon-Columbia
4. Chicago's Greatest Hits-Columbia
5. Wish You Were Here-Pink Floyd-Columbia
6. Windsong-John Denver-RCA
7. The Who By Numbers-The Who-MCA
8. Tryin' To Get The Feeling-Barry Manilow-Arista
9. Toys In The Attic-Aerosmith-Columbia
10. The Hissing Of Summer Lawns-Joni Mitchell-Asylum
11. Breakaway-Art Garfunkel-Columbia
12. Wind On The Water-David Crosby/Graham Nash-ABC
13. Fleetwood Mac-Reprise
14. Love To Love You Baby-Donna Summer-Oasis
15. Honey-Ohio Players-Mercury
16. K.C. & The Sunshine Band, TK
17. Save Me-Silver Convention-Midland Int'l
18. History-America's Greatest Hits-WB
19. Zuma-Neil Young-Reprise
20. Midnight Lightning-Jimi Hendrix-Reprise
21. Let's Do It Again-The Staple Singers-Curtom
22. One Of These Nights-Eagles-Asylum
23. Alive-Kiss-Casablanca
24. Family Reunion-The O'Jays-PI
25. Boogie Down U.S.A.-People's Choice-TSOP

5	1.	That's The Way I Like It-K.C. & The Sunshine Band-TK
6	2.	Feelings-Morris Albert-RCA
3	3.	Sky High-Jigsaw-Chelsea
4	4.	Fly, Robin, Fly-Silver Convention-Midland Int'l
2	5.	The Way I Want To Touch You-Captain & Tennille-A & M
1	6.	Who Loves You-Four Seasons-WB
9	7.	Winners And Losers-Hamilton, Joe Frank & Reynolds-Playboy
11	8.	Let's Do It Again-The Staple Singers-Curtom
16	9.	Fox On The Run-Sweet-Capitol
7	10.	Island Girl-Elton John-MCA
10	11.	Nights On Broadway-Bee Gees-RSO
15	12.	Saturday Night-Bay City Rollers-Arista
18	13.	I Write The Songs-Barry Manilow-Arista
8	14.	SOS-Abba-Atlantic
17	15.	You Sexy Thing-Hot Chocolate-Big Tree
13	16.	Venus And Mars Rock Show-Paul McCartney-Capitol
24	17.	My Little Town-Simon And Garfunkel-Columbia
12	18.	Miracles-Jefferson Starship-Grunt
14	19.	Bad Blood-Neil Sedaka-Rocket
22	20.	The Last Game Of The Season-David Geddes-Big Tree
27	21.	Love Rollercoaster-Ohio Players-Mercury
19	22.	Heat Wave-Linda Ronstadt-Asylum
29	23.	Theme From Mahogany-Diana Ross-Motown
28	24.	Convoy-C.W. McCall-MGM
20	25.	Lyin' Eyes-The Eagles-Asylum
31	26.	Rock And Roll All Nite-Kiss-Casablanca
32	27.	Walk Away From Love-David Ruffin-Motown
34	28.	Country Boy-Glen Campbell-Capitol
21	29.	I'm On Fire-5000 Volts-Philips
26	30.	Secret Love-Freddy Fender-ABC
30	31.	I'm Sorry-John Denver-RCA
23	32.	Welcome To My Nightmare-Alice Cooper-Atlantic
25	33.	Operator-The Manhattan Transfer-Atlantic
33	34.	Eighteen With A Bullet-Pete Wingfield-Island
35	35.	I Want'a Do Something Freaky To You-Leon Haywood-20th Century
36	36.	Games People Play-Spinners-

		Atlantic
37	37.	Lady Blue-Leon Russell-Shelter
38	38.	This Will Be-Natalie Cole-Capitol
39	39.	Fame-David Bowie-RCA
40	40.	It Only Takes A Minute-Taveres-Capitol

SUPER CFL SURVEY DEC 20
75

wcfl wcfl wcfl wcfl wcfl wcfl wcfl wcfl wcfl wcfl wcfl wcfl wcfl wcfl

SUPER ALBUMS

1. Chicago's Greatest Hits-Columbia
2. Red Octopus-Jefferson Starship-Grunt
3. Still Crazy After All These Years-Paul Simon-Columbia
4. Rock Of The Westies-Elton John-MCA
5. Toys In The Attic-Aerosmith-Columbia
6. Wish You Were Here-Pink Floyd-Columbia
7. History-America's Greatest Hits-WB
8. The Who By Numbers-The Who-MCA
9. The Hissing Of Summer Lawns-Joni Mitchell-Asylum
10. Tryin' To Get The Feeling-Barry Manilow-Arista
11. Love To Love You Baby-Donna Summer-Oasis
12. K.C. & The Sunshine Band-TK
13. Windsong-John Denver-RCA
14. Save Me-Silver Convention-Midland Int'l
15. Gratitude-Earth, Wind & Fire-Columbia
16. Wind On The Water-David Crosby/Graham Nash-ABC
17. Fleetwood Mac-Reprise
18. Family Reunion-The O'Jays-PI
19. Feelings-Morris Albert-RCA
20. Minstrel In The Gallery-Jethro Tull-Chrysalis
21. Greatest Hits-Seals & Crofts-WB
22. Honey-Ohio Players-Mercury
23. Zuma-Neil Young-Reprise
24. Breakaway-Art Garfunkel-Columbia
25. Wake Up Everybody-Harold Melvin & The Blue Notes-PI

2	1.	Feelings-Morris Albert-RCA
3	2.	Sky High-Jigsaw-Chelsea
1	3.	That's The Way I Like It-K.C. & The Sunshine Band-TK
7	4.	Winners And Losers-Hamilton, Joe Frank & Reynolds-Playboy
9	5.	Fox On The Run-Sweet-Capitol
8	6.	Let's Do It Again-The Staple Singers-Curtom
4	7.	Fly, Robin, Fly-Silver Convention-Midland Int'l
12	8.	Saturday Night-Bay City Rollers-Arista
13	9.	I Write The Songs-Barry Manilow-Arista
5	10.	The Way I Want To Touch You-Captain & Tennille-A & M
24	11.	Convoy-C.W. McCall-MGM
6	12.	Who Loves You-Four Seasons-WB
17	13.	My Little Town-Simon And Garfunkel-Columbia
15	14.	You Sexy Thing-Hot Chocolate-Big Tree
10	15.	Island Girl-Elton John-MCA
21	16.	Love Rollercoaster-Ohio Players-Mercury
16	17.	Venus And Mars Rock Show-Paul McCartney-Capitol
23	18.	Theme From Mahogany-Diana Ross-Motown
11	19.	Nights On Broadway-Bee Gees-RSO
14	20.	SOS-Abba-Atlantic
26	21.	Rock And Roll All Nite-Kiss-Casablanca
18	22.	Miracles-Jefferson Starship-Grunt
28	23.	Country Boy-Glen Campbell-Capitol
27	24.	Walk Away From Love-David Ruffin-Motown
--	25.	I Love Music-O'Jays-PI
20	26.	The Last Game Of The Season-David Geddes-Big Tree
19	27.	Bad Blood-Neil Sedaka-Rocket
--	28.	Evil Woman-Electric Light Orchestra-UA
--	29.	Breaking Up Is Hard To Do-Neil Sedaka-Rocket
22	30.	Heat Wave-Linda Ronstadt-Asylum
--	31.	All By Myself-Eric Carmen-Arista
25	32.	Lyin' Eyes-The Eagles-Asylum
30	33.	Secret Love-Freddy Fender-ABC
29	34.	I'm On Fire-5000 Volts-Philips
32	35.	Welcome To My Nightmare-Alice Cooper-Atlantic
33	36.	Operator-The Manhattan Transfer-Atlantic
36	37.	Games People Play-Spinners-Atlantic
34	38.	Eighteen With A Bullet-Pete

Wingfield-Island

31 39. I'm Sorry-John Denver-RCA

35 40. I Want'a Do Something Freaky To You-Leon Haywood-20th Century

SUPER CFL SURVEY DEC 27
75

wcfl wcfl wcfl wcfl wcfl wcfl wcfl wcfl wcfl wcfl wcfl wcfl wcfl wcfl

SUPER ALBUMS

1. Chicago's Greatest Hits-Columbia
2. Still Crazy After All These Years-Paul Simon-Columbia
3. Rock Of The Westies-Elton John-MCA
4. Red Octopus-Jefferson Starship-Grunt
5. History-America's Greatest hits-WB
6. Tryin' To Get The Feeling-Barry Manilow-Arista
7. Wish You Were Here-Pink Floyd-Columbia
8. Toys In The Attic-Aerosmith-Columbia
9. Love To Love You Baby-Donna Summer-Oasis
10. The Hissing Of Summer Lawns-Joni Mitchell-Asylum
11. The Who By Numbers-The Who-MCA
12. Family Reunion-The O'Jays-PI
13. Gratitude-Earth, Wind & Fire-Columbia
14. Windsong-John Denver-RCA
15. Breakaway-Art Garfunkel-Columbia
16. Numbers-Cat Stevens-A & M
17. Fleetwood Mac-Reprise
18. K.C. & The Sunshine Band-TK
19. Wind On The Water-David Crosby/Graham Nash-ABC
20. Wake Up Everybody-Harold Melvin & The Blue Notes-PI
21. Seals & Crofts' Greatest Hits-WB
22. Let's Do It Again-The Staple Singers-Curtom
23. Alive-Kiss-Casablanca
24. Boogie Down U.S.A.-People's Choice-TSOP
25. Save Me-Silver Convention-Midland Int'l

5	1.	Fox On The Fun-Sweet-Capitol
4	2.	Winners And Losers-Hamilton, Joe Frank & Reynolds-Playboy
3	3.	
1	4.	Feelings-Morris Albert-RCA
8	5.	Saturday Night-Bay City Rollers-Arista
6	6.	Let's Do It Again-The Staple Singers-Curtom
9	7.	I Write The Songs-Barry Manilow-Arista
11	8.	Convoy, C.W. McCall-MGM
2	9.	Sky High-Jigsaw-Chelsea
13	10.	My Little Town-Simon And Garfunkel-Columbia
14	11.	You Sexy Thing-Hot Chocolate-Big Tree
16	12.	Love Rollercoaster-Ohio Players-Mercury
7	13.	Fly, Robin, Fly-Silver Convention-Midland Int'l
18	14.	Theme From Mahogany-Diana Ross-Motown
12	15.	Who Loves You-Four Seasons-WB
10	16.	The Way I Want To Touch You-Captain & Tennille-A & M
21	17.	Rock And Roll All Nite-Kiss-Casablanca
23	18.	Country Boy-Glen Campbell-Capitol
19	19.	Nights On Broadway-Bee Gees-RSO
25	20.	I Love Music-O'Jays-PI
24	21.	Walk Away From Love-David Ruffin-Motown
20	22.	SOS-Abba-Atlantic
15	23.	Island Girl-Elton John-MCA
28	24.	Evil Woman-Electric Light Orchestra-UA
17	25.	Venus And Mars Rock Show-Paul McCartney-Capitol
29	26.	Breaking Up Is Hard To Do-Neil Sedaka-Rocket
--	27.	Sing A Song-Earth, Wind & Fire-Columbia
31	28.	All By Myself-Eric Carmen-Arista
--	29.	Love To Love You Baby-Donna Summer-Oasis
22	30.	Miracles-Jefferson Starship-Grunt
--	31.	December, 1963 (Oh, What A Night)-The Four Seasons-WB
--	32.	Fanny-Bee Gees-RSO
27	33.	Bad Blood-Neil Sedaka-Rocket
30	34.	Heat Wave-Linda Ronstadt-Asylum
26	35.	The Last Game Of The Season-David Geddes-Big Tree
32	36.	Lyin' Eyes-The Eagles-Asylum
33	37.	Secret Love-Freddy Fender-ABC
37	38.	Games People Play-Spinners-Atlantic

36	39.	Operator-The Manhattan Transfer-Atlantic
34	40.	I'm On Fire-5000 Volts-Philips

1976

SUPER CFL HITS OF '75

wcfl wcfl wcfl wcfl wcfl wcfl wcfl wcfl wcfl wcfl wcfl

1.	Love Will Keep Us Together	The Captain & Tennille	A & M
2.	Jive Talkin'	Bee Gees	RSO
3.	Have You Never Been Mellow	Olivia Newton-John	MCA
4.	Rhinestone Cowboy	Glen Campbell	Capitol
5.	Philadelphia Freedom	The Elton John Band	MCA
6.	Bad Blood	Neil Sedaka	Rocket
7.	The Hustle	Van McCoy	Avco
8.	Listen To What The Man Said	Paul McCartney	Capitol
9.	Fallin' In Love	Hamilton, Joe Frank & Reynolds	Playboy
10.	Get Down Tonight	K.C. & The Sunshine Band	TK
11.	Magic	Pilot	EMI
12.	Games People Play	Spinners	Atlantic
13.	Why Can't We Be Friends?	War	United Artists
14.	Another Somebody Done Somebody Wrong Song	B.J. Thomas	ABC
15.	Lady Marmalade	Labelle	Epic
16.	Ballroom Blitz	Sweet	Capitol
17.	My Eyes Adored You	Frankie Valli	Private Stock
18.	Someone Saved My Life Tonight	Elton John	MCA
19.	Mandy	Barry Manilow	Bell
20.	Black Water	The Doobie Brothers	Warner Brothers
21.	Fame	David Bowie	RCA
22.	Best Of My Love	The Eagles	Asylum
23.	Island Girl	Elton John	MCA
24.	Wildfire	Michael Murphey	Epic
25.	Please Mr. Postman	Carpenters	A & M
26.	Sister Golden Hair	America	Warner Brothers
27.	Lucy In The Sky With Diamonds	Elton John	MCA
28.	Who Loves You	Four Seasons	Warner Brothers
29.	Killer Queen	Queen	Elektra
30.	At Seventeen	Janis Ian	Columbia
31.	Lady	Styx	Wooden Nickel
32.	Miracles	Jefferson Starship	Grunt
33.	Laughter In The Rain	Neil Sedaka	MCA
34.	Lovin' You		
35.	Tush	Minnie Riperton	Epic
36.	Dance With Me	ZZ Top	London
37.	You Are So Beautiful	Orleans	Asylum
38.	It's A Miracle	Joe Cocker	A & M
39.	I'm Sorry	Barry Manilow	Arista
40.	Feelings	John Denver	RCA
		Morris Albert	RCA

SUPER CFL SURVEY JAN 17

wcfl wcfl wcfl wcfl wcfl wcfl wcfl wcfl wcfl wcfl wcfl wcfl wcfl wcfl

SUPER ALBUMS

1. Chicago's Greatest Hits-Columbia
2. Still Crazy After All These Years-Paul Simon-Columbia
3. History-America's Greatest Hits-WB
4. Gratitude-Earth, Wind & Fire-Columbia
5. Toys In The Attic-Aerosmith-Columbia
6. Tryin' To Get The Feeling-Barry Manilow-Arista
7. Love To Love You Baby-Donna Summer-Oasis
8. Red Octopus-Jefferson Starship-Grunt
9. Rock Of The Westies-Elton John-MCA
10. Wish You Were Here-Pink Floyd-Columbia
11. The Who By Numbers-The Who-MCA
12. Alive-Kiss-Casablanca
13. The Hissing Of Summer Lawns-Joni Mitchell-Asylum
14. Wake Up Everybody-Harold Melvin & The Blue Notes-PI
15. K.C. & The Sunshine Band-TK
16. Windsong-John Denver-RCA
17. Fleetwood Mac-Reprise
18. Family Reunion-The O'Jays-PI
19. Seals & Crofts' Greatest Hits-WB
20. Breakaway-Art Garfunkel-Columbia
21. Black Bear Road-C.W. McCall-MGM
22. Equinox-Styx-A & M
23. A Night At The Opera-Queen-Elektra
24. Wind On The Water-David Crosby/Graham Nash-ABC
25. Face The Music-The Electric Light Orchestra-UA

8	1.	Convoy-C.W. McCall-MGM
5	2.	Saturday Night-Bay City Rollers-Arista
7	3.	I Write The Songs-Barry Manilow-Arista
1	4.	Fox On The Run-Sweet-Capitol
2	5.	Winners And Losers-Hamilton, Joe Frank & Reynolds-Playboy
12	6.	Love Rollercoaster-Ohio Players-Mercury
14	7.	Theme From Mahogany-Diana Ross-Motown
10	8.	My Little Town-Simon And Garfunkel-Columbia
11	9.	You Sexy Thing-Hot Chocolate-Big Tree
3	10.	That's The Way I Like It-K.C. & The Sunshine Band-TK
20	11.	I Love Music-O'Jays-PI
17	12.	Rock And Roll All Nite-Kiss-Casablanca
18	13.	Country Boy-Glen Campbell-Capitol
21	14.	Walk Away From Love-David Ruffin-Motown
26	15.	Breaking Up Is Hard To Do-Neil Sedaka-Rocket
24	16.	Evil Woman-Electric Light Orchestra-UA
4	17.	Feelings-Morris Albert-RCA
6	18.	Let's Do It Again-The Staple Singers-Curtom
29	19.	Love To Love you Baby-Donna Summer-Oasis
9	20.	Sky High-Jigsaw-Chelsea
27	21.	Sing A Song-Earth, Wind & Fire-Columbia
13	22.	Fly, Robin, Fly-Silver Convention-Midland Int'l
28	23.	All By Myself-Eric Carmen-Arista
--	24.	Over My Head-Fleetwood Mac-Reprise
31	25.	December, 1963 (Oh, What A Night)-The Four Seasons-WB
--	26.	Fly Away-John Denver-RCA
15	27.	Who Loves You-Four Seasons-WB
32	28.	Fanny-Bee Gees-RSO
--	29.	Squeeze Box-The Who-MCA
--	30.	50 Ways To Leave Your Lover-Paul Simon-Columbia
16	31.	The Way I Want To Touch You-Captain & Tennille-A & M
--	32.	Yesterday's Hero-John Paul Young-Ariola
--	33.	Take It To The Limit-Eagles-Asylum
19	34.	Nights On Broadway-Bee Gees-RSO
--	35.	Paloma Blanca-George Baker Selection-WB

33	36.	Bad Blood-Neil Sedaka-Rocket
23	37.	Island Girl-Elton John-MCA
30	38.	Miracles-Jefferson Starship-Grunt
22	39.	SOS-Abba-Atlantic
25	40.	Venus And Mars Rock Show-Paul McCartney-Capitol

SUPER CFL SURVEY JAN 24

wcfl wcfl wcfl wcfl wcfl wcfl wcfl wcfl wcfl wcfl wcfl wcfl wcfl wcfl

SUPER ALBUMS

1. Chicago's Greatest Hits-Columbia
2. Still Crazy After All These Years-Paul Simon-Columbia
3. Gratitude-Earth, Wind & Fire-Columbia
4. History-America's Greatest Hits-WB
5. Tryin' To Get The Feeling-Barry Manilow-Arista
6. Alive-Kiss-Casablanca
7. Love To Love You Baby-Donna Summer-Oasis
8. Toys In The Attic-Aerosmith-Columbia
9. Red Octopus-Jefferson Starship-Grunt
10. Wish You Were Here-Pink Floyd-Columbia
11. Wake Up Everybody- Harold Melvin & The Blue Notes-PI
12. The Hissing Of Summer Lawns-Joni Mitchell-Asylum
13. Family Reunion-The O'Jays-PI
14. The Who By Numbers-The Who-MCA
15. Fleetwood Mac-Reprise
16. Rock Of The Westies-Elton John-MCA
17. Windsong-John Denver-RCA
18. Breakaway-Art Garfunkel-Columbia
19. Greatest Hits-Helen Reddy-Capitol
20. K.C. & The Sunshine Band-TK
21. Head On-Bachman-Turner Overdrive-Mercury
22. Black Bear Road-C.W. McCall-MGM
23. A Night At The Opera-Queen-Elektra
24. Wind On The Water-David Crosby/Graham Nash-ABC
25. Equinox-Styx-A & M

1	1.	Convoy-C.W. McCall-MGM
3	2.	I Write The Songs-Barry Manilow-Arista
6	3.	Love Rollercoaster-Ohio Players-Mercury
2	4.	Saturday Night-Bay City Rollers-Arista
7	5.	Theme From Mahogany-Diana Ross-Motown
8	6.	My Little Town-Simon And Garfunkel-Columbia
4	7.	Fox On The Run-Sweet-Capitol
12	8.	Rock And Roll All Nite-Kiss-Casablanca
11	9.	I Love Music-O'Jays-PI
19	10.	Love To Love You Baby-Donna Summer-Oasis
5	11.	Winners And Losers-Hamilton, Joe Frank & Reynolds-Playboy
16	12.	Evil Woman-Electric Light Orchestra-UA
14	13.	Walk Away From Love-David Ruffin-Motown
15	14.	Breaking Up Is Hard To Do-Neil Sedaka-Rocket
10	15.	That's The Way I Like It-K.C. & The Sunshine Band-TK
21	16.	Sing A Song-Earth, Wind & Fire-Columbia
9	17.	You Sexy Thing-Hot Chocolate-Big Tree
13	18.	Country Boy-Glen Campbell-Capitol
24	19.	Over My Head-Fleetwood Mac-Reprise
26	20.	Fly Away-John Denver-RCA
23	21.	All By Myself-Eric Carmen-Arista
25	22.	December, 1963 (Oh, What A Night)-The Four Seasons-WB
30	23.	50 Ways To Leave Your Lover-Paul Simon-Columbia
29	24.	Squeeze Box-The Who-MCA
28	25.	Fanny-Bee Gees-RSO
--	26.	Times Of Your Life-Paul Anka-UA
--	27.	Love Machine-The Miracles-Tamla
17	28.	Feelings-Morris Albert-RCA
20	29.	Sky High-Jigsaw-Chelsea
--	30.	Theme From S.W.A.T.-Rhythm Heritage-ABC
33	31.	Take It To The Limit-Eagles-Asylum
32	32.	Yesterday's Here-John Paul Young-Ariola
35	33.	Paloma Blanca-George Baker Selection-WB
22	34.	Fly, Robin, Fly-Silver Convention-Midland Int'l

18	35.	Let's Do It Again-The Staple Singers-Curtom
--	36.	Can The Can-Suzi Quatro-Big Tree
31	37.	The Way I Want To Touch You-Captain & Tennille-A & M
34	38.	Nights On Broadway-Bee Gees-RSO
27	39.	Who Loves You-Four Seasons-WB
36	40.	Bad Blood-Neil Sedaka-Rocket

SUPER CFL SURVEY JAN 31

wcfl wcfl wcfl wcfl wcfl wcfl wcfl wcfl wcfl wcfl wcfl wcfl wcfl wcfl

SUPER ALBUMS

1. Chicago's Greatest Hits-Columbia
2. Still Crazy After All These Years-Paul Simon-Columbia
3. Gratitude-Earth, Wind & Fire-Columbia
4. History-America's Greatest Hits-WB
5. Desire-Bob Dylan-Columbia
6. Love To Love You Baby-Donna Summer-Oasis
7. Tryin' To Get The Feeling-Barry Manilow-Arista
8. Toys In The Attic-Aerosmith-Columbia
9. The Hissing Of Summer Lawns-Joni Mitchell-Asylum
10. Alive-Kiss-Casablanca
11. Family Reunion-O'Jays-PI
12. Fleetwood Mac-Reprise
13. The Who By Numbers-The Who-MCA
14. "M.U."-The Best Of Jethro Tull-Chrysalis
15. Wake Up Everybody- Harold Melvin & The Blue Notes-PI
16. Wish You Were Here-Pink Floyd-Columbia
17. Equinox-Styx-A & M
18. Rock Of The Westies-Elton John-MCA
19. Red Octopus-Jefferson Starship-Grunt
20. Elite Hotel-Emmylou Harris-Reprise
21. Black Bear Road-C.W. McCall-MGM
22. Windsong-John Denver-RCA
23. Face The Music-The Electric Light Orchestra-UA
24. Breakaway-Art Garfunkel-Columbia
25. Helen Reddy's Greatest Hits-Capitol

3	1.	Love Rollercoaster-Ohio Players-Mercury
2	2.	I Write The Songs-Barry Manilow-Arista
5	3.	Theme From Mahogany-Diana Ross-Motown
1	4.	Convoy-C.W. McCall-MGM
10	5.	Love To Love You Baby-Donna Summer-Oasis
8	6.	Rock And Roll All Nite-Kiss-Casablanca
14	7.	Breaking Up Is Hard To Do-Neil Sedaka-Rocket
9	8.	I Love Music-O'Jays-PI
12	9.	Evil Woman-Electric Light Orchestra-UA
4	10.	Saturday Night-Bay City Rollers-Arista
6	11.	My Little Town-Simon And Garfunkel-Columbia
7	12.	Fox On The Run-Sweet-Capitol
11	13.	Winners And Losers-Hamilton, Joe Frank & Reynolds-Playboy
16	14.	Sing A Song-Earth, Wind & Fire-Columbia
13	15.	Walk Away From Love-David Ruffin-Motown
19	16.	Over My Head-Fleetwood Mac-Reprise
22	17.	December, 1963 (Oh, What A Night)-The Four Seasons-WB
21	18.	All By Myself-Eric Carmen-Arista
23	19.	50 Ways To Leave Your Lover-Paul Simon-Columbia
25	20.	Fanny-Bee Gees-RSO
15	21.	That's The Way I Like It-K.C. & The Sunshine Band-TK
20	22.	Fly Away-John Denver-RCA
26	23.	Times Of Your Life-Paul Anka-UA
24	24.	Squeeze Box-The Who-MCA
27	25.	Love Machine-The Miracles-Tamla
31	26.	Take It To The Limit-Eagles-Asylum
17	27.	You Sexy Thing-Hot Chocolate-Big Tree
30	28.	Theme From S.W.A.T.-Rhythm Heritage-ABC
18	29.	Country Boy-Glen Campbell-Capitol
33	30.	Paloma Blanca-George Baker Selection-WB
--	31.	Only Sixteen-Dr. Hook-Capitol
--	32.	Dream Weaver-Gary Wright-WB
28	33.	Feelings-Morris Albert-RCA
36	34.	Can The Can-Suzi Quatro-Big Tree
--	35.	Renegade-Michael Murphey-Epic
34	36.	Fly, Robin, Fly-Silver Convention-Midland Int'l

29	37.	Sky High-Jigsaw-Chelsea
35	38.	Let's Do It Again-The Staple Singers-Curtom
37	39.	The Way I Want To Touch You-Captain & Tennille-A & M
32	40.	Yesterday's Here-John Paul Young-Ariolal

SUPER CFL SURVEY FEB 7

wcfl wcfl wcfl wcfl wcfl wcfl wcfl wcfl wcfl wcfl wcfl wcfl wcfl wcfl

SUPER ALBUMS

1. Chicago's Greatest Hits-Columbia
2. Still Crazy After All These Years-Paul Simon-Columbia
3. Gratitude-Earth, Wind & Fire-Columbia
4. Desire-Bob Dylan-Columbia
5. Tryin' To Get The Feeling-Barry Manilow-Arista
6. Love To Love You Baby-Donna Summer-Oasis
7. History-America's Greatest Hits-WB
8. Toys In The Attic-Aerosmith-Columbia
9. Alive-Kiss-Casablanca
10. The Who By Numbers-The Who-MCA
11. "M.U."-The Best Of Jethro Tull-Chrysalis
12. Wish You Were Here-Pink Floyd-Columbia
13. Fleetwood Mac-Reprise
14. The Hissing Of Summer Lawns-Joni Mitchell-Asylum
15. Red Octopus-Jefferson Starship-Grunt
16. Wake Up Everybody-Harold Melvin & The Blue Notes-PI
17. Family Reunion-O'Jays-PI
18. City Life-The Blackbyrds-Fantasy
19. Head On-Bachman-Turner Overdrive-Mercury
20. Black Bear Road-C.W. McCall-MGM
21. Elite Hotel-Emmylou Harris-Reprise
22. Breakaway-Art Garfunkel-Columbia
23. K.C. & The Sunshine Band-TK
24. Helen Reddy's Greatest Hits-Capitol
25. Equinox-Styx-A & M

2	1.	I Write The Songs-Barry Manilow-Arista
5	2.	Love To Love You Baby-Donna Summer-Oasis
1	3.	Love Rollercoaster-Ohio Players-Mercury
6	4.	Rock And Roll All Nite-Kiss-Casablanca
9	5.	Evil Woman-Electric Light Orchestra-UA
7	6.	Breaking Up Is Hard To Do-Neil Sedaka-Rocket
8	7.	I Love Music-O'Jays-PI
3	8.	Theme From Mahogany-Diana Ross-Motown
19	9.	50 Ways To Leave Your Lover-Paul Simon-Columbia
4	10.	Convoy-C.W. McCall-MGM
16	11.	Over My Head-Fleetwood Mac-Reprise
10	12.	Saturday Night-Bay City Rollers-Arista
17	13.	December, 1963 (Oh, What A Night)-The Four Seasons-WB
14	14.	Sing A Song-Earth, Wind & Fire-Columbia
18	15.	All By Myself-Eric Carmen-Arista
11	16.	My Little Town-Simon And Garfunkel-Columbia
20	17.	Fanny-Bee Gees-RSO
24	18.	Squeeze Box-The Who-MCA
23	19.	Times Of Your Life-Paul Anka-UA
25	20.	Love Machine-The Miracles-Tamla
12	21.	Fox On The Run-Sweet-Capitol
15	22.	Walk Away From Love-David Ruffin-Motown
26	23.	Take It To The Limit-Eagles-Asylum
13	24.	Winners And Losers-Hamilton, Joe Frank & Reynolds-Playboy
28	25.	Theme From S.W.A.T.-Rhythm Heritage-ABC
22	26.	Fly Away-John Denver-RCA
32	27.	Dream Weaver-Gary Wright-WB
31	28.	Fly Away-John Denver-RCA
30	29.	Paloma Blanca-George Baker Selection-WB
--	30.	Lonely Night (Angel Face)-Captain And Tennille-A & M
--	31.	Love Is The Drug-Roxy Music-Atco
34	32.	Can The Can-Suzi Quatro-Big Tree
35	33.	Renegade-Michael Murphey-Epic
--	34.	Dream On-Aerosmith-Columbia
21	35.	That's The Way I Like It-K.C. & The Sunshine Band-TK
27	36.	You Sexy Thing-Hot Chocolate-Big Tree
33	37.	Feelings-Morris Albert-RCA

29	38.	Country Boy-Glen Campbell-Capitol
37	39.	Sky High-Jigsaw-Chelsea
36	40.	Fly, Robin, Fly-Silver Convention-Midland Int'l

SUPER CFL SURVEY FEB 14

wcfl wcfl wcfl wcfl wcfl wcfl wcfl wcfl wcfl wcfl wcfl wcfl wcfl wcfl

SUPER ALBUMS

1. Desire-Bob Dylan-Columbia
2. Still Crazy After All These Years-Paul Simon-Columbia
3. Gratitude-Earth, Wind & Fire-Columbia
4. Chicago's Greatest Hits-Columbia
5. "M.U."-The Best Of Jethro Tull-Chrysalis
6. History-America's Greatest Hits-WB
7. Alive-Kiss-Casablanca
8. Toys In The Attic-Aerosmith-Columbia
9. Tryin' To Get The Feeling-Barry Manilow-Arista
10. Wake Up Everybody-Harold Melvin & The Blue Notes-PI
11. Fleetwood Mac-Reprise
12. Love To Love You Baby-Donna Summer-Oasis
13. Rufus Featuring Chaka Khan-ABC
14. The Who By Numbers-The Who-MCA
15. Family Reunion-O'Jays-PI
16. Wish You Were Here-Pink Floyd-Columbia
17. Helen Reddy's Greatest Hits-Capitol
18. City Life-The Blackbyrds-Fantasy
19. The Hissing Of Summer Lawns-Joni Mitchell-Asylum
20. Places And Spaces-Donald Byrd-Blue Note
21. Aftertones-Janis Ian-Columbia
22. Equinox-Styx-A & M
23. Elite Hotel-Emmylou Harris-Reprise
24. Black Bear Road-C.W. McCall-MGM
25. Face The Music-The Electric Light Orchestra-UA

9	1.	50 Ways To Leave Your Lover-Paul Simon-Columbia
4	2.	Rock And Roll All Nite-Kiss-Casablanca
6	3.	Breaking Up Is Hard To Do-Neil Sedaka-Rocket
5	4.	Evil Woman-Electric Light Orchestra-UA
2	5.	Love To Love You Baby-Donna Summer-Oasis
13	6.	December,1963 (Oh, What A Night)-The Four Seasons-WB
1	7.	I Write The Songs-Barry Manilow-Arista
7	8.	I Love Music-O'Jays-PI
11	9.	Over My Head-Fleetwood Mac-Reprise
17	10.	Fanny-Bee Gees-RSO
15	11.	All By Myself-Eric Carmen-Arista
3	12.	Love Rollercoaster-Ohio Players-Mercury
8	13.	Theme From Mahogany-Diana Ross-Motown
18	14.	Squeeze Box-The Who-MCA
10	15.	Convoy-C.W. McCall-MGM
14	16.	Sing A Song-Earth, Wind & Fire-Columbia
20	17.	Love Machine-The Miracles-Tamia
25	18.	Theme From S.W.A.T.-Rhythm Heritage-ABC
23	19.	Take It To The Limit-Eagles-Asylum
19	20.	Times Of Your Life-Paul Anka-UA
12	21.	Saturday Night-Bay City Rollers-Arista
27	22.	Dream Weaver-Gary Wright-WB
21	23.	Fox On The Run-Sweet-Capitol
28	24.	Only Sixteen-Dr. Hook-Capitol
30	25.	Lonely Night (Angel Face)-Captain And Tennille-A & M
16	26.	My Little Town-Simon And Garfunkel-Columbia
34	27.	Dream On-Aerosmith-Columbia
22	28.	Walk Away From Love-David Ruffin-Motown
31	29.	Love Is The Drug-Roxy Music-Atco
32	30.	Can The Can-Suzi Quatro-Big Tree
33	31.	Renegade-Michael Murphey-Epic
--	32.	Money Honey-Bay City Rollers-Arista
--	33.	Bohemian Rhapsody-Queen-Elektra
29	34.	Paloma Blanca-George Baker Selection-WB
--	35.	Action-Sweet-Capitol
24	36.	Winners And Losers-Hamilton, Joe Frank & Reynolds-Playboy
26	37.	Fly Away-John Denver-RCA

37	38.	Feelings-Morris Albert-RCA
35	39.	That's The Way I Like It-K.C. & The Sunshine Band-TK
40	40.	Fly, Robin, Fly-Silver Convention-Midland Int'l

SUPER CFL SURVEY FEB 21

wcfl wcfl wcfl wcfl wcfl wcfl wcfl wcfl wcfl wcfl wcfl wcfl wcfl wcfl

1	1.	50 Ways To Leave Your Lover-Paul Simon-Columbia
3	2.	Breaking Up Is Hard To Do-Neil Sedaka-Rocket
6	3.	December, 1963 (Oh, What a Night)-The Four Seasons-WB
4	4.	Evil Woman-Electric Light Orchestra-UA
2	5.	Rock And Roll All Nite-Kiss-Casablanca
11	6.	All By Myself-Eric Carmen-Arista
10	7.	Fanny-Bee Gees-RSO
9	8.	Over My Head-Fleetwood Mac-Reprise
14	9.	Squeeze Box-The Who-MCA
18	10.	Theme From S.W.A.T.-Rhythm Heritage-ABC
5	11.	Love To Love You Baby-Donna Summer-Oasis
8	12.	I Love Music-O'Jays-PI
17	13.	Love Machine-The Miracles-Tamia
7	14.	I Write The Songs-Barry Manilow-Arista
22	15.	Dream Weaver-Gary Wright-WB
12	16.	Love Rollercoaster-Ohio Players-Mercury
19	17.	Take It To The Limit-Eagles-Asylum
13	18.	Theme From Mahogany-Diana Ross-Motown
25	19.	Lonely Night (Angel Face)-Captain And Tennille-A & M
27	20.	Dream On-Aerosmith-Columbia
24	21.	Only Sixteen-Dr. Hook-Capitol
16	22.	Sing A Song-Earth, Wind & Fire-Columbia
20	23.	Times Of Your Life-Paul Anka-UA
15	24.	Convoy-C.W. McCall-MGM
29	25.	Love Is The Drug-Roxy Music-Atco
21	26.	Saturday Night-Bay City Rollers-Arista
31	27.	Renegade-Michael Murphey-Epic
32	28.	Money Honey-Bay City Rollers-Arista
23	29.	Fox On The Run-Sweet-Capitol
--	30.	Love Hurts-Nazareth-A & M
33	31.	Bohemian Rhapsody-Queen-Elektra
--	32.	Junk Food Junkie-Larry Groce-WB
35	33.	Action-Sweet-Capitol
--	34.	Deep Purple-Donny And Marie Osmond-MGM
--	35.	Slow Ride-Foghat-Bearsville
--	36.	Sweet Thing-Rufus Featuring Chaka Khan-ABC
--	37.	Right Back Where We Started From-Maxine Nightingale-UA
--	38.	I Do, I Do, I Do, I Do, I Do-Abba-Atlantic
26	39.	My Little Town-Simon And Garfunkel-Columbia
28	40.	Walk Away From Love-David Ruffin-Motown

SUPER ALBUMS

1. Desire-Bob Dylan-Columbia
2. Still Crazy After All These Years-Paul Simon-Columbia
3. Gratitude-Earth, Wind & Fire-Columbia
4. Chicago's Greatest Hits-Columbia
5. "M.U."-The Best Of Jethro Tull-Chrysalis
6. Alive-Kiss-Casablanca

Toys In The Attic-Aerosmith-Columbia

8. Tryin' To Get The Feeling-Barry Manilow-Arista
9. Rufus Featuring Chaka Khan-ABC
10. Fleetwood Mac-Reprise
11. History-America's Greatest Hits-WB
12. Wake Up Everybody-Harold Melvin & The Blue Notes-PI
13. Love To Love You Baby-Donna Summer-Oasis
14. Family Reunion-O'Jays-PI
15. City Life-The Blackbyrds-Fantasy
16. Run With The Pack-Bad Company-Swan Song
17. Songs For The New Depression-Bette Midler-Atlantic
18. Aftertones-Janis Ian-Columbia
19. Frampton Comes Alive-Peter Frampton-A & M
20. The Who By Numbers-The Who-MCA
21. Native Sons-Loggins & Messina-Columbia
22. Breakaway-Art Garfunkel-Columbia
23. Equinox-Styx-A & M
24. Station To Station-David Bowie-RCA
25. Places And Spaces-Donald Byrd-Blue Note

SUPER CFL

APRIL FOOL SURVEY

week ending April 1, 1975

1. Eres Tu-Somos Nosotros-Que dice?
2. The Joker-William Simon-Czar
3. Show And Tell-John Dean-Spill
4. The Ward's Prayer-Sister Mary Elephant-Packyderm
5. Goodbye Yellow Brick Road-Chicago Department of Streets And Sanitation-Asphalt
6. Never, Never Gonna Give Ya Up-R. Nixon-Leader
7. Loves Theme-Joe Love-Mine
8. You're 16-Humbert Humbert-Tired
9. If I Were A Carpentersville-Karen Carpenter and David Seville-Duo
10. Free Ride-The RTA-Dreamer
11. The Way We Were-S. Agnew & R. Nixon-Bygone
12. The Canadians-Byron McBeaver-C'est toi?
13. Spiders And Snakes-Lincoln Park-Zoo
14. Mockingfurd-Carly Simon & Wally Cox-Mismatch
15. TSOT (The Sound of Toronto)-LSMFT-McHenry International
16. Space Race-CCCP & USA-Splash
17. Sunshine On My Shoulders-A Streaker-Everywhere
18. First Kiss-Tuesday-Essex
19. Sexy Mama-Oedipus-Delphi
20. Dee Dee Dee Dee Dee (Mindblower)-Trembling Rocks-Trembling Rocks
21. Teenage Lament '24-Rudy Valee-Bygone
22. Paper Roses-Euell Gibbons-Munch
23. Who's In The Strawberry Patch With Sally?-Sally's Father-Shotgun
24. Last Time I Saw Him-The FBI-Search
25. Smokin' In The Girl's Room-The Boys-Men's Lib
26. Right Place, Wrong Time-The Plumbers-White House
27. Just You 'N' Me-B. Riggs & B. J. King-Showdown
28. Tryin' To Hold On To My Woman-S. Bono-Share
29. We've Only Just Begun (Again)-Harold Wilson-Rerun

30, Jet-Bennie & The Jets-Redundant

APPENDIX

Debut	Peak	Title	Artist	Pos.	Wks.
3/11/71	3/11/71	Pearl	Janis Joplin	1(3)	8
3/11/71	3/11/71	Chicago III	Chicago	2(2)	5
3/11/71	3/11/71	Tumbleweed Connection	Elton John	3	1
3/11/71	3/11/71	All Things Must Pass	George Harrison	4	1
3/11/71	4/1/71	Jesus Christ Superstar	Various Artists	1(7)	17
3/11/71	3/18/71	The Cry Of Love	Jimi Hendrix	3	7
3/11/71	4/22/71	Tea For The Tillerman	Cat Stevens	3(2)	11
3/11/71	3/18/71	Emerson, Lake & Palmer	Emerson, Lake & Palmer	4	10
3/11/71	3/25/71	If I Could Only Remember My Name	David Crosby	2(3)	8
3/11/71	3/18/71	If You Could Read My Mind	Gordon Lightfoot	8	2
3/18/71	4/1/71	Paranoid	Black Sabbath	4	3
3/18/71	3/18/71	Manna	Bread	10	1
3/25/71	4/29/71	Moving Toward Happiness	Mason Proffit	5	6
3/25/71	4/8/71	Abraxas	Santana	4	3
4/8/71	4/22/71	Love Story	Andy Williams	8	3
4/15/71	4/15/71	Woodstock II	Various Artists	2(2)	4
4/15/71	5/6/71	Up To Date	Partridge Family	3	6
4/29/71	4/29/71	4 Way Street	Crosby, Stills, Nash & Young	2(3)	7
4/29/71	6/10/71	Tapestry	Carole King	1(11)	32
4/29/71	5/6/71	Golden Biscuits	Three Dog Night	6	6
5/6/71	7/8/71	Mud Slide Slim	James Taylor	2(3)	19
5/13/71	5/27/71	Aqualung	Jethro Tull	2	9
5/13/71	5/20/71	Sticky Fingers	Rolling Stones	1(3)	15
5/20/71	5/20/71	L.A. Woman	Doors	8	1
5/20/71	5/27/71	Survival	Grand Funk Railroad	8	4
5/27/71	6/10/71	Carly Simon	Carly Simon	6	5
6/3/71	8/5/71	Ram	Paul & Linda McCartney	1(2)	23
6/10/71	8/26/71	Carpenters	Carpenters	1	25
6/17/71	6/17/71	Songs For Beginners	Graham Nash	7	4
6/17/71	7/15/71	Live At Fillmore West	Aretha Franklin	6	5
7/1/71	8/5/71	What's Going On	Marvin Gaye	6	8
7/1/71	9/23/71	Every Picture Tells A Story	Rod Stewart	1(4)	18
7/8/71	7/8/71	Chase	Chase	9	1
7/8/71	7/22/71	Tarkus	Emerson, Lake & Palmer	8	3
7/15/71	8/12/71	Blood, Sweat & Tears #4	Blood, Sweat & Tears	5	7
7/22/71	8/5/71	Stephen Stills #2	Stephen Stills	2	9
7/29/71	8/5/71	Blue	Joni Mitchell	9	3
8/19/71	9/2/71	Who's Next	Who	5	13
8/26/71	9/16/71	Every Good Boy Deserves Favor	Moody Blues	1	11

WCFL Albums 1971-76

Debut	Peak	Title	Artist	Pos.	Wks.
9/2/71	11/25/71	"Shaft" Sound Track	Isaac Hayes	2	18
9/9/71	9/16/71	Master Of Reality	Black Sabbath	6	5
9/16/71	9/16/71	Sound Magazine	Partridge Family	10	2
9/23/71	10/7/71	Bark	Jefferson Airplane	5	4
9/30/71	10/21/71	Imagine	John Lennon	1(4)	12
10/14/71	11/4/71	Santana	Santana	1(2)	15
10/14/71	12/2/71	Teaser & The Firecat	Cat Stevens	1	17
11/11/71	11/18/71	After All This Time	Bonnie Koloc	8	4
11/18/71	12/16/71	There's A Riot Going On	Sly & The Family Stone	3	8
11/18/71	12/9/71	Chicago At Carnegie Hall	Chicago	1(5)	13
12/2/71	12/23/71	Led Zeppelin	Led Zeppelin	2(2)	13
12/9/71	12/23/71	E Pluribus Funk	Grand Funk Railroad	8	2
12/9/71	12/9/71	Stones	Neil Diamond	10	2
12/16/71	12/30/71	Carole King Music	Carole King	2(2)	17
12/16/71	1/27/72	American Pie	Don McLean	1(7)	17
12/23/71	12/30/71	Wings Wild Life	Paul & Linda McCartney	6	6
1/13/72	1/20/72	The Concert For Bangla Desh	Various Artists	1	9
1/13/72	2/17/72	Hot Rocks	Rolling Stones	2	12

1/20/72	3/9/72	The Low Spark Of High Heeled Boys	Traffic	5	9
1/27/72	2/10/72	Pictures At An Exhibition	Emerson, Lake & Palmer	6	5
2/10/72	3/16/72	Baby I'm A Want You	Bread	3	15
2/17/72	4/6/72	Paul Simon	Paul Simon	4	13
2/17/72	3/2/72	Fragile	Yes	2(3)	18
2/24/72	2/24/72	A Nod Is As Good As A Wink	Faces	10	1
3/2/72	3/16/72	Harvest	Neil Young	1(3)	20
3/9/72	3/23/72	Nilsson Schmilsson	Nilsson	4	13
3/16/72	4/6/72	America	America	1(3)	17
3/30/72	4/27/72	First Take	Roberta Flack	1(5)	15
4/13/72	5/11/72	Eat A Peach	Allman Brothers Band	5	12
4/27/72	5/25/72	Smokin'	Humble Pie	7	8
5/4/72	6/15/72	Graham Nash	David Crosby	4	10
5/18/71	6/1/72	Thick As A Brick	Jethro Tull	1(8)	16
6/1/72	6/29/72	Manassas	Stephen Stills	5	7
6/8/72	6/15/72	Exile On Main St.	Rolling Stones	2(5)	11
6/22/72	7/29/72	Honky Chateau	Elton John	1(3)	18
6/22/72	7/6/72	Joplin In Concert	Janis Joplin	4	6
7/6/72	7/6/72	Roberta Flack	Donny Hathaway	9	1
7/15/72	7/29/72	Elvis As Recorded At Madison Square Garden	Elvis Presley	3	7
7/15/72	8/5/72	School's Out	Alice Cooper	3	11

WCFL Albums 1971-76

Debut	Peak	Title	Artist	Pos.	Wks.
7/15/72	8/5/72	Simon & Garfunkel's	Simon & Garfunkel	2(3)	14
7/15/72	8/26/72	A Song For You	Carpenters	4	12
7/22/72	7/22/72	Procol Harem Live	Procol Harem	7	4
7/22/72	9/30/72	Moods	Neil Diamond	5	14
8/5/72	8/19/72	Chicago V	Chicago	1(7)	16
8/19/72	10/14/72	Trilogy	Emerson, Lake & Palmer	1	18
8/19/72	9/23/72	Never A Dull Moment	Rod Stewart	4	15
8/26/72	10/7/72	Big Bambu	Cheech & Chong	1	12
9/16/72	10/7/72	Carney	Leon Russell	6	6
9/30/72	10/21/72	Days Of Future Past	Moody Blues	1(5)	11
10/7/72	11/25/72	Superfly	Curtis Mayfield	3(2)	15
10/14/72	11/11/72	Close To The Edge	Yes	5	6
10/28/72	11/25/72	Catch Bull At Four	Cat Stevens	1	18
10/28/72	11/18/72	Black Sabbath, Vol. 4	Black Sabbath	4	8
11/11/72	12/2/72	Caravanserai	Santana	7	8
11/18/72	12/2/72	Seventh Sojourn	Moody Blues	1(5)	17
11/18/72	12/23/72	Rhymes & Reasons	Carole King	4	12
11/25/72	1/13/72	Living In The Past	Jethro Tull	2(5)	13
12/9/72	12/16/72	Homecoming	America	4	12
12/16/72	12/30/72	Summer Breeze	Seals & Crofts	5	6
12/23/72	1/13/73	No Secrets	Carly Simon	1(9)	15
12/23/72	1/20/73	Tommy	Various Artists	3	9
1/13/73	1/13/73	All Directions	Temptations	10	1
1/20/73	3/3/73	The World Is A Ghetto	War	5	8
1/27/73	2/10/73	One Man Dog	James Taylor	7	2
2/3/73	2/3/73	Who Do We Think We Are	Deep Purple	9	1
2/10/73	3/3/73	Shoot Out At The Fantasy	Traffic	4	11
2/10/73	2/24/73	Rocky Mountain High	John Denver	2	10
2/17/73	3/3/73	Don't Shoot Me I'm Only The Piano Player	Elton John	2(2)	12
2/24/73	3/17/73	Deliverance Sound Track	Deliverance	1(3)	9
3/3/73	4/21/73	Aloha From Hawaii	Elvis Presley	5	12
3/3/73	3/31/73	Prelude	Deodato	6	4
4/10/73	4/7/73	They Only Come Out At Night	Edgar Winter Group	1	17
3/24/73	3/3/73	Billion Dollar Babies	Alice Cooper	2(5)	17
3/24/73	4/14/73	The Dark Side Of The Moon	Pink Floyd	1(15)	30
4/14/73	4/14/73	Eat It	Humble Pie	5	1
4/21/73	5/12/73	Houses Of The Holy	Led Zeppelin	2(3)	29
4/21/73	4/28/73	The Best Of Bread	Bread	6	11
4/21/73	5/12/73	Moving Waves	Focus	4	9
4/28/73	6/2/73	The Beatles/1967-1970	Beatles	3(2)	11
4/28/73	5/26/73	The Beatles/1962-1966	Beatles	6	8

WCFL Albums 1971-76

Debut	Peak	Title	Artist	Pos.	Wks.
5/12/73	8/18/73	Diamond Girl	Seals & Crofts	3(3)	18
5/19/73	5/19/73	Ooh-La-La	Faces	10	1
5/26/73	6/16/73	Red Rose Speedway	Paul McCartney	3(3)	10
6/2/73	7/14/73	There Goes Rhymin' Simon	Paul Simon	4	13
6/23/73	6/23/73	Living In The Material	George Harrison	2(4)	9
6/30/73	7/14/73	Made In Japan	Deep Purple	3(2)	13
7/7/73	9/29/73	The Captain & Me	Doobie Brothers	5	21
7/7/73	7/7/73	Yessongs	Yes	10	1
7/21/73	7/28/73	Chicago VI	Chicago	1(8)	1
7/21/73	7/28/73	Love Devotion Surrender	Santana & McLaughlin	8	6
8/4/73	8/11/73	A Passion Play	Jethro Tull	2(4)	13
8/11/73	8/18/73	Foreigner	Cat Stevens	4	12
8/25/73	9/8/73	We're An American Band	Grand Funk	2(3)	13
9/1/73	9/22/73	Brothers And Sisters	Allman Brothers Band	1(4)	17
9/15/73	10/6/73	Los Cochinos @#!!*	Cheech Y Chong	2(5)	16
9/22/73	10/6/73	Killing Me Softly	Roberta Flack	8	8
9/22/73	10/27/73	Let's Get It On	Marvin Gaye	4	10
9/22/73	9/29/73	Deliver The Word	War	10	6
9/22/73	9/22/73	American Graffiti Soundtrack	Various Artists	10	17
9/22/73	10/20/73	Innervisions	Stevie Wonder	8	10
9/22/73	11/3/73	3 + 3	Isley Brothers	11	4
9/22/73	9/22/73	Fresh	Sly & The Family Stone	19	1
9/22/73	9/29/73	Long Hard Climb	Helen Reddy	15	3
9/29/73	9/29/73	Sufficiently Breathless	Captain Beyond	20	2
10/6/73	10/20/73	Goats Head Soup	Rolling Stones	1(5)	12
10/6/73	10/20/73	Angel Clare	Art Garfunkel	7	9
10/6/73	10/6/73	Rainbow Concert	Eric Clapton	11	5
10/13/73	11/3/73	Sweet Freedom	Uriah Heep	4	1
10/13/73	11/17/73	The Smoker You Drink, The Player You Get	Joe Walsh	10	9
10/20/73	11/24/73	Goodbye Yellow Brick Road	Elton John`1	1(5)	41
10/20/73	10/20/73	Over-Nite Sensation	Mothers	18	1
10/20/73	12/8/73	Life And Times	Jim Croce	6	14
11/3/73	11/17/73	Time Fades Away	Neil Young	12	4
11/3/73	12/7/73	Jonathan Livingston Seagull	Neil Diamond	2(2)	13
11/3/73	12/8/73	Ol' Blue Eyes Is Back	Frank Sinatra	11	7
11/3/73	11/10/73	Wake Of The Flood	Grateful Dead	10	6
11/10/73	12/15/73	Quadrophenia	The Who	2	16
11/17/73	11/17/73	Full Sail	Loggins & Messina	11	3
11/17/73	11/24/73	Hat Trick	America	13	4
11/17/73	1/19/74	You Don't Mess Around	Jim Croce	1(5)	21

WCFL Albums 1971-76

Debut	Peak	Title	Artist	Pos.	Wks.
11/24/73	12/8/73	Ringo	Ringo Starr	3	12
11/24/73	12/8/73	Mind Games	John Lennon	7	6
12/1/73	12/8/73	Laid Back	Gregg Allman	12	11
12/1/73	12/15/73	The Joker	Steve Miller Band`	6	12
12/8/73	12/29/73	The Singles	Carpenters	2(2)	12
12/8/73	12/8/73	Welcome	Santana	17	3
12/8/73	12/22/73	Live Dates	Wishbone Ash	18	2
12/15/73	12/29/73	I Got A Name	Jim Croce	1(2)	13
12/15/73	2/9/74	Brain Salad Surgery	Emerson, Lake & Palmer	2(2)	17
12/15/73	12/29/73	Muscle Of Love	Alice Cooper	9	10
12/15/73	12/29/73	Bette Midler	Bette Midler	11	7
12/15/73	19/29/73	The Serpent Is Rising	Styx	15	2
12/22/73	1/12/74	Bright White	Shawn Phillips	15	3
12/29/73	1/12/74	"Billy Jack"	Sound Track	11	5
12/29/73	6/8/74	Band On the Run	Paul McCartney	1(5)	36
12/29/73	12/29/73	"Sunshine"	Sound Track	19	1
12/29/73	1/26/74	Dylan	Bob Dylan	16	4
1/12/74	1/12/74	Big Bambu	Cheech & Chong	14	3
1/19/74	3/2/74	John Denver's Greatest Hits	John Denver	1(4)	28
1/19/74	2/16/74	Under The Influence Of Love Unlimited	Love Unlimited	7	9

1/19/74	1/19/74	Stone Gon'	Barry White	19	1
2/4/74	2/23/74	Sabbath, Bloody Sabbath	Black Sabbath	2	14
2/9/74	3/2/74	Hotcakes	Carly Simon	3(3)	18
2/9/74	3/23/74	Tales From Topographic Oceans	Yes	6	10
2/9/74	3/2/74	Planet Waves	Bob Dylan	6	8
2/16/74	2/23/74	Court And Spark	Joni Mitchell	1	24
2/16/74	2/16/74	"The Way We Were"	Sound Track	13	5
2/23/74	3/9/74	The Way We Were	Barbra Streisand	1(2)	10
3/2/74	4/13/74	Tubular Bells	Mike Oldfield	1(2)	15
3/2/74	3/16/74	Love Is The Message	MFSB	17	2
3/9/74	3/30/74	Burn	Deep Purple	6	10
3/9/74	3/30/74	Unborn Child	Seals And Crofts	7	14
3/9/74	3/23/74	All American Boy	Rick Derringer	18	2
3/16/74	3/16/74	Saints & Sinners	Johnny Winter	13	4
3/16/74	3/16/74	Rock N Roll Animal	Lou Reed	19	2
3/16/74	3/30/74	Ship Ahoy	O'Jays	15	2
3/23/74	4/6/74	What Were Once Vices Are Now Habits	Dobbie Brothers	4	17
30/30/74	4/27/74	"The Sting"	Sound Track	1(6)	19
3/30/74	4/13/74	Live Rhymin'	Paul Simon	18	3
4/6/74	4/6/74	Chicago VII	Chicago	3(4)	17

WCFL Albums-1971-76

Debut	Peak	Title	Artist	Pos.	Wks.
4/6/74	6/8/74	Shinin' On	Grand Funk	5	17
4/13/74	6/1/74	Buddha And The Chocolate	Cat Stevens	3(2)	18
4/27/74	8/3/84	Bachman-Turner Overdrive II	Bachman-Turner Overdrive	3(4)	29
4/27/74	6/1/74	Apostrophe'	Frank Zappa	14	12
4/27/74	6/1/74	"Maria Muldaur"	Maria Muldaur	10	12
5/4/74	5/4/74	Behind Closed Doors	Charlie Rich	12	2
5/4/74	5/4/74	Starless And Bible Black	King Crimson	16	4
5/4/74	6/29/74	Bridge Of Sighs	Robin Trower	7	23
5/11/74	6/1/74	Road Food	Guess Who	7	10
5/11/74	5/25/74	Secret Treaties	Blue Oyster Cult	17	2
5/25/74	6/8/74	On Stage	Loggins And Messina	17	6
6/1/74	7/6/74	Sundown	Gordon Lightfoot	1	15
6/1/74	6/1/74	Best Of Bread Volume II	Bread	18	2
6/8/74	6/29/74	Tres Hombres	ZZ Top	3(3)	13
6/8/74	6/8/74	Queen II	Queen	18	4
6/15/74	7/6/74	"Claudine" Sound Track	Gladys Knight & The Pips	6	13
6/15/74	6/15/74	Open Our Eyes	Earth, Wind & Fire	17	2
6/15/74	7/13/74	+'Justments	Bill Withers	14	10
6/15/74	6/15/74	Mighty Love	The Spinners	21	1
6/15/74	6/29/74	Street Lady	Donald Byrd	20	4
6/15/74	7/20/74	Head Hunters	Herbie Hancock	17	9
6/22/74	6/22/74	Diamond Dogs	David Bowie	16	10
6/22/74	6/29/74	Journey To The Center Of	Rick Wakeman	9	12
6/22/74	6/22/74	Sweet Exorcist	Curtis Mayfield	20	5
6/22/74	8/17/74	Body Heat	Quincy Jones	14	14
7/6/74	7/20/74	Caribou	Elton John	1(6)	26
7/6/74	7/27/74	Back Home Again	John Denver	2(5)	31
7/20/74	7/20/74	Mars Hotel	Grateful Dead	10	10
7/20/74	7/20/74	On The Border	Eagles	25	1
7/27/74	8/31/74	461 Ocean Boulevard	Eric Clapton	1(5)	16
7/27/74	8/10/74	Moontan	Golden Earring	5	10
8/3/74	4/26/74	Perfect Angel	Minnie Riperton	4	32
8/3/74	8/10/74	The Souther, Hillman, Furay Band	Souther, Hillman, Furay	14	8
8/3/74	8/24/74	On The Beach	Neil Young	6	11
8/3/74	8/3/74	Blue Magic	Blue Magic	24	1
8/10/74	8/31/74	You're Gonna Love Yourself In The Morning	Bonnie Koloc	18	5
8/10/74	8/24/74	Walking Man	James Taylor	24	2
8/17/74	9/14/74	Fulfillingness' First Finale	Stevie Wonder	3(2)	18
8/24/74	9/7/74	Rags To Rufus	Rufus	5	9
8/24/74	8/31/74	"Marvin Gaye Live"	Marvin Gaye	12	12

WCFL Albums-1971-76

Debut	Peak	Title	Artist	Pos.	Wks.
8/24/74	9/7/74	"Santana's Greatest Hits"	Santana	21	4
8/24/74	9/24/74	Holiday	America	12	6
8/31/74	10/12/74	"Bad Company"	Bad Company	2(5)	20
9/7/74	10/5/74	Not Fragile	Bachman-Turner Overdrive	1(8)	27
9/7/74	9/7/74	More, More, More	Latimore	11	5
9/7/74	9/7/74	Highway Call	Richard Betts	13	4
9/7/74	10/5/74	Anka	Paul Anka	12	9
9/14/74	9/28/74	Skin Tight	Ohio Players	8	13
9/14/74	9/28/74	One	Bob James	18	8
9/14/74	10/12/74	Can't Get Enough	Barry White	5	14
9/21/74	11/2/74	Welcome Back	Emerson, Lake & Palmer	5	10
9/21/74	10/26/74	So Far	Crosby, Stills, Nash & Young	6	11
9/21/74	10/19/74	Endless Summer	Beach Boys	5	18
9/28/74	11/2/74	If You Love Me Let Me Know	Olivia Newton-John	3	11
10/5/74	10/26/74	When The Eagle Flies	Traffic	11	11
10/5/74	10/5/74	Live It Up	Isley Brothers	17	6
10/5/74	11/9/74	Second Helping	Lynyrd Skynyrd	15	8
10/5/74	11/16/74	Wrap Around Joy	Carole King	2	13
10/5/74	10/5/74	Thrust	Herbie Hancock	25	1
10/12/74	11/9/74	Photographs And Memories	Jim Croce	3	18
10/12/74	10/12/74	All In Love Is Fair	Nancy Wilson	25	1
10/19/74	10/19/74	Roxy & Elsewhere	Zappa/Mothers	21	1
10/19/74	11/16/74	Walls & Bridges	John Lennon	10	11
10/19/74	11/2/74	Cheech & Chong's Wedding Album	Cheech & Chong	8	11
10/26/74	10/26/74	Late For The Sky	Jackson Browne	22	1
11/2/74	11/23/74	War Child	Jethro Tull	2(3)	22
11/2/74	11/2/74	Smiler	Rod Stewart	22	2
11/2/74	11/9/74	Odds & Sods	The Who	19	4
11/9/74	11/23/74	It's Only Rock 'N Roll	Rolling Stones	4	10
11/9/74	11/30/74	Serenade	Neil Diamond	3(2)	13
11/16/74	11/30/74	David Live	David Bowie	21	3
11/23/74	11/30/74	Mother Lode	Loggins And Messina	14	15
11/23/74	11/23/74	Borboletta	Santana	22	3
11/23/74	11/30/74	Greatest Hits	Elton John	1(12)	27
11/30/74	12/21/74	This Is The Moody Blues	Moody Blues	2(5)	13
11/30/74	11/30/74	Heavy	Stylistics	19	1
11/30/74	1/11/75	Butterfly	Barbra Streisand	5	11
12/7/74	1/18/75	Al Green Explores Your Mind	Al Green	11	9

WCFL Albums 1971-76

Debut	Peak	Title	Artist	Pos.	Wks
12/7/74	2/15/75	Fire	Ohio Players	2	19
12/14/74	1/11/75	Verities And Balderdash	Harry Chapin	4	8
12/14/74	12/28/74	Together For The First Time	B.B. King & Bobby Bland	15	2
12/14/74	1/11/75	Caught Up	Millie Jackson	19	5
12/14/74	12/14/74	Goodnight Vienna	Ringo Starr	23	2
12/21/74	2/1/75	Miles Of Aisles	Joni Mitchell	8	11
12/21/74	12/21/74	Stormbringer	Deep Purple	20	2
12/21/74	1/18/75	Melodies Of Love	Bobby Vinton	7	12
12/28/74	2/15/75	New And Improved	Spinners	9	15
12/28/74	1/25/75	Dark Horse	George Harrison	4	9
1/11/75	2/15/75	Relayer	Yes	7	9
1/11/75	1/11/75	Dragon Fly	Jefferson Starship	23	1
1/11/75	2/22/75	Souvenirs	Dan Fogelberg	12	11
1/18/75	2/8/75	Styx II	Styx	2(2)	12
1/18/75	3/1/75	Heart Like A Wheel	Linda Ronstadt	2	14
1/18/75	3/8/75	Average White Band	Average White Band	1	19
1/18/75	1/18/75	Light Of Worlds	Kool & the Gang	25	1
1/25/75	2/15/75	Joy To The World	Three Dog Night	12	7
1/25/75	2/22/75	So What	Joe Walsh	11	12
1/25/75	2/1/75	White Gold	Love Unlimited Orchestra	20	8
2/1/75	2/15/75	Barry Manilow II	Barry Manilow	8	10
2/15/75	3/1/75	Blood On The Tracks	Bob Dylan	1(2)	15
2/15/75	2/15/75	Never Can Say Goodbye	Gloria Gaynor	23	3

2/22/75	3/15/75	Do It	B.T. Express`	18	8
2/22/75	2/22/75	All The Girls In The World Beware!!!	Grand Funk	22	1
3/1/75	3/29/75	Have You Never Been Mellow	Olivia Newton-John	2(2)	15
3/8/75	4/5/75	For Earth Below	Robin Trower	6	12
3/8/75	3/8/75	Cold On The Shoulder	Gordon Lightfoot	15	7
3/8/75	5/10/75	Autobahn	Kraftwerk	4	15
3/8/75	4/5/75	"Phoebe Snow"	Phoebe Snow	15	7
3/15/75	3/22/75	Physical Graffiti	Led Zeppelin	1(8)	24
3/15/75	4/5/75	Rufusized	Rufus	10	9
3/15/75	4/12/75	"An Evening With John Denver	John Denver	9	14
3/22/75	3/22/75	Rock 'N Roll	John Lennon	21	2
3/22/75	4/5/75	Nightbirds	Labelle	18	8
3/22/75	5/31/75	That's The Way Of The World	Earth, Wind & Fire	2(5)	27
3/29/75	5/3/75	Crash Landing	Jimi Hendrix	3	12
3/29/75	3/29/75	Sun Goddess	Ramsey Lewis	20	2
3/29/75	6/14/75	Welcome To My Nightmare	Alice Cooper	14	6

WCFL Albums 1971-76

Debut	Peak	Title	Artist	Pos.	Wks.
3/29/75	3/29/75	Yesterdays	Yes	24	1
4/5/75	4/5/75	Young Americans	David Bowie	14	3
4/5/75	4/5/75	On Your Feet Or On Your Knees	Blue Oyster Cult	25	1
4/12/75	5/10/75	Blue Jays	Justin Haywood & John Lodge	5	11
4/12/75	4/26/75	"Funny Lady"	Original Sound Track	18	3
4/19/75	5/17/75	Chicago VIII	Chicago	1(4)	20
4/19/75	4/19/75	There's One In Every Crowd	Eric Clapton	14	4
4/19/75	5/17/75	"Tommy"	Original Sound Track	4	13
4/19/75	5/24/75	Blow By Blow	Jeff Beck	7	18
4/19/75	4/19/75	Nuthin' Fancy	Lynyrd Skynyrd	20	3
4/19/75	5/24/75	Katy Lied	Steely Dan	18	5
4/26/75	5/31/75	Straight Shooter	Bad Company	6	10
4/26/75	5/31/75	Hearts	America	5	15
5/3/75	5/24/75	To Be True	Harold Melvin/Blue Notes	11	11
5/3/75	5/10/75	Just Another Way To Say I Love You	Barry White	16	6
5/3/75	5/3/75	Judith	Judy Collins	25	1
5/10/75	6/21/75	Playing Possum	Carly Simon	7	11
5/17/75	7/12/75	Mister Magic	Grover Washington, Jr.	11	12
5/17/75	5/24/75	A Song For You	Temptations	17	7
5/17/75	5/17/75	The Myths And Legends	Rick Wakeman	23	2
5/24/75	6/14/75	Stampede	Doobie Brothers	4	11
5/24/75	5/24/75	Crime Of The Century	Supertramp	15	2
5/24/75	9/20/75	Fandango	ZZ Top	9	9
5/31/75	7/12/75	Disco Baby	Van McCoy	4	14
5/31/75	5/31/75	Just A Boy	Leo Sayer	22	1
5/31/75	6/28/75	"The Original Sound Track	10CC	14	11
5.31.75	7/19/75	Four Wheel Drive	Bachman-Turner Overdrive	8	13
6/7/75	6/14/75	Captain Fantastic	Elton John	1(17)	25
6/7/75	6/28/75	Survival	O'Jays	11	14
6/14/75	7/12/75	Lake Shore Drive	Aliotta, Haynes & Jeremiah	14	7
6/14/75	6/21/75	A Quiet Storm	Smokey Robinson	22	4
6/21/75	7/5/75	Venus And Mars	Paul McCartney & Wings	2(7)	19
6/21/75	6/21/75	Diamonds & Rust	Joan Baez	17	4
6/21/75	8/30/75	Gorilla	James Taylor	5	19
6/21/75	10/18/75	Toys In The Attic	Aerosmith	3(2)	26
6/28/75	6/28/75	Metamorphosis	Rolling Stones	17	1
6/28/75	8/2/75	Spirit Of America	Beach Boys	11	10
7/5/75	8/30/75	One Of These Nights	Eagles	2(2)	24

WCFL Albums 1971-76

Debut	Peak	Title	Artist	Pos.	Wks
7/5/75	8/23/75	The Heat Is On	The Isley Brothers	2(2)	21

7/5/75	7/5/75	Horizon	Carpenters	14	4
7/5/75	8/2/75	Love Will Keep Us Together	Captain & Tenille	3	14
7/5/75	7/19/75	Blue Sky-Night Thunder	Michael Murphy	21	2
7/12/75	8/9/75	Made In The Shade	Rolling Stones	15	6
7/12/75	8/2/75	Stills	Stephen Stills	6	11
7/19/75	8/9/75	Cut The Cake	Average White Band	17	10
7/26/75	7/26/75	Tonight's The Night	Neil Young	18	3
7/26/75	9/6/75	Greatest Hits	Cat Stevens	8	13
7/26/75	8/2/75	One Size Fits All	Frank Zappa	19	2
8/2/75	11/15/75	Red Octopus	Jefferson Starship	1(2)	26
8/2/75	8/2/75	Latimore III	Latimore	21	1
8/9/75	8/23/75	The Basement Tapes	Bob Dylan & The Band	13	6
8/9/75	8/9/75	Moving Violation	Jackson 5	23	1
8/16/75	8/30/75	Why Can't We Be Friends?	War	12	9
8/16/75	2/21/76	Fleetwood Mac	Fleetwood Mac	10	26
8/23/75	9/6/75	Steppin'	Pointer Sisters	15	5
8/30/75	8/27/75	Pick Of The Litter	Spinners	11	13
8/30/75	8/30/75	Insides Out	Hamilton Bohannon	18	2
9/6/75	9/27/75	Is It Something I Said?	Richard Pryor	6	10
9/6/75	10/11/75	Sabotage	Black Sabbath	13	9
9/6/75	11/1/75	Honey	Ohio Players	7	16
9/13/75	9/13/75	Blues For Allah	Grateful Dead	16	4
9/13/75	9/13/75	Main Course	Bee Gees	24	1
9/20/75	11/1/75	Born To Run	Bruce Springsteen	5	11
9/20/75	10/18/75	Between The Lines	Janis Ian	18	6
9/20/75	9/20/75	Rhinestone Cowboy	Glen Campbell	23	2
9/27/75	10/11/75	Win, Lose Or Draw	Allman Brothers Band	7	8
9/27/75	9/27/75	Atlantic Crossing	Rod Stewart	17	2
9/27/75	9/27/75	E.C. Was Here	Eric Clapton	25	1
10/4/75	10/11/75	Wish You Were Here	Pink Floyd	1(5)	18
10/4/75	11/8/75	Minstrell In The Gallery	Jethro Tull	5	11
10/4/75	12/20/75	K.C. & The Sunshine Band	K.C. & The Sunshine Band	12	16
10/11/75	11/1/75	Windsong	John Denver	3(3)	15
10/11/75	11/8/75	Prisoner In Disguise	Linda Ronstadt	9	9
10/11/75	12/6/75	Save Me	Silver Convention	7	12
10/11/75	11/8/75	Mellow Madness	Quincy Jones	14	8
10/18/75	11/1/75	Extra Texture	George Harrison	8	6
10/25/75	11/22/75	Wind On The Water	David Crosby/Graham Nash	12	9
10/25/75	11/29/75	Boggie Down U.S.A.	Peoples Choice	19	8
11/1/75	11/8/75	The Who By Numbers	The Who	6	15

WCFL Albums 1971-76

Debut	Peak	Title	Artist	Pos.	Wks.
11/1/75	1/24/76	Alive	Kiss	6	11
11/1/75	12/27/75	Still Crazy After All These	Paul Simon	2(7)	15
11/8/75	11/29/75	Rock Of The Westies	Elton John	1(3)	11
11/8/75	11/29/75	Breakaway	Art Garfunkel	9	13
11/15/75	1/24/76	Trying To Get The Feeling	Barry Manilow	5	13
11/22/75	12/6/75	Let's Do It Again	Staple Singers	12	5
11/22/75	11/29/75	Al Green Is Love	Al Green	23	2
11/29/75	1/31/76	Love To Love You Baby	Donna Summer	6	11
12/6/75	12/20/75	Chicago's Greatest Hits	Chicago	1(6)	10
12/6/75	1/17/76	History-America's Greatest Hits	America	3	10
12/6/75	1/17/76	Seal's & Crofts Greatest Hits	Seals & Crofts	19	4
12/6/75	12/13/75	Zuma	Neil Young	19	3
12/6/75	12/20/75	Feelings	Morris Albert	19	2
12/13/75	12/20/75	The Hissing Of Summer Lawns	Joni Mitchell	9	8
12/13/75	12/13/75	Midnight Lightning	Jimi Hendrix	20	1
12/13/75	1/31/76	Family Reunion	O'Jays	11	9
12/20/75	1/24/76	Gratitude	Earth, Wind & Fire	3(5)	8
12/20/75	2/14/76	Wake Up Everybody	Harold Melvin/Blue Notes	10	8
12/27/75	12/27/75	Numbers	Cat Stevens	16	1
1/17/76	2/7/76	Black Bear Road	C.W.McCall	20	5
1/17/76	1/31/76	Equinox	Styx	17	6
1/17/76	1/17/76	A Night At The Opera	Queen	23	2
1/17/76	1/31/76	Face The Music	Electric Light Orchestra	23	3

1/24/76	2/14/76	Helen Reddy's Greatest Hits	Helen Reddy	17	4
1/24/76	2/7/76	Head On	Bachman-Turner Overdrive	19	2
1/31/76	2/14/76	Desire	Bob Dylan	1(2)	4
1/31/76	2/14/76	"M.U." The Best Of Jethro Tull	Jethro Tull	5	4
1/31/76	1/31/76	Elite Hotel	Emmylou Harris	20	3
2/7/76	2/21/76	City Life	Blackbyrds	15	3
2/14/76	2/21/76	Rufus Featuring Chaka Khan	Rufus	9	2
2/14/76	2/14/76	Places And Spaces	Donald Byrd	20	2
2/14/76	2/21/76	Aftertones	Janis Ian	18	2
2/21/76	2/21/76	Run With The Pack	Bad Company	16	1
2/21/76	2/21/76	Songs For The New Depression	Bette Midler	17	1
2/21/76	2/21/76	Frampton Comes Alive	Peter Frampton	19	1
2/21/76	2/21/76	Native Sons	Loggins & Messina	21	1
2/21/76	2/21/76	Station To Station	David Bowie	24	1

Made in United States
North Haven, CT
10 November 2026

22106130R00454